Fourth Australasia and South-East Asia Structural Engineering and Construction Conference (ASEA-SEC-4)

Streamlining Information Transfer Between Construction and Structure Engineering

Brisbane, Australia
3-5 December 2018

Editors:

Jim Shiau
Vanissorn Vimonsatit
Amarjit Singh
Siamak Yazdani

ISBN: 978-1-5108-9998-8

Printed from e-media with permission by:

Curran Associates, Inc.
57 Morehouse Lane
Red Hook, NY 12571

Some format issues inherent in the e-media version may also appear in this print version.

ISEC PRESS
PO Box 1334
Fargo, North Dakota 58107
USA

isec.sec@gmail.com

Additional copies of this publication are available from:

Curran Associates, Inc.
57 Morehouse Lane
Red Hook, NY 12571 USA
Phone: 845-758-0400
Fax: 845-758-2633
Email: curran@proceedings.com
Web: www.proceedings.com

TABLE OF CONTENTS

ARCHITECTURE & ARCHITECTURAL ENGINEERING

CONSTRUCTION

COST AND PROJECT MANAGEMENT

CONCRETE STRUCTURES AND NONLINEAR BEHAVIOR

Preface

Having conjoined efforts by academicians and industry professionals is necessary for the progress of the construction and built environment towards more efficient and cost-effective outcomes. This idea forms the basis in providing an avenue for sharing and exchanging ideas among academics and practitioners.

In recent years, the active research areas in civil engineering and construction are, for example, synthesis and design of advanced construction materials, development of low-carbon blended cement, high strength and high performance concrete, durable steel, and advanced pavement and foundation materials. Other active areas are the development of advanced analytical methods for structures and infrastructure, construction techniques, project cost management, and planning towards digitalisation, construction safety, health monitoring and forensic engineering.

The fourth Australasia and South East Asia Conference in Structural Engineering and Construction, ASEA-SEC-04, was held in Brisbane, Queensland, Australia, during 3 - 5 December 2018. ASEA-SEC conferences are part of the international conference series organised by the International Structural Engineering and Construction Society, ISEC, USA. ASEA-SEC-4 was hosted by the University of Southern Queensland, with the theme *Streamlining Information Transfer between Construction and Structural Engineering*.

The primary aim of the conference was to create a forum for academics, researchers and practitioners to discuss advances and innovations in construction and civil engineering fields.

Registered participants came from 20 countries. Papers in these proceedings cover a wide range of topics including architecture, architectural engineering, construction, cost and project management, construction safety, delay analysis, and dispute resolution, geotechnical and foundation Engineering, pavements and infrastructure, properties of concrete materials, concrete structures and nonlinear behaviour, prestressed concrete and bridge structures, and building structures and vibration.

Many authors and co-authors of the papers served as peer-reviewers in the abstract and manuscript review process. We are thankful for the cooperation and timely support provided by all the reviewers.

The conference venue and facilities supported by the University of Southern Queensland, Springfield campus are gratefully acknowledged. Finally, we thank all the participants, the technical committee members, and all our peers who participated and/or supported us, which led to the success of the conference.

Jim Shiau, *University of Southern Queensland, Australia*
Vanissorn Vimonsatit, *Curtin University, Australia*
Amarjit Singh, *University of Hawaii at Manoa, USA*
Siamak Yazdani, *Heyer Engineering, USA*
Editors

Acknowledgements

Host
University of Southern Queensland, Brisbane, Australia

Committees

Organizing Committee
Amarjit Singh, *President, ISEC Society*
University of Hawaii at Manoa, USA
Frank Yazdani, *Secretary, ISEC Society*
Heyer Engineering, USA
Vanissorn Vimonsatit, *Curtin University, Australia*

Technical Committee
Syed Faiz Ahmed, *Lievense Arabia Consulting Engineers, Saudi Arabia*
Thiru Aravinthan, *University of Southern Queensland, Australia*
Farhad Aslani, *The University of Western Australia, Australia*
Sai On Cheung, *City University of Hong Kong, Hong Kong*
Amr Eldieb, *United Arab Emirates University, UAE*
Chayanon Hansapinyo, *Chiang Mai University, Thailand*
Takashi Hara, *Tokuyama College of Technology, Japan*
Ali Karrech, *The University of Western Australia, Australia*
Karu Karunasena, *University of Southern Queensland, Australia*
Shukla Sanjay Kumar, *Edith Cowan University, Australia*
Hieng Ho Lau, *Curtin University, Malaysia*
Hyuk Lee, *Curtin University, Australia*
Allan Manalo, *University of Southern Queensland, Australia*
Aman Mwafy, *United Arab Emirates University, UAE*
Kate Nguyen, *The University of Melbourne, Australia*
Nazar K. Oukaili, *University of Baghdad, Iraq*
Chiew Sing Ping, *Singapore Institute of Technology, Singapore*
Vivian Tam, *Western Sydney University, Australia*
Brian Uy, *University of New South Wales, Australia*
Valentina Villa, *Politecnico di Torino, Italy*
Weerakaset Suanpaga, *Kasetsart University, Bangkok, Thailand*
Weerachart Tangchirapat, *King Mongkut's University of Technology Thonburi, Thailand*
Andrew Whyte, *Curtin University, Australia*
Kenneth Yiu, *University of Auckland, New Zealand*

Local Organizing Committee
Jim Shiau, *University of Southern Queensland, Australia* ***(Conference Chair)***
Vanissorn Vimonsatit, *Curtin University, Australia* ***(Conference Co-Chair)***
Weena Lokuge, *University of Southern Queensland, Australia*
Sourish Banerjee, *University of Southern Queensland, Australia*
Ali Mirzaghorbanali, *University of Southern Queensland, Australia*
Habib Alehossein, *University of Southern Queensland, Australia*
Andreas Nataatmadja, *University of Southern Queensland, Australia*
Soma Somasundaraswaran, *University of Southern Queensland, Australia*
Amirhossein Heravi, *University of Southern Queensland, Australia*
David Thorpe, *University of Southern Queensland, Australia*
Jo Devine, *University of Southern Queensland, Australia*
Nateque Mahmood, *University of Southern Queensland, Australia*
Trevor Drysdale, *University of Southern Queensland, Australia*
Vasantha Abeysekera, *University of Southern Queensland, Australia*

List of Reviewers

Naveed Anwar
Muhamad Abduh
Sarah Aboulfotouh
Sohrab Afrasiabi
Douglas Aghimien
Fadhil Al-Asadi
John Aliu
Saud Alotaibi
Hossein Askarinejad
Prem Pal Bansal
Bernhard Bauer
Baris Binici
Xiang Chen
Amila Dissanayake
Mohammed El-Gendy
Abiy Goshu
Edwin Harrer
Mohammad Hassan
Masaaki Hiroe
Yoshitaka Ishikawa
Ryo Isuda
Weiliang Jin
Haeyoung Kim
Hyuk Lee
Motlatso Mabeba
Constant Mahame
Nokulunga Mashwama
Mahmood MD Tahir
Yoichi Mimura
Panagiotis Mitropoulos
Mitsuhiro Miyamoto
Mohd Amiza Mohamed
Rajesh Pathak
Liaqat Qureshi
Amarjit Singh
John Smallwood
Xi Song
Ann Stewart
Weerakaset Suanpaga
Kosuke Takahashi
Fumiya Takami
Noriko Takiyama
Chipozya Tembo-Silungwe
Afshin Turk
Ageliki Valavanoglou
Valentina Villa
Vanissorn Vimonsatit
Jeong-Hun Won
Siamak Yazdani
Isamu Yoshitake

Streamlining Information Transfer between Construction and Structural Engineering
Edited by Shiau, J., Vimonsatit, V., Yazdani, S., and Singh, A.

ISBN: 978-0-9960437-7-9

HISTORIC BUILDINGS COMPARISONS OF PATHOLOGIES AND INTERVENTIONS: EXAMPLES OF REUSE

SILVIA MAZZETTO

School of Architecture and Design, Lebanese American University, Byblos, Lebanon

In the recent years, many Gulf States have emerged because of their significant urban growth to contemporary architecture. In the same time, the growing number of conservation projects shows the raising need to enforce appreciation of the architectural heritage, to enhance and regenerate the historical memory of the counties. This paper offers a comparison between some recent conservative projects in Qatar conducted to some significant residential religious and commercial buildings, analyzing the pathologies that affected the structures, the reasons of deterioration before the restorative interventions, the comparison of methods and techniques that were adopted during the works and the proposals for reuse. Analyses of the data have permitted to classify the building pathologies and the conducted interventions and could be applied to other conservation projects in the future.

Keywords: Traditional materials, Restoration, Deterioration, Conservation, Architectural heritage.

1 INTRODUCTION

This paper presents an analysis of some conservation works completed in Qatar with the aim of describing and comparing them concerning their various deteriorations, the materials, and tools used for construction and the several proposals for re-use (Carbonara 2012, Feilden 2003, Muñoz Viñas 2004, Jokilehto 1999).

Our analysis of the restoration interventions was supported by site surveys, data collection and the literature review related to the Qatari heritage conservative approach. The data have described the state of deterioration of the structures before restoration, and then the cause of such deterioration, using the ICOMOS-ISCS glossary (ICOMOS 2008). The original materials used for construction were analysed and described with the aim of providing an overview of the Qatari heritage conservation in and enhancing the value of the local identity, and emphasizing the importance of preserving the past through a relationship built over the historical continuity (Salama and Wiedman 2013, Jaidah and Bourennane 2009, Mazzetto and Petruccioli 2017, Mazzetto 2006, Brandi 2005, Al-Kholaifi 2006).

2 MANAGEMENT OF QATARI HERITAGE

In Qatar, the legislation that protects architectural heritage was defined only 1980 to preserve the architectural heritage that evidence of the local history and the past civilizations (Bianchi and Tonner 2013, Carter 2012).

In 2005 the government set up the two leading institutions the Qatar Museum Authority (QMA) and the Private Engineering Office (PEO) which are responsible for safeguarding the state's heritage. The task of both the institutions is to preserve the local heritage and manage buildings of historical value, both within Qatar and internationally, including architectural restoration and regeneration of urban sites, focusing on the integration of the old into the new construction projects.

Although the Qatari tradition in managing architectural heritage is not so long, there are some specific institutions tools and laws, which were established to manage the conservative interventions of historical buildings, promoting good practice examples of management in compliance with international norms.

Recently an IT platform, QNHER (the Qatar National Historic Environment Record, have been developed with the aim of managing and cataloging the collaborative and shared researches between local institutions.

3 ANALYSIS OF SOME ADAPTIVE RE-USE INTERVENTIONS READING

3.1 Souk Waqif

Thanks to the recent restrictions imposed by the Qatari institutions and regulations, some rehabilitative interventions have recently been launched to control the widespread demolition phenomenon and revitalize the abandoned urban fabric by reusing the areas as new entertainment places.

This began with the refurbishment intervention of Souq Waqif (2004-2008), built about 100 years ago and located in the historic center of Doha. The conservative and typological restoration project started in 2004 and was directed by the Private Engineering Office (PEO).

The restoration work, completed in 2008, has reduced the deterioration of the historical buildings, conserving all of the ancient structures constructed before 1950, and restoring the more original parts, following their architectural typology.

The traditional construction method for residential buildings was adequately maintained, using load-bearing walls done by using sun-dried bricks, coral stone, wooden beams and the gypsum mortar.

The more recent materials, such as corrugated metal sheet, were removed since they were interfering with the ancient typology.

The Souq Waqif adaptive reuse intervention constitutes of an urban regenerative project that affected a large part of the historic city center and is currently reused as a new entertainment place with many commercial areas, art exhibitions, theatre performances, and sporting events.

3.2 Al Wakrah Souq

The adaptive reuse project of Al Wakrah was completed in 2015, under the direction of the Private Engineering Office (PEO) and was financed by the Emir. The historical fabric of Al Wakrah fisherman village, located close to the ancient port, was regenerated after being abandoned for many years. The area was then transformed into the new Souq of Wakrah, through the reconstruction of many collapsed buildings, the adaptation of the properties, the removal of damaged elements and materials, and the management of new commercial units (Figure 1).

The restoration of Al Wakrah village and its reuse was a small action, with limited repercussions on the adjacent territory, which was not able to trigger physical-environmental rehabilitative results in the surrounding areas. In Al Wakrah currently are located new activities and functions, and still the real cultural, and social values are tangible and well preserved.

Figure 1. Al Wakrah Souq. Public spaces for the new entertainment activities. (Source: Author)

3.3 Al Dakhira Mosque

The neglected and abandoned Al Dakhira Mosque located in the north coast of Qatar was structurally restored in 2015 under the direction of the Private Engineering Office and is currently being used as new cultural and religious center.

Before the restoration works, the ancient structure was affected by the characteristic deterioration due to the wet saline environments, causing permanent problems of rising damp on the external walls. The plastered surfaces were almost detached due to the Saline efflorescence, and the structural stability of the building was damaged by the deteriorated foundations affected by high level of salinity percentage of the ground.

The external plasters were restored entirely; the wooden roofs were consolidated by using traditional materials.

With the aim of reusing the old mosque as a new religious center, the new aluminum doors were fitted, along with the systems upgrading.

3.4 The Heritage Houses Quarter at Msheireb

The residential Heritage Houses (2006-2015) conservation project, located in Msheireb Downtown Doha was completed in 2015 under the supervision of the Private Engineering Office PEO. The four historic residential buildings that date from the early 20th century (Figure 2), exemplified the local building tradition and were integrally preserved, recovering all the original construction materials and replacing only a few parts that were missing.

Today the Heritage Houses are reused as new museums of the Msheireb Quarter's history and included some exhibitions about the residential project and the history of slavery in the region. The national institution of Msheireb Museums, responsible for the safeguard of the Heritage Houses district has been established with the aim to protect and promote the value of the preserved buildings.

Figure 2. Mohammed bin Jassim House. (Source: Author)

4 RESULTS

The approach to restoration taken for each intervention was related to the government authorities in charge for the works. The completed projects are mainly divided into two main categories: the architectural and the urban. Most of the interventions had been conducted using traditional materials like wooden beams, blocks stone, and mud mortar for plasters.

The same construction methods were usually adopted in similar conditions, such as the mud mortars, the sun-dried bricks, the "danchal" wood beams used as lintels, and the wooden “marazim” gargoyles.

Because of the repetitive use of materials and tools of construction, the reasons of deterioration were usually analogous and mainly due to the systematic abandonment of buildings, leading to common collapses of the foundations.

All the interventions intended to maintain the existing materials as found, without alterations, but in some cases, because of the advanced level of deterioration, substantial structural interventions were carried out combining traditional and contemporary materials such as reinforced concrete, cement mortar, and concrete blocks. In the case of commercial and entertainment reuses, the restoration projects allowed for typological reproduction, requested by the new uses.

5 CONCLUSION

All restoration works in Qatar were conducted with the aim of preserving the existing architectural heritage, to transfer their value to the future generations.

Although many recent management tools exist, there is a lack of interaction between the various governmental institutions involved in the conservation of Qatari Heritage, to standardize the conservative methods throughout the country and promote the application of new approaches in the field of the conservation.

The use of natural materials should be imposed, avoiding contemporary products, which in some cases are responsible for the increasing deterioration of old materials.

This paper attempts to document the recent architectural heritage conservation projects in Qatar, with the aim of promoting a standardized methodology for conservation that should avoid the frequent use of a case-by-case approach, integrating the individual interventions as parts of an organic vision concerning good conservation practices.

References

Al-Kholaifi, M. J., The Traditional Architecture in Qatar, *National Council for Culture, Art and Heritage, Museums and Antiquities Department*, Qatar, Doha, 2006.

Bianchi A., and Tonner T., *Data Standards, Documentation and Responses to Cultural Heritage Management in Qatar,* 18th International Conference on Cultural Heritage and New Technologies CHNT 18, Vienna, 2013.

Brandi, C., *Il Restauro: Teoria e Pratica 1939-1986*, Rome, Editori Riuniti, Italy 2005.

Carbonara, G., An Italian Contribution to Architectural Restoration, *Frontiers of Architectural Research*, 1, 2-9, 2012.

Carter, R., *Sea of Pearls. Seven Thousand Years of the Industry that Shaped the Gulf.* London, Arabian Publishing, 2012.

Feilden, B., *Conservation of Historic Buildings*, Rutledge, United Kingdom, 2003.

ICOMOS, *Illustrated Glossary on Stone Deterioration Patterns,* ICOMOS International Scientific Committee for Stone, Rome, Italy, 2008.

Jaidah, I. M., and Bourennane, M., *The History of Qatar Architecture (1800-1950),* Skira, Italy, 2009.

Jokilehto, J., *A History of Architectural Conservation*, Elsevier, Butterworth-Heinemann, 1999.

Mazzetto, S., *Beni Militari Come Attori Protagonisti, in Cavallino–Treporti,* Projects for an Idea for a Park P. Grandinetti, S. Mazzetto, V. Skabar (eds.), Il Poligrafo, Venice, Italy, 50-51, 2006.

Mazzetto S., and Petruccioli A., Methods and Techniques Used in Significant Restoration Projects in Qatar, *Studies in Conservation*, Routledge, 2017.

Muñoz Viñas, S., *Contemporary Theory of Conservation*, Elsevier, Butterworth-Heinemann, 2004.

Salama, A. M., and Wiedman F., *Demystifying Doha: on Architecture and Urbanism in an Emerging City,* Ashgate, Farnham, England, 43-45, 2013.

Streamlining Information Transfer between Construction and Structural Engineering
Edited by Shiau, J., Vimonsatit, V., Yazdani, S., and Singh, A.

ISBN: 978-0-9960437-7-9

PATHOLOGIES AND INTERVENTIONS FOR DEFENSIVE STRUCTURES: THE SAFEGUARD OF ARCHITECTURAL HERITAGE

SILVIA MAZZETTO

School of Architecture and Design, Lebanese American University, Byblos, Lebanon

Affecting the direction of the rapid urbanization due to the oil discovery and exportation, in the Gulf Regions, there is a perpetual dispute, between the construction of a new, modern identity and the advocacy of traditional architecture as a possible way to establish new relations with the local history and culture. In the struggle for the definition of a new architectural identity in the emerging Arab cities, this paper focuses on the action needed to preserve architectural heritage, with the aim of analyzing and comparing some significant examples of pathologies that affected the existing defensive structure in Qatar and the restoration works that were recently completed. The analysis of data and results made it possible to compare the different pathologies and the conducted interventions with the aim of defining an approach that is consistent and add value to the Arab culture in matters of conservation and re-use.

Keywords: Reuse, Deterioration, Conservation, Restoration, Historical heritage.

1 INTRODUCTION

The amount of conservation and restoration projects recently ultimate in Qatar show the growing interest in the local architectural heritage and construction traditions, as a way of enhancing the knowledge of the memory. There is a need to discover and improve appreciation of the local techniques and materials to preserve.

This paper presents some conservation projects recently completed for historical defensive structures in Qatar, analyzing the collected data and comparing the results to offer an overview of the current status of their conservation, the reasons of their deterioration before restoration; the methods and the constructions materials adopted for the intervention; and the diverse proposals for reusing the restored buildings. The research intends to compare and classify the various interventions to apply similar materials and techniques to other restoration projects in future.

2 LITERATURE REVIEW

The recent growth of many Gulf States due to the discovery and export of oil (1960-1970), has produced massive investments but have also produced a negative impact on the safeguard of the architectural heritage. This rapid urban development has altered the old face of the historical monuments of Doha, in particular of traditional defensive buildings (forts, towers, defensive residential complex) that are currently abandoned and have lost their original function, not being anymore used for defensive purposes. Due to the deleterious status of abandonment, the defensive structures are gravely damaged, and close to the final collapse, the original materials have been stolen in the last decades, causing the total loss of their original identity.

Additionally, the local administrations of the Gulf Regions (Carter 2012) have given little attention to the fast demolition of historical defensive buildings, mainly due to the absence of any laws on the heritage safeguarding until 1980. Since then the interest in conservation has reawakened with the launch of many restoration projects such as urban regenerations (Souk Waqif 2004-2008) and architectural refurbishments (Heritage Houses in Msheireb Downtown Doha 2008-2017) in Qatar, re-using the neglected buildings and urban area as new entertainment places (Mazzetto and Petruccioli 2017) (Salama and Wiedman 2013, Jaidah and Bourennane 2009). The increasing interest for the conservation of the heritage has also led to passion in cataloguing the existing remains of the historic defensive buildings, which are currently in very poor conditions of preservation such as archaeological findings, fortifications, towers, defensive residential buildings, the remains of shipwrecks, and wells (Bianchi and Tonner 2013, Brandi 2005, Mazzetto 2006, Carbonara 2012, Feilden 2003, Muñoz Viñas 2004, Jokilehto 1999). The aim is to promote and enhance the Qatari architectural heritage, to safeguard and transfer them over to future generations.

3 RESTORATION PROJECTS OF DEFENSIVE STRUCTURES

The Qatari fortified structures were usually located close to the ancient towns and villages. In fact, in the past, the main reason for building fortified structures, such as forts or watchtowers, was to protect the groundwater wells of the villages because of their importance for the survival of the local population. They were also built to protect the people from incursions and attacks.

3.1 The Al Zubarah Fort

The Al Zubarah Fort is located on the north of Qatar, and it was built in 1938 as a defensive military structure, acting initially as a coastguard station. It is part of a complicated defensive system composed of towers for surveillance situated throughout the territory, along the main roads to defend the inhabited centers.

The squared plan of The Al Zubarah Fort (Figure 1) is regular, composed of three cylindrical and one squared towers at the corners with high defensive walls constructed with local stones. The tops of the towers have machicolations, and soldiers used the inner rooms.

Figure 1. Al Zubarah Fort, the restoration works recently completed (Source: Author).

The restoration works were completed in 2015 by the Qatari governmental institutions and included the reconstruction of the roofs, which consist of wooden poles protected by bitumen and a layer of sun-dried bricks. The plasters of the ceiling on the rooms, deteriorated because of erosion and the critical weather conditions, were also restored. The defensive perimeter walls, built from coral stone and limestone were restored by complex structural interventions to solve the damages caused by the weight over the foundations. The internal partitions were consolidated, by using cement mixture to enhance the structural stability. Many missing elements were refurbished including portions of the wooden gargoyles.

In the courtyard is located an ancient well, initially used for the drinking water supply, that was restored during the last intervention.

The fort is currently reused as a museum, where it exhibits the recently discovered archaeological findings from the Al Zubarah site. The place is including a new visitor route for visitors to discover the history of the fort and its well.

3.2 The Barzan Towers

The Barzan Towers are located along the Doha Expressway and were constructed at the beginning of the 20th century to protect the ancient wells. The towers were part of the southern side of the Doha's defensive historical system, against the Ottoman army.

Figure 2. The Barzan Towers in Doha. The restoration works recently completed (Source: Author).

In 1910 the private owner of the towers carried out some reconstruction and restoration works, later on, in 2003, the towers were consolidated, and then in 2015, the last restoration works were completed by the Qatari local governmental institutions. The two towers have a simple rectangular plan; the defensive perimeter walls were built with coral and limestone masonry blocks bonded with mud. The corner towers were originally constructed using traditional materials such as coral and limestone blocks, mud gypsum and plasters, decorated with wooden rainwater gargoyles ("marazim") and wooden lintels ("danchal") usually protected with a bituminous layer ("basgijl")

The completed restoration works included a structural consolidation of the towers' foundation, which had failed because of the weight. The structural stability of one tower was strengthened at the corner by increasing the wall thickness. The deteriorated plasters were reinstated; the external and internal architectural elements; doors and windows, the stairs, and internal walls were all refurbished, and the building services were fully upgraded (Figure 2).

The defensive function of the towers was transformed into a museum that is open to visitors although there isn't yet a thematic visitor route or any specific service for tourists.

3.3 The Al Thagab and Al Rakayat Forts

The Al Thagab and Al Rakayat Forts are located in the northern side of Qatar. They are typical desert forts constructed in the 19th century to defend from incursions the wells that have been in use from 1911 onwards.

Al Thagab and Al Rakayat are both rectangular in plan. The central courtyards were used by the people to store goods in case of attacks, and in both the forts there are the traditional deep wells located in the corner of the perimeter defensive walls.

The analysis of the two forts is similar since they both were structurally consolidated and reconstructed at the end of the 1990s under the direction of the local governmental institutions. The methods for construction of the walls were, in both the cases, done by layers of coral stones and limestone blocks bonded with mud mortar and gypsum plaster, reaching a great thickness to provide both protection and thermal insulation.

Nowadays still the perimeter walls of both the forts consist of alternated layers of coral rock and limestone bocks, having both regular and irregular shapes. Wooden beams construct the structures of the roofs, currently in a good state of conservation in Al Taghab fort (Figure 3) while in Al Rakayat fort they are damaged by deep cracks which generated the failure of extended portions of the ceiling as a consequence of the building's abandonment in the last decades.

Figure 3. Al Thagab Fort, the external walls are constructed with layers of alternated coral rock and limestone blocks having a regular shape (Source: Author).

In Al Thagab fort, during the last restoration works, the upper level of the external wall was reconstructed using regular blocks of limestone, which are easy to identify if compared with the original lower part, more irregular and fragmented. In both the forts the interventions included newly reconstructions of the perimeter walls due to their structural instability, such walls were left visible in al Taghab fort while being covered with the final plaster in Al Rekyak fort. The cracks still distinguishable in the external walls are probably the result of structural settlement as a consequence of the consolidation and reconstruction works. In both the forts the oldest, an original layer of the wall is located at the bottom of the structures and it is composed of alternated coral rock and limestone fragments consolidated with composite mixed mortar mixture. This layer was consolidated during the last intervention using mud and cement mortar.

In both the forts, the extensive structural reconstructions have partly altered the original historical configurations. Neither the plans nor the elevations appear to have preserved their original design, and after the reconstruction works carried at the beginning of the 19th century, the layout of the original plan have been modified.

Both the forts are private and although they are open and accessible for visits they do not offer any special touristic tour or particular service for the visitors.

4 RESULTS

Analysis of the completed restoration projects to the Qatari defensive structures show the methods and techniques used during the interventions, permitting a comparison between different approaches more or less preservationist. In fact, all the defensive structures that are currently used as museums were more preserved, with the intention of maintaining the existing materials as found, without alteration to exhibit publically.

Because of the typical use of traditional materials (mud, sundried bricks, coral rock limestone blocks, wooden poles) the analyzed causes of deterioration were always similar and mainly due to the severe weather conditions or the change of use of the buildings, which have caused abandonment and lack of maintenance generating collapses in foundations, capillary rising damp in the masonry, severe deterioration if the plasters.

The different types of restoration works carried out can mainly be divided into two categories, such as an artificial approach for substantial structural interventions and a natural one in case of conservative interventions finalized to more preservationist results.

The different restoration approaches were also due to the various government authorities responsible for the different interventions. It was also found that the methods adopted by each governmental institution have often been isolated, and there is a need to improve the coordination and management of the Qatari heritage safeguard.

5 CONCLUSION

In Doha, a new emergent global city, despite its fast urbanization and modernization, there is a growing need to safeguard the historical heritage.

This need to preserve the cultural traditions of Doha is strongly related to the search for a new architectural identity that could emphasize the value of diversity against the conventionality of the global economic interests.

The research presents some recent restoration projects carried out for the ancient fortified Qatari structures, showing how these adaptive reuse interventions can provide one valid alternative for the growth of the city, respecting and preserving the existing architectural culture and traditions.

However, there is still a lack of open dialogue between the involved institutions and the different administrations. The absence of a general coordinating plan for all the works may be leading to future problems in safeguarding and managing the heritage safeguard.

References

Bianchi, A., and Tonner, T., *Data Standards, Documentation and Responses to Cultural Heritage Management in Qatar,* 18th International Conference on Cultural Heritage and New Technologies CHNT 18, Vienna, 2013.

Brandi, C., *Il Restauro: Teoria e Pratica 1939-1986*, Rome, Editori Riuniti, Italy 2005.

Carbonara, G., An Italian Contribution to Architectural Restoration, *Frontiers of Architectural Research*, 1, 2-9, 2012.

Carter, R., *Sea of Pearls. Seven Thousand Years of the Industry that Shaped the Gulf.* London: Arabian Publishing, 2012.
Feilden, B., *Conservation of Historic Buildings*, United Kingdom: Rutledge. 2003.
Jaidah, I. M., and Bourennane, M., *The History of Qatar Architecture (1800-1950),* Skira, Italy, 2009.
Jokilehto, J., *A History of Architectural Conservation*, Elsevier: Butterworth-Heinemann, 1999.
Mazzetto S., and Petruccioli A., Methods and Techniques Used in Significant Restoration Projects in Qatar, *Studies in Conservation*, Routledge, 2017.
Mazzetto, S., *Beni Militari Come Attori Protagonisti, in Cavallino–Treporti*, Progetti per un idea di Parco P. Grandinetti, S. Mazzetto, V. Skabar (eds.), Il Poligrafo, Venice, Italy, 50-51, 2006.
Muñoz Viñas, S., *Contemporary Theory of Conservation*, Elsevier Butterworth-Heinemann, 2004.
Salama, A. M., and Wiedman F., *Demystifying Doha: on Architecture and Urbanism in an Emerging City,* Farnham, England: Ashgate, 43-45, 2013.

Streamlining Information Transfer between Construction and Structural Engineering
Edited by Shiau, J., Vimonsatit, V., Yazdani, S., and Singh, A.

ISBN: 978-0-9960437-7-9

BUILDING INFORMATION MODELING (BIM) ADOPTION AND IMPLEMENTATION: INTERACTION BETWEEN BIM USERS AND NON-BIM USERS

NGOC QUYET LE, MICHAEL ER, and SHANKAR SANKARAN

School of Built Environment, Faculty of Design Architecture and Building, University of Technology Sydney, NSW, Australia

The research introduces the combination of Diffusion of Innovation Theory and Activity Theory to investigate the process of adoption and implementation of BIM in the construction industry in Vietnam. Data was collected from three large main contractors as they are considered as innovation leading organizations. Qualitative research was employed using semi-structured interviews to analyze respondents' perspectives of their daily BIM activities. Main informants include senior managers, BIM team members and site staff. Key findings were the factors affecting the decision of BIM adoption of senior managers (i.e., non-BIM users), the factors affecting BIM implementation of site staff (i.e., BIM users), and the contradictions emerging when senior managers mandate to use BIM, and employ a BIM team as the change agent for instruction and collaboration on the construction site.

Keywords: Diffusion of innovation theory, Activity theory, Senior managers, Site staff, BIM team.

1 INTRODUCTION

The construction industry has a historically low increase of productivity when compared with manufacturing industries (Sveikauskas *et al.* 2016). BIM has been introduced as a way of addressing this problem; but even while the percentages of BIM adopters are reportedly increasing around the world (McGraw-Hill 2014), and particularly in Vietnam (Le *et al.* 2017), the level of BIM implementation is quite low. A key reason is not the failure of introducing innovation but the failure of adopting its implementation, which means that the failure to gain targeted employees' implementing the innovation by advancing their skills, using a consistent approach, and commitment to adopt the innovation (Klein and Knight 2005).

BIM research, in general, has used the term "adoption" and "implementation" interchangeably. Klein and Knight (2005) however argued that "Innovation adoption is the decision to use an innovation. Innovation implementation, in contrast, is the transition period during which [individuals] ideally become increasingly skillful, consistent, and committed in their use of an innovation." Further, while the factors affecting BIM adoption and implementation have been well addressed at individual level, such as BIM professionals; investigation of the factors emerging from the interplay between BIM users (e.g., site staff, sub-contractors, and designers) and non-BIM users (e.g., senior managers, clients, and the government agents) within projects using BIM is lacking.

This paper aims to address that gap by proposing a theoretical framework combining Diffusion of Innovation Theory (DOIT) and Activity Theory (AT) to investigate the process of adoption and implementation in a holistic manner. By doing this, the contradictions that emerge from the interactions between BIM users and non-BIM users in the Vietnamese construction industry can also be explored.

2 THEORETICAL FRAMEWORK

2.1 Diffusion of Innovation Theory (DOIT)

DOIT (Rogers 2003) seeks to explain how innovations are accepted and adopted by social groups. DOIT identified key variables that significantly affect the rate of adoption of an innovation, namely: innovation characteristics, characteristics of decision making unit (i.e., target of innovation adoption), prior knowledge, and environmental characteristics.

Innovation characteristics include five elements which are relative advantage (i.e., the degree to which the innovation is perceived to be superior to current practice), compatibility (i.e., the degree to which the innovation is consistent with the existing values, past experiences and needs of potential adopters), complexity (i.e., the degree to which an innovation is perceived as difficult to understand and use), trialability (i.e., the degree to which an innovation can be tested before permanent adoption), and observability (i.e., the degree to which the results of an innovation are visible to others).

For characteristics of decision making unit, the socio-economic status has a great influence on the possibility of adoption. For example, people having high position in an organization (e.g. senior managers) would likely influence the adoption of followers such as the employees (Wejnert 2002). Further, the behavior of using mass media and internet, and being members of social groups could speed up the adoption rate. Other personal variables, such as formal education and innovativeness are indicators for early adopters.

Another factor affecting the probability of adoption are prior conditions existing in the organization before the innovation is introduced. Rogers (2003) noted that the previous practice provides a basis of familiar by which an innovation can be interpreted, thus decreasing its uncertainty. In addition, an innovation can be perceived as compatible if it meets a felt need and as incompatible if it does not address a need or a problem of a potential adopter. According to Kee (2017), an innovation should be in alignment with the cultural norms of an organization in order to successfully diffuse.

Social environment is also a critical factor affecting the awareness of the need of innovation adoption. The main feature of social environment is that it is not under the direct control of the organization but has a positive effect on human attitude towards innovation adoption (Buć and Divjak 2016). For example, a government's mandate to adopt innovation can be perceived as a social influence. Furthermore, the innovation has to be accepted by the society (e.g., meeting cultural values) before the actual innovation diffusion can take place. As for instance, innovation diffusion does spread more rapidly in "individualistic cultures" because of the social status rewards associated with innovation in that culture (Tolba and Mourad 2011).

Rogers's DOIT is one of the most used theories for studying adoption of information technologies (IT), and understanding how IT innovations spread within and between communities (Lyytinen and Damsgaard 2001). However, this theory has been criticized to be simplified to focus solely on a new product or innovation, disregarding the complex societal, cultural, economic and other factors that determine how the product is adopted into society (Al-Mamary *et al.* 2016). Particularly, DOIT's scholars are found to often collaborate with manufacturers, the government agents, or senior managers to enable the innovation diffusion within a community (e.g., construction industry) by business strategies or policies (Chile 2017). These approaches are focusing on "top-down" diffusion of innovations, and do not guarantee long-term success due to neglecting the actual innovation practices of lower positions such as site engineers (Ayodele 2012). The limitation of DOIT could be compensated by supplementing the investigation of innovation using another theory such as AT. According to Engeström (1987), AT presents a holistic and ecological perspective on human activity, providing the means of studying human actions and interactions with tools (e.g., technologies) within a historical, cultural and environmental context.

2.2 Activity Theory (AT)

This study employed Engeström's AT (1987) as a lens to interpret the activity systems of adopting and implementing BIM in main contractors. Three key elements of AT are Subjects (actors engaged in the activity), Tools (instruments used in the activity) and Objects (the targets of the activity). The theory proposes that the work activity is mediated by previous perceptions and behaviors (a historical cultural background of actors) and motivated by objects that take the form of tools as a medium of action in order to obtain expected Outcomes.

However, outcomes of an activity system are not always expected results, but possibly unexpected results that were transformed from the contradictions emerging when the AT system's elements interact with each other (Plakitsi 2013). For example, a contradiction such as a breakdown in the activity, where a tool is used inappropriately or in an unanticipated manner by the subjects. Such contradictions, however, should be seen as a source of change and development rather than problems or conflicts (Engeström 2001). The identification of contradictions in an activity system could help actors to focus their efforts on the root causes of tensions, to make proper decisions and take actions to adopt change.

Other three elements of AT are related to social factors – namely rules (cultural norms and regulation governing the performance of an activity), community (environment or social context in which the activity is being carried out), and division of labor (hierarchical structure of activity – roles and responsibilities of actors in the activity system).

2.3 A Theoretical Framework Combining DOIT and AT

Figure 1 shows how DOIT and AT can be organized in a combined theoretical framework. The framework distinguishes the activity of making adoption decision of non-BIM users, and the activity of implementing an innovation of BIM users. These two activities are integrated but not assimilated. Also, it is argued that BIM adoption and implementation are occurring in mandatory settings (Kumar *et al.* 2017), and Vietnam is not an exception (Le *et al.* 2017).

This framework adopts the basic elements of AT which are: Subject – Tools – Object – Outcomes, and adds main features of DOIT on the elements of "Subject" and "Tools". For example, technological characteristics, personal characteristics, and prior knowledge. Some key elements of AT are reorganized to match the condition of the mandate. Non-BIM users (usually senior managers) are affected by environmental factors whereas BIM users (usually employees such as site engineers) are under the influence of organizational factors. Although the employees have no opportunities to make adoption decisions at the initial stage, they could re-evaluate "what an innovation means to them" during the implementation stage, and after that determining their behaviors/responses at post-adoption stage.

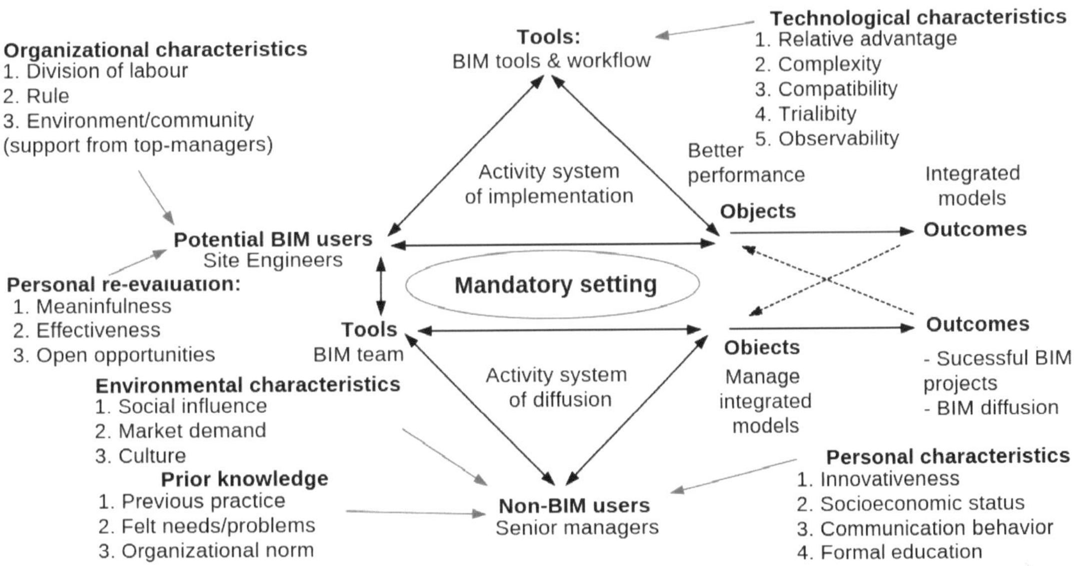

Figure 1. The combined framework of DOIT and AT.

The outcome of each activity system can impact the object of the other system and vice versa. For example, if the number of BIM adopters is accelerating, the site engineers would likely increase their awareness and intention to adopt BIM because they see the new job opportunities as well as competitive pressures on the labor market. On the other hand, if the site engineers create useful integrated models that can reduce issues such as costly re-works, and increase quality of building, then potentially senior managers might enhance the support and commitment.

3 APPLYING THE COMBINED FRAMEWORK FOR THE CONSTRUCTION STAGE

Due to the limited space, this paper only investigates the application of the combined framework of DOIT and AT during the construction stage, focusing on the factors affecting senior managers' decisions of BIM adoption, and information transfer among BIM team and site engineers.

3.1 Data Collection Method

The data used in this study was collected through semi- structured interviews of three main contractors in Vietnam with key informants are senior managers, BIM team, and site engineers. To maintain confidentiality, names of companies and respondents were put into codes (Table 1).

Table 1. Respondents' profile.

Contractors	Senior managers	BIM members	Site Engineers
C1	R1C1 (10+/0)	R3C1 (5+/3+)	R5C1 (15+/0)
	R2C1 (10+/0)	R4C1 (7+/3+)	R6C1 (7+/0)
C2	R1C2 (30+/0)	R3C2 (7+/5+)	R5C2 (20+/0)
	R2C2 (8+/5+)	R4C2 (3+/3+)	R6C2 (5+/0)
C3	R1C3 (20+/0)	R3C3 (20+/5+)	R5C3 (5+/0)
	R2C3 (10+/5+)	R4C3 (3+/2+)	R6C3 (10+/0)
Total	6 respondents	6 respondents	6 respondents

Code: Respondent (R)- Contractor (C)- (Industrial/BIM professional experience - years)

3.2 Findings and Discussions

3.2.1 *Factors affecting decision of BIM adoption of senior managers*

Through the interviews, the factor of "felt need" to innovation is more significant than environmental characteristics or personal characteristics. Despite low government, client, and competitive pressures on BIM adoption, main contractors are surpassing design professionals in term of BIM investment and application. This is because Vietnamese construction market has slowly shifted from a fully documented project to Design and Build (DB) based project delivery. As senior managers (R1C1, R2C2, and R3C3) said "The ability to implement technical solutions designed to achieve efficiency and aesthetics as well as reasonable costs were seen as the reason for the clients to contract us to design and build their entire projects." Therefore, large main contractors have been developing their in-house BIM experts and infrastructures for internal document management and quality control of project lifecycle. Their aims are to build trust with clients, and prepare for long-term plan of DB market in future.

3.2.2 *Factors affecting BIM implementation of site people*

The factors of "relative advantage", and "complexity" appear to be more influential in the level of BIM implementation within people based on site. Firstly, the majority of site personnel (e.g., R5C1, R6C1, R5C2, R6C2, R6C3) admitted that, "contractors only implement 3D BIM at the pre-construction stage for clash detection, and visualization purposes." This is due to the unavailability of information content developed for use at the actual construction stage such as 4D BIM (time management), and 5D BIM (cost management). While 4D BIM requires the constant collaboration of multiple participants such as sub-contractors, and designers to update the integrated model regarding change orders, specifications, and shop-drawings; 5D BIM obliges all parties to be transparent in the project budget. In the case study BIM was not a mandatory component of the contract which equated inconsistent commitment, and use of BIM through the construction project life-cycle; 4D BIM cannot be fully utilized. Additionally, the phenomenon of "lobbying for project" appears culturally as a part of business activity (e.g., winning tenders) in Vietnamese construction industry, thus, the interest groups such as main contractors and owners may be unwilling to share cost information with 5D BIM due to legal concerns.

Secondly, the use of BIM onsite is not flexible and speedy when addressing the urgent tasks or unexpected issues. Slow loading and editing of Revit files frustrated site staff. Further, the designs were often late in being handed over from the BIM team to the construction site, and in some cases the work at the building site had to start without ready- made designs to meet the programmed deadline. In other words, the developed models were not able to capture the "as-built" conditions, roadblocking extended usage of the 3D model to the construction stage.

3.2.3 *Contradictions emerging within the interaction between senior managers and site staff*

To increase the adoption rate of BIM, senior managers sent a qualified BIM team to site for onsite training and collaboration. This BIM team can be seen as a tool, serving as change agents in adoption activity (see Figure 1). However, the interactions between Senior managers – BIM team – Site staff created a contradiction that impeded the adoptability of BIM, particularly, the contradiction regarding "division of labor" issue. The BIM team had a responsibility to help site people with information management and visualization, and simultaneously report site activities related BIM uses to senior managers. The role of BIM team was negatively perceived as "double agent" by site people. Most BIM team members (e.g., R3C1, R4C1, R4C2, R3C3, R4C3) acknowledged that "we feel isolated as site people watch out us and are reluctant to support and share information for model update." As the result, senior managers failed to build "trust" with site staff despite supporting them with BIM experts.

4 CONCLUSIONS AND FUTURE WORK

This research proposed a framework combining DOIT and AT with the potential of holistically analyzing the process of BIM adoption and implementation across multiple parties during project life-cycle. Using this framework, not only the factors affecting BIM adoption and implementation of BIM users (e.g., relative advantage, and complexity) and non-BIM users (e.g., felt need) were individually identified; but the factors emerging from the interaction between them (e.g., contradictions in division of labor) were also explored. The future work will extend the range of interviews, with additional participants of BIM users (e.g., sub-contractors, and designers), and non-BIM users (e.g., clients, and the governments agents) to provide a complete view of their interactions in BIM based projects from the design stage to hand-over stage.

References

Al-Mamary, Y. H., Al-nashmi, M., Hassan, Y. A. G., and Shamsuddin, A., A Critical Review of Models and Theories in Field of Individual Acceptance of Technology, *International Journal of Hybrid Information Technology,* 9(6), 143–158, 2016.

Ayodele, A., A Critical Analysis of Diffusion of Innovation Theory, 2012. Retrieved from https://odinakadotnet.wordpress.com/2012/09/22/a-critical-analysis-of-diffusion-of-innovation-theory/ on April, 2018.

Buć, S., and Divjak, B., *Environmental Factors in the Diffusion of Innovation Model: Diffusion of e-Learning in a Higher Education*, Central European Conference on Information and Intelligent Systems, Varazdin, Croatia, September 21-23, 2016.

Chile, P., Some of the Disadvantages of Using the Diffusion Theory, 2017. Retrieved from https://bizfluent.com/info-8633492-disadvantages-using-diffusion-theory.html on April, 2018.

Engeström, Y., Expansive Learning at Work: Toward an Activity Theoretical Reconceptualization, *Journal of Education and Work*, 14(1), 133–156, 2001.

Engeström, Y., *Learning by Expanding,* Orienta-Konsultit Oy, Helsinki, 1987.

Kee, K. F., Adoption and Diffusion, In *The International Encyclopedia of Organizational Communication*, Scott, C., and Lewis, L. (eds.), 1-14, John Wiley & Sons, Inc., USA, 2017.

Klein, K.J. and Knight, A.P., Innovation Implementation: Overcoming the Challenge, *Current Directions in Psychological Science,* 14(5), 243-246, 2005.

Kumar, A., Gupta, M. P., and Ilavarasan, P. V, *Advances in Smart Cities: Smarter People, Governance, and Solutions*, CRC Press, USA, 2017.

Le, N. Q., Er, M., and Sankaran, S., *The Implementation of Building Information Modelling (BIM) in Construction Industry: Case Studies in Vietnam,* 6th International Conference on Civil Engineering (ICCEN 2017), Queensland, Australia, November 20-22, 2017.

Lyytinen, K. and Damsgaard, J., *What's Wrong with the Diffusion of Innovation Theory?* 4th Working Conference on Diffusing Software Product and Process Innovations, Banff, Canada, April, 2001.

McGraw-Hill, *The Business Value of BIM for Construction in Global Markets: How Contractors Around the World are Driving Innovation with Building Information Modeling, in Smart Market Report*, McGraw Hill Construction, USA, 2014.

Plakitsi, K., Activity Theory in Formal and Informal Science Education. In *Activity Theory in Formal and Informal Science Education: Research Dialogs,* Plakitsi, K. (eds.), 1–15, Sense Publishers, 2013.

Rogers, E. M., *Diffusion of Innovations* (5th ed.), The Free Press, New York, 2003.

Sveikauskas, L., Rowe, S., Mildenberger, J., Price, J. and Young, A., Productivity Growth in Construction, *Journal of Construction Engineering and Management*, 142(10), 04016045, 2016.

Tolba, A. H., and Mourad, M., Individual and Cultural Factors Affecting Diffusion of Innovation, *Journal of International Business & Cultural Studies*, 5, 1–16, 2011.

Wejnert, B., Integrating Models of Diffusion of Innovations: A Conceptual Framework, *Annual Review of Sociology*, 28(1), 297–326, 2002.

Streamlining Information Transfer between Construction and Structural Engineering
Edited by Shiau, J., Vimonsatit, V., Yazdani, S., and Singh, A.

ISBN: 978-0-9960437-7-9

MAPPING THE LITHOCERAMIC CLADDINGS CONSERVATION STATUS IN A BIM ENVIRONMENT

VALENTINA VILLA, MARIKA MANGOSIO, and EMILIA GARDA

Dept of Structural, Geotechnical and Building Eng., Politecnico di Torino, Torino, Italy

Italian Modern Movement architecture is characterized above all by the research and use of innovative, experimental and autarkic materials. Lithoceramics is one of the most interesting and widespread cladding materials and represents the Italian reinterpretation of klinker, a very popular ceramic material in Germany and in the Netherlands. Despite the excellent technical performances of the ceramic cladding, increasingly frequent detachment episodes make maintenance interventions necessary on this building stock. Starting from a brief illustration of the characteristics of lithoceramics and of the most significant applications, the paper intends to present a BIM-based digitization methodology for mapping the conservation status of this façade cladding technology, through exemplification on a significant case study. The BIM model contains all the information necessary for the mapping of the state of conservation, for the recovery and maintenance activities of the ceramic elements. The proposed methodology allows a faster and more efficient visualization of the present faults and proposes a targeted intervention system, in relation to the detected fault. This approach contributes to protecting the historical connotation of this building heritage, favoring restoration and cleaning, reducing the time and, consequently, costs of maintenance.

Keywords: Built heritage, Building envelope, Klinker, Modern movement architecture, Italy.

1 INTRODUCTION

Italian Modern Movement architecture is characterized, on the one hand, by the original use of traditional materials, but above all by the research and use of innovative, experimental and autarkic materials (Cupelloni 2017). The lithoceramics is one of the most interesting cladding materials and represents the Italian reinterpretation of klinker, a very popular ceramic material in Germany and in the Netherlands (Minnucci 1933). It is a very compact ceramic product with a high percentage of glass phase, obtained by wire drawing and subjected to a firing cycle at a temperature of 1200-1280°C. The versatility of shape, colors, and finishing of the cladding elements makes the building external surfaces vibrating, colorful and iridescent. The use of this 'modern skin' covers a wide time period, from the '30s to the end of the '60s and is extremely widespread in Italy (Mangosio 2006).

The excellent technical performance of this cladding solution has generally ensured a very limited maintenance activity over time. In recent years, frequent detachment episodes of individual façade elements or entire portions of cladding have, however, necessitated maintenance or recovery interventions in many buildings. The detachment is generally due to the failure of the connection between the cladding and the layer support. This failure can be caused by the loss of the mechanical characteristics of the mortar or by excessive tangential stresses between the cladding and the wall support (absence of open joint between facing elements, the absence of structural expansion joint).

The lack of critical awareness about the architectural and technological value of this material has often led to interventions of complete replacement of the coating, causing the irretrievable loss of a singular heritage of technical culture.

The study intends to provide designers and operators with a tool to selectively guide the intervention and to keep technical information and operational prescriptions useful for subsequent maintenance activities.

2 METHODOLOGY

The progressive digitization of the historical heritage is implementing an information management process that involves the collection and management of data through the BIM model. All conservative maintenance interventions, therefore, need to have an information model, which meets the specific needs of operators and scholars working on buildings, in particular, those dealt with in this research, i.e. buildings with lithoceramic cladding. For a correct setting of the BIM model, the proposed methodology identifies the first phase of identification of the most common pathologies refererred to this type of cladding.

A first subdivision can be made in relation to the types of fault (De Freitas 2013), that is detachment or lifting of surface elements or sections, surface degradation, detachment of fragments (Figure 1). As regards surface degradation, there are five types of faults: a) efflorescence, or superficial deposit of salts present in the mixing water of the tiles or the laying mortar; b) proliferation of micro-organisms and moulds on the surface due to stagnant water; c) chemical and physical aggression of the environment on the surface treatment of the tile; d) urban pollution deposit in relation to particular conformations of the façade and to its orientation; e) cracking of the surface treatment due to internal cohesion states. Conservative maintenance interventions may include two main interventions: a) replacement of damaged elements and consolidation of the layer support, b) cleaning.

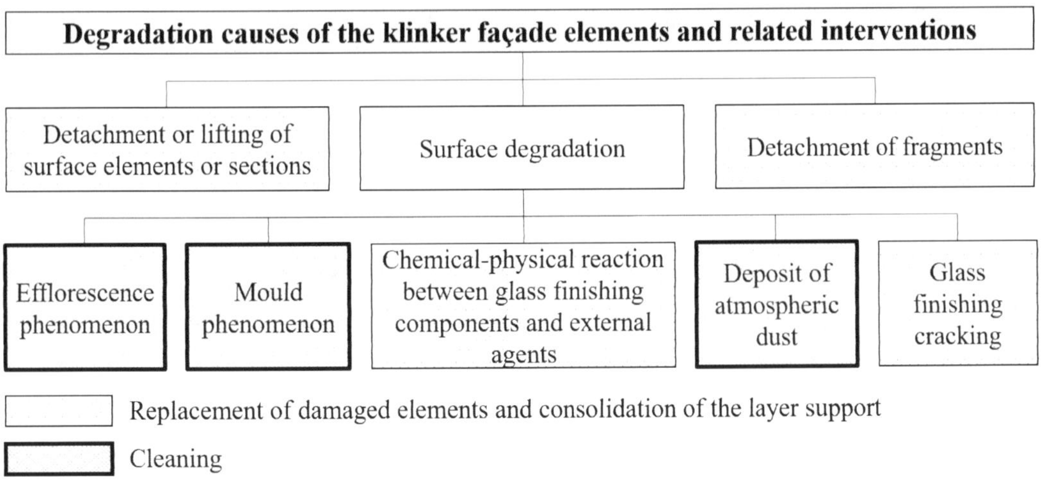

Figure 1. Degradation causes (Data processing by authors).

3 CASE STUDY

The case study is an apartment and office complex built in Turin between 1957 and 1959 for the real estate company ARVA by engineers Domenico Morelli and Felice Bardelli (Bagliani 1993). The ARVA House consists of two buildings separated by a courtyard-garden: the main building fronts onto Corso Marconi and is ten floors high, while the minor building on Via Sant'Anselmo - specific object of our study - is six floors high (Figure 2). The ground floor is intended for offices, while the upper floors have one flat per floor. A corner bow-window enlarges the living room space.

The secondary building is entirely covered in rectangular elements in ivory colored klinker (9.5 x 19.5 cm), supplied by Piccinelli Ceramics in Bergamo, which are arranged vertically and placed with open joint, so as the joint itself is able to absorb to any differential movement of the cladding due to the thermal expansion of the masonry support. The standard element has three posterior dovetail ridges for a better adherence to the bedding mortar.

In the original project the cladding equipment, the clamping device to the masonry and the resolution of every critical and singular technological crux are analyzed and illustrated up to the maximum detail scale. The care for the technological detail by the designers reveals the search for a lasting solution over time.

The critical points of the cladding, such as the edges, resolved continuously without an interposed joint, or the folds on the intrados of the cantilevered floor, defined with special angular elements, are approached with great care. The window outlines are obtained by subtraction from the modular structure of the cladding and the openings are defined with special pieces in L shape.

Figure 2. View of the ARVA House façades from the street (Ph. M. Mangosio).

4 RESULTS

Applying the proposed methodology to the case study presented in the previous paragraph, an in-depth research on the historical project documents was carried out, studying, in particular, the construction details and the technological solutions proposed by the designers (an extract is shown in Figure 3a).

In fact, the modularity of the material required a very accurate study of the singular points (angles, spouts, vaults, etc.) to then define the vertical trend of the more regular part of the façade (Figure 3b).

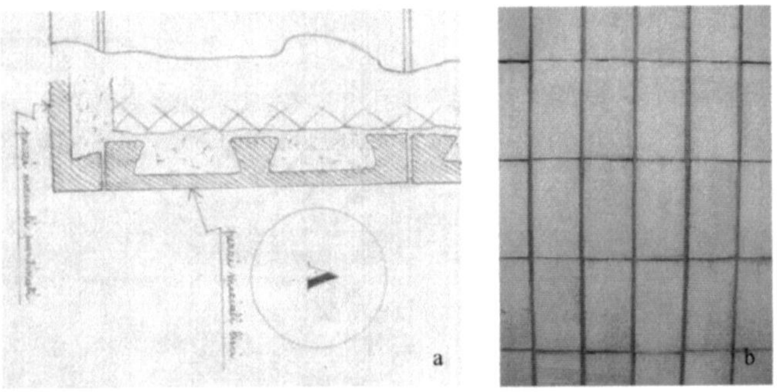

Figure 3. a) Details of the façade klinker element (Politecnico di Torino DIST, Laboratory of History and Cultural Heritage, MRL_23B17ARVA_83) – b) Detail of the regular part of façade (Ph. M. Mangosio).

Following the project indications, the family model of the "Klinker" façade element was created. In the geometric modeling phase, the dimensions were inserted as object parameters (Figure 4).

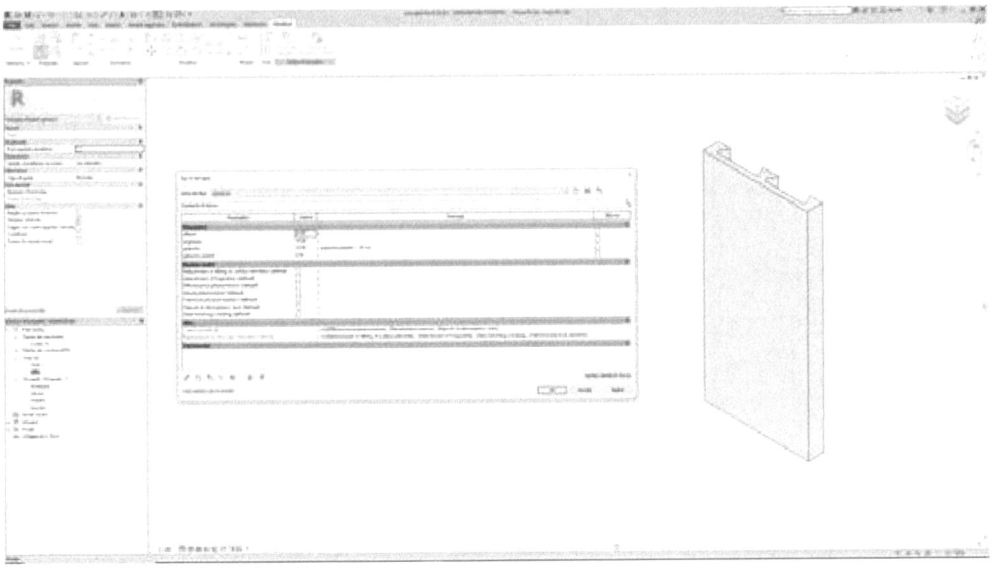

Figure 4. 3D view of the Klinker family model with parameters of the degradation state of the elements (Data processing by authors).

This operation makes the use of these elements more flexible even for projects that have a similar technology but with slightly different geometries. The BIM model was then associated with the information component, i.e. all the parameters that must be completed in order to correctly define the state of conservation.

Figure 5. D. Morelli, F. Bardelli, Façade drawing of the secondary building, 1957-1959 (Politecnico di Torino DIST, Laboratory of History and Cultural Heritage, MRL_23B17).

By visualizing the façade's projections, referring to the original drawing (Figure 5) and associating a color to each fault, the areas on which work must be immediately identified and, in relation to the degradation causes, the intervention mode is automatically defined (Figure 6).

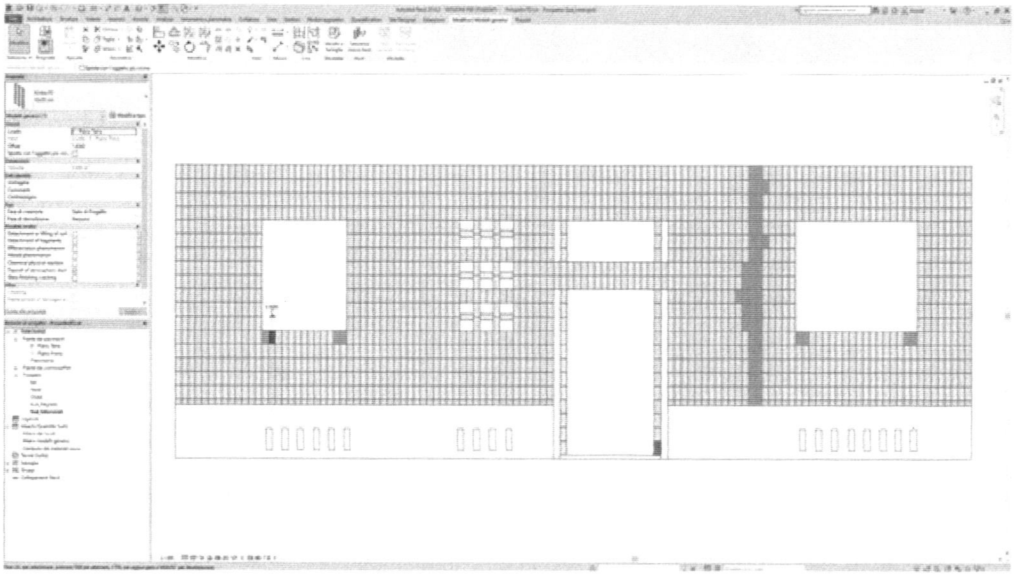

Figure 6. View of the elevation of the BIM model in which it is highlighted the state of maintenance of the façade cladding and the suggested intervention mode. In green, cleaning; in red, replacement of damaged elements and consolidation of the layer support (Data processing by authors).

5 CONCLUSIONS

In Italy, the litoceramic cladding not only connotes numerous architectures of the Modern Movement but also numerous buildings of lesser authors, of recognized design value, created between the Thirties and the Sixties in many Italian cities. This architectural heritage is not protected by Italian legislation but must be preserved as well as the great historical architecture.

The recent regulations which favor the improvement of energy performance, unfortunately, do not take into account the cultural, historical and technological value of these buildings, with the result that often the façades are covered with external wall insulation systems or, in the worst case, demolished and replaced with ventilated façades or with anonymous coatings.

The proposed methodology contributes to the digitalization of the historical heritage, assisting, with innovative technological tools, processes, and mapping of the degradation already in use. The BIM-based data collection allows storing a large amount of data, linking information on the state of conservation and degradation to each coded element, on how to restore/replace elements and on the cleaning products to be used.

This type of approach allows to keep information about the building history, the cladding building technique and the characteristics of the material, to intervene in a coherent and conscious way, especially in the case of the replacement of the elements, to evaluate and perform only the processings actually necessary, to contain maintenance costs and finally to keep memory of the intervention for the benefit of future maintenance.

In the next few years it will be necessary to intervene substantially on this heritage for reasons of obsolescence: to this end, as a next step of the research, the system here described could be linked to the databases currently used by companies for the routine maintenance of buildings.

References

Bagliani, D. (ed.), *Domenico Morelli Ingegnere Architetto*, Toso, Torino, 1993.

Cupelloni, L. (ed.), *Materiali del Moderno. Campo, Temi e Modi Del Progetto di Riqualificazione,* Gangemi, Roma, 2017.

De Freitas, V.P. (ed.), A State-of-the-Art Report on Building Pathology. CIB – W086 Building Pathology, Publication n. 393, CIB, Rotterdam, 2013. Retrieved from: http://site.cibworld.nl/dl/publications/pub_393.pdf

Mangosio, M., *La Nascita e l'evoluzione Dell'industria Italiana Della Litoceramica e i Suoi Riflessi Nella Cultura Del Costruire, in Storia dell'Ingegneria.* Atti del 1° Convegno Nazionale, Buccaro, A., Fabbricatore, G. and Papa L.M. (eds.), 575-578, Cuzzolin, Napoli, 2006.

Minnucci, G., *La Litoceramica 'Italklinker'*, Architettura, 4, 265-268, 1933.

Streamlining Information Transfer between Construction and Structural Engineering
Edited by Shiau, J., Vimonsatit, V., Yazdani, S., and Singh, A.

ISBN: 978-0-9960437-7-9

PRESTRESSED CONCRETE TECHNOLOGY IN UNDERDEVELOPED COUNTRIES

SEVGI LOKCE[1] and ERSAN GOKSU[2]

[1]*Department of Architecture, Atılım University, Ankara, Turkey*
[2]*MEG Technical Consultants, Marly, Switzerland*

Prestressed Concrete Technology developed by the middle of the last century depending and the increase in material qualities steel and as well as concrete. Compared to the conventional reinforced concrete the P.C technology is not widely distributed and applied in all over the world. Knowledge, Materials, Equipment and so called 'Patents' are still under the monopoly of some Industrial Countries. Process to develop the P.C. technology in underdeveloped countries even in developed countries is a special issue and will be treated in this paper. Research and to develop 'Know-How' i.e. Procuration, Projecting, Consulting, Execution and the Monitoring as well as methodical approach of Marketing will be discussed with concrete examples and case studies. In due time the mistakes and failures which prevents the development of this High Building technology will be emphasized to have the attention of the investors and the concerning Engineers and Architects.

Keywords: Patents, Anchorages, High strength steel, Flowchart.

1 INTRODUCTION

Prestressed concrete systems can be defined as creating a compression zone of a structural member, in the region where tensional stresses are expected to appear, so that whole section under total load combination receives a stress level less than concrete compressive strength. Special attention has to be paid to avoid the stress increases in the section above allowable ones. The prestressed concrete is not the only instance of prestressing. In daily life prestressing is used to carry books before the development of prestressed concrete as seen in Figure 1.

A prestressed concrete structure is different from a conventional reinforced concrete structure due to the application of an initial load on the structure prior to its use. The initial load or 'prestress' is applied to enable the structure to counteract the tensional stresses arising during its service period. Generally, there are two types of prestressed application in the technology field. The first one is Pretensioning System, which is produced in factories, in prestressing bed and the second one is Posttensioning System which are mostly in-situ applications. In this paper only one of the two types Prestressed Concrete applications "Posttensioning System" will be discussed.

1.1 Advantages of Prestressed Concrete

Prestressed concrete systems have brought lots of advantages above all as a construction material, as well as in construction technology.

Figure 1. Man carrying books using prestressing.

From the construction material point of view, using prestressed concrete reduces the slab and beam thickness. Reduced slab thickness enables additional floors to be built for the same allowable building height. Prestressed concrete also allows crossing large spans. This is a reflection of freedom of design concept proposed by Le Corbusier, example shown in Figure 2. This feature improves freedom of design. The use of high strength steel reduces the mild reinforcement requirement. Therefore, cost and construction time of the projects can be reduced. As a construction technology prestressed concrete improves structural systems by producing monolithic structural elements, reducing dead loads and limiting displacements, prestressed concrete designs have also highly controllable construction sequences, which allows practical execution.

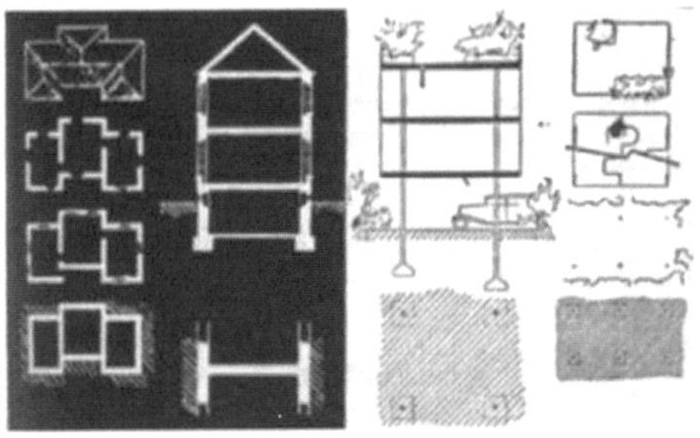

Figure 2. Freedom of design concept of Le Corbusier adopted from Robertson (1940).

2 PRESTRESSED CONCRETE TECHNOLOGY KNOW-HOW

The International Federation for Prestressing (FIP), a professional organization in Europe was established in 1952. Foundation of FIP is the beginning of the spreading of "Prestressed Concrete Technology" all over the World including developed as well as underdeveloped countries under license and know-how certified by FIP. Those companies, who have their "patents" certified by FIP, kept their know-how delicately.

It is apparent that Know-How is a result of a research and development operation. Not only the development of main parts of "Prestressed Concrete Technology" but also methodology in the structural calculations and the methods of application are the main issues in this stage.

3 PRESTRESSED CONCRETE TECHNOLOGY IN UNDERDEVELOPED COUNTRIES

Ersan Goksu who is the co-writer has 50 years of experience in prestressed concrete projects in both developed and underdeveloped countries. Based on his observations, main causes of limited prestressed concrete use in developing and less developed countries can be explained in the following topics;

3.1 Lack of Information

New technology often increases the volume of the economy since it creates opportunities for investments. However, not paying attention to research and development activities in concrete engineering results in limited usage of these technologies in developing countries (Goksu 2004). In general, utilization of prestressed concrete technology in the structures, brings us, the reduction in mild reinforcement, formwork system, saving and economic gain from construction time, totally about 30% of total investment cost.

Utility, defined in accordance with the definition of "freedom of design" especially in high building applications, with a span of 4.00 m to 6.00 m. is increased in recent years due to the quality of concrete up to 8.00 m. Although with prestressed concrete technology up to 18.00 m the beam-free flooring openings can be attained (Goksu 2005). The inexperience of the designer during the preliminary project, especially because of the "fear reinforcement" that he/she will use, and the deficiencies caused by his inability to know the calculation methods, face the investor in the form that the structure will be suddenly uneconomical. Since the new technology conveyed to the investor, the more expensive alternative investors often reject the new technology.

3.2 Lack of Promotion

Companies who own know-how do not care enough about the promotion in these underdeveloped countries as far as the spreading of the know-how, with desire to protect their monopoly.

Fear resulting from lack of information leads to attitudes and the introduction of negative arguments to support this negativity (Studer *et al.* 1999).

3.3 Conservatism of Know-How Owners

Both the principles, the advantages and benefits arising from projecting and manufacturing and applications of this technology, and in due course protection of technology "patent" system by the FIP, prevent the spreading of the information abroad. Therefore, companies with the know-how increase their profits, whereas designers and contractors in underdeveloped countries could not generate feasible options due to the patent limitations.

Case: Institutions who have the Know-how executing the projects abroad by carrying out their own projects in accordance with their own interests, and so-called know-how transfer does not come to question. Thus, the system so called "hard-ware", is launched according to their profitability rates (Goksu 2004).

In addition, the ratio of mild steel, which is required to be used in parallel with the posttensioning system, to the materials in question – prestressed concrete (PC) materials-, provides companies the possibility to sell more or more profitable amount of PC materials. Even very similar to the material amounts – mild steel and PC – for a structure in two countries, could result in very different costs because of big unit price differences.

In order to increase the cost of the PC materials in a flat slab of a building project, to propose the bonded posttensioning instead of unbonded posttensioning is another case of an engineering trick.

Similarly, instead of using the type of anchors to be used in the project, using a large number of smaller anchors is a separate manufacturer tactic.

3.4 Tender System

Unit price method in tender systems hinders the application of new technologies, which proposes more economical alternatives to the implementer of a project.

In other words, the tendency to escape from the new technologies, which bring enormous amount of economy in the investments will be rejected as a result of the profits in the cost of structure will be completely owned by the owner and will not provide the contractor with a profit. Corruption is also a parameter that supports this tender system, which prevents the development of new technologies. In such kind of tender systems, it is clear that the principle, much material, occurs in the form of high price and high corruption sum.

FIDIC conditions prevent this procedure, especially the silver book, design and construct bidding method face difficulties in such auctions. In this type of tender systems, the alternative project proposition chance will be closed to the contractors.

The lack of optimization studies and Technology Research almost eliminate the development of 'Prestressed Concrete technology' in underdeveloped countries. On the other hand, the main problem of underdeveloped countries is the lack of cost analysis and control. Cost optimization and searching for economical solutions are duties of technical staff and engineers (Goksu 2005).

4 FRAMEWORK FOR DEVELOPMENT OF PRESTRESSED CONCRETE TECHNOLOGY IN UNDERDEVELOPED COUNTRIES

Developing a post-tensioning system needs a team work between civil, metallurgical and mechanical engineers, technicians and draftsmen in order to be able to attain the success. An example multidisciplinary organization chart is presented in Figure 3.

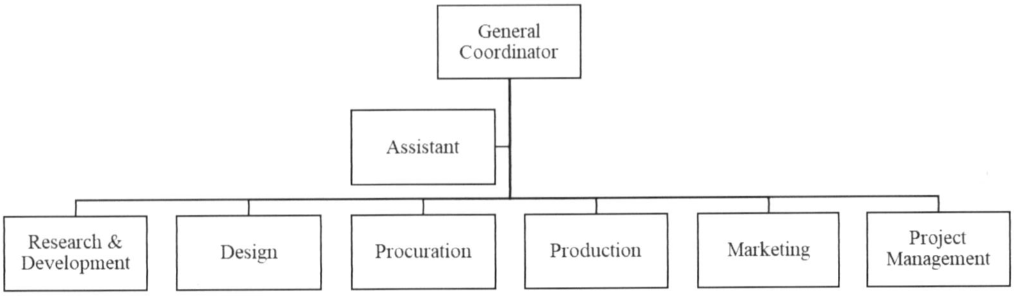

Figure 3. Example organization chart for successful multidisciplinary prestressed concrete technology development team.

4.1 Development of Prestressing Systems –Patents & Technology

All prestressing systems that are proposed for use in Switzerland shall be submitted in the following form to expedite approval of the system or systems. The following checklist includes the minimum required information necessary for approval by the EMPA - Switzerland (Eidgenössische Materialprüfungs und Forschungsanstalt /The Swiss Federal Laboratories for Materials Science and Technology) for a new or modified post-tensioning system:

1. Description

This part includes current product description literature of the system or systems being proposed. Prior listing of the system including specific details of projects, where it has been used. Complete records of tests run on the system and explanation of how seating loss is to be controlled and measured.

*2. **Hardware***

In the hardware section anchor head, bearing plate, wedges or nuts, and trumpet are described with detailed drawings. Mill certificates and quality control documents are shared.

*3. **Calculation***

Calculations for stresses behind bearing plate at service load after losses and stresses behind bearing plate at 95% specified ultimate tensile strength are shown in this section. Maximum bending stress in bearing plate of 95% specified ultimate tensile strength calculations are presented.

*4. **System***

Detailed drawings of the anchorage system, jacking system as well as duct and grouting details are shown in this section. Complete information on grouting procedures and equipment to be used are explained. Description of how system components are protected from physical damage and corrosion, description of tendon repair or replacement occur and description of how qualified technical assistance provided in the field for the contractor performing the work are detailed.

Therefore, development of the anchorages and their complementary parts, anchor head, anchor plate, spirals, trumpets, grouting bores and ducks – sheaths and static and dynamic analysis and design of those part and verification of their conformity of standards, norms, rules and regulations should be performed. All the devices i.e., all part of prestressing segments should be manufactured according to the corresponding standards (FIP) and fulfill the prescribed rules and regulations with conceivable certificates patents and quality testes necessary.

4.2 Presentation and Marketing

Presentation of this technology includes multiple tasks beginning from the preliminary design such as; consulting the other engineers, architects and design offices, evaluation of bill of quantity, preparation of bidding documents, comparison with the other alternatives and presenting the economy of the system, and convincing the client on this technology till to the execution plans and details. Presentations should also have basis on structural calculations, stability verifications, detail studies in the domain of the mission.

Marketing is another principal task to make system and the technology familiar with state institutions, private investors, architects, and especially engineers and the students. Common efforts with professional organizations to attract attention of the leaders through courses, seminars, colloquiums, and meetings are invaluable for this mission. Marketing also includes preparation of the technical documents, publications and all kind of advertisements is covered in his duty range.

Marketing situation in underdeveloped countries is completely different from the industrial countries as far as the state structures and private investments concerned:

- State Structures: Mostly projected in abroad by foreign engineering groups and approved by the local officials and executed without taking into account of the economy.
- Private Structures: Local engineers do not have the necessary experiences to attack the projects and to propose the alternative technologies and solutions. Prestressing

technology does not only need well qualified materials but it needs quality engineers and workmanship as well.

Marketing and project consultants of the firm have to take as well the corresponding responsibility with the partner engineering group and/or architects against the owner and laws. As a result, the local engineers are afraid to proposing alternative solutions to the owner because of the risks and the amount of projecting and planning tasks. On the other hand, they make the projects without the support of specialists and make designs in very safe side without taking any risks. In both ways solutions become uneconomical and as a result, prestressing know-how could not be cultivated.

5 CONCLUSION

Prestressed concrete technology can be very beneficial for underdeveloped countries by increasing structural quality and by providing economically optimized options. For successful integration of research and development into construction site, private and state organizations should pool their efforts. By promotion campaigns and supportive legislations know-how of prestressed concrete could be more widespread.

References

Goksu, E. and Jungo, A., Export von Schweizer Bautechnologie, *Freiburger Nachrichten,* Freiburg, Germany, October 2004.

Goksu, E. and Kibar, A., 'Ardgermeli Beton Teknolojisi - *Sheraton Kongre ve Kültür Merkezi'nde Hotel' Nurol Dünyası,* Ankara, Turkey, March 2005.

Robertson, H., ARCHITECTURE 1919-1939, *Journal of the Royal Society of Arts*, *88*(4557), 449-463, 1940.

Studer, J., Badoux, M., Bertogg, M., Goksu, E., Isler, P., Lestuzzi, P., and Zwicky, P., Erdbeben in der Westtürkei vom 17. August 1999, *Schweizer Ingenieur und Architekt*, (43), 1999.

Streamlining Information Transfer between Construction and Structural Engineering
Edited by Shiau, J., Vimonsatit, V., Yazdani, S., and Singh, A.

ISBN: 978-0-9960437-7-9

STRUCTURAL AND GEOTECHNICAL DESIGN OF A HIGH-RISE BUILDING USING A 3D BIM GENERATED MODEL

AHMED ROMAIH, SAMA TAHA, MINA NAGUIB, SALMA IBRAHIM, YOUSSEF HALABI, EZZELDIN SAYED-AHMED, SAFWAN KHEDR, and EZZAT FAHMY

Dept of Construction Engineering, The American University in Cairo, Cairo, Egypt

This paper presents the work performed by an undergraduate student group at the American University in Cairo as a first phase of their graduation project that mimics a state-of-the-art industrial experience in structural and geotechnical design of a high-rise building using building information model (BIM). A 3-D BIM was developed for the structure and used to generate an analytical model which is exported to a structural numerical analysis program, followed by a full design of the building elements and foundation. The considered building is 218 m high, in which the architectural design requires a successive 2-degrees twist per floor through the building height, resulting in a total twist of 80 degrees between the first and last floors. The architect also retracted the slabs every six stories from a group of columns, leaving 24 m-height laterally unsupported peripheral columns. As such, the vertical and horizontal building irregularities present challenges in the structural modeling and design and requires thorough analyses, particularly for seismic and wind considerations. Due to the high water table at the building site and the existence of a 12.5 underground basement, a special dewatering technique was proposed, along with the full tanking design consideration of the building basement.

Keywords: Building information modeling, Dewatering, Numerical modeling, Seismic analysis, Tall buildings.

1 PROJECT OVERVIEW

The tower is 218 m high in which the architectural design requires a successive 2-degrees twist per floor through the building height resulting in a total twist of 80 degrees between the first and last floors. Furthermore, the architect also retracted the slabs every 6 stories from a group of columns leaving 24 m-height laterally unsupported edge columns. As such, the vertical and horizontal building irregularities present significant challenges in the structural design and requires thorough structural analyses, particularly for seismic and wind considerations, as well as the construction sequence. Due to the high water table at the building site and the existence of a 12.5 underground basement, special dewatering technique was proposed along with the "full-tanking" design consideration of the building basement.

Building information modeling (BIM) is a digital representation of the structural composition and function. For the past decades, conventional Civil Engineering relied on 2D drawings only, such as elevations, plans, and sections, and each engineering discipline was working in its own field. Therefore, extensive coordination problems arose when the designs from different disciplines were integrated, which impacted the project's design time, cost and even quality. It is evident that BIM eliminates such dilemma and allows for a live-interdisciplinary coordination between the different design disciplines, therefore minimizing the impact on time, cost and quality.

In this work, BIM for the building was first built on Revit, then, an analytical model was developed. This analytical model was imported to ETABS for numerical analysis of the tower. Beside gravity loads, seismic and wind loads were considered in the structural analysis of the building according to ASCE/SEI 7-16 (2016) provisions. The tower was designed as a reinforced concrete framed structure with shear walls according to ACI 318M-14 (2014) provisions. The structural design was performed using ETABS as well as SAFE; the later was used to design the floor flat slabs of the tower. Geo5 was also used to design the retaining walls in the basement considering the soil properties and the groundwater table. Moreover, additional software such as STAAD and PCA columns were also employed to validate ETABS' outcomes. Finally, a 4D simulation model that shows the detailed schedule and construction process of the tower was performed on Navisworks.

After many design iterations, all structural members of the tower were designed and the students submitted complete structural drawings, a 3D BIM along with the 3D structural numerical model and full calculation sheets. The project was challenging and mimicking a typical industrial application, which serves to prepare the students for their post-graduate career in structural analysis and design.

2 STRUCTURAL MODELLING AND ANALYSIS

A reinforced concrete structural system was chosen and shown in Figure 1 is the generation and development of the model for the structure using Revit.

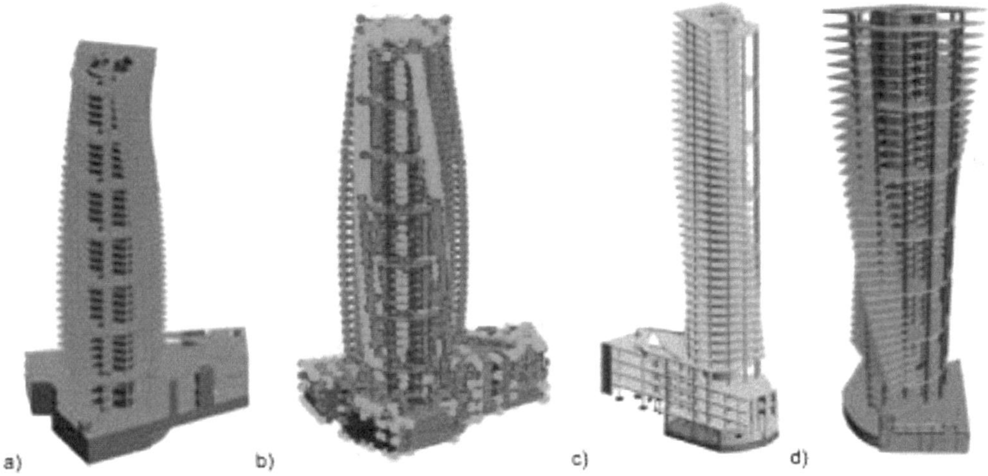

Figure 1. 3D model development. a) Initial BIM, b) Analytical model, c) ETABS model, d) Updated BIM.

The chosen structural system was composed of flat slabs supported by reinforced concrete columns and shear walls. Then, an analytical model was produced from the BIM: this analytical model needed significant adjustment to fit the numerical analysis requirements. This process by itself presents a challenge to undergraduate students and requires building knowledge in both modeling and analysis phases. The adjusted analytical model was exported to ETABS, where the building was analyzed under the effects of gravity and lateral loads. The analysis was performed according to ACI 318M-14 (2014) with all loads and load combinations satisfying the requirements of ASCE/SEI 7-16 (2016).

For seismic analysis, the building type, the vertical and horizontal irregularities, the seismic zone requirements, the risk category, the importance factor, etc. were all considered in the numerical modeling of the building using ETABS. Wind load with all its combinations was also analyzed. The base shear, overturning moment and distribution of forces per floor were obtained and checked.

Modal analysis was performed using Modal participation mass ratios as the results in Table 1 demonstrate and the time period of each mode was calculated where internal forces were obtained based on the SRSS technique.

Table 1. Modal participation mass ratios.

TABLE: Modal Participating Mass Ratios

Case	Mode	Period sec	UX	UY	UZ	Sum U	Sum U	Sum U	RX	RY	RZ	Sum R	Sum R	Sum R
Modal	1	5.773	0.5421	0.0228	0	0.5421	0.0228	0	0.0098	0.3549	0.0433	0.0098	0.3549	0.0433
Modal	2	4.383	0.0516	0.4468	0	0.5937	0.4697	0	0.2996	0.0352	0.074	0.3094	0.39	0.1173
Modal	3	2.811	0.0248	0.125	0	0.6185	0.5947	0	0.1083	0.0085	0.464	0.4177	0.3985	0.5812
Modal	4	1.362	0.0795	0.0585	0	0.698	0.6532	0	0.0605	0.0972	0.0033	0.4782	0.4957	0.5846
Modal	5	1.156	0.0702	0.1184	0	0.7682	0.7717	0	0.0881	0.0672	0.000008285	0.5663	0.563	0.5846
Modal	6	0.834	0.0146	0.0141	0	0.7828	0.7857	0	0.0182	0.024	0.091	0.5845	0.587	0.6756
Modal	7	0.603	0.0209	0.0172	0	0.8037	0.8029	0	0.0257	0.028	0.0125	0.6102	0.615	0.6881
Modal	8	0.487	0.0262	0.0267	0	0.8299	0.8297	0	0.0398	0.0361	0.0004	0.6501	0.6511	0.6885
Modal	9	0.434	0.0057	0.0044	0	0.8356	0.8341	0	0.0091	0.0094	0.0225	0.6591	0.6605	0.711
Modal	10	0.37	0.00003172	0	0	0.8356	0.8341	0	7.994E-07	0.00004876	0.000006204	0.6591	0.6605	0.711
Modal	11	0.369	0.0001	0.000007454	0	0.8357	0.8341	0	0.0000109	0.0002	0.000007071	0.6592	0.6607	0.711
Modal	12	0.365	0.00000643	0.00001494	0	0.8357	0.8341	0	0.00002553	0.000008906	0.000007375	0.6592	0.6607	0.711

The maximum period of the building for the first mode was found to be 5.8 seconds. The first mode of the building was a translational mode in the x-direction. The second mode was also a translational mode but in the y-direction, while the third mode was a torsional mode: the chosen a central reinforced concrete shear walls delayed this torsional effect. Mass participation for each mode was checked and the first 20 modes were considered in the analysis.

3 GEOTECHNICAL ANALYSIS AND DESIGN

Based on the geotechnical investigation performed on the building site, different dewatering methods were scrutinized as the ground water level was found to be 8.5 m below ground level. A polymer plug was chosen to counterweight the uplift pressure and to prevent water from seepage through the foundation. Full tanking was also adopted for the basement in order to act after the lifetime of the plug when it decays. The basement's retaining walls were analyzed and designed using Geo5; struts were designed to act as their support for large distances.

As illustrated in Figure 2, which is a cross-section for the foundation system details for the model, isolated footings were adopted for the three-floor podium columns (above the water level) while raft foundation with retaining walls was chosen under the tower (below the water level).

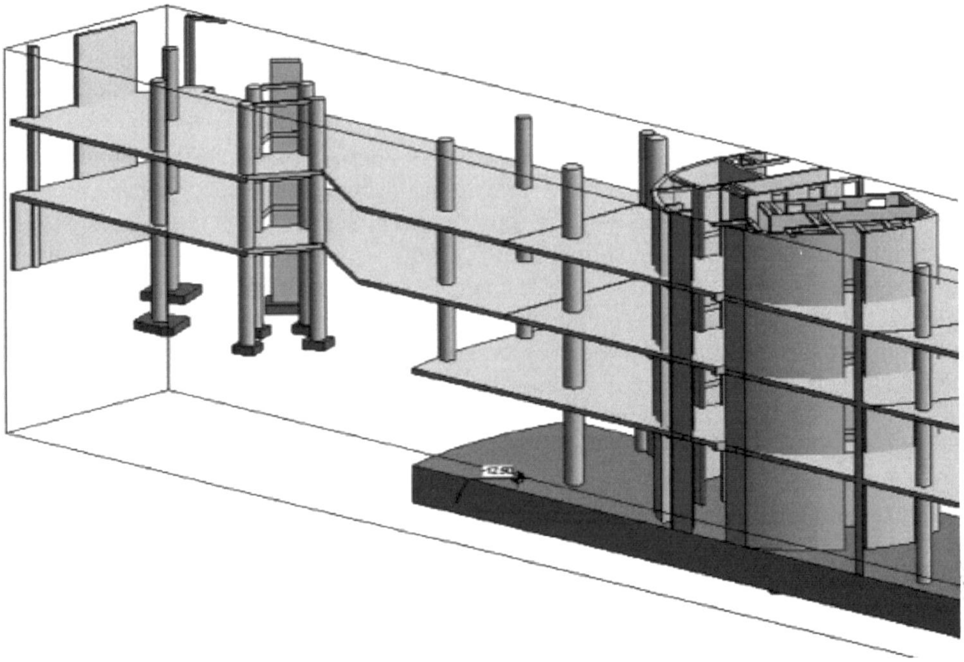

Figure 2. Section of the foundation systems.

The analysis and design of the raft was done using SAFE, considering the soil modulus of subgrade reaction, maintaining the differential settlement within acceptable limits as per the results shown in Figure 3.

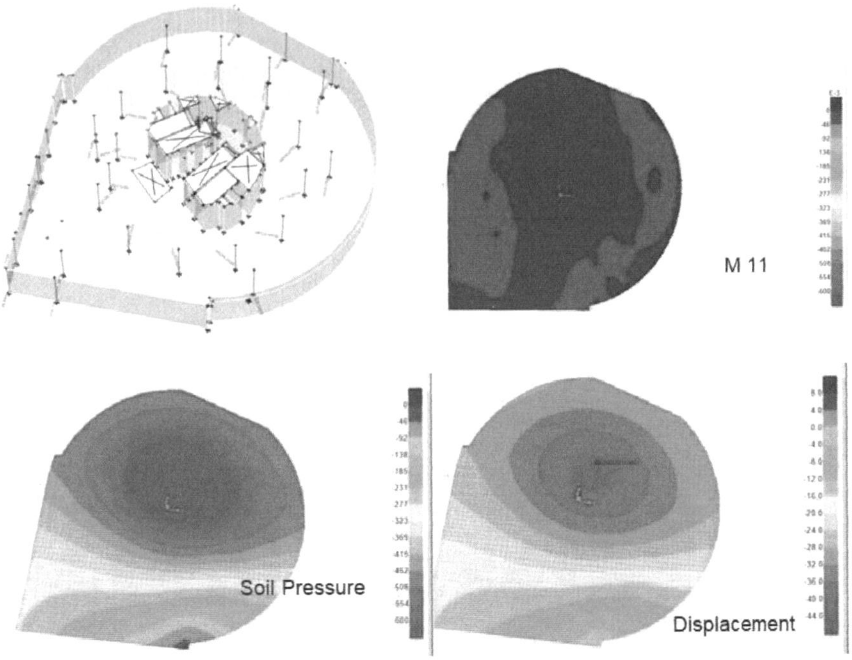

Figure 3. Analysis and results of tower raft.

4 STRUCTURAL DESIGN

Columns were designed on ETABS according to ACI 318M-14 (2014) provisions. To validate the design performed by ETABS, random samples of the columns were chosen and designed on PCA Columns; another software specifically used for RC columns design. The results showed that, the design performed by ETABS was close, to that performed by PCA Columns. Some of the columns in the atrium gardens (Figure 1) had an unsupported length of 24 m. This led to design these columns with a 1.7 m diameter circular cross-sections; a huge section but required for column stability and for minimizing the second order (p-d) effect. Therefore, an alternative design was also proposed where these columns were designed as composite steel-concrete sections (with steel hot-rolled section embedded inside a concrete circular section). This design allowed the column cross section to be reduced from 1.7 m to 1.0 m; therefore, increasing the area that can be used by the client and of course reducing the cost of construction.

The shear walls were designed for bending moments and shear forces using ETABS and the walls boundary zone was checked where reinforcement was added according to the ACI 318M-14 (2014) provisions. Displayed in Figure 4 is the reinforcement drawn according to the design. These walls serve to reduce the torsional effect, which arises from the severe horizontal and vertical irregularity of the structures imposed by the architectural design.

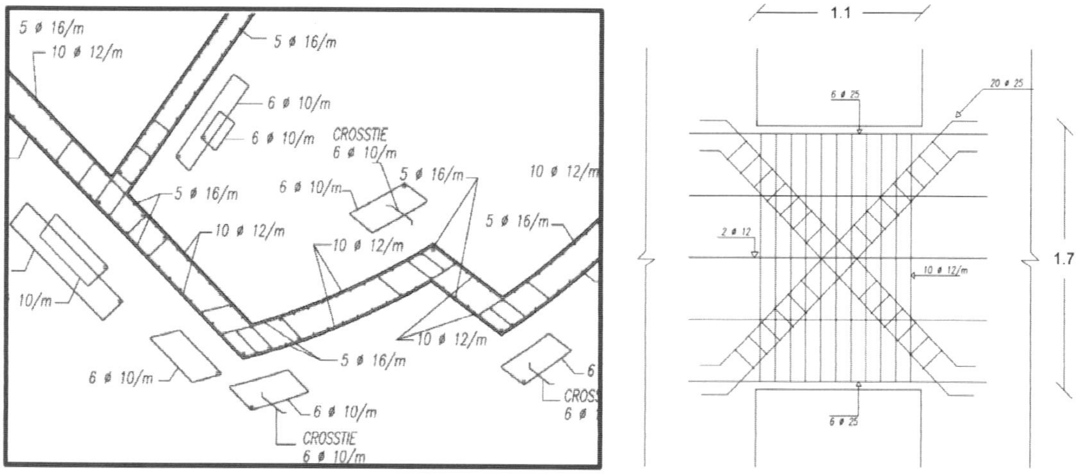

Figure 4. Core wall and link / coupling beam reinforcement.

Beams were designed using the 3D model on ETABS according to ACI 318M-14 (2014) provisions. However, to validate this design, random beam samples were chosen and manually designed. This check proved that the design performed by ETABS is both safe and economic. Almost all the beams were used to connect the shear walls in the core area in order to add additional lateral stiffness to the building in resisting lateral loads. As such, these beams were manually designed as coupling beams based on ACI318M-14 (2014) provisions and their reinforcement is shown in Figure 4.

The slabs were exported from ETABS to SAFE to be analyzed and designed as flat slabs. The short- and long-term deflections were checked against the limits specified by ACI 318M-14 (2014) provisions. Due to the presence of a 9 m cantilever slabs at three tips of the building floors as they twist by 2 degrees (Figure 3), the slab thickness was increased to make sure the cantilever long-term deflection is within the acceptable limits of ACI 318M-14 (2014). As illustrated in Figure 5, a typical mesh was added, and additional reinforcement was needed at certain areas. For the podium, a waffle slab system was designed due to their huge spans and area.

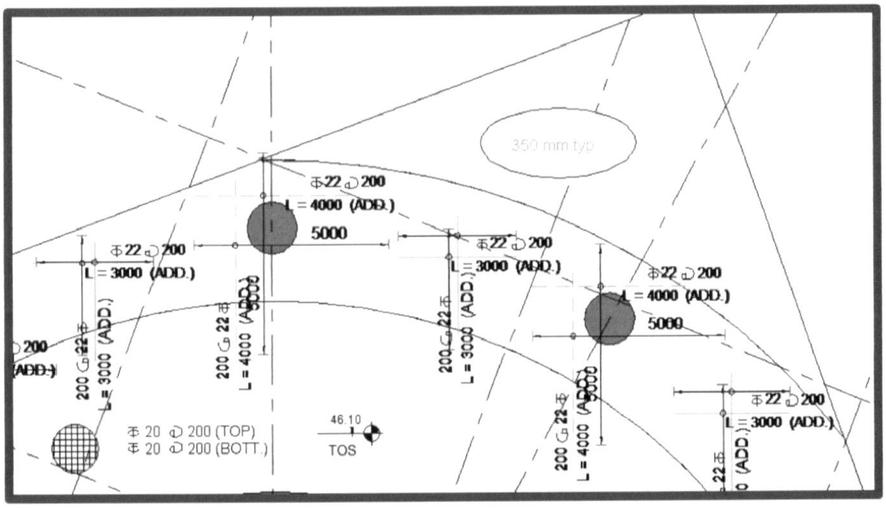

Figure 5. Tower slab reinforcement.

5 CONCLUSIONS

Significant learning outcomes and lessons were gained/ learned by the students in this capstone course. Full 3D BIM and analytical model, construction methods for the substructure, 3D numerical analysis, calculation sheets, and design development structural drawings were delivered. Using different software was a real challenge as moving a model or data from one software to the other took a time and efforts to adjust these models between software. This gave the students knowledge and prepared them for their professional career. Designing different structural elements such as coupling beams, waffled slabs in the podium, composite column for long column, and reinforced concrete raft foundation on elastic support added another edge to the students in this capstone course.

References

ACI 318M-14, *Building Code Requirements for Structural Concrete (ACI 318-14) and Commentary (ACI 318R-14)*, American Concrete Institute, Committee 318, Farmington Hills, MI, USA, 2014.

ASCE/SEI 7-16, *Minimum Design Loads and Associated Criteria for Buildings and Other Structures*, American Society of Civil Engineers, Reston, VA, USA, 2016.

Streamlining Information Transfer between Construction and Structural Engineering
Edited by Shiau, J., Vimonsatit, V., Yazdani, S., and Singh, A.

ISBN: 978-0-9960437-7-9

A CONCEPTUAL FRAMEWORK FOR IDENTIFYING CONSTRUCTION ACTIVITY INTENSITY IN HARSH WEATHER CONDITIONS

AMMAR S.M. MOOHIALDIN[1], FIONA LAMARI[1], MARC MISKA[1], and BAMBANG T. SUHARIADI[2]

[1]*Science and Engineering Faculty, Queensland University of Technology, Brisbane, Australia.*
[2]*Dept of Construction Engineering and Management, King Fahd University of Petroleum and Minerals, Dhahran, Saudi Arabia.*

Harsh Weather Conditions (HWC) are environmental hazards that directly affect construction worker's health and safety, with a corresponding negative impact on their productivity. Climate predictions predict worsening scenarios of hot and humid weather conditions extending into regions not previously subject to HWC. The predictions pose a significant potential risk and elements of current practices in the construction industry seriously compromise workers' safety and productivity. HWC have been shown to have impacts on individuals and influence working environments which taken together lead to inhibiting the achievement of practical, safe and productive practices. This research aims to highlight the need to automatically identify the construction activity intensity in HWC, with proposing a conceptual framework of computerized vision analysis (CVA) system. This framework includes construction activity matrices that classify steel, formwork, and concreate work into macro and micro movements, with including the engaged materials, tools, and equipment. This system is expected to help in measuring the impacts of HWC on construction workers' safety and its implications on their productivity by identifying the activity intensity level and worker's idle time at the crew levels.

Keywords: HWC, CVA, Safety, Productivity.

1 BACKGROUND

The construction projects include various types of the activities that are conducted in heavy intensive work and many risk factors can affect the delivery of the projects such as management and finical related factors, site conditions, worker's accidents and injuries, poor worker's productivity, weather conditions. The workers are the key success factors of the construction activities as most of the construction work depends on their physical effort. The environmental factors such as hot and humid weather conditions have been found to be one of the most influential factors in the construction work environment that influence the workers' safety and negatively impact their productivity. Thus, this research will try to improve the worker's safety with respect the effects of the hot and humid weather conditions that can positively improve their productivity and consequently the delivery of the construction projects.

1.1 Heat-Related Injuries and Fatalities

Eighty nine fatal injuries caused by the impact of HWC were recorded in the construction industry between 2011–2016 in the United States (U.S. Bureau of Labor Statistics 2016). In Japan, yearly average heat-related deaths between 1991 and 2001 was around 14 workers and 194 heat cases of absenteeism were reported in 2001 (Inaba and Mirbod 2007), which increased dramatically and reached 47 deaths in 2010 (Yi *et al.* 2016). In Hong Kong, the statistics for 2007-2011 reported 32 injures and 11 deaths that were caused by heat stress (Li *et al.* 2016), and in 2012, there were 28% of workers who experienced heat stroke symptoms (Yi *et al.* 2016). Between 1900–2011, general statistical records in Australia showed 55% of the natural-related factors fatalities were due to the effects of HWC (Zander *et al.* 2015). Between 1844-2010, it was also estimated the fatalities in Australia due to effects of HWC by 5332 deaths (Coates *et al.* 2014). The Office of Industrial Relations in Queensland (personal communication, May 9, 2018) reported that between the period 2010 and 2017, 53 hear-related claims were recorded within the construction workers. Four claims were fatal or serious whereas 49 claims were identified as non-serious claims. Furthermore, according to the industry heat map index of the Queensland, construction industry and Construction workers have the highest heat-related harm index (3) in a scale from low harm (1) to high harm (3) (Worksafe Queensland 2017). These alarming statistics of heat-related illnesses and fatalities have been considered as driven to investigate the actual impact of hot and humid weather conditions on the construction worker safety.

1.2 Construction Worker's Productivity Under HWC

A combination of physically demanding work and HWC can disrupt workers' productivity and overall performance (Li *et al.* 2016) and this has been directly linked to workers' health and safety (H&S) (Choudhry 2017, Shirali *et al.* 2016, Cheng *et al.* 2012). The effects of hot and humid weather conditions have extended to workers' productivity. For instance, construction organizations have faced challenges in dealing with workers' productivity losses due to hot and humid weather conditions, which have been found to be the main reason for 64% of the variation in productivity levels (Koehn and Brown 1985). Local indications in Australia have shown significant losses in workers' productivity due to HWC (Zander *et al.* 2015, Srinavin and Mohamed 2003). In Queensland, hot and humid weather conditions have been found to be the most influential factors on workers productivity (Conor and Panuwatwanich, 2013). This could explain the incurred productivity-related losses in this industry due to HWC, for example, Zander *et al.* (2015) estimated the worker productivity losses at $6.2 billion as a result of harsh weather in Australia during the period 2013–2014, the results of the Singh *et al.* (2013) study showed that Australian companies may lose one-third of their workers' productivity in HWC, Yi and Chan (2017) found that the crew productivity time is decreased by 0.33% when the Wet Bulb Globe Temperature (WBGT) increase by 1°C. Therefore, not only worker's safety but their productivity are both considered important in understanding to what extent the HWC influences the construction industry. Accordingly, it is important to apply an integrated approach including safety and productivity to assess the effects of HWC which can help to achieve aligned improvements in worker's safety and productivity (Choudhry 2017, Cheng *et al.* 2013, Miller and Bates 2007).

2 THE CONTEXT OF THIS RESEARCH

Previous studies in different regions in the world have investigated the problem of HWC in the construction industry by focusing on the individual factors of the construction workers that have manifested a wide range of variation (Yang and Chan 2017, Gatti *et al.* 2014), made generalization of the research outcomes difficult (Hunt 2011), and ignored the regional variations (Yang and Chan 2017, Tymvios *et al.* 2016, Dutta *et al.* 2015, Bates *et al.* 2009). Moreover, the literature has highlighted significant limitations in the applied methods of data collection (Rowlinson *et al.* 2014) in particular, that the intrusive sensing technology can hinder the workers' activities or create discomfort (e.g., sensors that weigh 1.5 kg; Wong *et al.* 2014, Chan *et al.* 2012), cannot persuade the employees and management in real-time monitoring, and require special physiological status monitoring (PSM) sensors and knowledge in physiology (Lee and Migliaccio 2016, Lee *et al.* 2015, Liang *et al.* 2011). There was also limited information about the applicability of such technologies including its associated costs and the social and culture influences of adopting continuous monitoring systems based on PSM (Gatti *et al.* 2013, Zhou *et al.* 2013). It is, therefore, imperative that research methods use sophisticated measuring devices and monitoring systems that have small and practical designs (Gatti *et al.* 2013, Russell-Smith and Lepech 2011, Miller and Bates 2007, Abdelhamid and Everett 2002).

Hence, the research problem that will be addressed in this study is that the existing research raises the question that how the effects of HWC can be measured in the construction site with non-intrusive and real-time measurements to enhance a proactive decision making that ensure safe work without compromising worker's productivity. There is also a lack of research to automate work-related factors, in particular, activity intensity under HWC, especially in Queensland.

3 THE RESEARCH AIM AND QUESTIONS

The aim of this research is to investigate the extent to which HWC impact on construction workers' safety and how HWC may influence crews' productivity levels while considering the influence of activity intensity levels of a limited range of activities at the crew level. This study proposes to use a quantitative approach involving real-time measurements of environmental conditions and automated activity intensity identification with crews' idle time and the utilization of computerized vision analysis (CVA). Real data synchronization will be employed as an input to a proactive framework that supports the site management with safe and productive decisions. To achieve the stated aim, this research addresses the following main questions:

- What is the potential applicability of an automated activity intensity identification system under HWC?
- To what extent do the various types of construction activity intensities impact on workers' safety?
- What is the potential application of CVA technology in supporting proactive decisions to maintain both safety and productivity under HWC?

4 RESEARCH METHOD

The measurements of HWC impacts on construction workers ideally depend on collecting information of the individual factors of the workers such HR, BR, and oxygen uptake level. Previous research has extensively employed this type of information in estimating the activity intensity levels which have been linked to the weather measurements such as degree of temperature, humidity and WBGT. However, the literature suggests that such measurements to be impractical when considering the real site applications due to intrusiveness of the PSM sensors; a wide range of variations in the collected information due to the effects of the worker's individual factors; and the time-, cost- and effort-consuming method of attaching sensors to the body of each site worker in different working shifts. Therefore, this research proposes a framework of AII under HWC to achieve a non-intrusive system for monitoring the worker's safety and productivity under HWC. This framework is expected to measure the worker's activities (activity, typical movements, engaged materials, tools, equipment and workstation), their idle time and weather parameters (degree of temperature, humidity and WBGT) in a real-time manner. This approach will employ the CVA technology of real-time video records to measure worker's activities. Thus, the system needs to collect real-time data from the video records and a local weather station.

4.1 Research Design

The research literature review included the understanding of the impacts of the HWC and the nature of the targeted construction activities. Based on that the research problem and aim have been formulated. To achieve the research aim, the research method will include three phases: (1) phase one, activity intensity matrix and CVA system preparation; (2) phase two will be designated to conduct a pilot measurement; and (3) phase three includes site measurements.

4.1.1 *Framework development*

This research proposes to develop an automated AII under HWC. This framework will enable the understanding and the identification of the nature of the construction activities, their associated intensities and the links between worker's safety and idle time under HWC. This framework can help to facilitate proactive decision-making for both safety and productivity under HWC. The AII formwork will be built on the concept of providing a massive amount of valuable information of the worker activities in non-intrusive and real-time manner. Thus, the data acquisition component of this framework will include the application of 2D sit cameras. The second component of the framework will include two layers of quantitative data analysis: (1) CVA application; and (2) data synchronization, activity intensity identification, WBGT thresholds and worker's idle time optimization. The last component of the framework will report the data analysis results to the decision makers via summarized detailed information of the worker's activity intensity levels compared to the WBGT thresholds and optimized idle time, as Figure 1 shows.

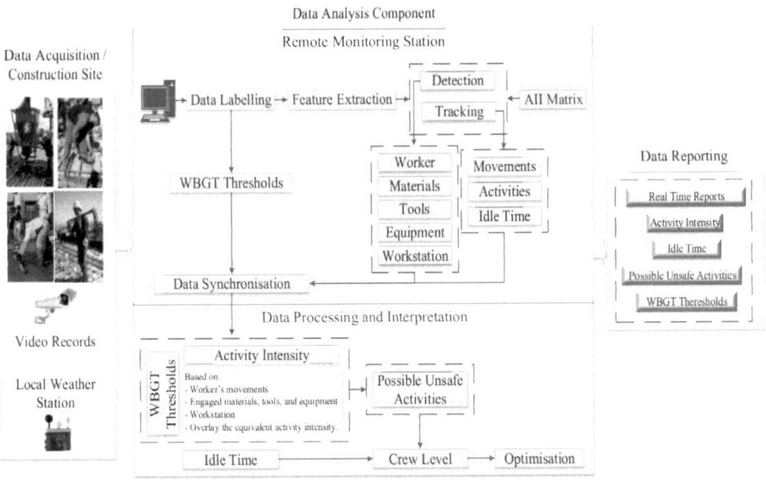

Figure 1. The proposed conceptual framework of AII under HWC.

5 CONCLUSIONS AND RECOMMENDATIONS

Literature review shows that construction projects are seriously encountered loss in workers due to injuries and fatalities as well as the increase of their idle time under HWC. The current state of the art related to HWC generally applied intrusive sensors to measure the effects of HWC and extensively emphasized on the individual factors of the works. However, the debate as to what the characteristics of the performed manual tasks and work-related factors that can help to improve safe work environment under HWC, becomes a pertinent question. Therefore, the aim of this research is to investigate the extent to which HWC impact on construction workers' safety and how HWC may influence crews' productivity levels while considering the influences of activity intensity levels of a limited range of activities at the crew level. This research proposes to develop an automated AII framework to achieve non-intrusiveness in site measurements under HWC. This framework is expected to develop an automated, non-invasive monitoring system that provides supports real-time data collection.

References

Abdelhamid, T. S., and Everett, J. G., Physiological Demands during Construction Work. *Journal of Construction Engineering and Management,* 128, 427–437, 2002.

Bates, G. P., Miller, V. S., and Joubert, D. M., Hydration Status of Expatriate Manual Workers During Summer in the Middle East, *Annals of Occupational Hygiene,* 54, 137–143, 2009.

Chan, A. P. C., Yi, W., Wong, D. P., Yam, M. C. H., and Chan, D. W. M., Determining an Optimal Recovery Time for Construction Rebar Workers After Working to Exhaustion in a Hot and Humid Environment. *Building and Environment*, 58, 163–171, 2012.

Cheng, E. W., Ryan, N., and Kelly, S., Exploring the Perceived Influence of Safety Management Practices on Project Performance in the Construction Industry. *Safety Science*, 50, 363–369, 2012.

Cheng, T., Migliaccio, G. C., Teizer, J., and Gatti, U. C., Data Fusion of Real-Time Location Sensing and Physiological Status Monitoring for Ergonomics Analysis of Construction Workers, *Journal of Computing in Civil Engineering,* 27, 320–335, 2013.

Choudhry, R. M., Achieving Safety and Productivity in Construction Projects, *Journal of Civil Engineering and Management*, 23, 311–318, 2017.

Coates, L., Haynes, K., O'Brien, J., McAneney, J., and de Oliveira, F.D., Exploring 167 Years of Vulnerability: An Examination of Extreme Heat Events in Australia 1844–2010, *Environmental Science & Policy*, 42, 33–44, 2014.

Conor, O., and Panuwatwanich, K., *The Impact of Fatigue on Labour Productivity: Case Study of Dam Construction Project in Queensland*, 4th International Conference on Engineering, Project, and Production Management From EPPM, 993–1005. 2013.

Dutta, P., Rajiva, A., Andhare, D., Azhar, G., Tiwari, A., and Sheffield, P., Ahmedabad., Heat and Climate Study Group, Perceived Heat Stress and Health Effects on Construction Workers, *Indian Journal Occupat'onal and Environmental Medicine.*, 19, 151, 2015.

Gatti, U., Migliaccio, G., Bogus, S. M., Priyadarshini, S., and Scharrer, A., Using Workforce's Physiological Strain Monitoring to Enhance Social Sustainability of Construction, *Journal of Architectural Engineering.*, 19, 179–185, 2013.

Gatti, U. C., Migliaccio, G. C., Bogus, S. M., and Schneider, S., An Exploratory Study of The Relationship Between Construction Workforce Physical Strain and Task Level Productivity, *Construction Management and Economics*, 32, 548–564, 2014.

Hunt, A. P., *Heat Strain, Hydration Status, and Symptoms of Heat Illness in Surface Mine Workers*. .Sc. (with Distinction), 2011.

Inaba, R., and Mirbod, S. M., Comparison of Subjective Symptoms and Hot Prevention Measures in Summer between Traffic Control Workers and Construction Workers in Japan, *Industrial Health,* 45, 91–99, 2007.

Koehn, E., and Brown, G., Climatic Effects on Construction, *Journal of Construction Engineering and Management,* 111, 129–137, 1985.

Lee, W. and Migliaccio, G. C., Physiological Cost of Concrete Construction Activities, *Construction*

Innovation, 16, 281–306, 2016.

Lee, W., Migliaccio, G. C., Lin, K. Y., and Russo, F., *Lessons Learned from Using Bio-and Environmental Sensing in Construction: A Field Implementation*, 5th International/11th Construction Specialty Conference, Vancouver, British Columbia, 270–280. 2015.

Li, X., Chow, K. H., Zhu, Y., and Lin, Y., Evaluating The Impacts of High-Temperature Outdoor Working Environments on Construction Labor Productivity in China: A Case Study of Rebar Workers, *Building and Environment*, 95, 42–52, 2016.

Liang, C., Zheng, G., Zhu, N., Tian, Z., Lu, S., and Chen, Y., A New Environmental Heat Stress Index for Indoor Hot and Humid Environments Based on Cox Regression, *Building and Environment*, 46, 2472–2479, 2011.

Miller, V. S., and Bates, G. P., The Thermal Work Limit Is a Simple Reliable Heat Index for the Protection of Workers in Thermally Stressful Environments, *Annals of Occupational Hygiene*, 51, 553–561, 2007.

Rowlinson, S., YunyanJia, A., Li, B., and ChuanjingJu, C., Management of Climatic Heat Stress Risk in Construction: A Review of Practices, Methodologies, and Future Research, *Accidents; Analysis and Prevention*, 66, 187–198, 2014.

Russell-Smith, S., and Lepech, M., Dynamic Life Cycle Assessment of Building Design and Retrofit Processes, *Journal of Computing in Civil Engineering,* Reston, VA, 760–767, 2011.

Shirali, G., Savari, R., Ahmadiangali, K., and Salehi, V., The Effect of Safety Costs on Productivity and Quality: A Case Study of Five Steel Companies in Ahvaz, *Jundishapur Journal of Health Science,* 9, 0–5, 2016.

Singh, S., Hanna, E. G., and Kjellstrom, T., Working in Australia's Heat: Health Promotion Concerns for Health and Productivity, *Heal. Promot. Int.,* 30, 239–250, 2013.

Srinavin, K., and Mohamed, S., Thermal Environment and Construction Workers' Productivity: Some Evidence from Thailand, *Building and Environment,* 38, 339–345, 2003.

Tymvios, N., Behm, M., Jia, Y., and Johnson, K., Heat Stress in the U.S . Construction Industry, *CIB WBC 2016,* Tampere, Finland, 2016.

U.S. Bureau of Labor Statistics, 2016, Occupational Injuries / Illnesses and Fatal Injuries Profiles Occupational Injuries and Illnesses and Fatal Injuries. Retrieved from https://data.bls.gov/gqt/RequestData on March 22, 2018.

Wong, D. P., Chung, J. W., Chan, A. P.-C., Wong, F. K., and Yi, W., Comparing the Physiological and Perceptual Responses of Construction Workers (Bar Benders and Bar Fixers) In a Hot Environment, *Applied Ergonomics*, 45, 1705–1711, 2014.

Worksafe Queensland, 2017, Industry Heat Map. Retrieved from https://www.worksafe.qld.gov.au/forms-and-resources/statistics/harm-index-heat-maps-of-work-related-injuries/industry-heat-map on June 26, 2017.

Yang, Y., and Chan, A. P., Heat Stress Intervention Research in Construction: Gaps and Recommendations, *Industrial Health,* 55, 201–209, 2017.

Yi, W., and Chan, A., Effects of Heat Stress on Construction Labor Productivity in Hong Kong: A Case Study of Rebar Workers, *International Journal of Environmental Research and Public Health*, 14, 1055, 2017.

Yi, W., Chan, A. P. C., Wang, X., and Wang, J., Development of an Early-Warning System for Site Work in Hot and Humid Environments: A Case Study, *Automation in Construction*, 62, 101–113, 2016.

Zander, K. K., Botzen, W. J. W., Oppermann, E., Kjellstrom, T., and Garnett, S.T., Heat Stress Causes Substantial Labour Productivity Loss in Australia, *Nature Climate Change,* 5, 647–651, 2015.

Zhou, Z., Irizarry, J., and Li, Q., Applying Advanced Technology to Improve Safety Management in The Construction Industry: A Literature Review, *Construction Management and Economics*, 31, 606–622, 2013.

Streamlining Information Transfer between Construction and Structural Engineering
Edited by Shiau, J., Vimonsatit, V., Yazdani, S., and Singh, A.

ISBN: 978-0-9960437-7-9

DESIGN AND PLANNING OF SCAFFOLDING SYSTEM

TAKASHI HARA[1] and KATSUKIYO SHIMOMURA[2]

[1]*Dept of Civil Engineering and Architecture, National Institute of Technology, Tokuyama College, Shunan, Japan*
[2]*Miwa Consultants Co. Ltd., Iwakuni, Japan*

In this paper, the automatic design method of scaffold was proposed. This method was programmed using web based PHP program under the safety standard and the designer's know-how. The size and the numerical quantity of the scaffold were calculated considering the relation between the scaffolding condition and the constructing structure. Three dimensional positions and the numbers of the individual scaffold materials were calculated and recorded in the data server simultaneously. Then, three dimensional structural drawings were made automatically using the AutoCAD with VBA by using the previously designed material positions. Therefore, 3D drawings were done automatically. From the proposed design method, the precise quantities of scaffold materials were obtained. Also, the drawing of the scaffold was done automatically without any laborious drawing by hands.

Keywords: Drawing, Web, Cloud, CAD, PHP.

1 INTRODUCTION

Scaffold systems in the construction site play an important role for a safety and a stable construction of structures. There are many kinds of scaffold systems corresponding to the structural shape and constructional condition. The design of the scaffold has been done under the safety standard by Ministry of Health, Labor and Welfare (2010) and Temporary Works Forum (2014) as well as the designer's knowledge. These standards specify the performance and safety requirements of the scaffold and introduce the general structural design methods of scaffoldings.
In the example application, Figure 1 shows the hanging scaffold which is constructed to maintain the pipelines racked on the frames. Many kinds of materials are required to construct the hanging scaffold as shown in Figure 1. In designing the scaffold, the positions, the size and the quantities of all materials must be calculated. The design and the planning of the scaffolding system are quite laborious. Especially, the quantity surveying and cost estimation of the scaffold materials such as jacks, transoms, ledgers, etc, have been laborious. Therefore, these works have been empirically done based on the total volume or the total surface area of the structure. The discrepancy between design and actual quantities of the scaffold materials occurred frequently and brought the incorrect supply, the incorrect schedule and the unsafe condition.

In this paper, the automatic design and drawing methods of scaffold were proposed. The method was presented for the sample hanging scaffold and the applicability for other scaffold system was also presented.

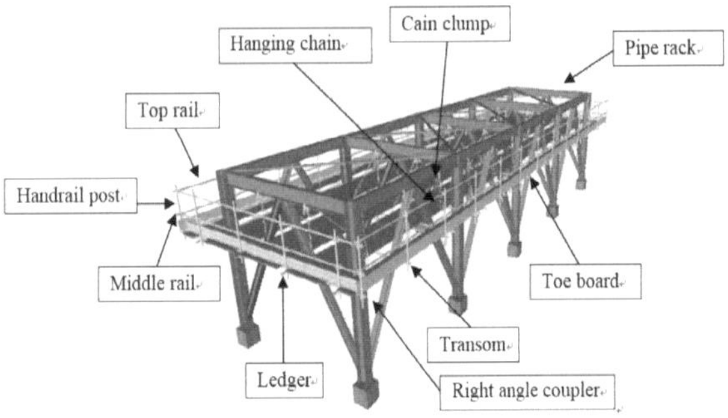

Figure 1. Example of hanging scaffold.

2 DESIGN PROCEDURE

2.1 System Environment

The system is composed of the web server, the file server and the client computer or PDA. The web and the file servers can be placed on the clouding system. The material quantity surveying and cost estimation will be done outside the office and these data will be stored on the storage inside the office for the later use.

2.2 Quantity Surveying Method

The quantity surveying method is represented using the example of the hanging scaffold shown in Figure 1. The calculation system is running on the web with PHP program (Cowburn 2018).

2.2.1 *Web data input*

Figure 2 shows the web input page. Firstly, the user should input the geometric conditions of the constructing structure such as the size of pipeline racks. Rack length and rack width are 40m and 10m, respectively. The column distance is 8m. The setting position is 0m. Then, the scaffold design conditions should be defined. The ledger distance is initially defined as 1.8m. The transom arrangement is selected (8th raw in Figure 2). In this example, 4 transoms are arranged between rack column. The longest steel pipe length is selected as 6m in this example. Five patterns of aisle can be selected (see Figure 3). In this example 4 aisles are selected. In each aisle, two types of plank can be selected. In this example, 3 planks are selected. The calculation with PHP program (Cowburn 2018) is done. Each calculation method is shown below.

2.2.2 *Origin of the coordinate system in the calculation*

After pushing the "Execute" button, all materials are calculated sequentially based on the origin of the coordinate system that is the reference point of the system. In this system, the origin is defined as the top center of the beam placed at the edge of the rack (Figure 3). All positions of the composed materials are calculated based on this point. The axis x is defined along the rack axis from the origin and the axis z is defined as the vertical direction. The axis y is defined perpendicular to x-z plane. The material positions and the quantities are calculated considering the geometric conditions of the materials. The relations between materials are defined under the Ministry of Health, Labor and Welfare (2010) specifying the safety and the strength requirements of the scaffold. Positions and the length of each material are evaluated below.

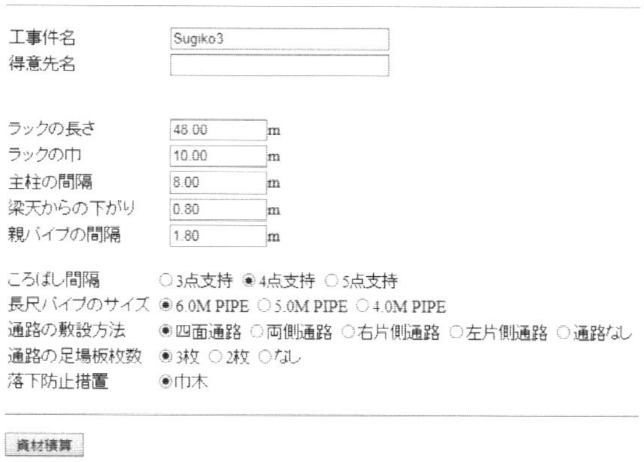

1. Name of project
2. Name of client

3. Length of rack
4. Width of rack
5. Distance of columns of rack
6. Setting position of hanging scaffold
7. Distance of ledgers

8. Method of transoms arrangement
9. Pipe length
10. Type of aisles
11. Plank rows
12. Toe board

13. Execute

Figure 2. Input data on WEB.

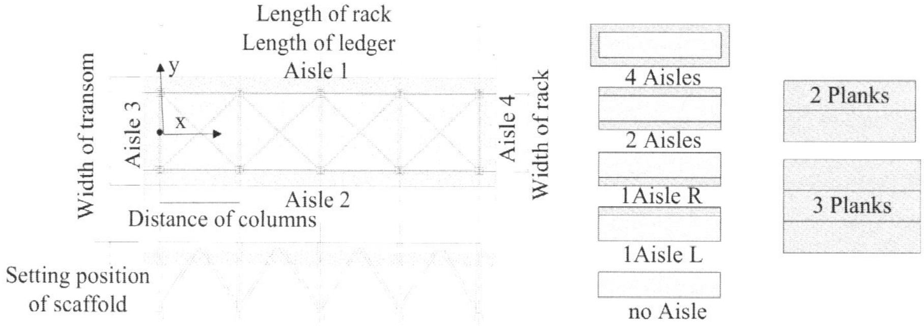

Figure 3. Rack and scaffold.

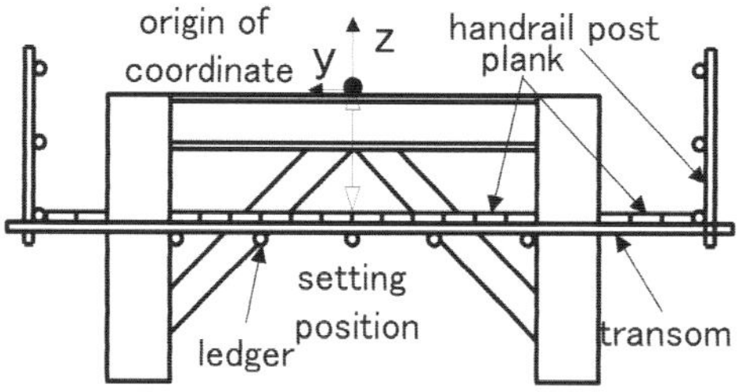

Figure 4. Side view of hanging scaffold.

2.2.3 *Position and length of ledgers*

The ledgers are placed within the rack columns. The depth is the distance between the origin of the coordinate system and the working platform on the planks. The distance between the ledgers is defined on the web (7th raw in Figure 2). Considering the width of the frame, each ledger is placed at regular interval. The position of the ledger is under the transom below the planks (Figure 4). The required length of the ledger is evaluated considering the geometrical relation of other materials (see Figure 3). The length of the ledger is the summation of the required length, that is the rack length and the aisle width, and the extra length of pipe defined in the Ministry of Health, Labor and Welfare standard (2010).

2.2.4 *Position and length of transoms*

The first transom is placed at 100mm from the end of the legers depend on Ministry of Health, Labor and Welfare (2010). These are placed on the ledgers. Then other transoms are placed repeatedly depending on the arrangement of the transoms (8th raw of Figure 2). The distance between transoms is the quotient of the column distance and the number of transoms (8th raw of Figure 2). The length of each transom is calculated considering the rack width and the aisle width as well as the extra length of pipe. The height of transom is calculated adding 76mm to the height of ledger. 76mm is the distance between right angle couplers. Then, each coordinate of the transoms is defined as the crossing points between legers and transoms.

2.2.5 *Positions and numbers of planks*

Planks are placed from the center of the rack beam (the origin) to both columns of the rack. The coordinate and the number of the planks are calculated from the origin. The planks are also placed on the aisles defined on the web (11th raw of Figure 2).

2.2.6 *Position, length and numbers of handrails and their post*

Additional pipes are placed at both ends of the transoms. Then, handrail posts are arranged at the intersection of these pipes and the transoms. The length of the post is 150mm. The top rails and the mid rails are placed on the handrail posts. These numbers are calculated and the coordinate of handrails and the handrail posts are also calculated.

2.2.7 *Position and the length of toe boards*

The coordinate of the toe boards is evaluated considering the position of hand rail post.

2.2.8 *Positions and numbers of the hanging chain and the chain clump*

The coordinate and the quantities of the hanging chain and the chain clump are evaluated in regular interval.

2.2.9 *Positions and number of right angle couplers*

The position and the number of the right angle couplers are calculated at each intersection of the steel pipes.

2.2.10 *Storage of the coordinate data and the quantities of the materials*

After evaluating the geometric relations of the materials based on the standards to assure the safety and the strength of them, the quantity of each material is accounted.

The steel pipes used for scaffold have several cut lengths. Combining several sizes of the pipe lengths, the quantities of pipes are calculated. Table 1 shows the calculation results from the input data (Figure 2). Also, the coordinate data for all materials are stored in the data file. These data are used for drawing and for managing the materials.

3 DRAWING THE SCAFFOLDING

Using the estimated data, the drawings will be done on each terminal (PC or PDA) by using the CAD system and can confirm the estimation data. Also, the quantities shown in Table 1 will be presented on each terminal.

Table 1. Evaluated quantities after calculation.

Material	Quantity
Steel pipe 6(m)	214
Steel pipe 2.5(m)	13
Steel pipe 1(m)	55
Steel pipe 1.5(m)	116
Coupler	214
Plank 4.0m	540
Plank 3.0m	0
Plank 2.0m	45
Plank 1.5m	0
Plank 1.0m	45
Toe board 1.8m	72
Right angle coupler	807
Loop chain	52
Hanging chain	120
Chain clump	120

The drawing of the scaffold will be done by using AutoCAD. Each material of unit length is registered in the CAD. Using the position and the length of each material, the drawing is automatically done by the program coded by Visual Basic for Application (VBA). The results are shown in Figure 5(a). From the total view, the arrangement of the materials is confirmed. If the design is not suitable, the designer can go back to the web input system (Figure 2) and can reconfirm the result. Also, as the drawing is done on the cad system, it is easy to change the viewpoint and to zoon in detail (Figure 5(b)).

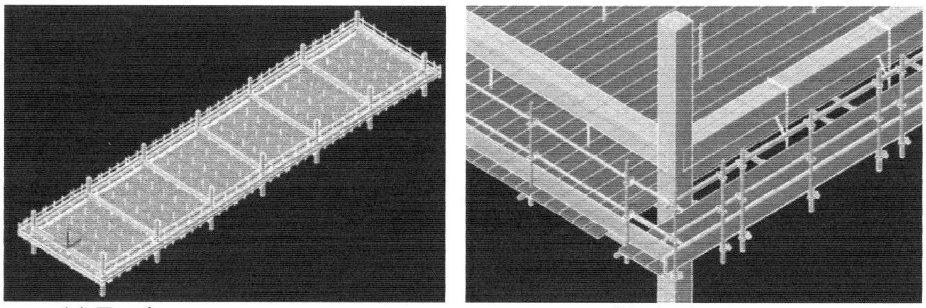

(a) Total arrangement (b) At the corner

Figure 5. Total view of the hanging scaffold.

4 APPLICATION TO OTHER SCAFFOLD SYSTEM

The design system for hanging scaffold is easy to apply to the other type of scaffold. Figure 6 and Figure 7 show the drawings applied to the building structure and the tank maintenance, respectively. These scaffolding systems require a lot of materials and each material is registered as the block data. It is laborious work to account the materials and to make the drawing by hand. However, in this system, only the outline of the building and the gap between building and the scaffold are input data on the web. As the same manor of the previous chapter, the web system calculates the positions, the size and the quantity of each material. Then the drawing system is completely done and the quantities of the materials are obtained in detail. This type of scaffolding has not flexibility of material choice. Therefore, the additional pipes are installed. In Figure 7, there is the gap between column materials in the scaffold system as shown as the circle because the distance between tank and the scaffold should be kept and the width of a material is constant. In such case, the additional materials to fill the gap are designed after drawings.

However, as the proposed system recognizes the coordinates of entire materials, it is easy to add the positions and the numbers of the steel pipes and the right angle couplers.

Figure 6. Scaffold for the building.

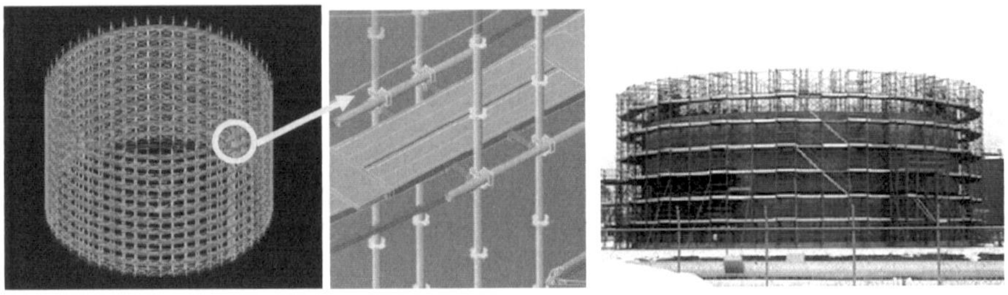

Figure 7. Scaffold for the tank maintenance.

From the accounting and drawing, it is easy to communicate the designers and the workers using the 3D drawings. Also, once the designed data are stored in the storage server on the clouding, it is easy to manage the supply of the materials and the other construction managements.

5 CONCLUSIONS

In this paper, the scaffold design system using the web calculation and CAD drawing was presented. Hanging scaffold was selected as the example. The design of the scaffold has been quite difficult and laborious. For example, when the hanging scaffold shown in Figure 5 was designed by the experienced engineer, it took about a couple of hours to a half day. Also, it took about the same time or more to draw the design drawing. However, the proposed design system will help the engineers' work. Also, it is easy to recalculate or redrawing quickly using this system. The proposed design system could show the design data on the familiar CAD system and could account the precise quantities automatically.

The design system was applied to the hanging scaffold example. Then the applicability of other type of scaffold was presented. Also, it is easy to manage the supply of the materials and the other construction control because the designed data are stored in the storage server on the clouding. The methodology of three dimensional CAD drawing enables us to apply to BIM and CIM software.

References

Cowburn P., *PHP Documentation*, Zend Con & Open Enterprise, U.S.A. 2018.
Ministry of Health, Labor and Welfare, *Occupational Safety and Health Regulations*, MHLW, Japan, 2010.
Temporary Works Forum, *The Use of European Standards for Temporary Works Design*, U.K., 2014.

Streamlining Information Transfer between Construction and Structural Engineering
Edited by Shiau, J., Vimonsatit, V., Yazdani, S., and Singh, A.

ISBN: 978-0-9960437-7-9

MATHEMATICAL MODEL AND SIMULATION OF VEHICLE PLATOON OVERTAKING OTHER VEHICLE

EISUKE KITA[1] and HIROKI SAKAMOTO[2]

[1]*Graduate School of Informatics, Nagoya University, Nagoya, Japan*
[2]*Graduate School of Information Science, Nagoya University, Nagoya, Japan*

A vehicle platoon is the method of grouping vehicles into platoon and driving synchronously. The vehicles in the platoon can travel safely in very small vehicle distance with the help of the mechanical and electric systems and thus, the traffic flow can be increased without any additional road construction. In this study, the interaction behavior between the vehicle platoon of three vehicles and the other vehicle is discussed. When the platoon of three vehicles overtakes the other vehicle, the vehicles in the platoon change the lane, overtake the preceding slow-travelling vehicle and changes the lane again. The vehicle velocity is controlled according to the vehicle following model (Bierley 1963, Chandler *et. al.* 1958, Gazis *et. al.* 1961, Helly 1959). The control models of vehicle velocity and behavior are defined in the mathematical model and discussed in the experiments of the robot vehicle. The results show that the model can make the platoon overtakes the other vehicle safely. In the near future, the model should be improved in order to enhance the safety and the efficiency of the vehicle platoon.

Keywords: Vehicle velocity, Vehicle following model, LEGO MINDSTORM.

1 INTRODUCTION

In vehicle platoon, vehicles travel as a coordinated array with small vehicle head distance by the help of the electric and mechanical control system. It is considered as the important technique for increasing the traffic volume safely. The most important issue of the vehicle-platoon control system is to keep safe and adequate head distance between vehicles even when some vehicles in the platoon show down suddenly. Simply, it is most important to keep the vehicle platoon stable. For this purpose, authors studied the velocity control model of the vehicles in the platoon (Kita *et. al.* 2014, 2016). When, however, the vehicle platoon travels along the highway and inter-city road network, the control of vehicle platoon becomes more complicated. One of such situations is the overtaking of the platoon. When the platoon of a few vehicles overtakes the other vehicle, each vehicle in the platoon departs from the group control of the platoon and then, overtaken the vehicle.

The aim of this study is to present the control model of the vehicle velocity in the platoon. The model is formulated according to the vehicle following model in case that the platoon of three vehicles overtakes the other vehicle. The validity of the control model is discussed in the computer simulation and the experiment of LEGO MINDSTORM (LEGO).

The remaining part of this paper is organized as follows. In section 2, the model is formulated by means of vehicle following model. In section 3, the model parameters are determined in the computer simulation. In section 4, the model with designed parameters is applied for the experiments of LEGO MINDSTORM. In section 5, the conclusion is summarized again.

2 MATHEMATICAL MODEL OF VELOCITY CONTROL

2.1 Traffic Situation

One vehicle platoon and two single vehicles travel in the same direction (Figure 1). The platoon is composed of three vehicles which can travel individually. One of two single vehicles travels in from of the platoon in the same direction as the platoon at the slower speed than the platoon. The other travels along the other lane. Three vehicles in the platoon are named as the lead, the first following and the second following vehicles, respectively.

The behavior of the vehicles in the platoon is summarized as follows.

1) The lead vehicle of the platoon determines whether the platoon can overtake its preceding travelling vehicle.
2) When it is decided that the platoon cannot overtake, the lead vehicle slow down to the velocity of the frontal vehicle.
3) When it is decided that the platoon can overtake, each vehicle of the platoon changes the lane, overtake the frontal vehicle and change the lane again.

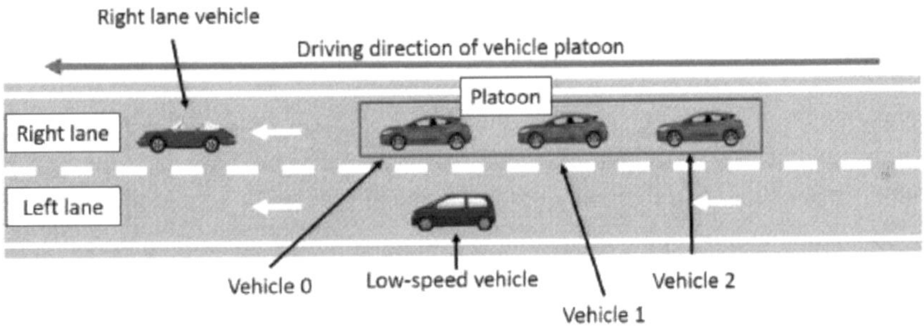

Figure 1. Traffic situation composing of one vehicle platoon and two single travelling vehicles.

2.2 Velocity Control Model

The vehicle platoon is composed of three vehicles. Their velocity is controlled by the vehicle following model, which controls the vehicle velocity according to the mathematical model with the information from the frontal vehicles. Chandler model is adopted in this study. The vehicle acceleration rate is defined as the linear function of the velocity distance of the vehicle and its frontal vehicle as follows.

$$\ddot{x}_1(t+\Delta t) = \alpha\left(\dot{x}_{n-1}(t) - \dot{x}_n(t)\right) \quad (n = 1,2) \tag{1}$$

The variables $x_n(t)$ and $x_{n-1}(t)$ are the positions of the vehicle n and its nearest preceding vehicle $(n-1)$, respectively. The terms $\dot{x}_{n-1}(t) - \dot{x}_n(t)$ denotes the vehicle velocity difference between the vehicles $(n-1)$ and n. The parameters α is the sensitivity parameter.

2.3 Overtaking Condition

Each vehicle in the platoon overtakes the frontal vehicle by turns. The overtaking situation is shown in Figure 2. The vehicle P will overtake the vehicle L in order to avoid the collision with the vehicle R.

Assume that the positions of the vehicles L, P and R are referred as $x_L(t)$, $x_P(t)$ and $x_R(t)$ at the time t, respectively. The following condition is held for them.

$$x_L(t) > x_P(t) > x_R(t) \tag{2}$$

The velocities of the vehicles L, P and R are as follows.

$$v_L(t) > 0, v_P(t) > 0, v_R(t) > 0 \tag{3}$$

The distance between the vehicle P and the vehicle R is given as follows.

$$x_P(t) - x_R(t) - \frac{l_P + l_R}{2} \tag{4}$$

Assume that the passing time of the vehicle P over the vehicle L is Δt. If the vehicle R doesn't come into collision with the vehicle P, the following condition has to be satisfied.

$$\left(x_P(t) - x_R(t) - \frac{l_P + l_R}{2}\right) + v_P(t)\Delta t \geq v_R(t)\Delta t$$

$$x_P(t) - x_R(t) \geq \{v_R(t) - v_P(t)\}\Delta t + \frac{l_P + l_R}{2} \tag{5}$$

When the vehicle P starts to overtake the vehicle L, the distance between the vehicle L and the vehicle P is $x_L(t) - x_P(t)$. The following relation is held.

$$v_P(t)\Delta t = v_L(t)\Delta t + x_L(t) - x_P(t) + \frac{l_L + l_P}{2} \tag{6}$$

The following equation is obtained from equations (5) and (6).

$$x_P(t) - x_R(t) \geq \frac{v_R(t) - v_P(t)}{v_P(t) - v_L(t)}\left\{x_L(t) - x_P(t) + \frac{l_L + l_P}{2}\right\} + \frac{l_P + l_R}{2} \tag{7}$$

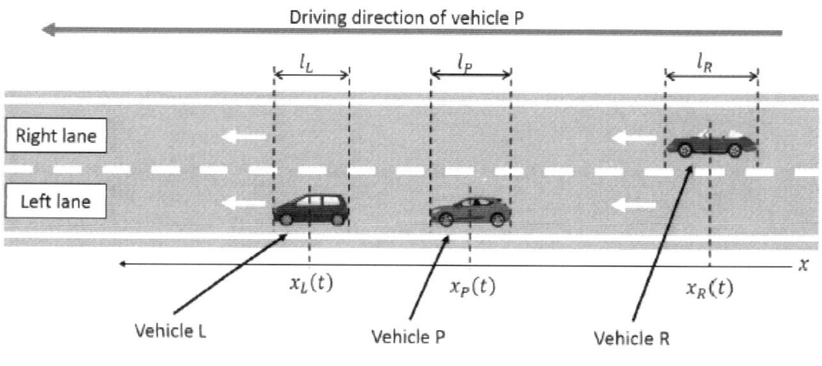

Figure 2. The vehicle p overtakes the vehicle L for avoiding the collision with the vehicle R.

Table 1. Initial position and velocity of vehicles.

Vehicle	Initial velocity (cm/s)	Initial position (cm)
Vehicle 0	11	0
Vehicle 1	11	-33
Vehicle 2	11	-66
Low-speed vehicle	7.5	48
Right lane vehicle	16.5	-33

3 SIMULATION AND EXPERIMENTAL RESULTS

Along the road, one vehicle platoon and two single vehicles travel in the same direction. The vehicle platoon is composed of three vehicles, which are named as Vehicle 0, Vehicle 1 and Vehicle 2. One single vehicle, "the low-speed vehicle", travels along the same road at the slower speed than the vehicle platoon. Another single vehicle, "the right lane vehicle", travels along the other lane. Their initial position and the initial velocity are listed on Table 1. The vehicle 0, the vehicle 1 and the vehicle 2 of the platoon, the low-speed vehicle and the right lane vehicle locate at 0 time-step at 0 cm, -33 cm, -66 cm, 48 cm and -33 cm, respectively. The low-speed vehicle and the right lane vehicle travel at the constant velocity of 7.5 cm/s and 16.5 cm/s, respectively. The platoon starts at the velocity of 11 cm/s.

The numerical simulation results are shown in Figure 3. The platoon follows the preceding travelling vehicle from 24 to 67 time-steps and at 68 time-step, starts to overtake the preceding travelling vehicle. The vehicle 2, which is the last vehicle of the platoon, overtakes the preceding vehicle at the 163 time-step.

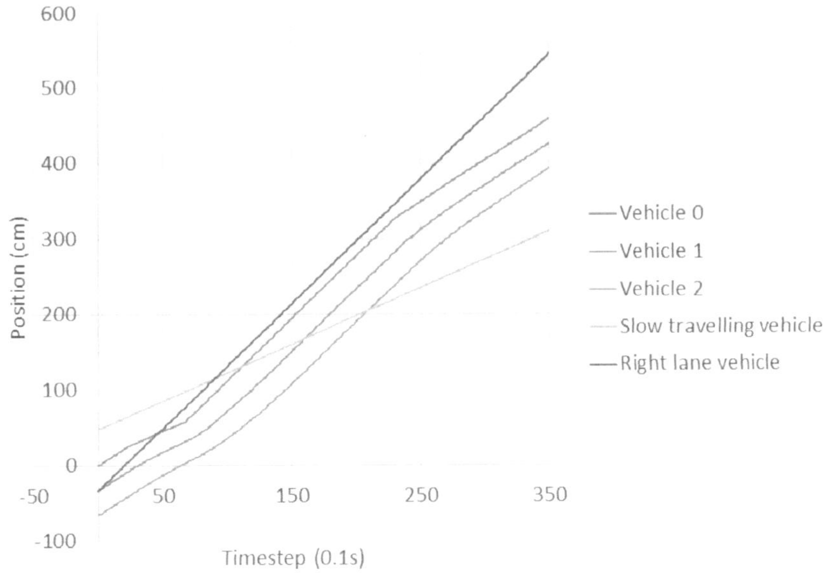

Figure 3. Trajectory of vehicle positions in simulation.

The velocity control model is used for the vehicles in the experimenal result of LEGO MINDSTORMS EV3. The experiment is illustrated in Figure 4. The trajectroy of the vehicle positions is shown in Figure 5. The low-speed and the right lane vehicles travel at the constant speed, just like as the siumulation in Figure 3. The trajectories of the vehicles in the platoon are also similar to them in the siumulation in Figure 3. The differences between the vehicles in the platoon in Figures 3 and 5 are caused by the delay time of the velocity control model and the noise included in the input data such as the sensor information, the lane changing of vehicles, and so on.

Figure 4. Experimental status.

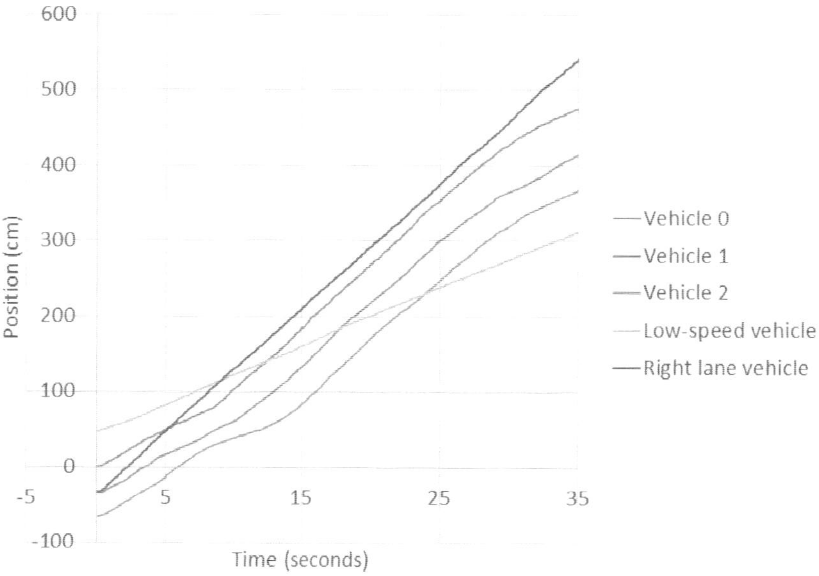

Figure 5. Trajectory of vehicle positions in experiment of LEGO MINDSTORM.

4 CONCLUSIONS

This paper describes the velocity control model of the vehicles in the platoon, when the platoon avoids the vehicle travelling on the other lane and overtakes the preceding traveling vehicle. The model is defined as the Chandler model which controls the vehicle velocity according to the velocity difference between the vehicle and its preceding vehicles. The overtaking condition which the vehicles in the platoon can overtake the preceding travelling vehicle is defined from the positions and velocities of the vehicles and their distances.

The model was applied for the velocity control of the vehicles in the computer simulation and the experiments of LEGO MINDSTORMS EX3. The experimental result was compared with the computer simulation result. The vehicle behavior in the experimental result is agreed relatively well with the vehicle behavior in the computer simulation result although the slight difference between the computer simulation and the experimental results is observed. It is assumed that the difference is caused by the delay time of the velocity control model and the noise included in the input data such as the sensor information, the lane changing of vehicles, and so on.

Acknowledgments

This work was supported by JSPS KAKENHI Grant Number 15K05760.

References

Bierley, R. L., Investigation of an Intervehicle Spacing Display, *Highway Research Record*, 25, 58-75, 1963.

Chandler, R. E., Herman, R., and Montroll, E. W., Traffic Dynamics: Studies in Car Following, *Operations Research*, 6(2), 165-184, 1958.

Gazis, D. C., Herman, R., and Rothery, R. W., Non-linear Follow-the-Leader Models of Traffic Flow, *Operations Research*, 9, 545-567, 1961.

Helly, W., Simulation of Bottlenecks in Single-Lane Traffic Flow, *Theory of Traffic Flow*, 207-238, 1959.

Kita, E., Asahina, K., Ushida, C., Wakita, Y., and Tamaki, T., *Vehicle Platoon Simulation Based on Multi-Leader Vehicle Following Model*, Proceedings of The 9th International Conference on Engineering Computational Technology, CD-ROM, 2014.

Kita, E., Takaue, H., Yamada, M., Sakamoto, H., *Simulation of Vehicle Separated from Vehicle Platoon,* 24th International Congress of Theoretical and Applied Mechanics (ICTAM 2016*)*, 2016.

Streamlining Information Transfer between Construction and Structural Engineering
Edited by Shiau, J., Vimonsatit, V., Yazdani, S., and Singh, A.

ISBN: 978-0-9960437-7-9

BUSINESS SURVIVAL RATE IN CONSTRUCTION INDUSTRY IN RELATION TO OTHER INDUSTRIES: A COMPARATIVE ANALYSIS

IFTE CHOUDHURY

Dept of Construction Science, Texas A&M University, College Station, USA

Construction industry is one of the largest industrial sector in the United States that employs close to ten million people and makes a high contribution to the growth of the country's economy. In spite of the huge impact that the industry has on the US economy, construction businesses have a hard time surviving in the market, with construction companies having the least survival rate among all the industries. Five-year survival rate of construction companies is one of the lowest compared to other industries. This study aims at providing evidence that the construction industry suffers the most as compared to the other industries in terms of business survival rate. A General Linear Model was used for statistical analysis. Results show a significant difference between the construction industry and other industries providing evidence that the construction industry businesses have the least survival rate.

Keywords: Business failure, Business entry rate, Construction sector, General linear model, US economy.

1 INTRODUCTION

Business survival rate is the proportion of companies in an industry remaining operational at the end of a period of time. It can be measured on a yearly basis by finding out the difference between yearly entry and failure rates of enterprises in an industry.

Failure in business is considered to be a state of insolvency which is an event when a company is unable to meet its liabilities (Clusel and Lagarde 2011). Frederikslust (1978) defines failure of a business as an inability to meet its obligations. Rate of failures vary from industry to industry and also from business to business in the same industry.

The failure of businesses has a significant impact on the country's economy. Business failure in the construction industry, which historically accounts for about ten percent to the national gross domestic product and employs about ten million workers (Nunnally 2011), must have a significant impact on the economy of the nation.

The purpose of this study was to make an analysis of companies surviving in construction sector, on a yearly basis, compared with other industrial sectors. Another objective was to find out whether there is a relationship between the rates of emergence of new companies and survival on a yearly basis in nine different sectors of industry in the United States. The sectors of industry chosen were (1) agriculture, forestry, and fishing (AGR); (2) construction (CON); (3) manufacturing (MAN); (4) mining (MIN); (5) transportation, communication and public utilities (TCU); (6) wholesale trade (WHO); (7) retail trade (RET); (8) finance, insurance, and real estate (FIRE); and (9) services (SRV).

2 METHODOLOGY

2.1 Study Population

The study population was nine different industrial sectors in the United States with a longitudinal database of 38 years from 1977 to 2014. The sample size was 342. Unit of analysis was the business entity or company in an industrial sector.

2.2 Data Collection

Data related to business attributes of nine different industrial sectors in the United States was collected from the database of the United States Census Bureau (n.d.). The data for the study consisted of all business related information of the industrial sectors for 38 years from 1977 to 2014. The information included number of establishments in operation at the beginning of a year, number of new entries per year, and number of business exits per year.

2.3 Variables

The following variables were used in the study:

2.3.1 *Entry rate (ENTRATE)*

This variable identifies the rate at which new businesses emerge and start operations in a particular industrial sector. It was operationalized as a yearly rate as follows:

$$\left(\frac{\text{No.of new companies entering in year } i}{\text{Total no.of companies in operation in year } i}\right) * 100 \tag{1}$$

2.3.2 *Survival rate (SURVRATE)*

This variable is the difference between yearly entry and exit rates of companies in a particular industrial sector. It was operationalized as a yearly rate as follows:

$$\left(\frac{\text{No.of new companies entering in year } i - \text{No.of companies exiting in year } i}{\text{Total no.of companies in operation in year } i}\right) * 100 \tag{2}$$

2.3.3 *Sector (SECTOR)*

This is a categorical variable that classifies companies by the industry under which the establishment is categorized. The industries were classified as follows: (1) Agriculture, forestry, and fishing (AGR); (2) Construction (CON); (3) Manufacturing (MAN); (4) Mining (MIN); (5) Transportation, communication and public utilities (TCU); (6) Wholesale trade (WHO); (7) Retail trade (RET); (8) Finance, insurance, and real estate (FIRE); and (9) Services (SRV).

3 ANALYSIS AND RESULTS

The data was analyzed using General Linear Model (GLM). It is an analysis of variance procedure in which the calculations are performed using a least squares regression approach to describe the statistical relationship between one or more predictors and a continuous response variable. Predictors can be factors and covariates. The design may either be balanced or unbalanced. GLM can perform multiple comparisons among or between factor level means to find significant differences. The following model was used for analysis:

$$SURVRATE = \beta_0 + \beta_1 SECTOR + \beta_2 ENTRATE + \varepsilon \quad (3)$$

where (1) ENTRATE = the rate at which new businesses emerge and start operations in a particular industrial sector, (2) SECTOR = categorical variable classifying companies by industrial sectors under which the establishment is categorized, (3) SURVRATE = the difference between yearly entry and exit rates of companies in a particular industrial sector, (4) β_0 = intercept, (5) β_1 and β_2 = regression coefficients, and (6) ε = error term.

Results of the analysis are shown in Tables 1, 2, and 3.

Table 1. Results of general linear model analysis.

Source	F-value	Significance (p-value)
Corrected Model[a]	204.660	0.000
Intercept	276.742	0.000
ENTRATE	1239.314	0.000
SECTOR	29.325	0.000

a. $R^2 = 0.847$ (Adjusted $R^2 = 0.843$)

Table 2. Parameter estimates (t-values).

Parameter	Coefficient	t-value	Significance (p-value)
Intercept	-4.166	-13.620	0.000
ENTRATE	.858	35.204	0.000
SECTOR = AGR	-1.038	-3.583	0.000
SECTOR = CON	-3.168	-11.439	0.000
SECTOR = FIRE	.359	1.364	0.173
SECTOR = MAN	-.766	-2.958	0.003
SECTOR = MIN	-1.282	-4.873	0.000
SECTOR = RET	-1.630	-6.198	0.000
SECTOR = SRV	-.510	-1.927	0.055
SECTOR = TCU	-.864	-3.157	0.002
SECTOR = WHO	0[a]	.	.

a. This parameter is set to zero because it is redundant.

Table 3. Pairwise comparison of SURVRATE.

Comparison	Mean difference	Significance (p-value)
CON vs AGR	-2.130*	0.000
CON vs FIRE	-3.527*	0.000
CON vs MAN	-2.402*	0.000
CON vs MIN	-1.886*	0.000
CON vs RET	-1.538*	0.000
CON vs SRV	-2.658*	0.000
CON vs TCU	-2.304*	0.000
CON vs WHO	-3.168*	0.000

* The mean difference is significant at the .05 level.

The model has been derived from empirical data. Predictive efficacy of this particular model was found to be quite high with an R^2 of 0.847, and an adjusted R^2 of 0.843. This means that about 84 percent of the variances in survival rates are explained by the variables included in the model. Figure 1 shows how close the data is fitted to the regression line.

The F-value of the model was found to be 204.660, which is statistically significant at less than the 0.0001 level. It indicates that the model, as a whole, accounts quite well for the behavior of the predictor variables.

ENTRATE, as hypothesized, was found to have a statistically significant positive relationship with the dependent variable SURVRATE at a level of significance of less than 0.0001 (Figure 1). When the data is segregated for the construction industry, the regression line seems to have a much closer fit (Figure 2).

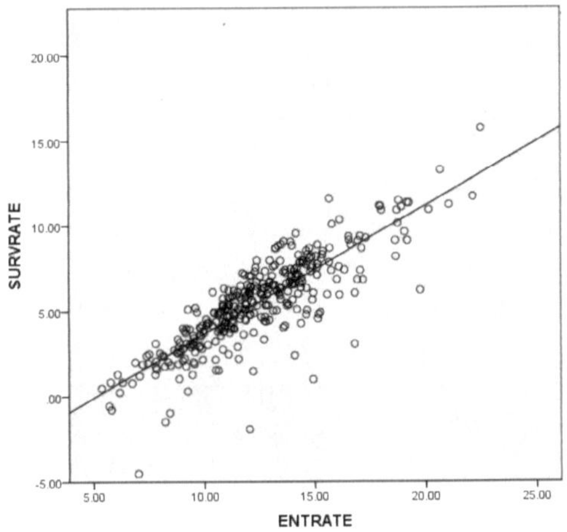

Figure 1. Relationship between entry and survival rates in all industrial sectors.

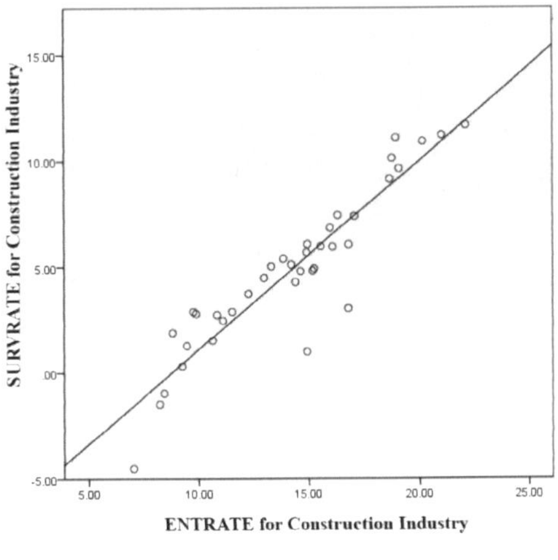

Figure 2. Relationship between entry and survival rates only for construction industry.

SECTOR also had a statistically significant relationship with SURVRATE at the level of significance less than 0.0001. All industrial sectors, except FIRE and SRV, were found to have statistically significant relationship with SURVRATE. The mean SURVRATE in construction industry was found to be the lowest among all sectors.

4 DISCUSSIONS

Survival rate of businesses, as found in this study, is a function the emergence of new enterprises and industrial sectors. But the factors that trigger business failure were beyond the scope of this research.

Kangari (1988) stated long back that "the construction industry has characteristics that sharply distinguish it from other sectors of the economy. It is fragmented, very sensitive to economic cycles, and highly competitive because of the large number of firms and relative ease of entry (p. 173)." This discussion highlights some of the probable determinants of survival rate specific to the construction industry.

According to Kisner and Fosbroke (1994), the construction industry is plagued with injuries and hazards. The construction industry has the highest fatal injuries and workdays lost among all the industries (Dong *et al.* 2014). Because of higher degrees of exposures to occupational hazards in construction industry compared to other sectors, construction industry confronts an increased level of risks associated with safety. At times, it may lead to business closures.

Another risk that sets the construction industry apart from other industries is the seasonal slowdown that is witnessed in extremely hot or cold seasons. According to Koehn and Brown (1985), works in extreme weather conditions (either cold or hot) are often stalled or slackened in most of the United States. Therefore, seasonal sluggishness is one of the risks that can have an enormous impact on the survival of construction businesses.

The risk associated with change orders and project delays are commonly occurring risks in the construction industry. Change orders, leading to delays, may sometimes have a negative impact on the productivity and efficiency of a project (Moselhi *et al.* 2005). A study by Srdic and Selih (2015) indicates that delays in projects can often lead to additional costs, conflicts and litigation. It often renders businesses vulnerable, making them struggle for survival.

Poor quality of work leads to increased work hours for labors and equipment, resulting in increased overall expenses for a contractor (Love *et al.* 2016). Survival of companies may be at stake if quality of work is not maintained.

One last thought about the low survival rate of businesses in construction industry is related to innovations. One of the main reasons for low productivity and quality of products in comparison with other industries is believed to be a lack of innovation (Yusof *et al.* 2014). Existing empirical studies consistently find a positive relationship between innovative activity and company survival (Yusof *et al.* 2014, Buddelmeyer *et al.* 2006). Successful innovations have been found to have clear positive effects on survival in other industries.

5 CONCLUSIONS

This study was conducted to find out whether survival rates of businesses are lower than those in other industrial sectors in the United States. The objectives of the study were twofold: (1) to make a comparative analysis between construction sector and other industrial sectors with respect to survival rates on a yearly basis and (2) to find out whether there is a relationship between the rates of emergence of new companies and survival in nine different sectors of industry in the United States. The data, obtained from the United States Census Bureau, was analyzed using a General Linear Model. The spread of the data incorporated a time period of 38 years from 1977 to 2014. The findings provide evidence that the construction industry has the least business survival rate among the nine industrial sectors taken into consideration for this study.

The scope of the study did not include the analysis of the actual determinants of business survival rate in the construction industry. Some conjectures were made on the probable causes that might have an effect on business failure. The factors include risks associated with safety, injuries, and fatalities; seasonal slowdown; changes in scope of work; time overrun; quality of work; labor productivity; and technical innovations. Future studies related to business survival in construction industry may include some or all of these factors to find out the correlates of survival rate in the industry.

References

Buddelmeyer, H., Jensen, P. H., and Webster, E., *Innovation and The Determinants of Firm Survival*, Discussion paper no. 2386. Bonn, Germany: Institute for the Study of Labor, October 2006.

Clusel, S., and Lagarde, D., Reducing the Risks Faced by Small Businesses: The Lifecycle Concept, *Advances in Safety, Reliability and Risk Management, ESREL 2011,* 280, 2011.

Dong, X., Largay, J. A., Wang, X., and Windau, J. A., Fatalities in The Construction Industry: Findings From a Revision of The BLS Occupational Injury and Illness Classification System, *Monthly Labor Review*, US Bureau of Labor Statistics, Washington, DC, July 2014.

Frederikslust, R.A.I. van, *Predictability of Corporate Failure: Models for Prediction of Corporate Failure and For Evaluation of Debt Capacity*, Springer, NY, 1978.

Kangari, R., Business Failure in Construction Industry, *Journal of Construction Engineering and Management*, 114(2), 172-190, 1988.

Kisner, S. M., and Fosbroke, D. E., Injury Hazards in The Construction Industry, *Journal of Occupational and Environmental Medicine*, 36(2), 137-143, 1994.

Koehn, E. and Brown, G., Climatic Effects on Construction, *Journal of Construction Engineering and Management*, 111(2), 129-137, 1985.

Love, P. E., Teo, P., Morrison, J., and Grove, M., Quality and Safety in Construction: Creating a No-Harm Environment, *Journal of Construction Engineering and Management*, 142(8), 1-10, 2016.

Moselhi, O., Assem, I., and El-Rayes, K., Change Orders Impact on Labor Productivity, *Journal of Construction Engineering and Management*, 131(3), 354-359, 2005.

Nunnally, S. W., *Construction Methods and Management*: 8th Edition, Prentice Hall, Columbus, OH, 2011.

Srdić, A. and Šelih, J., Delays in Construction Projects: Causes and Mitigation. Organization, *Technology & Management in Construction: An International Journal*, 7(3), 1383-1389, 2015.

United States Census Bureau, Business Dynamics Statistics, n.d. Retrieved from https://www.census.gov/ces/dataproducts/bds/data_firm.html on March 1, 2017.

Yusof, N. A., Kamal, M. E., Seng, L. K., and Iranmanesh, M., Are Innovations Being Created or Adopted in the Construction Industry? Exploring Innovation in the Construction Industry, *SAGE Open,* July-September, 1-9, 2014.

Streamlining Information Transfer between Construction and Structural Engineering
Edited by Shiau, J., Vimonsatit, V., Yazdani, S., and Singh, A.

ISBN: 978-0-9960437-7-9

DESIGN FEATURES FOR IMPROVED CONSTRUCTABILITY OF MARINE PIER PROJECTS

PANAGIOTIS MITROPOULOS and VICTOR TAJIMA

Dept of Civil, Construction, and Environmental Engineering, San Diego State University, San Diego, USA

Marine construction projects face significant challenges due to their physical environment as well as safety and environmental regulations. A well-engineered design with constructability in mind is essential for the successful construction of such projects. This study investigated the design features that created significant constructability benefits on the Pier 12 Replacement Project at San Diego, California. The US $82 Million fixed price design-build project involved the construction of a new 1,500-foot pier that could fulfill the utility needs of multiple stationed ships. The following design features were important in improving the constructability of the new pier: (1) Selection of plumb piles; (2) Design of uniform and longer pile spacing; (3) Increased depth of pile dowel tubes; (4) Use of T-headed pile dowels; (5) Avoiding the use of pile caps; (6) Minimizing the fluctuations of deck soffit elevations; (7) Minimizing the expansion joints; (8) Minimizing the variety of rebar sizes; and (9) Increasing the depth of concrete rebar cover.

Keywords: Marine construction, Design-build, Piles, Dowels, Concrete, Deck.

1 INTRODUCTION

Construction efficiency and costs are heavily dependent on the quality of the design. The Construction Industry Institute (CII) defined constructability as "the optimum use of construction knowledge and experience in planning design, procurement, and field operations to achieve overall project objectives" (CII 1993). It is the "potential 'bridge' between the design and construction disciplines. Applying constructability concepts at the early stage of construction projects reduces the cost of design and construction and improves the project schedule, design coordination and worker productivity (CII 1993, Jergeas and Van der Put 2001, Pulaski and Horman 2005, Ruby 2008, Song *et al.* 2009, Raviv *et al.* 2012, Kifokeris and Xenidis 2017). According to the CII (1993), constructability efforts save on average, 4.3% on project cost and 7.5% on schedule. These savings have been shown to represent a 10:1 return on investment in the constructability effort.

Marine construction projects take place over and under the water, in order to build structures such as wharves, piers, terminals, bridges, marinas and outfalls. The marine environment, the design requirements, and the strict safety and environmental regulations, create significant challenges for such projects. A well-engineered design with constructability kept in mind is essential for successful construction of such project. The objective of this study was to identify design features that created significant constructability benefits on the Pier 12 Replacement

Project at San Diego, California. This design-built project was performed by the same contractor and designer, who designed and built similar piers (Pier 10 and 11 replacement projects) at the same naval base. As a result of the lessons learned, the design engineers incorporated several changes in the design of the new Pier 12 to improve constructability. This investigation focused on two major operations; pile driving and concrete construction, which included building and setting the formwork, fabricating rebar layouts, and pouring concrete.

2 BACKGROUND

Previous research has identified several constructability principles: (1) Design simplicity, which involves the minimization of the number of materials, sizes and components (Zolfagharian and Irizarry 2017), and the simplification of construction details (Kuo and Wium 2014). (2) Standardization and repetition. The repetition of layout (floor grids, dimensions of elements, and floor height) benefits the learning curve and increases productivity. Standardization of material simplifies procurement and material management and allows volume purchase discounts (O'Connor *et al.* 1987, Zolfagharian and Irizarry 2017). (3) Preassembled components typically improve productivity, parallel sequencing of activities, increase safety, improve quality control, reduce the need for scaffolding (O'Connor *et al.* 1987), and reduced exposure to weather conditions. (4) Consideration of interfaces where joining of materials are difficult (Kuo and Wium 2014). (5) Design solutions that can reduce sensitivity to adverse weather conditions, e.g. by reducing the quality sensitive work conducted outdoors (O'Connor *et al.* 1987). (6) Enhanced accessibility of manpower, materials, and equipment (O'Connor *et al.* 1987, Zolfagharian and Irizarry 2017).

Construction researchers have developed guidelines for the constructability of reinforced concrete structures (Fischer 1991), steel structures (Ruby 2008), façades (Horn 2015), formwork (Jarkas 2010, Jiang *et al.* 2015), and commercial buildings (Zolfagharian and Irizarry 2017). With regards to constructability of marine pier structures, the use of large precast components has been used to accelerate pier construction, saving time on formwork and concrete curing (Rosner *et. al* 2010, Jaradat 2013).

3 METHOD

This investigation used the case study method at the project level. Data were collected through interviews with the project personnel (design engineers, project manager, estimator, quality control and field personnel) and extensive field observations. One of the authors was working as a project engineer on the project and had access to the personnel and operations. The design engineers and project management staff had worked previously on several similar piers (original Pier 12, Pier 10, etc.). The engineers and contractor discussed the lessons learned from previous pier projects and explained the construction advantages of the selected design features. Interviews with craft employees (laborers, pile drivers, surveyors, and foremen) discussed the difficulties and problems that field encountered. Finally, extensive observations and documentation of the construction activities recorded the details of the operations examined.

4 PROJECT DESCRIPTION

The Pier 12 Replacement project involved the construction of a new pier 1,500 feet long, and 117 feet wide that could fulfill the utility needs of multiple stationed ships. The cost was US $82 Million, and the schedule was 25 months. The delivery method was design-build with fixed price contract. The major operations included: (1) Dredging of approximately 130,000 cubic yards around the pier area; (2) Demolition of the existing 1,460-foot Pier 12, (3) Driving 512 24-inch octagonal precast prestressed concrete piles configured into 64 bents (rows) of piles that contain eight piles; (4) Construction of the new Pier 12; (5) Construction of one guard tower; and (6) Installation of 56 100-ton mooring bollards; and (7) Driving 204 concrete fender piles. The construction of the new pier required (1) Construction of the 24 inch thick main deck surrounded by a 22.5-foot utility trench making the entire pier deck approximately 175,000 square feet and 15,000 cubic yards of concrete; (2) Construction of a utility system consisting of a plumbing network that fulfills the ship's needs for fresh water, oily waste, and sewage; and (3) Installation of the electrical package. The investigation focused on the pile driving and concrete operations.

The pile driving operation included the following activities: (1) Receive and stack the concrete piles; (2) Load the piles onto a 7,500 square foot flat barge using a derrick barge crane; (3) Drive the piles in the bay floor using the derrick barge crane, and a D-80 hammer; (5) Cut the top of the piles at the proper elevation (or pour additional length if needed), and (6) set the dowels. The formwork and rebar operation included the following activities: (1) Set friction collars onto concrete piles; (2) Set I-beams on top of the friction collars; (3) Set prefabricated formwork panels and parallam beams on top of I-beams to elevation; (4) Set sidewall formwork panels and lateral supports; and (5) Install reinforcement bars and electrical conduits. Concrete placing included the primary pour for the 72-foot long deck section over four center piles, and secondary pour for the utility trench sections over the outer four piles. Finally, the formwork panels were stripped by removing the friction collars while the I-beams and formwork panels were secured to the concrete deck with stripping cables, and then cutting the stripping cables, dropping the I-beams and panels into the water.

5 RESULTS: CONSTRUCTABILITY OPPORTUNITIES

The design-build contractor identified nine design decisions that were significant for the constructability of Pier 12.

5.1 Plumb Instead of Battered Piles

The use of plumb piles provided several benefits: (1) *Fewer piles*. Pier 12 required 512 plumb piles, as opposed to same size Pier 10 that called out for 362 battered piles and 334 plumb piles totaling 696 piles. This was 184 fewer piles to be transported, handled and driven. (2) *Reduced cycle time*. When driving the battered piles of Pier 10, the crane had to relocate and re-anchor up to eight times for every two bents of piles it drove. In contrast, for plum piles of Pier 12, the crane had to relocate and re-anchor five times for every two bents of piles driven. This decrease in the number of moves drastically reduced the cycle time and increased productivity. On Pier 10, an average of 8 battered piles were driven per day, while on Pier 12 an average 11 plumb piles were driven per day. (3) *Lower pile driving cost*. Considering the cost for materials, labor and equipment, using plumb pile over battered piles saved the project about $4,500 per pile and a total of $828,000. (4) *Faster installation of friction collars*. Since battered piles are set at an angle, the workers must carefully set the friction collars and then use an alignment collar in order to evenly set the I-beams above it. The use of plumb piles eliminated the need for alignment collars and made installation of squeeze collars faster. This change saved money with equipment and labor cost adding up to about $125,000. (5) *Improved safety*. Installing battered piles involved higher safety risk due to chipping off, rubbing on gate, and risk of ears on hammer breaking.

5.2 Pile Spacing

The spacing between pile bents determines the length of the deck panels. Larger spacing increases the length of formwork and makes formwork easier to set, but it increases the concrete load on the form, which is in turn requires more reinforcement bars, as well as more friction collars. Thus, to determine an economical length of the panels, the contractor considered the trade-off between efficiency of formwork installation against the increased friction collars and rebar required. For Pier 12, the designer considered three spacing options and selected a standard bent-to-bent span distance of 24 feet except for the last two spans (10 and 16 feet). A longer and a shorter span were found to be less economical. The uniform distance resulted in uniform size deck panels (48 feet long). This reduces potential mistakes due to a variety of panel sizes. Standard spacing also allows reuse of formwork panels down the pier, which reduces waste of formwork material.

5.3 Depth of Pile Dowel Tubes

Each precast pile required eight rebar dowels to tie into the reinforcement bars in the deck of the pier. The spec required that these dowels were embedded 6 feet in the pile. For this, each pile had at the top eight corrugated metal tubes of 2 inch diameter for the dowels to be embedded and grouted into place. On the Pier 12 project, the contractor ordered the piles with dowel tubes of 16 feet. This way, a pile could be cut up to 10 feet (in order to be at the correct elevation) without requiring extra drilling to embed the dowels. This "lesson" was learned on a previous project, where the contractor used dowel tubes 10 feet deep. However, a large portion of the piles had to be cut by more than 4 feet, and the contractor had to perform extensive drilling to achieve embedment of 6 feet. The labor and equipment time and cost to drill more embedment depth Is much greater than just filling the extra embedment tube length with more grout. The 16 feet depth increased the tolerance for the variation in pile depth.

5.4 T-headed Instead of Angled Pile Dowels

The original design called for 90-degree angled pile dowels. However, because the piles are not at the perfect location, the angled dowels were often out of position and did not meet the required minimum cover depth of three inches. This led to numerous rebar dowels needing to be bent to make the needed 3-inch cover. Bending these rebar dowels out in the field was not an easy task and cost the crew great deal of time. Other dowels had to be cut and retrofitted with a T-head using a lenton rebar splice. After the contractor identified the problem, they changed the design to T- headed dowels. These dowels were more expensive and needed a three-week lead-time but drastically reduced the need to rework the dowels, as they provided larger tolerance to meet the rebar cover requirements.

5.5 Deck Design Without Pile Caps

Pile caps and pile cap beams are needed when a thinner deck is thinner, or a precast deck is used. In those cases, the pile caps are formed and poured before the deck, adding another step to the operation. Pile caps are also labor intensive and require extensive formwork, thus increasing material and production costs. Pile caps also pose an easy target for deterioration and corrosion. In Pier 12, there were no separate pile caps; the required reinforcement was part of the thick deck. The main deck was 24 inches thick, but at the along the sides of the pier the utility trench the depth of the trench reached 4. 5 feet. With the thick deck, the pile cap was incorporated in the deck soffit, which eliminated the need for separate formwork and pouring.

5.6 Flush Deck Soffit

In Pier 12 Replacement, the design of the deck soffit was flush. Eliminating fluctuations in the elevation of the deck soffit greatly simplified the formwork fabrication and rebar installation. This was a change from the old Pier 12, which had four different soffit profiles. The changing elevations across the bent as well as from bent to bent have a substantial impact on the labor costs for formwork and rebar fabrication and installation. Having the experience of the old pier 12, the designer and contractor simplified the design of the deck soffit for the new pier 12. Minimizing the fluctuations of the deck soffit elevations at Pier 12 created significant savings, as the fabrication of the deck formwork was a lot simpler and faster. The saving in labor and time (without accounting for potential errors and rework) exceeded the additional cost of material.

5.7 Minimum Number of Expansion Joints

The installation of expansion joints is very labor intensive and requires extensive formwork. The Pier 10 replacement project required three construction joints with base isolators along the structure. The installation of the expansion joint and base isolators required a total of 650 man-hours and 6 weeks duration. In addition, the expansion joints in the middle of the pier also prevent the reuse of the deck panels and formwork. After considering the lessons from Pier 10, the designer and contractor decided to install at Pier 12 only one construction joint with base isolator at the base of the pier since the use of plumb piles gave the structure more flexibility from internal and external forces.

5.8 Low Variety of Rebar Sizes

In Pier 12, the designer also tried to keep the complexity of rebar configurations at minimum, by standardizing rebar sizes for each work area of the pier and using few rebar sizes at each area. The majority of the concrete deck used #6 and #8 size rebar, the utility trenches used #5 and #10 rebar, the precast electrical vaults used #4 and #5 rebar, the precast trench lids used #4 and #9 rebar. The simplicity and standardization of the rebar resulted in faster installation and fewer mistakes when installing rebar cages.

5.9 Depth of Rebar Cover

On the Pier 12 project, the design specified 3 ½ -inch concrete cover with a DCI corrosion-inhibiting admixture to inhibit rebar corrosion. The reason was to avoid the difficulties faced on Pier 10, which used epoxy- coated rebar and a standard 3-inch concrete cover. Since the reinforcement bars in Pier 10 had a cover less than 3 ½ -inches, all bars had to be coated with epoxy due to the corrosive marine environment. The epoxy coat was applied by the manufacturer prior to shipping. The contractor identified the following disadvantages of the coated rebar: (1) it was about $1 per pound more expensive than non-epoxied rebar; (2) it required very careful handling - if the epoxy was chipped or damaged, the bars had to be patched with epoxy; (3) the crew found it difficult walking on the rebar due to the slippery conditions. The crew also reported difficulties clearing debris from the rebar when working with a 3 inches cover.

On Pier 12, the design specified 3-½ inch cover with a corrosion inhibiting concrete admixture. This required more concrete but eliminated the need for epoxied rebar and all the associated difficulties. It also allowed the ironworkers to bend any rebar on site if needed, since damaging the epoxy was not an issue and made easier to clear out debris in the rebar.

6 CONCLUSIONS

The study identified nine constructability improvements that the designer and builder incorporated in the design of Pier 12 replacement project based on their experience with previous similar piers. Most of these practices aimed at creating a simple and standardized deck design. The *pile spacing*, the *flush soffit*, the *avoidance of pile caps*, and the *minimum use of expansion joints* reduced the variety of formwork panels, simplified formwork fabrication and installation and increased the reuse of the panels. The trade-off was increased quantity of concrete.

The *rebar design*, and the 3 ½ inch *rebar cover* reduced the difficulties of the rebar activity. The trade-off was again increased quantity of concrete. The *plumb piles* and *pile spacing* reduced the scope, complexity and difficulties of piling driving and friction collar installation. The trade-off was an increased flexibility of the structure, which in this case was a desired attribute. Finally, two other features—the increased *depth of dowel embedment*, and the use of *T-headed dowels*, demonstrate the importance of design details that provide adequate tolerance and minimize the costs of adjustments to the variability of field conditions, such as extra drilling, and modifying the dowels, which generate significant additional costs.

References

CII, *Constructability Implementation Guide, Special Publication 34–1,* Construction Industry Institute, University of Texas at Austin, Austin, Texas, 1993.

Fischer, M., *Constructibility Input to Preliminary Design of Reinforced Concrete Structures, Technical Rep. No. 64,* Center for Integrated Facility Engineering, Stanford Univ., Stanford, USA, 1991.

Horn, A., *Integrating Constructability into Conceptual Structural Design and Optimization,* MS Thesis, Massachusetts Institute of Technology, Massachusetts, 2015.

Jaradat, O., *Rapid Bridge Construction Methods Applied to Piers and Wharves,* Ports 2013, 1275-1284, 2013.

Jarkas, A. M., Buildability Factors Affecting Formwork Labour Productivity of Building Floors, *Canadian J.ournal of Civil Engineering*, 37, 1383–1394, 2010.

Jergeas, G. and Van der Put, J., Benefits of Constructability on Construction Projects, *Journal Construction Engineering and Management*, 127(2), 281-290, 2001.

Jiang, L., Leicht, R. M., and Messner, J. I., *Towards Automated Constructability Checking: A Case Study of Aligning Design Information with Formwork Decisions,* Computing in Civil Eng., 531-539, 2015.

Kifokeris, D. and Xenidis, Y., Constructability: Outline of Past, Present, and Future Research, *Journal Construction Engineering and Management*, 143(8) 1-13, 2017.

Kuo, V. and Wium, J., The Management of Constructability Knowledge in The Building Industry Through Lessons Learnt Programme, J. *of the South African Institution of Civil Eng.*, 56(1), 20-27, 2014.

O'Connor, J.T., Rusch, S.E., and Schulz, M.J., Constructability Concepts for Engineering and Procurement, *Journal Construction Engineering and Management*, 113(2), 235–248, 1987.

Pulaski, M. and Horman, M., Organizing Constructability Knowledge for Design, *Journal Construction Engineering and Management*, 131(8), 911–919, 2005.

Raviv, G., Shapira, A. F., and Sacks, R., Relationships Between Methods for Constructability Analysis during Design and Constructability Failures in Projects, in *Construction Challenges in a Flat World*, ASCE, Cai, H., Kandil, A., Hastak, M., and Dunston, P. S. (eds), 515-524, W Lafayette, Indiana, 2012.

Rosner, C., Shafer, T., and Byres, R., *Rapid Pier Delivery Using Precast Concrete Components,* Ports 2010, 1098-1107, 2010.

Ruby, D. I., *Constructability of Structural Steel Buildings*, AISC, Chicago, 2008.

Song, L., Mohamed, Y., and AbouRizk, S. M., Early Contractor Involvement in Design and Its Impact on Construction Schedule Performance, *Journal of Management in Engineering*, 25(1), 12–20, 2009.

Zolfagharian, S. and Irizarry, J., Constructability Assessment Model for Commercial Building Designs in the United States, *Journal Construction Engineering and Management*, 143(8), 2017.

Streamlining Information Transfer between Construction and Structural Engineering
Edited by Shiau, J., Vimonsatit, V., Yazdani, S., and Singh, A.

ISBN: 978-0-9960437-7-9

DIGITALISATION IN CONSTRUCTION INDUSTRY: CONSTRUCTION PROFESSIONALS PERSPECTIVE

DOUGLAS AGHIMIEN, CLINTON AIGBAVBOA, AYODEJI OKE, and NTEBOHENG KOLOKO

Sustainable Human Settlement and Construction Research Centre, Faculty of Engineering and the Built Environment, University of Johannesburg, Johannesburg, South Africa

This study presents the result of an assessment of digitalisation within the South African construction industry from the construction professionals' perspective. The study adopted a quantitative approach through a questionnaire survey carried out among construction professionals in Gauteng Province, South Africa. Appropriate descriptive statistical tools were used in analysing the data gathered. The study revealed that the major stages of construction where digitalisation is evident are the design phase, and the feasibility phase. The major construction processes where it is mostly visible are, construction cost control, cost planning, preliminary cost estimate and building system analysis. In addition, the most significant benefits to be derived from digitalisation in the industry includes time-saving in construction projects delivery, increase productivity, increase speed of work, increase document quality, speeding up of response time, and simpler working methods. This study contributes to the body of knowledge as it reveals the happenings surrounding digitalisation within the South African construction industry; an aspect that seems to be lacking research attention.

Keywords: Digital technologies, Digital transformation, Digital tools, ICT, Technological advancement, Industry4.0.

1 INTRODUCTION

Digitalisation has become a popular concept around the world today due to its ability to create efficiency in operations, effectiveness, and provide new opportunities. Industries such as banking, manufacturing, and retailing have all come to understand the benefit of digitalisation and have keyed into the future by using it as a new approach for ensuring competitive gain and efficiency (Osunsanmi *et al.* 2018). While these other industries have adopted the use of this obviously beneficial development, the construction industry is yet to fully embrace it in the delivery of its services. This failure to harness the inherent benefit of digitalisation has significantly affected the industry, especially in developing countries, as no major transformation is evident within their construction sector (Castagnino *et al.* 2016). The case is no different with South Africa, as most construction services are still being delivered using traditional delivery methods. This has adversely affected the growth of the industry, the quality of services rendered, and the satisfaction thereof. Halim (2010) have earlier noted that there is no industry that can effectively succeed in solving its growth problems unless its goals and strategies towards achieving these goals are rooted in the realm of knowledge creation, and driven by digital technology.

Ibem and Laryea (2014) submitted that the use of digital technologies has only recently become a common idea among construction participants, because of the attention garnered over time by BIM (Building Information Modelling). Therefore, if construction services are to be delivered with less effort and higher accuracy, then digitalisation among construction participants must change from being just an "idea" to becoming a "norm". There is the need for the industry to change from just being aware and partially adopting digitalisation through BIM, to fully implementing it in all facets of its activities.

This study recognises the fact that there are different published studies that have been done on digital technology, and information and communication technology (ICT) in construction in countries around the world (El-Mashaleh 2007, Ibem and Laryea 2014, Oladapo 2007, Peansupap and Walker 2015) just to mention a few. Benefits of the use of these technological advancements have been revealed. However, there has been a paucity of information in literature as regards digitalisation within the South African construction industry. This industry just like its counterparts in every other country plays a vital role in the economic development of South Africa. Its proper functioning and alignment with current happenings within the construction world is necessary for it to stay relevant and serve its purpose in the socio-economic development of the country. It is based on this knowledge that this study examined digitalisation in the delivery of construction projects in the South African construction industry as well as its perceived benefits. The study contributes to the body of knowledge as it reveals the happenings surrounding digitalisation within the South African construction industry; an aspect that seems to be lacking research attention.

2 OVERVIEW OF DIGITALISATION

Rouse (2017) posited that digitalisation is the process of organising and transforming information into a distinct data set that is digital in nature. This transformed information becomes the binary data that is understandable and can be processed by computers and other devices with computing capacity. According to Berger (2016) digitalisation is more about working with tools and practices based on information and communication technology. Castagnino *et al.* (2016) observed that digitalisation can be crucial in the transformation of the three major life-cycle phases of construction projects; that is the design and engineering phase, construction phase and operation phase. Adopting digital technologies for the various construction processes within these different construction life-cycle phases will largely reduce uncertainty and promote satisfactory construction products. The advantages of utilising digital technologies as part of development in construction are great. In fact, Hashim *et al.* (2013) identified the benefits of using digital technology in procuring construction projects to include among others, improvement in process quality, adequate savings in construction cost, adequate clients and participants satisfaction, increased responsiveness and productivity, expansion in market, and project delivered in the most effective manner. Staub-French and Fischer (2017) noted that using digital technologies go a long way in distinguishing clashes in design, and facilitates outlines of schedule in the construction procedure. Aside from this, it was also discovered that less rework and fewer change orders are recorded when digitalisation is adopted. These rework are in most cases costly and occur due to countless encounter that is undiscovered until they are experienced in the construction phase. Above all, the adoption of digitalisation gives better cost control.

3 RESEARCH METHODOLOGY

This study adopted a survey approach and quantitative data were gathered from construction professionals in Gauteng province, South Africa, most of who have considerable years of experience within the construction industry. These construction professionals are Architects, Construction managers, Engineers, General Contractors and Quantity Surveyors. These professionals are directly involved in the delivery of construction project, hence their selection for the study. The instrument for data collection from these professionals was a questionnaire. The questionnaire was employed because of its ease of use and its ability to cover a wide range of participants (Tan and Yeoh 2011). Sixty (60) questionnaires were conveniently distributed to the identified construction professionals, out of which 52 were retrieved and deemed fit for analysis. Data gathered were analysed using percentage for background information of the respondents. Mean Item score (MIS) was used to rank the use of digital technology in the different construction processes and stages, and its perceived benefits. Kruskal-Wallis H-Test; a non-parametric test used in ascertaining the significant difference in the perception of three or more categories of respondents, was employed in determining consistency in the opinion of the different professionals sampled. Cronbach's alpha values of 0.716, 0.735 and 0.874 were also derived for the use of digitalisation in the construction process, construction stages, and the perceived benefits of digitalisation. This result implies that the instrument used is reliable as they tend closer to 1.0

4 FINDINGS AND DISCUSSIONS

4.1 Digitalisation in the Construction Industry

In determining the state of digitalisation within the South African construction industry, different construction stages were presented to the respondent and they were asked to rate these construction stages base on the level of digital technology used in them using a Likert scale of 1 to 5. Result in Table 1 shows the ranking of these stages with their associated significant p-value derived from Kruskal-Wallis H-test conducted. From the table, it is evident that all the stages have a p-value of above 0.05. This implies that at 95% confidence level, there is a statistically significant relationship in the view of the different construction professionals as regards the adoption of digitalisation in all the construction stages. A look at the table also shows that all the assessed construction stages have a mean value of well above the average of 3.0. This implies that to a considerable extent, digital technologies are being used within the industry for the identified construction stages. Digitalisation in the South African construction industry is more evident at the design phase and the feasibility phase as mean values of 4.40 and 4.12 were derived for both stages.

Table 1. Adoption of digitalisation for different construction stages.

Construction Stages	Mean	Rank	Chi-Sq.	Sig.
Design Phase	4.40	1	5.961	0.241
Feasibility Phase	4.12	2	3.514	0.476
Maintenance Phase	3.88	3	6.326	0.176
Construction Phase	3.85	4	7.537	0.110
Operation Phase	3.79	5	1.925	0.749
Decommissioning Phase	3.71	6	1.493	0.828

Certain construction processes were also identified from the review of related literature and respondent were asked to rate these processes base on the level of digital technology used in them. Result in Table 2 shows that all the processes have a p-value of above 0.05. This implies that there is a statistically significant relationship in the view of the different construction professionals as regards the adoption of digitalisation in all the assessed construction processes.

In addition, a cursory look at the table shows that all the processes assessed have a mean value of above average of 3.0. This implies that to a considerable extent digital technologies are being used within the industry for the identified construction processes.

The key construction process where digital technology is mostly used is in construction cost control. This ranked the highest with a mean value of 4.13. Cost planning, preliminary cost estimate, and building system analysis followed this, while the least aspect is 3D/4D coordination, and maintenance schedule. This result implies that digital technologies geared towards cost reduction and controls especially during designing are mostly being adopted. Relating the earlier result, which shows that digitalisation, is common at the design and feasibility stages, with this result, it is shown that while designing using digital technology, the cost of the project is also being considered. Digital technology is used to carry out preliminary estimate at the feasibility stage so as to determine the possible cost of construction, while at the designing phase, it is used to plan and control cost alongside the design in line with the clients' budget.

Table 2. Adoption of digitalisation for different construction processes.

Construction processes	Mean	Rank	Chi-Sq.	Sig.
Cost control	4.13	1	5.913	0.206
Cost Planning	3.79	2	8.241	0.083
Preliminary cost estimating	3.77	3	4.642	0.326
Building System Analysis	3.71	4	3.688	0.450
Existing condition Modelling	3.62	5	3.400	0.493
Procurement Stages	3.60	6	3.693	0.449
Design Authoring	3.54	7	2.009	0.734
Conflict resolution	3.44	8	1.516	0.824
Structural analysis	3.44	8	3.830	0.429
3D/4D Co-ordination	3.38	10	3.876	0.423
Maintenance Schedule	3.33	11	0.707	0.950

4.3 Perceived Benefits of Digitalisation

Result in Table 3 shows the ranking of the different benefits identified from literature with their associated significant p-value. From the table, it is evident that all the benefits have a p-value of above 0.05. This implies that at 95% confidence level, there is a statistically significant relationship in the view of the different construction professionals as regards the benefits of digitalisation. A look at the table shows that all the identified benefits also have their mean value to be above the average of 3.0. This implies that to a considerable extent with the digitalisation of the industry come the identified benefits.

Table 3. Perceived benefits of digitalisation in the construction industry.

Benefits	Mean	Rank	Chi-Sq.	Sig.
Saves time	4.37	1	1.604	0.808
Increases Productivity	4.18	2	3.512	0.476
Increases speed of work	4.17	3	1.462	0.833
Increases document quality	4.15	4	2.988	0.560
Speeds up response times	4.06	5	2.110	0.716
Simpler working methods	4.06	5	1.020	0.907
More accurate documentation	3.98	7	1.490	0.828
Reduces degree of difficulty	3.96	8	1.687	0.793
Reduces construction error	3.83	9	1.903	0.754
Proportion of new work	3.81	10	1.862	0.761

The most significant benefits to be derived from the digitalisation of the industry includes time saving, increase productivity, increase speed of work, increase document quality, speeding up of response time, and simpler working methods. The least benefits of digitalisation are, reduction of construction error, and the promotion of new work. Despite the fact that these benefits were ranked the least, they have a mean value of well above the average of 3.0, which means that they can still be achieved with the adoption of digitalisation of the industry.

4.4 Discussion

Findings of the study reveal that digitalisation in the South African AEC industry is more evident at the design phase and the feasibility phase than the decommissioning phase of construction projects. The major construction processes where digitalisation is mostly visible are, construction cost control, cost planning, preliminary cost estimate and building system analysis. It is, however, less evident in the maintenance schedule. This means that more attention is given to digitalisation in the cost of construction, while the use of digitalisation in maintenance schedule requires more attention if effective maintenance of construction works is to be achieved. Contrary to Castagnino *et al.* (2016) submission that digitalisation can be crucial in the transformation of the three major life-cycle phases of construction projects, findings of the study shows that the South African construction industry has only adopted digitalisation in the early phase of construction (feasibility and design phase). This digitalisation is mostly in the cost and design aspect, which is similar to happenings in other Middle East/African countries as observed by Jung and Lee (2015). Findings of this study further confirm Jung and Lee (2015) submission that the adoption of digitalisation in maintenance schedule in most African countries is low, as this aspect of construction was rated the least area where digitalisation is being used in South Africa.

Findings also revealed that the most significant benefits to be derived from the digitalisation of the AEC industry includes time-saving in construction projects delivery, increase productivity, increase the speed of work, increase document quality, speeding up of response time, and simpler working methods. This finding further confirms Hashim *et al.* (2013) submission that some of the benefits of using digital technologies in the delivery of construction projects include, improvement in process quality, adequate savings in construction cost, adequate clients and participants' satisfaction, increased responsiveness and productivity, expansion in market, and project delivered in the most effective manner. Findings also agrees with Staub-French and Fischer (2017) submission that digitalisation goes a long way in distinguishing clashes in design, and also facilitates outlines of schedule in the construction procedure. Also, less rework and fewer change orders are recorded when digitalisation is adopted.

5 CONCLUSION

Based on the findings, the study concludes that the major stages of construction where digitalisation is evident are the design phase, and the feasibility phase. The major construction processes where digitalisation is mostly visible are, construction cost control, cost planning, preliminary cost estimate and building system analysis. It is, however, less evident in the maintenance schedule. In addition, the most significant benefits to be derived from the digitalisation in the industry includes time-saving in construction projects delivery, increase productivity, increase the speed of work, increase document quality, speeding up of response time, and simpler working methods.

The study, therefore, recommends the need for further adoption of digital technology in other aspects of construction aside feasibility and designing. This will go a long way in achieving the holistic benefits inherent in digitalisation. Also, more focus is needed in the digitalisation of construction processes such as maintenance schedule. The findings of this study will go a long way in assisting construction professionals and stakeholders in understanding the inherent benefit in digitalisation. However, though result reveals that there is the considerable use of digital technology in some aspect of the construction processes within the South African construction industry, care must be taken in generalising the result as it is based on the perception of construction professionals. Herein lies the limitation of the study. Therefore, further study is needed to effectively determine the practical adoption of digitalisation within the industry.

References

Berger, R., *Digitization in the Construction Industry*, Think Act, 1-16, 2016.

Castagnino, S., Rothballer, C., and Gerbert, P., What's the Future of the Construction Industry? World Economic Forum, 2016. Retrieved from https://www.weforum.org/agenda/2016/04/building-in-the-fourth-industrial-revolution/ on October 2018.

El-Mashaleh, M. S., Benchmarking Information Technology Utilization in The Construction Industry in Jordan, *Journal of Information Technology in Construction,* 12(19), 279-291, 2007.

Halim, H. C., *Assessment of The Application of Information and Communication Technology in Real Estate Practice (A Case Study of Lagos Metropolis)*, MSc. Thesis, Department of Estate Management, University of Nigeria, Enugu, 2010.

Hashim, N., Said, I., and Idris, N. H., Exploring E-Procurement Value for Construction Companies in Malaysia, *Procedia Technology*, 9, 836–845, 2013.

Ibem E. O. and Laryea S., Survey of Digital Technologies in Procurement of Construction Projects', *Automation in Construction*, 46, 11-21, 2014.

Jung, W., and Lee, G., The Status of BIM Adoption on Six Continents, World Academy of Science, *Engineering and Technology International Journal of Civil and Environmental Engineering,* 9(5), 512-516, 2015.

Oladapo, A. A., An Investigation into the Use of ICT in The Nigerian Construction Industry, *Journal of Information Technology in Construction*, 12, 261-277, 2007.

Osunsanmi, T. O., Aigbavboa, C. O., and Oke, A. E., Construction 4.0: The Future of South Africa Construction Industry, World Academy of Science, *Engineering and Technology International Journal of Civil and Environmental Engineering*, 12, (3), 206-212, 2018.

Peansupap, V. and Walker D. H., Factors Enabling Information and Communication Technology Diffusion and Actual Implementation in Construction Organisations, *Journal of Information Technology in Construction*, 10(4), 193-218, 2015.

Rouse, M., *Mean of Digitalisation*, 2017. Retrieved from http://whatis.techtarget.com/definition/digitization on October 2018.

Staub-French, S., and Fischer, M., *Enabling Concurrent Engineering Through 4d Cad*, In Anumba, C. J., John M. K., and Decelle, A. F. C., Concurrent Engineering in Construction Projects, 201-217, London, New York, Taylor Francis, 2017.

Tan, C. K., and Yeoh, K. C., *A Study on the Use of Measurement Software in the Preparation of Bills of Quantities Among Malaysian Quantity Surveying Firms.* i9th International Conference on ICT and Knowledge Engineering, Bangkok, Thailand, Jan 12-13, 2011.

Streamlining Information Transfer between Construction and Structural Engineering
Edited by Shiau, J., Vimonsatit, V., Yazdani, S., and Singh, A.

ISBN: 978-0-9960437-7-9

CONCEPTUAL BUSINESS MODELS OF STRATEGIC LOGISTIC SERVICES FOR SMALL-SIZED CONTRACTORS

MUHAMAD ABDUH

Faculty of Civil and Environmental Engineering, Institut Teknologi Bandung, Bandung, Indonesia

The new Indonesian Construction Service Act has explicitly introduced the importance of supply chains for construction project delivery and sustainable supports of small-sized contractors to the Indonesian construction industry's efficiency. Most of the construction cost comes from the material cost to be delivered to the project site that needs proper logistic management. For large and medium-sized contractors, their capacities and projects values enable them to access closer tier of the supply chains; to the producers of strategic materials with more efficient transactions. This is not the case for small-sized contractors, therefore, there should be a strategic logistic service provided for helping them. This paper presents a development of a conceptual business model of a strategic logistic service, especially the third-party logistic service, for small-sized contractors in a small coverage area of operation. Seven business models were identified with the preliminarily identified potential to be the model for the region of Greater Bandung, Indonesia. The models were derived based on the origin entity that could transform or be created as a new business or extended business. The conceptual business models will be used for further research to develop proper and feasible business models of strategic logistic services for small-size contractors in the Greater Bandung area.

Keywords: Distribution, Material, Supply chain, Transportation, Warehouse.

1 INTRODUCTION

Indonesia is currently building its infrastructure intensively, and at the same time, a new Construction Services Act was launched in 2017 that attempts to correct the lack of previous legislation, so that infrastructure development in Indonesia would be supported by good quality construction services. Explicitly, the importance of construction supply chains to support the implementation of construction services with the fulfillment of domestic material, equipment adequacy, and the support of the state to the development of a small construction services business entity to support the construction services has been emphasized.

As in other countries, this construction supply chains will support contractors, especially in the supply of construction materials. Each commodity has its own supply chains with its own structure, behavior, and performance. One of the important things that determine the materials' shipping cost to the location of the construction project is how well the contractor can manage its supply chains, including the logistics system. For large and medium-sized contractors, the capacity and value of the projects they often work on allow them to gain further upstream access of strategic commodities with economically valuable volumes, so that the supply chains they get will be short, the cost will be low, and it increases their competitiveness.

However, this does not occur to small-sized contractors because they do not get the same opportunity so that the material logistics costs become relatively more expensive per unit of each commodity. Hence, construction products from small contractors are relatively more expensive. It has an impact on the users of the construction product, namely the community; people would find it difficult to have good quality buildings at affordable prices.

In the logistics system, there are currently many strategic logistics services that try to provide logistics services more efficiently and effectively by offering shorter access from users to producers. Some of these services are like third-party logistics (3PL or TPL) and fourth party logistics (4PL), which have been developed in many countries with convincing success rates, especially for general commodities and manufacturing. The strategic logistics services system has also been developed in the construction industry, but it is usually carried out to support large contractors (Ekeskär and Rudberg 2015). However, more potentials are identified that the services would be directed to serve SMEs (Dybskaya and Vinogradov 2018).

2 SMALL-SIZED CONTRACTORS IN INDONESIA

The government's alignment with small-sized contractors cannot be separated from their roles in the implementation of construction services and in maintaining the stability and economic development of the country. Although the number of workers and the value of sales are small, their presence in the community is very significant in both to support the needs of the community in building procurement and as a cooperation partner in a supply chains managed by large and medium-sized contractors.

In Indonesia, based on 2013 data from the National Construction Services Development Board or LPJKN, there were 128,570 registered contractors; with 108,626 or 85% are small-sized contractors, 17,511 or 14% are medium-sized contractors, and large-sized contractors only 2,433 or 2%. The number of small-sized contractors, which is more than 80%, only absorbs less than 20% of construction works, while large and medium-sized contractors absorb most of the construction work in terms of value.

Thus, the level of small-zed contractors' competition is very high with relatively small construction value. If a small-sized contractor wants to grow, they must compete well in absorbing the construction market share of 20% and they need to try to take a large portion of the construction projects which carried out by large and medium-sized contractors, by becoming a subcontractor or partner of the contractor. The empowerment of small-sized contractors through the supply chain management mechanism owned by large and medium-sized contractors is believed to be able to increase the potential of developing small contractors themselves.

On the other hand, even the capacity of the small-sized contractor still depends on the supply chains that it makes or has because the value of the construction that has been done is relatively small in terms of quantity. The small-sized contractor does not get quantity benefits in the form of price discounts, a certainty of quantity, the certainty of delivery, and in negotiating the price of construction materials with members of the supply chains. Thus, the commodity prices obtained by small contractors per unit quantity are relatively more expensive compared to the prices set by large and medium contractors who have the advantage of the quantity of construction work. It is all because the access to producers is shorter; in the other words, the more parties involved in the supply chains, the more expensive the commodity will be (Abduh *et al.* 2012).

3 CONSTRUCTION SUPPLY CHAINS IN INDONESIA

In line with the understanding of supply chains in the context of manufacturing, the context of supply chains in construction are defined as a process of a set of activities to change raw materials to become the finished products e.g. roads, buildings, and planning services, which are used by service users without admitting the organizational boundaries. The supply chain in construction has different characteristics from the supply chain in manufacturing, such as the temporary nature of a construction project that produces a unique product with project team members who are constantly changing from one project to another, and by always moving locations. Regardless, all supply chains that are inherently in the market will be invited to come to the project site within a period, and it is the challenge. Thus, the supply chain management of construction projects is very important.

There are four roles of supply chain management (SCM) (Vrijhoef and Koskela 2000). The first role is related to clarifying the relationship between the supply chain and the location of the construction project. The information shared between supply chains and construction projects are very important. The second role of SCM is to load supplies from the supply chain smoothly so that it will be in accordance with the needs at the project site. The third role of SCM is to move as much activity as possible on the construction project site to the supply chain. This is closely related to product quality improvement, such as pre-fabrication of structural components. Lastly, the role of SCM is to incorporate the supply chain and construction project locations in real terms, and this means combining the three previous SCM roles. Roles 1 to 3, often known as the construction project logistics system, while the 4th role is the meaning of the SCM itself.

In Indonesia, studies on construction supply chains existed in the framework of the application of lean construction in Indonesia in 2005. In the early stage, studies were carried out to get a more complete form of the construction supply chains pattern. Furthermore, various initial studies were followed up with studies that led to efficient supply chain management methods. One of the main studies that have been done was the relationship between various parties involved in the implementation of construction, namely owners, contractors, sub-contractors, suppliers, and others. An efficient relationship has been studied based on the description of the relationships in construction projects, which in the end can be identified by various suitable methods in the construction supply chain management. However, the construction supply chains cost in Indonesia is not clear and has not been well developed and detailed enough. It happens because the companies do not feel like they need to break down the cost structure in detail, to control the supply-chain financing. In addition, the components of steel material purchase costs are significant, while transportation and inventory costs do not significantly affect the total cost (Abduh *et al.* 2012).

For construction industry level, the supply chain studies collaborate with the government to ensure the fulfillment and sustainability of strategic commodities, and the supply chain structure of large, medium and small contractors, and to create a solid construction business. For several national strategic commodities, such as reinforcing steel, steel profiles, and cement, their supply chain structures and the behavior of its supply chain members have been defined. Yet, the structure and channel of the supply chain of these strategic commodities can only be optimized by large contractors, while small and medium contractors must be empowered by larger contractors and the government with affirmative policy (Abduh and Pribadi 2014).

4 STRATEGIC LOGISTICS SERVICES FOR SMALL-SIZED CONTRACTORS

Nowadays, there has been a lot of development in manufacturing and general commodities. A strategic logistics service (SLS), such as third-party logistics (TPL or 3PL) or even fourth party logistics (4PL) connects the needs of efficient logistics between the upstream and downstream of a supply chain (Aguezzoul 2014). TPL is a logistics service that is held by third parties with the consideration that the company can focus on its core business. Usually, TPL covers warehouse services as well as transportation services. Furthermore, TPL would provide integration services with a supply chain management system (SCM), so that it is sometimes referred to the third-party supply chain management or 3PSCM and sometimes also called as a supply chain management service provider (SCMSP). The 4PL is a further development of 3PL, an integrator who provides a complete supply chain management service. There are many benefits that the TPL could offer and the competitive advantages for doing so, such as operating cost reduction, invested capital reduction, flexibility increase, customer service level improvement, and innovation capability improvement (Marchet *et al.* 2018).

It seems that the development in strategic logistics services (SLS) will continue to expand in line with the increasing globalization and business process efficiency (Saglietto 2013). Based on the first role in the Vrijhoef and Koskela (2000) model, the TPL can play a role as an unclear interface in the supply chain construction management (Ekeskär and Rudberg 2015). In some developed countries, the existence of strategic logistic service, especially TPL, has been applied in construction, but it is all to support large-sized contractors (Ekeskär *et al.* 2016). Whereas in fact, large and medium-sized contractors can manage their own supply chains, while small-sized contractors, even from the beginning, do not get the benefit from the existence of a construction supply chain. However, the potential of TPL to serve the SMEs (Dybskaya and Vinogradov 2018) has been identified.

Based on the previous study, small-sized contractors in Indonesia must be empowered by utilizing the existence of construction supply chains (Abduh and Pribadi 2014). There is a lack of access of small-sized contractors to available supply chains so that they cannot be compared to large and medium-sized contractors in the cement supply chain. Based on these considerations, it is necessary to see the possibility of developing a strategic logistics service system (SLS) to support small contractors, because they do not have good access to the construction supply chain to get their strategic commodities.

5 CONCEPTUAL MODELS

Based on the needs as previously explained, a general conceptual model of TPL for construction projects is shown in Figure 1. For that reason, to develop the TPL system for construction projects and specifically to support small-sized contractors in Indonesia, there must be a study of

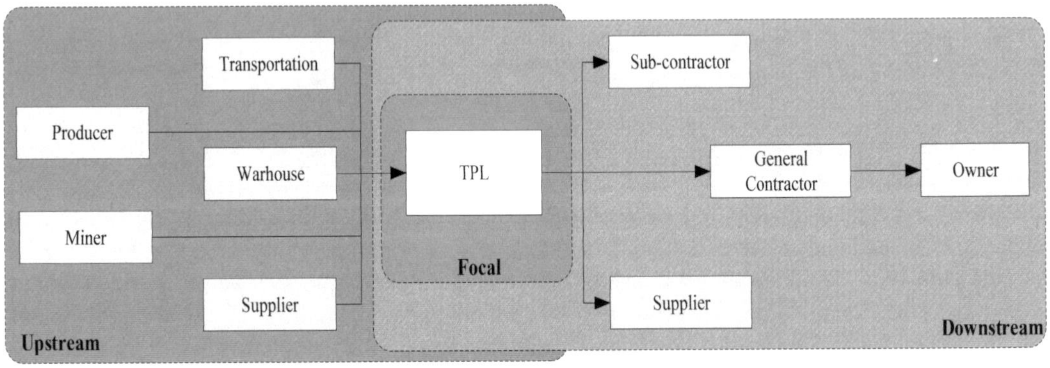

Figure 1. TPL conceptual model for construction projects.

the 3 domains in the construction supply chain with the TPL interface, i.e., the upstream, focal and downstream domains.

The study in the upstream domain is related to obtaining a complete figure of the structure and behavior of the supply chain construction's strategic commodities in a range of study areas, including the warehouse service market and transportation services, as a support. For the downstream, the need for construction in a range of study areas, as well as the capacity and the competition level of the small contractor must also be seen. While in the focal domains, the assessment of entity development, business modeling, and overall system integration must be carried out. The results of these studies will open opportunities for a new business model in the logistics field, such as strategic logistics services for small contractors. It can later be done in the form of community service by developing a start-up business.

For that purpose, a multi-year research to develop the TPL system for a small contractor in Greater Bandung area, covering Bandung city, Cimahi city, Bandung regency and West Bandung regency. In this area, there are 2,216 small-sized contractors, or constituted the majority (83.7%) of the contracting population in the Greater Bandung that were engaged in residential building as much as 25, 2%, in the field of civil engineering buildings 58.3%, and in the field of special buildings (other) 16.5%, with the value of construction in around IDR 66 trillion.

Based on the previous study on potential parties existed in Greater Bandung area, there are several possibilities how the TPL could be developed with focusing only to major commodities (Table 1) for small-sized contractors that are operating in residential and civil engineering buildings. In general, the business of TPL could be developed as a new business that is developed from scratch and as an extended business from an existing business that could be originated from upstream as well as downstream of the supply chain. A new business could be developed by the local government (public), a private company, or as a PPP project. In the downstream of the supply chain, the existing contractors, sub-contractors, or suppliers could extend their business to set up TPL service. While in the upstream, a transportation service provider, as well as warehouse service provider, could also extend their businesses to include other services as a TPL provider. However, every possible set-up could have advantages and disadvantages that will determine its potentials and feasibility to be TPL business models in the Greater Bandung area for small-sized contractors.

Table 1. Major commodities for small-sized contractors.

Residential Buildings	Civil Engineering Buildings
Brick	Aggregate
Aggregate	Clay
Timber	Stone
Cement	Cement
Glass	Rebar
Rebar	Asphalt
Precast concrete	Precast concrete

Based on the distinctive 3PL selection criteria identified in order to build a strategic outsourcing relationship (Marchet *et al.* 2018), the preliminarily identified potential of each business model of TPL based on the origin entity is shown in Table 2. The selection criteria used in the table are related to: geographical spread offered by the TPL (Geography); reputation, reliability, customer service level, customer satisfaction and contract fulfilment that TPL could achieve (Performance); information sharing and trust that TPL could commit (Trust); and the availability of appropriate physical facilities and equipment (Facility). The potential level of each business model depends on the existing management capability, technology, competition, and networks of supply chains. Therefore, for each business model as depicted in Table 2, the management of TPL should include strategic planning, inventory management, transportation, capacity planning, and information technology (Gunasekaran and Ngai 2003).

Table 2. Potential business model of TPL for small-sized contractors.

Business Model		TPL Selection Criteria			
		Geography	**Performance**	**Trust**	**Facility**
New Business	Public	◎	○	◎	◎
	Private	○	○	○	○
	PPP	●	○	●	●
Extended Business	Contractor/Sub-contractor	○	○	○	○
	Supplier	●	⊙	⊙	◎
	Transportation Service	◎	◎	◎	⊙
	Warehouse Service	⊙	◎	◎	⊙

Legend: ○: none; ●: small; ◎ medium; ⊙: high.

6 CONCLUSION

This paper conveys the needs of strategic logistics service, especially the third party logistic (TPL), to empower small contractors in the Greater Bandung area, Indonesia. Moreover, seven business models were introduced to be further assessed their potentials and feasibilities to be the TPL models that suit for small-sized contractors in the area. It is expected that the results of the research can provide opportunities for Indonesian construction industries using the new business model and the supply chain management approach, specifically to empower small contractors.

References

Abduh, M., and Pribadi, K.S., *Harmonizing the Indonesian Construction Resources Supply Chain*, in *The 20th Asia Construct Conference*, Hong Kong, 13-14 November 2014.

Abduh, M., Soemardi, B. W., and Wirahadikusumah, R. D., Indonesian Construction Supply Chains Cost Structure and Factors: A Case Study Of Two Projects, *Journal of Civil Engineering and Management*, *18*(2), 2012.

Aguezzoul, A., The Third-Party Logistics Selection: A Review Of The Literature, *Omega*, 69–78, 2014.

Dybskaya, V. V. and Vinogradov, A. B., Promising Directions For The Logistics Service Providers Development On The Russian Market In Times Of Recession, *Transport and Telecommunication*, *19*(2), 151–163, 2018.

Ekeskär, A. and Rudberg, M., *Third-Party Logistics In Construction: Perspectives From Suppliers And Transport Providers*, 22nd Annual International Euroma Conference, 1–11, August 2015.

Ekeskär, A., Rudberg, M., and Ekesk, A., Third-Party Logistics In Construction: The Case Of A Large Hospital Project, *Construction Management and Economics*, (34), 174–191, 2016.

Gunasekaran, A. and Ngai, E. W. T., The Successful Management Of A Small Logistics Company, *International Journal of Physical Distribution and Logistics Management*, *33*(9), 825–842, 2003.

Marchet, G., Melacini, M., Perotti, S., and Sassi, C., Types Of Logistics Outsourcing And Related Impact On The 3PL Buying Process: Empirical Evidence, *International Journal Of Logistics Systems And Management*, *30*(2), 139, 2018.

Saglietto, L., Towards A Classification of Fourth Party Logistics (4PL), *Universal Journal of Industrial And Business Management*, *1*(3), 104–116, 2013.

Vrijhoef, R. and Koskela, L., The Four Roles of Supply Chain Management in Construction, *European Journal of Purchasing and Supply Management*, *6*(3–4), 169–178, 2000.

Streamlining Information Transfer between Construction and Structural Engineering
Edited by Shiau, J., Vimonsatit, V., Yazdani, S., and Singh, A.

ISBN: 978-0-9960437-7-9

THE ROLE OF EFFECTIVE COLLABORATION FOR SUCCESSFUL DELIVERY OF CONSTRUCTION PROJECTS

CONSTANT MAHAME, THEONESTE BIGIRIMANA, and MARIE JUDITH KUNDWA

Dept of Civil Engineering Construction and Management, Lanzhou Jiaotong University, Lanzhou, China

This paper is written with the purpose of assessing the role of effective collaboration for successful delivery of construction projects. Generally, in construction industry, to come up with the successful project delivery; the construction management team has to work against any factor that may lead to cost overruns, project delays, poor quality delivery and high level of variation. Construction projects present extreme varieties and occur in different size, shape and quality. Undertaking any construction project is a process that involves many bodies and organizations. Effective collaboration between the various parties is crucial for successful completion of both design and construction. Based on a case study of a commercial building project, this work discusses how project managers can use collaboration approach to solve project challenges that may arise during execution such as high level of project variations, endless time extension and disputes between parties to the contract. Collaborative Approach in Problem Solving helps the integration of knowledge and gain of experience, to reduce payment delay, achieve the proper evaluation and control of the schedule of activities and to motivate the contractor to achieve effectiveness and efficiency. The effective collaboration between the various parties and the early involvement of the contractor are crucial for successful completion of both design and construction and the early integration of knowledge.

Keywords: Collaborative working, Construction management, Contractual relations, Project management, Proper project delivery, Contractual parties.

1 INTRODUCTION

1.1 Background

The construction industry is an economic investment and its relationship with economic development is well posited. Many studies have highlighted the significant contribution of the construction industry to national economic development (Myers 2013). Among the major economic sectors, the importance of the construction industry is unique regardless of whether the country is underdeveloped, developing or developed (Olanrewaju and Abdul-Aziz 2015). Generally, in construction industry, to come up with the successful project delivery; the construction management team has to work against any factor that may lead to cost overruns, project delays, poor quality delivery and high level of variation. Undertaking any construction project is a process that may involve many bodies and organizations. Effective collaboration between the various parties is crucial for successful completion of both design and construction.

In general, these parties can be referred to as the 'client', the 'design team' and the 'construction group'.

As said by Lahdenperä (2012), in recent years the construction industry has experienced a development towards new collaborative project arrangements, driven by dissatisfaction with the inbuilt opportunism and adversarial relations in traditional project delivery forms like design-bid-build and design-build. Executing construction projects with a design-bid-build approach creates a separation between the design and construction processes (Forbes and Ahmed 2010). This separation is shown to hinder integration of knowledge between the technical consultant and the contractor, as well as diminishing the opportunity for contractors to have influence on the design decision (Song *et al.* 2009). Introducing collaboration and engaging the different project parties at an earlier stage of the construction project process is one way to include this construction knowledge and create better relationships between the client, contractor and consultant (Song *et al.* 2009, Mosey 2009, Walker and Lloyd-Walker 2015).

1.2 Collaborative Management

According to Collaborative Management Tutorial (2018), Collaborative management can be defined as a collection of various management techniques that enlighten a sense of unity and teamwork among managers, supervisors, and the employees within a business organization. It can be broadly seen as the act of working together as a team to accomplish a common goal within a given timeframe.

The concept behind collaborative management style is to: allow managers to combine their strengths with the strengths of their team, make it possible to collectively overcome any weaknesses found among the team members, and enhance the efficiency and productivity of all the company.

1.3 Collaborative Approach in Problem Solving

One of the important objectives of the collaboration on any project is to solve disputes and taking effective and efficient decisions.

As applied to the organization, the collaborative problem solving model sets forth two major tenets: First, the problems are well understood in terms of the cause, the issue, and the bugs present (Rather than pointing out mistakes of the team members); second, these problems are best addressed by mutual discussions where everybody speaks and everybody listens (rather than showing superiority and trying to dominate other members).

Collaboratively resolving problems isn't necessarily all that complicated, but it's something for which the leaders need to have experience as well as patience, so it can take a while for all involved members to feel comfortable and keep their views open.

1.4 Collaborative Decision Making

The term collaborative decision making refers to processes in which autonomous or semi-autonomous actors interact through formal and informal negotiation, jointly creating rules and structures governing their relationships and ways to act or decide on the issues that brought them together" to achieve mutually beneficial interactions (Thomson *et al.* 2007).

It is very important to note that on any construction project, the management is as successful and powerful as its decision making based on the contract arrangement agreed-on for the particular project.

As mentioned earlier, executing construction projects with a traditional approach creates a separation between the design and construction processes. This separation is shown to hinder integration of knowledge between the technical consultant and the contractor, as well as diminishing the opportunity for contractors to get involved on the design decision.

The construction projects are highly subjected to changes or variations that need to be early assessed in a way to minimize them and/or solved in collaboration by all parties to contract. Any collaboration between project contractual parties helps to prevent delay of decision making and long-time disputes, helps to build good relationship and to achieve the timely delivery of the project. This needs to be a shared concept; in understanding and acting process to all contractual parties to the construction management team as whole.

1.5 Collaborative Project Delivery Forms

Partnering: The basics of partnering can be associated to a number of hard elements such as formal integrative mechanisms like workshops for team-building and common goals, financial incentive systems, dispute resolutions procedures and continuous improvements programmes (Bresnen and Marshall 2002, Nyström 2005, Eriksson and Westerberg 2011, and Bygballe *et al.* 2010). Additionally, as both a prerequisite and result of these integrative mechanisms trust and mutual understanding has an important place in partnering arrangements (Bresnen and Marshall 2002, Nyström 2005). Complementally, optional procedures like early involvement of contractors in concurrent engineering, joint project office and joint IT-tools can be adopted for increased partnering in a specific project (Eriksson and Westerberg 2011).

Two more distinct partnering approaches can be defined, namely project partnering and strategic partnering. In project partnering the different project parties agree on working collaboratively in a specific project (Rigby *et al.* 2009). Strategic partnering, on the other hand, is when a client (or contractor in that sense) works with a limited set of suppliers through a number of projects, with an intention to improve relations and performance over time (Rigby *et al.* 2009).

Relational contracting with early contractor involvement: The early contractor involvement (ECI) is a form of collaboration that originates from the ideas of partnering but with the goal to involve the contractor at an earlier stage through formal contractual mechanisms (Mosey 2009). It is a two-stage project delivery approach where the contractor is procured for two separate phases with open books. Figure 1 presents an overview of ECI compared to design-build and design-bid-build delivery forms from a construction project life-cycle perspective.

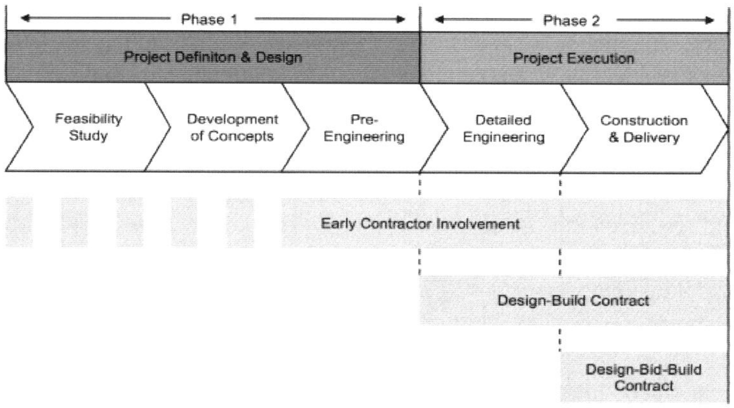

Figure 1. Construction project delivery activities in relation to different contract forms.

The point of time at which the contractor is awarded in an ECI project can vary depending on the project. The time of involvement ranges from the initial idea phase to the pre-engineering phase (Walker and Lloyd-Walker 2015). The range of work for the contractor, and the consultants, therefore, depends on the time of involvement of the contractor.

1.6 Opportunities and Challenges of Collaboration on Project Delivery

There is not a small number of advantages associated with collaboration on for delivery of any construction projects, including- Helping to develop a strong mind-set of best-for-project, building long-term relationship, achieving high performance and increased commitment in the project delivery, integration of knowledge and gain of experience, achieved constructability, avoidance of opportunistic behaviour, it enforces project parties to work together sharing both pain and gain; etc.

However, there are also the challenges associated to lack of proper integration of collaboration during project delivery, it includes: High level of project variations, increased number of disputes, cost overruns, project delays, poor quality delivered, long-term rivalry, etc.

2 METHODOLOGY

The Commercial Building Project mentioned in this article as an illustrative case study, provides a clear image about challenges of lack of collaboration during project delivery and advantages of adopting collaborative work for the best-for-project. This case study focused primarily on the execution stage, where complexity and novelty of processes are highest.

It is important to note that, one of authors was directly involved in the project management team as consultant quantity surveyor; which gives credit to genuineness of information and data analysis to be provided.

3 THE COMMERCIAL BUILDING PROJECT

3.1 Background

The new commercial building project is located in Rwanda, it comprises of six blocks with "U-shaped form; four blocks and two blocks have two and three floors respectively. As for the title, it is built for commercial activities. The building covers the surface area of 3,395m2, and in total is 22, 650m2 with a total cost of around two million dollars. The contractual approach adopted on this project was design-bid-build or traditional method of contract.

The client, a private woodwork cooperative, had the main objective of achieving low total operating costs over the lifecycle of the building. This included construction costs, costs for technical operation and maintenance of the building.

3.2 Project Challenges

Generally, this project suffered three major problems namely high level of variations, high cost-overruns and endless time extension resulting from poor integration of constructability principles and lack proper collaboration approach. Other causes included; poor project study, faulty, ambiguous, or defective working drawings; incomplete specifications; shortage of specialists in the consultant/design firm and lack of trust during project execution between parties on the contract.

According to Project Financial Appraisal As at 31st December 2013, the mentioned problems led to:

1) The contract price rise from 2,047,672$ to 2,891,010$. It equals the rise of construction cost is 41% (forty one percent) or 843,339$. It is beyond allowable contract additional cost according to Rwanda Public Procurement Authority (RPPA) conditions of contract, which can only allow 15% rise for public projects. Fortunately, the client was a private developer.
2) *Disputes between all parties to the contract*:
   - The client accused the contractor to delay the project as opposed to the contract agreement.
   - The contractor also was accusing the design/consultant firm to be responsible of problems related to a big amount of variations occurred during execution resulting from undetailed, ambiguous, or defective working drawings; incomplete specifications and late of making decision to any available change.
   - The consultant used to accuse the contractor to intentionally change the figure numbers of works executed during preparation of interim payments in order to obtain high payment.
3) *Endless extension of time*. Initially the project completion agreed-on in the contract was supposed to take place on 31 October 2013.

For the first time, in the site meeting of 24th September 2013 the contractor started to propose about time extension after realizing that the completion date agreed on in the contract will not be met. In the following site meetings both parties agreed to shift the completion time on 31st December 2013.

Latterly the works were not finished as on 31st December, negotiation agreed to place completion time on 28th February 2014, but the objective couldn't be met still. And finally, the project got completed by October.

3.3 How Collaboration Helped in Problem Solving

Despites challenges that occurred during execution of the Commercial Building Project rooting from improper planning and preparation of the project; there was a need to come up with a common problem solving approach namely the collaborative approach in problem solving. The following are important measures that were agreed-on by all parties for achieving that collaborative problem solving:

1) After realizing the high level of project variations in terms of design, cost and schedule, all parties to the contract (client, consultant and contractor) came to agreement of working together to re-work on the project financial appraisal. It as such decided in the way of integrating different knowledge to effectively and efficiently conduct the activity.
2) The second agreement was that at the end of each month of the contractual period, quantity surveyors of both the consultant and contractor parties had to work together to measure executed works to be paid in the interim payments. It properly replaced the former method whereby contractor quantifiers were the only supposed to measure the periodical executed work. That was such modified in order to prevent multiple checking of measurements by the consultant, which used to delay the approval of interim payments.
3) After facing the above-mentioned problems; the new collaborative agreement instructed the meeting minutes has to be checked and modified if necessary directly at the end of meeting and got signed by all parties to contract at the same meeting location. It is important to note here that the site meetings schedule changed from unplanned (or the client or consultant instructs it) to planned schedule, at least once a week every Friday. This helped the evaluation and control of the schedule of activities.

4) The client agreed on providing bonuses to the contractor if he timely completes works based on the new project schedule. This motivated the contractor to raise commitment and achieve the high performance.

4 DISCUSSION AND CONCLUSION

The purpose of this article was to assess the role of effective collaboration for proper delivery of the construction project with an illustrative case to clearly draw the understanding of the literature. The collaboration model chosen for the Commercial Building Project was the scheduled meetings, working together by professionals from both parties during preparation of project financial appraisal interim payments, and finally provision of bonuses by the client. Even if the project was completed with all issues between parties not settled, however the adopted collaborative approach helped the integration of knowledge and gain of experience, reduced payment delay, achieve the proper evaluation and control of the schedule of activities and motivated the contractor to achieve high performance and efficiently complete the project.

Collaborative project delivery presents many advantages than challenges on any construction projects, especially when it is well organized and established from the early phase of the project. That is from project planning phase to project execution.

The effective collaboration between the various parties and the early involvement of the contractor are crucial for successful completion of both design and construction and the early integration of knowledge.

References

Bresnen, M., and Marshall, N., The Engineering or Evolution of Co-Operation? A Tale of Two Partnering Projects, *International Journal of Project Management*, 20(7), 497-50, 2002.

Bygballe, L. E., Jahre, M., and Swärd, A., Partnering Relationships in Construction: A Literature Review, *Journal of Purchasing and Supply Management*, 16(4), 239-25, 2010.

Collaborative Management Tutorial, retrieved from https://www.tutorialspoint.com/collaborative_management/collaborative_management_introduction.htm on October 2018.

Eriksson, P. E., and Westerberg, M., Effects of Cooperative Procurement Procedures on Construction Project Performance: A Conceptual Framework, *International Journal of Project Management*, 29(2), 197-208, 2011.

Forbes, L. H., and Ahmed, S. M., *Modern Construction: Lean Project Delivery and Integrated Practices*, CRC Press, Boca Rato, 2010.

Lahdenperä, P., Making Sense of the Multi-Party Contractual Arrangements of Project Partnering, Project Alliancing and Integrated Project Delivery, *Construction Management and Economics,* 30(1), 57-79, 2012.

Mosey, D., *Early Contractor Involvement in Building Procurement: Contracts, Partnering and Project Management*, Wiley-Blackwell, Chichester, 2009.

Myers, D., *Construction Economics: A New Approach*, 3rd ed., New York: Routledge, 2013.

Nyström, J., The Definition of Partnering as A Wittgenstein Family-Resemblance Concept, *Construction Management and Economics*, 23(5), 473-48, 2005.

Olanrewaju, A. L., and Abdul-Aziz, A.-R., *Building Maintenance Processes and Practices, An Overview of the Construction Industry*. DOI 10.1007/978-981-287-263-0_2, Springer, Singapore, 2015.

Rigby, J., Courtney, R., Lowe, D., Klepfisch, G., Larsen, J. N., and Lahdenperä, P., *Study on Voluntary Arrangements for Collaborative Working in the Field of Construction Services*, Brussels: DG ENTR, European Commission, 2009.

Song, L., Mohamed, Y., and AbouRizk, S. M., Early Contractor Involvement in Design and Its Impact on Construction Schedule Performance, *Journal of Management in Engineering*, 25(1), 12-20, 2009.

Walker, D. H. T., and Lloyd-Walker, B. M., *Collaborative Project Procurement Arrangements*, Project Management Institute, Newtown Square, PA, 2015.

Streamlining Information Transfer between Construction and Structural Engineering
Edited by Shiau, J., Vimonsatit, V., Yazdani, S., and Singh, A.

ISBN: 978-0-9960437-7-9

PREQUALIFICATION METHOD STATEMENT IN CONSTRUCTION PROCUREMENT

BENEDETTA BOSSI and NATALE MASSIMO MINOTTI

Dept of A.B.C., Politecnico di Milano, Milan, Italy

The current competitive environment within the construction sector is increasingly difficult and challenging, leading all the main players to increase their focus and efforts upon procurement strategies and policies. Therefore, procurement practice is important in order to effectively perform and close out the construction process and meet stakeholders' needs. Starting from a deep analysis of procurement processes framed within the project life cycle, the research proposed the prequalification phase as a selection strategy necessary to identify the contractor. In order to select the project delivery system most suitable to the customer and the project to be developed, an operating procedure has been proposed, allowing who deals with procurement better ruling the purchasing processes. In the proposed procedure, each activity is defined through a general overview, the execution process and its tools. The outputs of the study are also sample documents and model forms that can be easily utilized assisting the procurement phase. Thus, since the procurement process has assumed increasing importance in the strategic planning of the construction industry, the aim of the research is to provide a prequalification method necessary to select the most suitable contractor before the tender, in order to reduce any risks during the entire construction process.

Keywords: Building sector, Tender, Contractor, Supplier, Delivering, Project management.

1 INTRODUCTION

Procurement process has been defined as a strategy to satisfy clients' development and/or operational needs with respect to the provision of constructed facilities for a discrete life cycle (Lenard and Mohsini 1998). Thus procurement is the acquisition of the necessary project resources to realize a constructed facility and can be referred to a process of combining these necessary resources together (Rowlinson and McDermott 1999).

In a construction context which is increasingly competitive, excellence in the procurement process is essential in order to generate savings, reduce risks and optimize costs and times in any project, thus procurement process is an area of vital importance and interest to organizations responsible for delivering project outcomes (Alias *et al.* 2012).

Even though many studies have investigated different aspects of procurement, such as selection process, influence of procurement choice on the project performance and contractual mechanism of procurement alternatives (Rahmani and Maqsood 2017), there is a lack of studies focusing on supplier prequalification phase as significant part of the procurement process.

As underlined by de Araújo *et al.* (2017), managers should pay special attention to two phases of the project procurement process: (1) supplier selection and (2) supplier evaluation. Therefore, selecting the right supplier for a construction project, as well as evaluating this supplier's performance while the contract is being implemented, plays an important role in ensuring a positive outcome.

As a matter of fact, what highly influences the procurement processes is what methods are used to select and evaluate suppliers. Also Lambropoulos (2007) highlights the importance of choosing an appropriate procurement method and asserts that a "best overall" award method does not exist, but the most appropriate method depends on the specific conditions for each particular project.

Since the offers analysis and negotiation process in a construction bid is often long and complicated, only a supplier preselection can make the bid process more efficient and reduce the risk of a short-list with unsuitable suppliers. Therefore, we present ongoing research on procurement process that proposes the prequalification phase as a selection strategy necessary to identify a bidder list of suitable contractors before the tender.

In the first part of the paper we report a synthesis of the analysis carried out on the standard procurement process used in the construction sector. In the second part we propose a prequalification operating procedure with its specific activities and deliverables.

2 PROCUREMENT PROCESS ANALYSIS

Procurement is for a company one of the biggest costs to manage, meaning that the products and services provided by the main subcontractors typically represent a high percentage on the total cost of the construction project, accounting for as much as 80% in some industries (Neuhaus *et al.* 2015). Thus, the relationship with suppliers is a crucial aspect of the construction business. Any error or delay caused by a subcontractor has a direct and serious impact on the efficiency and profitability of the work. This is why the contract system and the procurement management can lead to a very different result on a project's success.

2.1 Procurement Across the Project

Procurement process occurs in each phase of a project. The process is not composed by discrete components with well-defined interfaces, but these components overlap and interact throughout the project (Project Management Institute 2008). Starting from the PMBOK©Guide, for each of the six macro phases of the project, procurement key activities and key deliverables have been analyzed (see Table 1).

2.2 Procurement Process and Its Operating Procedure

The procurement process in construction has been summarized in the following six stages: (1) Supplier requirements identification, (2) Bidder qualification, (3) Invitation to make an offer, (4) Bid evaluation, (5) Negotiation and (6) Contract awarding. Inside the Bidder Qualification a Procurement Operating Procedure is necessary and it has been divided into four process groups and their activities and deliverables: (1) Procurement initiating and strategic planning processes, (2) Prequalification, (3) Supplier selection and contract awarding and (4) Monitoring and closing.

The Prequalification process does not take place after the strategy planning, but it is conducted simultaneously at the initiating processes after the identification of the project requirements. Table 2 represents key activities and key deliverables identified for the Prequalification process.

During the process potential bidders are requested to submit information on their companies in order for the client to select the companies that will participate to the tender. It consists of the verification of certain formal requirements, such as adequate proof of technical capability (also through on-going site inspections), prior experience, financial and human resources, equipment and other physical facilities, managerial capacity and reliability necessary to carry out the project.

Table 1. Procurement key activities and key deliverables in each single project phase.

KEY ACTIVITIES	KEY DELIVERABLES
Procurement at the Feasibility and Conceptual Phase	
Market Investigation	Make or Buy decisions
Obtain key supplier input	Supplier Prequalification
Conduct Make or Buy analysis	Preliminary Cost Estimate
Procurement at the Preliminary Engineering Phase	
Finalize purchase orders and contract terms and conditions	Procurement Management Plan
Prepare documentation to procure critical and long lead equipment/ materials	Budget Cost Estimate
	Tender opening for long lead equipment/ materials
Procurement at the Detailed Design Phase	
Prepare purchase order documentation	Tender opening for equipment, materials and construction services
Procure long lead equipment/ Materials	
Procurement at the Construction and Start-Up Phase	
Procure equipment, materials and construction services	Procurement /contract documents
Procurement at the Closeout Phase	
Close purchase orders and contract accounts	Close out report

Table 2. Prequalification process key activities and key deliverables.

KEY ACTIVITIES	KEY DELIVERABLES
Define chartered prequalification team, roles and responsibility	Responsibility assignment matrix (RAM)
Conduct market investigation	Prequalification Plan
Define prequalification strategic planning	Categories of work and purchasing item
Define prequalification criteria	Define prequalification process timing by steps
Set up prequalification documents	Prequalification score grid
Solicit interest from potential bidders	Prequalified bidder short list
Applicants evaluation	Prequalification final report
Close prequalification	

Based on the prequalification results, a bidder list is drawn up of the companies eligible to compete. This is an opportunity to ensure that in the final bid evaluation, the contracting authority is able to compare "apples to apples" rather than "apples to oranges" (Diop 2017).

Starting from Prequalification process key activities and key deliverables identified, below we describe the prequalification operating procedure proposed.

3 PREQUALIFICATION PROCESS

The prequalification objective is to enter into an agreement with a contractor who possesses the necessary technical skills, resources and financial backing to give the client the best possible chance of the project being completed within the required time, cost and quality standards.

Thus, the prequalification procedure allows to select potential contractors and suppliers so that, during the bid analysis, the client consultant could evaluate offers focusing on the economic and technical issues, avoiding the second time analysis of those subjects already validated. Besides, prequalification reduces time dedicated to the offers analysis, saving time necessary for the negotiation phase.

3.1 Prequalification Operating Procedure

The prequalification operating procedure proposed is characterized by various activities, in Figure 1 are shown the links between them. Each activity has been identified by a Roman numeral and the most significant are described in the following paragraphs. The left column contains major factors that constitute the input of some activities; in the middle, tool & techniques are the documents that can be used performing the prequalification. The right-hand column lists the main documents or documentable outcomes of the activities.

3.1.1 *Prequalification strategic planning*

Prequalification requires a strategic planning phase to develop the appropriate selection strategy and respect the times dictated by the tendering process and also the project schedule. The first assessment to do would be to understand if the qualification is performed because the client is looking for potential bidders with reference to a particular project, or whether it is a general qualification carried out regardless of the type of items to be purchased but that allows the company to create its own database of qualified contractors. Anyhow for each category of work, in which it was decided to divide the project, or each item to purchase, the procurement manager must decide the prequalification approach and develop a work plan. Thus, the study has developed a model form "Prequalification strategy" and another "Prequalification plan".

3.1.2 *Prequalification criteria and documents*

To conduct an evaluation of possible candidates, prequalification criteria have been fixed and first divided and organized in the following groups: general criteria, financial criteria, technical capabilities, past performance and contract-specific criteria. Prequalification criteria have to be set up at the same time as the prequalification documents are prepared and the criteria should not be changed during the prequalification evaluation process.

The prequalification documentation depends on the type of contract that has been selected, the complexity of the item to purchase and the project complexity. They are prepared by the Procurement Department with assistance for the legal, administrative and technical issues. Prequalification documents include: prequalification letter, instructions to applicants, technical documents, prequalification document list and prequalification submission form.

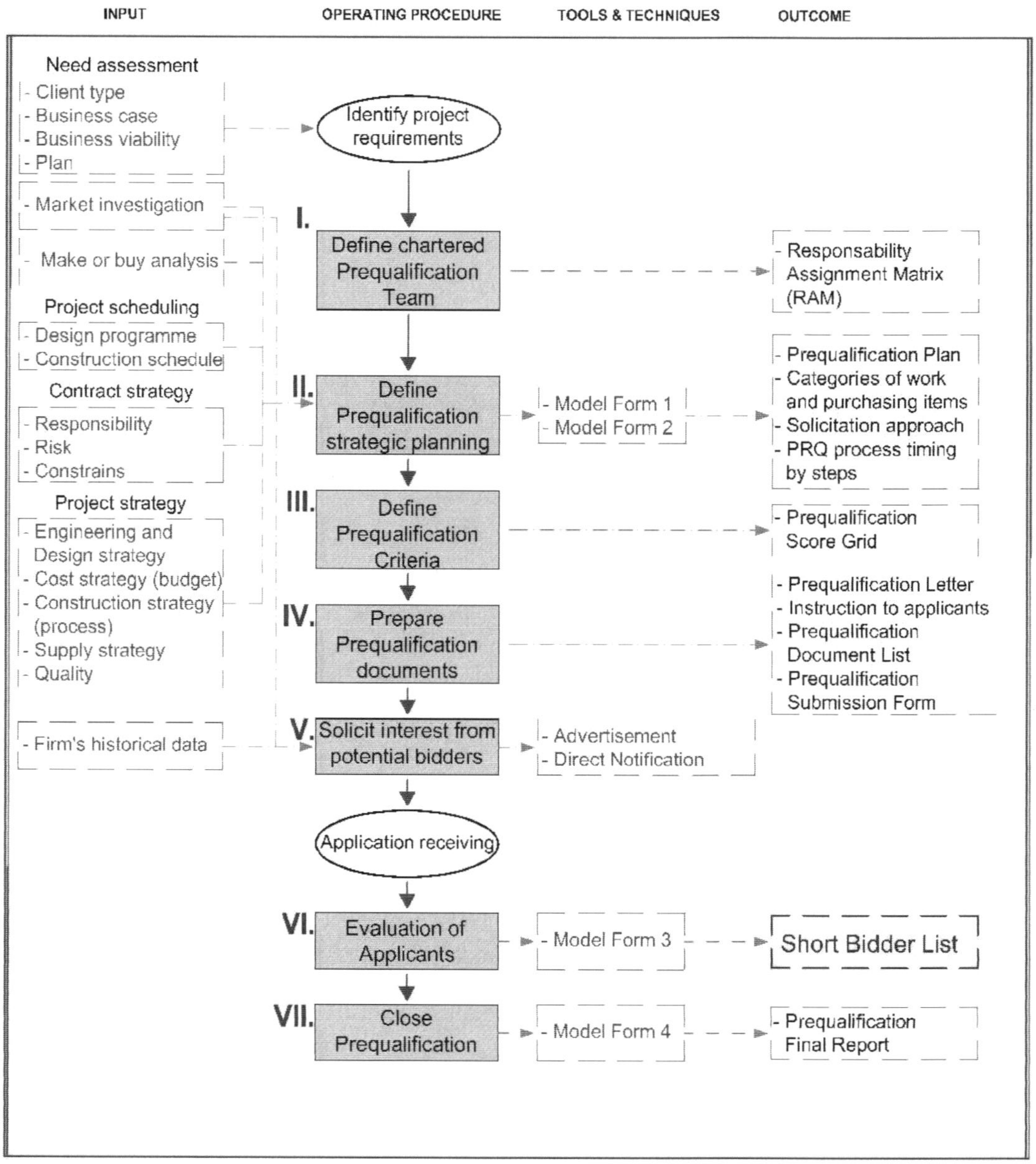

Figure 1. Prequalification process.

3.1.3 *Evaluation of applicants*

The evaluation procedure has been divided in two stages according to the complexity of the qualification requirements. The first step is a Preliminary Examination to assess the formality of the documents required, the completeness of application and the eligibility of the application, for each bidder, utilizing the pass-or-fail method. In the second place, a qualitative evaluation is employed to examine whether submitted documents comply with the qualification requirements by following the criteria set up. The result of the prequalification procedure is the selection of prequalified applicants and the recommendation to the client. If the client approves the bidder list becomes part of the Procurement Plan.

4 CONCLUSION

The goal of the procurement phase is to secure bidders of each bid package who are qualified, competitive, interested in the work, and capable of doing the work within the time, cost and quality requirements. The principles and procedures outlined are applicable to the procurement of goods, services and works. In particular, the study proposes the prequalification phase as a selection strategy necessary to identify a bidder list of suitable contractors before the tender, in order to save time during the offers analysis. The research work is going on focusing on the analysis and identification of the prequalification criteria based on categories of work or items to purchase. Furthermore, in order to validate the operative procedure proposed, a significant application is necessary in procurement process, identifying similar tender case studies.

References

Alias, R. H., Noor, N. M. M., Selamat, A., Saman, M. Y. M., and Abdullah, M. L., Decision Making Model for Electronic Tender Evaluation (ETE) Using Fuzzy AHP with Extent Analysis Method, *International Journal of Digital Content Technology and its Applications,* 6 (22), 79-89, 2012.

de Araújo, M. C. B., Alencar, L. H., and de Miranda Mota, C. M., Project Procurement Management: A Structured Literature Review, *International Journal of Project Management,* Elsevier, 35, 353-377, January, 2017.

Diop, M., United Nations Development Programme. Implementation-Tendering&Procurement. 2017. Retrieved from http://pppue.undp.2margraf.com/en/16_2.htm on September 16, 2018.

Lambropoulos, S., The Use of Time and Cost Utility for Construction Contract Award Under European Union Legislation, *Building and Environment*, Elsevier, 42(1), 452-463, January, 2007.

Lenard, D., and Mohsini, R., *Recommendations from the Organisational Workshop*, CIB W-92 Procurement – The Way Forward. The University of Montreal, Davidson C. H. and T. A. Meguid, CIB, 1: 79-81, May 18-22, 1998.

Neuhaus, K., Umbeck, T., Schmitz, A., and Gerhard, R., Why Procurement Matters, January 22, 2015. Retrieved from http://www.bain.com/publications/articles/why-procurement-matters-video.aspx on September 16, 2018.

Project Management Institute, *A Guide to the Project Management Body of Knowledge (PMBOK© Guide)*. Fourth ed. Pennsylvania: Project Management Institute, Inc., 2008.

Rahmani, F., and Maqsood, T., An Overview of Construction Procurement Methods in Australia, *Engineering, Construction and Architectural Management*, Emerald Publishing Limited, 24(4), 593-609, 2017.

Rowlinson, S., and McDermott, P., *Procurement System: A Guide to Best Practice*, Spon Press, London, 1999.

Streamlining Information Transfer between Construction and Structural Engineering
Edited by Shiau, J., Vimonsatit, V., Yazdani, S., and Singh, A.

ISBN: 978-0-9960437-7-9

TENDER EVALUATION: MCDM TECHNIQUES

SOMYA AGRAWAL, KEYUR A. KAPADIA, and RAJIV GUPTA

Dept of Civil Engineering, BITS Pilani, Pilani, India

The decision of tender evaluation and allotment is one of the important step of any construction project. There has been a trend away in tender evaluation process from considering quantitative parameters ("lowest price win", "maximum benefit win") to other qualitative parameters (past performance, time of completion). This paper reviewed past and recent literature in order to identify the current tendering process and the various techniques that have been applied previously. The study utilized different MCDM techniques for evaluation of tendering process. Multi-criteria decision making research has become one of the main areas of research for dealing complex and critical decision problems. This paper outlines the use of various systematic decision making tool for the tender selection. Qualitative and quantitative criteria are compared, and the uncertainty associated with the selection of right contractor is highly reduced. By using a case study the tender evaluation is fully investigated applying various MCDM techniques. Results from the different methods are compared and fuzzy extend analysis, fuzzy TOPSIS showed slight difference in results from the other methods.

Keywords: Multi-attribute utility theory, Multiple-regression, AHP, TOPSIS, Fuzzy AHP, Fuzzy TOPSIS.

1 INTRODUCTION

Tender selection and its evaluation process are a critical issue in construction management (Halil 2007). The identification of best contractor for the work is a challenge i.e. performance during challenging situation, Acceptable rates, Quality of work etc. the contractor selection process has received considerable attention to Management and client. Tender evaluation does not involve only quantitative criteria such as rate and time of completion but also qualitative parameters are to be considered such as quality of work, experience, efficiency of contractors etc. (Sun 2010). In most of the cases of contractor selection, turns out to be a bit complex process if both qualitative and quantitative parameters are to be considered. Recently, MCDM techniques have acquired considerable attention in evaluation of tenders, procurement process, bidding and other tenders which take care of both types of parameters. Further the methods commonly practised in many countries is the best value method, called MEAT (Most Economically Advantageous Tendering) (Ramon and Cristóbal 2012), and was further improved by using the weights obtained from AHP (Zhu K., 1983). Previously, Hatush and Skitmore (1998) applied Multi-attribute utility function for evaluation of tender and was further used in case study (Zhu *et al.* 1983). Also the Fuzzy set theory was applied by Nyugen (1985) for first time in evaluation of tender in 1984. Further, Holt (1998) described various methods that can be applied in evaluation of tender i.e. Bespoke's Approach, Multi- attribute analysis, Multi-attribute Utility theory, Multiple-regression, Fuzzy Sets, Custer analysis (Nyugen 1985). The benefits and disadvantages of the described methods were studied. But none of the method was applied or used in any case studies (Nyugen 1985). In

2012, DEA Approach was used for evaluation of tenders, in this Falagario *et al.* (2012) says Cross-Efficiency DEA can be used to remove the uncertainty Among DMU's (Decision Making Unit). The Method was evaluated by a Case study of Italian public procurement where LW (Linear Weighing) method was used and the result obtained were same for the first three cases and changed for the others. The result varied a lot from fuzzy AHP (Halil 2007). Wong (2004) used Logistic regression approach and the performance of the contractor was predict from the previous data. The regression equation was used and the present rating according to the tender details were taken as the input.(Falagario *et al.* 2012). In 2007 Farisah Mohammad Ali said that instead of evaluating the just on basis of minimum rate a cut-off rate should be there, tender below which should be directly rejected, the cut-off rate was decided on the basis of the statistical method (Wang *et al.* 2013). Further Sun (2010) discussed Fuzzy AHP and Fuzzy TOPSIS in evaluation of the tender and applied on a case study(Gupta *et al.* 2016).

This paper studied the effectiveness of various MCDM methods i.e. Bespoke's Approach (Nyugen 1985), Multi-attribute analysis (Wong 2004), Multi-attribute Utility theory (Ramon and Cristóbal 2012), Multiple-regression, AHP (Hatush and Skitmore 1998), TOPSIS, Fuzzy AHP (Chang 1996, Aggarwal *et al.* 2013), Fuzzy TOPSIS (Holt 1998) for evaluation of tender and further some of the methods are applied on a case study of Aakash Ganga project tender in BITS-Pilani. The above methods are not discussed in detail due to paucity of space and directly applied to a real case and results are compared.

2 METHODS AND CASE STUDY

This study has considered a tender floated for "Aakash Ganga", a rain water harvesting project to assist villages in rural India that have little access to clean water. In this project, every homeowner with adequate size of roof top is asked to lease the rights to harvest their rooftop rainwater. So, this is based on public utility model. Then this rain water is collected into underground storage reservoir through various conduits interconnected together. Aakash Ganga IT Network (ITN) aims to store, capture and disseminate information for development, design and scale up of Aakash Ganga network. The four agencies Tender1, Tender2, Tender3, and Tender4 have sent proposals for development of the online database required for the Aakash Ganga Rainwater Harvesting Network. The study includes application of six multi-criteria decision making techniques. Here both qualitative (experience) and quantitative (cost) parameters are considered. The major performance attributes are the criteria for selection of alternative (Tendering agencies) for allocating tender (Gupta *et al.* 2016). Criteria: C1 = economic aspect; C2 = Time for host on server; C3 = Database; C4 = software; C5 = experience.

2.1 AHP

The pair wise comparison matrix for the Aakash Ganga project was formed by experts. Tables 1 to 6 indicate the Saaty's weightage with respect to criterion.

Table 1. Pair wise Comparison of Criteria (Single Dimension).

	C1	C2	C3	C4	C5
C1	1	3	5	5	7
C2	1/3	1	3	3	5
C3	1/5	1/3	1	1	3
C4	1/5	1/3	1	1	3
C5	1/7	1/5	1/3	1/3	1

Table 2. Pair wise comparison of Tenders w.r.t. criteria economic aspect.

Economic Aspect	Tender 1	Tender 2	Tender 3	Tender 4
Tender 1	1	3	3	1/3
Tender 2	1/3	1	1	1/5
Tender 3	1/3	1	1	1/5
Tender 4	3	5	5	1

Table 3. Pair wise comparison matrix of Tenders w.r.t. criteria time of host.

Time for Host	Tender 1	Tender 2	Tender 3	Tender 4
Tender 1	1	1/7	3	1
Tender 2	7	1	9	7
Tender 3	1/3	1/9	1	1/3
Tender 4	1	1/7	3	1

Table 4. Pair wise comparison of matrix w.r.t. criteria database.

Database	Tender 1	Tender 2	Tender 3	Tender 4
Tender 1	1	1	2	2
Tender 2	1	1	2	2
Tender 3	1/2	1/2	1	1
Tender 4	1/2	1/2	1	1

Table 5. Pair wise comparison matrix w.r.t. criteria software.

Software	Tender 1	Tender 2	Tender 3	Tender 4
Tender 1	1	1	3	1
Tender 2	1	1	3	1
Tender 3	1/3	1/3	1	1/3
Tender 4	1	1	3	1

Table 6. Pair wise comparison of Tenders w.r.t criteria Experience.

Experience	Tender 1	Tender 2	Tender 3	Tender 4
Tender 1	1	1/3	1/6	1/3
Tender 2	3	1	1/3	1
Tender 3	6	3	1	3
Tender 4	3	1	1/3	1

The final result was obtained by following the AHP procedure and shown in concluding table.

2.2 TOPSIS

The real values were taken for the economic aspect while the other values were given rating from 0-9 based on their performance in individual criteria. The economic aspect was to be minimized while other was to be maximized. In order to carry out maximization for all the criteria the economic aspect was multiplied by -1. The normalized matrix for the Real matrix shown in Table 7.

From the above normalized matrix the distance from the positive ideal point and negative ideal point were obtained and the final ranking were given. The weights used for the TOPSIS were obtained from Fuzzy AHP. The final ranking is shown in Table 8.

Table 7. Real value matrix tender to criteria.

	C1	C2	C3	C4	C5
Tender 1	-8,48,000	3	2	3	1
Tender 2	-15,20,000	9	2	3	3
Tender 3	-15,46,350	1	1	1	6
Tender 4	-4,75,000	3	1	3	3

Table 8. Normalized matrix tender to criteria.

	C1	C2	C3	C4	C5
Tender 1	-0.19319	0.1875	0.333	0.3	0.0769
Tender 2	-0.34629	0.5625	0.333	0.3	0.230
Tender 3	-0.35229	0.0625	0.1667	0.1	0.4615
Tender 4	-0.10821	0.1875	0.1667	0.3	0.230

2.3 Fuzzy AHP

The linguistic pair wise rating matrix was formed for the case study and the evaluation was done using both the Fuzzy AHP approaches i.e. Buckley and Chang method. The modified Chang approach has been used for the rating purpose. The pair wise comparison matrix for fuzzy AHP is shown below in Tables 9 to 14. Linguistic terms are indicated in table: Equal/ Equally Important / Equally weak (E/EI/EW); Weakly important/ Weakly weak(WI/WW); Slightly important/ Slightly Weak (SL/SM); Moderately important /Moderately weak(MI/MW); Fairly important / Fairly Weak (FI/FW); Strongly plus Important / strongly plus weak (SP/SS); Very important/ Very weak (VI/(VW); Very- Very Important / Very-very weak (VV/VS); Absolutely important/ Absolutely weak (AI/SW).

Table 9. Pair wise comparison matrix for criteria.

	C1	C2	C3	C4	C5
C1	E	SI	FI	FI	VI
C2	SW	E	SI	SI	FI
C3	FW	SW	E	EI	SI
C4	FW	SW	EW	E	SI
C5	VW	FW	SW	SW	E

Table 10. Pair wise comparison matrix for tender w.r.t. criteria economic aspect.

Economic aspect	Tender 1	Tender 2	Tender 3	Tender 4
Tender 1	E	SI	SI	SW
Tender 2	SW	E	EI	FW
Tender 3	SW	EW	E	FW
Tender 4	SI	FI	FI	E

Table 11. Pair wise comparison matrix for tender w.r.t. criteria Time of host.

Time of host	Tender 1	Tender 2	Tender 3	Tender 4
Tender 1	E	VW	SI	EI
Tender 2	VI	E	AI	VI
Tender 3	SW	AW	E	SW
Tender 4	EW	VW	SI	E

Table 12. Pair wise comparison matrix for tender w.r.t. criteria Software.

Software	Tender 1	Tender 2	Tender 3	Tender 4
Tender 1	E	EI	SI	EI
Tender 2	EW	EI	SI	EI
Tender 3	SW	SW	E	SW
Tender 4	EW	EW	SI	EI

Table 13. Pair wise comparison matrix for tender w.r.t. criteria Experience.

Experience	Tender 1	Tender 2	Tender 3	Tender 4
Tender 1	E	SW	SW	SW
Tender 2	SI	E	SW	EI
Tender 3	SI	SI	E	SI
Tender 4	SI	EW	SW	E

Table 14. Pair wise comparison matrix for Tender w.r.t. criteria Database.

Database	Tender 1	Tender 2	Tender 3	Tender 4
Tender 1	E	EI	SI	SI
Tender 2	EW	E	SI	SI
Tender 3	SW	SW	E	EI
Tender 4	SW	SW	EW	E

Further the weights were obtained. Final ranking and the weights obtained using both Chang and Buckley approaches are shown in concluding table.

2.4 Fuzzy TOPSIS

The fuzzy TOPSIS matrix was formed by rating the alternative w.r.t. criteria in linguistic terms. The matrix formed for evaluation is given in Table 15. The weights obtained from Fuzzy AHP Buckley's approach were used for calculation.

Table 15. Comparison matrix for Tender w.r.t. criteria Degree of Satisfaction.

	Fuzzy TOPSIS	
Alternatives	Degree of satisfaction	Rank
Tender 1	0.608	1
Tender 2	0.4814	3
Tender 3	0.409	4
Tender 4	0.5443	2

2.5 Result Comparison

The final results are shown in Table 16.

Table 16. Comparison of results obtained by different methods.

Alter-native	AHP	TOPSIS	MCUT	Fuzzy AHP (Buckley)	Fuzzy AHP (Ext. analysis)	Fuzzy TOPSIS
Tender 1	2	2	2	2	3	1
Tender 2	3	3	3	3	1	3
Tender 3	4	4	4	4	4	4
Tender 4	1	1	1	1	2	2

3 CONCLUSION

Tender evaluation being one of the most crucial steps of any construction management process, various MCDM techniques are applied to select the optimum tender based on qualitative and quantitative parameters. The results obtained do not vary in most of the techniques. Tender 4 comes out to be ranked first in most cases. Further, in the multi criteria utility method, it will be more helpful if case individual functions are prepared for individual criteria. Fuzzy AHP Buckley's and Chang's approach provide different ranking.

References

Aggarwal, R., Singh, S., and Ahp, A. C., AHP and Extent Fuzzy AHP Approach for Prioritization of Performance Measurement Attributes, *International Journal of Industrial and Manufacturing Engineering* Vol:7, No:1, 6–11, 2013.

Chang, D., *Applications of The Extent Analysis Method on Fuzzy AHP*, European Journal of Operational ResearchVolume 95, Issue 3, Pages 649-655, December 1996.

Falagario, F., Sciancalepore, F., Costantino, N., and Pietroforte, R., Using a DEA-Cross Efficiency Approach in Public Procurement Tenders, *European Journal of Operational Research*, 218(2), 523–529, 2012.

Gupta, R., Kondepati, L. K., and Gupta, A., Tender Evaluation by Visual Decision Support System, International *Journal of Computing, Communications & Instrumentation Engineering*, Vol. 3, Issue 1, 137–143, 2016.

Halil, F. M., Contractor's Perception of the Use of Statistical Approach in the Tender Evaluation at the Public Works Department, Malaysia, *American Journal of Applied Sciences*, 4. 10.3844/ajassp.2007.1084.1089.

Hatush, Z., and Skitmore, M., Contractor Selection Using Multicriteria Utility Theory: An Additive Model, *Building and Environment*, 33(97), 105–115, 1998.

Holt, G. D., Which Contractor Selection Methodology?, *Internatinal Journal of Project Management*, 6(3), 153–164, 1998.

Nguyen,V.U., Tender Evaluation by Fuzzy Sets, *Journal of Construction Engineering and Management*, 111(3) , 231-243, September, 1985.

Ramón J. and Cristóbal S., Contractor Selection Using Multicriteria Decision-Making Methods, *Journal of Construction Engineering and Management* , 138(6), 751-758, June, 2012.

Sun, C., A performance evaluation model by integrating fuzzy AHP and fuzzy TOPSIS methods, *Expert Systems with Application*, 37(12), 7745–7754, 2010.

Wong, C. H., Contractor Performance Prediction Model for the United Kingdom Construction Contractor: Study of Logistic Regression Approach, *Journal of Construction Engineering and Managemen*t, Vol. 130, (5), 691–698, October 2004.

Wang, W. C., Yu, W., Yang, I. T., Lin, C. C., Lee, M. T., and Cheng, Y. Y., Applying the AHP to Support The Best-Value Contractor Selection – Lessons Learned from Two Case Studies in Taiwan, *Journal of Civil Engineering and Management*, 19(1), 24–36, 2013.

Zhu, K., Shang, J., and Yang, S.,The Triangular Fuzzy AHP : Fallacy of the Popular Extent Analysis Method, 1–22,. *SSRN Electronic Journal*, 2012.

Streamlining Information Transfer between Construction and Structural Engineering
Edited by Shiau, J., Vimonsatit, V., Yazdani, S., and Singh, A.

ISBN: 978-0-9960437-7-9

EDUCATION: A CHANGE AGENT FOR PROJECT FAILURE?

ANN STEWART[1], AYMAN NASSIF[2], and ANDREW STEWART OBE[1]

[1]*Faculty of Business and Law, University of Portsmouth, Portsmouth, UK*
[2]*School of Civil Engineering & Construction, University of Portsmouth, Portsmouth, UK*

This research paper seeks to critically assess the role of education in addressing the skills gap in managerial competences and resilience to project failure within the project management profession. In order to achieve this it will address issues relating to why projects fail. To conduct this research an unobtrusive approach is undertaken through the use of content analysis. Content analysis is used to examine the qualitative data drawn from 4 case studies on project failure. It will use a thematic approach and coding to identify emergent patterns and trends. This in turn will help to capture and identify lessons learnt that could be adopted to create a new model of educational and leadership competences for project managers. By encouraging more critical reflection in training and educational development programs, it is proposed that this will empower project managers towards a better understanding of their corporate social responsibility within their profession. This paper concludes that failure to properly educate project managers has serious implications on their professional identity and credibility as practitioners, as well as for future pedagogical practice.

Keywords: Training, Lessons learnt, Resilience, Reflective practice, Competences.

1 INTRODUCTION

A search of the literature in a quest to establish the reason for project failure gives no single answer. The reasons for project failure are diverse, ranging from the use of fixed price contracts as a single point of failure (Jørgensen *et al.* 2017), changes within the execution stage which can adversely affect project performance (Chen 2015), failure to address stakeholder complexity in large cultural building projects (Mok *et al.* 2017), or blamed on deficient management (Sage *et al.* 2014) to name but a few. As projects are unique and have a degree of complexity within the environment in which they operate (Bakhshi *et al.* 2016), this opens up the ongoing debate of what then is project success and what is project failure? Are these terms mutually exclusive? To answer these questions Alami (2016) suggests that project success is based on parameters. If the parameters are accomplished then the project can be considered as successful. However, failure to meet objectives within the defined parameters could ultimately lead to an unsuccessful project outcome and project failure. As far as Alami (2016) is concerned project success is about survival and how to manage change whilst trying to keep the "eco-system" balanced. Thus the outcome of a project will largely depend on the complexity of a project's "ecosystem" within which it operates.

In view of the recent project failure in Miami, USA, in March 2018, where a pedestrian bridge under construction commissioned by Florida International University collapsed killing 6 people and injuring at least 9 others, this research aims to critically assess the role of education as a moral change agent to improve professional practice, in seeking to address the ongoing debate within the discipline of project management as to why projects fail. It is therefore hypothesized that there is a need for project management professionals to continuously reflect and critique their own practice, so that they would be better trained and equipped to deal with challenging circumstances that affect their professionalism, credibility and the safety of others. There are two research objectives. The first being to carry out an in-depth analysis of four different construction projects that hit the media within the past 5 decades over criticisms of project failure, in order to identify emergent patterns and trends. The second objective is to capture and identify lessons learnt from the case studies that could be adopted to create a new model of educational and leadership competencies for project managers.

2 RESEARCH METHODOLOGY

This research explores four case studies within the construction sector that are well known, as shown in Table 1. The reason for their selection is due to the fact that they have attracted widespread media attention as failed projects.

Table 1. Overview of case studies.

Case Study	What happened?
Grenfell Tower Block (London, UK) 14 June 2017	Fire broke out in a 4th floor flat of a 24 storey building and spread rapidly. It set the exterior cladding alight, which turned out to be flammable material with no clear evidence of pre-testing. 72 lives were lost, possibly more in the carnage. The cost of refurbishment in 2016 was £8.6 M
Florida Bridge (USA) 15 March 2018	The bridge collapsed while being assembled across a busy highway killing 6 people. Two days before the catastrophe a crack was detected by an engineer who reported it as not a safety concern to the Department of Transport. The message was picked up on the day of the disaster when a meeting was convened less than 5 hours before the bridge collapsed. Cost of the bridge USD 9.4 M with USD 2.4 M over budget.
Kutai Kartanegara bridge (Indonesia) 27 Nov 2011	The bridge collapsed while undergoing maintenance work killing 12 people and injuring dozens more. Poor quality of material was detected as well and criticisms were addressed against the working methods of the maintenance contractors. The cost of the bridge was USD 7.4 M.
West Gate Bridge (Melbourne, Australia) 14 Oct 1970	Whilst undergoing construction, the bridge collapsed on 14 October 1970. Thirty five construction workers lost their lives as they were trying to resolve a technical issue over a 4" camber through what was criticized as poor judgement and serious miscalculations. At the time of construction, the cost of the bridge was AUD 202 M.

The impact and cause of the failure will be assessed against 10 factors, for example, cultural, ethical, technological and legal; and 10 behaviours which includes driven to deliver, courage to challenge and accountability. These factors were carefully selected for the purpose of the research to give deeper insight and richer data sets for the case studies. An unobtrusive approach was used for this research using content analysis. Content analysis is used to examine the qualitative data from the case studies using a thematic approach and coding to enable the researchers to make inferences and draw meaningful conclusions (Graneheim *et al.* 2017). Inferential coding was selected. This method was chosen as a research technique to enable observations of content to be captured and codified using a coding sheet containing up to 10 sub categories of analysis in order to provide more in-depth, critical and objective data for analysis. Before the coding took place it was necessary to undertake a wide review of the case studies by consulting as many sources as possible ranging from the social media, professional Institutes, practitioner and government websites, books and journal articles, in order to gain an holistic and more objective understanding of what took place, to reduce research bias. Table 1 provides a brief synopsis of the four case studies used in this research.

Having compiled Table 1 the next step was to review the case studies and source materials again and this time with 3 coders (authors) to systematically examine the factors that could have contributed to the project failure in each individual case study. Each factor was ranked using a scale of 0 to 10 where 0 represents no evidence available to suggest there was an issue, 1 suggests very little evidence, 5 represents a reasonable amount of evidence exists through to a score of 10 which indicates a very high amount of evidence exists. Three coders were chosen to ensure more validity and reliability of the data and to reduce subjectivity through individual bias. To help with the coding, definitions of each factor were established, to increase reliability and to help achieve a common frame of reference (Graneheim *et. al* 2017). This then took the coding to a different level which was more evaluative. Adopting an evaluative approach to coding helps to assess the quality of the content and then make a judgement. This judgement is then recorded using an ordinal rating which can help to capture behaviours. The coders then compared individual rankings and agreed the overall scores which are presented in Table 2 below. The two categories examined are "cause of project failure" and "impact of project failure." The maximum score for each sub category is a total of 40 points.

Table 2. Ranking of causes and impacts of failure, content analysis.

Criterion	**Ranking of cause of failure**					**Ranking of impact of failure**				
	CS1	**CS2**	**CS3**	**CS4**	**Total**	**CS1**	**CS2**	**CS3**	**CS4**	**Total**
1. Political	4	3	1	2	10	9	7	6	7	29
2. Legal	9	8	4	7	28	9	9	8	8	34
3. Economic (Financial)	10	4	6	2	22	9	7	8	7	31
4. Societal	6	4	3	2	15	9	9	7	8	33
5. Cultural	7	5	6	6	24	8	5	5	7	25
6. Environmental	4	4	4	1	13	4	6	6	4	20
7. Technological	9	9	8	8	34	9	9	6	8	32
8. Ethical (corruption)	8	3	8	2	21	8	2	6	3	19
9. Educational/Training	7	9	9	8	33	8	8	8	8	32
10. Lessons identified	10	8	7	8	33	9	9	9	8	35

3 RESEARCH FINDINGS

This section of the paper will discuss the overall results presented in Table 2 to identify the key factors that caused the projects to fail and to critically assess their impact.

Figure 1 illustrates that the three most significant sub categories, when combining the causes and impact of project failure are technological, educational/training and lessons identified. The least significant are environmental and ethical. The significant variances between causes and impact of failure are in the subcategories of political and societal. Taken individually, the highest impact of failure are lessons identified and legal issues. In contrast the highest causes of failure were technological, education/training and lessons identified. A discussion on the significant subcategories findings will now be evaluated.

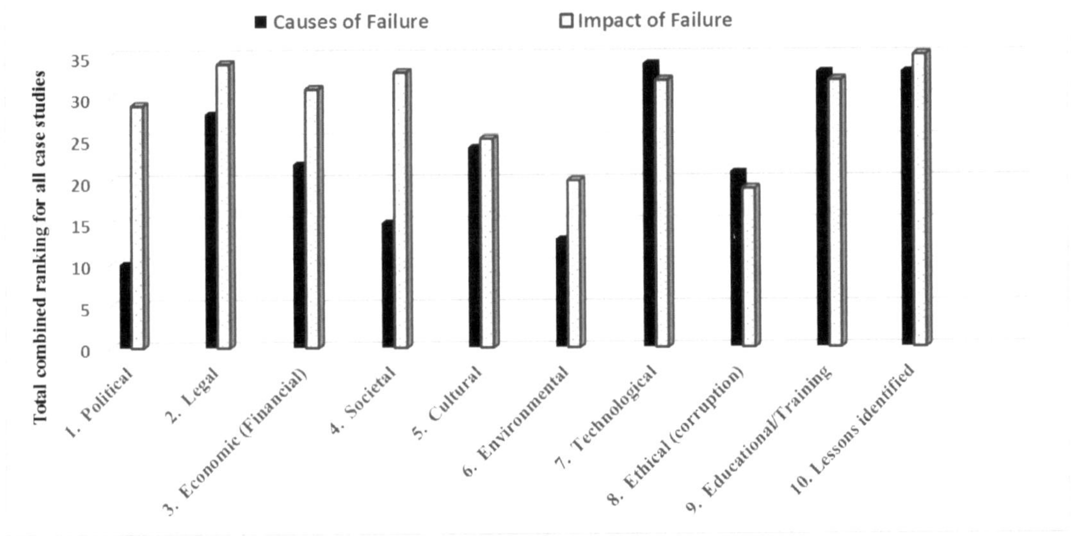

Figure 1. Comparison between causes and impacts of project failures.

3.1 Technical Issues

The technological issues revealed in the case study analysis were due to the following observations, failure to test building equipment and materials and implement an overall strategy for fire safety evacuation (e.g. Grenfell, cladding); overcomplicated concrete mix (Florida Bridge); poor quality material which adversely affected the structure and lack of understanding and knowledge of the structure (Indonesian Bridge, nuts and bolts), poor judgement and serious miscalculations (West Gate Bridge). These issues identified were all preventable and highlights the need for more critical reflection to learn from failure (Mann 2008).

3.2 Political and Legal Issues

Politically there was a lack of leadership in decision-making by the government over the banning of combustible materials which meant there was little respect for compliance from industry (e.g. the Grenfell, cladding). Legal aspects include breach of fire safety regulations with direct criticisms addressed to the fire service for failure to implement PLAN B evacuation policy (Grenfell) and compliance with health and safety legislation (in all four case studies). All the case studies involve manslaughter offences either individually or corporately and highlights the need for tighter regulation (Bridle and Sims 2009) and governance.

3.3 Economic (Financial Issues)

Evidence suggests that there are several compromises in the case studies to reduce costs either in the short term or long-term as in the Florida Bridge example of self-cleaning concrete whilst this was an innovative concept it was largely untested. Other examples include cheaper products being used in both the cladding (Grenfell) and cheaper material such as the bolts and nuts and lack of regular inspection checks (Indonesia Bridge). This suggests financial compromises have been made but it does not take into account the economic loss when things go wrong (Wong *et al.* 2005).

3.4 Societal and Environmental Issues

There is an emergent trend that society trusts the "Brand name "of the recognised Designers and Construction companies. This is validated from the case studies. Therefore nobody questioned the Designers competence and thus blindly accepted their work without question even when flaws were detected as in the cases of the West Gate and Florida bridge designs. This then challenges the behaviour and thinking of society as a whole and suggests that the failure can be attributed to societal failure (Brady 2014, 2015, 2016).

3.5 Cultural and Ethical Issues

Industry wide there appears to be a general acceptance of non-compliance with regulations, codes and safety standards especially where they were not evident (Grenfell, Indonesia Bridge, West Gate and Florida Bridge). When faults were suspected there was a general lack of interest to follow up with concerns (Grenfell, West Fate and Indonesia Bridge) or where this was evident (Florida Bridge) it was not diligently pursued. This shows a lack of corporate social responsibility and accountability (Mok *et al.* 2017, Victoria 1971).

3.6 Educational and Training Issues

There was a lack of training for complex fire scenarios and guidance for the Public on the "stay put" policy (Grenfell). There was also a paucity of training on the importance of adherence to regulations, codes and standards as evident in all 4 case studies. There needs to be more education in the importance of mutual interdisciplinary respect between the technical stakeholders, e.g. Designers, Engineers, Construction specialists, Consultants and regulatory bodies. Education of project managers must include these aspects and competences to understand and interpret complex structures and their performance through rigorous testing methods before materials are approved as safe for construction projects (Terwel *et al.* 2014, Collings 2008).

4 CONCLUSIONS

In response to the research aims and objectives this paper has identified that there is no single factor that can be taken in isolation for project failure. Although there are lessons to be learned from each case study, history repeats itself and therefore there appears to be little learning actually taking place. Therefore this paper concludes that it is not enough to focus on regulated behaviour and the "academisation" of professionals through simply equipping them with tools and techniques and the technical 'know-how' to manage projects and teams. It has to be reinforced through sustained education. Thus it is argued that education must play a central role to act as a moral change agent. Evidence from this research demonstrates that failure to properly educate project managers has serious implications on their professional identity and credibility as practitioners. This can be alleviated by way of encouraging all PMs to engage in continuous learning and education through Continuing Professional Development (CPD) programs and become members of the professional body such as the Association of Project Managers (APM, UK), or the Project Management Institute (PMI), which gives international recognition. For aspiring PMs they should be encouraged to obtain professional qualifications and membership in project management in order to practice. This paper further concludes that professionals should be continually tested against the professional standards and behavioural competences as set out by their respective professional institutions. In this respect they will remain current with emergent knowledge, professional standards and legislation in an attempt to be held more accountable for their corporate social responsibility not only to their stakeholders but also to the wider society. It is recommended that following further research, supported by empirical evidence, a new model of educational and leadership competencies for project managers could be developed.

References

Alami, A., Why do Projects Fail?. *PM World Journal*. 5, 1-9, 2016.

Bakhshi, J., Ireland, V., and Gorod, A., Clarifying the Project Complexity Construct: Past, Present and Future, *International Journal of Project Management*, 34(7), 1199-1213, 2016.

Brady, S. P., Learning from Structural Failure: The Challenges and Opportunities, *Proceedings of the Institution of Civil Engineers - Forensic Engineering*, 167(1), 10–15. https://doi.org/10.1680/feng.13.00018, 2014.

Brady, S., When Structural Failure Reflects Societal Failure, *The Structural Engineer: Journal of the Institution of Structural Engineer*, 93(12), 30-31, 2015.

Brady, S., West Gate Bridge Collapse–The Story of the Box Girders, *The Structural Engineer: Journal of the Institution of Structural Engineer*, 94(10), 26-28, 2016.

Bridle, R. J., and Sims, F., The Effect of Bridge Failures on UK Technical Policy and Practice, *Proceedings of the ICE - Engineering History and Heritage*, 162(1), 39–49. https://doi.org/10.1680/ehh2009.162.1.39, 2009.

Chen, H. L., Performance Measurement and the Prediction of Capital Project Failure, *International Journal of Project Management*, 33(6), 1393-1404, 2015.

Collings, D., Lessons from Historical Bridge Failures, *Proceedings of the Institution of Civil Engineers - Civil Engineering*, 161(6), 20–27. https://doi.org/10.1680/cien.2008.161.6.20, 2008.

Graneheim, U. H., Lindgren, B. M., & Lundman, B. Methodological challenges in qualitative content analysis: A discussion paper. *Nurse education today*, *56*, 29-34, 2017..

Jørgensen, M., Mohagheghi, P., and Grimstad, S., Direct and Indirect Connections Between Type of Contract and Software Project Outcome, *International Journal of Project Management*, 35(8), 1573-1586. https://doi.org/10.1016/j.ijproman.2017.09.003, 2017.

Mann, A., Learning from Failures at the Interface, *Proceedings of the Institution of Civil Engineers - Civil Engineering*, 161(6), 8–15. https://doi.org/10.1680/cien.2008.161.6.8, 2008.

Mok, K. Y., Shen, G. Q., and Yang, R. J., Addressing Stakeholder Complexity and Major Pitfalls in Large Cultural Building Projects, *International Journal of Project Management*, 35(3), 463-478, 2017.

Sage, D., Dainty, A., and Brookes, N., A Critical Argument in Favor of Theoretical Pluralism: Project Failure and the Many and Varied Limitations of Project Management, *International Journal of Project Management*, 32(4), 544-555, 2014.

Terwel, K., Boot, W., and Nelisse, M., Structural Unsafety Revealed by Failure Databases, *Proceedings of the Institution of Civil Engineers - Forensic Engineering*, 167(1), 16–26. https://doi.org/10.1680/feng.13.00019, 2014.

Victoria. Royal Commission into the Failure of the West Gate Bridge, Barber EH. Report of Royal Commission Into the Failure of the West Gate Bridge. Government Press; 1971.

Wong, S. M., Onof, C. J., and Hobbs, R. E., Models for Evaluating the Costs of Bridge Failure, *Proceedings of the Institution of Civil Engineers - Bridge Engineering*, 158(3), 117–128. https://doi.org/10.1680/bren.2005.158.3.117, 2005.

Streamlining Information Transfer between Construction and Structural Engineering
Edited by Shiau, J., Vimonsatit, V., Yazdani, S., and Singh, A.

ISBN: 978-0-9960437-7-9

ATTRACTING AND IDENTIFYING HIGH PERFORMING PROJECT MANAGERS

KRISTEN HURTADO, ANUSREE SASEENDRAN, JOHN SAVICKY, and KENNETH SULLIVAN

School of Sustainable Engineering and the Built Environment, Arizona State University, Tempe, USA

Construction project managers (PMs) are critical to the execution of successful construction projects, conducting and orchestrating the intricacies of dynamic and complex projects. A large state university was challenged with attracting and retaining PMs in their construction department during a period of rapid university growth and departmental re-organization. The university tried current models for selecting construction PM firms largely based on commodities-based procurement and/or low bid structures and was unsuccessful. A new model was developed that considered and analyzed both the capabilities and qualifications of the individual PMs. The importance of individuals being able to identify their unique capabilities, be accountable for their performance, and operate in a transparent environment are critical concepts within this model. The new model also created an environment of organizational transparency, requiring PMs to measure their performance and the performance of their projects. The model resulted in an overwhelming amount of high quality PMs seeking to join the university. The university initially sought a specific skills set for new PMs, but revised their criteria and future selection processes for hiring future PMs based on the results of this model. Analysis of the PM capabilities, qualifications, and performance are shared as well as lessons learned to refine the model. This model can also be used to identify high performing individuals in other positions or disciplines.

Keywords: Procurement, Performance-based qualification, Expertise-driven, Organizational transparency.

1 INTRODUCTION

In project management literature, project success is a well-researched area of study. Effective project management is one of the factors that has been increasingly linked to the success of a project (Spalek 2005; Papke-Shields *et al.* 2010, Cooke-Davies and Arzymanow 2003, Din *et al.* 2011). Studies that have focused specifically on the construction industry have also demonstrated similar findings. For instance, Yang *et al.* (2011) found that increase in levels of leadership of Project Managers (PMs) enhances team member relations, thereby improving the chances of project success. Chan *et al.* (2004) proposed a conceptual framework of factors affecting project success in construction, in which one of the five main contributing factors to success was project management. Relationship management, which is an important part of a PM's role, was found to have a significant effect on the performance of construction projects (Meng 2012).

In spite of the overwhelming evidence demonstrating the importance of the role a PM plays in the success of a construction project, sophisticated practices to identify and hire PMs whose capabilities align with that of the project requirements do not currently exist, particularly in the public sector. Lavigna (2002) claimed that "at every level of government, HRM/personnel systems are being criticized as inflexible, slow, rule-bound, and user-*un*friendly". In a study that compared hiring practices in the public and private sectors, Osoian and Zaharie (2014) found that advertising job openings is more complicated in the public sector and it does not attract as many direct applications from candidates as in the private sector. In addition, Glassdoor Economic Research Report (Chamberlain 2018) found that the length of the average interview process in the United States had increased by approximately 77% from 2010 to 2014.

1.1 Traditional Hiring Formats

When an organization has a need for professionals such as PMs, one of two alternatives is generally pursued: 1) hiring professionals internally, or 2) hiring an external consultant to supply the professionals as needed. The steps involved in internal hiring, particularly in a public setting, is time-intensive and takes approximately four to five months, as it involves writing job description, advertising the opening, reviewing the resumes, interviews and finalizing the award (Osoian and Zaharie 2014). Furthermore, barriers in the laws and how agencies use these laws may hinder initiation of demotion or removal actions if the hired PM is performing poorly (McPhie 2010). Hiring external consultants to outsource the work is the more uncommon approach of the two. This method awards the contract to a single firm based on their qualifications, rather than the capabilities and expertise of the individuals who will be providing the service. This poses a risk of the organization receiving a poorly performing PM from a well-qualified consultant firm. Expertise-driven hiring mitigates this risk by evaluating specific individuals instead. Hiring a singly form usually follows a low bid approach, where the vendor with the lowest proposed cost is selected. However, this approach has been linked to inconsistency in performance, quality and reliability (Assaf and Al-Hejji 2006).

In response to the challenges identified, the researchers developed a new model of hiring that takes into account both the capabilities and qualifications of the individual PMs being hired. The following sections discuss the new hiring model that was proposed by the researchers, and details of a case study where the model was utilized to hire PMs.

2 EXPERTISE-DRIVEN HIRING FRAMEWORK

Expertise-driven hiring mitigates the long duration taken to hire internally by outsourcing the services to an external organization. The researchers estimated a total time of eight weeks between advertising the job and making the award which makes the process 58% faster than traditional hiring. It involves a sequential list of items that needs to be completed by the organization seeking to identify high performing project managers (Figure 1). The first step of outsourcing PM services is needs assessment, which involves evaluating the current state and potential future needs of the organization. This is key to providing definition to the needs of the organization and communicating pain points, if any, thereby providing visibility into the organization and a chance to ascertain how they can add value to PMs who are proposing.

Following this, a request for qualifications (RFQ) is issued to engage consultants to supply one or more personnel to provide PM services. This is a crucial point of departure from traditional solutions to soliciting professional services where a consultant is requested to propose on behalf of all the personnel in the firm and the lowest bidder is commonly selected. Expertise-driven hiring mitigates the risks associated with hiring a single firm that may have varying capabilities among their PMs by evaluating specific individuals who will be hired instead. Additionally, expertise-driven hiring utilizes the five filters of best value procurement in order to ensure that the PMs being hired provide maximum value in terms of money, risk, and quality of work (Sullivan 2010). In expertise-driven hiring, a consultant is allowed to propose more than one qualified PM for the services being requested, although a separate proposal is required for each PM being proposed. Each individual PM responding to the RFQ is required to submit the following documents, in addition to basic information such as resume and references:

- Capability Statement – to illustrate the PM's unique capabilities that sets them apart
- Service Approach – to allow the PM to create a roadmap that describes the PM's vision
- Risks and Challenges – to allow the PM to describe any potential issues that may occur
- Measurements and Metrics – to provide a summary of how the PM will document and track their performance on this service, and
- Cost – yearly and hourly cost or fee for informational purposes until negotiation period.

The next step in the process is evaluation of proposals. An evaluation committee that consists of representatives from the major stakeholders at the organization evaluates the proposals blindly after removing any personally identifiable information to ensure fairness. This is followed by a shortlisting of PMs and interviews, and a clarification period with the shortlisted PMs. During this period, the organization negotiates and enters into contracts with the top-rated PMs.

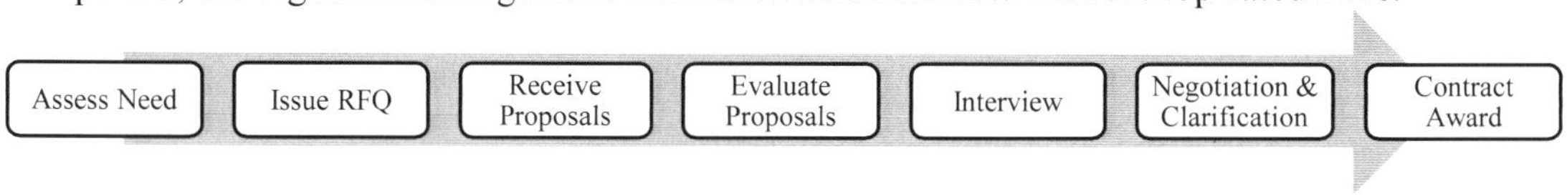

Figure 1. Expertise driven hiring framework.

Expertise-driven hiring ensures that an organization does not need to make long-term commitments to who is being hired as compared to internal hiring. The organization can set up yearly contracts with an option for renewal contingent on the performance of who is being hired.

3 CASE STUDY – BACKGROUND

A large public university situated in southwestern United States is the focus of this study. The university has an estimated student population of 22,000 and 1,000 faculty. At the time of this study, the university was facing a unique set of challenges fueled by the tremendous growth in student enrollment and faculty onboarding. The Office of Architects and Engineers (A&E) at the university manages the programming, design and construction of new and remodeled capital projects, and is responsible for an active project load of over $1.1 billion had recently undergone several changes, including a restructuring and combining of three departments – A&E, Construction, and Facilities. Additionally, its director was looking to retire and the department would then have to be managed by one of the senior PMs. In total, the restructured A&E department had a staff of seven PMs (from senior to entry-level) that were managing roughly 500 projects. This left roughly 236 projects ($26M) that had not been assigned a PM. Furthermore, additional projects were being requested by the university's stakeholders on a daily basis and there was a growing perception that A&E was unable keep up with the increasing demand. The A&E department also predicted a rapid growth in the requested number of projects per year over the next five years.

In order to keep up with the increasing demand, the A&E office was seeking to hire up to ten PMs on staff. However, the department estimated that the traditional hiring process would take four to five months to perform, which was considerably longer than they could afford considering the immediate need. Therefore, they decided to forego the traditional process and adopt an expertise-driven hiring process proposed by the researchers as an alternative.

4 FINDINGS AND DISCUSSION

The university implemented expertise-driven hiring in order to address an immediate need on campus and identify high-performing PMs. The A&E office performed a needs assessment of the university, which revealed that the 236 approved projects that had not been assigned PMs could be broadly classified as Heating, Ventilation and Air-Conditioning (HVAC, 40%), classroom/laboratory (18%), small/basic renovations (9%), and others (33%). It was also estimated that the university would undertake an average of 492 projects each year for the next five years valuing a total of $718 million. These assessments helped the university determine the areas wherein the PMs being evaluated would need competency.

4.1 Proposal Submittal and Evaluation

The advertisement of the university's need through an RFQ generated a much higher level of interest among PMs than anticipated by the university and the university received proposals from 42 applicants. Three addenda were issued to answer questions from potential proposers, as a result of which the interviews were delayed by two weeks. Five evaluators assessed the proposals, based on a 1, 5, 10 scale, 10 being exceptional, and 1 being poor.

After the scores were calculated, the researchers conducted a *point difference analysis* of the 42 responses for the purpose of identifying which proposers would proceed to the interview phase. Only the 14th and 22nd highest ranked PMs had a large enough point difference from the next highest ranked to be defensible. To avoid the possibility of excluding talented PMs, the university decided to proceed with 22 PMs. However, three of the 22 were offered other opportunities that were more desirable; thereby reducing the shortlisted pool to 19.

4.2 Interviews, Clarification, and Contract Award

The shortlisted candidates underwent 40 minute interviews, after which the evaluators scored them. These scores were combined with the proposal scores to obtain a total score for a PM. Among the top nine PMs, eight were from different organizations indicating the value of expertise-driven hiring, since onboarding a single consultant would have resulted in one to two high quality PMs being used (assuming the same firms would have been selected, which is very unlikely). Further, of the top nine PM firms, only one was familiar to the university. The expertise-driven process attracted new firms to the university that had not previously worked with the university or had chosen not to participate in the past. This process thus successfully mitigated the risk of low responses that is associated with traditional hiring. At the end of the process, the university hired three PMs for their immediate needs and maintained a list of shortlisted candidates in the event of a future need.

4.2.1 Analysis of the applicants

The university received 42 proposals from 21 consultants. 83% of the applicants declared a bachelor's degree, and 43% of the applicants were A/E licensed. However, the shortlisted applicants also had a similar background (85% bachelor's degree, 55% A/E licensed), demonstrating that a degree or an A/E accreditation is not a differentiating factor for performance. Among the shortlisted candidates, 86% indicated competency in only one of the three areas that the university was interested in, thereby denoting their specialization in a field.

10 of the 21 firms proposed only one candidate. Six firms proposed two candidates each, three firms proposed three candidates, one firm proposed four, and the last one proposed seven candidates. The following table indicates the number of applicants each firm proposed, and the number of applicants from each firm that made it to the top nine.

Table 1. Number of applicants per consultant firm.

Consultant Firm	Number of Applicants	Number of Top Nine Applicants
Firm 1	4	0
Firm 2	3	0
Firm 3	7	2
Firm 4	2	0
Firm 5	2	0
Firm 6	3	0
Firm 7	3	1
Firm 8	2	0
Firm 9	2	1
Firm 10	2	1
Firm 11	2	1

4.3 Satisfaction with Expertise-Driven Hiring

Shortly after the contracts were awarded, the key stakeholders in the University were asked to complete a short anonymous survey about the new process. Two executive level personnel who were involved in the expertise-driven hiring process at the university at all stages completed the survey. Of the responses recorded, the average overall satisfaction with the new process was a 7.5 out of 10. When the responders were asked to rate their level of comfort with unique methods of the process, they rated an 8 or above for all sections but one. The lowest rated criteria was the interview process at 6.5. The researchers hypothesized that this was due to the university-caused delay in interviews by a month as well as the significant change that this process represented, with some participants being highly resistant to any changes of the traditional process.

The university has since renewed all of the current PM contracts due to their measured performance on completed projects for the university. As the university's needs change, they will continue to have the flexibility to choose to renew or terminate any of these PM contracts on a yearly basis. This is useful for public organizations, as internally hired personnel are subject to certain labor laws and regulations.

4.4 Lessons Learned

A procurement of "people" presents its own unique challenges and differences compared to that of a typical procurement of a firm for goods or services. Initially, the credentials of a PM (having a bachelor's degree, etc.) weighted heavily against their ranking. PMs were expected initially to have at least a bachelor's degree, and ideally further credentials. Later, after narrowing the original pool to a shortlisted pool, it became evident that background credentials and level of education did not correlate with a PM's final score or ranking. One suggestion moving forward with similar procurements was to reduce the focus on a PM's degree and credentials.

From the start of scope drafting through contract award, the number of PMs to be hired was unclear to the university. By the time the university had decided to proceed with the top three ranked PMs, two of the highest-ranking PMs had accepted other offers. Additionally, moving forward, it should be made clear initially to PMs who they should contact with any questions. In the interview process, it became evident that PMs had several questions regarding their project load, responsibilities and the university policies and procedures. Providing them with this information up front would increase transparency and help retain vendor interest.

5 CONCLUSION

Although identifying high performing project managers is known to be crucial to the success of a project, existing hiring practices are time-consuming and inefficient, particularly in the public sector. Further, the traditional process does not always attract the highest quality experts, since individual companies may have fewer experts among their teams. The researchers therefore proposed an expertise-driven hiring framework, where the project management service is outsourced to individual PMs. A case study where this model was used to identify and hire PMs revealed that the new process is faster, attracts higher interest from well-qualified and experienced PMs from different organizations, and results in a high satisfaction among the hiring personnel. As organizations look to hiring high-performing PMs to better manage their portfolio of projects, this research will encourage them to rethink their current hiring process in favor of identifying expertise among individual PMs. Future research should focus on testing this model at other organizations – both public and private, to assess whether the results of this study can be duplicated in those settings as well.

References

Assaf, S. A. and Al-Hejji, S., Causes of Delay in Large Construction Projects, *International Journal of Project Management*, *24*(4), 349-357, 2006.

Chamberlain, A., Why is Hiring Taking Longer? New Insights from Glassdoor Data, Glassdoor.com, June 2015. Retrieved from https://www.glassdoor.com/research/app/uploads/sites/2/2015/06/GD_Report_3.pdf on October 2018.

Chan, A. P., Scott, D., and Chan, A. P., Factors Affecting the Success of a Construction Project, *Journal of Construction Engineering and Management*, 130(1), 153-155, 2004.

Cooke-Davies, T. J. and Arzymanow, A., The Maturity of Project Management in Different Industries: An Investigation into Variations between Project Management Models, *International Journal of Project Management*, 21(6), 471-478, 2003.

Din, S., Abd-Hamid, Z., and Bryde, D. J., ISO 9000 Certification and Construction Project Performance: The Malaysian Experience. *International Journal of Project Management,* 29(8), 1044-1056, 2011.

Lavigna, R. J., Best Practices in Public-Sector Human Resources: Wisconsin State Government, *Human Resource Management: Published in Cooperation with the School of Business Administration, The University of Michigan and in alliance with the Society of Human Resources Management*, 41(3), 369-384, 2002.

McPhie, N. A., (Ed.), *Addressing Poor Performers and the Law*. DIANE Publishing, 2010.

Meng, X., The Effect of Relationship Management on Project Performance in Construction, *International Journal of Project Management*, 30(2), 188-198, 2012.

Osoian, C. and Zaharie, M., Recruitment for Competencies in Public and Private Sectors, *Transylvanian Review of Administrative Sciences*, 10(41), 129-145, 2014.

Papke-Shields, K. E., Beise, C., and Quan, J., Do Project Managers Practice What They Preach, and Does It Matter to Project Success? *International Journal of Project Management,* 28(7), 650-662, 2010.

Spalek, S., *Critical Success Factors in Project Management: To Fail Or Not to Fail, That is the Question!*. Project Management Institute, 2005.

Sullivan, K. T., Quality Management Programs in The Construction Industry: Best Value Compared with Other Methodologies, *Journal of Management in Engineering*, *27*(4), 210-219, 2010.

Yang, L. R., Huang, C. F., and Wu, K. S., The Association Among Project Manager's Leadership Style, Teamwork and Project Success, *International Journal of Project Management*, 29(3), 258-267, 2011.

Streamlining Information Transfer between Construction and Structural Engineering
Edited by Shiau, J., Vimonsatit, V., Yazdani, S., and Singh, A.

ISBN: 978-0-9960437-7-9

COOPERATIVE PROJECT DEVELOPMENT – INFLUENCE OF PARTICIPANTS ON SUCCESS FACTORS OF REAL ESTATE PROJECTS

EDWIN HARRER, BERNHARD BAUER, and GOTTFRIED MAUERHOFER

Institute for Construction Management and Economics, Technical University Graz, Graz, Austria

Successful project development is characterized by a multitude of risks and high uncertainties. Particularly for the correct assessment of the input data, such as market and location factors, as well as the right estimation of the requirements of the user, high competences are demanded by the project developer. Many decisions, which are required for a successful project development, therefore have to be made in very early stages of the development process. Based on this it has been postulated that an early involvement of stakeholders in the process of project development has influence on the project success. In this context there was executed a quantitative survey to examine competences of planners, contractors, brokers and operators to influence deducted success factors of project development. Findings revealed that the early integration of the participants influence the project success strongly. Particularly planners have a very high influence on object based factors like building design and layout of the ground plan. For optimizing the market and location factors also an integration of brokers and operators is essential. For developers it is recommended that they are informed of the possibilities of influencing the different success factors by an integration of the main participants. So the security for project relevant decisions in the early stage of the development process can be provided.

Keywords: Partnering, Integration of participants, Value of properties, Project success, Decision-making, Development process, Early involvement.

1 INTRODUCTION

1.1 Initial Situation

Until the end of the 20th Century at most European real estate markets, the demand exceeded the supply of properties. The increasing saturation of the real estate market in Europe led -- and still leads -- to a situation that buildings have to be adapted more and more to the requirements of the users and their increasing expectations (Brauer 2013). To provide market-conformity and valuable buildings it is necessary to give further incentives - beyond the basic expectations of the users - to buy or rent the corresponding property. This can only be achieved by an exact assessment of these expectations as well as a high flexibility for adapting the building to the different subjective demands of the users.

Furthermore, the rapid technological progress and the associated adaptations of standards and regulations, lead to a wealth of required disciplines and knowledge in real estate project development. The number of project participants and the complexity of the projects are increasing rapidly, resulting in higher coordination efforts and increasing project risks. Clients and builders who usually have little or no expertise in the execution of construction projects are thus often overwhelmed. The consequence is a massive transfer of risks[1] from the builders towards the contractors of building projects (Kochendörfer 2008). Due to the high level of competition at the construction market, it is often not possible to price in these risks in building contracts or control them by the contractors itself. So, conflicts during the project realization phase already are bound to occur (Mosey 2009).

In comparison to the product manufacturing industry the construction sector traditionally is characterized by relatively low profit margins. Despite adjustments in the procurement system, for real estate projects the lowest price is still the most important factor awarding a building contract. There are existing procurement systems which consider other factors to find the best offer but they regularly are arranged in a way that also the price represents the crucial criteria. To be able to operate economically entrepreneurs are forced either to optimize their process efficiency (get the highest achievable result with the least effort as possible) or improve their income by an aggressive claim-management. This is a quite common phenomenon during the execution of building projects. The consequences of this strategy is, that the success of projects massively can be compromised. Lack of quality in planning and execution, cost and deadline overruns or conflicts during the realization of building projects are almost inevitable with this form of project execution (Girmscheid 2014).

The trend of digitalization, which has captured all economic sectors has also profound effects on the construction and real estate industry. Big-data, virtual reality, augmented reality or building information modeling are only some keywords mentioned here. By integrating these tools in a reasonable way into the project development process, they can improve the accuracy of forecasting market data, enable interdisciplinary planning on a virtual building model or allow an early beginning of project-marketing.

1.2 Cooperative Project-Development

The developments in the construction and real estate industry mentioned above do not claim to be complete but are some distinct indicators that there is a demand to react on. The increasing implementation of forms of partnering in construction management, incentive-based contract models as well as the application of lean management are indications that the market participants are trying to react on the changed conditions (Heidemann 2011, ÖBV - Österreichische Bautechnik Vereinigung 2013). When looking at the whole lifecycle of the project development process, starting with the project initiation up to a new development or demolition of the building, these aspects are often not accounted sufficiently, or they are deliberately ignored in favor to a short-term yield optimization or a fast realization of the project.

The traditional process model of project development presupposes a sequential integration of the participants. That means the integration of them is based on the project progress. In case of success-relevant decisions, which often have to be made in early phases, the participants are not yet involved in the project. Competencies which may be relevant for the further course of the project therefore are not available. His affects especially the security of forecasted data as for example the specification of user requirements, the estimation of loan and buying prices, the determination of durable costs and deadlines or the improvement of the qualities of the project. Optimizations of the project are certainly possible and necessary at later phases, but with the ongoing progression of the project the influenceability for major adaptions decrease fast and the costs for them increase rapidly (IG Lebenszyklus Hochbau 2014).

---

[1]For example planning and coordination risks, arrangement of cost-caps or lump sums for insufficient detailed projects, etc.

Not only cooperative project management at planning and execution, also the concept of cooperative project development is characterized by the early involvement and the parallel processing through various participants. For this the participants not only come from the spheres of planning and execution, they are included from all phases of the project development process (Fröch 2013, Harrer and Mauerhofer 2017). Consequently operators, users, real estate brokers, financiers or investors, etc. are integrated in the decision-making process at an early stage. So therefore the different participants are already involved during the phase of project conception where the feasibility and efficiency of the project are checked and the main decisions for the ongoing project are made.

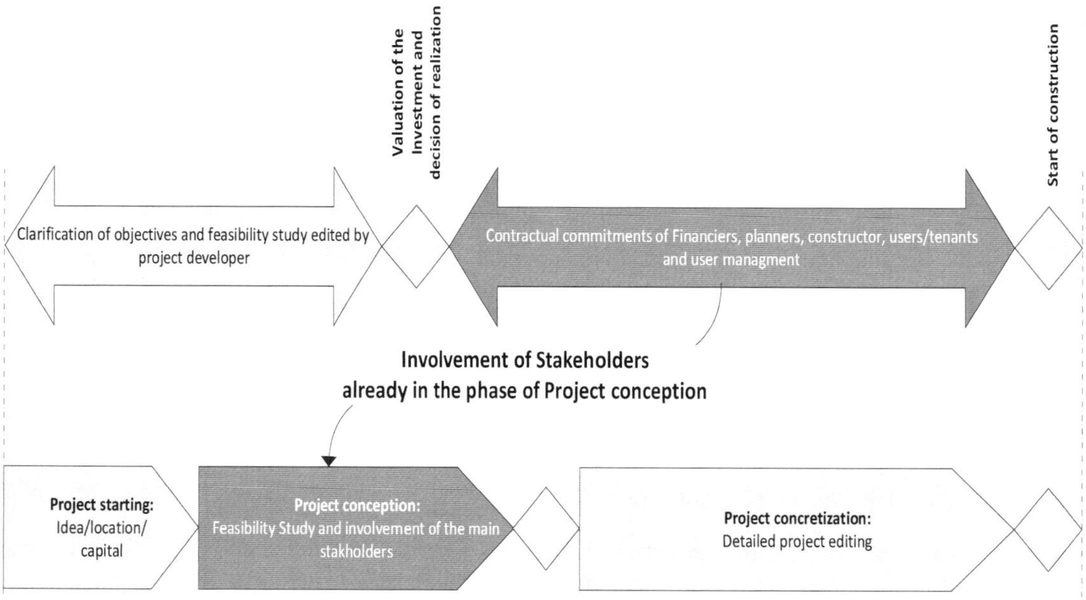

Figure 1. Integration of stakeholders in the phase of project-conception (Harrer 2016).

Based on these considerations it was postulated that project decisions which are regularly made by the project developer can be influenced by the early involvement of the respective actors. The aim of the research is to quantify the influence from essential participants on different success factors of the development process, in order to be able to give project developers an assistance to optimize the value of their project by applying a cooperative process model.

2 METHODOLOGY

Considering the market value of real estate properties as the basis for its recoverability, it can be assumed that all factors affecting its value are reflected by it (Eser 2009). That means, that the willingness of a tenant or investor to pay the appropriate rental or purchase price only will be given if all decision relevant parameters are involved in this price. This approach is also pursued by Fröch (2013) in his work, in which he demonstrates the optimization of project development by integration selected parameters of sustainability. In order to quantify the influence of the participants, based on the parameters to calculate the market value of properties, nine key success factors were defined. These are *location, market, building design, layout of ground plan, outdoor areas, building quality, flexibility of using, interior design and comfort.*

By using a standardized questionnaire, 680 experts were invited to estimate the possibilities of incorporating the competences of planners, contractors, brokers and operators to these factors. 83 questionnaires could be used for the evaluation, reflecting a responding rate of approx. 12%.

To each factor were assigned three to four task areas of the project development process on which the competences of the participants could be estimated by using a five-digit Likert scale. To quantify the answers a score from zero points (no competence) to four points (high competence) was applied.

3 RESULTS

Figure 2 vividly illustrates the possibility of bringing in the competences of the participants for the nine defined success factors of real estate properties.

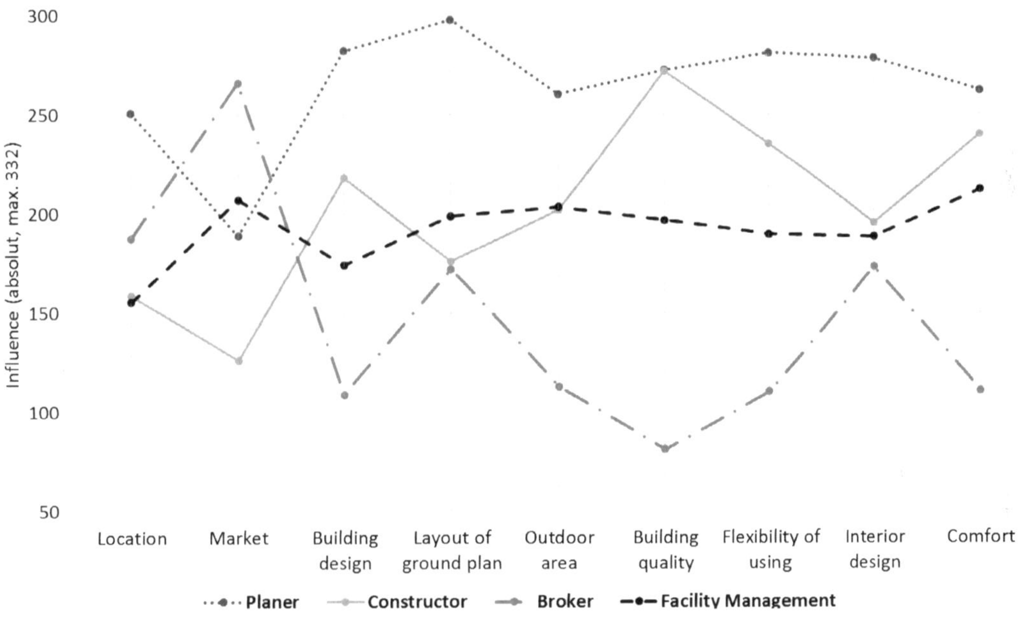

Figure 2. Influence of project participants on success factors of project development.

The integration of *planners* has the highest impact on almost all factors. In particular the object-dependent factors such as building design, layout of ground plan, interior design, etc. can be optimized by planners. The reasons for this are, on the one hand, the high level of competence in the planning of the building, but also the independence of planners to estimate costs and deadlines as well as the securing of building quality during the execution phase.

The early involvement of the *constructor* has the most impact to improve the quality of the building. By setting specific incentives, conflicts between client and constructor can be reduced. The security of predicted costs and target dates also can be increased by the involvement of the constructor in the early phase of project development.

The involvement of the *broker* is mainly important to optimize the assessment of market and location factors. High competencies in forecasting regional developments, the assessment of rental income or retail prices as well as the information on current projects which are standing in competition are crucial task areas which can be affected.

The effects of integrating the *facility management* is estimated in the middle of the defined success factors. This is mainly because of the high experience required for estimating the costs for operating buildings as well as the assessment of durability and lifetimes of used materials. Also, relatively high competences for defining adequate user requirements are attributed to the Facility Management.

4 CONCLUSION AND FURTHER RESEARCH

The results of the conducted expert survey shows the possibilities of the participants to influence the nine derived success factors in the early stage of project development. For an optimization of the individual factors it is obvious which participants should be involved in the project development process early. For the project developer so it is possible to decide which participants are important to improve the success of the project. One of the main key findings is the high influence of planners to mainly all deducted success factors. So it is absolutely essential to implement the planner early in the development process. Also, the integration of the broker is very important for a correct assessment of the market factors like the estimation of user requirements or the regional development.

The next step of the research is to observe the influence of the nine deducted success factors on the determination of the market value. So it is possible to show directly the relative influence of the different participants on the value-determining parameters. An independent part of the further research is to analyze different project organization forms how the participants can be integrated into the project development process. Which project types are suitable for a cooperative processing, what aspects are important for the implementation and so far, an early integration of the participants is practicable.

References

Brauer, K.-U., *Grundlagen der Immobilienwirtschaft,* Recht - Steuern - Marketing - Finanzierung - Bestandsmanagement - Projektentwicklung, Springer Fachmedien, Wiesbaden, 2013.

Eser, B., *Erzielung nachhaltig hoher Büroimmobilienwerte,* Ein Entscheidungsmodell für die Planungsoptimierung, Gabler Verlag / GWV Fachverlage GmbH, Wiesbaden, 2009.

Fröch, G., *Probabilistische Bewertung und systematische Optimierung von Projektentwicklungen unter Berücksichtigung von Nachhaltigkeitskriterien,* Bauwirtschaft und Projektmanagement, University Press, Innsbruck, 2013.

Girmscheid, G., *Projektabwicklung in der Bauwirtschaft-prozessorientiert,* Wege zur Win-Win-Situation für Auftraggeber und Auftragnehmer, Springer, Berlin, Heidelberg, 2014.

Harrer, E. *Value Oriented Project Development - Effects of an early Integration of Stakeholders to the Value of Properties,* First European and Mediterranean Structural Engineering and Construction Conference, EURO-MED-SEC-1, Istanbul, Turkey, May 24-29, 2016.

Harrer, E. and Mauerhofer, G. *Partnerschaftliche Projektentwicklung - Möglichkeiten kooperativer Projektabwicklungsstrategien und deren Auswirkungen auf den Wert von Immobilien,* Festschrift zum 60. Geburtstag von Univ.-Prof. Dr.-Ing. Christoph Motzko, Darmstadt, 2017.

Heidemann, A., *Kooperative Projektabwicklung im Bauwesen unter der Berücksichtigung von Lean-Prinzipien - Entwicklung eines Lean-Projektabwicklungssystems,* Internationale Untersuchungen im Hinblick auf die Umsetzung und Anwendbarkeit in Deutschland, KIT Scientific Publishing, Karlsruhe, 2011.

IG Lebenszyklus Hochbau, *Der Weg zum lebenszyklusorientierten Hochbau,* Leitfaden für Bauherrn sowie Vertreter der Bau- und Immobilienbranche, 2014.

Kochendörfer, B. *Erfordernis des Partnering Angesichts der Deutschen Marktverhältnisse,* Partnering in der Bau- und Immobilienwirtschaft, Projektmanagement- und Vertragsstandards in Deutschland, Stuttgart, 2008.

Mosey, D., *Early Contractor Involvement in Building Procurement,* Contracts, Partnering and Project Management, Wiley- Blackwell, Chichester, U.K, 2009.

ÖBV - Österreichische Bautechnik Vereinigung, *Kooperative Projektabwicklung - Empfehlungen zur erfolgreichen Umsetzung komplexer Bauvorhaben*. Merkblatt, Wien, 2013.

Streamlining Information Transfer between Construction and Structural Engineering
Edited by Shiau, J., Vimonsatit, V., Yazdani, S., and Singh, A.

ISBN: 978-0-9960437-7-9

ROLE OF FINANCIAL CAPABILITY IN MANAGING IMPACTS OF FOREX FLUCTUATIONS ON CONSTRUCTION BUSINESS

MOHD AMIZAN BIN MOHAMED[1,2], MELISSA TEO[2], STEPHEN KAJEWSKI[2], and BAMBANG TRIGUNARSYAH[2]

[1]*Faculty of Civil Engineering, UiTM, Shah Alam, Malaysia*
[2]*School of Civil Engineering and Built Environment, Science and Engineering Faculty Queensland University of Technology, Brisbane, Australia*

Many reasons have led construction organizations to venture into international projects in the past years. Despite offering opportunities for business expansion, risks from foreign exchange (FOREX) fluctuations are some of the challenges that need addressing. These fluctuations have affected the construction business at company level and project levels. Most of the impacts are related to financial issues. Thus, coordinating strategic plans to deal and mitigate the impacts of FOREX fluctuations is crucial. A qualitative research was conducted in response to this. Data collection for this research involved interviewing 14 respondents who represented Malaysian construction organizations and actively involves in international construction projects. The respondents ranged from Chief Executive Officer (CEO), Chief Financial Officer (CFO), financial and project managers. NVivo Version 11.0 was used in managing the transcribed interview data. The data were analyzed via content analysis. Results revealed that FOREX fluctuations give impacts on the financial performance of construction businesses. This study also suggested mitigation actions that could be implemented to manage these issues. The most important finding is that Financial Capability (FC) is relevant and essential to overcome those issues. Finally, a framework of FC in managing impacts of FOREX fluctuations on construction business is proposed.

Keywords: Foreign exchange (FOREX), Financial capabilities (FC), Business performance.

1 INTRODUCTION

The Foreign exchange (FOREX) is a decentralised market and is considered to be the world's leading trading area for currencies and for investors, consequently, it is important for them to predict any trends and fluctuations in the world currencies (Talebi *et al.* 2014). The same concept also can be applied to the companies that are involved in foreign exchange trading. Foreign Exchange Trading (FX) has been around for over four decades, but it started when the first 'paper' money in the world was used (Fabian *et al.* 2017). When there are any fluctuations in the foreign exchange, the risk will rise which may cause a disturbance in a business's operation (Ehrlich *et al.* 2012). Based on this unpredictability, FOREX may have a huge impact on world activities.

An international construction business will also face some difficulties due to the FOREX fluctuations (Mohamed *et al.* 2015). If the FOREX fluctuates, the financial condition of the construction company might be affected (Yean *et al.* 2006). Previous studies on the FOREX fluctuations only focused at the project level but not at the construction company level (Mohamed *et al.* 2015). Consequently, the construction company may face dire consequences if they have poor management skills in dealing with the impact of FOREX fluctuations (Ehrlich *et al.* 2012). Based on the above research, it shows that any fluctuations of FOREX may cause disturbance and major problems to a construction business. Thus, it is a significant benefit to the construction companies if they can manage the FOREX fluctuations to sustain their business performance.

2 LITERATURE REVIEW

The questions are: what are the impacts of FOREX fluctuations on construction businesses? How can these companies manage the impacts of FOREX fluctuation in their business? And how can the financial capability within the business be used to manage those impacts?

2.1 Impacts of Forex Fluctuations on Construction Business

A company dealing with FOREX maybe trading with several foreign currencies, so when any fluctuations happen a risk may arise (Ehrlich *et al.* 2012). Any construction companies that operate their business overseas may also be affected. For example, the costs overrun in a project may happen because the price of raw materials increases due to the FOREX fluctuations (Fidan *et al.* 2011). In addition, other problems may arise in an international construction business such as a delay in projects, confusion in financial management, lack of competitiveness, critical abandonment and ultimately, insolvency and bankruptcy. Hence, it is important for construction companies to manage the impact of FOREX fluctuations to protect their profit margins and sustainability and also their business performance.

2.2 Managing the Impacts of Forex Fluctuations

FOREX fluctuations are most likely the cause for the most common problems at project level such as cost overruns and delay. However, these fluctuations may also have an effect at the company level such as the financial stability and performance. In other words, fluctuations of the market can affect the profit and losses (P & L) of the company so it can be seen as an impact at company level too. Hence, it is not appropriate to just only manage the impacts at the project level because the effects are also felt at the company level. The appropriate actions need to be taken in order to mitigate the impacts of FOREX fluctuations to guarantee the sustainability of a construction business performance (Ehrlich *et al.* 2012, Mohamed *et al.* 2015). This study will focus only on one capability which is Financial Capability (FC).

2.3 Financial Capability

'Financial Capability' term was used by Xiao and O'Neill (2016) where some important elements in finance that need to be considered by an international construction company such as bonds, cash flow, and the transaction of money (Jung *et al.* 2012), which can be affected and directly related to FOREX fluctuations.

The financial capability of a construction company can be used to improve its current business and to also mitigate the impacts of FOREX fluctuations (Jung *et al.* 2012). Financial capability is an important aspect where it helps construction companies to maintain the expected cash flow and profit margin (Ahn *et al.* 2009). There are many aspects of financial capability, and one of the main ones being able to negotiate a loan with a bank to lock in a fixed rate for the required currency or borrowing in order to purchase the materials early before their projects start.

3 METHODOLOGY

The data for this study was collected through face to face interviews with 14 respondents from Malaysian construction companies. These companies fulfilled the study criteria before interview. The criteria were: (a) The company has some overseas projects, (b) The company is an active construction business and (c) The company has 3 years experienced in the field.

The interviews were conducted on the top management of the company including; Chief Executive Officer (CEO), Chief Financial Officer (CFO), and financial and project managers. The data was then analysed by using a software that supports qualitative research called 'NVivo version 11'.

4 RESULTS

The impacts of FOREX fluctuations on the business performance is divided into 2 categories which are; project level and company level.

4.1 Impacts of FOREX Fluctuations Toward Financial Capability

The study identified several impacts of FOREX fluctuations that related to the financial capability of a construction company. The next stage of the research was to categorise the findings into 6 sub-factors (As Table 1). The ref value in Table 1 means that the subfactor was mentioned by a respondent in the interviews.

Table 1. Impacts of FOREX fluctuations towards financial capability.

Factor	Sub-Factors	Ref	% of ref	Ranking
Impacts of FOREX fluctuations	Translation	119	20.20	1
	Transaction	118	20.03	2
	Profit	113	19.19	3
	Economic	88	14.94	4
	Cash Loss	81	13.75	5
	Non Profit	70	11.88	6
	Total	**589**	**100**	

4.2 Mitigation Action in Term of Financial Capability

The study uncovered a lot of actions that can mitigate a lack of financial capability as mentioned by interviewees stated in Table 2.

One of them is by a hedging method. According to Ahn *et al.* (2009), it is possible to mitigate the exchange rate and currency conversion risk by having dual currency contracts and hedging tools, these were found to be effective in China. The best way to understand hedging is to think of it as insurance. Researchers such as Ehrlich *et al.* (2012) state that the hedging method can provide a level of protection by limiting the possibility of financial repercussions on the business. Hence, the construction company can usually have an agreement or negotiate a contract with the parties they are involved with. The construction companies can hedge with their suppliers or clients with an agreement related to FOREX fluctuations.

In addition, some construction companies mitigate the impacts of FOREX fluctuations by buying a property in the foreign country; this is sometimes seen as a flexible option. The company will buy the property in the foreign country because the price of the property can often increase and alleviate the impact of FOREX fluctuations. Nonetheless, this method depends on the regulations and rules of the foreign government of the construction that they are involved in.

Table 2. Mitigation actions in term of financial capability.

Method	Description
Hedging	• Make an agreement with the suppliers or the banks to lock the price of the materials or loan. Companies can make an agreement on fixed rate loan contract with lending banks in order to reduce the impacts of FOREX fluctuations (Wang et al., 2008).
Buying property	• Instead of bring back the money to the origin country, the construction companies can buy a property at the foreign country, and make a profit out of it and overcome the shortfall. Some investors are relying on sales of property to overcome their shortfall (Atkinson et al., 2007).
Open separate account	• Open separate account is considered as a strategic investment in the foreign market. By having a separated account the FOREX involved in the future will be reduce hence reduce the risk of FOREX fluctuations. Cash and investment policy; match sources and utilization of funding; and reserved cash (retained earnings) is considered as one of the risk management in term of financial capability at the international markets (Wethyavivorn et al., 2009)

4.3 Financial Capability

However, the mitigating actions in term of financial capability are unable to be utilized by the construction business companies. Hence it is important to study the financial capability of the construction company. Financial capability is considered a conversion factor necessary for an individual or even organization to use a resource, money, into various financial goods (Allmark and Machaczek 2015).

Table 3 is the list of subfactors of financial capability. Based on the interviews, the financial capability of the construction company can be divided into several sub-factors, which are financial management, asset, material, labour and loan.

Financial management was observed as the most important and crucial aspect of the financial capability of the construction company. It scored the highest ranking followed by asset, materials, labor and loan.

These findings can help the construction company in managing their finances in the future. However, the interviewees also mentioned several other aspects related to financial capabilities such as assets, material, labor and the loan associated with the project.

Table 3. Subfactors of the Financial Capability.

Factor	Sub-Factors	Ref	% of ref	Ranking
Financial Capability	Financial Management	488	44.00	1
	Asset	219	19.75	2
	Material	160	14.43	3
	Labour	147	13.26	4
	Loan	95	8.57	5
	Total	**1109**	**100**	

4.4 Proposed Framework of Financial Capability

A framework to manage the impacts of FOREX fluctuations is proposed in Figure 1. The proposed framework shows that the impact of the fluctuations of FOREX could be seen at two (2) different level in the construction industry; namely, project and company level, both contributing to the overall company.

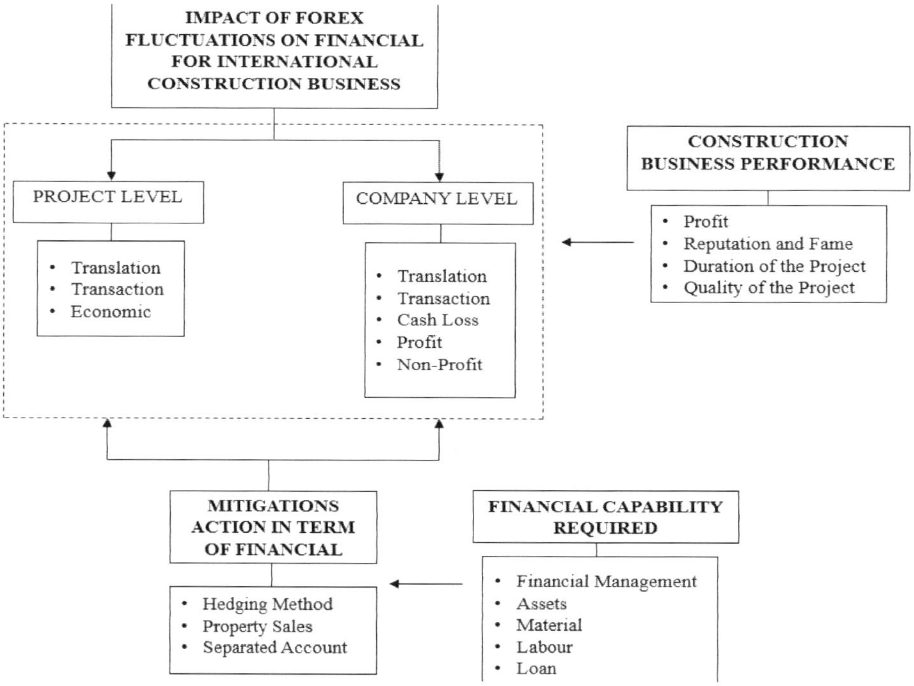

Figure 1. Proposed framework of the financial capability towards the impacts of FOREX fluctuations on construction business performance.

The important part of this proposed framework is highlighting the mitigation actions that need to be carried out to resolve the issues raised from the FOREX fluctuations. This framework strongly addresses the role of financial capability in applying the mitigation actions to manage those impacts of FOREX fluctuations.

The proposed framework is developed based on the perspective of Malaysian construction companies who were involved with projects overseas. This framework can be a guide for the construction companies from different countries as well. However, it might be different in terms of impacts and mitigation actions locally, but the key point is that financial capability has a role in implementing the mitigation actions.

5 CONCLUSIONS

The objective and aim of the study were achieved. The impacts of FOREX fluctuations on a construction business performance from a financial capability perspective were identified. The mitigating actions in term of financial capability were also identified. Based on the literature review and analysis from the respondents a financial capability framework was proposed and can be used as a guideline for a construction company when they are involved with the overseas market. Moreover, the important capabilities for the construction business company when they are in an international market have also been discussed. This study has focused on the financial capability of construction companies. The literature reviews and data analysis were conducted for a deep discussion on this capability. Thus, the objectives and aims of the study were achieved.

References

Ackermann, F., Pohl, W., and Schmedders, K., Optimal and Naive Diversification in Currency Markets. Management Science 63(10):3347-3360. https://doi.org/10.1287/mnsc.2016.2497, 2017.

Ahn, Y. H., Holley, P., and Kang, J. S., Risk Management of Exchange Rates in International Construction, *International Journal of Construction Education and Research*, 5(1), 24-44. https://doi.org/10.1080/15578770902717550, 2009.

Allmark, P., and Machaczek, K., Financial Capability, Health and Disability. *BMC Public Health,* 15(1), 243. https://doi.org/10.1186/s12889-015-1589-5, 2015.

Ehrlich, M., Woodward, D., and Tiong, R., A State- of- Practice Survey on Managing FX Exposure in Project Companies, Construction Companies and SMEs, *Journal of Financial Management of Property and Construction,* 17(1), 29-48. https://doi.org/10.1108/13664381211211037, 2012.

Fidan, G., Dikmen, I., Tanyer, A. M., and Birgonul, M. T., Ontology for Relating Risk and Vulnerability to Cost Overrun in International Projects, *Journal of Computing in Civil Engineering*, 25(4), 302-315. https://doi.org/10.1061/(ASCE)CP.1943-5487.0000090, 2011.

Jung, W., Han, S. H., Asce, M., Koo, B. and Jang, W., Which Strategies are More Effective for International Contractors during Boom and Recession Periods?, 28(3), 281–290. https://doi.org/10.1061/(ASCE)ME.1943-5479.0000087, 2012.

Mohamed, M. A., Bin, Teo, M., Kajewski, S., and Trigunarsyah, B., A Relationship of Managing Impacts of FOREX Fluctuations and Organizational Capabilities in Construction Business. Proceedings of the 6th International Conference on Construction Engineering and Project Management 11-14 October 2015, Busan, Korea, (October), 477–480, 2015.

Talebi, H., Hoang, W., and Gavrilova, M. L., Multi-Scale Foreign Exchange Rates Ensemble for Classification of Trends in Forex Market, *Procedia Computer Science*, 29, 2065–2075. https://doi.org/10.1016/j.procs.2014.05.190, 2014.

Xiao, J. J., and O'Neill, B., Consumer Financial Education and Financial Capability, *International Journal of Consumer Studies*, *40*(6), 712-721, 2016.

Streamlining Information Transfer between Construction and Structural Engineering
Edited by Shiau, J., Vimonsatit, V., Yazdani, S., and Singh, A.

ISBN: 978-0-9960437-7-9

CHALLENGES OF FRONT-END LOADING IN CONSTRUCTION PROJECT DELIVERY

DOUGLAS AGHIMIEN, CLINTON AIGBAVBOA, AYODEJI OKE, and MATLEKO SETATI

Faculty of Engineering and the Built Environment, University of Johannesburg, Johannesburg, South Africa

Adopting a good pre-project planning or front-end loading method during the early stage of construction is important to the success of the project. However, most construction projects have failed as a result of poor project planning at their early stage. Based on this knowledge, this study assessed the challenges facing the use of front-end loading. The study adopted a quantitative survey approach and questionnaire was used to harness information on the objective of the study from construction professionals in Gauteng Province, South Africa. Data gathered were analysed using statistical tools such as percentage, mean score and standard deviation. Findings revealed that the major challenges affecting the use of front-end loading are inability to identify importance of the process, unreliable information during early project stages, insufficient time to thoroughly carry out the front-end loading process, indecisiveness or lack of knowledge by the client, and lack of structured project team during inception phases of the project. It is therefore recommended that the construction clients be made aware of the importance of front-end loading, while project team members should be introduced to the project from the early stage of the project.

Keywords: Front-end planning, Pre-project planning, Project Performance, Project success.

1 INTRODUCTION

Project Management refers to the application of acquired knowledge and the use of tools, techniques, and skills for project activities with the aim of meeting the project requirements (Project Management Institute 2012). With proper project management, project objectives are met through the integration and application of various processes such as planning, execution, monitoring and controlling, and closing. Planning has proven to be crucial in the successful delivery of any project. Planning makes it easy to understand how the project team intends to execute the work in a systematic manner, which will facilitate achievement of the project objectives. Furthermore, planning also improves ease of identification of the project risks. It is important to establish the manner in which the activities to be completed will be carried out as the process will indicate whether it will be possible, given the available resources. Wang and Gibson (2012) noted that the decisions made during the early planning phases of projects have a significant bearing on the final project performance.

Project performance has overtime been evaluated using a number of factors which may be sought after individually or combined. The timely completion, compliance with the quality requirements, compliance with the targeted budget and the client's satisfaction are the factors, which are used to determine project success. Unfortunately, construction project delivery in developing countries, South Africa included, has failed in most cases to deliver construction projects that perform to the set performance criteria. Most construction projects are delivered way over budget and far beyond the stated completion time. This is not including the frequent client and end-users dissatisfaction, failure to meet health and safety standards and sustainability criteria. In South Africa, Emuze (2011) observed that it is rare for the dream of construction clients to come to reality due to the poor performance of the construction industry. There is no doubt that many of these construction projects executed in these developing countries failed to reach their full potential as a result of inadequate pre-project planning also known as front-end loading, during conceptualisation (Morris 2005).

As a result of the complexity of construction project and the need to deliver projects successfully, it has become important to implement a process which will facilitate effective decision making to prepare for the engineering and construction process which follows pre-planning (Sherif and Price 1999). Andersson and Rosenberg (2012) and Gibson and Pappas (2003) have noted that front-end loading has a significant contribution to project success. It is essential where meeting project objectives are concerned. According to Shlopak *et al.* (2014), front-end loading also referred to as pre-project planning, front-end planning, programming, schematic design or design development, is regarded as the foundation that needs to be created to ensure predictable and efficient project delivery. According to the Construction Industry Development Board (CIDB) (2016), the adoption of a robust front-end loading is crucial to the successful delivery of any project. However, in most construction projects the front-end loading phase is the phase where minimal or no emphasis is invested (Construction Industry Institute, 2012). Moreover, Edkins *et al.* (2013) have noted that the front-end of project management practice is very often poorly understood and thus, inconsistent from project to project and also between industries. The practice is consequently confused, and this is as a result of lacking effective, clear and adequate guidance concerning it. Front-end loading is the most important point in the lifecycle of a project as the decisions, which are made during this phase of the project, have impact on the final project schedule and cost which are adopted as performance indicators on most construction projects (Son *et al.* 2015).

In defining front-end loading, although there are variations in the definitions, they ultimately allude to the same notion. The definitions all highlight the importance of great planning at the very front-end of the project and also the importance of keeping the planning in this phase consistent from initial endorsement up until a decision has been made to continue with the physical construction work (Williams and Samset 2010). Morris (2011) simply described front-end loading as the preliminary/emergence phase of the project. Saputelli *et al.* (2008) described front-end loading as a methodology whose main objective is focused on capital project planning. According to the Construction Industry Institute (2012), front-end loading is a process, which is aimed at attaining maximum project success through the gathering of information with regards to any risk that may be associated with the execution of the project.

Merrow (2011) emphasised that the front-end loading process continues until complete authorisation of the viability of the project has been established and capital has been fully granted. According to Laufer and Tucker (2006), the planning phase of construction projects seeks to identify whether or not it is viable to commit any resources to the project. The planning of construction projects consumes a significant amount of resources and it, therefore, becomes concerning when construction projects fail to reach their full success potential. The construction planning process identifies goals, which need to be achieved and through efficient planning techniques implemented, aims at satisfying these goals. In addition, Wang and Gibson (2012) emphasised that the decisions made by the project team members during this early stage of the project are of great significance and may determine the outcome of the project. Green and Perry (2008) noted that the level of front-end loading and project performance are directly proportional and the relationship can be examined through the overall project costs. Integration of the project times minimises later changes to the project and these minimal changes are significant because they determine the overall costs, improve adherence to the project schedule plan, and also improve the operability.

Considering the poor delivery of construction projects with respect to set project objectives, and the obvious importance of project pre-planning, it is therefore important to critically evaluate the challenges of front-end loading in the delivery of construction projects in South Africa. This is done with a view of improving construction project delivery in the country through adequate planning process.

2 METHODOLOGY

This study assessed the challenges of front-end loading in the delivery of construction projects in South Africa. The study was conducted in Gauteng province based on the availability of a large number of construction organisations and projects within the province. A quantitative survey approach was adopted for the study with information gathered from construction professionals directly involved in the delivery of construction projects within the study area. These professionals include architects, construction managers, civil and structural engineers, project managers and the quantity surveyors. Questionnaire was adopted as the instrument for data collection due to its ease of usage and the ability to cover a wide range of respondents (Tan 2011). The questionnaire used was designed in two sections and was administered to the identified professionals based on their convenience and willingness to partake in the survey. The first section of the questionnaire sort answers to questions relating to the background of the respondents, while the second section addressed the objectives of the research, which was to assess the challenges facing the proper implementation of front-end loading on construction projects. A total of 60 questionnaires were conveniently distributed, with 44 deemed fit for analysis upon return. The remaining 16 were dropped due to inaccurately filling and omission of answers to important questions. This 44 analysed questionnaire represents 73% of the distributed questionnaire and was considered adequate for analysis.

In analysing the data gathered, information on the respondent's background was analysed using percentage, while mean score and standard deviation (SD) were used to analysed data gathered the challenges of the use of front-end loading. These challenges were ranked from the highest mean score down to the lowest. However, where two variables have the same mean score the variable with the lowest SD was ranked first as suggested by Field (2005). The reliability of the questionnaire used was tested using the Cronbach alpha test. The Cronbach alpha test gives a measure of the inward consistency of a test or scale and it ranges between 0 and 1 (Tavakol and Dennick 2011). The greater the score the more reliable the generated scale is, however, the indicated score of 0.7 and above is said to be an acceptable reliability coefficient but lower thresholds are sometimes used in the literature (Santos 1999). Cronbach's alpha value of 0.825 was derived. Since these alpha values are above the recommended 0.7 it, therefore, follows that the adopted research instrument is reliable.

3 FINDINGS AND DISCUSSIONS

3.1 Background Information

The analysis of the background information of the respondents shows that the most prominent professionals were the Project Managers (27.3%) followed by the Quantity Surveyors (25.0%) and Construction Managers (22.7%). The least represented were the Engineers (13.6%) and Architects (11.4%). In terms of highest level of education attained, findings revealed that most respondents (34%) held bachelor's degree, followed by honours degree (29.5%) and a national diploma (22.7%). In addition, the average years of respondents working experience within the built environment was assessed. The result revealed that the highest range of working experience is between 11 and 20 years (34.1%), this is followed by 6-10 years (31.8%), and 1-5 years (29.5%). Only a few respondents (4.5%) have a working experience of over 20 years. From the result, it is evident that the identified population for the study is adequately represented, with a considerable amount of education attained to be able to understand and interpret the questions posed in the study. Also, there is a considerable high amount of years of working experience among the respondents. Thus, the answers received from the respondents can be relied upon as these would have been given based on the understanding of happenings within the South African construction industry.

3.2 Challenges of Front-End Loading in the South African Construction Industry

The challenges of the use of front-end loading in the South African construction industry was assessed by presenting respondents with some list of challenges identified from the review of literature. Respondents were asked to rank these challenges based on their level of significance using a scale of 1 to 5, with 5 being very high, and 1 being not significant. The result in Table 1 shows the ranking of these challenges as perceived by the respondents. From the Table, it is evident that all the 10 assessed challenges have a mean value of above average of 3.0. This means that to a considerable extent, respondents believe that these challenges are affecting the use of front-end loading in the study area. Chief of these challenges are inability to identify importance of the process with a mean score of 4.09, unreliable information during early project stages with a mean score of 4.07, insufficient time to thoroughly carry out the front-end loading process with a mean score of 3.89, indecisiveness or lack of knowledge by the client with a mean score of 3.89, and lack of structured project team during inception phases of the project with a mean score of 3.86.

Table 1. Challenges of Front-End Loading.

Challenges	Mean	SD	Rank
Inability to identify the importance of the process	4.09	1.273	1
Unreliable information during the early project stages	4.07	1.228	2
Insufficient time to thoroughly carry out the front-end loading process	3.89	1.224	3
Indecisiveness or lack of knowledge by the client	3.89	1.146	4
Lack of structured project team during inception phases of the project	3.86	1.047	5
Insufficient budget to carry out the front-end loading process	3.70	1.231	6
Insufficient knowledge on the process by the project team	3.70	1.407	7
The uncertainty of project schedule	3.59	1.317	8
The uncertainty of project budget	3.55	1.130	9
The uncertainty of project delivery method	3.41	1.263	10

The dynamic and complex nature of construction projects plays a part in increased project risk and ultimately promotes automatic project failure. However, with sufficient planning the risk may be reduced and lead the projects towards better performance. In addition, great expertise and broad understanding of construction are required to lead the construction project as it facilitates improved project performance. The findings imply that project managers need to use their expert knowledge to advise both their clients and the project team on the importance of the front-end loading process. Moreover, they need to utilise their role to guide the clients in applying processes, which will facilitate the best possible project outcome. Findings of this study further corroborate Cleland and Ireland (2010) submission that front-end loading is not adequately carried out because most clients fail to identify the importance of the process and very quickly wants to see progress in the project. The findings also affirm Cooper (2008) assertions that construction projects fail because of the lack of planning in the pre-project phases, poor leadership, and unreliable data. Consequently, the Independent Project Analysis (2002) stated that it is important to develop a structured team and manage the transition of the team members during the different phases of the project. Findings of this study show that as a result of lack of structured project team, most projects fail to employ proper front-end loading.

4 CONCLUSION

Based on the findings of the study, it is concluded that the major challenges affecting the use of front-end loading are inability to identify importance of the process, unreliable information during early project stages, insufficient time to thoroughly carry out the front-end loading process, indecisiveness or lack of knowledge by the client, and lack of structured project team during inception phases of the project. Thus, there is the need to increase clients' knowledge of the process of front-end loading and their understanding of its importance. Introduction of project team members to the project from the early stage is also crucial, as this will give a good view and understanding of the project from the outset and also help improve communication within the project team.

Findings of this study contribute to the body of existing knowledge as it shows the major challenges facing the use of front-end loading for construction projects. The findings will go a long way in assisting construction participants in understanding the need for adopting this process in the delivery of successful construction projects, and the challenges they need to avoid in order to effectively adopt this system. Although this study contributes significantly to the body of literature emanating from South Africa, its findings cannot be generalised as it was conducted in a single province within the country. Thus, further studies can be conducted in order regions of the country in order to get a wider view of the topic and also to compare result.

References

Andersson, C., and Rosenberg, L., *The Preconstruction Planning Process from a Site Manager Perspective.* Chalmers University of Technology, 2012.

Cleland, D., and Ireland, L., *Project Manager's Portable Handbook* 3rd ed., McGraw-Hill, New York, 2010.

Construction Industry Development Board, *Construction Industry Development Board,* 2016. Retrieved from https://registers.cidb.org.za/PublicContractors/Reports?RenderReport=trueandType=COWand provinceid=49e86b26-80c6-e111-a22d00155d022301andgradeid=1andownership=0andgender=0and youth=0andeffectivedate=2016/05/07%2012:18:18%20AM.

Construction Industry Institute, *CII Best Practices Guide: Improving Project Performance,* United States of America, 2012.

Cooper, R.G., The Stage-Gate Idea-to Launch Process-Update, What's New and NexGen Systems, *Journal of Product Innovation Management,* 25(3), 213-232, 2008.

Edkins, A., Geraldi, J., and Smith, A., Exploring the Front-End of Project Management, *Engineering Project Organization Journal,* 3(2), 71-85, 2013.

Emuze, F. A., *Performance Improvement in South African Construction.,* PhD Thesis in Construction Management Submitted to the Faculty of Engineering, The Built Environment and Information Technology at the Nelson Mandela Metropolitan University, 2011.

Field, A., Discovering Statistics, Using SPSS for Windows, Sage Publications, London, 2005.

Gibson, E., and Pappas, M. P., *Starting Smart: Key Practices for Developing Scopes of Work for Facility Projects*, Washington, DC: National Academies Press, 2003.

Green, D. W., and Perry, H. R., Capital Project Execution and Analysis" in *Perry's Chemical Engineers' Handbook,* 8th edn, McGraw-Hill, New York, 2008.

Independent Project Analysis, HPT impact, *Hydrocarbon Processing,* 23, 2002.
Laufer, A., and Tucker, R. L., Is Construction Project Planning Really Doing Its Job? A Critical Examination of Focus, Role and Process, *Construction Management and Economics,* 5(3), 243-256, 2006.
Merrow, E. W., *Industrial Megaprojects: Concepts, Strategies, and Practices for Success*, Wiley, USA, 2011.
Morris, P., *Managing the Front-End: How Project Managers Shape Business Strategy and Manage Project Definition,* PMI Global Congress Proceedings. Edingburgh, Scotland, 2005.
Morris, P., Managing the Front-End; Back to Beginning, *Project Perspectives,* (33), 4-8, 2011.
Project Management Institute, What is Project Management?, 2012. Retrieved from https://www.pmi.org/about/learn-about-pmi/what-is-project-management on June 23, 2018
Santos, J. R. A., Cronbach's alpha: A Tool for Assessing the Reliability of Scales. *Journal of Extension, 37(2),* 1–5, 1999.
Saputelli, L., Hull, R., and Alfonzo, A., *Front End Loading Provides Foundation for Smarter Project Execution. Oil and Gas Financial Journal*, 5(7): Special Report, 2008.
Sherif, M. A., and Price, A. D. F., *A Framework for Preproject Planning*. Proceedings 15th Annual ARCOM Conference, 15-17 September, Liverpool, UK, Association of Researchers in Construction Management, 2, 435-444, 1999.
Shlopak, M., Emblemsvag, J. and Oterhals, O., Front End Loading as an Integral Part of the Project Execution Model in Lean Ship Building, *Proceedings of the 22nd Annual Conference of the International Group for Lean Construction*, Akademika forlag, Oslo, 2014.
Son, H., Lee, S., and Kim, C., An Empirical Investigation of Key Pre-Project Planning Practices Affecting the Cost Performance of Green Building, *Procedia Engineering,* 11(8), 37-41, 2015.
Tan, W. C. K., *Practical Research Methods,* Pearson Custom, Singapore, 2011.
Tavakol, M., and Dennick, R., Making Sense of Cronbach's Alpha, *International Journal of Medical Education,* 2(1), 53-55, 2011.
Wang, Y., and Gibson Jr, E., A Study of Pre-Project Planning snd Project Success Using Anns and Regression Models, *Automation in Construction*. 19(3), 341-346, 2012.
Williams, T., and Samset, K., Issues in the Front-End Decision Making on Projects, *Project Management Journal,* 41(2), 38-49, 2010.

Streamlining Information Transfer between Construction and Structural Engineering
Edited by Shiau, J., Vimonsatit, V., Yazdani, S., and Singh, A.

ISBN: 978-0-9960437-7-9

SCHEDULE RISK ANALYSIS OF RAILWAY PROJECTS USING MONTE CARLO SIMULATION FOR IMPROVED PROJECT MANAGEMENT

MOTLATSO MABEBA[1], JAN HARM C PRETORIUS[1], and LEON PRETORIUS[2]

[1]*Postgraduate School of Engineering Management, University of Johannesburg, Johannesburg, South Africa*
[2]*Graduate School of Technology Management, University of Pretoria, Pretoria, South Africa*

Railways have been used throughout history for the transportation of goods. Even though the inception of rail transport improved civilization, due to its inefficiencies, road transport is at present dominating the freight and logistics industry. Company A, which has the largest market share in the rail freight business, has embarked on projects to improve rail efficiencies by moving higher volumes of freight timeously. Most of the projects embarked on by Company A have failed largely due to the poor planning of the projects in the feasibility stages. Most of the planning schedules are overoptimistic and unrealistic making them unreliable and difficult to track. The scope of this study was to investigate the way in which planning schedules of Company A are developed by undertaking a schedule risk analysis on one of the planning schedules titled 'Design of railway exchange yard' and using Monte Carlo simulation to validate the schedule. If projects of Company A can be planned better, using schedule risk analysis, projects can become more successful and completed within the required time frame.

Keywords: Critical path method, Schedule sensitivity index, PERT.

1 INTRODUCTION

Railways have been of great significance in the progression of civilization and in the transportation industry at large. The inception of the railway improved the efficiency of the logistics and transportation industry as goods were able to be moved from one place to another in higher volumes. Steam-powered locomotives were first developed in 1825 with the first railway line opened for public use in Darlington (Chandra and Agarwal 2013). Modern railways include electrical and diesel-powered locomotives. The use of railways for transportation of goods has grown steadily over time but road transport has had a substantially faster growth rate and has a larger market share in the transportation and logistics industry. The market share of railways in freight business is only between 15-25%, which is low compared to the capital cost of its infrastructure. In 2013, it was documented that for general freight business, road transport accounted for 221 billion tonne-kilometers of freight delivery in South Africa while rail accounted for only 39 million tonne-kilometers (Scott 2007).

A company (Company A) in Southern Africa has the largest market share in the railway industry regarding freight business. Due to the declining growth of railway use in the freight business, Company A has embarked on major projects in the past 5 years with the aim of transferring freight volumes from road to rail. The types of projects that have been implemented by Company A include the following:

- Upgrading track infrastructure to run faster trains and thus increase volumes;
- Purchasing trains that require less maintenance, and thus less downtime of trains;
- Construction of new railway lines to run more routes and thus increase overall volumes;
- Running longer trains to transport higher freight volumes.

The formative planning stages of these projects implemented by Company A have been thoroughly neglected and have resulted in overrunning project schedules leading to late project delivery. Most of the projects overran the budget substantially due to the late completion of the projects. The schedules for the projects are unrealistic and thus projects are not able to realize benefits at the planned dates. The planning schedules do not incorporate risk and uncertainty, which leads to project failure because most deadlines are not met.

The focus of this report was to analyze a planning schedule which was developed by Company A in March 2015 and to use Monte Carlo simulation to validate the activity durations and completion dates. The planning schedule used in this study was titled 'Design of a railway exchange yard' and the Critical Path Method (CPM) was used to develop this schedule. A schedule risk analysis was done on the existing planning schedule (Design of a railway exchange yard) for Company A. Schedule risk analysis is a scheduling technique which is used to determine activity durations of a schedule by incorporating risk and uncertainty and determining whether the completion date estimated is accurate with a certain level of confidence (Tao *et al.* 2017).

2 LITERATURE REVIEW

A schedule or project plan is a major tool, which is used in project management to monitor, manage and control project activities. One of the key factors that determine whether a project is successful is when it is completed on time (Patel *et al.* 2013). A schedule breaks down the deliverables of a project into activities with durations to determine the total duration required to deliver or complete the project. Project schedules are commonly based on a Work Breakdown Structure (WBS) which groups work into manageable portions (PMBOK® 2013). One of the critical project management concepts used as a framework for managing projects, is the project management triple constraint triangle. Quality forms an integral part of the constraint triangle. An extension or restriction of one of the constraints, may impact on the performance of the other constraint and thus that of the entire project (Mokoena *et al.* 2013).

2.1 Monte Carlo Methods

Monte Carlo methods also known as Monte Carlo experiments are a form of computational algorithms that use repeated random sampling to obtain a numerical result. Monte Carlo methods are used to provide approximate solutions in several complex, physical as well as mathematical problems and have formally been used since the 1940s (Walker 2016). Monte Carlo simulation, also known as probability simulation is a mathematical technique, which is used to model risk and uncertainty in a quantitative analysis. Monte Carlo simulation is used widely in project management, statistics, finance, insurance, research and development and in engineering. Simulations were first started in the 1930s by Enrico Fermi and have since gained popularity in modeling different physical and conceptual systems (El-Reedy 2016). Monte Carlo simulations are also used in the validation of project schedules.

2.2 Schedule Risk Analysis

Uncertainty is a very big part of projects and project schedules are no exception. Risks in project schedules reduce the probability that a project will be completed within the specified time. For projects to be completed on time, uncertainty needs to be considered during schedule development. Monte Carlo simulations are a very effective tool in determining the impact of the identified project risks on the project duration and thus form a critical part in schedule risk analysis (Martinelli and Milosevic 2016). Projects sometimes experience technical challenges and schedule risk analysis addresses the probability that the planned completion date will be met with a certain level of confidence and determines the effect of activities slipping onto the schedule's critical path (GAO Cost Estimating and Assessment Guide 2010). The following are the steps followed to perform schedule risk analysis: 1) preparation of project schedule, 2) identification of schedule risk sources, 4) estimation of 3-point activity duration ranges-assignment of probabilities, 5) Monte Carlo simulation.

2.3 Simulation of Schedule

Once the schedule has been prepared and developed using the CPM, risks identified and activity ranges as well as distributions have been determined, Monte Carlo simulation can be run to determine with a certain level of confidence if the estimated completion date is realistic. The purpose of a simulation is to replicate real-life problems using a set of defined assumptions. This method makes it easy to control risk within a project schedule and is also a good decision-making tool (Wang and Huang 2009). The Monte Carlo simulation for schedule risk analysis answers three critical questions in project management and in schedule development namely: 1) Is the completion date of the project, or the duration provided to complete the project realistic? 2) Is the duration provided, the most-likely duration? 3) How much more time can be allocated as contingency to reduce the overrun risk exposure?

3 METHODOLOGY

To perform schedule risk analysis on the 'Design of a railway exchange yard' schedule, the following steps were taken:

- A schedule was prepared using the CPM in Primavera P6 software using the 'Design of a railway exchange yard' specification. Three questionnaires were then developed and distributed together with the schedule once the schedule was completed.
- A questionnaire was conducted using SurveyMonkey to identify the 10 key activities in the schedule with the highest risk source. Two questionnaires were distributed together with the 'Design of a railway exchange yard' specification. The purpose of the specification was to give context to duration estimation.
- Questionnaire 2 was titled: Three-point Estimation. The purpose of this questionnaire was to use the three-point duration estimation technique to estimate activity durations of the 10 activities identified as having the highest risk source in the schedule.
- Questionnaire 3 (Activity Distribution). The purpose was to identify the distribution functions, which would best model each of the 10 risky activities in the schedule.
- Once the three-point duration estimation and distribution had been identified, a Risk Analysis Software and a Monte Carlo simulation was run.
- The output from the Monte Carlo simulation included the following: 1) probability distribution, with the expected completion date, 2) a tornado chart which provided sensitivity of the results.

4 RESULTS AND ANALYSIS OF RESULTS

The activities of 'Design of a railway exchange yard' schedule was developed by the planner of Company A. The schedule commenced on 9 March 2015 and the completion date was 11 October 2016. All relationships were FS (Finish to Start). The questionnaires were distributed to the project team from Company A with the 'Design of railway exchange yard' schedule in order to validate the 'Design of a railway exchange yard' schedule. The sample sizes used for each of the questionnaires can be seen in Table 1. Therefore, the relevant project team selected included: 1) planners of various technical backgrounds, 2) project managers, and 3) design engineers.

Table 1. Sample size for questionnaires 1-3.

	Planners	Project M/Construction Managers	Design Engineers	Total
Q1	10	15	10	35
Q2	10	10	10	30
Q3	15	10	10	35

4.1 Questionnaire 1 (Risky Project Activities)

The purpose of Questionnaire 1 was to identify activities in the schedule that had the highest source of risk according to the project team. All the activities in the 'Design of railway exchange yard' schedule was inputted as part of Questionnaire 1 using SurveyMonkey. As can be seen from Table 2, Perway and Architectural designs are the activities that have the highest risk in the schedule. This might be because these activities are the first to commence and if delayed, they delay the start of other engineering designs such as structural, electrical and mechanical.

Table 2. Results of risky activities selected by project team.

Activities	%
Project initiation-business case development	1.86
Procurement of survey	3.18
Procurement of Geotech	3.45
Site survey	*9.02*
Geotechnical fieldwork and report	*11.94*
Perway design	*14.85*
Drainage design	*6.10*
Architectural design	*14.06*
Mechanical design	*6.37*
Structural design	4.77
Health and safety management plan	2.65
Environmental: Water use license	*5.57*
Municipal submissions	*7.96*
Quality management plan	3.98
Risk management plan	3.71

4.2 Questionnaire 2 and 3

The purpose of Questionnaire 2 was to determine the activity durations of the identified risky activities using the three-point estimation. The optimistic, most-likely and pessimistic durations for each activity were provided and the average duration (in weeks) for each activity was calculated. The purpose of Questionnaire 3 was to determine the distribution that best fits the risky activities identified. The project team selected the distributions that best represented the activities and the results were provided in graphs and tables as percentages of the total number of respondents.

4.3 Monte Carlo Simulation Results

To perform a Monte Carlo simulation, the 'Design of a railway exchange yard' schedule was inputted into Primavera Risk Analysis Software. As per Questionnaire 1 and Questionnaire 2, the optimistic, most-likely and pessimistic durations were inputted into the software together with the preferred distribution functions. All other activities were inputted with the most-likely duration. 1000 Monte Carlo iterations were performed on the schedule. Table 3 summarizes the risky activities together with durations (In Weeks) as well as distributions.

Table 3. Summary of activity durations (In Weeks) and distributions.

Activities	**Optimistic**	**Most likely**	**Pessimistic**	**Distribution**
Site survey	6	9	12	Beta Pert
Geotechnical fieldwork and report	8	10	14	Beta Pert
Perway design	4	5	8	Triangle
Drainage design	4	5	8	Tringle
Architectural design	6	8	10	Beta Pert
Mechanical design	3	6	8	Triangle
Environmental: Water use license	52	52	52	Uniform
Municipal submissions	17	24	33	Beta Pert

Figure 1 shows the probability distribution for all (frequency histogram) the possible dates that the project can be completed. Using this graph, one can determine with a certain confidence level when the project will be completed. From the graph, the completion date with a 50% confidence level is 7/10/2016. The completion date with an 80% confidence level is 7/11/2016. The completion date with 100% completion level is 14/12/2016. The original completion date of the schedule before simulation was 11/10/2016. This shows that the original schedule was optimistic and unrealistic and did not account for risk and uncertainty.

Figure 2 displays the activities that have the highest risk in the project schedule and thus the activities that have the greatest effect on the overall schedule. The tornado chart is based on the metric Schedule Risk Index (SSI). The following activities have the highest sensitivity and overall highest influence on the schedule: 1) application of water use license 2) geotechnical field work and report 3) procurement of geotechnical studies.

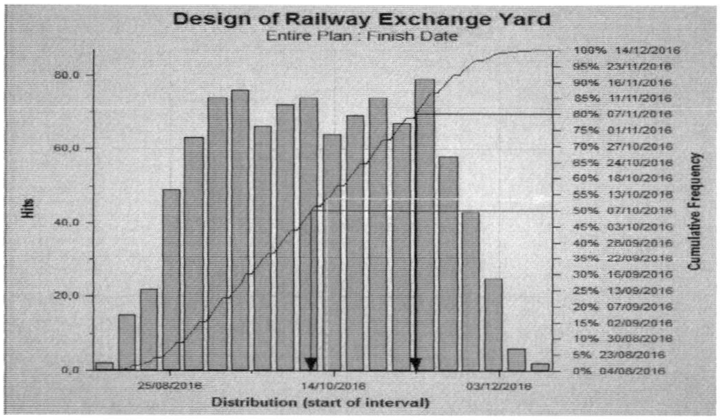

Figure 1. Probability distribution.

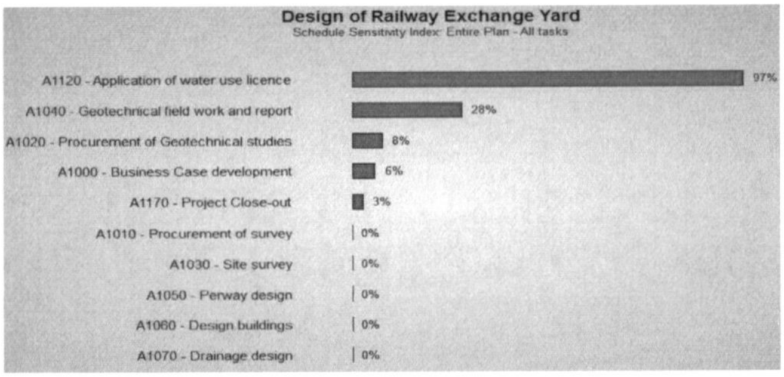

Figure 2. Schedule Sensitivity Index: Entire plan.

5 CONCLUSIONS

This study has illustrated the importance of risk identification in the development of project schedules. The schedule risk analysis performed on the schedule provided guidance as to which activities had the highest risk source and which activities had the potential to delay the entire project. The schedule risk analysis provided a confidence level as to when the project would be completed, which gives the project team comfort. As per the results, the projected completion date based on the Monte Carlo simulation was 14/12/2016, which is two months later than the original estimated date. This proves that the original schedule was overoptimistic and unrealistic. It is recommenced, that Company A uses schedule risk analysis to develop all planning schedules. Schedule risk analysis accounts for risk and uncertainty and once mitigation measures are put in place, the schedule can be managed with ease. Company A will thus be able to plan for the execution of projects at the correct time and benefits can be realized as planned.

References

Chandra, S., and Agarwal, M. M., *Railway Engineering.* (2nd Edition), Oxford: Oxford University Press, 2013.

El-Reedy, M. A., *Project Management in the Oil and Gas Industry.* New Jersey: John Wiley & Sons, 2016.

GAO Cost Estimating and Assessment Guide, *Best Practices for Developing and Managing Capital Program Costs*, (GAO-09-3SP), U.S. Government Accountability Office, 2010.

Martinelli, R. and Milosevic, D., *Project Management Toolbox.* (2nd Edition), New Jersey: John Wiley & Sons, 2016.

Mokoena, T. S., Pretorius, J. H. C., and Wyngaard, C. J. V., *Triple Constraint Considerations in The Management of Construction Projects.* Industrial Engineering and Engineering Management (IEEM), 2013 IEEE International Conference, 2013.

Patel, A., Bosela, P. A., and Delatte, N. J., 1976 Montreal Olympics: Case Study of Project Management Failure, *Journal of Performance of Constructed Facilities*, 27(3), 362-369, 2013.

PMBOK®, *A Guide to the Project Management Body of Knowledge* (PMBOK® Guide), Newtown Square: Project Management Institute, 2013.

Scott, S., *Market Analysis: The South African Transport and Logistics Industry - An Industry in Turmoil*, Cape Town: Smart Procurement, 2007.

Tao, L., Wu, D., Lui, S., and Lambert, J., Schedule Risk Analysis for New-Product Development: The GERT Method Extended by A Characteristic Function. *Reliability Engineering & System Safety*, 167, 464-473, 2017.

Walker, D. J., *Computational Physics - An Introduction. Mercury Learning and Information.* Mercury Learning and Information, 2016.

Wang, X. X., and Huang, J. W., *Risk Analysis of Construction Schedule Based on Monte Carlo Simulation.* Computer Network and Multimedia Technology, CNMT 2009 International Symposium, 2009.

Streamlining Information Transfer between Construction and Structural Engineering
Edited by Shiau, J., Vimonsatit, V., Yazdani, S., and Singh, A.

ISBN: 978-0-9960437-7-9

MANAGEMENT RELATED CRITICAL SUCCESS FACTORS FOR THE ADMINISTRATION OF BUILDING PROJECTS

OSEI YAW SAFO-KANTANKA, CLINTON OHIS AIGBAVBOA, and WELLINGTON DIDIBHUKU THWALA

Sustainable Human Settlement and Construction Research Centre, Faculty of Engineering and the Built Environment, University of Johannesburg, Johannesburg, South Africa

The failure of building projects or works is becoming endemic in the global stage, of which Ghana is of no exception. This study identified project manager related critical success factors that affect the success of building projects in Technical Universities in Ghana. Five technical universities were randomly selected for the study. The study selected 120 respondents using a purposive sampling technique. A self-administered questionnaire was used to collect data from the respondents. Descriptive data analysis methods such as mean, standard deviation, standard error was used to analyze the data. The analysis revealed that the project manager related critical success factors for construction projects in technical universities in Ghana included: the competence of the project manager; the experience of the project manager; the quality, cost and time commitment of the project manager; the project manager's authority to take decisions; and the leadership skills of the project manager. The conclusion from the study is that project managers play a very crucial role towards the success of building projects in the technical universities in Ghana.

Keywords: Competence, Experience, Quality, Cost and time commitment, Decisions, Leadership skills.

1 INTRODUCTION

In Ghana, like the global stage, abandonment has been common for construction projects for years or decades. This failure of building projects or works is becoming endemic in the Ghanaian construction industry. It has been established that contractors and project owners are jointly and severally responsible for the rise in non-completion of building projects within the agreed contract period (Ofori 2012). To ensure a successful project, it is important to understand the requirements of projects right from the beginning and put in place an appropriate project plan, which offers the right direction to project managers and their teams and accordingly execute the project (Alias *et al.* 2014). This is because; Ghazali *et al.* (2017) and Dosumu *and* Onukwube (2013) assert that a successful project is the one that is completed within time and budget. Quality, cost, time, users' satisfaction, professionals' commitment and attainment of organizational goals have been recognised as essential elements of project success (Jha and Iyer 2006). In addition, the success of a project and project critical success factors (CSFs) is regarded as one of the main ways to improve the effective delivery of a project (Almarri and Hijleh 2017). The successful delivery of building projects is largely dependent on how the projects are controlled and managed (Young and Mustaffa 2012). The critical success factors (CSFs) are very supportive to a useful decision making Young and Mustaffa (2012).

The mandates of the polytechnics have been strengthened and expanded by the parliament of Ghana under a new Act, the Polytechnic Act 2007 (Act 745) to provide a wide range of science disciplines and applied arts at sub-degree, degree and postgraduate degree levels. Currently, eight polytechnics have been upgraded to technical university status, which is expected to result in monumental building projects hence the need to ensure the success of such projects.

However, it is mostly reported that the building construction projects in Ghana are confronted with many challenges that results to poor delivery of building projects. This normally leads to calamities, disputes, non-payments or delays in the payment of contract sums. It is therefore important to avoid these challenges by identifying the factors that affect the delivery of building projects in Ghana. Project managers are the vehicles towards the delivery of building projects. Understanding the project managers related factors that affect the construction of building projects is critical towards finding solutions to this challenge. There are however dearth of empirical evidence on the various project mamanger related factors that affect the success of building projects in Ghana. As a result, this research examines the Critical Success Factors (CSFs) for the delivery of technical universities building projects.

2 LITERATURE REVIEW

2.1 Project Success Factors

CSFs in construction are factors that are significant towards the success of an activity (Han *et al.* 2012). There are a number of studies on CSF for a successful project, however there is lack of consensus among these studies on the criteria for determining the success of a project and the factors that influence these successes. Majority of the studies only concentrates on the critical project attributes that are unique to green building projects like the early involvement of the project team members and integrated project execution methods that are mostly not regarded as a significant success factor for the improvement of project performance with regards to time, cost and quality by majority of previous studies (Chan *et al.* 2004).

Han *et al.* (2012) identified 15 factors that are critical to construction project success in Malaysia and grouped them under seven key headings. The results suggest a strong consistency in perception between respondents in recognizing the significance of human related factors such as competence, commitment, communication and cooperation towards the success of a construction project. They further recommended that more emphasis be placed on improving the human related factors in order to ensure the successful implementation of a construction project in future. Furthermore, it is stressed that in order to improve project performance and industry, there is a need to look into industry specific strategies and research to be carried out within the local context and business operating environment. Dosumu and Onukwube (2013) also identified 18 CSF in construction industry in South Africa and grouped them according to four models: comfort, competence, communication and commitment. The findings suggest that both project managers and contractors strongly support the CSF as significant in achieving project success.

Pinto and Slevin (1988), Silva *et al.* (2016) also identified various CSFs that affect the management of projects and these factors include client and consultant related factors, construction industry related factors, political, economic, legal and social environment related factors etc. Project CSFs identified by these different researchers stated above can be categorised into two broad types/categories, which are having little or no control/influence on the success factors as to the manager having full or a considerable degree of control/influence. Eleven success factors, on which project manager has a considerable level of control were identified as presented in Table 1.

Table 1. The contractor and project manager related factors.

Project Manager related Critical Success Factors	Authors Listed in References
Project Manager's competence	4, 5, 7, 10, 12, 13
Project Manager's experience	3, 4, 7, 8, 12, 14
Project Manager's authority to take day-to-day decisions	3, 5, 7, 12, 13
Project Manager's authority to take financial decision, selecting key team members, etc.	7, 10, 12, 13
Technical capability of project manager	5, 6, 12, 13
Leadership skills of project manager	4, 7, 8, 10, 12
Organizing skills of project manager	10, 11, 13,
Project manager's commitment to meet quality, cost & time	3, 8, 10, 14
Project manager's early & continued involvement in project	4, 10, 13, 14
Project manager's adaptability to changes in project plan	3, 5, 10, 12, 13
Project manager's ability to delegate authority	4, 10, 13

3 RESEARCH METHODOLOGY

Both descriptive and survey methods were adopted for the study. The population comprised five technical universities from Ghana. Accra Technical University in Greater Accra region; Kumasi Technical University in Ashanti region, Sunyani Technical University in Brong Ahafo Region, Koforidua Technical University in Eastern Region and Tamale Technical University in Northern Region, which were chosen by using simple random sampling technique.

The respondents of the study comprised consultants, project officers, procurement staff, other university authorities and construction firms involved in the execution of projects in the selected technical universities in Ghana between 2010 and 2017. Another major criterion in selecting the consultants was based on experience and scope of the work-experience of more than 5 years for project worth GHS 200,000.00 (USD 50,000.00) and above. One hundred and twenty (120) respondents were purposively selected for the study. Self-administered questionnaires were used to collect data from the respondents. The questionnaire consists of closed-ended questions. The closed-ended items were mainly made up of Likert scale questions. The researcher used a digital recording device, which ensured that participant responses to questions accurately reflected in the data transcript to facilitate data analysis. Descriptive data analysis methods such as mean, standard deviation, and standard error were used to analyze the data. Similarly, the quantitative aspect of the data was analyzed using descriptive statistics through the SPSS software (23.0).

4 DATA ANALYSIS AND RESULTS

Table 2. presents the results on the project managers related factors that affect the success of building projects in Technical Universities in Ghana.

A total of eight project manager related CSFs were identified. Table 2 shows that the 'competence of the project manager' was the highest ranked project manager related CSF to building projects. This response item obtained a mean rating of 4.33 and standard deviation of 0.955. In addition, the 'experience of the project manager' (mean = 4.15 and SD = 0.903) was the second ranked project manager related CSF to building projects. The results further indicate that the 'quality, cost and time of the project manager' (mean = 4.10 and SD = 0.856) was the fourth ranked project manager related factor for building projects at Technical Universities in Ghana. The 'project manager's authority to take decisions' and the 'leadership skill of the project manager' were ranked fourth and fifth respectively.

Table 2. Project manager related CSF for building projects.

Project Manager Related CSF for Building Projects	N	Mean	Std. Deviation	Std. Error Mean	Sig. (1-tailed)	Rank
Project Manager's competence	94	4.33	.955	.098	.000	1
Project Manager's experience	94	4.15	.903	.093	.000	2
Quality, cost & time of project manager	94	4.10	.856	.088	.000	3
Authority to take financial decisions	94	4.02	1.005	.104	.000	4
Leadership skills of project manager	94	3.96	.879	.091	.000	5
Organizing skills of project manager	94	3.88	.866	.089	.000	6
PM's adaptability to make changes in project plan	94	3.85	.867	.089	.000	7
Motivating skills of project manager	94	3.74	.879	.091	.004	8

It can be obtained that the 'project manager's authority to take decisions' and the 'leadership skill of the project manager' obtained mean ratings of 4.02and 3.96 respectively. Table 2 further shows that the sixth ranked project manager related CSF for building projects at Technical Universities in Ghana was 'organizing skills of the project manager' (mean = 3.88 and SD = 0.866). In addition, the 'project manager's adaptability to make changes in project plan' with a mean of 3.85 and standard deviation of 0.867 was the seventh ranked project manager related CSF. The results presented in Table 2 further indicates that the 'motivation skill of the project manager' (mean = 3.74 and SD = 0.879) was the eighth ranked.

The results revealed that the 'competence of the project manager' was the highest ranked project manager related CSF to building projects. This confirms the findings of earlier studies such as Han *et al.* (2012), Pinto and Slevin (1988). In addition, the 'experience of the project manager' was the second ranked project manager related CSF to building projects. Then again, this result is a confirmation of the findings of Jha and Iyer (2006) and Ghazali *et al.* (2017). These authors reported that the success of a project is largely influenced by the experience and level of competence of project managers.

These results are also in consonance with the findings of (Ofori 2014, Han *et al.* 2012, Ghazali *et al.* 2017) who found that the ability of the project manager to take decisions relating to finance and project team members is critical to a project's success. In addition, the 'project manager's adaptability to make changes in a project plan's and his organizing skills was ranked quiet high for project manager related CSF.

The results further revealed that the 'motivation skill of the project manager' was also important to a project manager's CSF, which obtained a mean rating of more than 3.7. In addition, the 'average' mean value of the project manager related CSF for building projects was 4.00, which is more than 3.5. It is clear here that project manager related CSFs are very important for building projects of Technical Universities in Ghana.

To determine whether the population considered a particular CSF as important or not, a one sample t-test was carried out. Table 3 showing rankings of various project related CSF was also examined. In addition to the mean ratings, the level of significance of the various CSF is presented. A critical factor, variable or response item with a p-value of less than .05 indicates that such critical factor is significant. The results revealed that the 'competence of the project manager' was ranked highest among the rest of the project CSF. This was followed by 'contractors cash flow', 'project monitoring', 'experience of project manager', 'effectiveness of project supervision' and quality, 'cost and time of project manager', which were ranked second, third, fourth, fifth and sixth respectively. In addition, the p-value of each of these CSF was .000. This suggests that these projects CSF were very important for the building projects.

Table 3. T-test summary of rankings of various project critical success factors.

T-test rankings of various project CSF	Mean	SD	Rank	Sig. (1-tailed)
PM Related Factors: Project Manager's competence	4.33	.955	1	.000
Contractor Related Factors: Contractor's cash flow	4.30	1.014	2	.000
PM Related Factors: Project Manager's experience	4.15	.903	3	.000
PM Related Factors: Quality, cost & time of PM	4.10	.856	4	.000
Contractor Related Factors: Contractor's experience	4.10	1.108	5	.000
PM Related Factors: Authority to take financial decisions	4.02	1.005	6	.000
Client Related Factors: Timely decision on project variations	4.00	1.037	7	.000
Project Related Factors: Stakeholders Communication system	3.97	.978	8	.000
PM Related Factors: Leadership skills of project manager	3.96	.879	9	.000
PM Related Factors: Organizing skills of project manager	3.88	.866	10	.000
Client Related Factors: Clear definition of project scope	3.87	1.080	11	. 0005
Project Related Factors: Stakeholders Coordination effectiveness	3.86	1.043	12	. 0005
PM Related Factors: Manager's adaptability to make changes	3.85	.867	13	.000
Project Related Factors: Effective site quality assurance program	3.84	1.009	14	. 0005
Contractor Related Factors: Site management activities efficiency	3.77	.921	15	.003
PM Related Factors: Motivating skills of project manager	3.74	.879	16	.004

4.1 Implications of the Findings

These results suggest that project managers play critical roles towards the success of building projects in Technical Universities in Ghana. These results imply that contractors or clients must appoint experienced and knowledgeable individuals to manage their projects. Indeed, a competent and experience project manager will have the foresight to anticipate problems and put measures in place to mitigate such problems. In addition, it is important to note that project managers have scope of authority defined by the contractor or clients and this scope should be clear and unambiguous. The scope should be based on experience and competence of the project managers and must possess motivation skills. In this way, they must have the ability to motivate themselves and others. This is critically important because construction projects are always met with uncertainties and challenges and a project manager should be able to 'whip' the workers and other parties to the project along. The implication of this result is that project managers must abreast themselves with emerging technologies that can be used to ensure the success of projects.

5 CONCLUSIONS

The study concludes that a project managers CSF is crucial to the successful implementation of Technical University projects. The Project manager's CSF's must include: the competence, the experience, appreciate and commitment to a higher sense of quality, cost and time. The project manager's authority to take decisions and the leadership skills should not be over looked in addition to his organizing skills, adaptability to or make changes in project plan and motivation skill of the manager. It is thus concluded that project manager's play crucial role towards the success of a building projects in the Technical Universities in Ghana. It is recommended that the government of Ghana and the technical universities authorities identify reliable Project managers with such skills and also resource adequately before a project is started to ensure projects success but further studies should look into CSF for effective stakeholder participation or engagement.

References

Alias, Z., Zawawi, E. M. A., Yusof, K., and Aris, N. M., Determining CSF of Project Management Practice: A Conceptual Framework, *Procedia - Social and Behavioral Sciences,* 153, 61–69, 2014.

Almarri, K. and Hijleh, B., Critical Success Factors for Public Private Partnerships in the UAE Construction Industry-A Comparative Analysis between the UAE and the UK, *Journal of Engineering, Project, and Production Management,* 7(1), 21-32, 2017.

Chan, A. P. C., Scott, D., and Chan, A. P. L., Factors Affecting the Success of a Construction Project, *Journal of Construction Engineering and Management,* 130, 153-155, 2004.

Dosumu, O. S. and Onukwube, H. N., Analysis of Project Success Criteria in the Nigerian Construction Industry, *International Journal of Sustainable Construction Engineering and Technology*, 4(1), 31-47, 2013.

Ghazali, F. M., Rashid, S. A., and Sadullah, A. M., The CSF for Public-Private Partnership Highway Construction Project in Malaysia, *Journal of Engineering and Technology* (JET), 8(1), 69-84, 2017.

Han, W. S., Yusof, A. M., Ismail, S., and Aun, N. C., Reviewing The Notions of Construction Project Success. *International Journal of Business and Management*, 7(1), 90-101, 2012.

Jha, K. N. and Iyer, K. C., Critical Factors Affecting Quality Performance in Construction Projects, *Total Quality Management*, 17(9), 1155-1170, 2006.

Ofori, G., New Perspectives on Construction in Developing Countries, Spon Press, Abingdon, 1-15, 2012.

Ofori, D. F., Project Management Practices and Critical Success Factors–A Developing Country Perspective, *International Journal of Business and Management*, 8(21), 14-31, 2014.

Pinto, J. K. and Slevin, D. P., *Critical Success Factors Across the Project Life Cycle*, Project Management Institute, 1988.

Silva, S. K., Warnakulasuriya, B. N. F., and Arachchige, B. J. H., Critical Success Factors. En Route for Success of Construction Projects, *International Journal of Business and Social Science,* 7(3), 27- 37, 2016.

Young, C. Y. and Mustaffa, E. N., Analysis of Factors Critical to Construction Project Success in Malaysia, *Engineering, Construction and Architectural Management,* 19(5), 543-556, 2012.

Streamlining Information Transfer between Construction and Structural Engineering
Edited by Shiau, J., Vimonsatit, V., Yazdani, S., and Singh, A.

ISBN: 978-0-9960437-7-9

CHALLENGES FACED BY PUBLIC-PRIVATE PARTNERSHIP IN IMPROVING SERVICE DELIVERY

NOKULUNGA XOLILE MASHWAMA, CLINTON AIGBAVBOA, and WELLINGTON DIDIBHUKU THWALA

Dept of Construction Management and Quantity Surveying, University of Johannesburg, Johannesburg, South Africa

Partnerships, of any kind, come with challenges and obstacles that might be a threat to the success of the vision they have. PPPs are faced with a lot of challenges when it comes to infrastructure delivery in South Africa and across the globe. Hence, the study was conducted to determine the challenges faced by Public Private Partnerships (PPPs) in the South Africa. The data in this paper was obtained from secondary and primary sources. 90 structured questionnaires were distributed to construction stakeholders in the public and private sectors, who were involved in the PPP's projects in South Africa in the Gauteng province. 80 Questionnaires were brought back and were valid for use. The study revealed that: PPPs are inflexible; there is limited trust between public and private sector professionals; lack of sufficient time to implement PPPs; reduced control of public assets; conflict of interest between private and public sector; public acceptability; PPPs are not integrated into sector planning, lack of resources dedicated to fostering PPPs, a policy bias toward traditional public procurement and against PPPs, lack of fiscal imperative to use PPPs and policy on PPP changes with changes in political leadership, etc. The research has revealed that PPPs are lacking at the highest levels of leadership especially at governmental level. Also, PPPs must be favored in general, not in specific circumstances.

Keywords: Construction industry, Infrastructure, South Africa.

1 INTRODUCTION

Infrastructure investment plays a pivotal role in short term and long term of economic growth and job creation in a country. Furthermore, it brings greater economic returns on investment than any other forms of capital expenditure (Chunn 2014, Rogerson 2010). The South African National Development Plan has priorities like investment in infrastructure to grow the economy while reducing unemployment and inequality (Budget review 2018). The development plan states that the infrastructure investment as a percentage of GDP needs to increase to 30% in 2030 (Budget Review 2018). Furthermore, in 2016 the Gross Domestic Product (GDP) was sitting at about 19.5%, which proves that the government of South Africa has made progress towards this goal. Despite the good plan in place, South Africa is one of the countries that are experiencing backlog in infrastructural service delivery and this has a negative impact to the economy and the residents of the country (Budget Review 2018).

The South African National Development Plan (SANDP) has been doing their best to come up with solutions of reducing the service delivery backlogs by introducing PPPs (Department

Finance 2000). However, PPPs is not the answer to infrastructural service backlog but it can only reduce the challenge faced by the government (Greve and Hodge 2005). The government cannot fund infrastructure alone, hence, crowding in private sector investment could bring along financing needed to meet the national development plans target. In addition, it will contribute to improved decision making and more rigorous assessment and accountability in infrastructure project (Budget Review 2018). There has been a considerably more research work conducted, both locally and internationally, on the nature and form of PPPs. However, there has not been any sufficient investigation on the challenges faced by Public Private Partnership in meeting the infrastructural needs of South Africa.

2 DEFINITION OF PUBLIC PRIVATE PARTNERSHIP

Available literature shows that there is no common definition for the concept of public private partnerships, as it is usually defined in relation to the various forms it assumes. Several definitions of PPPs have been used by different scholars, governments and international organizations (Asmati 2010). PPP is defined in South African law, and adopted by the Treasury Regulations 2000 based on the three essential elements (Nyagwachi and Smallwood 2006); 1) A contractual arrangement whereby a private party performs a departmental function on behalf of a national or provincial department for a specified time, 2) Substantial risk transfer to the private party, 3) A schedule of outcome-based financial rewards, derived either from service tariffs or user charges, from a departmental or other budget, or from a combination of these sources.

3 CHALLENGES OR OBSTACLES FACED BY PPPS IN SOUTH AFRICA AND AROUND THE WORLD

Despite the success of the PPP projects in South Africa, the number of new project transaction has declined over the past six years, decreasing from an estimated R10.7 billion in 2011/12 to R 5 billion in 2017/18. Mainly as a result of challenges were faced by PPPs in South Africa. Every partnership of any kind comes with challenges and obstacles that might be a threat to success of the vision they have. Review literature shows that PPPs are faced with a lot of challenges when it comes to its implementation to deliver services in South Africa and across the globe. This paper will only discuss few challenges that were found in the existing literature (Mashwama *et al.* 2017b, Kupke *et al.* 2004, Grimsey and Lewis 2004, Manchidi and Merifield 2001, Nyagwachi and Smallwood 2006).

3.1 Lack of Highest Level Policy Direction

PPPs are lacking at the highest levels of leadership at government level. The management of PPPs from government level does not provide clearer and more predictable intent on when and why to do PPPs in general, and in specific sectors.

3.2 Lack of Consistent Political Resolve

Reviewed literature states that political leadership shows inconsistent commitment to PPPs. According to Nyagwachi and Smallwood (2004) commitment to PPPs suffers from the fact that policy on PPPs changes with changes in political leadership. This fact was based on the research that was conducted by SPAID and it showed that frequent leadership changes within the PPP Unit had to some extent harmed the development of the policy environment for PPPs in South Africa.

3.3 Lack of Resources Dedicated to Fostering Public Private Partnership

The resources to do all that is needed to be done to foster successful PPPs are lacking in SA (Jamali and Sumali 2004).

3.4 Policy Bias Toward Traditional Public Procurement

In South Africa, like in other countries, traditional public procurement of infrastructure services is the default choice (Budget Review 2018). Line departments must therefore first decide they want to consider a PPP for a given project and then prove that a PPP would provide more value-for-money than traditional government procurement.

According to Babatunde *et al.* (2012), Houghton (2011), Chunn (2014) and Chib *et al.* (2009), this are the additional challenges faced around the globe:

- Regulatory environment;
- Lack of information;
- Project development;
- Lack of institutional capacity;
- Financing availability.

4 RESEARCH METHODOLOGY

4.1 Research Approach and Design

This study adopted a quantitative approach, as the purpose was to investigate the challenges faced by PPPs in improving service delivery in Gauteng South Africa. Quantitative research is based on the measurement of quantity or amount. It is applicable to phenomena that can be expressed in terms of quantity (Mashwama *et al.* 2017a). A well-structured questionnaire was distributed to different construction companies in Gauteng Province, amongst construction professionals such as civil engineers, project managers, directors, quantity surveyors, construction managers and contractors who were involved in the PPP projects.

The questionnaire was sent via e-mails, some were delivered. 90 Questionnaires were distributed and 80 came completed and eligible to use and reflects 89 % response rate. The study was conducted from reliable scholarly sources such as articles, journals, books, publications, websites and site experience on the field. The study was limited to Gauteng Province and the target population was the professionals who have been involved with PPP project in Gauteng. The reason why we only involving professionals who were and are actively involved in the PPP project so we can get direct views on the challenges faced by the PPPs in improving service delivery. Random sampling was adopted and used for study.

4.2 5- Point Linkert Scale and MIS

5- point Linkert scale was adopted for the study, which gave a wider range of possible scores and increase statistical analyses that are available to the researcher (Mashwama *et al.* 2017a). The 5 point scales were transformed to mean item score abbreviated as (MIS) for each of the challenges faced by PPPs whilst implementing projects. The computation of the mean item score (MIS) was calculated from the total of all weighted responses and then relating it to the total responses on a particular aspect. The formula in Eq. (1) is used to rank the challenges facing roads construction projects based on frequency of occurrences as identified by participants.

$$MIS = \frac{1n1 + 2n2 + 3n3 + 4n4 + 5n5}{\sum N} \qquad (1)$$

where;

n1 = *number of respondents for strongly disagree*
n2 = *number of respondents for disagree*
n3 = *number of respondents for neutral*
n4 = *number of respondents for agree*
n5 = *number of respondents for strongly agree*
N = *Total number of respondents*

5 FINDINGS AND DISCUSSIONS

5.1 Challenges Faced by PPPs in South Africa

Table 1. Obstacles faced by PPP.

Obstacles in the PPP's	Mean	Std	Rank
Limited trust between public and private sector professionals	4,90	0,995	1
Conflict of interest between private and public sector	4,89	0,985	2
Reduced control of public assets	4,83	0,958	3
Lack of sufficient time to implement PPPs	4,81	1,208	4
PPPs are inflexible in terms of administrative burden associated with altering the contract	4,78	1,147	5
PPPs are inflexible in general	4,76	0,964	6
A policy bias toward traditional public procurement and against PPPs	4,74	0,924	7
A lack of fiscal imperative to use PPPs	4,71	1,047	8
Policy on PPPs changes with changes in political leadership	4,56	0,940	9
Lack of competition in the industry since big companies are used for partnerships	4,38	0,976	10
Lack of confidence or authority to make critical decision in managing PPPs	4,15	0,853	11
Lack of consistent political resolve	4,11	0,983	12
Public acceptability	4,05	1,048	13
PPPs are integrated into sector planning	3,95	1,109	14
Lack of resources dedicated to fostering PPPs	3,82	0,980	15
Lack of financing availability	3,75	1,024	16
PPPs lack high policy direction and political resolve	3,72	0,924	17
A lack of resources within the PPP unit to promote PPPs	3,63	1,147	18
Lack of capacity in implementing PPP agencies	3,59	1,135	19
Lack of institutional capacity for PPPs	3,53	1,020	20

Respondents were asked to rank the challenges faced by PPPs in South Africa. The respondent ranked first limited trust between public and private sector professionals with an MIS=4,90 and STD=0,995; followed by conflict of interest between private and public sector and was ranked 2nd with an MIS=4,89 and STD= 0,985; Reduced control of public assets was ranked 3rd with an MIS= 4,83 and STD= 0,958; Lack of sufficient time to implement PPPs was ranked 4th with an MIS=4,81 and STD=1,208; PPPs are inflexible in terms of administrative burden associated with altering the contract was ranked 5th with an MIS=4,78 and STD=1,147; PPPs are inflexible in general was ranked 6th with an MIS=4,76 and STD=0,964; A policy bias toward traditional public procurement and against PPP was ranked 7th with an MIS= 4,74 and STD=0,924; A lack of fiscal imperative to use PPPs was ranked 8th with an MIS=4,71and 1,047; Policy on PPPs changes with changes in political leadership was ranked 9th with an MIS=4,56 and STD=0,940; Lacks of competition in the industry since big companies are used for partnerships was ranked 10th with an MIS=4,38 and STD=0,976; Lack of confidence or authority to make critical decision in managing PPPs was ranked 11th with MIS=4,15 and STD=0,853; Lack of consistent political resolve was ranked 12th with MIS= 4,11 and STD= 0,983;Public acceptability was ranked 13th with MIS=4,05 and STD=1,048; PPPs are integrated into sector planning was ranked 14th with MIS= 3,95 and STD=1,109; Lack of resources dedicated to fostering PPPs was ranked 15th with MIS=3,82 and STD=0,980; Lack of financing availability with MIS= 3,75 and STD=1,024; PPPs lack high policy direction and political resolve with a MIS= 3,72 and STD= 0,924; A lack of resources within the PPP unit to promote PPPs with MIS=3,63 and a STD= 1,147; Lack of capacity in implementing PPP agencies was ranked 19th with MIS= 3,59 and a STD=1,135; lack of institutional capacity for PPPs was ranked last with MIS=3,53 and STD=1,020.

6 CONCLUSIONS

The findings in the study presented the challenges faced by PPPs in South Africa. In addition, limited trust between public and private sector; conflict of interest between private and public sector; reduced control of public assets and lack of sufficient time to implement PPPs were ranked the highest. The findings revealed that the community at large needs more awareness of PPPs structure, policies. Moreover, the government needs to increase PPP project pipeline, policies don't have to change due to political shifting or leadership. The government needs to streamline the implementation PPP to reduce the time it takes to complete project planning and ease the regulatory regime so that it is less cumbersome for both the private and public sector.

7 RECOMMENDATIONS

The concept of PPP is relatively new in the functioning of the SA economy, and therefore it will require further research, which will seek to closely examine on how the PPP initiatives can be entrenched as a reliable mode of delivering effective service delivery. Further research should also be conducted in assessing the approach and attitude of government institutions towards the PPPs, and thereby measuring the involvement of PPPs in the broader government infrastructure development projects. If indeed PPPs are perceived as an effective and efficient means of delivery infrastructure, it will be interesting to see as to what extent PPPs are involved in the overall government infrastructure development program.

References

Asmati, A., *Is Public Private Partnership an Effective Instrument to Implement SDIs?*, FIG Congress Facing the Challenges-Building Capacity Held In Sydney. Conducted by Library Association in Australia. Sydney, Library Association, 2010.

Babatunde, O. S., Opawole, A., and Olusegun, A., Critical Success Factors in Public-Private Partnership on Infrastructure Delivery in Nigeria, *Journal of Facilities Management,* 10(3), 212-225, 2012.

Budget review, *Public- Private Partnerships*, 153-158, South Africa, 2018.

Chib, Y., Ibbs, W. C., and Kwak, H. Y., Towards a Comprehensive Understanding of Public Private Partnerships for Infrastructure Development. *Journal of California Management Review*, 51(2):51-76, 2009.

Chunn S., *Key Complex Issues Impacting Public Private Partnerships for Transportation Renewal Projects in the United States*, MSc Thesis, Master of Science at Colorado State University Fort Collins, Colorado, 2014.

Department of Finance, *A Strategic Framework for Delivering Public Services Through Public-Private Partnerships*, Republic of South Africa, 2000.

Greve, C., and Hodge, G., *The Challenge of Public-Private Partnerships Learning from International Experiences*, Edward Elgar, *UK*, 2005.

Grimsey, D., and Lewis, K. M., *Public Private Partnerships the Worldwide Revolution in Infrastructure Provision and Project Finance*, Edward Elgar, *UK*, 2004.

Houghton, J., Negotiating the Global and Local Development: Evaluating Development Though Public-Private Partnerships in Durban, South Arica, *Journal of Urban Forum*, 22, 75-93, 2011.

Jamali, D., and Sumali, S., Success and Failure Mechanisms of Private-Public Partnerships in Developing Countries, *The International Journal of Public Sector Management*, 17(5), 414-430, 2004.

Kupke, S., Latteman, C., and Stieglitz, S., Impact of PPPs to broadband Diffusion in Europe, *Transforming Government People, Process and Policies*, 3(4), 355-374, 2009.

Manchidi, E. M., and Merifield, A., Public-Private Partnerships: Public Infrastructure Investment and Prospects for Economic Growth in South Africa, *Empowerment Through Economic Transformation*, Edited by Khosa, M.M. Durban, African Millennium Press, 2001.

Mashwama, N. X, Aigbavboa, C., and Thwala, D., An Assessment of the Critical Success Factor for the Reduction of Cost of Poor Quality in Construction Projects in Swaziland, *Procedia Engineering*. 196: 447-453, 2017a.

Mashwama, N. X, Aigbaboa, C., and Thwala, D., *A Theoretical Assessment of the Challenges of Public Private Partnership In Improving Infrastructure Service Delivery in Swaziland*, Proceedings of the 4th International Conference on Infrastructure Development and Investment Strategies for Africa, DII-2017. 30 August -1September 2017. 299-239, Livingstone, Zambia, 2017b.

Nyagwachi, J. N., and Smallwood, J. J., *South African Public Private Partnership Projects a Systematic Model for Planning and Implementation*, Department of Construction Nelson Mandela Metropolitan University, Port Elizabeth, 2006.

Rogerson, C. M., In Search of Public Sector-Private Sector Partnerships for Local Economic Development in South Africa, *Journal of Urban Forum*, 21, 441-456, 2010.

Streamlining Information Transfer between Construction and Structural Engineering
Edited by Shiau, J., Vimonsatit, V., Yazdani, S., and Singh, A.

ISBN: 978-0-9960437-7-9

PRACTICAL VALIDATION MEASUREMENTS OF A PHYSIOLOGICAL STATUS MONITORING SENSOR IN REAL CONSTRUCTION ACTIVITIES

AMMAR S.M. MOOHIALDIN[1], BAMBANG T. SUHARIADI[2], and MOHSIN K. SIDDIQUI[3]

[1]*Science and Engineering Faculty, Queensland University of Technology, Brisbane, Australia*
[2]*Dept of Construction Engineering and Management, King Fahd University of Petroleum and Minerals, Dhahran, Saudi Arabia*
[3]*College of Engineering, Qatar University, Al-Doha, Qatar*

The construction industry is one of the most hazardous industry and construction worker's health and safety in hot and humid weather conditions become a topic of a great interest for researchers and practitioners. Various physiological monitoring sensors have been employed to measure the physiological impacts of such weather conditions. However, the ability of these sensors to deal with the dynamic nature of the construction industry still under controversial arguments and there is a need to justify their applicability in real working conditions. This paper aims to investigate the accuracy of an off-the-shelf physiological monitoring sensor (Zephyr BioHarnessTM 3) in a real working environment when considering the impact of hot and humid weather conditions. Fifteen different construction workers participated in three site measurements, with monitoring two physiological parameters heart rate (HR) and breathing rate (BR). A non-parametric statistical test (Mann-Whitney) was applied to identify whether there is a significant difference between the medians of live data in the remote monitoring station and saved data of the sensor internal memories. The results showed that there are significant differences between live data and saved data. These differences were highly noticed in the third site measurements, as it included working in confined areas and trenches. The accuracy of live data is highly affected by types of activities and appropriate positions of the wireless connection devices (ECHO gate and repeater).

Keywords: Zephyr BioHarnessTM 3, Health and safety related on-site data collection, Heart rate, Breathing rate.

1 BACKGROUND

There are several technologies can be used for Physiological Status Monitoring (PSM) purposes. According to Alam and Ben Hamida (2014), there are 41 different wearable sensors available in the market for medical applications, 77 for fitness applications, 117 for lifestyle monitoring and 26 for entertainment.

"Zephyr" (Annapolis, Maryland, USA) is a leading global provider of healthcare products, and it is recognized innovator in patient monitoring and respiratory care devices. This technology is directed toward physiological status monitoring in training applications and in high-stress environmental conditions. In addition, it was successfully applied in some studies that are cooperated with Fire Department, NASA Ames Research Centre, National Guard Civil Support

Teams, and multiple US Special Forces contracts. Most of these studies supported and validated the application of Zephyr™ technology under extreme operating environmental conditions (Zephyranywhere (About Zephyr) 2015). The main tool applied for PSM is BioHarness™ 3, which is available in different forms based on the way of contact with the human body (compression shirt, chest strap, and holder). In addition, there are another tools are provided by Zephyr™ such as HxM™ BT, which mostly used for fitness monitoring. Zephyr website addressed several applications of their technology in various disciplines, such as sports, medical and other applications in addition to some published journals.

Different studies addressed the PSM applications by using Zephyr technology for different purposes, however few of them demonstrate the applicability, reliability, and accuracy of this technology in a real construction environment. Zephyr technology was recommended in several studies as a valid tool for PSM of construction workers. To illustrate, Gatti *et al.* (2011) assessed the reliability and accuracy of collected data of HR and acceleration by Zephyr technology during recording dynamic activities of construction workers. They concluded that Zephyr technology is a recommended technology for construction safety applications. A further application for Zephyr technology ("BioHarness BT") was addressed by Gatti *et al.* (2012a) for monitoring HR of construction workers in a lab-setting experiment, to identify physical strain level based on HR recorded data. Based on the result of Gatti *et al.* (2012b), Zephyr technology is considered as an effective technology for monitoring health and safety conditions of workers in the construction industry. It accurately and effectively measures HR in a dynamic nature of their tasks as well as their environmental conditions, without any interruption or hindrances. Furthermore, Migliaccio *et al.* (2012) were able to assess the physiological and ergonomic status of bending activities of construction workers where the authors incorporated the UWB PSM recorded data of construction workers by using Zephyr sensors. The authors concluded that Zephyr technology is a reliable tool for monitoring construction activities remotely. This conclusion was also supported by Cheng *et al.* (2013a and 2013b).

In another study, Gatti *et al.* (2014a) utilized Zephyr technology in construction physical strain assessment. They recommended HR, relative HR, and BR as effective measures for physical strain. Furthermore, Gatti *et al.* (2014b) validated Zephyr technology (BioHarness BT 1) as an effective tool for monitoring the physiological and ergonomic status of construction workers in both static and dynamic activities. The authors argued that Zephyr technology provides valid records for physiological status (HR) in both dynamic and static activities.

Another field of Zephyr technology application in the construction industry was addressed in Lee and Migliaccio (2014) study, where the authors utilized Zephyr BioHarness™ 3 in assessing the impact of hard weather conditions on construction workers' health and physiological conditions. Recently, Lee *et al.* (2015) addressed that, Zephyr BioHarness™ 3 chest belt is the most accurate and applicable tool for monitoring physiological status of construction workers while they are performing their tasks without any interruption or inconvenience. It is an important characteristic for effective PSM tools to contain a warning system for detecting vital signs of the subjects under study (Buller and Karis, 2007). This character is available in Zephyr BioHarness™ 3 technology, where it gives warning signs through producing different warning sounds and lights.

A further investigation in PSM technology was addressed in Gatti *et al.* (2012b) study such that the authors evaluated three different technologies for PSM applications "Zephyr BioHarness (BH), Zephyr HxM, and Hidalgo EQ-01". These technologies have different dimensions, weights, data transaction methods and measured variables. The result of Gatti *et al.* (2012b) comparison reveals to that, BH and EQ-01 technology provide reliable and effective tools for monitoring physiological status through measuring HR and BR of construction workers without any interruptions to their tasks. The most important factor for getting a successful application of PSM technology are size, applicability, and reliability of the utilized sensors, and it should not hinder or interrupt workers performing the required tasks and activities. Another important issue is that, there is not any study addressed the application of PSM technology in Saudi construction market, and only one study was conducted in Arabian Gulf Region (UAE) using "Polar S720i" sensors.

Therefore, there is a higher need to investigate the argument that, Zephyr technology provides accurate, reliable and valid real-time records for construction workers while they are performing their normal tasks, without creating any interruption or discomfort especially under extremely hot and humid weather conditions in Saudi Arabia.

2 RESEARCH METHOD

Real construction site measurements were held in three different construction sites in Dhahran, Saudi Arabia. The participation was totally voluntary, and the participants were divided into groups according to their job, working shift, the day of recording. There was not any interruption to the participants' activities such that their physiological status was recorded simultaneously with passing time by using Zephyr technology. The two different sit measurements were conducted in the first construction in which steel structure installation and filter preparation for an AC station. The third site measurements included two different activities (shoveling and formwork) in order to prepare foundations for a parking building. All measurements were conducted under extremely hot and humid weather condition such that the highest degree temperature and relative humidity in the first and second site measurements are 48°C/19% and 47°C/20%, respectively. These measurements were conducted during morning shifts with considering three different sessions (outdoors measurements, indoors measurements, and resting measurements). Third measurements were conducted during night shifts where relative humidity was higher than the morning shifts. The highest degree and relative humidity in the third site measurements are 33°C and 40%, respectively. In each measurement, five participants and five sensors were employed in the measurements, which means fifteen participants in all measurements.

Zephyr sensors are able to provide wide options of physiological status reports, which will be employed in this study, to test the proposed hypothesis. Zephyr sensors can provide hourly records for different parameters and vital signs. In this study, only HR and BR recorded, and live data will be tested. Zephyr BioHarness technology provides two different types of software for monitoring the participants and providing warning signs in dangerous situations. The available software is OmniSense Analysis and OmniSense Live. The OmniSense Live assists continuous monitoring purposes and provides warning signs for all participants at the same time. The other software, OmniSense Analysis provides a graphical presentation of the monitored subjects and it supports analysis purposes.

Zephyr BioHarnessTM 3 provides one ECHO gate and one ECHO repeater to be used in the conducted measurements. Each one is able to cover up to 300 yards range i.e. 600 yards totally. Therefore, participants should be within this range and ECHO repeater should be placed in 300 yards far from ECHO Gateway.

After each recording session, the participants were asked to answer the following questions based on their feeling and perception:

- Is wearing BioHarness Belt makes you perform your activities without any discomfort?
- Do you prefer that this belt was designed to be worn on wrist or shoulder?

3 DATA ANALYSIS

The results of the proposed two questions show that one of the participants declared that wearing Zephyr belts create any discomfort. Nine of participants preferred Zephyr belts to be manufactured in such way that it possible to worn in wrist or shoulder and only six of them did not.

Zephyr sensors are wireless sensor has the ability to monitor physical remotely. The wireless connection ECHO Gate covers a circle of 300-yards diameter circle. The repeater also covers a circle of 300-yards diameter circle which means Zephyr system can cover 600-yards totally. Based on this, the participants in construction site measurements moves within this range. However, the signal of the Zephyr sensors was missed several times during recording sessions. The sensors have the ability to keep the records in an internal memory in addition to the live records that are sent to the software simultaneously. It is important to identify whether there is a significant difference between the live recorded data (HR and BR) and the data that is saved in the internal memory of Zephyr sensors. In this research, Mann-Whitney test is applied to identify these differences with considering significant level $\alpha = 0.05$. The proposed null and alternatives hypothesis are:

- H_0: There is not a significant difference among the medians of live and saved records of both HR and BR.
- H_1: There is a significant difference among the medians of live and saved records of both HR and BR.

The results of the proposed tests are summarized in Table 1. Mann-Whitney test results shown in Table 1 indicate to some differences between live and recorded data. In the first construction site measurements, only two significant differences were identified in HR data and one significant difference was identified in BR data. The second construction site measurements included three significant differences in HR data and two in BR data where in the third construction site measurements included the highest number of significant differences in HR and BR data, which are four significant differences in both categories.

Table 1. Results of Mann-Whitney test for recorded and live data comparison.

Subject	HR/BR	1st Measurement		2nd Measurement		3rd Measurement	
		P-Value	Decision	P-Value	Decision	P-Value	Decision
1	HR	0.000	Significant	0.000	Significant	0.0337	Significant
	BR	0.000	Significant	0.000	Significant	0.000	Significant
2	HR	0.000	Significant	0.3894	Insignificant	0.000	Significant
	BR	1.000	Insignificant	0.5581	Insignificant	0.9856	Insignificant
3	HR	0.0931	Insignificant	0.0053	Significant	0.000	Significant
	BR	0.6886	Insignificant	0.9409	Insignificant	0.000	Significant
4	HR	0.0545	Insignificant	0.000	Significant	0.000	Significant
	BR	0.4283	Insignificant	0.000	Significant	0.000	Significant
5	HR	0.7878	Insignificant	0.4991	Insignificant	1.000	Insignificant
	BR	0.3744	Insignificant	0.8093	Insignificant	0.000	Significant

4 RESULTS AND DISCUSSION

The results of the proposed paper show that Zephyr BioHarness Belts are suitable to be used in construction industry to monitor construction workers' physiological status remotely under extremely hot and humid weather condition. Zephyr BioHarness Belts do not create any discomfort or interrupt the workers' activities. In addition, most of the participants preferred that Zephyr belts to be manufactured in such way that it possible to be worn in wrist or shoulder rather than chest belts.

The results of the Mann-Whitney test show that there are significant differences between recorded and live HR and BR data. However, the estimated differences between the recorded data and live data do not exceed the value of one in both HR and BR in the first and second measurements which is a good indication to a good level of accuracy in the recorded and live data of HR and BR. On the other hand, the maximum differences between the recorded and live BR data reach to the value of eight in the third measurements. In addition, these differences were not so much effective especially in the first and second site measurements. In these measurements, the construction site under study did not include any confounded areas or working in trenches. In the third measurements the location under study included working in trenches, which may create problems in the wireless connection.

5 CONCLUSIONS AND RECOMMENDATIONS

Different innovative technologies have been successfully applied to support construction workers' health and safety. However, there was not any research that addressed validity test of these technologies in Saudi construction market. This paper is aiming to investigate the accuracy of the recorded data that is provided by Zephyr BioHarnessTM 3 sensors with considering extremely hot and humid weather conditions in Saudi Arabia. Three different construction site measurements were conducted in order to test the proposed hypothesis. Two of these measurements were taking place on the same construction site where the third one was conducted in different site measurements. The results of the proposed paper show that Zephyr BioHarness Belts are suitable to be used in construction industry to monitor construction workers' physiological status remotely under extremely hot and humid weather condition. It is preferable to redesign Zephyr belts in such way that it possible to be worn in wrist or shoulder rather than chest belts. The accuracy of the recorded data is highly affected by the construction site such that the results show some significant differences between recorded and live data in the third site measurements. In these measurements, construction workers performed their tasks in trenches, which may create problems in the wireless connection. Zephyr technology is recommended to be applied in construction safety and productivity applications. In order to improve wireless connection, it is recommended to select appropriate positions for the ECHO gate and one ECHO repeater in the construction site.

Acknowledgements

The authors would like to thank King Fahd University of Petroleum and Minerals and Saudi National Science, Technology and Innovation Plan (NSTIP) for their support that made this research possible.

References

Alam, M. M., and Ben Hamida, E., Advances in Wearable Sensor Technology and its Applications in Mobile Workforce's Health Monitoring and Safety Management, *SPE Middle East Health, Safety, Environment & Sustainable Development Conference and Exhibition*, Society of Petroleum Engineers. Doha, Qatar. 2014.

Buller, M. J., and Karis, A. J., Heart Rate and Respiratory Rate During Fitness Training, Obstacle Course Runs, and Mass Casualty Simulations-Performance of Two Warfighter Physiological Status Monitoring Systems (No. USARIEM-TR-T07-10). *Natick, MA: Army Research Institute of Environmental Medicine, Biophysics and Biomedical Division*. 2007.

Cheng, T., Migliaccio, G. C., Teizer, J., and Gatti, U. C., Data Fusion of Real-Time Location Sensing and Physiological Status Monitoring for Ergonomics Analysis of Construction Workers, *Journal of Computing in Civil Engineering*, 27(3), 320-335, 2013a.

Cheng, T., Teizer, J., Migliaccio, G. C., and Gatti, U. C., Automated Task-Level Activity Analysis Through Fusion of Real Time Location Sensors and Worker's Thoracic Posture Data, *Automation in Construction*, 29, 24-39, 2013b.

Gatti, U. C., Migliaccio, G. C., and Schneider, S., Wearable physiological status monitors for measuring and evaluating worker's physical strain: preliminary validation, *International Workshop on Computing in Civil Engineering 2011*, 194-201, Miami, Florida, United States. doi.org/10.1061/41182(416)24, 2011.

Gatti, U., Migliaccio, G., Bogus, S., and Schneider, S., Using Wearable Physiological Status Monitors for Analyzing the Physical Strain-Productivity Relationship for Construction Tasks, *International Conference on Computing in Civil Engineering 2012*, 577-585. Clearwater Beach, Florida, United States. 2012a.

Gatti, U., Migliaccio, G., Bogus, S. M., Priyadarshini, S., and Scharrer, A., Using Workforce's Physiological Strain Monitoring to Enhance Social Sustainability of Construction, *Journal of Architectural Engineering*, 19(3), 179-185. 2012b.

Gatti, U. C., Migliaccio, G. C., Bogus, S. M., and Schneider, S., An Exploratory Study of the Relationship Between Construction Workforce Physical Strain and Task Level Productivity, *Construction Management and Economics*, 32(6), 548-564. 2014a.

Gatti, U. C., Schneider, S., and Migliaccio, G. C., Physiological Condition Monitoring of Construction Workers, *Automation in Construction*, 44, 227-233. 2014b.

Lee, W., and Migliaccio, G. C., Field Use of Physiological Status Monitoring (PSM) to Identify Construction Workers' Physiologically Acceptable Bounds and Heart Rate Zones, *International Conference on Computing in Civil and Building Engineering*, Orlando, Florida, United States. doi.org/10.1061/9780784413616.129, June 23-25, 2014.

Lee, W., Migliaccio, G. C., Lin, K. Y., and Russo, F., Lessons learned from using bio-and environmental sensing in construction: A field implementation, *in Proceedings of 5th International/11th Construction Specialty Conference*, Vancouver, British Columbia. 2015.

Migliaccio, G., Teizer, J., Cheng, T., and Gatti, U., Automatic Identification of Unsafe Bending Behavior of Construction Workers Using Real-Time Location Sensing and Physiological Status Monitoring, *in Proceedings of Construction Research Congress*, West Lafayette, Indiana, United States, 633-642. doi.org/10.1061/9780784412329.064, May 21-23, 2012.

Zephyranywhere (About Zephyr) (Zephyr™), retrieved from http://www.zephyranywhere.com/about-zephyr on September 13, 2015.

Streamlining Information Transfer between Construction and Structural Engineering
Edited by Shiau, J., Vimonsatit, V., Yazdani, S., and Singh, A.

ISBN: 978-0-9960437-7-9

ASSIGNMENT OF RESPONSIBILITY AND ROLE TO CONSTRUCTION CLIENT IN OCCUPATIONAL SAFETY AND HEALTH SYSTEM

JEONG-HUN WON[1], NAM-GWON JANG[1], SANG-SEOM KO[2], and TAE-SANG KIM[3]

[1]*Dept of Safety Engineering, Chungbuk National University, Cheongju, South Korea*
[2]*Eco-Institute, Hanwha Group, Seoul, South Korea*
[3]*Consumer Product Business Division, Korea Conformity Laboratories, Seoul, South Korea*

This study investigated the client's duty and role to prevent construction accidents by participating construction client into the occupational safety and health management system. Since the fatal accidents of the construction sites in Korea are higher than other OECD countries and approximately 500 workers annually have died on construction sites, various efforts have been performed to prevent construction accidents. Recently, in Korea, the client's participation in the safety and health management system has been focused as a fundamental method to prevent serious accidents. To examine the necessity of client's participation, the survey was performed for construction participants. And, survey results and existing studies about client's role were analyzed to assign responsibility and role to the client. Results showed that safety managers and supervisors were strongly agreed for imposing client's duty (average scores of 4.373 and 4.107 in likert 5point scale) while clients showed the lowest agreement (average score of 3.153). In addition, the detailed role was proposed in the whole project stage, that is plan, design, and construction stage. Client can be a key player in reducing the fatal construction accidents by doing the accurate roles proposed in this study.

Keywords: Participation, Fatal accident, Survey, Construction accidents.

1 INTRODUCTION

The statistics for industrial accidents in Korea shows that the accident rate and fatality rate of construction industry are higher than those of manufactural industry. For the accident rate, the construction industry showed an increasing trend recently while the overall industry is continuously decreasing (Jang *et al.* 2017). In 2016, the total number of deaths in the construction sites in Korea was 554, which accounts for 31.18% of all deaths from industrial accidents.

The role and responsibility for workers' safety and health in Korea construction industry has been focused on construction contractors who employ the workers (Won *et al.* 2015). Most of duties about the construction workers' safety and health are imposed on the contractors while, for the clients, there is no specified duty even though they contribute to the accident by ordering incorrect requirements.

In order to induce construction client's role and responsibility in construction safety and health management system, this study examined the necessity of construction client's duty and suggested the client's role for plan, design, and construction stage. The survey was performed to investigate the method of imposing client's duty and various literatures were reviewed to suggest the client's role.

2 ACCIDENT STATISTICS IN KOREA CONSTRUCTION SITE

The annual variation of the fatality rate was shown in Figure 1 in order to compare the accident trend of construction and manufacture industries (MOEL 2007-2016). Fatality rate, which represents death rates per 10,000 people, showed a decreasing trend in the manufacturing industry. However, it seems that the fatality rate in construction industry is starting to increase recently. The rate for the construction industry was the highest at 2.21 in 2013, and thereafter, it drastically declined by 1.47 times from the rate in 2013, to 1.47 in 2015. However, in 2016, the fatality rate in the construction industry increased again and was 1.83 times higher than that in manufacturing industry. In addition, the number of deaths in Korea construction sites was tabulated from 2011 to 2016 in Table 1. The average number of deaths is 536.

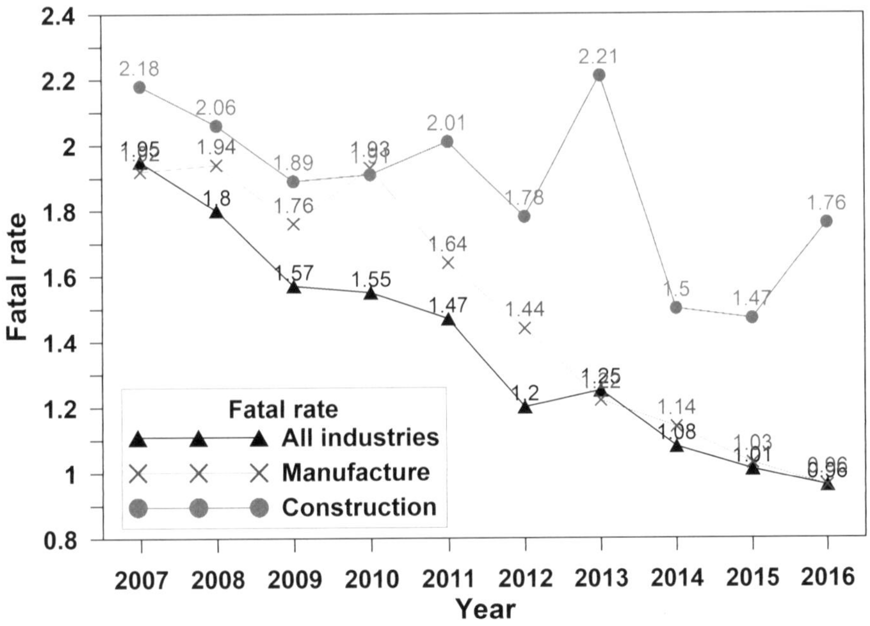

Figure 1. Annual change of industrial fatality rate in Korea.

Table 1. Number of deaths in Korean construction sites.

Year	2011	2012	2013	2014	2015	2016
Number of death	621	496	567	486	493	554

3 SURVEY RESULTS

Survey was conducted to examine the necessity for inducing the client's role and responsibility in construction safety and health management. Various stakeholders such as clients, safety managers, construction managers, and supervisors were included in the survey. The numbers of clients, safety managers, construction managers, supervisors, other experts were 179, 119, 63, 28, 7, respectively.

Using the likert 5-point scale (1-Not at all helpful, 2-Not very helpful, 3-Natural, 4-Somewhat helpful, 5-Very Helpful), the effect of imposing the client's duty in the construction safety and health management was surveyed (Table 2). The average score of total respondents was 3.724. The clients showed the lowest agreement (average score 3.153) since they should have a mandatory duty for construction safety and health management. For the effects of prevention of construction accidents by means of imposing client's duty, safety managers and supervisors were strongly agreed. The average scores were 4.373 and 4.107, respectively.

Table 2. Average score for imposing construction clients' duty according to stakeholders.

Classification	Client	Safety manager	Construction manager	Supervisor	Total
Likert 5 point average score (standard deviation)	3.153 (1.173)	4.373 (0.985)	3.794 (1.194)	4.107 (1.166)	3.724 (1.241)

To induce the clients' duty in the law for construction safety and health management, 6 options were inquired as follows.

- Option 1: Client should appoint the safety and health specialist who carries out client's role and responsibility in the site in order to perform for client's role and responsibility.
- Option 2: Client's duty is included in the range of contractors' duty (The clients should have unique and direct role and responsibility similar to contractors.)
- Option 3: Client should employ exclusive safety and health supervisor in order to perform for client's role and responsibility.
- Option 4: Client should appoint a main contractor who will responsible for client's role and responsibility.
- Option 5: Do not impose client' duty (same to current system).
- Option 6: Others

For the 6 options, option 3 was most favorable, and then, option 1 was selected (Table 3). For only client's response, option 3 using supervisors, was dominantly. However, for the safety managers and construction managers, option 1 and option 2 was mainly preferred. Supervisors preferred option 1 and option 3.

Table 3. Selection for 6 options about method for imposing client's duty.

Classification	Option selection (%)					
	1	2	3	4	5	6
Client	18.7	4.3	49.7	8.0	18.7	0.5
Safety manager	34.1	32.6	20.7	6.7	4.4	1.5
Construction manager	27.4	25.8	17.7	9.7	14.5	4.8
Supervisor	37.5	15.6	31.3	6.3	9.4	0.0
Total	27.1	17.9	33.6	7.5	12.5	1.4

4 CLIENTS' ROLE IN PLAN, DESIGN AND CONSTRUCTION STAGE

Domestic construction process and guidance, foreign law and safety systems, and various researches were studied to identify the clients' role through the whole construction process (EU 1992, HSE 2015, Health and Safety Authority 2009, Huang and Hinze 2006, ILO 1992,

Smallwood 1998). Then, experts about construction safety and health discussed about the clients' role according to the construction stage. Finally, the clients' role was suggested in plan, design, construction stages as shown in Figure 2.

In the plan stage, clients should identify hazards controlled intensively and establish the approximate risk reduction plan. In the design stage, clients should control design process to carry out DFS (design for safety) by a competent designer. In the construction process, clients should select a competent contractor in the safety and health management. And, clients should check and audit the construction site whether the contractor implement the workplace safety and health plan or not.

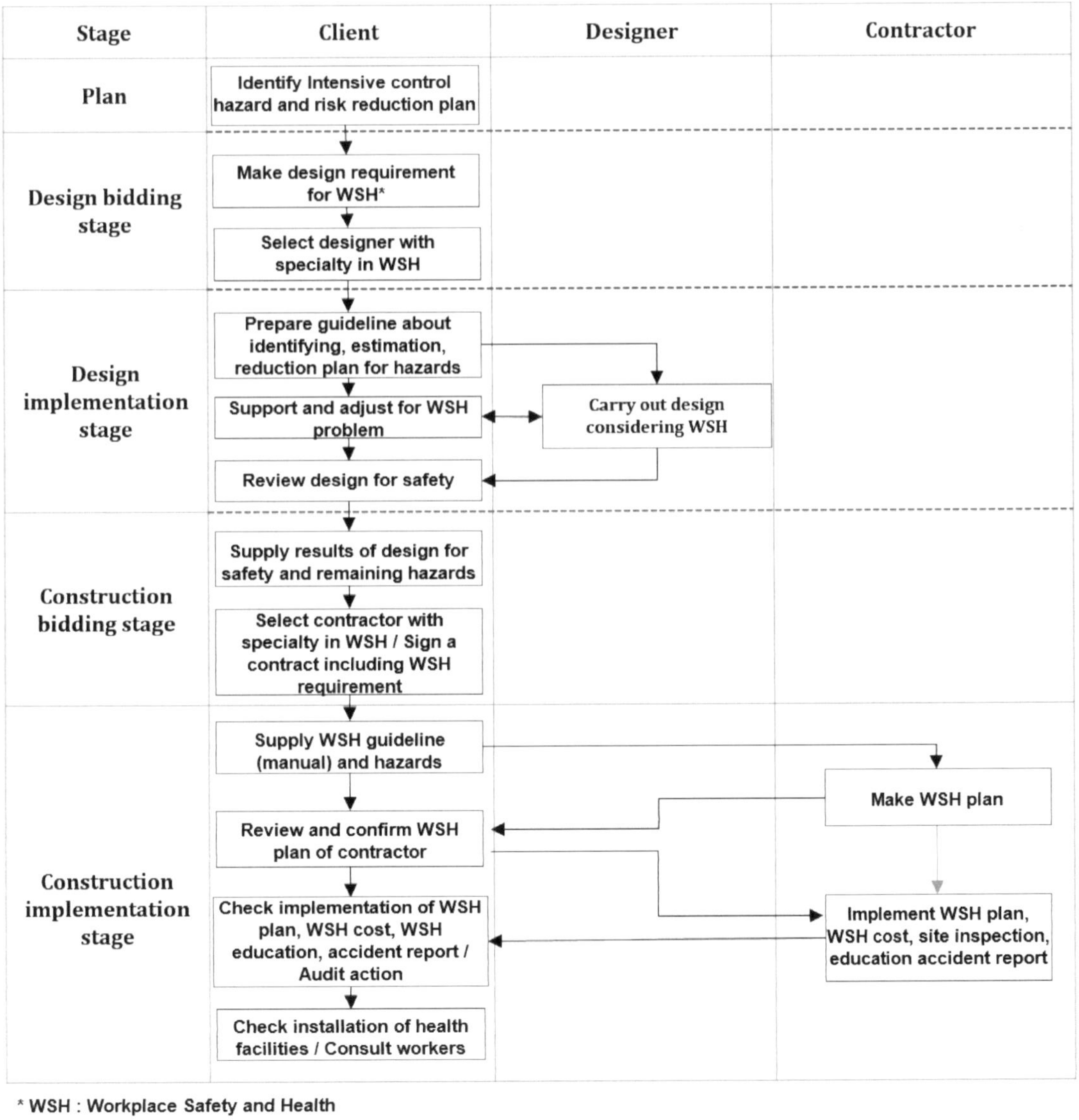

Figure 2. Suggestion of clients' role according to construction stage.

5 CONCLUSION

The method for imposing construction client's duty in Korea construction safety and health manage system was surveyed and the client's role was suggested in the whole construction process. From the survey, it was found that safety managers and supervisors were strongly agreed to impose the client's duty legally. For the imposing method for client's duty in a law, safety managers and construction managers preferred to appoint a safety and health specialist instead of the client or to assign a client direct role and responsibility, while the client wanted to give client's role and responsibility to the supervisor. The client can be a key player and mediator in the construction safety and health management system by performing the appropriate role suggested in this study according to plan, design, and construction stage

Acknowledgments

This study has been supported partially by the expert education project in disaster management of Ministry of the Interior and Safety.

References

EU, *Council Directive 92/57/EEC*, 1992.

HSE, *The Construction (Design and Management) Regulations 2015*.

Health and Safety Authority, *Clients in Construction – Best Practice Guidance*, 2009.

ILO, *Safety and Health in Construction - Code of Practice*, 1992.

Huang, Z., and Hinze, J., Owner's Role in Construction Safety, *Journal of Construction Engineering and Management*, 132(2), 164-173, Feburary 2006.

Jang, N. G., Song, B. C., and Won, J. H., *Suggestion for Client Oriented Safety and Health Management System in Korea Construction Site,* Asia Pacific Symposium on Safety 2017, Kitakyushu, Japan. November 30-December 1, 2017.

MOEL *Occupational Injuries and Illnesses Statistics,* Ministry of Employment and Labor in Korea, 2007-2016.

Smallwood, J., Client Influence on Contractor Health and Safety in South Africa, Building Research & Information, 26(3), 181-189, 1998.

Won, J. H., Park, H. K., Lim, S. J., and Park, Y. K., Investigation of Construction Work Participants' Recognition for Assigning Safety and Health Management Responsibility to Client, *Journal of Korean Society of Safety*, 30(5), 59-66, October, 2015.

Streamlining Information Transfer between Construction and Structural Engineering
Edited by Shiau, J., Vimonsatit, V., Yazdani, S., and Singh, A.

ISBN: 978-0-9960437-7-9

CASE STUDIES: DISPUTE RESOLUTIONS SELECTION IN CONSTRUCTION

AMARJIT SINGH and XI SONG

Dept of Civil Engineering, University of Hawaii at Manoa, Honolulu, USA

Disputes simply cannot be prevented all the time in the construction industry. As such, common dispute resolution systems such as negotiation, mediation, arbitration and litigation are widely used. Compared to litigation, negotiation and mediation have high popularity owing to their low cost and faster duration to dispensation. However, proper selection of dispute resolution techniques becomes very critical in real business practice. In this article, a numerical method using pairwise comparisons and the analytic hierarchy process are used to discover the most appropriate method evaluated against specific selection factors. Eight major selection factors are used, and an importance scale is assigned to weigh each factor, followed by a pairwise comparison. The rationale for selection of factors is explained. Final scores of the effectiveness of each dispute resolution technique are calculated for six cases that are meaningful in the construction industry, thereby providing a practical guideline and system to disputants for selecting a suitable dispute resolution method for their specific case.

Keywords: Alternate dispute resolution (ADR), Negotiation, Mediation, Arbitration, Litigation, Pairwise comparison, Analytic hierarchy process.

1 INTRODUCTION

Disputes commonly happen in construction projects due to underprepared contracts, uncleared scope of work, insufficient planning, budget issue, and communication problems, among many other (Cheung and Pang 2013). It is very important to address and properly resolve the disputes to ensure a successful construction project. However, the complexity and uncertainty of construction projects led to the rising popularity of alternative dispute resolution (ADR) techniques, in contrast to litigation. ADR methods are those such as mediation, negotiation, and arbitration. In this article, six case studies are presented to illustrate the selection of an appropriate dispute solution strategy.

2 METHODOLOGY OF DISPUTE RESOLUTION SELECTION

Four major dispute resolution techniques that are widely used in construction projects are negotiation, mediation, arbitration and litigation. Among these four techniques, negotiation is the most common method in the construction industry, further reporting that more than 70% of disputes are resolved using negotiation (Tam 1998). Mediation is performed by a third-party mediator, if negotiation cannot provide a satisfactory solution. Arbitration is a method where disputants agree to seek solutions by the decision of an arbitrator, where the solution may be legally enforced. Litigation is considered as the ultimate legal method for settling controversies or disputes between and among persons, organizations, and governments. It is obvious that when disputes occur, disputants need to decide on which resolution to pursue. In this article, a criterion-based rating method is utilized.

2.1 Criteria for Dispute Resolution Selection

Cheung and Suen (2002) did a study by sending a survey to 13 professionals to identify five most important criteria for selecting a construction dispute resolution strategy from a list of 16. Based on the voting results, the five most important criteria were overall duration, relative cost, flexibility in issues and strategy, confidentiality, and preservation of the relationship. Other minor considerations included binding decision and enforcement, degree of control by parties and degree of control by third-party neutral, etc. In the construction industry, different entities will put ahead their own interests and preferences to choose the criteria to evaluate. In terms of having a quantitative rating system, the criteria weighting system is developed.

2.2 Criteria Raw Points

The scores in the study by Cheung and Suen (2002) to identify important criteria were converted to raw points as summarized in Table 1. Higher scores represent higher suitability and appropriateness. Raw points act as base parameter in the entire selection procedure. In the next step, the weighting evaluation is done by using the analytic hierarchy process (AHP).

Table 1. Raw points for the dispute resolutions strategies (Data source: Cheung and Suen, 2002).

Attribute designation	Decision parameter	Dispute resolution strategies			
		Negotiation	Mediation	Arbitration	Litigation
A	Overall duration	9	8	2	1
B	Relative cost	8	8	2	1
C	Flexibility	10	9	5	1
D	Confidentiality	8	8	9	2
E	Preservation of relationship	9	8	3	1
F	Binding decision and enforcement	1	2	10	10
G	Degree of control by parties	10	9	5	3
H	Degree of control by third party neutral	1	3	9	10

2.3 Weighting Method

The weights assigned to selected criteria in the construction dispute case can be somewhat subjective. Therefore, a systematic weighting method is necessary. Hence, the analytical hierarchy process (AHP) is adopted to ascertain the priority ratings of those weights, and also to ascertain whether the weights have been assigned with reason, without bias. The AHP process employs pairwise comparison techniques in which the disputants must make judgments on the relative standings of the criteria, and it is guided by an importance scale shown in Table 2.

2.4 Selection Procedure

When there is a dispute in a construction project, disputants can use the following step-wise procedure to identify the appropriate resolution:

1. Evaluate the dispute and review the criteria for selection;
2. Select major criteria that apply to the dispute and conduct pairwise comparisons;
3. Apply the results from pairwise comparisons to AHP to obtain normalized weights for each selected criterion;
4. Get the raw points of each selected criterion for every dispute resolution from Table 1;
5. Calculate the weighted scores by multiplying the raw points by the normalized weights;
6. Get the total weighted score for each dispute resolution, and compare the results;
7. Select the appropriate dispute resolution based on total weighted score. Usually the highest score represents the most favorable choice.

Table 2. Importance scale.

Level of preference	Numerical points	Reciprocal for diagonal
Equal	1	1
Moderate	3	0.33
Strong	6	0.17
Very Strong	9	0.11

3 CASE STUDIES

In the construction industry, disputes usually occur between or among partners, collaborators, owners and contractors, etc. Based on the different roles and requirements on a construction project, and to illustrate the selection of a dispute resolution technique, the following six cases are studied:

Case I: Large enterprise/corporation
Case II: Small business owner
Case III: General contractor
Case IV: Subcontractor
Case V: Disputant who wants to maintain a good business relationship
Case VI: Disputant who is in a high-confidentiality environment

Table 3. Selection criteria for six cases.

Attribute designation	Decision parameter	Case I	Case II	Case III	Case IV	Case V	Case VI
A	Overall duration	X	X	X	X	X	X
B	Relative cost	X	X	X	X	X	X
C	Flexibility				X		
D	Confidentiality	X		X			X
E	Preservation of relationship	X		X		X	
F	Binding decision and enforcement						
G	Degree of control by parties			X			
H	Degree of control by third party neutral						

When a construction dispute arises, the involved parties must evaluate all listed criteria and select the relevant ones to reflect their business interests and aspirations. Table 3 shows the different choices of criteria for the six cases based on the relevance they have to the specific case

(Haugen and Singh 2014). It is noticeable that all the cases considered overall duration and relative cost as their dispute resolution factors. From Table 1, it is obvious that negotiation and mediation have the highest points in these two criteria. Furthermore, negotiation or mediation may be decided as the best practice for dispute resolution even without conducting the selection procedure. However, the detailed selection procedure is still necessary to provide a quantitative guidance for the disputants to take final decisions.

Subsequently, the weights for each criterion are calculated using pairwise comparison as shown in Tables 4 to 9. Attribute designations A to H are assigned to each criterion. Selected criteria are paired, and level of preference is assigned to each pair using the numerical scale shown in Table 2. For example, in Case I, E (preservation of relationship) has strong preference over D (confidentiality). Therefore, the scale number "6" is filled in the corresponding cell; while 0.17 is filled in the reciprocal cell. Furthermore, the geometric mean and normalized weight are calculated (Saaty 1980).

Table 4. Pairwise and AHP matrix for Case I.

Criteria	Pairwise/AHP points				Geometric mean	Normalized weight
	E	D	A	B		
E	1	6	3	0.17	1.32	0.20
D	0.17	1	0.17	0.11	0.24	0.04
A	0.33	6	1	0.17	0.76	0.12
B	6	9	6	1	4.24	0.65
Sum	-	-	-	-	6.55	1.00

Table 5. Pairwise and AHP matrix for Case II.

Criteria	Pairwise/AHP points		Geometric mean	Normalized weight
	A	B		
A	1	0.33	0.58	0.25
B	3	1	1.73	0.75
Sum	-	-	2.31	1.00

Table 6. Pairwise and AHP matrix for Case III.

Criteria	Pairwise/AHP points					Geometric mean	Normalized weight
	E	D	A	B	G		
E	1	6	6	0.17	6	2.05	0.24
D	0.17	1	0.33	0.11	6	0.52	0.06
A	0.17	3	1	0.17	3	0.76	0.09
B	6	9	6	1	9	4.93	0.58
G	0.17	0.17	0.33	0.11	1	0.25	0.03
Sum	-	-	-	-	-	8.51	1.00

Table 7. Pairwise and AHP matrix for Case IV.

Criteria	Pairwise/AHP points			Geometric mean	Normalized weight
	A	B	C		
A	1	0.11	0.33	0.33	0.06
B	9	1	9	4.33	0.81
C	3	0.11	1	0.69	0.13
Sum	-	-	-	5.35	1.00

Table 8. Pairwise and AHP matrix for Case V.

Criteria	Pairwise/AHP points			Geometric mean	Normalized weight
	E	A	B		
E	1	9	6	3.78	0.75
A	0.11	1	0.17	0.26	0.05
B	0.17	6	1	1.00	0.20
Sum	-	-	-	5.04	1.00

Table 9. Pairwise and AHP matrix for Case VI.

Criteria	Pairwise/AHP points			Geometric mean	Normalized weight
	D	B	A		
D	1	9	9	4.33	0.79
B	0.11	1	6	0.87	0.16
A	0.11	0.17	1	0.26	0.05
Sum	-	-	-	5.46	1.00

Table 10. Case scores summary.

Case/Criteria		AHP Scores	Negotiation		Mediation		Arbitration		Litigation	
			Raw Points	Score	Raw Points	Score	Raw Points	Score	Raw Points	Score
I	E	0.20	9	1.81	8	1.61	3	0.60	1	0.20
	D	0.04	8	0.29	8	0.29	9	0.32	2	0.07
	A	0.12	9	1.04	8	0.93	2	0.23	1	0.12
	B	0.65	8	5.18	8	5.18	2	1.29	1	0.65
	Total			**8.32**		**8.00**		**2.45**		**1.04**
II	A	0.25	9	2.25	8	2.00	2	0.5	1	0.25
	B	0.75	8	6	8	6.00	2	1.5	1	0.75
	Total			**8.25**		**8.00**		**2**		**1**
III	E	0.24	9	2.17	8	1.93	3	0.72	1	0.24
	D	0.06	8	0.49	8	0.49	9	0.55	2	0.12
	A	0.09	9	0.80	8	0.71	2	0.18	1	0.09
	B	0.58	8	4.64	8	4.64	2	1.16	1	0.58
	G	0.03	10	0.30	9	0.27	5	0.15	2	0.06
	Total			**8.39**		**8.03**		**2.76**		**1.09**
IV	A	0.06	9	0.56	8	0.50	2	0.12	1	0.06
	B	0.81	8	6.47	8	6.47	2	1.62	1	0.81
	C	0.13	10	1.30	9	1.17	5	0.65	1	0.13
	Total			**8.32**		**8.13**		**2.39**		**1.00**
V	E	0.75	9	6.74	8	5.99	2	1.50	1	0.75
	A	0.05	8	0.42	8	0.42	2	0.10	1	0.05
	B	0.20	10	1.98	9	1.78	5	0.99	1	0.20
	Total			**9.15**		**8.20**		**2.59**		**1.00**
VI	D	0.79	8	6.33	8	6.33	9	7.13	2	1.58
	B	0.16	8	1.28	8	1.28	2	0.32	1	0.16
	A	0.05	9	0.44	8	0.39	2	0.10	1	0.05
	Total			**8.05**		**8.00**		**7.54**		**1.79**

Table 10 summarizes the final dispute resolution selection for the six cases, the numerical results show that the most preferred resolution for all six cases is negotiation, followed by mediation; least preferable option is litigation. However, in real construction disputes, negotiation may not always work, in which case the disputants must select other resolutions based on their situation.

Consistency Ratio (CR) is used to verify the reasonability of the pairwise comparison and weights. CR is the ratio between CI and RI, where a reference random index (RI) is provided for different sizes of an AHP matrix. The Consistency Index (CI) is calculated using Equation 1 (Saaty, 1980), in which, λ_{max} is the largest eigenvalue for the *n* size AHP matrix.

$$CI = \frac{\lambda_{max} - n}{n - 1} \tag{1}$$

For instance, if the acceptable consistency ratio is 15%, Case I yields a consistency ratio of 14.4%. Being less than 15%, the assignment of weights is considered reasonable.

4 DISCUSSION AND CONCLUSION

Among the six cases, Case I to Case IV considered overall duration and relative cost as the first two priorities due to the nature of the business. However, if different companies or parties will make distinctive decisions in pairwise comparison, the results may vary. Case V and Case VI represent particular requirements that may occur in the construction industry. As such, the resolution procedure recommended negotiation as the first option. Owners should also consider using alternate resolution based on their own market position and business attribute. It is worth noticing that in Case VI, the score of arbitration is 7.54, only 0.5 less than mediation. It is indicated that when confidentiality has high priority in the business, choosing arbitration as the dispute resolution should be in the construction contract.

By conducting six hypothetical case studies, this paper presented a numerical method for dispute resolution selection, based on eight measures such as overall duration, relative cost, flexibility, confidentiality, etc. Among these four resolution strategies, negotiation and mediation were discovered to be the most favorable solutions. Furthermore, based on the different demands in dispute events, eight major criteria provided a guideline for disputants to convert their subjective opinion into a systemic rating system by using pairwise comparison and AHP. The following observations are made to these six cases:

- Overall duration and cost are considered primarily in all six cases.
- Because of the time consuming and expensive litigation process, ADR methods are preferred over litigation.
- Negotiation was the most preferred, followed closely by mediation, and then by arbitration. Litigation is the least preferred by a factor of more than 4 times compared to negotiation.

References

Cheung, S. and Pang, K. H. Y., Anatomy of Construction Disputes, *Journal of Construction Engineering and Management,* 139. 15-23. 10.1061/(ASCE)CO.1943-7862.0000532, 2013.

Cheung, S. O. and Suen, H., A., Multi-Attribute Utility Model for Dispute Resolution Strategy Selection, *Construction Management and Eco*nomics, 20, 557-568, 2002.

Haugen, T. and Singh, A., Dispute Resolution Strategy Selection, *Journal of Legal Affairs and Dispute Resolution in Engineering and Construction*, ASCE, online 2014.

Saaty, T. L., *The Analytic Hierarchy Process: Planning, Priority Setting, Resource Allocation*, McGraw-Hill, New York, 1980.

Tam, P. K., *Alternative Dispute Resolution Effectiveness and Its Acceptability to the Construction Industry of the HKSAR*, MSc Thesis, Polytechnic Univ., Hong Kong, 1998.

Streamlining Information Transfer between Construction and Structural Engineering
Edited by Shiau, J., Vimonsatit, V., Yazdani, S., and Singh, A.

ISBN: 978-0-9960437-7-9

A GRAPHICAL USER INTERFACE (GUI) FOR FORENSIC ANALYSIS OF DELAY AND DISRUPTION

AGELIKI VALAVANOGLOU[1], DANIJEL REBOLJ[2], and DETLEF HECK[1]

[1]*Institute of Construction Management and Economics, Technical University of Graz, Graz, Austria*

[2]*Faculty of Civil Engineering, Transportation Engineering and Architecture, University of Maribor, Maribor, Slovenia*

Disrupting events are common in construction and can have a significant impact on the successful completion of a project. Disputes concerning disruptive events often arise because no party wants to accept responsibility for the extra cost incurred. The resolution of this kind of disputes lies within the clear demonstration of causation, liability and the quantification of the disrupting event. Purpose of this research is to explore the potential of Building Information Modeling (BIM) in assisting forensic delay and disruption analysis supported by a Graphical User Interface (GUI). An extensive literature review highlighted the shortcomings of current practices and identified the reconstruction of events through incomplete and unstructured documentation as one of the primary challenges faced by the analysis expert. The findings of the literature review formed the basis for the development of a GUI designed to incorporate all the necessary information for the identification of the causal link of events, the liability and the calculation of damages. The usability of the GUI prototype was tested on a case study, indicating an increase in the overall efficiency and reduction of time spent by the forensic analyst in the retrieval of relevant information.

Keywords: Construction claims, Building information modeling, Documentation, Dispute prevention, Dispute resolution, Information management.

1 INTRODUCTION

Construction projects can be influenced by several factors such as the numerous participating parties, weather conditions, risks associated with the construction process as well as unforeseeable events which can impact the successful completion of a project (Chan *et al.* 2004). The effects of disrupting events can be manifold and can lead to delay to activities or to the overall completion date as well as increase in cost (Rosenfeld 2014). The successful resolution of claims regarding delay and disruption are based on the clear demonstration of causation, entitlement and quantification (Williams *et al.* 2013). One of the primary challenges encountered in analysis of delay and disruption is the reconstruction of events through incomplete as-built documentation (Hammad and Alkass 2000).

The emergence of technology has greatly improved the process of project documentation in the last decades, but the construction industry has been slow to adapt, especially concerning construction progress documentation (Stowe *et al.* 2014). Apart from Project Management software, Building Information Modeling (BIM) has emerged to revolutionize the construction industry. It has opened up opportunities for optimization in all phases of construction, and research has already identified the potential of BIM in dispute prevention and dispute resolution (Greenwald 2013) as well as to support forensic analysis of delay and disruption (Gibbs *et al.* 2012).

The objective of this study is to develop a prototype GUI linking together all necessary information stored in BIM and Project Management software, to be used as factual evidence supporting a claim. The aim is to alleviate the difficulty of information retrieval as well as to facilitate the forensic analysis. The prototype was tested on a simple case study to determine its usability and identify possible limitations.

2 DISRUPTION CLAIMS AND BIM TECHNOLOGY

The success of forensic analysis both for delay and for disruption is based on the quality of the project progress documentation available. Retrieval of all relevant documentation is the first step in claim preparation, not only to gain insight into the events leading to delay and disruption, but also for the comparison of the as-planned (AP) and the as-built (AB), their discrepancies and the analysis of the causal link between events and their primary and secondary effects. The documentation requirements for the impact assessment of disrupting events include, accuracy through continuous updates during the construction process, systematic structure, high level of detail and comprehensive composition (Reister 2007).

Proper contemporaneous project documentation can be very time consuming (Vidogah and Ndekurgi 1997) and is often regarded as unnecessary or superfluous. In the last decades the progress of Information Technology has alleviated the onerous task of contemporaneous project documentation, but non-existent, erroneous, or lacking information still rank highest as obstacles in delay analysis methods (Braimah and Ndekurgi 2009). As delay and disruption analysis has to be supported through a plethora of evidence it is very paper-intensive (Gibbs et al. 2017) with the elementary information retrieved from disruption notifications, documentation of performance change, as-planned and as-built construction schedules, allocation of resources, daily site reports, photos, consultation meeting reports, correspondence and request for information and respond lists (Trauner 2009, Vygen 2011).

BIM technology offers novel opportunities of integrated and interoperable processes in the construction lifecycle providing a centralized virtual information environment (Eastmann *et al.* 2011). Research and practice have already recognized the potential of BIM in easing the access to coordinated contemporaneous project information through multiple dimensions (Chou and Yang 2017). Studies focusing on assisting delay claims through BIM have illustrated that 4D Models, linking building components from the 3D model to the construction schedule (Kensek 2014) and visualizing the construction sequence by comparing the actual versus the planned construction process can support the presentation of the analysis and findings (Pickavance 2007). Coyne (2008) outlined the integration of 4D tools into typical schedule analysis steps. Guevremont and Hammad (2018) illustrated the benefits of identification, visualization, quantification, and responsibility assignment of delay events by identifying the resulting spatiotemporal conflicts in 4D simulations. Burr (2016) states that benefits gained through the implementation of BIM can only be obtained through the premise of best practice project management such as qualitative construction documentation. Gibbs *et al.* (2017) conducted a study presenting computer-generated exhibits (CGE), such as videos of virtual construction sequences, highlighting the importance of "displaying only the facts" as well as accompanying the presentation with "a narrative from the CGE creator". The study also emphasized that there is still an element of distrust in this kind of evidence presentation methods and the accuracy of the supporting facts. Chou and Young (2017) conducted an extensive literature review on whether a BIM-based approach can aid the mitigation of problems encountered in schedule delay analysis and concluded that although it can greatly benefit information collection there is a lack of focus on the presentation of analytical results.

There have been attempts to make retrieval of information more user-friendly by increasing efficient and effective distribution of information amongst different BIM users. Lee *et al.* (2011) proposed a Web-based interactive GUI to support acquisition of required information with 2D, 3D views, and 4D simulations, options for data search as well as the location of the construction site on Google Maps.

3 GUI FOR RETRIEVAL OF INFORMATION FOR DELAY AND DISRUPTION ANALYSIS

In the analysis of delay and disruption particular focus is placed on the comparison of the AP versus the AB schedule linked to the allocated resources for the visualization and identification of discrepancies in time and cost. Figure 1 illustrates the system schema behind the design of the GUI, incorporating information regarding construction progress provided by digital on- and off-site documentation stored in Project Management software with additional attributes and metadata of the construction process. The BIM system provides the AP and the AB model with the construction schedule and the respective cost. APIs (Application Programming Interfaces) enable automated data extraction from the respective systems.

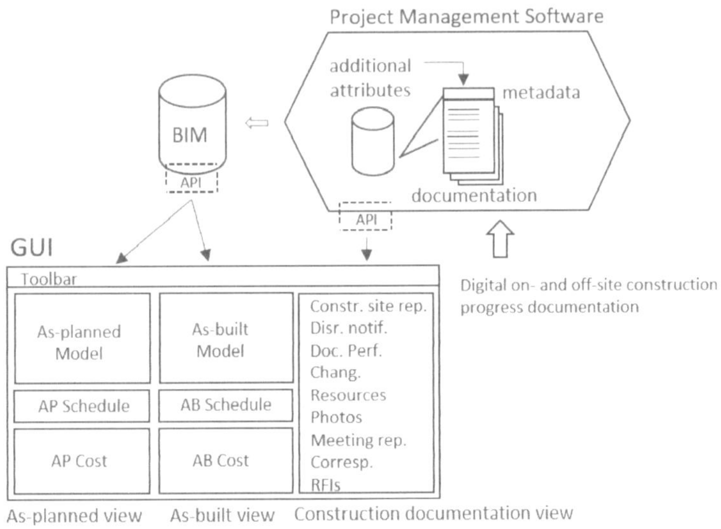

Figure 1. System schema of developed GUI.

Figure 2 shows the developed GUI prototype, which was tested on a simple case study to demonstrate the functionalities, intended to increase the overall efficiency and reduction of time spent by the forensic analyst in the retrieval of relevant information. In the GUI the expert is able to contemporaneously visualize the AP and the AB construction process, analyze the AP and AB schedules and costs, as well as retrieve information from progress documentation such as, disruption notifications, documentation of performance change, construction schedule, daily construction reports, photographical documentation, consultation meeting reports, relevant correspondence and requests for information submittal and respond lists.

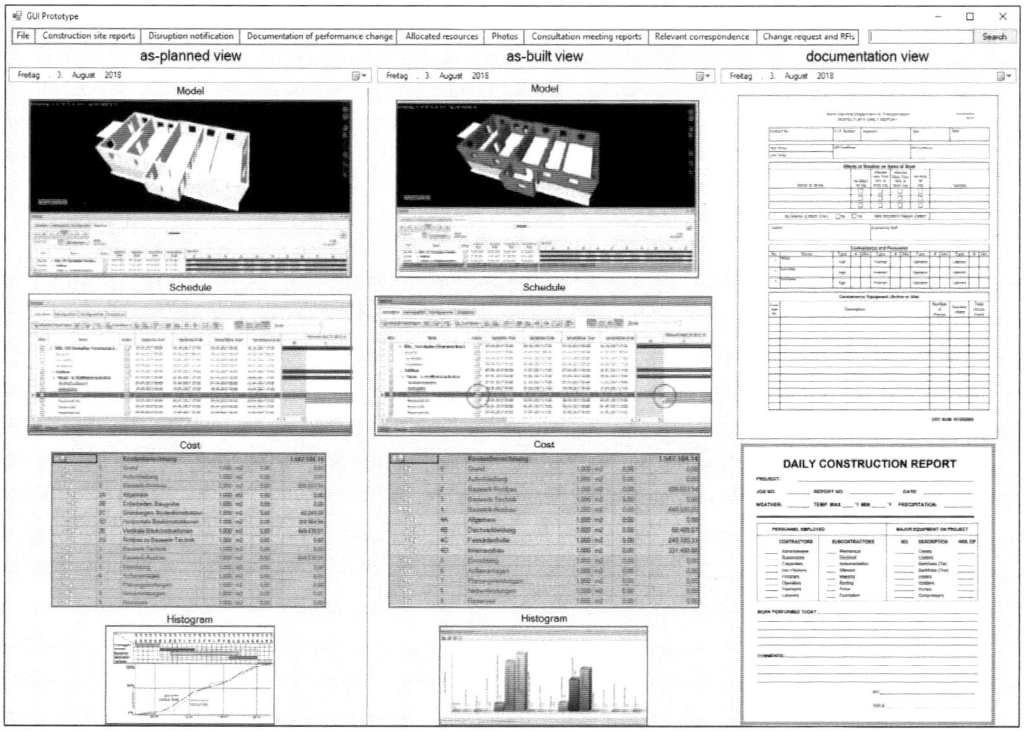

Figure 2. GUI displaying information necessary for delay and disruption analysis.

3.1 Case Study

The applicability of the developed GUI was tested on a theoretical case study of a two-story residential building. Forensic analysis was called to settle a dispute regarding a claim made by the contractor requesting compensation for additional cost of construction site overhead, labour and equipment. The contractor's claim stated that due to differing site conditions the foundation excavation works were extended by five days, affecting masonry work on the critical path. The contractor was granted an extension of time (EOT) but compensation only for three out of the five days. The client argued that the contractor was able to accelerate the works after the delay, resulting in only three days delay to the project completion date.

Figure 3 illustrates the steps in the application of the GUI for the reconstruction of events by the forensic analyst. The contractor's disruption notification stated that the schedule was disrupted due to differing site conditions claiming an EOT of five days and additional costs for site office overhead, 2 extra labourers dispatched to the site, and an additional excavator. Contemporaneous AP and AB visualization as well as photo documentation provided an overview of the effect of the delay on downstream activities and the completion date of the project. Causal link between events was determined with a schedule delay analysis method and extra cost incurred by the affected resources. Construction progress documentation, AB schedule and cost calculation showed that the contractor was able to speed up downstream activities by dispatching two extra labourers to the construction site. The delay to the completion date was reduced from five to three days.

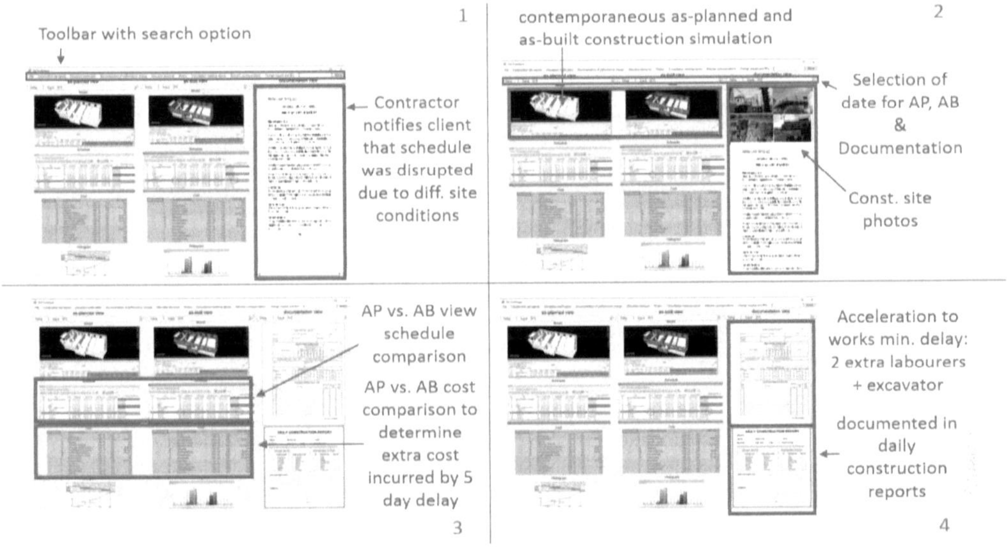

Figure 3. Case study information retrieval for forensic analysis of disrupting event.

The forensic analysis concluded that the contractor was entitled to the EOT and the extra cost incurred by the two additional laborers, excavator, and site overhead, but only for the three days delay to the completion date.

4 CONCLUSIONS

There are various triggers that can cause delay and disruption in a construction project, some attributable to the owner, others attributable to the client or in case of true concurrency to both. Effects of delay and disruption can be diverse and must be individually identified and analyzed.

The developed GUI prototype was designed to integrate information to support the analysis and be used as factual evidence, aiming to alleviate the difficulty of information retrieval as well as the distrust in new presentation methods. The prototype was tested on a simple case study indicating an increase in the overall efficiency and reduction of time spent by the forensic analyst in the retrieval of relevant information. It is important to note that the forensic analysis expert continues to play a significant role in the determination of the accuracy of the provided information and the subsequent schedule analysis, identification of the causal events, and appointment of liability.

Further research is necessary in order to test the developed GUI on larger more complex projects and conduct a usability study with experts to identify possible limitations.

References

Burr, A. (ed.), *Delay and Disruption in Construction Contracts*, Routledge, Abington, 2016.

Braimah, N., and Ndekurgi, I., Consultants' Perceptions on Construction Delay Analysis Methodologies, *Journal of Construction Engineering and Management*, 135, 1279-1288, 2009.

Chan, A., Scott, D., and Chan, A., Factor Affecting the Success of a Construction Project, *Journal of Construction Engineering and Management*, 130(1), 153-155, February 2004.

Chou, H., and Yang, J., *Preliminary Evaluation of BIM-based Approaches for Schedule Delay Analysis,* IOP Conf. Series, Materials Science and Engineering, Volume 245, WMCAUS, 1-8, Prague, Czech Republic, 12–16 June 2017.

Coyne, K., *Leveraging the Power of 4D Models for Analyzing and Presenting CPM Schedule Delay Analysis*, 52nd Annual Meeting of the Association for the Advancement of Cost Engineering, 1-10, Morgantown, USA 2008.

Eastmann, Ch., Teicholz, P., Sacks, R., and Liston, K., *BIM Handbook, A Guide to Building Information Modeling for Owners, Managers, Designers, Engineers, and Contractors*, John Wiley & Sons, New Jersey, USA, 2011.

Gibbs, D., Emmit, S., Ruikar, K., and Lord, W., *An Investigation into Weather Building Information Modelling (BIM) Can Assist with Construction Delay Claims*, Proceedings of the First UK Academic Conference on BIM, 36 – 44, Northumbria University, 5-7 September 2012.

Gibbs, D., Lord, W., Emmit, S., and Ruikar, K., Interactive Exhibit to Assist with Understanding Project Delays, *Journal of Legal Affairs and Dispute Resolution in Engineering and Construction*, 9(1), 1-12, February 2017.

Greenwald, N., A Creative Proposal for Dispute Systems Design for Construction Projects Employing BIM, *Journal of Legal Affairs and Dispute Resolution in Engineering and Construction*, 5(1), 2-5, February 2013.

Guevremont, M., and Hammad, A., Visualization of Delay Claim Analysis Using 4D Simulation, *Journal of Legal Affairs and Dispute Resolution in. Engineering and Construction*, 10(3), 1-8, 2018.

Hammad, M., and Alkass, S., *A Web-Based Construction Project Document Information Center in Support of Claims Preparation,* Eighth International Conference on Computing in Civil and Building Engineering, 844-851, Stanford, August 14-16, 2000.

Kensek, K., *Building Information Modeling*, Routledge, Abingdon, 2014.

Lee, B., Chuang, T., and Wu, I., *Design and Implementation of an Interactive GUI for BIM Systems*, CONVR2011, 11th International Conference on Construction Applications of Virtual Reality, Bargstädt, H., and Ailland, K. (eds.), 106-116, Weimar, November 3-4, 2011.

Pickavance, K., Using Advanced Forensic Animations to Resolve Complex Disruption Claims, Society of Construction Law, 2007. Retrieved from https://www.scl.org.uk/papers/using-advanced-forensic-animations-resolve-complex-disruption-claims on October 11, 2016.

Reister, D., *Nachträge im Bauvertrag,* Werner, Neuwied, 2007.

Rosenfeld, Y., Root-Cause Analysis of Construction-Cost Overruns, *Journal of Construction Engineering Management*, 140(1), 1-10, 2014.

Stowe, K., Zhang, S., Teizer, J., and Jaselskis, E., Capturing the Return on Investment of All-In Building Information Modeling: Structured Approach, *Practice Periodical on Structural Design and Construction*, 20(1), 1-5, 2015.

Trauner, T., *Construction Delays, Documenting Causes, Winning Claims, and Recovering Costs*, Elsevier, Amsterdam, 2009.

Vidogah, W., and Ndekurgi, I., Improving Management of Claims: Contractor's Perspective, *Journal of Management in Engineering*, 13(5), 37-44, 1997.

Williams, T., Ackermann, F., and Eden, C., Structuring A Delay and Disruption Claim: An Application of Cause-Mapping and System Dynamics, *European Journal of Operational Research*, Elsevier, 148, 192-204, 2003.

Streamlining Information Transfer between Construction and Structural Engineering
Edited by Shiau, J., Vimonsatit, V., Yazdani, S., and Singh, A.

ISBN: 978-0-9960437-7-9

DESIGN AND BUILD – A DELAY REDUCTION METHOD FOR CONSTRUCTION PROJECTS

KURT CROSSMAN[1], INNOCENT MUSONDA[2], ANNLIZE L MARNEWICK[1], and JAN HARM C PRETORIUS[1]

[1]*Postgraduate School of Engineering Management, University of Johannesburg, Johannesburg, South Africa*
[2]*Construction Management and Quantity Surveying, University of Johannesburg, Johannesburg, South Africa*

The paper reports on a study of the impact of traditional and design & build delivery methods to ameliorate schedule delays in the South African construction projects. A case study approach of two similar projects was adopted for the study. Data was compiled from document analysis and semi-structured interviews. Selection of projects was purposive and convenience based. A comparative analysis of the two projects was performed on delays to inform on which delivery method ameliorated better the impact of schedule delays. Findings were that design and build marginally reduced the impact of consultant related delays. Little impact was however found for client, contractor and design change related delays. Most construction projects suffer delays, which inevitably increases the contract sums. Reducing the impact of delays is therefore critical. Choice of a delivery method offers an opportunity to reduce the delays and therefore it is valuable to understand which delivery method offered better results. Findings underscore the importance of seeking alternative methods to improve performance. The design and build method seemed to perform better than the traditional method on delay reduction. The implication is that although clients prefer "safe" methods, alternative delivery methods offered better results.

Keywords: Project delivery method, Traditional method, Impact of project of delays, Prevalence of project delays.

1 INTRODUCTION

The construction of large public and private sector water treatment facilities is largely implemented through a project delivery method named design-bid-build, otherwise known as the traditional method. Under this method, the management of the contracting parties and design of the facility is performed by engineering consultants or architects and the construction by a chosen contractor. However, these contractors are kept at a distance hampering synergy and innovation (Davis *et al.* 2008).

The traditional method of project delivery is still used as it is considered to obtain the lowest cost for the client and protect their interests with the expected guarantee of a sound engineering job (Love *et al.* 2014). Despite this the benefit to the client is lower than expected (Ng 2005) due to the consultants' lack of specialized skills and poor cooperation with contractors, ultimately leading to cost overruns caused largely by delays and rework. This has resulted in the decrease in popularity of traditional fixed-price tendering through an engineering consultant (Oyegoke *et al.* 2009), due to competition from more modern and successful design & build delivery method.

According to Chan *et al.* (2011), the design & build project delivery method allows a selected engineering contractor complete control of the design and construction process. One entity is responsible for both the design and construction of the project, allowing the project owner to sign only one contract, thus maintaining a single point of responsibility. This allows the contractor to apply their expertise and determine practical constructability problems early on, usually only found at an advanced stage when rework is expensive. Design & build delivery methods harness the expertise and experience of the selected contractor which maximizes client value by reducing delays, fostering innovation and reducing the project schedule. As a result, literature by Konchar and Sanvido (1997), Hale *et al.* (2009) and Ibbs *et al.* (2003) all found that design and build methods are beginning to produce better results than traditional methods globally.

The success of this method in developed countries has led to extraordinary growth in its adoption. However, the adoption of the design & build method in South Africa was low according to Grobler and Pretorius (2002). Further research is required to determine, whether local role players have the ability implement design & build effectively and reap its benefits.

1.1 Pre-Tender Decisions

The owner of a construction project has several important decisions to make before being put out to tender. Three such decisions which need to be made as early on as possible in a project according to the Design Build Institution of America (2015) are the; project delivery method, procurement method and contract type. Proper consideration of each decision is required to match the project characteristics, as these are some of the most important factors contributing to project success (Kumaraswamy and Dissanayaka 2001). The project delivery method defines the roles and responsibilities of the parties involved and is one of the most important decisions the project owner must make. According to the Construction Management Association of America (2012) there are four types of project delivery methods namely; Design-bid-build or traditional methods, construction management at risk, design & build and Integrated project delivery method. Before a construction project can go out to tender the project owner must make the above-mentioned important decisions which has the potential to impact greatly on the project success. The project owner must carefully consider the characteristic of the construction project and choose the best suited project delivery, procurement and contracting method.

1.2 Project Delivery Methods

In South Africa, the dominant project delivery method used to implement construction projects is the traditional method followed by the design & build delivery method according to Grobler and Pretorius (2002). The traditional delivery method, also known as design-bid-build is still the most commonly used method to procure construction projects (Mahdi and Alreshaid 2005). Under this method, the project is typically split into separate sequential phases of design development, tender, contract award and construction phases. The fundamental principle of the traditional method is the signing of two or more contracts by the owner, one with the designer and another with the contractor (Davis *et al.* 2008). The contractor is responsible for building the project as designed and the designer is responsible for designing the project to the client's specifications and also runs the project as the project owners' representative (Mahdi and Alreshaid 2005). This is intended to ensure a quality project, which meets the client's needs and is used on most public works projects. The clearly defined roles and responsibilities make the contractual arranges easy to manage, however these relationships have been shown to become adversarial when rework and change orders are required, leading to cost and time overrun.

Design & build can be summarized as a delivery method where the responsibility of both construction and design falls under one entity. The owner of the project can then sign only contract, allowing all responsibility to be contained to single point (Mosey 1998). Under design & build, the complete design and construction is not always completed solely by the contracted entity, different forms exist, however in all cases the adversarial relationship experienced between design and construction teams found in traditional methods is mitigated. The design & build delivery method has numerous advantages largely as a result of the single point of responsibility. These advantages include: a shortened project duration (life Cycle), establishing more accurate cost, reducing cost, increasing innovation and reduced claims.

Table 1. Project delivery method schedule growth comparison.

		Schedule growth	
Author	**Project Type**	**Design & build (%)**	**Traditional (%)**
Shane *et al.* (2013)	Roads	23	20
Shrestha *et al.* (2012)	Highways	-4.2	4.8
Ibbs *et al.* (2003)	Buildings	4.1	6.5
Konchar and Sanvido (1997)	Industrial buildings	0	5
Hale *et al.* (2009)	Military facilities	11.4	13.8

The success of this method in developed countries as illustrated in Table 1 has led to extraordinary growth in its adoption. However, the adoption of the design & build method in South Africa is low according to Grobler and Pretorius (2002). Further research is required to determine whether local role players can implement design & build effectively and reap its benefits.

2 RESEARCH METHODOLOGY

The aim of this research was to determine how the project delivery method used on a construction project impacts on delays, thus allowing the researcher to determine whether the design and build project delivery method can reduce delays on construction projects when compared with the traditional method, as seen internationally.

This research made use of two similar case studies, which differed as far as possible by the project delivery method used only. The case study method was chosen for its ability to determine operational links over a period of time instead of simply frequencies of specific occurrences. It provides insights into organizations and gives an in-depth understanding of the issue being studied. Multiple sources of information were used for analysis, which included documentation and interviews. Documentation relating to delays was gathered for each case and entered into a case study database. The project contract managers of each case were interviewed in semi-structured manner, using a questionnaire as a guideline. Responses were transcribed and coded, allowing the generation of themes, which were compared with documentation allowing in depth analysis of each case. A comparison was then performed between cases and conclusions made from the findings presented.

2.1 Case Description

Case study 1 covered the capacity upgrade to a waste water treatment works which is managed by the government institution “ABC”. The institution required a 50ml/day conventional activated sludge extension to the existing plant in the form of an additional module to treat the increasing load. The project was implemented using the traditional project delivery method resulting in the signing of five separate contracts between consultants and contractors. A competitive bidding process based on a supplied bill of quantities was used by the client. The selected contractor was awarded the contract at a cost of R118 million and expected duration of 110 weeks.

Case study 2 was performed on the capacity upgrade to waste water treatment works (WWTW), which is managed by the government institution "ABC". In 2014 the organization identified the need to upgrade the WWTW by 5ml/d within 2 years to cope with the expanding population of the area. However, due to space constraints on the plant, an upgrade using the treatment method in operation was not possible. A specialized solution was needed to fit the requirement. The researchers' organization was chosen as the only bidder able to meet the project requirements using their proprietary technology.

The design & build project delivery method was therefore used as the researchers organization held the technology rights. The researcher's organizations consultant was employed by the client in order to develop the tender specification. Once the contract was awarded, the consultant then returned under the employ of the researcher's organization, otherwise known as novation design & build. The researcher's organization was the sole supplier thus the price was negotiated to a guaranteed maximum price of R25 million with an expected completion time of 72 weeks.

3 ANALYSIS AND COMPARISON

Project documentation was gathered for each case and entered into a case study database. Interviews were then conducted and transcribed. The transcriptions were then coded using CAQDAS software and case specific themes developed. Analysis and comparison between cases was then possible using the gathered documentation, themes and literature.

The Pre-tender decisions taken by the project owner are presented below and compared to findings in literature. It is clear that the decisions made by the project owner followed exactly that suggested by literature. Both cases made pre-tender decisions as suggested by literature and thus one can conclude that the environment was correct for both projects to succeed. Each case study's delays were then compared in order to determine the length and cause of delays, which were present for each case study. Delays affecting the project timeline were extensive in both case studies, however the causes under each case vary.

Table 2. Pre-Tender decision comparison.

	Case 1	**Literature**	**Case 2**	**Literature**
Project delivery method	Traditional	Traditional	Design & Build	Design & Build
Procurement method	Low Bid	Low Bid	Best Value	Best Value
Contract method	Fixed Price	Fixed Price	Guaranteed maximum price	Guaranteed maximum price

Table 3. Schedule growth comparison.

Case study	**Start date**	**Planned completion**	**Actual Completion**	**Overrun**	**Schedule growth**
Case 1	11/10/2012	16/11/2014	8/1/2016	625	81%
Case 2	2/5/2015	16/3/2016	incomplete	622 (at time of writing)	123%

Table 3 presents the schedule overrun experienced in each case study. Both case studies experienced schedule growth far in excess of that found in literature. In addition to the greater schedule growth when compared to literature, design & build was also found to experience larger schedule growth than the traditional method. These discrepancies with literature were then analyzed further in conjunction with interview data in order to determine the cause. A detailed breakdown was then produced and illustrated in Table 4.

Table 4. Project delivery method impact on delays.

Delay type	Case 1 (Traditional)	Case 2 (Design & build)	Design & build delay impact
Civil	Significant	Mild	Reduction
Variation Orders	Significant	Mild	Reduction
Consultant	Mild	Negligible	Reduction
Design Changes	Mild/Negligible	Mild/Negligible	No effect
Contractor	Mild/significant	Mild/significant	No effect
Client	Mild	Significant	Large increase

Negligible <= 1 week   Mild 0-3 Months   Significant > 3 Months

Table 4 presents the impact of the project delivery method on delays. It is clear that design and build was able to reduce the effect of civil delays. This was in part due to the fact the contractor was able to choose their own trusted sub-contractor to perform the work. The civil sub-contractor was under the direct employ of the design & build contractor. This meant the design & build contractor was able to apply considerable pressure to speed up work and reduce potential delays due to design changes. The impact of variation orders was reduced from significant to mild. This was due to the ability of design & build to provide a more complete design up front, resulting in fewer changes during the project. In addition, the need for formal documentation and approvals was reduced because the design & build contractor was fully responsible for the design. Approvals for changes from an external consultant were not required allowing changes to be made without approval delays. The consultant delays were reduced from mild to negligible, largely due to the elimination of the need for official RFI's. The design process was more fluid under the design & build method as the consultant was in-house to the design & build contractor reducing the need for formal documentation and increasing cooperation.

No change in impact on design changes was found between case 1 and case 2. The impact on delays due to the contractor was also found to have no impact across delivery methods. However, it is possible this was due to the same contractor being used in both case studies. Perhaps the single most important finding is the increase in delays experienced due to client related factors. Under the traditional project delivery method, a consultant is present to act as the client agent and make decisions on their behalf and manage the construction process. Under the design & build delivery method the client has no agent to support them in decision making and management of the project. The design & build contractor now has the ability to make a large percentage of these decisions for the client, however, the client is still required to make some decisions. This places an extra burden on the client's resources to make these decisions and plan accordingly.

Furthermore, significant delays were experienced due to the inability of the client to complete work within their scope on time. The costliest decision by the client was to not award a variation to the design & build contractor. This decision, along with having work in the client's scope removed the single point of responsibility shown to be design & builds greatest advantage. The result of these decisions led to the projects most significant delays and ultimately caused the project to run vastly over schedule.

4 CONCLUSIONS

The aim of the research was to determine how the project delivery method effects the prevalence and impact of delays with goal of proving and reduction though the design & build method. The results of this research found that design and build was able to reduce the impact of delays related to: Civil, Variation Orders and Consultant.

However, while design and build was able to reduce the impact of delays in three areas, it also increased the impact of delays related to the project client. Importantly, this increase in impact of client related delays was so extensive that it negated the reduction in impact experienced in the three areas. As a result, design and build experienced 42% more schedule growth than its traditional counterpart. The design and build delivery method does not necessarily reduce delays on construction projects, however, the design and build project studied was not able to maintain the single point of responsibility required for successful implementation. Therefore, further research is suggested using a properly implemented design and build project.

References

Chan, D. W., Chan, A. P., Lam, P. T., Yeung, J. F., and Chan, J. H., Risk Ranking and Analysis in Target Cost Contracts: Empirical Evidence from The Construction Industry, *International Journal of Project Management*, 29(6), 751-763, 2011.

Davis, P., Love, P., and Baccarini, D., Building Procurement Methods, *CRC Construction Innovation*, 8–10, June 2008.

Design Build Institution of America, Choosing a Project Delivery Method, A Design-Build Done Right Primer, 2–3, April 2015.

Grobler, K. and Pretorius, L., An Evaluation of Design-Build as Procurement Method for Building and Civil Engineering Projects in South Africa, *Journal of the South African Institution of Civil Engineering*, 44(2), 13-19, Jan 2002.

Hale, D. R., Shrestha, P. P., Gibson Jr, G. E., and Migliaccio, G. C. Empirical comparison of design/build and design/bid/build project delivery methods. *Journal of Construction Engineering and Management*, 135(7), 579–587. 2009.

Ibbs, C. W., Kwak, Y. H., Ng, T., and Odabasi, A. M., Project Delivery Systems and Project Change: Quantitative Analysis, *Journal of Construction Engineering and Management*, 129(4), 382-387, 2003.

Konchar, M. and Sanvido, V., Comparison of US project Delivery Systems, *Journal of Construction Engineering and Management*, 124(6), 435-444. 1998.

Kumaraswamy, M. M. and Dissanayaka, S. M., Developing a Decision Support System for Building Project Procurement, *Building and Environment*, 36(3), 337–349, 2001.

Love, E. P., O'Donoghue, D., Davis, R. P., and Smith, J., Procurement of Public Sector Facilities: Views of Early Contractor Involvement, *Facilities*, 32(9/10), 460-471, 2014.

Mahdi, I. M. and Alreshaid, K., Decision Support System for Selecting the Proper Project Delivery Method Using Analytical Hierarchy Process (AHP), *International Journal of Project Management*, 23(7), 564–572, 2005.

Mosey, D., *Design and Build in Action*, Chartridge Books Oxford, 1998.

Ng, S. T., Performance of Engineering Consultants in ISO 9000-based Quality Management Systems implementation, Engineering, *Construction and Architectural Management*, Emerald Group Publishing Limited, 12(6), 519–532, 2005.

Oyegoke, A. S., Dickinson, M., Khalfan, M. M., McDermott, P., and Rowlinson, S., Construction Project Procurement Routes: An In-Depth Critique, *International Journal of Managing Projects in Business*, 2(3), 338-354, 2009.

Shane, J., Strong, K., Gransberg, D., et al. Project Management Strategies for Complex Projects, Project R-10 final report, Strategic Highway Research Program 2, Transportation Research Board, National Academies, Washington, D.C., 199pp. 2013.

Shrestha, P. P., O'Connor, J. T., and Gibson, G. E., Jr. Performance comparison of large design-build and design-bid-build highway projects. *Journal of Construction Engineering and Management*, 138(1), 1-13. 2012.

Streamlining Information Transfer between Construction and Structural Engineering
Edited by Shiau, J., Vimonsatit, V., Yazdani, S., and Singh, A.

ISBN: 978-0-9960437-7-9

FINANCIAL PROVISION FOR CONSTRUCTION HEALTH AND SAFETY

ROXY MALAN[1] and JOHN SMALLWOOD[2]

[1]*Dept of Quantity Surveying, Nelson Mandela University, Port Elizabeth, South Africa*
[2]*Dept of Construction Management, Nelson Mandela University, Port Elizabeth, South Africa*

South African Health and Safety (H&S) legislation and the Construction Regulations require wide ranging H&S interventions on the part of all stakeholders including ensuring that adequate financial allowance is made for H&S. Given the afore mentioned, a study was conducted to determine the perceptions of general contractors (GCs) and quantity surveyors (QSs) with respect to financial provision for construction H&S. A survey of literature informed the development of the survey instrument in the form of a questionnaire, which was administered among GCs and QSs. The salient findings include: H&S preliminaries should be included in Bills of Quantities, and the inclusion of thirty-nine items therein is perceived as more than important, as opposed to less than important. It is concluded that a scientific approach through the inclusion of detailed H&S Preliminaries items will: "level the playing fields"; engender scientific calculation of the costs related thereto, and enable clients, with the assistance of QSs, to ensure that principal contractors (PCs) have made adequate financial provision for H&S in their tenders. Recommendations include: H&S preliminaries should be included in the BoQ based on the project specific H&S specification provided by the client, and a CHSA should be appointed in the early stages of projects to assist the QS with respect to interpreting the H&S specification when compiling H&S items for the BoQ.

Keywords: Contract documentation, Construction industry development board (CIDB), Risk assessment, Cost of accidents.

1 INTRODUCTION

The Construction Industry Development Board (CIDB) (2009) report *Construction Health & Safety Status & Recommendations* highlighted the considerable number of accidents, fatalities, and other injuries that occur in the South African construction industry. The report cited the high-level of non-compliance with H&S legislative requirements, which is indicative of a deficiency of effective management and supervision of H&S on construction sites as well as planning from the inception / conception of projects within the context of project management.

The CIDB's report indicates the disabling injury incidence rate (DIIR) to be 0.98 i.e. 0.98 disabling injuries per 100 workers, the all industry average being 0.78, and a fatality rate of 25.5 per 100 000 workers, which does not compare favorably with international rates (CIDB 2009). The Australian construction industry fatality rate is 3.29 for all construction (Safe Work Australia, 2017), and 1.94 in the United Kingdom (Health and Safety Executive (HSE) (2016). The severity rate (SR) in turn indicates the number of days lost due to accidents for every 1 000 hours worked. The South African construction industry SR 1.14 is the fourth highest, after fishing, mining, and transport, the all industry average being 0.59. Given that the average worker works 2000 hours per year, if the SR is multiplied by two, the average number of days lost per worker per year can be computed – the construction industry lost 2.28 working days per worker. This is equivalent to 1.0% of working time.

The CIDB (2009) refer to Smallwood's 2004 findings that the total cost of accidents (COA) could have been between 4.3% and 5.4%, based upon the value of construction work completed in South Africa. The key issue relative to the COA is that ultimately, clients incur the COA as the COA is included in contractors' cost structures in the form of indirect costs, as contractors do not disaggregate costs when preparing tenders.

Given the aforementioned, a study named *The Introduction of H&S Preliminaries in the Eastern Cape Province* was conducted among GCs and QSs to determine perceptions, problems, and practices, and the potential value of introducing H&S preliminaries in order to assist contractors in terms of making adequate financial provision for H&S.

2 REVIEW OF THE LITERATURE

2.1 Legislation and Regulations

The Construction Regulations (2014) of the Republic of South Africa state that the client must provide designers with an H&S specification, and the designers need to submit a report to the client before the client provides the H&S specification to the principal contractor (PC) when the project goes out to tender. This report should include, *inter alia*, all relevant H&S information with respect to the design of the relevant structure that may affect the pricing of the construction work. The client who is required to provide the PC with an H&S specification, is also required to ensure that the PC has made adequate financial allowance for H&S. The latter is cascaded down the supply chain, as the PC must ensure that potential contractors submitting tenders have made provision for the cost of health and safety measures during the construction process.

2.2 Challenges with Respect to Financial Provision for H&S

According to Smallwood (2011), due to competitive tendering, contractors need to limit costs, and consequently, the winning tender is unlikely to make adequate provision for H&S equipment, welfare facilities, and a healthy and safe working environment.

Smallwood and Emuze (2014) required respondents to a study to indicate their concurrence with respect to a range of statements. A mean score between 1.00 and 5.00 is recorded in parentheses after the statements: Competitive tendering without reference to H&S marginalizes H&S (4.09), contractors are afforded the opportunity to price H&S on an equitable basis (2.36), and contractors are afforded the opportunity to price items included in H&S specifications on an equitable basis (2.36).

2.3 Standard Contract Documentation

The most widely used form of contract for construction in South Africa is the Joint Building Contracts Committee (JBCC) Principal Building Agreement, the latest addition being Edition 6.1 – March 2014. Other forms of contracts include the General Conditions of Contract (GCC), International Federation of Consulting Engineers (FIDIC), and the New Engineering Contract (NEC). The GCC does not make any explicit reference to H&S, other than "reporting of accidents." The FIDIC and NEC contracts originated overseas, and therefore provide conflicting clauses in terms of the H&S legislation in South Africa.

According to the CIDB (2009), the JBCC does not make any explicit reference to H&S, but does refer to the need for parties to comply with laws and regulations that govern the work that needs to be executed. According to Smallwood and Emuze (2014), scope exists for the standard forms of contract to include more direct reference to construction H&S.

2.4 Form of Financial Provision for H&S

Wells and Hawkins (2009) state that to avoid misunderstanding of what is required and to facilitate the checking of contractors' financial provision for H&S, it is recommended that H&S items that can be separately priced be listed as prime cost items, provisional sums, or the use of another form of pricing mechanism. They provide examples of items that can be addressed in such a manner, namely the preparation and updating of an H&S plan; provision of temporary works such as scaffolding and hoarding, H&S Officer, H&S training, attendance of H&S Committee meetings, provision of welfare facilities, provision of personal protective equipment (PPE), and medical examinations. Wells and Hawkins (2009) also state that it is possible, and may be considered preferable, to take the cost of meeting the client's H&S requirements out of competition, by pre-pricing H&S items, and they cite the approach adopted in Hong Kong in 1996 under the 'Pay for Safety' scheme.

The study conducted by Smallwood and Emuze (2014) determined the concurrence with respect to a further range of statements as: contract document enabled financial provision for H&S promotes H&S (4.36); a detailed H&S section should be included in the Preliminaries (4.27), and a provisional sum should be provided for H&S in the preliminaries (3.64).

3 RESEARCH

3.1 Research Method and Sample Stratum

The quantitative component of the study reported on, targeted two sample strata, namely members of the Association of South African Quantity Surveyors (ASAQS) East Cape Chapter, QSs, and GC members of the East Cape Master Builders Association (ECMBA). 12 responses were received from the 77 ASAQS sample, which equates to a response rate of 15.6%. 14 responses were received from the 58 ECMBA sample, which equates to a response rate of 24.1%.

92.8% of the GC respondents were male, and 7.2% were female, while 91.6% of the QS respondents were male, and 8.4%m were female.

Most of the respondents are over 44 years of age – 61.6% of the GC respondents and 75% of the QS respondents.

In terms of qualifications, 53.8% of GC respondents possess a diploma, and 30.8% an Honors degree. 75% of QS respondents possess an Honors degree, and 25% a Masters or Doctoral degree.

69.2% of GC respondents, and 66.7% of QS respondents have been working in the industry for more than 18 years.

3.2 Research Findings

Respondents were required to indicate the extent to which they agreed with the introduction of H&S preliminaries in Bills of Quantities on a scale of 1 (strongly disagree) to 5 (strongly agree). The resultant GC mean score (MS) between 1.00 and 5.00, namely 4.43, indicates that the agreement is between strongly agree to agree / agree. The resultant QS MS, namely 3.75 indicates that the agreement is between neutral to agree / agree.

Table 2 presents the importance of the inclusion of 39 items in an H&S Preliminaries section to contractors and QSs in terms of MS based on percentage responses to a scale of 1 (not at all) to 7 (extremely). It is notable that all the MSs are > 4.00, the midpoint of the MS range, which indicates that their inclusion is more than important as opposed to less than important.

However, a summary of the MS ranges in the form of Table 1 provides further insight. MSs > 6.14 ≤ 7.00 indicate that the inclusion is between very to extremely / extremely important: contractors (20 No. = 51.3%); QSs (5 No. = 12.9%), and mean (9 No. = 23.1%). MSs > 5.28 ≤ 6.14 indicate that the inclusion is between more than important to very / very important: contractors (17 No. = 43.6%); QSs (22 No. = 56.4%), and mean (22 No. = 56.4%). MSs > 4.42 ≤ 5.28 indicate that the inclusion is between important to more than important / more than important: contractors (2 No. = 5.1 %); QSs (10 No. = 25.6 %), and mean (8 No. = 20.5%). MSs > 3.56 ≤ 4.42 indicate that the inclusion is between less than important to important / important: contractors (0 No. = 0.0%); QSs (2 No. = 5.1%), and mean (0 No. = 0.0%). Based upon the GCs rating, 37 of the items are in the upper two ranges, whereas in the case of the QSs, 27 are. Furthermore, the mean MS for the 39 items is 6.05 in the case of the GCs, and 5.48 in the case of the QSs.

Table 1. Summary of the perceived importance of the inclusion of items in an H&S Preliminaries section.

Range	**GCs**		**QSs**		**Mean**	
	No.	**%**	**No.**	**%**	**No.**	**%**
> 6.14 ≤ 7.00	20	51.3	5	12.9	9	23.1
> 5.28 ≤ 6.14	17	43.6	22	56.4	22	56.4
> 4.42 ≤ 5.28	2	5.1	10	25.6	8	20.5
> 3.56 ≤ 4.42	0	0.0	2	5.1	0	0.0
Total	39	100.0	39	100.0	39	100.0

It is now appropriate to discuss the individual items in Table 2. In terms of the MS range > 6.14 ≤ 7.00 (very to extremely / extremely important), which includes nine items, it is notable that first aid is ranked joint first with H&S plan. First aid is primarily a labor cost in that a first aider is invariably required on most projects. The development of a H&S plan is a time-consuming activity, and may require the engagement of a consultant. PPE (3rd) is a major cost, especially personal fall arrest equipment, and special respiratory equipment. Hoarding and / or public walkways (joint 4th), and guarding and barricading (joint 7th) are related, and invariably entail major cost. Storage for flammable goods – Regulation 27 (b) (joint 4th) and fire precautions (joint 7th) also entail major cost, and in the case of the latter, servicing. Risk assessment (6th) is an integral part of managing construction and construction H&S, however, contractors may engage consultants to assist.

The MS range > 5.28 ≤ 6.14 (important to very / very important) includes twenty-two items. Both temporary electrical installations (10th) and special scaffolding (11th) entail major cost. However, special scaffolding, engineering design and / or certification (joint 12th), suspended scaffolding (joint 14th), design of temporary works (joint 18th), and design of permanent structures (joint 24th) invariably require specialist designer contributions and thus entail major expense. Allowance for H&S Audits (joint twelfth) may have been ranked such due to contractors engaging consultants to assist, even though H&S audits are a client function. H&S Representatives (joint 14th) are elected from the workforce, and therefore when they conduct inspections, attend meetings, and participate in investigations. WCs (joint 14th) are a key welfare facility and a major cost, and are linked to WHBs (20th), and showers joint 30th. The H&S file (17th), which must be handed over to the client by the principal contractor or GC, entails a range of H&S

documentation, and therefore there are labor and printing costs related thereto. SWPs and method statements (joint 18th) may require contractors to engage consultants to assist. Access and catch platforms ranked joint 21st entail substantial cost. Signage (23rd), education and training (joint 24th), meetings (26th), and inspections (27th) are legal requirements. Medicals (28th) may be required due to varying circumstances, including regulations, and are a major cost item. A full-time H&S Officer (29th) has financial implications, and transport of workers (joint 30th) is an issue as inappropriate transport contributes to the high motor vehicle accident (MVA) fatality rate.

Table 2. Perceived importance of the inclusion of items in an H&S Preliminaries section.

Item	GCs		QSs		Mean	
	MS	Rank	MS	Rank	MS	Rank
First aid	6.77	1	6.44	1	6.61	1=
H&S plan	6.69	2	6.33	2	6.51	1=
Personal protective equipment (PPE)	6.62	3	6.33	2	6.48	3
Hoarding and / or public walkways	6.54	4	6.22	4	6.38	4=
Storage for flammable goods – Regulation 27 (b)	6.54	4	6.22	4	6.38	4=
Risk assessment	6.46	6	6.11	6	6.29	6
Fire precautions	6.46	6	6.00	7	6.23	7=
Guarding and barricading	6.46	6	6.00	7	6.23	7=
Scaffolding	6.46	6	5.89	9	6.18	9
Temporary electrical installations	6.38	10	5.78	10	6.08	10
Special scaffolding	6.31	11	5.78	10	6.05	11
Allowance for H&S Audits	6.31	11	5.67	12	5.99	12=
Engineering design and / or certification	6.31	11	5.67	12	5.99	12=
H&S Representatives	6.23	14	5.67	12	5.95	14=
Suspended scaffolding	6.23	14	5.67	12	5.95	14=
WCs	6.23	14	5.67	12	5.95	14=
H&S file	6.17	17	5.56	17	5.87	17
Design of temporary works	6.15	18	5.56	17	5.86	18=
SWPs and method statements	6.15	18	5.56	17	5.86	18=
WHBs	6.15	18	5.56	17	5.86	20
Access	6.08	21	5.56	17	5.82	21=
Catch platforms	6.08	21	5.56	17	5.82	21=
Signage	6.00	23	5.56	17	5.78	23
Design of permanent structures	6.00	23	5.33	24	5.67	24=
Education and training	6.00	23	5.33	24	5.67	24=
Meetings	5.85	27	5.33	24	5.59	26
Inspections	5.83	28	5.33	24	5.58	27
Medicals	5.82	29	5.22	28	5.52	28
Full-time H&S Officer	6.00	23	5.00	31	5.50	29
Showers	5.69	30	5.22	28	5.46	30=
Transport of workers	5.69	30	5.22	28	5.46	30=
Housekeeping	5.62	32	4.89	32	5.26	32=
Maintenance	5.62	32	4.89	32	5.26	32=
General administration	5.62	32	4.78	34	5.20	34
Storage	5.38	35	4.78	34	5.08	35
Environmental measurement	5.38	35	4.67	36	5.03	36
Mess room	5.31	37	4.67	36	4.99	37
Biological monitoring	5.23	38	4.33	38	4.78	38
Living accommodation	5.08	39	4.22	39	4.65	39
	6.05		5.48		5.76	

4 CONCLUSIONS

Based upon the extent to which GCs and QSs agree with the introduction of H&S preliminaries in Bills of Quantities, it can be concluded that the support therefore is not constrained to the contracting sector of the industry. Given the degree of importance of the inclusion of thirty-nine items in an H&S Preliminaries section it can be concluded that such a section should be comprehensive, and that a provisional sum or a single item will not suffice.

Furthermore, a scientific approach through the inclusion of detailed H&S Preliminaries items will: 'level the playing fields'; engender scientific calculation of the costs related thereto, and enable clients, with the assistance of QSs, to ensure that principal contractors (PCs) have made adequate financial provision for H&S in their tenders.

5 RECOMMENDATIONS

H&S preliminaries should be included in the BoQ based on the project specific H&S specification provided by the client. A CHSA should be appointed at the first stage (project initiation and briefing), or at the latest, the second stage (concept and feasibility) of a project to assist the QS with respect to interpreting the H&S specification when compiling H&S items for the BoQ. The relevant associations, and more specifically the JBCC committee, need to consider the inclusion of H&S preliminaries to keep abreast with regulations relating to H&S. Associations representing each profession, namely the MBAs and ASAQS, need to inform members of their responsibilities with respect to the Construction Regulations (2014).

Comprehensive research should be conducted by the ASAQS at national level to determine current QS practices relative to facilitating financial provision for H&S by contractors. Furthermore, such research should determine the percentage such financial provision constitutes of project value for the various sectors and types of construction taking a range of characteristics such as stories, and floor area per story into account.

References

Construction Industry Development Board (CIDB), *Construction Health & Safety in South Africa Status & Recommendations*, CIDB, Pretoria, The Republic of South Africa, 2009.

Health and Safety Executive (HSE), *Statistics on Fatal Injuries in the Workplace Great Britain 2016*, HSE, 2016.

Safe Work Australia, *Key Work Health and Safety Statistics Australia 2017*, Safe Work Australia, Canberra, Australia, 2017.

Smallwood, J. J., Financial Provision for Health and Safety (H&S) in Construction, in *Proceedings of the CIB W099 Conference Prevention: Means to the End of Safety and Health Injuries, Illness and Fatalities,* Washington D.C., USA, August 24-26, 2011.

Smallwood, J. J. and Emuze, F., Financial Provision for Construction Health and Safety (H&S), in *Proceedings of the Construction Research Congress*, 1881-1890, Atlanta, Georgia, USA, May 19-21, 2014.

The Construction Regulations, *Government Gazette No. 37305*, Pretoria, The Republic of South Africa, 2014.

Wells, J. and Hawkins, J., *Promoting Construction Health and Safety through Procurement: A Briefing Note for Developing Countries*, Engineers Against Poverty, Institution of Civil Engineers (ICE), London, 2009.

Streamlining Information Transfer between Construction and Structural Engineering
Edited by Shiau, J., Vimonsatit, V., Yazdani, S., and Singh, A.

ISBN: 978-0-9960437-7-9

PLANT, ROAD CONSTRUCTION, TRAFFIC, AND GENERAL HEALTH AND SAFETY: PERSPECTIVES OF GENERAL CONTRACTORS

JOHN SMALLWOOD[1] and LANCE WENTZEL[2]

[1]*Dept of Construction Management, Nelson Mandela University, Port Elizabeth, South Africa*
[2]*Dept of Construction Management and Quantity Surveying, Cape Peninsula University of Technology, Cape Town, South Africa*

Statistics and various studies indicate that plant, road construction, and traffic safety are major issues in terms of construction H&S. Furthermore, motor vehicle accidents (MVAs) during employment are the dominating cause of fatalities in South African construction. A study was conducted among general contractor (GC) members of a civil engineering sector employer association using a self-administered questionnaire. The salient findings include: plant, road construction, and traffic safety are important aspects in respondents' organizations; plant is the more important of the three aspects; traffic accidents account for a fair percentage of total accidents, and respondents' organizations use a range of such devices and equipment to realize plant, road construction, and traffic safety. Conclusions include: respondents' organizations are managing the three aspects, and H&S programs must focus thereon. Recommendations include that respondents' organizations need to focus on drug and other substance abuse testing and medicals, and that plant, road construction, and traffic safety must feature prominently in H&S programs.

Keywords: Motor vehicle accidents, Construction industry development board, Market and social influences, Injury prevention, Construction road signs.

1 INTRODUCTION

Motor vehicle accidents (MVAs) during employment are the dominating cause of fatalities in South African construction, contributing 47% of fatalities (Construction Industry Development Board (CIDB) 2009). In 2017, 43 / 69 (62.3%) of fatalities were attributable to MVAs during employment in construction (Federated Employers Assurance (FEM) 2018). "Struck by" accidents, which would include struck by plant, contributed 8.7% of fatalities. Furthermore, Sinclair (2010) states that although South Africa boasts a well-deserved reputation for having some of the finest roads on the continent, road construction and traffic safety remains a dominant problem within the South African and worldwide construction industry.

According to the Center for Construction Research and Training (CPWR) (2018), road construction site fatalities accounted for approximately 9% of all construction fatalities between 2003 and 2016 in the United States of America (USA), of which approximately 50% were due to being struck by a vehicle or mobile equipment.

However, the Health and Safety Executive (HSE) emphatically states that, "Construction vehicle accidents can be prevented by the effective management of transport operations throughout the construction process."

Given the aforementioned, a study named *Plant, Road Construction and Traffic Safety* was conducted among GCs to determine perceptions, problems, practices, and injury experience.

2 REVIEW OF THE LITERATURE

2.1 Road Construction and Traffic Safety

Sinclair (2010) postulates that road construction by its very nature represents some of the highest risk work environments of all. These are roads which demand a great deal from the driver - attention, consideration and compliance with traffic regulations in an unforgiving road context. The efforts to improve safety on roads ironically create road construction and traffic safety challenges in the short-term; this is the paradox that countries around the world are dealing with daily (Sinclair 2010). According to Arrive Alive (2012), road construction zones present a deadly hazard for workers, motorists, and pedestrians. This hazard is brought about by high speed limits, impatient drivers, and widespread traffic congestion. Furthermore, heat, driving stress, and long stretches of highway under construction – create a recipe for extreme driving hazards for motorists and road workers alike.

Construction zones are also called "cone zones" - those portions of the highways marked by cones, barrels, and signs where road construction is taking place (Arrive Alive 2012). Some work zones might have a speed-monitoring device to alert motorists of their speed prior to entering the work zone. Even though they are marked and signposted as areas where motorists must slow down and drive with extra caution, many drivers speed up to get through the construction area as quickly as possible, which in its totality produces poor road construction traffic safety (Arrive Alive 2012).

2.2 Factors That Influence Plant, Road Construction, and Traffic Safety

According to BOMEL Limited (2004), there are four major influences in terms of the causes of plant, road construction and traffic accidents within the construction industry. These are: direct influences, which directly influence the likelihood of an accident being caused; organizational influences, which influence and reflect the culture, procedures, and behavior promulgated by the organization; policy level influences, which reflect the expectations of the decision makers in the employers of those at risk and the organizations they interface with e.g. clients, suppliers, and subcontractors, and environmental influences, which cover the wider political, regulatory, market, and social influences which impact the policy influences.

The HSE (2009) contend that most construction transport accidents result from the inadequate segregation of pedestrians and vehicles. This can be mitigated by careful planning, particularly at the design stage, and by controlling vehicle operations during construction work. Inadequate planning and control are the root cause of many construction vehicle accidents, which often involve: vehicles or people being struck by their loads, particularly when reversing; vehicles striking services and obstructions; disregard of manufacturers' instructions for safe use; inadequate training of drivers and signalers, and unsafe loading and transportation of materials on vehicles.

2.3 Preventing Plant Related Accidents

The HSE (2009) advocate the following to prevent plant related incidents: detailed planning; hazard identification and risk assessment; public protection; establishment of separate pedestrian and vehicle routes; the deployment of signalers; dedicated loading and storage areas; information and warning; appropriate vehicle selection; vehicle inspection and maintenance; safe driving and work practices; circumspect loading of vehicles, and competent experienced drivers / operators, which are not under the influence of alcohol or drugs.

2.4 Preventing Injuries at Road Construction Sites

The CPWR (2018) advocate the following categories of solutions to prevent injuries at road construction sites: increased visibility; positive barriers between workers and traffic; warning systems for drivers; warning systems for workers; speed reduction systems; impact attenuators; engineering controls; separate workers on foot from plant and equipment; raise awareness, and equipment safety and rollover prevention.

3 RESEARCH

3.1 Research Method and Sample Stratum

A study named *Plant, Road Construction, and Traffic Safety* was conducted among general contractor (GC) members of a civil engineering sector employer association using a self-administered questionnaire. 55 Responses were included in the analysis of the data, which entailed the computation of frequencies, and a measure of central tendency in the form of a mean score (MS) to enable the comparison of responses to Likert type scale questions, and the ranking of variables related thereto.

3.2 Research Findings

Table 1 indicates the importance of plant, road construction, traffic, and general safety in respondents' organizations in terms of percentage responses to a scale of 1 (not important) to 5 (very important), and a mean score (MS) ranging between 1.00 and 5.00. It is notable that all the MSs are above the midpoint of 3.00, which indicates that in general, the respondents perceive plant, road construction, traffic, and general safety in to be more than important as opposed to less than important. It is notable that 3 / 4 (75%) MSs are $> 4.20 \leq 5.00$, which indicates that plant, general, and traffic safety are between more than important to very important / very important. Road construction safety's MS is marginally below the cut-point of the former range.

Table 1. The importance of plant, road construction, traffic, and general safety in respondents' organizations.

Type	Response (%)						MS	Rank
	Un-	Not ....Very						
	sure	1	2	3	4	5		
Plant	0.0	3.6	3.6	0.0	16.1	76.8	4.59	1
General	1.8	3.6	0.0	12.7	23.6	58.2	4.35	2
Traffic	1.8	7.1	0.0	12.5	14.3	64.3	4.31	3
Road construction	3.6	10.7	3.6	7.1	10.7	64.3	4.19	4

Table 2 indicates the extent to which plant, road construction, traffic, and general safety constitute a problem in respondents' organizations in terms of percentage responses to a scale of 1 (does not) to 5 (major), and a MS ranging between 1.00 and 5.00. It is notable that all the MSs are below the midpoint of 3.00, which indicates that in general, the respondents perceive plant, road construction, traffic, and general safety to constitute a minor as opposed to a major problem. However, in terms of ranges, all the MSs are > 1.80 ≤ 2.60, which indicates that they constitute between does not to less than a problem / less than a problem.

Table 2. The extent to which plant, road construction, traffic, and general safety constitute a problem in respondents' organizations.

Type	Response (%)						MS	Rank
	Don't know	Does not..Major						
		1	2	3	4	5		
Traffic	0.0	27.3	23.6	25.5	16.4	7.3	2.53	1
General	5.4	23.2	35.7	30.4	3.6	1.8	2.21	2
Road construction	3.6	33.9	30.4	17.9	7.1	7.1	2.20	3
Plant	0.0	25.0	44.6	25.0	0.0	5.4	2.16	4

Table 3 indicates the extent to which respondents agree with plant, road construction, and traffic safety related statements in terms of percentage responses to a scale of 1 (does not) to 5 (major), and a MS ranging between 1.00 and 5.00. It is notable that 4 / 6 (66.7%) MSs are > 3.00, which in general indicates agreement, and that 2 / 6 (33.3%) MSs are < 3.00, which in general indicates disagreement. However, in terms of ranges, 3 / 6 (50%) MSs are > 4.20 ≤ 5.00, which indicates that respondents agree to strongly agree / strongly agree that plant, road construction, and traffic safety are important components of H&S. In terms of accidents relative to the respective categories accounting for a fair percentage of total accidents, the agreement with respect to traffic is between neutral to agree / agree as the MS of 3.55 is > 3.40 ≤ 4.20. In the case of road construction, and plant operating, the agreement is between disagree to neutral / neutral as the MSs of 2.64, and 2.70 respectively are > 2.60 ≤ 3.40.

Table 3. Extent to which respondents agree with plant, road construction, and traffic safety related statements.

Statement	Response (%)						MS
	Don't know	Strongly disagree	Disagree	Neutral	Agree	Strongly agree	
Plant safety is an important component of H&S	0.0	0.0	1.8	0.0	36.4	61.8	4.58
Traffic safety is an important component of H&S	1.8	0.0	0.0	3.6	29.1	65.5	4.56
Road construction safety is an important component of H&S	1.8	0.0	0.0	9.1	29.1	60.0	4.45
Traffic accidents account for a fair percentage of total accidents	1.8	3.6	14.3	21.4	37.5	21.4	3.55
Plant operating accidents account for a fair percentage of total accidents	7.1	10.7	26.8	28.6	21.4	5.4	2.70
Road construction accidents account for a fair percentage of total accidents	10.7	8.9	28.6	25.0	21.4	5.4	2.64

In terms of the extent to which respondents' organizations conduct medicals, training, and drug and other substance abuse testing, it is notable that operator training (89.1%) is ranked first, followed by drug and other substance abuse testing (41.8%), operator in-employment medicals (31.5%), and operator pre-employment medicals (22.2%).

Table 4 indicates the extent to which various issues constitute a problem in respondents' organizations and on projects, in terms of percentage responses to a scale of 1 (not) to 5 (major), and a MS ranging from 1.00 to 5.00. It is notable that only 3 / 11 MSs are > 3.00, which indicates the extent is major as opposed to minor. It is notable that no MSs are > 4.20 ≤ 5.00 – a near major to a major / major extent. Only 1 / 11 (9.9%) MSs is > 3.40 ≤ 4.20, namely speeding by public during work on or adjacent to existing roads, which indicates the frequency is between some extent to a near major extent / near major extent. Only 3 / 11 (%) of the MSs are > 2.60 ≤ 3.40, which indicates that the extent is between near minor to some extent / some extent - poor road (public) surface conditions, poor road marking, and poor lighting. The remaining 7 / 11 MSs are 1.80 ≤ 2.60, which indicates the extent is between not to near minor / near minor.

Table 4. The extent to which issues constitute a problem in respondents' organizations and on projects.

Issue	Response (%)						MS	Rank
	Don't know	Not				Major		
		1	2	3	4	5		
Speeding by public during work on or adjacent to existing roads	0.0	3.6	3.6	14.3	35.7	42.9	4.11	1
Poor road (public) surface conditions	1.8	10.7	12.5	35.7	21.4	17.9	3.24	2
Poor road marking	1.8	14.3	19.6	21.4	28.6	14.3	3.09	3
Poor lighting	5.4	14.3	25.0	21.4	21.4	12.5	2.92	4
Inadequate operator / driver education	1.8	21.4	32.1	30.4	10.7	3.6	2.42	5
Operating / Driving while under the influence	3.6	33.9	30.4	25.0	3.6	3.6	2.10	6
Overloaded vehicles – Materials / Plant	1.8	37.5	30.4	26.8	1.8	1.8	1.98	7
Overloaded vehicles - Personnel	7.3	32.7	41.8	12.7	3.6	1.8	1.92	8
Operator / Driver fatigue	0.0	33.9	39.3	19.6	1.8	1.8	1.88	9
Unlicensed operators / drivers	0.0	50.0	19.6	23.2	7.1	0.0	1.87	10
Unroadworthy plant / vehicles	1.8	46.4	32.1	14.3	5.4	1.8	1.84	11

In terms of the extent of use of devices and equipment when respondents' organizations undertake any type of road construction, 7 / 8 devices / equipment were identified by more than 60% of respondents, 5 / 8 by more than 80% of respondents, and 2 / 8 (25%) by more than 90% of respondents. Construction road signs were identified by 100% of respondents, followed by chevron hazard plates (95.7%).

It is notable that in terms of accidents, MVAs recorded the highest mean (4.50) followed by general (3.12), plant operating (2.23), and road construction (1.48). However, in terms of injuries, general recorded the highest (1.51), followed by road construction (1.30), and jointly by MVAs and plant operating (0.98).

In terms of the extent to which respondents' organizations calculate various types of accident rates, disabling injuries (100%) predominates, followed by plant accidents per hours operated (88.2%), fatalities (65.3%), and traffic accidents per km travelled (62.7%).

4 CONCLUSIONS

The degree of importance of plant, traffic, and road construction safety relative to general H&S leads to the conclusion that they are important aspects in respondents' organizations, and in civil engineering construction H&S. This is underscored by the extent to which respondents agree that plant, road construction, and traffic safety are an important component of H&S. The relative importance of plant safety leads to the conclusion that plant is the more important of the three aspects.

The extent to which plant, road construction, traffic, and general safety constitute a problem in respondents' organizations leads to the conclusion that the respondents' organizations are managing such aspects.

The extent to which respondents agree that traffic accidents account for a fair percentage of total accidents underscores the findings of literature, and industry statistics. This leads to the conclusion that H&S programs must focus thereon.

The extent to which respondents' organizations conduct medicals, training, and drug and other substance abuse testing leads to the conclusion that they appreciate the importance of operator training, but need to focus on drug and other substance abuse testing and medicals.

The extent to which certain issues constitute a problem in respondents' organizations and on projects, namely speeding by the public during work on or adjacent to existing roads, poor road (public) surface conditions, poor road marking, and poor lighting, lead to the conclusion that traffic safety is an important issue. This is underscored by the literature.

The extent of use of devices and equipment when respondents' organizations undertake any type of road construction leads to the conclusion that they use a range of such devices and equipment to realize plant, road construction, and traffic safety, and general H&S.

The mean number of accidents and injuries that occurred in respondents' organizations highlights the importance of MVAs during employment as reflected in the literature, and the conclusion that traffic safety must feature prominently in H&S programs. The extent to which respondents' organizations calculate various types of accident rates, namely plant accidents per hours operated, and traffic accidents per km travelled, leads to the conclusion that they are measured because they are issues, which further motivates the importance of plant and traffic safety.

Future research should review the status quo relative to the issues reported on, and given a key finding, namely that traffic accidents account for a fair percentage of total accidents, traffic safety issues should be addressed in more detail.

References

Arrive Alive, *Road Safety Information, Education and Tips: Arrive Alive South Africa*, 2012. Retrieved from www.arrivealive.co.za on February 25, 2013.

BOMEL Limited, *Improving Health and Safety in Construction, Phase 2- Depth and Breadth: Volume 3 Construction Transport Accidents, Underlying Causes and Risk Control*, Health and Safety Executive, Norwich, United Kingdom, 2004.

Center for Construction Research and Training (CPWR), *Quarterly Data Report, Fatal Injuries at Road Construction Sites among Construction Workers*, Second Quarter, Washington, USA, 2018.

Construction Industry Development Board (CIDB), *Construction Health & Safety in South Africa Status & Recommendations*, CIDB, Pretoria, the Republic of South Africa, 2009.

Federated Employers Assurance (FEM), *Injuries per Injury Type for all MBA Regions from 2017 to 2018.* Retrieved from www.fem.co.za/Layer_SL/FEM_Home/FEM_Accident_Stats/FEM_Accident_Stats.htm on March 11, 2018.

Health and Safety Executive (HSE), *The Safe Use of Vehicles on Construction Sites*, HSE, Richmond, United Kingdom, 2009.

Sinclair, M., Vulnerability at Road Construction Sites, South Africa, in *Proceedings of the 29th Southern African Transport Conference*, 610-620, Pretoria, the Republic of South Africa, August, 16-19, 2010.

Streamlining Information Transfer between Construction and Structural Engineering
Edited by Shiau, J., Vimonsatit, V., Yazdani, S., and Singh, A.

ISBN: 978-0-9960437-7-9

ASSESSING LEVEL OF COMPLIANCE OF OCCUPATIONAL HEALTH AND SAFETY AMONG SMALL AND MEDIUM ENTERPRISES

NOKULUNGA MASHWAMA[1], DIDIBHUKU THWALA[2], and CLINTON AIGBAVBOA[2]

[1]*Dept of Construction Management and Quantity Surveying, University of Johannesburg, Johannesburg, South Africa*

[2]*Sustainable Human Settlement and Construction Research Centre, Faculty of Engineering and the Built Environment, University of Johannesburg, Johannesburg, South Africa*

Occupational Health and Safety (OH&S) is the responsibility of everyone at work, construction companies have to be aware that they are responsible for managing and improving issues related to H&S on construction sites. The data in this research was obtained from secondary and primary sources. 70 Questionnaires were dispersed to different companies in the construction industry, 42 were brought back and they were all valid and usable. This represented 60% response rate. The research revealed that the level of implementation was not practice well to the level of acceptance. Therefore, the study went on the assess the hindrance of the poor implementation among small medium enterprise's (SMEs) which were poor supervision, lack of regular audits and inspection, poor material and component, poor management commitment, lack of equipment and tools, lack of training and risk education, lack of communication, lack of skilled workforce, poor working condition, poor subcontractors involvement, lack of incentive for good performance, poor investigation and risk assessment. The research point out that safety and health of the workers is not negotiable, is all pervasive, increases productivity, and leads to better performance, improve the company image, reduced claims and accidents also reduce lost times spent on injuries and property loss. Emphases on the OHS level of implementation are greatly urged to be implemented non-stop to overcome the challenges facing the SME's contractors.

Keywords: Construction industry, Hindrances, SME, South Africa, Health and safety training.

1 INTRODUCTION

The primary objective of any occupational health and safety (OH&S) legislation is the prevention of accidents and their consequences in terms of injury, disablement and fatality, and ill health within the work environment (James *et al*. 2012). The achievement of this objective depends on good legislation supported by effective, sensible and accountable enforcement (Hsiao and Amstrong 2012). Worldwide, the construction industry contributes a large number of injuries (James *et al*. 2012). However, construction in developing countries, including South Africa, performs worse than construction in developed countries (Ramutloa 2007). Furthermore, there is a high level of non-compliance with H&S regulations in South Africa.

According to Mashwama *et al*. (2017a), the construction industry creates employment and serves as a means of generating income for many people and also as a means of application of technologies in various degrees. Furthermore, the sector engages about 240.000 workers in South Africa thus contributing significantly to the gross domestic product (GDP) (James *et al.* 2012). Nonetheless, the construction industry can be interlinked with the economy of a country and this often is an indicator of how healthy the state of the said economy is (James *et al.* 2011). Therefore, if the construction sector and the economy of a country are linked, it is necessary for the sector to effectively attend to the welfare of the workers in that industry (Brace and Gibb 2005). Moreover, the sector is associated with high risks particularly for workers nonetheless its importance in the social and economic development in South Africa (Brace and Gibb 2005, James *et al.* 2011). The construction industry has great number of injury and fatality among workers and this is substantiated by the (Chartered Institute of Building 2008) that many construction workers become unfit at one point or the other as a result of minor injuries or some kind of problems in their health.

Health and Safety (H&S) is the responsibility of everyone at work, construction companies have to be aware that they are responsible for managing and improving issues related to H&S on construction sites (Davies and Tomasin 2000). Adherence to regulations and legislation is not the only way to ensure safety on the working environment however it provides a basis for the employment and enforcement of good construction practices at minimal cost (Opaleye and Talukhaba 2014). Hence the study is to examine the level of implementation among SME's in-order to prevent future occurrences of poor implementation and discover the actual hindrance for SME's to implement. Therefore, health and safety is the responsibility of everyone at work on site.

2 OCCUPATIONAL HEALTH AND SAFETY IN SOUTH AFRICA

Worldwide construction industry contributes a large number of injuries (Department of Labour 2011). However, construction in developing countries, including South Africa, performs worse than construction in developed countries (Department of Labour 2011). Furthermore, there is a high level of non-compliance with H&S regulations in South Africa. According to Opaleye and Talukhaba (2014), poor performance in construction H&S is attributable to a lack of management commitment, inadequate supervision, and inadequate or lack of H&S training. Moreover, a lack of worker involvement, personal risk appreciation and work pressures also contribute to poor implementation of health and safety. Management and leadership at all levels are therefore important to improve construction H&S in South Africa. Leadership needs to manifest itself among all the stakeholders, commencing with clients and including project managers, designers, quantity surveyors, contractors, manufacturers and suppliers (Opaleye and Talukhaba 2014).

3 SMALL AND MEDUM ENTERPRISE (SME'S) CONTRACTORS IN SOUTH AFRICA

The growing importance of SMEs is clear, and SMEs currently account for about half of all people in formal employment in South Africa (Audet and Courteret 2012). SMEs in South Africa are alive, well and regard themselves as highly competitive. However, few SME owners give the government credit for their initiatives and efforts to promote SMEs (Fakoki 2010). According to Fakoki (2010), given the amount of resources being ploughed into SME development, the situation reflects poor communication rather than poor strategy. The survey also shows that less than half of the SME owners were positive about legislation, the impact of skills development programmes and the impact of BEE, the latter having more negative than positive scores (CIDB 2011). A low 12% positive rating was given for the impact of general government incentives for SMEs (Fakoki 2010, Lunsche 2010).

The lack of sufficient capital and credit is often a major handicap in the development of SMEs, especially in their growth stages (CIDB 2012, Jooste 2012). In developing countries worldwide, close to 95% of all SMEs have to rely solely on the personal resources and initiatives of the owners (Jooste 2012). The existence of a cheap source of labor, combined with the low socio-economic status of workers are major obstacles to improving OH&S as this limits workers' capacity to resist working under poor OH&S conditions on sites (Fakoki 2010, Lunsche 2010). Low skill levels linked with the adoption of labor intensive methods by the construction industry in South Africa and many other developing countries means managing OH&S can be challenging (Audet and Courteret 2012). It is clear that changing the attitudes of SMEs in construction must be handled to ensure workers' rights are not flagrantly abused in an industry setting where large numbers of the labor force are unskilled and largely uneducated (Audet and Courteret 2012). SMEs do not create better quality jobs than large enterprises, SMEs are usually created as a last resort and not as a first choice of employment and therefore have limited growth potential (Fakoki 2010).

4 METHODOLOGY

4.1 Research Approach and Design

Quantitative approach method was used to collect data for the study. A quantitative research is an enquiry into an identified problem, measured with numbers, and analysed using statistical techniques (Mashwama *et al.* 2017b). The goal is to determine whether the predictive generalizations of a theory hold true. Much research in the engineering and management sphere involve asking and obtaining answers to questions through conducting surveys of people by use of structured questionnaires, interviews and observations (Mashwama *et al.* 2017b). The study utilised questionnaires to collect data and answer the objective of the study which was to investigate the hindrances of implementation of health and safety among SME's in the Gauteng Province of South Africa. The study was carried out in Johannesburg area of South Africa, on small and medium enterprises (SME's) contractors. The target population were the owners and managers of SME's in Johannesburg Metropolitan Municipality. The reason for choosing owners and managers was that, they can offer useful information on the occupational health and safety challenges among SME's contractors in the construction industry of South Africa. 70 Questionnaires were distributed to owners and managers of SME's, 42 were brought back (15 owners and 27 managers). This reflected 60% response rate.

4.2 Analysis

In this study, the analysis employed simple statistical methodology, which is descriptive statistics (mean, mode, median, number, percentage, range, standard deviations). Respondent were required to respond to question based on the five point Likert scale. The mean item score was adopted to rank the factors from highest to lowest. The Mean Item Score (MIS) is expressed and calculated for each item as in Eq. (1):

$$MIS = \frac{1n1 + 2n2 + 3n3 + 4n4 + 5n5}{\sum N} \quad (1)$$

where:

$n1$ = *number of respondents for strongly disagree*
$n2$ = *number of respondents for disagree*

n3 = number of respondents for neutral
n4 = number of respondents for agree
n5 = number of respondents for strongly agree
N = Total number of respondents

5 FINDINGS AND DISCUSSIONS

5.1 Level of Implementation of OHS in SME's Contractors

Different construction sites were visited to check the level of compliance with OHS among the SME's, after inspectors issued notices to different site such as improvement notices, contravention notices and prohibition notices about 21% were non-compliant with OHS at about 10-20; 30-40 were non-compliant at 20%; followed by 20-30 were non-compliant at 18%; followed by 50-60 who were compliant at 16%, between 40-50 were non-compliant at 9%, 70-80 were compliant at 7%, 60-70 were compliant at 5%, 80-90 were compliant at 3% and between 90-100 were compliant at 1%.

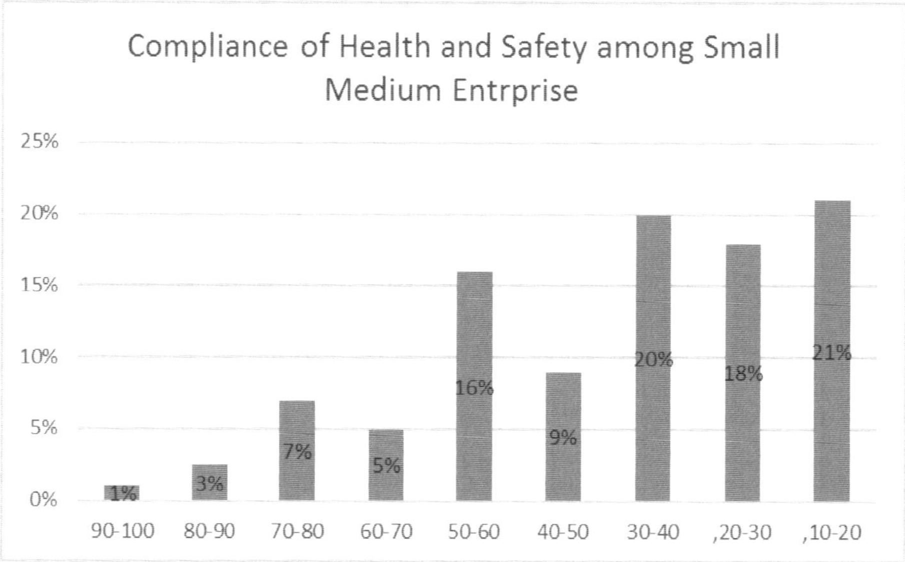

Figure 1. Compliance of OHS in SME's.

5.2 Hindrances of OHS Implementation by SME's

Respondent were asked to rank the factors that hinder the success implementation of OHS among SME's contractors. The responded rank lack of regular inspections and audits the highest with a MIS =4.40 and STD=0.964; poor management/supervision/information flow; Material and components was ranked 2 with a MIS=4.31 and STD=0.924 and 1.047 respectively; lack of management commitment was ranked 3 with MIS=4.29 and STD=0.970; Equipment and tools was ranked 4 with and MIS=4.21 and STD =0.976; lack of communication between workers was ranked 5th with MIS=4.17 and STD 0.853; poor employees involvement was ranked 6th with MIS=4.14 and STD=1.049; lack of training and risk education was ranked 7th, with MIS=4.12 and STD=1.109; lack of skilled workforce was ranked 8 with MIS=4.10 and STD=0.983; work area access was ranked 9th with MIS=4.02 and STD=1.024; poor work/jobsite conditions ranked 10th with MIS=3.98 and STD=0.924; subcontractors involvement was ranked 11 with MIS=3.95

and STD=1.147; lack of incentive for good performance was ranked 12 with MIS=3.93 and STD=1.135; investigations and risk assessments was ranked 13th with MIS=3.83 and STD=1.208.

Table 1. Implementation level.

Descriptive Statistics	MIS	Std. Deviation	Rank
Lack of regular inspections and audits	4.40	.964	1
Poor management/supervision / information flow	4.31	.924	2
Material and components	4.31	1.047	2
Lack of management commitment	4.29	.970	3
Equipment and tools	4.21	.976	4
Lack of good communication between workers	4.17	.853	5
Poor employees involvement	4.14	1.049	6
Lack of training and risk education	4.12	1.109	7
Lack of skilled workforce	4.10	.983	8
Work area access	4.02	1.024	9
Poor work/jobsite conditions	3.98	.924	10
Subcontractors involvement	3.95	1.147	11
Incentive for good performance	3.93	1.135	12
Investigations and risk assessments	3.83	1.208	13

6 CONCLUSIONS

The findings of this study presented that OHS is poorly implemented by the SME's contractors. Moreover, the factors that hinder the success implementation of OHS among SMEs were lack of communication, lack of management commitment, lack of regular inspections and audits; poor management/supervision/information flow; lack of management commitment; equipment and tools; lack of communication between workers; poor employee involvement; lack of training and risk education; lack of skilled workforce; poor work/jobsite conditions. Occupational health and safety of the workers is not negotiable, is all pervasive, increases productivity, and leads to better performance, improve the company image, reduced claims and accidents also reduce lost times spent on injuries.

7 RECOMMENDATIONS

The concept of OHS implementation is not yet well accepted and practices by the SME's, therefore, regular training must be provided by the government in order to emphases the critic of OHS and must be emphasized by all stakeholders involved in the construction industry, in-order to increase the awareness and the seriousness of the implementation of OHS.

References

Audet, J., and Courteret, P., Coaching the Entrepreneur, Features and Success Factors, *Journal of Small Business Enterprise and Development,* 19(1), 515-531, 2012.

Brace, C. L., and Gibb, A. G. F., A Health Management Process for the Construction Industry, in *Rethinking and Revitalizing Construction Safety, Health and Quality,* Haupt, T and Smallwood, J. (Eds.), *Port Elizabeth*, South Africa, 2005.

CIDB, SME's Development, 2011. Retrieved from www.info.gov.za/view/downlaodfileAction?id=68205. January 20, 2017.

Chartered Institute of Building. Health and Safety in the Construction Industry. CIOB Supports a Zero Tolerance Approach to Unsafe and Unhealthy Practices, 2008. Retrieved from http://www.hse.gov.uk/humanfactors/comah/common, 09 June 2017.

Davies, V. J., and Tomasin, K., *Construction Safety Handbook: Hardcover, 303* Kurasa, Kuchapishwa, Thomas Telford, 2000.

Department of Labour, Occupational Health and Safety (OH&S) Inspectorate in Relation to the Construction Industry, *Government Gazette* 539 (2010) 33176, 2011.

Fakoki, O., *Obstacles to the Growth of SME's in South Africa. A Principal Component Analysis Approach.* PhD Thesis, University of Forte Hare, South Africa, 2010.

Hsiao, H., and Armstrong, T. J., Preface to The Special Section on Occupational Fall Prevention and Protection: National Institute for Occupational Safety and Health (NIOSH), *Hum Factors 2012*, 54(3), 301-302, June 2012.

James, P. M., Braam Rust, A. A., and Kingman, L., The Well-Being of Workers in The South African Construction Industry: A Model for Employment Assistance, *African Journal of Business Management,* 6(4), 1553-1558, 2011.

James, P. M., Braam Rust, A. A., and Kingma, L., *The Well-Being of Workers in The South African Construction Industry: A Model for Employment Assistance,* Cape Peninsula University of Technology, 2012.

Jooste, C., South African Transition to Consolidated Budget, *South African Journal of Economics*, 80(2) 181-199, 2012.

Lunsche, S., *The Quite Death-Where to Next for South Africa Economic Policy*, Wits Business School University of Witwatersrand, Johannesburg, 2010.

Mashwama, N. X., Aigbavboa, C., and Thwala, D., An Assessment of the Critical Success Factor for The Reduction of Cost of Poor Quality in Construction Projects in Swaziland. *Procedia Engineering.* 196: 447-453, 2017a.

Mashwama, N. X., Aigbaboa, C., and Thwala, D., A Theoretical Assessment of the Challenges of Public Private Partnership in Improving Infrastructure Service Delivery in Swaziland, *Proceedings of the 4th International Conference on Infrastructure Development and Investment Strategies for Africa*, DII-2017. 30 August -1 September 2017. 299-239, Livingstone, Zambia, 2017b.

Opaleye, O. S., and Talukhaba, A. A., *Evaluating Causes of Workers Fatality on Construction Sites in The City of Tshwane Metropolitan Municipality.* In: Proceedings 8th Construction Industry Board (cidb) Postgraduate Conference. 10-11 February 2014, Laryea, S. and Ibem, E. (Eds), 281-292, University of the Witwatersrand, Johanneburg, South Africa, 2014.

Ramutloa, L., *Department of Labour Republic of South Africa, Amended Occupational Health and Safety Act.*, 2-27, 2007.

Streamlining Information Transfer between Construction and Structural Engineering
Edited by Shiau, J., Vimonsatit, V., Yazdani, S., and Singh, A.

ISBN: 978-0-9960437-7-9

IMPORTATION OF RELATIONAL PROJECT DELIVERY AGREEMENTS EXPLOITING THE BIM APPROACH IN LEGISLATION

GIUSEPPE MARTINO DI GIUDA[1], PAOLO ETTORE GIANA[1], VALENTINA VILLA[2], and SARA VALAGUZZA[3]

[1]*Dept of ABC, Politecnico di Milano, Milan, Italy*
[2]*Dept of DISEG, Politecnico di Torino, Torino, Italy*
[3]*Dept of Italian and Supranational Public Law, Università degli Studi di Milano, Milan, Italy*

The research proposes an overview of the importation of a Relational Project Delivery Agreement (RPDA) in Italian legislation and the related issues. The AEC sector fragmentation, caused by an increase of building complexity and a change in the industry structure, is demanding a collaborative approach to the project to allow the possibility of a holistic vision based on a BIM approach. This work provides an overview of RPDAs applications on different contexts, highlighting benefits and issues related, including litigation resolution processes. A special focus has been kept in an Italian context, where the stagnant construction market needs to be revolutionized, and better interactions among stakeholders are required. The research imports and adapts an alliance agreement, FAC-1 (Framework Alliance Contract), on Italian framework, tailoring the most important features of the model to the Italian legislation. The goal of this work is providing a methodology to validate a standard form of contract, aiming at an added value to construction sector. Further developments consist in guidelines for contract management and evaluation of the project behavior during different phases of the process.

Keywords: Collaboration, Multiparty, Relational contracts, Framework alliance.

1 INTRODUCTION

This research proposes an analysis of the fragmentation of the construction sector, due to the contractual tradition, that obliges the different actors to act in pursuit of their own interests, producing a decrease in re-work and in lack of process optimizations performances, as well as collaboration with the supply chain (Akyuz *et al.* 2014). In most countries, construction projects are carried out by means of standard contracts, this procedure is new on the Italian scene, still used to building custom contracts (McKinsey Global Institute 2017). In order to optimize the entire process, an analysis on the worldwide different types of contracts and their implications for the construction process was carried out. In this sense, the study of the state of the art of collaborative contracts, Relational Project Delivery Agreements (RPDAs), has highlighted their wider potential impact on the sector. In order to import and adapt a collaborative procurement model in Italian legislation, collaboration has developed among some Italian universities (University of Milan, Milan Polytechnic and University of Brescia) and Prof. David Mosey at

Kings College London. This alliance framework, considering the actual Italian procurement law, defines a new standard contract able to create collaborative interactions.

2 RELATIONAL PROJECT DELIVERY AGREEMENTS

Beside different contract approaches, there are different collaboration levels that could be set in a project (Lahdenperä 2012). The contract could be seen just as a promissory agreement among people recognized by the law, which sets the rules of interactions among the participants. According to Construction Leadership Council (CLC) (Construction Industry Council 2013) report on productivity, collaboration set in contractual stage, provides the basis for the Lean principles of reducing reworking and optimizing processes. The un-alignment of design team's goals and the site design work-flow goals create a waste, especially during the construction phase, resulting in time and budget overrun (Sacks *et al.* 2010). The lack of communication among team members produces (Tauriainen *et al.* 2016) misunderstanding resulting in an un-optimized building. All these circumstances are emphasized when collaboration doesn't involve all the participant to the project (Myerson 1999). In US, a new form of collaborative contracts is called Integrated Project Delivery (IPD) (AIACC 2014), it started as an alternative to the tradition Design-Bid-Built, in order to reduce complex project risks, and it imposed a mental shift in the fulfilling of the contract. In Europe Collaborative contracts are represented by Integrated Design, which does not involve linked contract to the framework. In this scenario, UK is the leading country that has applied and developed a RPDA for AEC industry. FAC-1, Framework Alliance Contract, is the first European contractual standard that is able to link more single set from a bi-lateral contract in a multi-party agreement. This approach is applicable in different scenarios and its methodology fits with BIM methodology and with collaborative system (Di Giuda *et al.* 2017a, Di Giuda *et al.* 2017b).

The European Directive 2014/24 recognizes the joint planning as one of the methods for building public works, the core principle of that partnership is that the design quality would be superior if the project is developed jointly. The Laws realized on January 28, 2016, Legislative Decree 50/2016, no. 11, (Italian Parliament 2016) has drastically limited the use of integrated tender based on technical design to "works or components characterized by a remarkable innovative or technological content". The reviewed Contracts Legislation has further restricted the joint design operation scope and execution contract to activities characterized by high technology or innovation elements that need the division of these phases. The value of a joint project development is superior when the project is produced by collaboration instead of a sequential development, so a joint alliance adoption impacts on time and costs.

3 METHODOLOGY

Productivity increase and time waste decrease are the RPDAs relevant advantages. The evolution of BIM methodology is imposing also a shift from traditional contract, which segregates different stakeholders, to a collaborative one. In this context, the BIM use helps in controlling and implementing the result of contractual evolution form during the whole project evolution. This research provides an analysis of the importation of a collaborative agreement in Italian legislation enhancing a BIM approach. A multidisciplinary group of researchers understood the possibility to import FAC-1 into Italian legislation. The FAC-1 was conceived and drafted by Prof. David Mosey PhD, Director of the Centre for Construction Law and Dispute Resolution of King's College in London and was subsequently translated and adapted to the Italian legal context by the University of Milan. To prove the importation legitimacy, four levels of control analyses were put in practice:

(i) First control: contractual model internal quality control in term of immediate comprehensibility of the terminologies used in the translation and the consistency of the contractual model with the reference market (Italian end European).
(ii) Second control: a semi-internal quality control developed in collaboration with Prof. David Mosey and based on a contract flux comparison between the two versions, explaining the modification of the contract to fit the Italian legislation.
(iii) Third control: external control that requires the help of some institutional entities, local authorities, universities, trade associations, representatives of the administrative judiciary and private operators. They provided some doubts and uncertainties, from different perspectives, which were analyzed and solved discrepancies.
(iv) Fourth control: second round of external control which took place during a round table discussion, where all the people, invited to discuss during the previous step to submit comments to the contract standard, attended the discussion. During this event, a few points of the standard were discussed, explained and commented for the regulatory bodies sensitivity, the coherence with Italian legislation and the market.
(v) The contributions received - further verified, also from the point of view of consistency with the original philosophy of the contractual model - were included in the final version of the document, which is therefore enriched by the experience of various leading players of the national panorama.

4 ADAPTATION OF FAC-1 ON ITALIAN LEGISLATION

The adaptation process of FAC-1 was developed by collaboration among people with different background and adapted the standard to Italian legislation, modifying its content in a drastic way in order to fit a different legislative standard. This research will provide a project and program management tool to clients, who would like to manage complex situations. According to the analysis conducted by OICE (Association of Organizations of Engineering and Technical-Economic Consulting), BIM projects are drastically increasing their importance in Italian market and for this reason, the framework will provide to it a solid base on which BIM methodology can fit the client requests. The Italian FAC-1 establishes a system regulates legal relations among several subjects involved in the implementation of one or more projects, defined as Programs, by linking several contracts with a view to encouraging collaboration and coordination of the various activities. Through the stipulation of FAC-1, parties undertake to work in a collaborative spirit and to carry out activities, called "Alliance Activities" – added value, site organization efficiency, BIM use - consistent with the aims identified by the Customer, which may be public or private.

4.1 FAC-1: Structure and Features

FAC-1 is structured as a Common Law Contract; and for this reason, it's useful starting from the General Conditions, containing the general rules, then moving on to the initial part.

The entire document is related to the Definitions, which states the way to read correctly the standard contracts. Consequently, the General Conditions, referred in the Cooperation Agreement, are the contract column. One of the main features of the contract is that it could be bended to the requests of the collaboration members just selecting the clauses that they want to apply. FAC-1 has pre-structured Annexes and Modules in order to set the requests made or the action carried by the collaboration. Through an information preset included in each request, each party structures the request in order to provide all the needed information.

Figure 1. Hierarchical relationship among FAC-1 documents.

The provisions contained in the General Conditions and Annexes may be departed from or modified in the FAC-1, which is composed of several Modules, specifying respectively:

- The Objectives (KPI), Success measures, Targets, Objectives and Bonuses (Schedule 1);
- The Timetable (Schedule 2);
- The Risk Register (Schedule 3);
- Awarding Procedures (Schedule 4);
- Contractual Models (Schedule 5);
- Legal requirements and special terms (Schedule 6);

FAC-1 is a contract that suits a general goal called Objectives, which will be accomplished by specific Objectives measured through Key Performance Indicator (KPI) (Jonsson *et al.* 2017). The contract can pursue both the aims of a Project or a Program. The collaboration is set among the Client, the Alliance Manager and the Collaboration Members who sign the FAC-1.

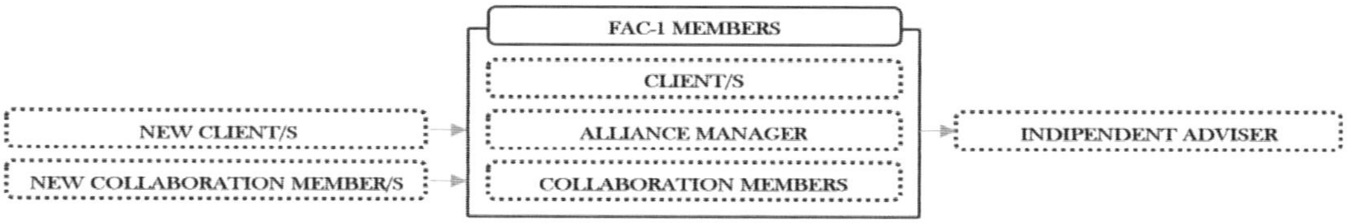

Figure 2. Structure of FAC-1.

The standard flexibility allows including more members to the collaboration both as a New Client and as a New Collaboration Member. To pursuit the Collaboration scope, a series of activities have to be completed and each member can propose a series of activities to the Client in order to optimize the project, their payment is at the discretion of the Alliance Manager and the Client. Sometimes the added value that is gathered from a new activity repay the effort extended in the action. Whether FAC-1 intervenes in a Program already in the design or execution phase, or where FAC-1 precedes the start of any activity related to a Program, the Collaboration Members are required to present their Collaboration Proposals, which may be limited to simple participation in meetings, drafting reports or sharing relevant information with other Collaboration Members, or they may consist in the provision of specific services. The Collaboration Activities execution takes place - according to the schedule in part two - on the basis of Requests formulated by the Client, indicating the contents to be implemented, the execution terms and the amount due for them, calculated on the basis of the Collaboration Price. Payment for Collaboration Activities shall be made in accordance with the conditions set out in

FAC-1, a specific payment policy states the workflow of the procedure to trace all the payment among all the parties and to set the fee for delays. The Alliance Manager in relation to the Purposes of Collaboration monitors the performances and the Objectives, based on Members' activities, are assessed by shared KPI. The profit sharing is one of the keys of contracts based on collaborative procurements. For instance, the partaking of the supply chain can optimize the process and produce a huge reduction of the costs, thanks to shared procedures.

4.2 Theoretical Examples of the Application of FAC-1

In order to demonstrate FAC-1 flexibility, two examples of its application are presented, both on side of public and private investors. If we deal with the public sector, multiple municipalities can decide to optimize the street lighting service by creating a single management system for them all. Through this alliance, parties can be joined in the contractual scheme by establishing a dialogue aimed at a single goal dictated by the Client. Thanks to a shared supply chain, single installation companies can participate by offering, for example, an implementation of their System's performance. At the same time, the various companies being companies in the sector can help the maintenance manager to choose ways that are more appropriate by improving the Final Service. Finally, customers can choose which of the improvement proposals to adopt.

While if we deal in the context of the private client, the general contractor of a work decides to establish a collaboration agreement with other participants in order to optimize the process in terms of work quality, integration between design and construction and safety. In this scenario the designer could offer, as an additional service, the implementation through BIM methodology so as to reduce the design inefficiencies and the process inconsistency, in this sense the Client could accept an extra cost in the face of fewer problems in the execution phase and final saving greater than the cost incurred. In this way, proposals made by the other collaboration members can enhance the process offering costs management reduction through energy improvements or flows optimization through schedule changes. These collaborations improve the final result, setting an exchange of information and reducing misunderstandings and costs.

5 DISCUSSION AND FURTHER DEVELOPMENT

FAC-1 is a flexible contractual model in which parties are given the opportunity to consider efficiencies in the supply chain that make the flow of information more transparent and reduce the overall cost of performance. The Client could use the standard in order to create a collaboration, legally valid, among their sub-consultants or sub-contractors. The higher level of transparency and increased responsibility, towards both the Client and other Collaboration Components, required by each private operator in the pursuit of collaboration, are counterbalanced by the economic compensation provided. FAC-1 is a contract that regulates and manages the inter-relations among different contracts and, namely, the relation among parties, which are not directly associated over a contract. Due to the Italian legislation, FAC-1 permits to connect different parties and to enhance collaboration among them, below some advantages achievable are shown:

- Different subjects' activities coordination with greater guarantees of results and with a reduction of unexpected interference, possible over-budget and overrun of time;
- Complex project efficient management with multiple subjects' contributions;
- Added value thanks to collaboration, in terms of work or service sustainability, site organization and working conditions efficiency, collaboration with the supply chain;
- All professionals' involvement in the collaboration in order to prevent or reduce the mistakes, which must be reported to Alliance Manager that improves project final quality;

- Promoting transparency in relation to the specific aims and objectives of collaboration;
- Risks prevention and management during execution;
- Litigation reduction by preventing and extra-judicially handling possible claims.

In addition, FAC-1 is intended to build a solid legal framework for the BIM use in the construction sector, through the following means of developing a positive interaction among different design teams and linking the various phases of construction of a work (Alwash *et al.* 2017). FAC-1 approach invites the participants to submit their Collaboration Proposals and enhances the professional expertise, thereby exploiting economies of scale and achieving cash or other benefits.

At this point, the research provides a new approach to the problem of the inefficiency of the sector, in fact the collaboration that RPDA established among stakeholders is difficult to achieve and, most of the time, it is unattainable in a traditional process, although the promised success (Ghassemi *et al.* 2011). The research aims at providing to Italian and European legislation a contractual model, which aligns AEC sector performances to other sectors that already use this type of agreement. The new contract improves the processes management decreasing the public administration burden often due to litigations caused by traditional contractual procedures (Valaguzza 2018). The research will produce to integrate collaboration in work procedures, guideline for the contract application both for private and public users. The "Adda Martesana" Municipality will use this contractual model as part of the project to build a middle school in Liscate (Italy).

References

AIACC, Integrated Project Delivery: An Updated Working Definition, 2014.

Akyuz, G. A., Gursoy, G., and Celebi, N., Supply Chain Collaboration: A Conceptual Maturity Model, *Industrial Engineering*, 2333–2335, 2014.

Alwash, A., Love, P. E. D., and Olatunji, O., Impact and Remedy of Legal Uncertainties in Building Information Modeling, *Journal of Legal Affairs and Dispute Resolution in Engineering and Construction*, 1–7, 2017.

Construction Industry Council, Building Information Cic / BIM Protocol. *Cic*, 1–15, 2013.

Di Giuda, G. M., Villa, V., and Giana, P. E., *Collaborative Contract with Building Information Modelling: Comparison Between USA and European Approach. ISEC-9*, Resilient Structures and Sustainable Construction, Valencia, Spain, 2017a.

Di Giuda, G. M., Villa, V., Giana, P. E., Tagliabue, L. C., and Ciribini, A. L. C., LEAN Construction Applied to a BIM Process: How to Control Point Attribution in Meat Tender Process, *Tema*, *3*(1), 35–44, 2017b.

Ghassemi, R., and Becerik-Gerber, B., Transitioning to Integrated Project Delivery: Potential Barriers and Lessons Learned, *Lean Construction Journal*, 32–52, 2011.

Italian Parliament. D.Lgs. 50/2016 "Codice dei contratti Pubblici" e s.m.i., 2016.

Jonsson, H., and Rudberg, M., KPIs for Measuring Performance of Production Systems for Residential Buildings, *Construction Innovation*, *17*(3), 381–403, 2017.

Lahdenperä, P., Making Sense of The Multi-Party Contractual Arrangements of Project Partnering, Project Alliancing and Integrated Project Delivery, *Construction Management and Economics*, *30*(1), 57–79, 2012.

McKinsey Global Institute, *Reinventing Construction: A Route to Higher Productivity*, 1–155, 2017.

Myerson, R., B., Nash Equilibrium and the History of Economic Theory, *Journal Economy Literature,* 36, 1067-1082, 1999.

Sacks, R., Radosavljevic, M., and Barak, R., Requirements for Building Information Modeling Based Lean Production Management Systems for Construction, *Automation in Construction*, 2010.

Tauriainen, M., Marttinen, P., Dave, B., and Koskela, L., The Effects of BIM and Lean Construction on Design Management Practices, *Procedia Engineering*, 2016.

Valaguzza, S., *Governare per contratto*. CCLM Editoriale Scientifica, Napoli. 2018

Streamlining Information Transfer between Construction and Structural Engineering
Edited by Shiau, J., Vimonsatit, V., Yazdani, S., and Singh, A.

ISBN: 978-0-9960437-7-9

GROUND STRENGTHENING METHOD BY MULTISTAGE LOCAL CONSOLIDATION

RYUNOSUKE MATSUNAGA and HARUYUKI YAMAMOTO

Development Technology, Hiroshima University IDEC, Higashi-Hiroshima, Japan

Authors are investigating a ground strengthening method, which effectively consolidates the construction area of requisite minimum in a short time by local consolidation for the soft ground. In this paper, Authors research that the effect of increasing the strength of soft ground by using the three-dimensional elasto-plastic finite element analysis when the area of local loading is gradually changed. First, it was found that the strength of the soft ground just below the local loading surface increased with a short time by the local consolidation and ultimate bearing capacity increases to 2.6 times as much as the initial bearing capacity by the multistage loading. Next, analysis was carried out in the case of the multistage loading with cyclic loading. In the reloading process, loading area is reduced to 1/3 and 1/6 of the loading area of the previous stage. In this case, the ultimate bearing capacity increases to 8.25 times as much as the initial bearing capacity. In conclusion, it was found that the strength of the soft ground just under the loading surface is considerably increased by multistage local loading with reducing the loading area.

Keywords: Improvement, Soft ground, 3-D elasto-plastic analysis, Bearing capacity.

1 INTRODUCTION

This research is to find an effective ground improvement method that is not a conventional method of consolidating and strengthening the whole area for the problematic soft ground but local consolidation can effectively consolidate the exact size necessary to construction in a short time. In this paper, Authors research that the strengthening effect of the soft ground by using three-dimensional elasto-plastic finite element analysis when the area of the local loading is gradually changed in multistage local consolidation. There are few documents that directly relate to this research.

2 PRELIMINARY ANALYSIS BY FINITE ELEMENT METHOD

As a preliminary analysis, in order to study the effectiveness of consolidation strengthening by local loading, governing equations based on Biot consolidation theory were formulated and three-dimensional linear consolidation analysis by finite element method was performed. Figure1 shows the analytical model and analytical constants of the ground subjected to local loading. For the boundary conditions related to displacement, the free end of the upper end face of the analysis region, the lower end face is fixed only in the vertical direction and the side face is fixed only in the horizontal direction. Boundary conditions related to permeability were drainage on the upper and lower surfaces, and undrained on the side. The length of one side of the analytical area/

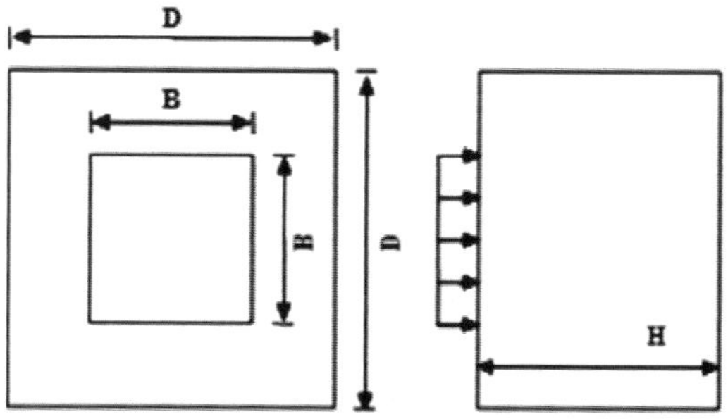

(Analytical Constant)
Lame's constant $\lambda = 1.0$, $\mu = 1.0$
Coefficient of consolidation Cv = 3.0

(Load)
Uniform distribution load p = 1.0

Figure 1. 3-D linear consolidation analytical model and analytical constants.

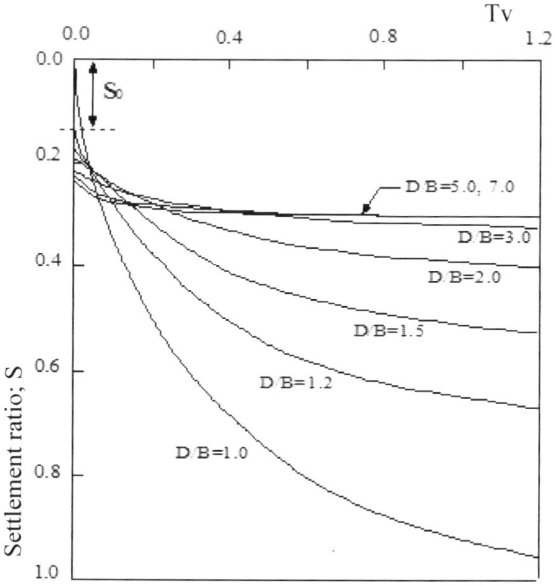

Figure 2. Relationship between settlement ratio and Tv.

length of one side of the loading area was set as D / B and this value was varied from 1.0 to 7.0 with depth of H, and the relationship between consolidation settlement and time was examined from the analytical result in each case. Figure 2 [1] shows the relationship between the settlement ratio S (settlement / settlement at U = 100% in one-dimensional consolidation) and the time coefficient Tv with respect to each of D / B = 1.0 to 7.0 (H / B = 3.75). As shown this figure, the settlement ratio decreases with increasing D / B and the settlement ratio converges to almost 0.3 when the D / B becomes 5 or more. Furthermore, in case of local loading, the consolidation subsidence progresses much faster than in case of one-dimensional consolidation, and the absolute value of consolidated settlement is also considerably small (for example, about 10% when D / B is 5.0 or more). This result gives favorable conditions that the main part of consolidation subsidence almost finishes during the construction period, the absolute value of the settlement is also small and the strength of the soft ground just under the local loading surface increases in a short period. As a second preliminary analysis, consolidation strengthening by multistage loading using the above-mentioned local loading (increase rate of C was set to $\Delta C = 0.3\ \Delta\bar{\sigma}_v$ [2], C; cohesion, $\bar{\sigma}_v$; vertical effective stress) is examined. Multistage loading is to gradually perform the next loading after being strengthened by the initial loading. With regard to the effectiveness of multistage loading, multistage loading was carried out with initial linear inner increment (the settlement is controlled within the linear range in one stage loading) and 90% increment (the loading is controlled up to 90% of the ultimate bearing stress of the previous stage) respectively using elasto-plastic incremental analysis by finite element method, and the strengthening effect was investigated. Under the plane strain condition, Figure 3 [3] shows the relationship between the load and the settlement / base width (Se / B) in case of the initial linear inner increment and the case of 90% increment. On both the initial linear inner increment and the 90% increment, the ultimate yield strength was increased, reaching 2.4 times and 2.6 times the initial ultimate bearing stress (qu = 5.14 C). From the results of this analysis, it was found that the strength of the soft ground just under the local loading surface increases, therefore it is possible to increase the ground strength without destroying the ground with multistage loading.

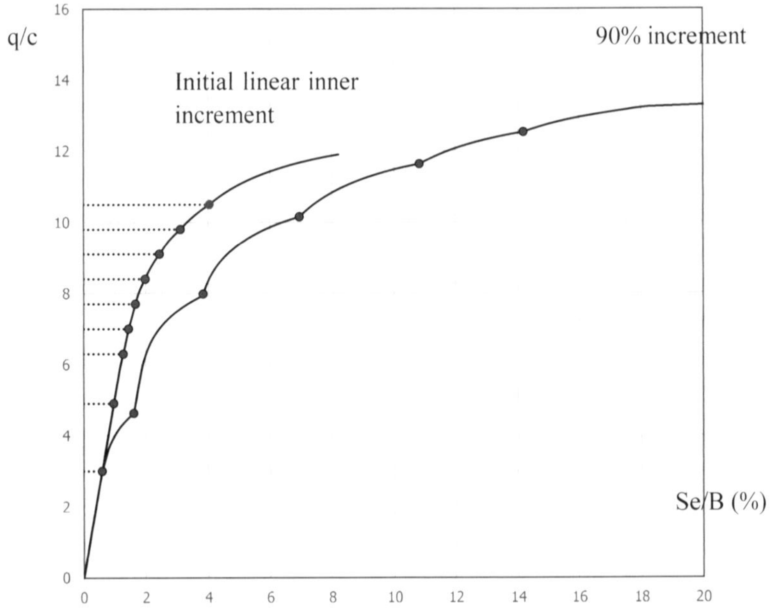

Figure 3. Relationship between the bearing capacity and settlement ratio in case of initial linear inner increment and 90% increment.

3 CONSIDERATION OF MULTISTAGE LOADING WITH CHANGE OF LOADING AREA

In this section, after carrying out the multistage loading as described above, authors examined the case of unloading once and reloading to the reducing area. Figure 4 shows this series of flows. The elasto-plastic increment analysis was also performed for this analysis. After multistage loading on the ground model, it was unloaded and reloaded to the loading area, which was reduced to 1/3 (or 1/6) of the loading area of the previous stage. Figure 5 shows the relationship between the load and the settlement / the foundation width in case where the loading area of the sixth stage is reduced to 1/3 (or 1/6) of the previous stage in the 90% increment. It can be seen that when the loading area is smaller than that of the previous stage, the ultimate bearing capacity greatly increases, and in case of the 1/6 loading area, it is 8.25 times as compared with the initial bearing capacity (qu = 5.14 C). This is because loading is applied to the central part of the loading area, which is sufficiently consolidated and strengthened by multistage loading with reducing of loading area. This shows the effectiveness of local consolidation strengthening by multistage loading that reduces the loading area in stages, giving efficient consolidation strengthening of the necessary area. As a construction example assumed at an actual construction site as shown Figure 6, it is conceivable to construct a 1/3 or 1/6 strip foundation on the soft ground which is consolidated and strengthened by multistage loading, further consolidation strengthening by building load is carried out and the bearing capacity can be greatly increased.

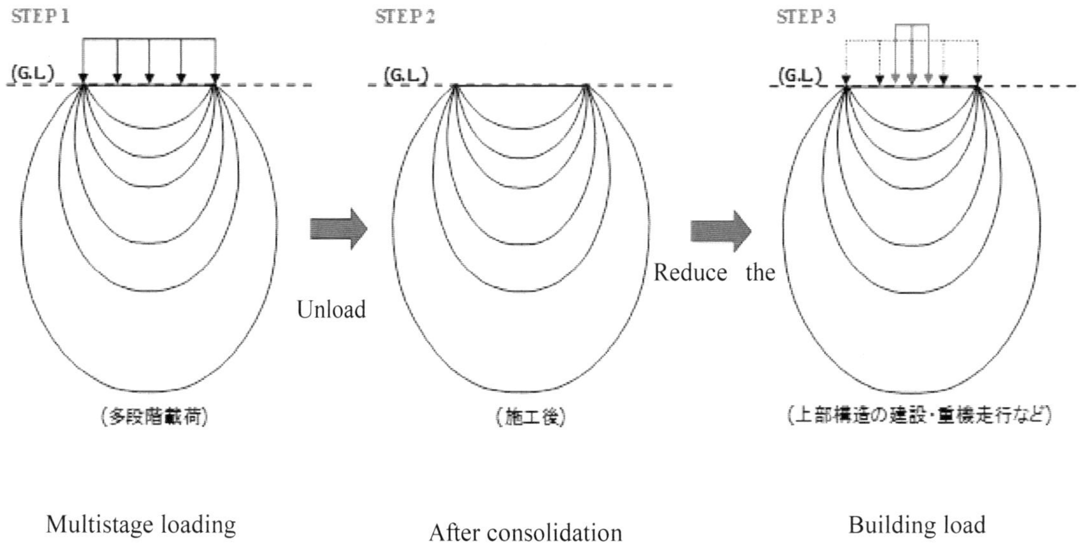

Figure 4. Construction image.

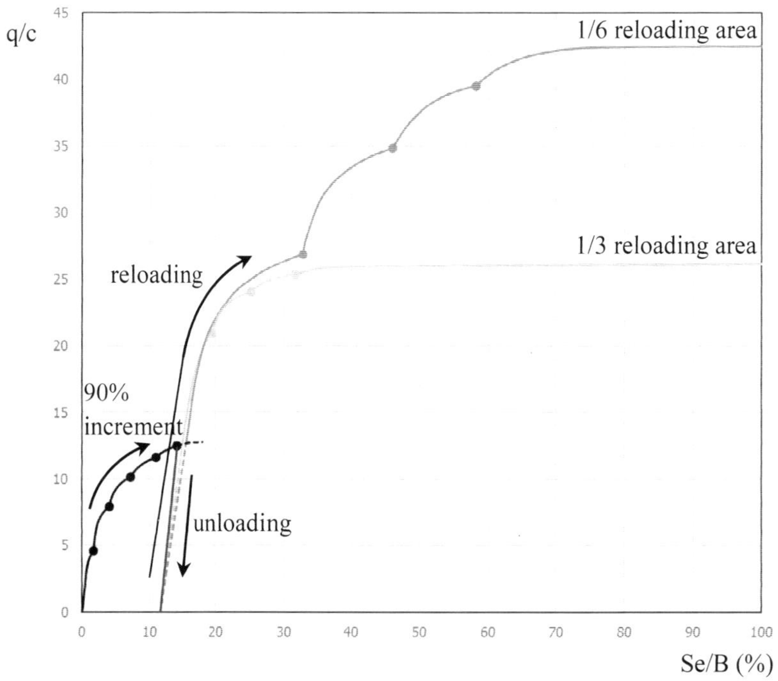

Figure 5. Load ~ Se/B relationship (90% increment, unloading and reloading to reducing area).

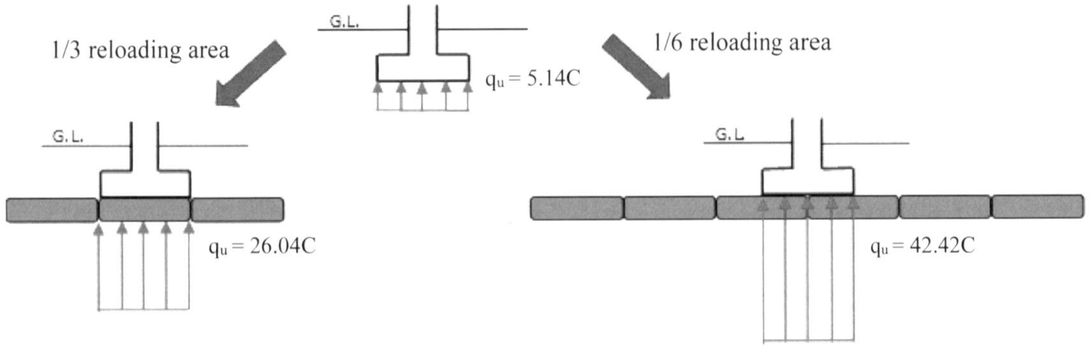

Figure 6. Improvement soft ground under the strip foundation.

4 CONCLUSION

In this study, authors investigated the effect of the increase in strength of soft ground when multistage loading on soft ground, unloading and reloading to the reduced area by three - dimensional elasto -plastic analysis. As a result, it was found that the strength of the soft ground immediately increased considerably under the loading surface. It was also found that effective and economical improved method can be developed by applying consolidation strengthening method of soft ground with suitable multistage loading and controlled loading area. As a future construction method, the building load aims to realize multistage loading by playing the role of reloading after the ground improvement by the preload method. When reloading, that area is required to be smaller than the preload area. In the future study, authors should conduct laboratory experiments on the contents of this research and it is necessary to compare the experimental results with the analytical results.

References

Yamamoto Haruyuki, Huang He, Matsuoka Hajime, *Study on Ground Strengthening by Local Consolidation (3-D Linear Consolidation Analysis)*, Proceedings of the Chugoku Branch of Japanese Architectural Institute, Vol. 34 (CD version), 73 -76, 2011. (in Japanese)

Matsuoka Hajime, *Basic Civil Engineering Series 15: Soil mechanics*, Mori Kita Publishing Co., Ltd., 2005. (in Japanese)

Yamamoto Haruyuki, Huang He, Matsuoka Hajime, *Ground Strengthening Method by Staged Local Consolidation (Increased Effect of Bearing Capacity)*, Proceedings of the Chugoku Branch of Japanese Architectural Institute, Vol. 35 (CD version), 97-100, March, 2012. (in Japanese)

Streamlining Information Transfer between Construction and Structural Engineering
Edited by Shiau, J., Vimonsatit, V., Yazdani, S., and Singh, A.

ISBN: 978-0-9960437-7-9

CRUSHING CHARACTERISTICS OF SOIL PARTICLE ON THE EFFECT OF STRESS PATH

KEIGO FUKUDA and HARUYUKI YAMAMOTO

Development Technology, Hiroshima University/IDEC, Higashi-Hiroshima, Japan

In previous studies, we have conducted tests under various conditions of stress for the particle crushing. Anyway, it is necessary to perform the crushing test that assumed stress conditions in ground under controlled each principal stresses individually, to find the mechanical characteristics of the soil with particle crushing. Therefore, the purpose of this study is to confirm the effect for particle crushing under various stress paths in the combination of principal stresses. We planned four combinations of principal stresses when operate the mean stress p and deviatoric stress q up to the Shear Failure Line (SFL) on the π-plane. First, we set tri-axial extension test as Lode angle θ=60° and tri-axial compression test as θ=0°, and set two of the remainder as θ=40°, 20°. The crushing tests are carried out using the high pressure true tri-axial compression apparatus under the planned stress paths. As a result, the progress of the particle crushing becomes active in order of θ=40°, 20°, 60°, 0°, and found out that deviatoric stress q has essential effect on the particle crushing.

Keywords: Mean stress, Deviatoric stress, Lode angle, High pressure, Plastic work, Relative breakage.

1 INTRODUCTION

Sand and gravels are granular materials, which composed of solid particles aggregated. Under the high pressure condition, particle crushing phenomenon occurs in which individual particles constituting the granular material are fractured. It is an important task to investigate the mechanical properties associated with particle crushing of sand under high pressure conditions because the mechanical properties of the soil particles change due to the change in the structure of the granules when particle crushing occurs. In the previous studies, the crushing experiments were conducted under the isotropic compression condition, the K_0 condition or the tri-axial compression condition, however, the load test should be conducted under conditions close to the actual ground stress conditions ($\sigma_1 \neq \sigma_2 \neq \sigma_3$). The difference in Relative breakage B_r should be examined when reaching the same stress state from various stress path. Then, we employed three kinds of experimental parameters; mean stress p, deviatoric stress q and Lode angle θ. The purpose of this study is to compare the effect of stress path and confirm the influence of stress path and deviatoric stress q on particle crushing. In addition, the volume reduction was discussed in consideration of the relation between mean stress p, deviatoric stress q and plastic volumetric strain ε_v^p.

2 OUTLINE OF TESTS

In this study, we used high pressure tri-axial compression apparatus (Yokura *et al.* 2015) to consider the effect of particle crushing under high pressure condition. Figure 1 shows the plan view of apparatus used in this study. As shown in Figure 1, a cubic test specimen with the shape of edge cutting 50 × 50 × 50 mm is installed in the center of this apparatus. The specimen was loaded by oil jack with maximum stress up to 200 MPa through loading plate. All weight of the loading device in the vertical direction are supported by the counterweight passed through the pulleys. Furthermore, in order to load using the reaction force in all directions of the apparatus, loads from the relative loading plate were devised to be equal. In addition, the movable part was supported by a ball bearing to move smoothly and without frictional resistance. In this study, Toyoura sand was used as the specimen in Japan. Toyoura sand is often used in various soil tests. By the general sieving test, the proportion by weight of particles having a particle diameter of 0.125 mm to 0.297 mm is large, and the weight ratio of particles of 0.074 mm or less is extremely small. The specimen was compacted to be initial relative density of 90% or more.

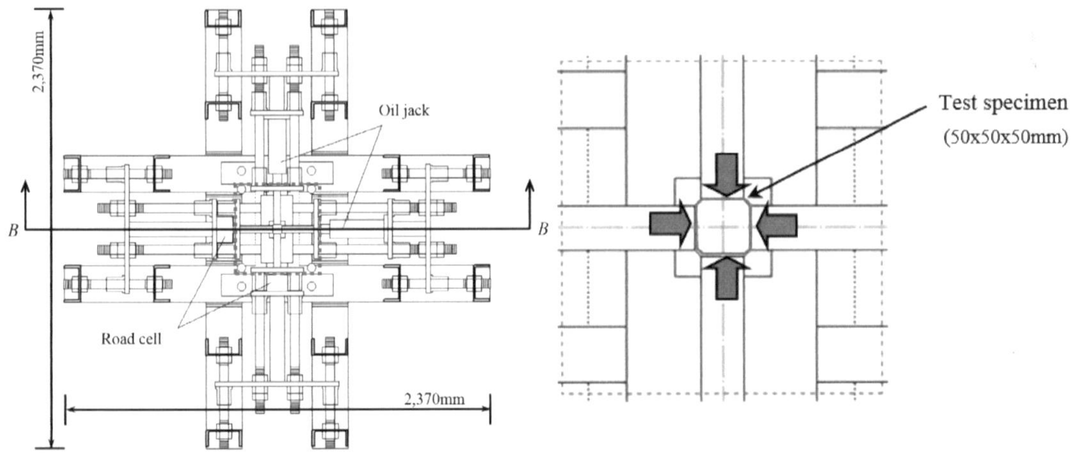

Figure 1. Plan view of high pressure tri-axial compression apparatus, enlarged view of dotted line portion.

Mean stress p, Deviatoric stress q and Lode angle θ are controlled in this study, additionally, it is necessary to make differences only in the stress path with the same mean stress p and deviatoric stress q. For that purpose, it is easy to understand by considering the three-dimensional principal stress space represented by σ_1, σ_2, σ_3. Therefore, experiments were conducted by setting a stress path on the π-plane. Figure 2 shows the stress path on the π-plane. The definitions of mean stress p and deviatoric stress q are shown below.

$$p = \frac{1}{3}(\sigma_1 + \sigma_2 + \sigma_3) \tag{1}$$

$$q = \frac{1}{\sqrt{2}}\sqrt{(\sigma_1 - \sigma_2)^2 + (\sigma_2 - \sigma_3)^2 + (\sigma_3 - \sigma_1)^2} \tag{2}$$

At the first step, we consider three π planes with mean stress p = 14.5, 30 and 50 MPa. Experiments were also conducted by stress paths with Lode angles θ = 0°, 20°, 40°, 60° for each deviatoric stress levels and the SFL condition, and the isotropic compression cases without deviatoric stress. In total, we conducted 51 cases of experiments as shown Figure 3.

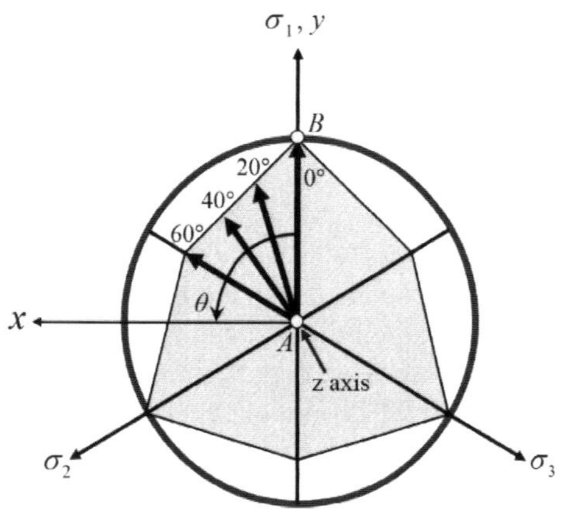

Figure 2. Stress path on the π-plane.

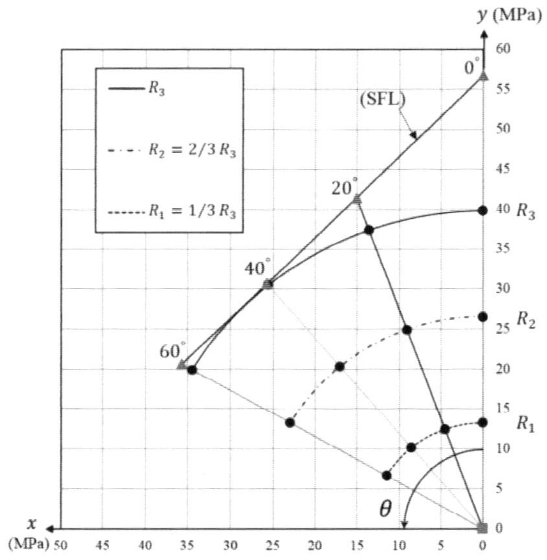

Figure 3. Explanation diagram of stress path in case of C test.

Then, the test of p = 14.5 MPa is called A test, the test of p = 30.0 MPa is B test and the test of p = 50.0 MPa is called C test. For example, experiment with the case of p = 14.5 MPa, the path of $\theta = 60°$ and folding back at the point of R_1 is expressed as "A-60°-R_1".

3 TEST RESULTS AND DISCUSSION

At the start, sieving test was conducted to determine the particle size after particle crushing. Volumetric strain $\varepsilon_v \sim$ p relationship and deviatoric strain $\varepsilon_d \sim$ q relationship are shown in Figure 4. The Plastic work W_p are calculated by adding up the area of each painted place (Roscoe *et al.* 1963). In addition, plastic volumetric strain and deviatoric strain, elastic volumetric strain and deviatoric strain are as shown in the figures. Since this plastic volumetric strain ε_v^p is largely related to the volume reduction of the sample, it is considered to be related to the Relative breakage B_r (Hardin 1985).

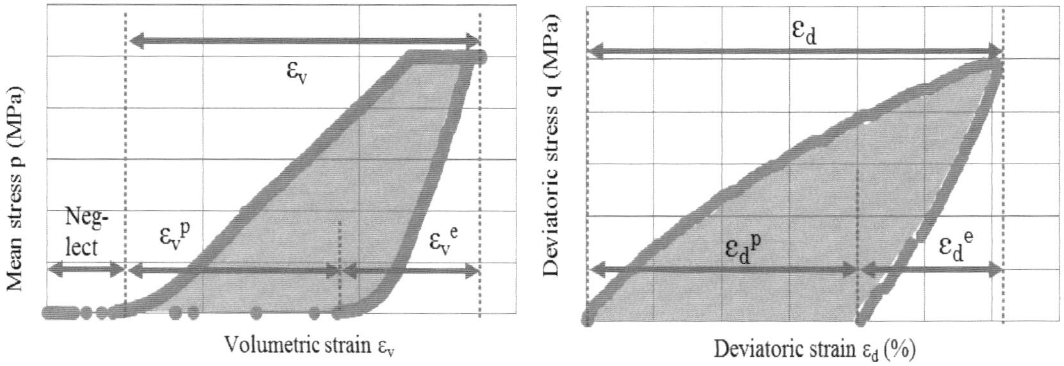

Figure 4. Definition of volumetric and deviatoric strain components.

Figure 5 shows the relationship between the Lode angles θ and B_r, and the relationship between the Lode angles θ and W_p obtained by the result of the sieving test. According to B_r and W_p obtained by the test, almost no difference due to the stress path of Lode angle θ was found. It is also understood that B_r and W_p fluctuate in proportion to the deviatoric stress q.

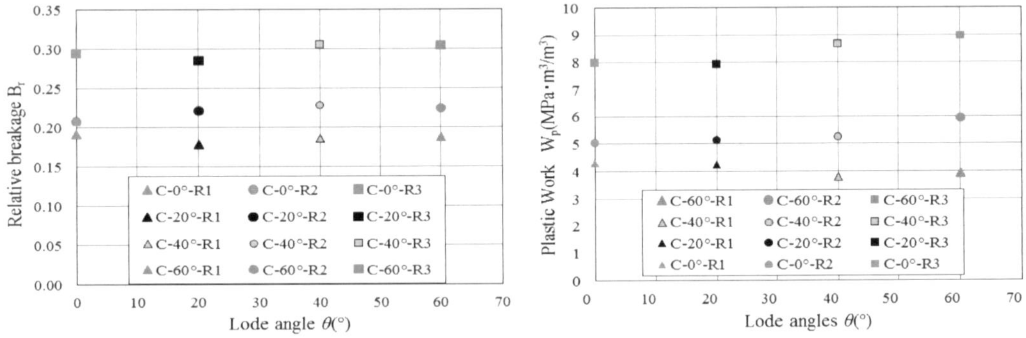

Figure 5. The Relative breakage B_r and the Plastic work W_p for Lode angle θ.

It is considered to the consume energy when particles break. Therefore, it can be inferred that there is a unique relationship between Relative breakage B_r and Plastic work W_p. Figure 6 shows the relationship between Plastic work W_p and Relative breakage B_r. The curve is convex downward as shown the figure. Also, since the increment of B_r gradually decreases, it seems that B_r has an upper limit. This is a proportional relationship as shown in Figure 7, the order of approximation is good (R^2=0.939).

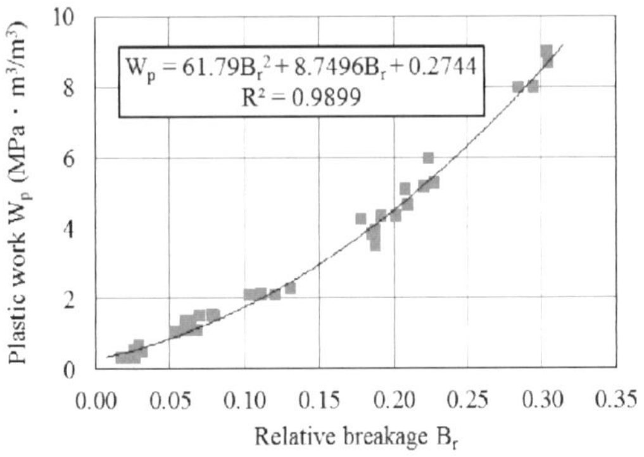

Figure 6. Relationship of $B_r - W_p$.

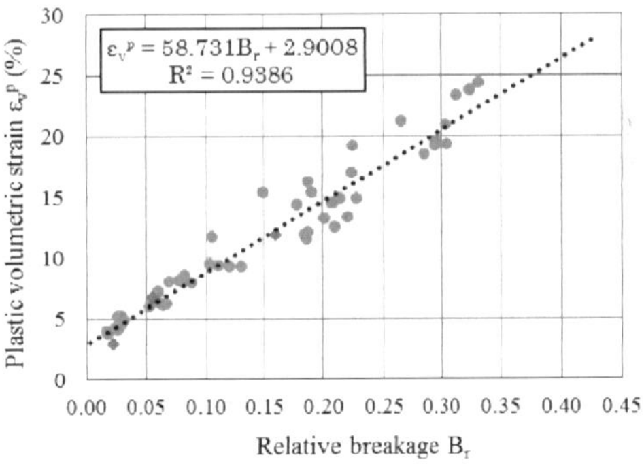

Figure 7. Relationship of $B_r - \varepsilon_v^p$.

4 EXPERIMENTAL FORMULA

In this chapter, the relationship between amount of particle crushing and volumetric strain ε_v was derived by experiments. The final goal is to be able to estimate the strain increment from the pressure on the soil particles by establishing these relationships. As shown in Figure 4, the volumetric strain ε_v is a sum of a plastic and elastic component. Therefore, each components are represented by mean stress p and deviatoric stress q. At first, it is explained the relationship between elastic volumetric strain ε_v^e and (p, q). According to the paper by (Nakai 1989), ε_v^e is expressed by the following Eq. (3).

$$\varepsilon_v^e = C_e\left[\left(\frac{p}{p_a}\right)^m\right] \quad \text{(In case of } p_0{=}0\text{)} \tag{3}$$

The unknown coefficients in Eq. (3) are C_e, m (pa is atmospheric pressure). C_e, m derived from the experimental data are shown in Figure 8 and 9. From these figures, it can be assumed that C_e, m are constants, therefore we regard the average value of experimental data as coefficients. In these Figures, rhombic marks are regarded as singularities with largely deviated from the average and are excluded when taking the average.

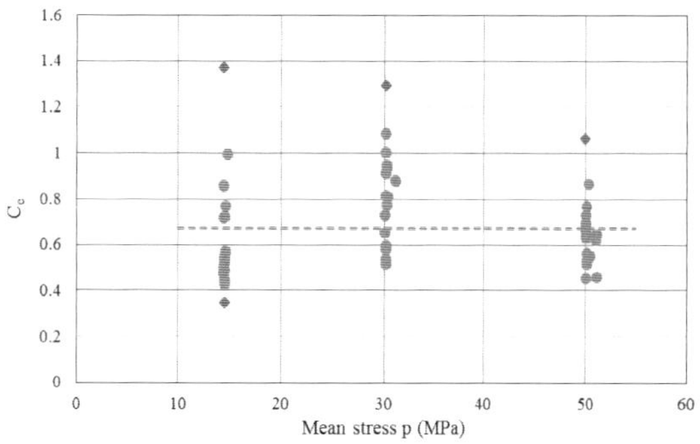

Figure 8. Relationship of p - C_e.

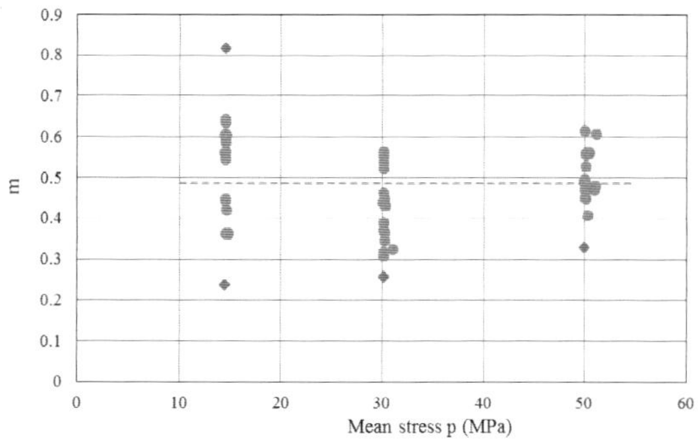

Figure 9. Relationship of p - m.

Figure 10 shows the relationship between the Relative breakage B_r and mean stress p. The increasing tendency of Relative breakage B_r gradually increases as the mean stress p increases in the case where the deviatoric stress q is small. On the other hand, we can be seen that the increasing tendency of Relative breakage B_r gradually decreases with increase in the mean stress p in the case where the deviatoric stress q is large. In addition, we can find that the influence on the increase of particle crushing is larger in the deviatoric stress q than the mean stress p. Based on the above tendency, an approximate expression of B_r - (p, q) relationship shown in the Eq. (4) and (5) are assumed. Also, stress path is not considered in this estimation. Figure 11 compares the estimated B_r with experimental data. Incidentally, B_r was obtained by substituting the coefficients A_0, r_c and c determined by the least-squares method into the Eq. (4) and (5). The order of approximation is good (R^2=0.974) by using this relational expression and Relative breakage B_r can be estimated from the combination of (p, q). Figure 12 is a three-dimensional image of this experimental formula. Experimental equations derived from the above calculations are shown in Table 1. Accordingly, the volumetric strain can be represented by the mean stress p and the deviatoric stress q.

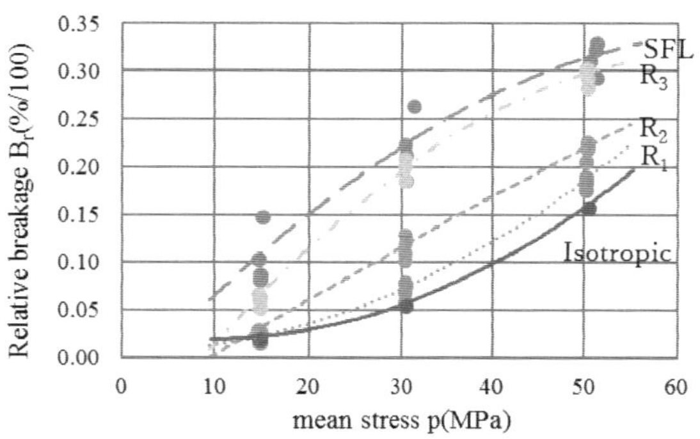

Figure 10. B_r - mean stress p.

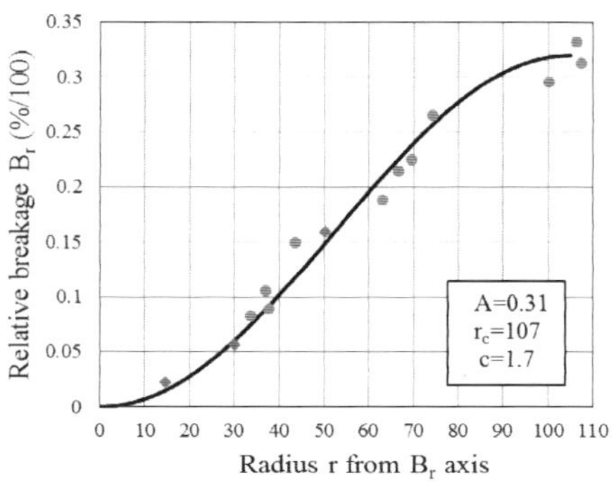

Figure 11. Estimation accuracy of B_r.

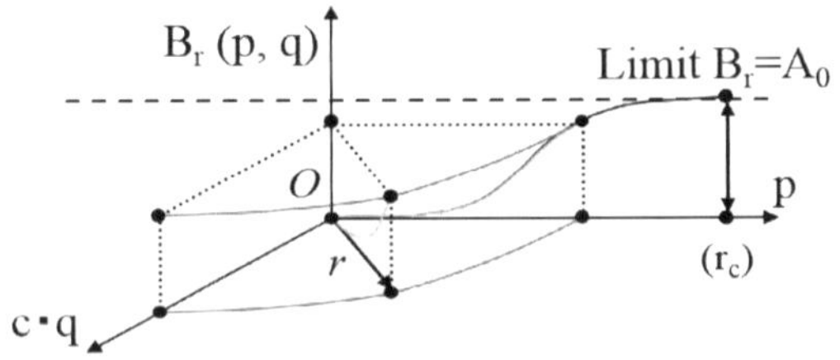

$$B_r = A \cdot \{1 - cos(^{\pi}/_2 \cdot {}^{r}/_{r_c})\} \tag{4}$$

$$r^2 = p^2 + (c \cdot q)^2,\ c\ ;\ \text{Magnification factor}\ \ c \geqq 1.0 \tag{5}$$

Figure 12. Conceptual diagram of B_r, p and q in three-dimensional space.

Table 1. List of estimated experiment formula.

Volumetric strain components	Coefficients
$\varepsilon_v^e = C_e \times (p/p_a)^m$	$C_e = 0.6725$ $m = 0.4863$ $p_a = 0.1$(MPa)
$\varepsilon_v^p = a_1 \times B_r(p, q) + b_1$	$a_1 = 58.731$ $b_1 = 2.9008$
$B_r = A \times \{1 - \cos(\pi/2 \times r/r_c)\}$ $r = \sqrt{p^2 + (c \times q)^2}$	A =0.31 $r_c = 107$(MPa) c = 1.7

5 CONCLUSION

From the experimental results of particle crushing tests, it was found that there is little influence of the Lode angle θ on the crushing property, and B_r, W_p increase in proportion to the deviatoric stress q. We derive an experimental formula to calculate the volumetric strain ε_v by mean stress p and deviatoric stress q. In addition, formulating the relation $\varepsilon_v^e \sim (p, q)$, $B_r \sim (p, q)$ and $\varepsilon_v^p \sim (p, q)$ were achieved success in the case of Toyoura sand. However, we have unfinished works to future studies as followings.

1) Applicability of the experimental formula for the different type of sand.
2) Examine the effect of different stress paths.
3) Experimental formula on the deviatoric strain ε_d.

Thereafter, these works will be confirmed by experiments; using different type of sand (e.g. weathering granite and Shirasu) and passing through different stress paths. Experimental equation of deviatoric strain ε_d will be estimated with same method as the case of volumetric strain ε_v.

References

Hardin, B. O., Crushing of Soil Particles, *Journal of Geotechnical Engineering,* 111(10), 1177-1192, 1985.

Nakai, T., An isotropic hardening elastoplastic model considering the stress path dependency in three-dimensional stresses, *Soils and Foundations,* 29(1), 119-139, 1989.

Roscoe, K. H., Schofield, A. N., and Thurairajar, A., Yielding of Clays in States Wetter than Critical, *Geotechnique,* 13(3), 211-240, 1963.

Yokura, K., Yamamoto, H., and Wu, Y., Crushing tests of soil particles by high pressure true tri-axial compression apparatus, *The 6th Japan-China Geotechnical Symposium,* 51-56, 2015.

Streamlining Information Transfer between Construction and Structural Engineering
Edited by Shiau, J., Vimonsatit, V., Yazdani, S., and Singh, A.

ISBN: 978-0-9960437-7-9

ENGINEERING CHARACTERISTICS AND EARTHQUAKE RESPONSE ANALYSIS OF GROUND

HARUYUKI YAMAMOTO[1] and MUNKHUNUR TOGTOKHBUYAN[2]

[1]*Development Technology, Hiroshima University / IDEC, Higashi-Hiroshima, Japan*
[2]*Civil Engineering and Architecture, Mongolian University of Science and Technology, Ulaanbaatar, Mongolia*

One-dimensional layered soil lumped mass ground response analysis was conducted for the representative site in Ulaanbaatar, Mongolia. The surficial geology of the site is predominantly composed of the gravely and sandy soil typical of this region in the central part of Ulaanbaatar. The natural period of soil profiles needs to be investigated under several circumstances. For example, these parameters-based study has indicated that damage due to earthquakes occurs when the natural periods, T_1 and T_2, of the ground are closer to that of a superstructure. Various computational procedures or methods have been proposed for this kind of the ground response analysis. In this paper, the numerical analysis method such as the lumped mass method within eigenvalue analysis is used to determine the natural periods of the ground. The ground surface, soil deposits, and bedrock are assumed to be horizontal. The soil deposits are subjected to shear deformation such as shear modulus, G, on the other hand, excitation of vibration could be a shear modulus on each layer. As well as to determine an engineering bedrock depth in the site, the methodology that is utilized in this paper is focused on the use of the correlation between SPT-N value and soil elastic Young`s modulus, E, in the soil profiles, and used over 100 boreholes data with SPT-N values in the vicinity of Ulaanbaatar.

Keywords: Natural periods, Lumped mass, Layered soil deposit, Elastic analysis.

1 INTRODUCTION

A large number of constructions have been built across the capital city Ulaanbaatar of Mongolia. Unfortunately, many believe buildings that are under construction in there suffer from a lack of thinking in a dynamic response analysis of the ground, that is based on either the engineering geology or seismology investigations and relates to the soil conditions. In the last several years, the fast growth in the number of available strong-motion records has allowed statistical studies of the characteristics of horizontal ground shaking, including the effect of local geology in many countries.

These studies have been purposed to obtain average values of peak ground acceleration, using different methods and data such as a basis for the comparison the Modified Mercalli Intensity Scale and several accelerograms for the various type of soil conditions on the site (Seed and Idriss 1967). These investigations and observations have also indicated that any buildings construct on deep or soft alluvium may be subjected to seismic forces several times larger than similar buildings on hard soil if the peak ground acceleration is the same in both cases. Therefore, the statistical analysis suggests that stiffness and depth of soil should be assumed in the seismic design of structures.

In Ulaanbaatar, it lacks data for seismic response analysis of the ground due to the data in the national code of Mongolia has been adopted from the Code of Russian Federation. To eliminate or solve this problem, we need to find various approaches for the dynamic ground response analysis which are based on not only accelerograms but also detailed geotechnical data, geology information and so on.

One of the primary representations of this assumption is determining the natural period, T, of a soil deposit. On the other hand, this characteristic site period depends on the stiffness and depth to the rock-like material of the soil profile at the site. Simple procedures for estimating natural periods of horizontal soil layers are given (Idriss and Seed 1968).

In this paper, a simple method such as the one-dimensional lumped mass solution for estimating T of multilayer the gravelly and the sandy soil profile with a linearly elastic is used in the representative site of Ulaanbaatar.

2 CORRELATION BETWEEN SPT-N VALUE AND DEFORMATION MODULUS E FOR THE GRAVELLY AND THE SANDY SOILS AROUND ULAANBAATAR

The first, to evaluate the ground response of soil deposit, an empirical equation of correlation between the blow count N value of Standard Penetration Test (SPT-N value) and elastic Young's modulus E of the gravelly and the sandy soils around Ulaanbaatar which was obtained. It helps to determine the bedrock depth in the study area. If SPT-N blow numbers count more than 50 in a borehole depth of strata, it can be considered to be bedrock depth.

Figure 1 shows the weak positive relationship between two parameters for various types of gravelly and sandy soils together, and comparison with the case of normal consolidated and over consolidated sandy soils in Japan. It is modified in the next attempt.

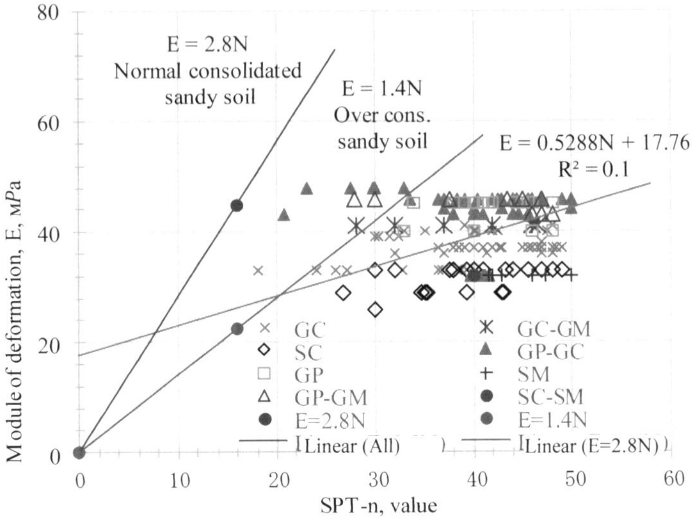

Figure 1. The correlation between SPT-N value and Modulus of Deformation E for Sandy and Gravelly soil. The blue and the brown lines are the cases of normal-consolidated and over-consolidated sandy soil respectively, in Japan.

Figure 2 presents the modified scatter plot of the positive relationship between two parameters such as SPT-N value and E, the value of the gravelly and sandy soils around Ulaanbaatar city. In this case, as shown in the Figure 1, all different N values of a type of soil corresponding to the same amount of E, values, that is integrated as follows:

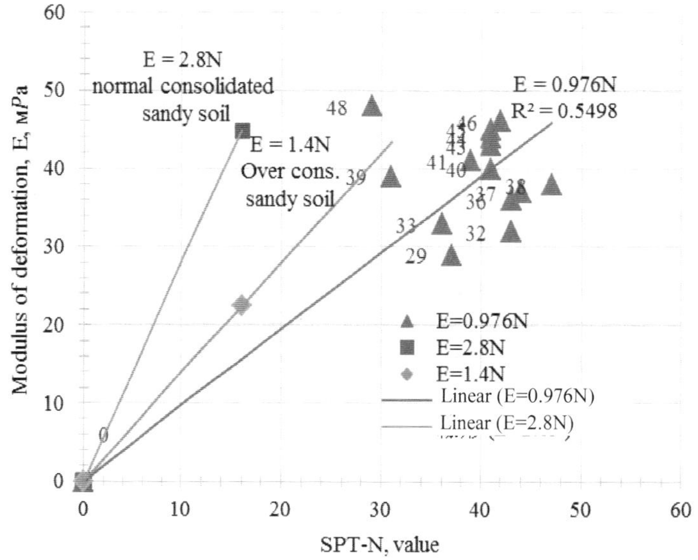

Figure 2. Modified scatter plot of the correlation between SPT-N value and Modulus of Deformation E for Sandy and Gravelly soil.

As a result of the correlation procedure, the empirical equation between SPT-N value and E value for gravelly and sandy soil was given by

$$E\ (MPa) = 0.976 \cdot N \tag{1}$$

A beneficial relationship that can provide the elastic properties from an in-situ test result it is given in Table 1, and the computed equation in this study is also compared with several countries in here.

Table 1. Correlations between SPT-N values and Elastic Young's Modulus of soils, the case of several countries.

<table>
<tr><th></th><th>Japan (MPa)</th><th>USA and Russia Es (KPa)</th><th>Mongolia (MPa)</th></tr>
<tr><td>Sand (normally consolidated)</td><td>Es = 1.4N</td><td>Es = 0.5(N+0.015)</td><td rowspan="5">E = 0.976N

It is found in case of all sandy and gravelly soils in Ulaanbaatar area</td></tr>
<tr><td>Sand (over consolidated)</td><td>Es = 2.8N</td><td>Es = 40 + 1.05N</td></tr>
<tr><td>Gravelly sand</td><td></td><td>IF, N > 15 Es = 0.6(N + 0.06) +2
IF, N < 15 Es= 0.6(N +0.06)</td></tr>
<tr><td>Clayey sand</td><td></td><td>Es = 0.32(N+0.015)</td></tr>
<tr><td>Silts, sandy silt, or clayey silt</td><td></td><td>Es = 0.3(N+0.006)</td></tr>
</table>

Totally over 429 boreholes without SPT investigation were carried out for the engineering geological investigation, and information has been accumulated in this study area. From the total boreholes, 132 boreholes, which were used in this research due to the majority of them drilled by shallow (around 6 meters' depth). Their engineering bedrock depth is specified by Eq. (1). For instance, in the case of 132 boreholes, if the SPT-N value is estimated by Eq. (1) is closer to the 50 in a stratum, which were assumed by the bedrock level for the site.

3 MODELING OF THE SOIL DEPOSIT

In many cases, the one-dimensional analysis gives larger displacement than the multidimensional analysis, when the ground layers are close to the horizontally layered. In this study, the one-dimensional coordinate system is chosen, where soil column can move freely under the one-dimensional analysis. Furthermore, the lumped mass idealization of the horizontal soil layers' model is created, as shown in Figure 3. In such a case, each layer including surface and bedrock boundary layers is assumed to be parallel with each other. Where m_i is a mass of layer i, h_i is a thickness of each corresponding layer i, k_i is spring constant, c_i and c_E are viscous and dashpot damping, during undamped free vibration both of them are null.

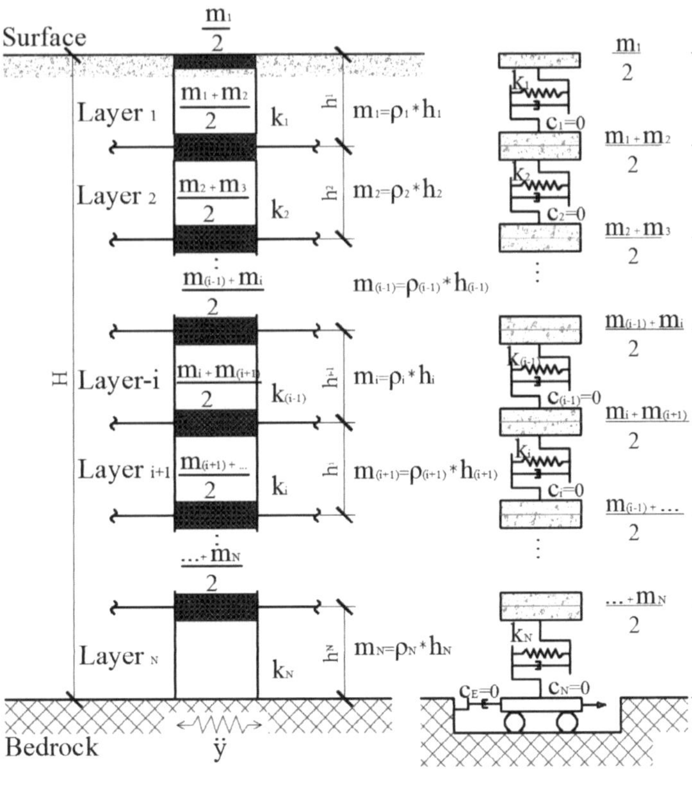

Figure 3. The lumped mass idealization of horizontal soil layers' model.

3.1 Numerical Analysis Method

For determining the natural period of the ground, the theory of the eigenvalue analysis is used. The undamped free vibration basic equation is given by;

$$[M]\{\ddot{y}\} + [k]\{y\} = 0 \tag{2}$$

Where, [M] is mass, or the diagonal matrix and [k] is the stiffness matrix, respectively.

Furthermore, the natural periods can be obtained by the general method. In this study, damping is neglected completely in ground response analysis, values of natural period T_1 and T_2 calculated by the Fortran programs for the based on the above general theory. As mentioned before one hundred thirty-two soil profiles were selected and their fundamental periods were estimated. For all spatial analysis, to create the counter map which corresponds with the results, ArcGIS has been used.

4 DISCUSSION OF RESULTS

According to the result of the counter map in Figure 4 and 5, the natural periods represents T_1=0.05-1.6 sec and T_2=0.01-0.57 sec respectively.

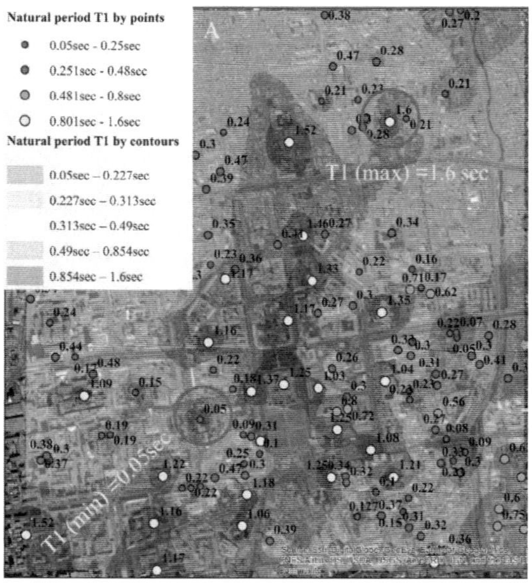

Figure 4. Natural period: T_1 distribution map.

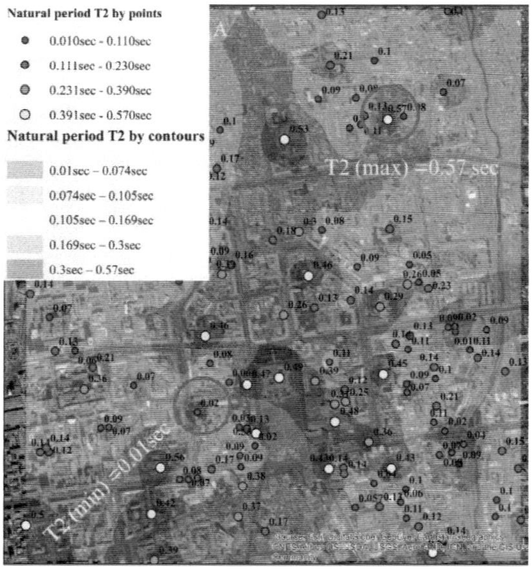

Figure 5. Natural period: T_2 distribution map.

These maps illustrate which area is risky or safe for the ground response due to the induced vibration, and so on. For example, as shown in Figure 4, two points are delineated by the red circle. One of them is related to the maximum natural period area and another presents the minimum natural period area. In such a case, the high-rise building in the maximum natural period area may be susceptible to the shock, but the small-rise building may show no damage.

As shown in Figure 6, the relationships between depth and natural period in a site are obtained. It could be useful for the study of geotechnical earthquake engineering and for the structural engineers in the future. If the special area of any object is interested in the apparent depth in the study area, the natural periods can be obtained using this chart.

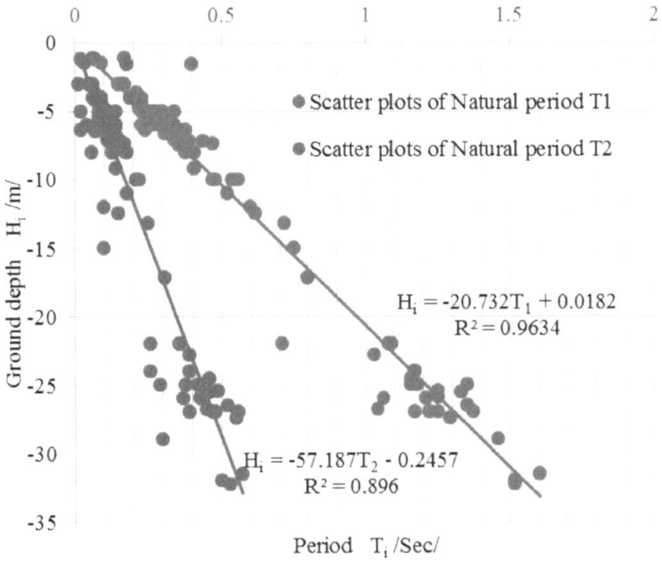

Figure 6. The relationship between ground depth and natural periods T_1 and T_2.

5 CONCLUSIONS

A simple method for estimating the natural periods of soil profiles with the determining engineering bedrock depth in the site has been presented. On the other hand, the linear elastic one-dimensional lumped mass ground response analysis was performed for the multilayered the gravelly and the sandy soil profiles. Input motion related to accelerograms-time history and damping was neglected, and horizontal excitation of motion was assumed by the shear stress of the ground. This study should be taken into account for the clay deposits around Ulaanbaatar city.

References

Idriss, I. M., and Seed, H. B., Seismic Response of Horizontal Soil Layers, *Journal of Soil Mechanics and Foundations Division,* 94, SM4,1003-1031, 1968.

Seed, H., and Idriss, I., Analysis of Soil Liquefaction: Niigata Earthquake, *Journal of Soil Mechanics and Foundations Division,* 93, SM3,83-108, 1967.

Streamlining Information Transfer between Construction and Structural Engineering
Edited by Shiau, J., Vimonsatit, V., Yazdani, S., and Singh, A.

ISBN: 978-0-9960437-7-9

AN EFFICIENT VOLUME REDUCTION APPROACH FOR WATER DREDGED SOIL BY ULTRASONIC WAVE

KOSUKE TAKAHASHI[1], SHUNJI UE[1], KEIJI KUWAJIMA[1], YASUSHI FUKUDA[1], and MITSUNORI OHUCHI[2]

[1]*Civil Engineering and Architecture, National Institute of Technology, Tokuyama College, Yamaguchi, Japan*
[2]*Yasuhara Environment Technology Co., Ltd., Yamaguchi, Japan*

Dredged soil caused by dredging work has brought the limits of the capacity of sediment disposal facilities in Japan. One way to solve this problem is considered to efficiently reduce the dredged soil and secure the capacity of the sediment disposal facilities. As a method of volume reduction, the technique utilizing ultrasonic vibration for compacting of soil particles has been studied so far. There are two purposes of this study. One is to clarify the volume reduction characteristic of water dredged soil by ultrasonic wave irradiation by changing the frequency or output voltage. The other is to take aim at 30% of the volume reduction. In the experimental method, the slurry of either sand or silt, its moisture content adjusted to 150%, was poured into each container. Ultrasonic wave was irradiated from the horizontal direction. After the experiment, the mass of the sample was measured. Relative density and volume reduction ratio were calculated through wet density, dry density, and void ratio. As a result, the highest volume reduction ratio of 27% was obtained when the ultrasonic wave of frequency 28kHz with output voltage 200V were irradiated in the silt. In conclusion, although we did not reach the target volume reduction of 30%, we were able to confirm the increase in density and the decrease in volume.

Keywords: Sediment disposal, Ultrasonic wave irradiation, Ultrasonic vibration, Dredging.

1 INTRODUCTION

Dredging is one of the most important work for constructing anchorages, channels, and underwater structures. Dredging work generates a lot of earth and sand. Part of one is used for landfill materials on the sea, but the rest is carried to sediment disposal facilities. In recent year, the limit of the capacity of sediment disposal facilities is near with the increase of dredged soil in Japan. Hence, we must reduce the volume of the dredged soil efficiently and secure the capacity of the sediment disposal facilities.

As a method of volume reduction, the technique utilizing ultrasonic vibration 20kHz for compacting of soil particles has been studied so far. Ouchi *et al.* (2016) conducted an experiment to irradiate the ultrasonic wave from the vertical direction to the soil tank of Toyoura sand. Thereby, they succeed in densification of the soil tank. However, this method cannot be applied to *fines* such as silt or clay. Fines are extremely small and light, so if ultrasonic wave are irradiated from above, fines will be blown up. This is called *convection phenomenon*.

In this study, ultrasonic wave is irradiated not from the vertical direction but from the horizontal direction of the soil tank. And we set two goals. One is to clarify the volume reduction characteristic of soil particles by ultrasonic wave irradiation by changing the frequency or output voltage. The other is to take aim at 30% of the volume reduction.

2 CONCEPT OF DENSIFICATION OF FINES BY ULTRASONIC WAVE

Figure 1 illustrates concepts of the behavior of fines which is irradiated ultrasonic wave from the horizontal direction. Focusing on the shape of the ultrasonic wave, there are two places where the wave is high and where the wave crosses at the center. The former is called an *anti-node*, and the latter is called a *node*. An anti-node has the effect of dispersing the fines, and a node has the effect of flocculating the fines. Therefore, fines move from an anti-node to a node and flocculate at a node, and densification of fines is promoted because of increasing self-weight of the fines.

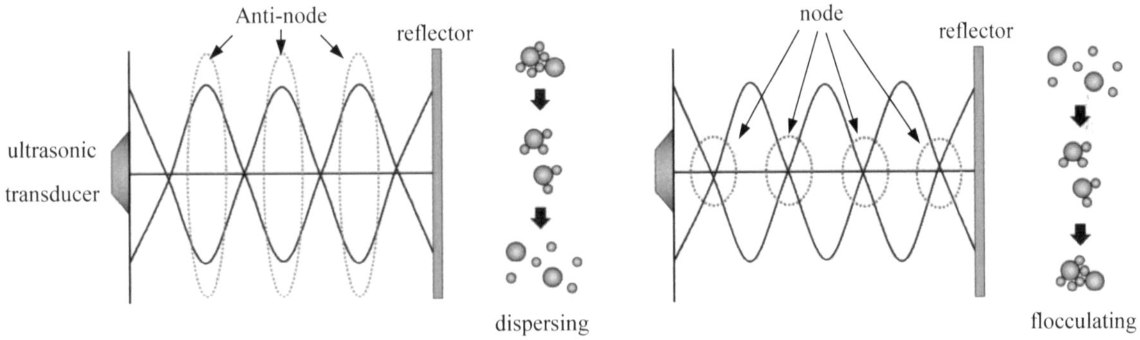

Figure 1. Mechanism of dispersion and flocculation of fines.

3 MATERIALS AND METHODS

3.1 Materials

In order to examine the relationship between the particle diameter and the volume reduction characteristics, the experiments were conducted with three kinds of samples. The samples, Toyoura sand, No.7 silica sand, and Silica powder, were used, all of which were mined in Yamaguchi Prefecture in Japan. Figure 2 shows their grading curves and photos, and Table 1 shows their physical properties.

Figure 2. Particle size distribution curves and images of experimental materials.

Table 1. Physical properties for experimental materials.

	Toyoura sand	No.7 silica sand	Silica powder
Specific gravity, G_s	2.637	2.604	2.590
Maximum void ratio, e_{max}	0.960	1.163	1.502
Minimum void ratio, e_{min}	0.640	0.655	0.668
D_{30} (mm)	0.2	0.1	0.006

3.2 Equipment

The equipment used in the experiment is as follows: ultrasonic transducer, container, ultrasonic transmitter, and digital oscilloscope. The ultrasonic wave is generated with the vibration of the piezoelectric ceramics in the ultrasonic transducer. An ultrasonic transmitter and an oscilloscope are connected to the ultrasonic transducer. The former is modified to output voltage and the latter is displayed the waveform. The frequencies used in the experiment are 28kHz, 40kHz and 60kHz, and the container size is designed to match each wavelength. Figure 3 shows the experimental equipment and detail of the ultrasonic transducer.

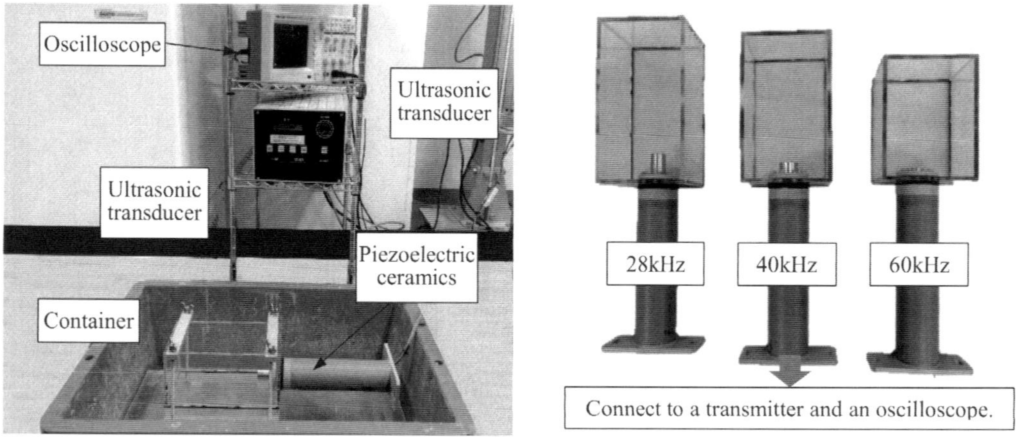

Figure 3. Construction of the experimental equipment and detail of the ultrasonic transducer.

3.3 Methods

The experimental conditions are listed in Table 2. The experiment was conducted three times under each condition. For comparison, we also conducted an unirradiated experiment.

Table 2. List for experimental conditions.

Frequency (kHz)	Output voltage (V)	Time to irradiate (min)	Number of experiments
0	0	30	3
28	100		
	200		
	300		
40	100		
	200		
	300		
60	100		
	200		
	300		

The experimental procedures are described below:

(1) the sample, its moisture content adjusted to 150%, is thoroughly stirred and then poured into a container.
(2) voltage is outputted by an ultrasonic transmitter, and the ultrasonic wave is directly irradiated to the sample for 30 minutes.
(3) after 30 minutes, the ultrasonic transmitter is stopped and the wet mass of the deposited sample, m, is measured.
(4) The sample is dried and then the dry mass, m_s, is measured.
(5) Wet density, ρ_t, dry density, ρ_d, and void ratio, e, are calculated.
(6) The relative density, D_r, and the volume reduction ratio, D_v, are calculated. The relative density is the compactness of a soil relative to that same soil under highly compacted and very uncompacted conditions. We defined the volume reduction ratio as the proportion of volume decrease by irradiating the ultrasonic wave. The relative density and the volume reduction ratio can be determined by the following equations:

$$D_r = \frac{e_{max} - e}{e_{max} - e_{min}} \times 100(\%) \tag{1}$$

$$D_v = \frac{e_0 - e}{1 + e_0} \times 100(\%) \tag{2}$$

Where e_{max} is the void ratio for the very loose (uncompacted) condition and e_{min} is the void ratio for the highly compacted condition. e_{max} and e_{min} are determined from the maximum density and minimum density tests. In addition, e_0 is the void ratio when the ultrasonic wave is unirradiated.

4 RESULTS AND DISCUSSION

We organized the values obtained in the experiment for each sample. Figure 4 to Figure 6 show the relationship between the output voltage and the relative density or the volume reduction ratio. The horizontal axis represents the output voltage, the vertical axis represents the values of the relative density or the volume reduction ratio. In addition, Figure 7 shows the relationship the particle size and the relative density or the volume reduction ratio. The horizontal axis represents the 30% cumulative average particle diameter of the samples, the vertical axis represents the values of the relative density or the volume reduction ratio. Graphs were color-coding these values under each frequency.

4.1 Toyoura Sand

Figure 4 shows the results in Toyoura sand. The relative density exceeds 70% at any frequency. This is a relatively high value, and the compaction using the ultrasonic wave can be said to be effective. On the other hand, the volume reduction ratio is almost the same as the unirradiated graph. It seems to be a low level. These results could be attributed to the particle size. We supposed that Toyoura sand has been already relatively tight before irradiating the ultrasonic wave. Hence the relative density becomes high and the volume reduction ratio becomes low.

4.2 No.7 Silica Sand

Figure 5 shows the results in No.7 silica sand. Comparison with Toyoura sand, the relative density was low and the volume reduction ratio was high. The volume reduction ratio peaked at frequency 28kHz with output voltage 200V.

4.3 Silica Powder

Figure 6 shows the results in Silica powder. The volume reduction ratio peaked at frequency 28kHz with output voltage 200V, and its value was 27.3%. There are many voids in Silica powder before irradiating the ultrasonic wave. Hence the volume reduction ratio of Silica powder becomes high.

4.4 Comparison In the 30% Cumulative Average Particle Diameter

Figure 7 shows the relationship between the 30% cumulative average particle diameter of samples and the relative density or volume reduction ratio, respectively. Regarding the horizontal axis, 0.2mm corresponds to Toyoura sand, 0.1mm corresponds to No.7 silica sand and 0.006mm corresponds to Silica powder. The relative density graphs are rising to the right and the volume reduction ratio graphs are dropping to the right, approximately. These trends are evident regarding the physical properties of the experimental materials. When the particle size decreases, the difference between the maximum void ratio and the minimum void ratio increases; That is, the results depend on the particle size. However, this study has not been clarified the behavior in the sample which the particle size is mixed. Therefore, it is necessary to conduct experiments and clarify the behavior of an actual dredged soil. These are challenges for the future.

Focusing on the frequency of the ultrasonic wave, when the frequency is 28kHz, the relative density and the volume reduction ratio is peaked overall. Hence, the frequency does not relate to the value of the relative density and the volume reduction ratio.

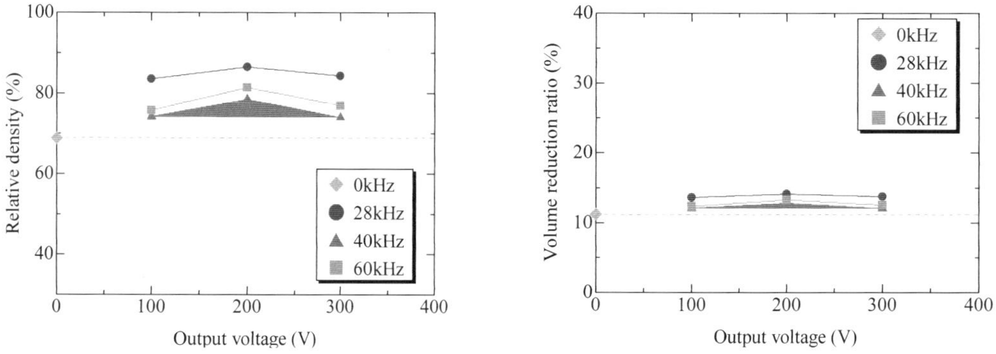

Figure 4. Experimental results in Toyoura sand.

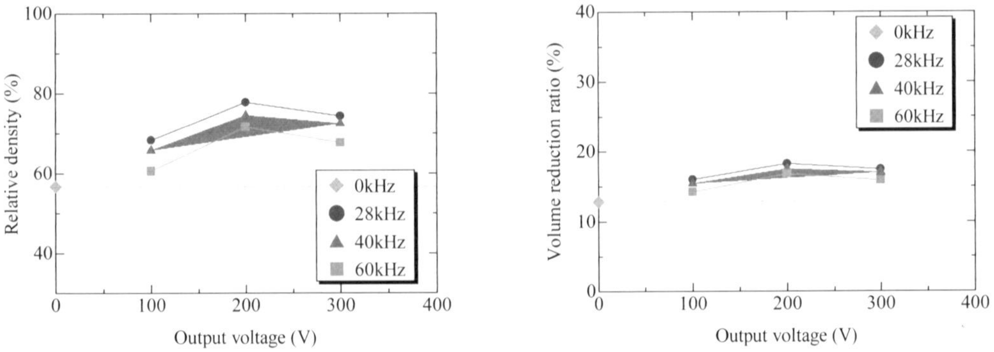

Figure 5. Experimental results in No.7 silica sand.

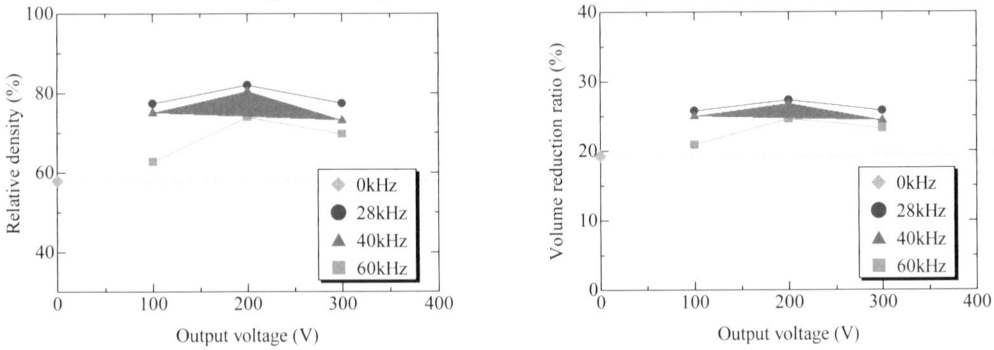

Figure 6. Experimental results in Silica powder.

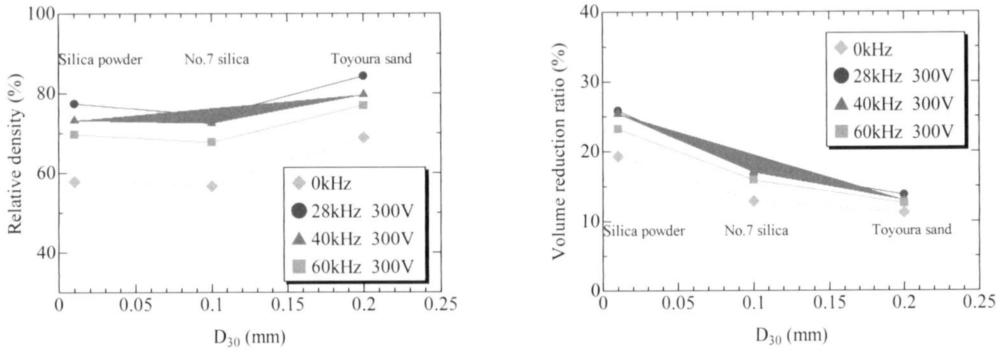

Figure 7. Comparison of the relationship between the 30% cumulative average particle diameter of samples and the relative density or volume reduction ratio.

5 CONCLUSIONS

The object of this study was to establish an efficient volume reduction method for sediment disposal facilities. From the results of the experiment on three kinds of samples with the ultrasonic wave, the following conclusions were derived:

- Irradiation of the ultrasonic wave from the horizontal direction was effective for increasing the density of fines.

- The results were extremely related to the particle size of the sample. The smaller the particle size was, the smaller the relative density was and the larger the volume reduction ratio was.
- In any of the samples, the volume reduction ratio was maximized at the ultrasonic wave of frequency 28kHz with output voltage 200V. Consequently, we couldn't find out a correlation between the particle size and the frequency.

References

Ohuchi M., Murakami T., Hyodo M., and Yoshimoto N., Fundamental Study on A Constructing Method of The High Density Sand Pile by Irradiation of The Ultrasonic Wave by the Practical Size, *Journal of Japan Society of Civil Engineers C*, Vol.66, No.4, 776-787, 2010.

Streamlining Information Transfer between Construction and Structural Engineering
Edited by Shiau, J., Vimonsatit, V., Yazdani, S., and Singh, A.

ISBN: 978-0-9960437-7-9

TURK SALTY MORTAR THROUGH ADDITION OF PLASTIC AND COTTON WASTE MATERIALS TO STABILIZE PROBLEMATIC COASTLINES

AFSHIN TURK[1], BAHRAM MOMBENI[2], SOHRAB AFRASIABI[3] and SHABNAM GHANAVATIZADEH[4]

[1]*Dept of Dam –Developing and Utilization, Ministry of Power, KWPA, Ahwaz, Iran*
[2]*Dept of Water Supply, Ministry of Power, KWPA, Ahwaz, Iran*
[3]*KTNSE, East South Water Company, KWPA, Ahwaz, Iran*
[4]*Dept of Biology, PNU University, East of Tehran, Iran*

Invention of Turk Salty Concrete is recommended for the GOTVAND Reservoir Dam, SW of Iran. This new type of concrete can be mixed with pure salty water to stabilize the salty domes. For the development of the mix, a ratio was defined as the proportion of the salty water mass to the mass of the materials used (clay, bentonite, cement type II). These proportions varied from 0.8 to 1.80 in the mix. The proposed mix presented a significant compressive strength when mechanically tested. The maximum bearing capacity was measured to be more than 10,500kpa. This is a significant capacity for salty cementation mixtures. In this paper, new mixture designs are reported developed by adding waste materials such as plastic bottles and cotton fibers. As the waste materials are detrimental to the environment and coastlines health, the inclusion of waste would have the added benefits of helping in the environmental issues. The results were found to be encouraging since it is shown that they improve durability, breaking stress and the modulus of toughness. Other research studies are currently underway examining the behavior of Turk Salt Mortar (TSM) by adding wasste materials such as date palm tree fibers. TSM can be used in deep injection grouting for protection of shorelines that come in contact with sea water.

Keywords: GOTVAND Dam, Fibers, Palm tree, Salty, Shorelines, Injection grouting, Clay, Bentonite.

1 INTRODUCTION

In ancient Egypt (Waltari 1945), bodies were mummified by filling them with pure salt to preserve them. Similar findings were discovered in the Chehrabad salt mines of Zanjan, Iran (Saltmen 1993). Similarly, the TSM mixes incorporate a huge amount of pure salt with clay and cement to form and possess rock-like properties (Turk 2007, 2008). TSM mix was shown to be a mix that could be used appropriately in dry or saturated conditions (Turk 2017). The mix is produced by the mixture of ultra-saturated salt water, Shushtar Ceramic Clay, Bentonite and Cement type II (Turk 2006, 2018; Damough 2011). This research study began in 2016 and was continued by the addition of waste materials. The new mix displays a different behavior in comparison with the original mixes. The broken, unreinforced samples show 30°-60° plane slope rupture as shown in Figure 1. When fibers were added, vertical failure cracks appeared in the broken samples that showed improvement in failure resistance (Figure 2). Figure 2 also shows failure patterns of samples when date palm tree fibers, cotton, and plastic shredded wastes were added.

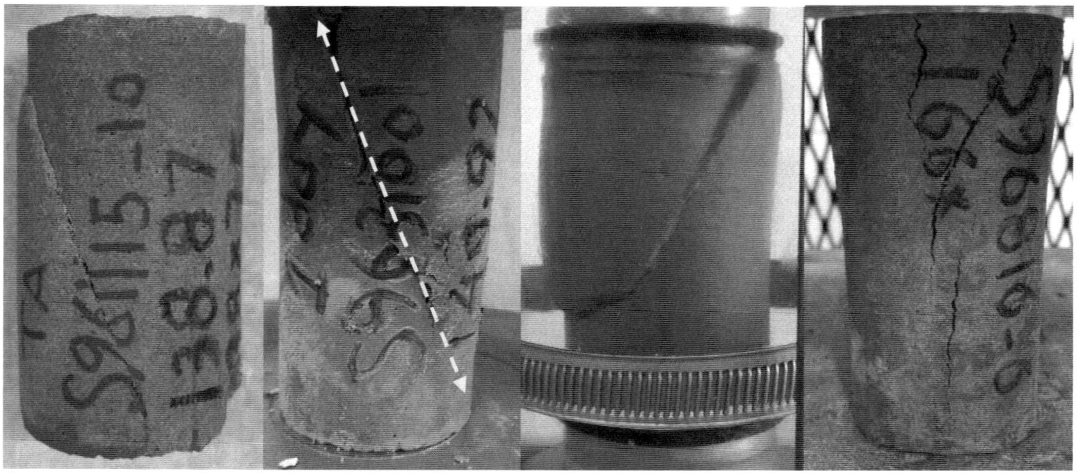

Figure 1. Pictures demonstrate mortar behavior at 30°-60° plane stress of non-fibers (KTNSE 2016, 2017).

Figure 2. Fibers: A) Palm Tree 1.5% B) Cotton 0.11% C, D) Useless Plastic 2.13% (KTNSE 2017, 2018).

A tri-axial UU test was conducted for sample S970216P-2 per BS standard 1377 Part 7, at two days age, with 2.1% weight ratio of plastic fibers at a cell pressure of 600kpa, strain 6.27%, and rapture stress 1'853kpa. Sample S970216C-2 was also tested per the BS standard at two days age with 1.54% cotton fibers, strain 1.81%, and rapture stress 7'240kps. It seems that the sample S970216P (plastic fibers) has more flexibility behavior than S970216C (cotton shards). In the initial days, flexibility of plastic fibers is 400% more than cotton fibers. This ratio decreases in the following days when the age of samples rise to 30days. This value is incredible and remarkable in contrast to the previous samples that were tested by TSM (Turk 2017). In Figure 2, fibers of plastic stick inside the TSM strongly, which acts such as a reinforced concrete. It is noted that the mortar is created by an ultra-salty mixture.

2 MATERIALS OF TSM

TSM is a mixture which occurs in the natural Anbar salt mine, and consists of water, Shushtar Ceramic Clay, Bentonite and Cement type II. The study aimed to stabilize the GOTVAND salty domes inside the fresh lake reservoir. TSM can be injected under pressure in deep formations (Gutiérrez 2016). Figure 3 explains the ratio of materials that should be mixed for TSM.

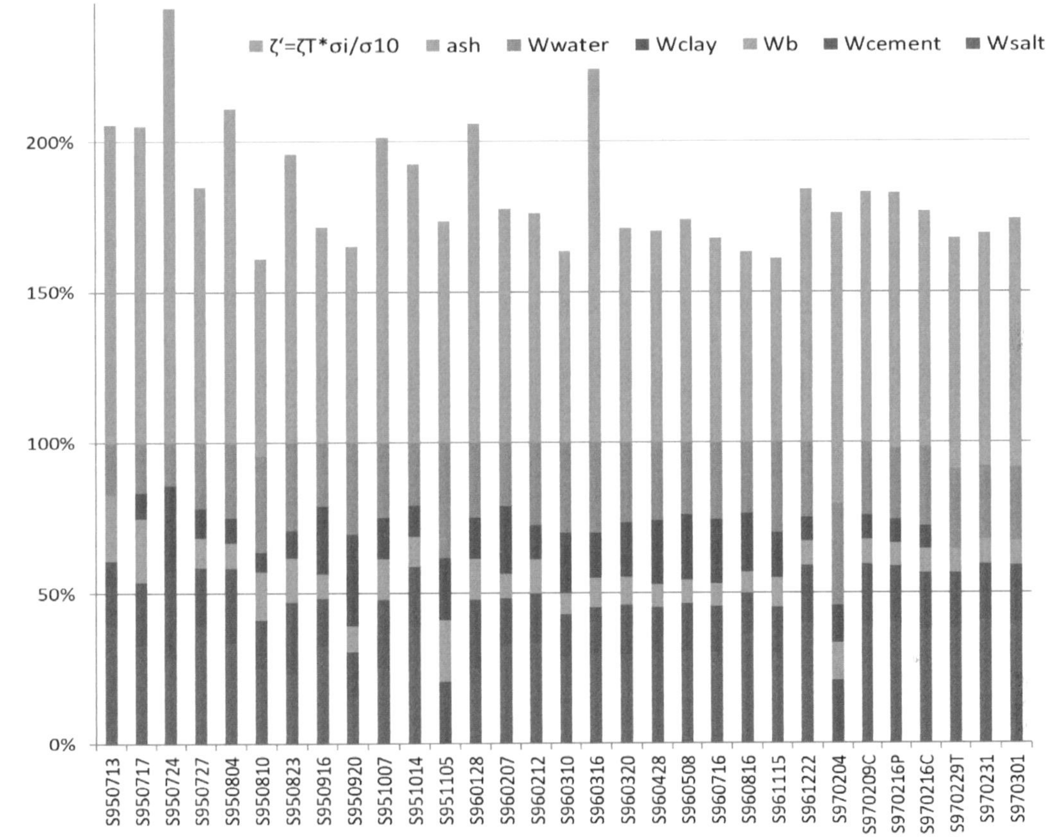

Figure 3. Materials will use in TSM Samples S(yy, mm, dd, ages, fiber, σ_Z) mixture design (Turk 2017).

2.1 Criterion of TSM Resistance and MTE

The maximum breaking stress was 13,800 kpa as registered by the compressive hydraulic jack. It belonged to Sample S950713 at 26 days on October 4th 2016. It can be selected as a relative criterion for all samples. Eq. (1) to Eq. (4) interpret the relative criterion (Turk 2018) and modulus of toughness energy per volume (Beer 2012). Samples have different cell pressures.

$$\zeta_{Turk}=\frac{W_{SaltWater}}{W_{Clay}+W_{Bentonite}+W_{Cement}} \tag{1}$$

$$\zeta'_{Turk}=\frac{\left(\sigma_{yymmdd0}\right)_i}{\sigma_{9507130}}\times\left(\zeta_{Turk}\right)_i \tag{2}$$

$$\left(U_{TSM}\right)_i = \int_0^{\varepsilon_{Ri}} \sigma_{Zi}.d\varepsilon_{Zi} \tag{3}$$

$$\left(\psi_{Turk}\right)_i = \frac{\left(U_{TSM}\right)_i}{\left(U_{TSM}\right)_\circ},\ i = 1, 2, 3, \ldots\ldots,\ \left(U_{TSM}\right)_\circ = \underbrace{\int_0^{\varepsilon_{R\circ}} \sigma_{Z\circ}.d\varepsilon_{Z\circ}}_{S970216P2-600kpa} \tag{4}$$

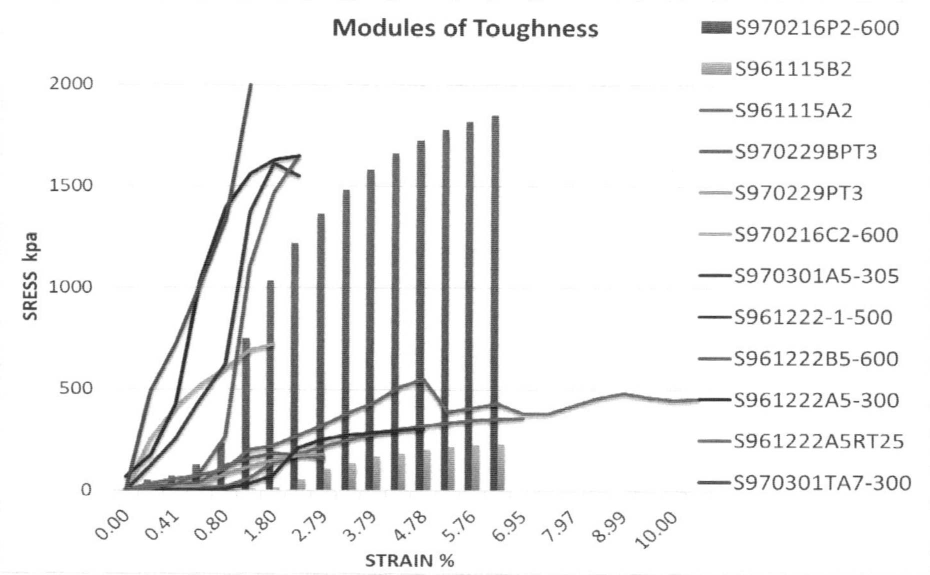

Figure 4. Modules of toughness energy respect to their σ_Z, σ_X and sample ages (KTN 2018).

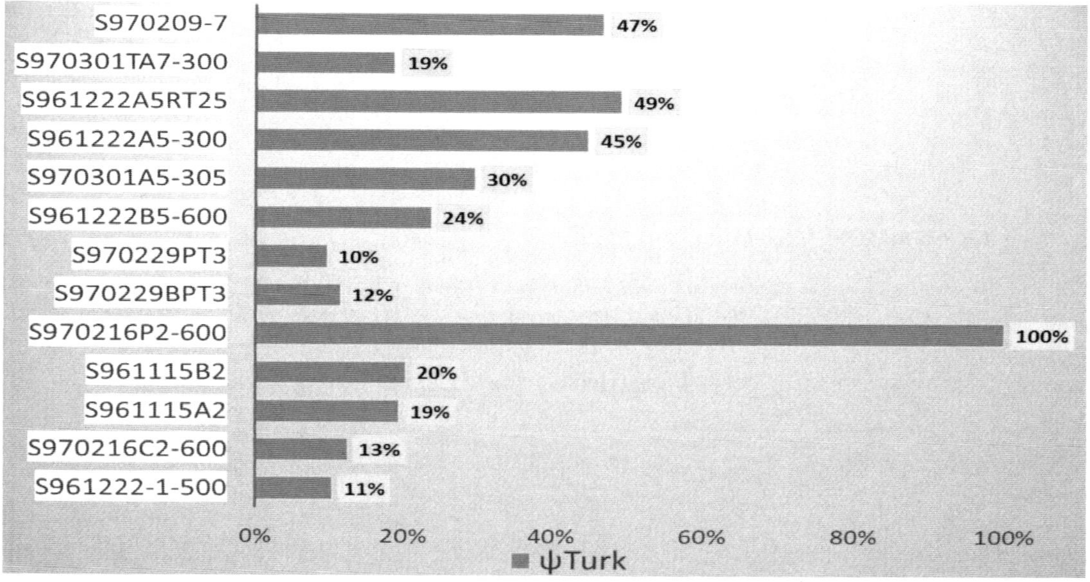

Figure 5. The value of ψ_{Turk}S970216P2-600kpa plastic fibers is the highest TSM (KTNSE 2017, 2018).

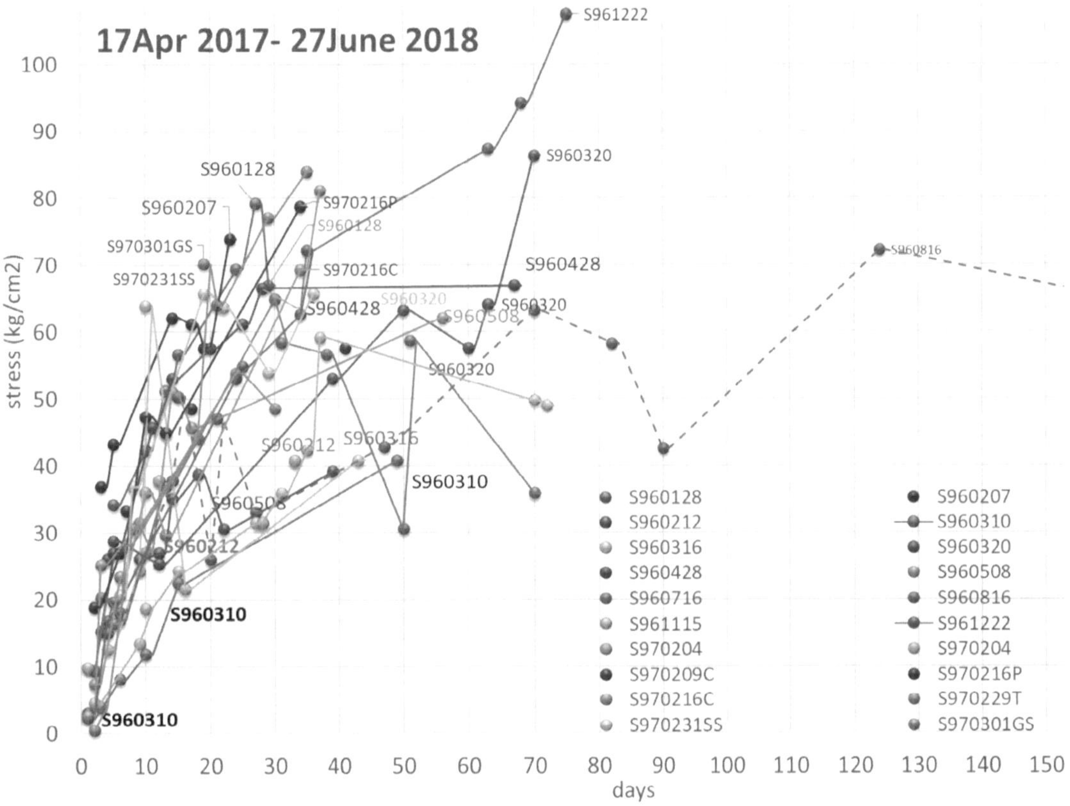

Figure 6. Breaking resistance of Samples registered from 17 April 2017 till 27 June 2018.

2.2 Criterion of TSM permeability

TSM is slightly different from the ACI 318 manual for concrete structural mixture design. In permeability testing, horizontal cross section is taken to obtain the percentage of water leakage. Eq. (5) to Eq. (7) and Table 1 explain Darcy's law of permeability using the cylindrical samples (Das 2010). Tri-Axial samples were tested by saturation rate, S97020920P100%, S970030135G and S970231S0.0% (60 min in 100^{oC} oven), in a pressure chamber with longitudinal sections.

$$Q=\frac{\Delta V}{\Delta t}=\frac{(W_1-W_\circ)}{\Delta t}=k.A.i=k\cdot(d\times h)\cdot\frac{\Delta H}{h} \quad (5)$$

$$k=\frac{Q}{d\cdot\Delta H}=\frac{0.0009}{3.8\times7'000}=2.9\times10^{-3}\ (cm/day) \quad (6)$$

$$\delta_{Turk}=\%\frac{\Delta W}{W_\circ} \quad (7)$$

Table 1. Permeability and water adsorption calculation, d=38mm, ζ_{Turk} =1.80, 1.86, 1.85 (KTNSE 2018).

Sample	Age (day)	Fiber %	W1 (gr)	W0 (gr)	Sr %	h (cm)	Δt (s)	Δh (cm)	Q (cm^3/s)	K (cm/day)	δ_{Turk} %
S970209P	20	0.11	139.44	138.63	100	8.7	900	7000	9.0E-4	2.9E-3	0.40
S970301G	5	0.0	156.05	155.29	35	7.9	1380	8366	5.5E-4	1.5E-3	0.49
S970231S	29	0.0	161.3	153.77	0.0	8.7	1800	7110	4.2E-3	1.34E-2	4.90

3 CONCLUSION

Salty Domes (HARZA 1968) are hazardous. Correspondingly, an alarm was conveyed to the government by NGOs and the local residents. The Ministry of Power declared that there is no feasible solution to stabilize the Salty formation of GOTVAND Lake. However, TSM is the only way to protect the fresh water against the impact of salts. Saltified TSM can be accepted, per the results of a pilot field using grouting operation inside the salty layers. Tentative samples proved the beneficial behavior of TSM by breaking resistance, durability in water saturation, and in elasticity behaviors. Unused plastic, cotton and palm fibers are combined with TSM cementation, which improves elasticity and strain behavior. Figure 4 and Eq. (4) show the ψ_{Turk} increasing in ratio with unused fibers, in contrast to non-fiber conditions. Unused fibers sharpen the σ-ϵ curve. Also, the toughness modulus increases during initial strain ($\epsilon < 2.0\%$). Modification behavior is clearly represented in Figure 5. Unfortunately, the unused plastic bottles increase day-to-day to destroy the shorelines and natural beauty of coasts. By the adoption of this method, onshore protective structures may be designed to save the dunes, estuary, ports, and more through the use of un-reinforced concrete. The TSM fibers axial test realized more than 8,000 kpa at 36days with ζ_{Turk}=1.85 proportion. Sample resistance grows till 100 days.

References

Beer, F. P. E., Johnston, R., Dewolf, jr. T., Mazurek, D. F., "Mechanic of Materials." *The McGraw-Hill Companies,* Inc., 6th edition, 2012.

Damough, N. A., Zarei, H. "Investigating of Dissolution Problems in Gachsaran Formation Salt Layers in GOTVAND Olia Dam Reservoir." Research KUD-84001, KWPA, Iran, 2011.

Das, B, M., "Principles of Geotechnical Engineering", *Cengage Learning Inc.,* USA, 2010.

KTNSE, "SE Water Supply, Act of 2016. Pub L. No. 951027." KWPA. 950708 IR.SE, September 2016.

KTNSE (Kavosh Taraz Niro Soil Engineering), Gotvand Test Act of 2016, research work, Laboratory Soil-Concrete, KTNSE–44890-SE, Pub L., No.951027 stat. 960119-K9631, May 2017.

KTNSE, Gotvand Test Act of 2018, research work, Laboratory Soil-Concrete, KTNSE–44890-SE, Pub L., No.1200/126238 stat. 970217-K9631, 2018.

Gutiérrez, F., and Lizaga, I., "Sinkholes Collapse Structures and Large Landslides in an Active Salt Dome Submerged by a Reservoir: The Unique Case of the Ambal Ridge in the Karun River, Zagros Mountains, Iran." Journal of Geomorphology, Elsevier, 254, 88–103, 2016.

HARZA, Consulting Engineering International, Dam Power Plant Appurtenances, Contract Documents 384-3, KWPA, August 1968.

Saltmen, National Museum of Iran, https://en.wikipedia.org/wiki/Saltmen on June 1993.

Turk, A., Stabilization of Pure Salty Formations of the GOTVAND Dam Lake and another Salty Drought Desert Regions through Invention of Turk Salty Mortar", Aerospace Division of ASCE, USA, 2018.

Turk, A. and Ghanavatizadeh, S. "Micro-pile root net theory to modify earthquake and liquefaction properties of soils based on biologic concepts, laboratory and field tests", presented at the Conference EARTH&SPACE 2006, Aerospace Division of ASCE, USA, 2006.

Turk, A., Ghanavatizadeh, S., Zaamari, A.A., and Kolahchi, A., "Regeneration of missed record data, vertical axes of Karkheh earth dam using cell pressure, mathematical aspects and SSM", Aerospace Division of ASCE, USA, 2008.

Turk, A., Salehi, F., Kolahchi, A., and Ghanavatizadeh, S., "Elasticity Module of Karkheh earth dam study & behavior through inclinometers & SSM", ISEC-4, Melbourne, Australia, 2007.

Turk, A. and Zaemeri, A. A., "Micro pile behaviors study using compressive (or tensile) pile load test and steps stiffness method", CSCE, Saskatoon, Canada, 2004.

Turk, A., Mombeni, B., Bahmaei, D., Khodabakhshi, H. R., Behdarvandi, A. M., and Ghanavatizadeh, S., "Coastal Protection of GOTVAND Reservoir DAM, EC Perfection of Salty Domes", ISEC-09, Valencia, Spain, 10.14455/ISEC.res.2017.61, 2017.

Waltari, M., *Sinuhe Egyptiläinen, WSOY Press,* Finland, 1945, Persian translation by Mansuri, Z., 1985.

Streamlining Information Transfer between Construction and Structural Engineering
Edited by Shiau, J., Vimonsatit, V., Yazdani, S., and Singh, A.

ISBN: 978-0-9960437-7-9

A SIMPLE CONSTITUTIVE MODEL FOR UNDRAINED SHEAR CHARACTERISTICS OF SANDY SOILS

YANG WU

School of Civil Engineering, Guangzhou University, Guangzhou, China

A vast amount of past experimental examinations reported that the internal peak angle of sand was jointly affected by the density and effective stress level. Several relationships were proposed between these elements. The dependence of dilatancy characteristics on the internal state of a granular material was examined and revealed. A simple constitutive model framework was established on a basis of several well-proven and experienced relationships for granular materials to simulate their undrained shear behavior. A basic hardening law connecting the varying tendency of the stress ratio with shear strain was employed. This model is capable of predicting the undrained monotonic stress-strain relationship of granular materials at different densities and various confining pressures. A series of parametric studies are conducted to investigate the susceptibility of the simulation results to the selected parameters. The simulation results also confirm the influential influences of dilatancy and deformability on the shear characteristics of granular materials at the critical state.

Keywords: Undrained shear behavior, Granular material, Stress-dilatancy, Constitutive relation, Critical state.

1 INTRODUCTION

The shear strength and deformation characteristics of granular soils subjected to shearing are of great interest in geomechanics and geoengineering practice (Bolton 1986). The sands during shearing tend to contract at loose state and show dilative tendency at dense state. Thus, the density state is regarded as an influential factor governing the shear response of sands. The mean stress level is also an important parameter affecting the shear response of sand.

In undrained conditions, flow failure of loose sand resulting from the steady state deformation following the unstable behavior is liable to occur. The steady state (Castro and Poulos 1977) is a state of deformation of soil without effective stress increment or decrement and with migration of pore water. The steady state is regarded as anonymous to the critical state in many experimental investigations (Been *et al*. 1991, Yamamuro and Lade 1998). The latter term is employed in the following section. Shear strength at the critical state was believed to be the undrained strength of granular soils with a flow failure. Flow deformation would continue infinitely once the outer force exceeds the undrained shear at the critical state. It was realized that the determination of shear strength at the critical state was more important than the decision of the stress triggering liquefaction during associated evaluations.

Some well-proven equations acquired from the drained triaxial tests are combined to formulate a simple constitutive model with the capacity of simulating the shear response of grain soils in this investigation. This study intends to provide a better understanding of the shear properties of granular materials with particular emphasis on the existence of critical state. The establishment of this simple model is also expected to be employed to evaluate the liquefaction phenomena of sand in laboratorial test.

2 CONSTITUTIVE MODEL

Some well-known equations associated with the strength and deformation characteristics of granular materials were summarized based on a large amount of experimental data in past investigations. Relevant relationships for this simple constitutive model are reviewed as below.

2.1 Peak Shear Strength Estimation

Past investigations revealed that the peak friction angle ϕ_p of granular materials was jointly governed by the levels of initial density and confining pressure (Bolton 1986, Gutierrez 2003, Lashkari 2009). A rise in density and a reduction in confining pressure enhance the peak friction angle ϕ_p of granular soils. The peak friction angle ϕ_p is composed of the frictional angle at the critical state ϕ_{cr} and the portion due to dilation behavior. The friction angle ϕ is correlated with the stress ratio η by the equation $\sin\phi=3\eta/(6+\eta)$ at any state during triaxial shearing. Analogous to this relationship, the peak stress ratio of granular materials η_p could be expressed using Eq. (1).

$$\eta_p = \eta_{cr} + CD_r \ln(\frac{p_{cr}}{p}) \tag{1}$$

where η_p is the peak stress ratio, η_{cr} is the stress ratio at the critical state, D_r is the relative density and C is a material parameter related to the dilatancy characteristics of soils. p_{cr} refers to the mean stress when the peak stress ratio reduces to the stress ratio at the critical state. The peak stress ratio η_p only emerges for the soil at the dense state. η_p approaches to η_{cr} at the critical state as the mean stress p is increased to be identical to p_{cr}. It is indicated that the peak stress ratio η_p is independent on the density and confining pressures once the mean stress p exceeds p_{cr}. Besides, p_{cr} is also greatly affected by the level of density.

2.2 Stress-dilatancy

Stress-dilation relationship is also an important mechanical characteristic to describe the shear response of soils. This relationship captures the evolution of strain increment in accompany with the stress ratio of granular materials in drained shearing test. Well prediction of the stress-dilatancy characteristic plays a major role in the usefulness of a constitutive model. Some classical expressions were proposed to characterize the stress-dilation behavior. The empirical equation in Cam-clay model (Roscoe *et al.* 1963) is written in Eq. (2).

$$D = M - \eta \tag{2}$$

where $D=d\varepsilon_{vd}/d\varepsilon_s$ is the rate of dilation, $d\varepsilon_{vd}$ is the volumetric strain increment due to dilation and $d\varepsilon_s$ is the shear strain increment. M represents the stress ratio at the critical state and is regarded as synonymous to η_{cr} in Eq. (1). It is noted that the summary of rate of dilation D and stress ratio η is a constant in Eq. (2). Li and Dafalias (2000) proposed a state-dependent stress-dilatancy form using the state parameter ψ as the state variable in Eq. (3).

$$D = d_o\left(e^{m\psi} - \frac{\eta}{M}\right) = \frac{d_o}{M}\left(Me^{m\psi} - \eta\right) \tag{3}$$

where m and d_o are two positive parameters. It is noticed that the Eq. (2) is the specific case of Eq. (3) when m is equal to zero and d_o is identical to M. At the critical state, the $\psi = 0$ and $\eta = M$ lead to a zero dilatancy. Thus, Eq. (3) is positively included in this simple model.

2.3 Distortion Hardening Law

The distortion relationship linking the development of stress ratio η with shear strain $d\varepsilon_s$ was proposed and well-proven in the hyperbolic stress-strain model (Duncan and Chang 1970). This simple hardening law assumes that granular soils are constantly approaching to the peak shear stress ratio η_p with increasing shear strain ε_s in Eq. (4).

$$\eta = \frac{\eta_p \varepsilon_s}{A + \varepsilon_s} \tag{4}$$

where A is a parameter associated with the initial shear modulus of soils.

2.4 Model Construction

Particular expressions reviewed above are combined and formulated to create a new simple constitutive model. The specific procedure to implement the calculation using the proposed simple model is explained in this section. The total volumetric strain increment $d\varepsilon_v$ is supposed to be composed of the portion due to consolidation $d\varepsilon_{vc}$ and the portion induced by dilation $d\varepsilon_{vd}$. The volumetric strain increment due to consolidation $d\varepsilon_{vc}$ in Eq. (5) could be calculated from the relationship between the mean stress p and the void ratio e during isotropic consolidation loading.

$$d\varepsilon_{vc} = \frac{\lambda}{1+e_o}\frac{dp}{p} \tag{5}$$

where λ is the slope of isotropic consolidation line, e_0 represents the initial void ratio of sand. The total volumetric strain increment $d\varepsilon_v$ could be acquired using Eqs. (3) and (5)

$$d\varepsilon_v = d\varepsilon_{vc} + d\varepsilon_{vd} = \frac{\lambda}{1+e_o}\frac{dp}{p} + \frac{d_o}{M}\left(Me^{m\psi} - \eta\right)d\varepsilon_s \tag{6}$$

For undrained condition, the total volumetric strain increment $d\varepsilon_v$ should be zero for a saturated sample when the pore fluid and soil particle compressibility are neglected. Thus, the expression in Eq. (6) is rearranged to display the mean stress increment dp,

$$dp = -\frac{p(1+e_o)}{\lambda}d_o\left(e^{m\psi} - \frac{\eta}{M}\right)d\varepsilon_s \tag{7}$$

Putting the Eqs. (4) and (1) into the Eq. (7), we further obtain the exact form of mean stress increment dp,

$$dp = -\frac{p(1+e_o)}{\lambda}d_o\left[e^{m\psi} - \frac{\varepsilon_s}{\varepsilon_s + A}\left(1 + \frac{CD_r}{M}\ln\frac{p_{cr}}{p}\right)\right]d\varepsilon_s \tag{8}$$

Eq. (8) descripts the variation in mean stress dp with the rise in shear strain increment $d\varepsilon_S$during undrained shearing in triaxial shearing test. The calculation implementation of dp is integrated numerically. Firstly, dp could be decided using Eq. (12) for a given increment $d\varepsilon_S$. Secondly, mean stress ($p \leftarrow p+dp$) and shear strain ($\varepsilon_S \leftarrow \varepsilon_S + d\varepsilon_S$) could be updated using superposition method. Then, peak stress ratio η_p could be re-determined using the new value of p and Eq. (1). The stress ratio η could also be updated using Eq. (8) to decide the new deviatoric stress q. Subsequently, the iteration calculation would continue with the same given increment $d\varepsilon_S$. Whole calculation would be completed once the desire level of shear strain is attained.

3 SIMULATION RESULTS

The simulation capacity of this simple constitutive model is demonstrated in this section. Figures 1-3 show the measured and predicted stress-strain response of Toyoura sand at loose, medium-dense and dense states at various confining pressures. In each figure, subtitle (a) describes the effective stress paths of sands and (b) shows the corresponding stress-strain curves of sands. The solid symbols are the measured results and the solid lines represent the predicted values. Verdugo and Ishihara (1996) conducted an extensive set of monotonic drained and undrained triaxial tests on Toyoura sand at a wide range of stresses and densities to determine the ultimate condition of sand. The initial void ratios of Toyoura sand are prepared as 0.907 (loose), 0.833 (medium-dense), 0.735 (dense). A set of high quality data was reported. The maximum and minimum void ratios of the Toyoura sand are 0.977 and 0.597, respectively.

The prediction values have a good agreement with the measured results of Toyoura sand subjected to undrained shearing in Figures 1-3. This constitutive model has capacity of describing the pure contractive behavior of loose samples at confining pressures of 1.0 and 2.0 MPa. This simple model is capable of predicting dilative behavior of medium-dense sand at confining pressures of 0.1 MPa and 1.0 MPa and contractive behavior of medium-dense sand at confining pressures of 2.0 MPa and 3.0 MPa in Figure 2. Both of predicted and measured results show that the deviatoric stresses of medium-dense sand at confining pressures of 0.1 MPa and 1.0 MPa monotonically increase with the shear strain progresses. However, medium-dense sand rapidly grains the peak shear strength and gradually loses the deviatoric stress with the rise in shear strain. For dense sand, simulated values are generally consistent with the observation results in Figure 3. The slight deviation of predicted result from the measured value at a confining pressure of 0.1 MPa in Figure 3 is probably resulted from the strong dilation estimation at low confining pressures by the selected stress-dilatancy relation in Eq. (7).

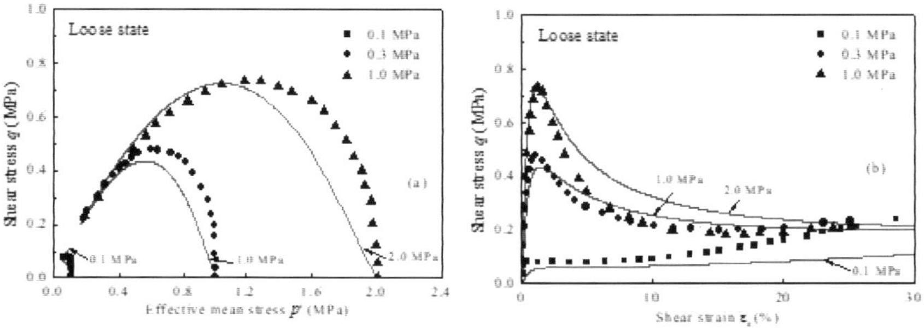

Figure 1. Predicted and measured stress-strain response of loose sand.

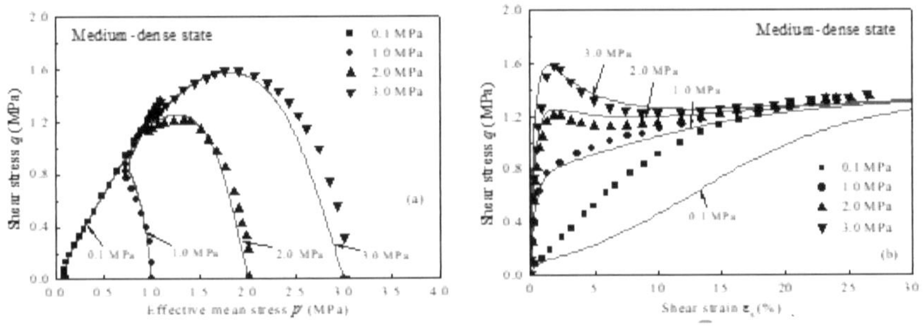

Figure 2. Predicted and measured stress-strain response of medium-dense sand.

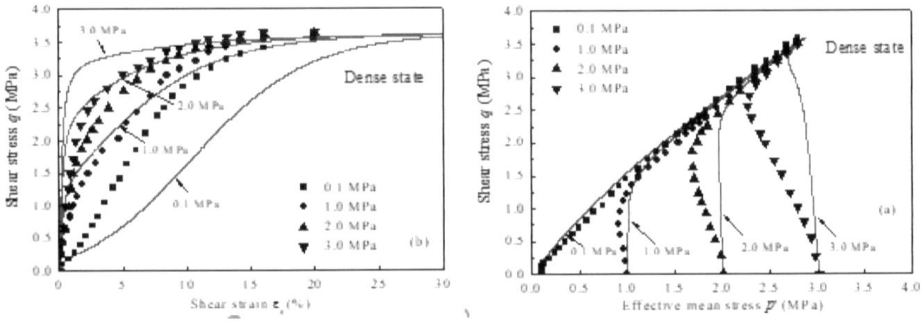

Figure 3. Predicted versus measured stress-strain response of dense sand.

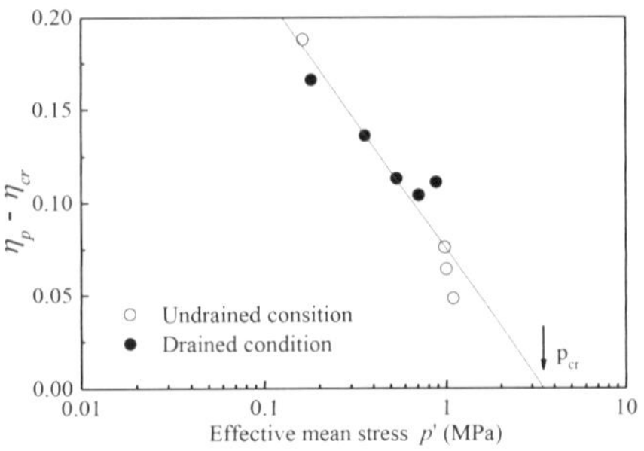

Figure 4. Determination of the mean stress at the critical state p_{cr}.

It is noted that the simulation results of sand specimens at different confining pressures and initial densities approach to the constant deviatoric stress q, which is the hint of the critical state as the axial strain progresses. It is indicated that the critical state could be attained at larger shear strains around 20% through the simulation using this simple constitutive model. Table 1 shows the set of simple model parameters for Toyoura sand at three densities. The determination method of the associated parameters is simply explained. The difference in the stress ratios at the peak state and the critical state are plotted against the effective mean stress to determine p_{cr} using Eq. (5) in Figure 4. It is seen that the rise in density increases the material parameter C and level of p_{cr}. With regard to the parameters related to the stress-dilatancy and the location of critical state line, they are independent on the density of sand.

Table 1. Constitutive model parameters for sand with different densities.

	C	D_r (%)	p_{cr} (MPa)	m	d_0	M	e_Γ	λ_{csl}	ξ	λ	A
Loose (e_0=0.907)	0.13	18.5	0.28	3.5	0.88	1.25	0.934	0.019	0.7	0.028	0.007
Medium-dense (e_0=0.833)	0.37	37.9	1.2	3.5	0.88	1.25	0.934	0.019	0.7	0.035	0.005
Dense (e_0=0.735)	0.47	63.7	3.0	3.5	0.88	1.25	0.934	0.019	0.7	0.050	0.002

4 CONCLUSIONS

A simple constitutive model was proposed based on some well-established relationships for sand obtained from drained shearing test to predict the undrain shear response and the static liquefaction of granular materials. The combined effects of density and mean stress level on the peak shear strength was considered and reflected in the relevant expressions. A rise in density and decline in the confining pressure lead to a marked dilation behavior of granular material. These features are reflected by the stress-dilatancy relationship considering the state of granular materials. The parameters of constitutive model can be determined form the triaxial shear test. The proposed simple constitutive model has the capacity of predicting density-dependent and stress-dependent undrained shear properties of granular material using a few parameters. The predicted results show good agreement with the measured values for sand at various densities and confining pressures.

References

Been, K., Jefferies, M. G., and Hachey, J., The Critical State of Sands, *Géotechnique*, ICE, 41(3), 365–381, March, 1991.

Bolton, M. D., The Strength and Dilatancy of Sands, *Géotechnique*, ICE, 36(1), 65–78. January, 1986.

Castro, G., and Poulos, S. J., Factors Affecting Liquefaction and Cyclic Mobility, *Journal of the Geotechnical Engineering Division*, ASCE, 103(6), 501–506, June, 1977.

Duncan, J. M., and Chang, C.-Y., Nonlinear Analysis of Stress and Strain in Soils, *Journal of the Soil Mechanics and Foundations Division*, ASCE, 96(5), 1629–1653. May, 1970.

Gutierrez, M. S., Modleing of the Steady State Response of Granular Soils, *Soils and Foundations*, Elsevier, 43(5), 93–105, May, 2003.

Lashkari, A., On the Modeling of the State Dependency of Granular Soils, *Computers and Geotechnics*, Elsevier, 36(7), 1237–1245, July, 2009.

Li, X. S., and Dafalias, Y. F., Dilatancy for Cohesionless Soils, *Géotechnique*, ICE, 50(4), 449–460, April, 2000.

Roscoe, K. H., Schofield, A. N., and Thurairajah, A., Yielding of Clays in States Wetter than Critical, *Géotechnique*, ICE, 13(3), 211–240, March,1963.

Verduguo, R., and Ishihara, K., The Steady State of Sandy Soils, *Soils and Foundations*, J-Stage, 36(2), 81–91. February, 1996.

Yamamuro, J. A., and Lade, P. V., Steady-State Concepts and Static Liquefaction of Silty Sands, *Journal of Geotechnical and Geoenvironmental Engineering*, ASCE, 124(9), 868–877, September, 1998.

Streamlining Information Transfer between Construction and Structural Engineering
Edited by Shiau, J., Vimonsatit, V., Yazdani, S., and Singh, A.

ISBN: 978-0-9960437-7-9

BIM-BASED APPROACH FOR SEISMIC RISK ANALYSIS

VALENTINA VILLA, MARCO DOMANESCHI, GIAN PAOLO CIMELLARO, and CARLO CALDERA

Dept of Structural, Geotechnical and Building Engineering, Politecnico di Torino, Torino, Italy

The Italian school buildings asset consists of over 40,000 units. The most (more than 60%) were built before the introduction of the national standard on school buildings and constructions in seismic areas. The present research aims to implement a methodology that consists in an informative modelling for seismic risk analysis. The objective of the activity is to provide the policy makers of a useful tool for screening the existing building stock, in order to define the priorities of intervention. The research is divided into several parts. First of all, the most recurrent building technologies have to be defined with respect to the year of construction and the structural characteristics. Furthermore, to assess the seismic risk, the seismic hazard has also to be analyzed. Next, a multiphase process of increasing complexity has to be defined, in which two approaches of seismic analysis are tested. The first one is related to "simple" construction technologies (e.g. frame structures) where information can be collected through visual screening during inspections and through the study of the existing documentation. The second one is a more refined approach that includes non-destructive testing on site and structural analysis. Both procedures lead to the definition of BIM models.

Keywords: School building, Structural monitoring, Numerical modelling, Simulations, Seismic vulnerability.

1 INTRODUCTION

The potential of Building Information Modeling for seismic risk analysis is closely linked to BIM as a relational database of buildings information. Started from '70 with Eastman *et al.* (2011), nowadays BIM is a methodology that is changing dramatically the AEC sector, thank to innovation in IT technology. There are many and many advantages using BIM in project and construction process: can support planning and cost estimation, management of design changes, visualization and simulation of design ideas, construction management, and building lifecycle control. Many recent researchers analyzed also BIM for Facility Management (FM) and for existing buildings (Volk *et al.* 2014, Becerik-Gerber *et al.* 2012), and some of them are focused on the use of sensors for monitoring and control of parameters for building uses (Dong *et al.* 2014). BIM represent the future of building management because is a database, which contains all the information about structural and non-structural components vulnerability (FEMA 2006). Some research has also started to define some ways in which Building Information Modelling (BIM) could assist in the assessment and mitigation of seismic risk. (Welch *et al.* 2014).

2 SCHOOL BUILDING STOCK IN ITALY

The school building heritage consists of more than 40,000 buildings, largely constructed before the introduction of the technical regulations on school building (DM 18/2/1975) and before the settlement of special requirements standards for seismic areas (Law No. 64/1974) (Tagliabue and Villa 2017). Indeed, almost 2/3 of schools were built before 1974 and before that date, there was no technical standard in force that guided the design, defining at least the performance aspects.

Accordingly, most of the existing schools were built without static verification and with a level of seismic safety not in line with current standards. Nowadays, the seismic safety is established by define the seismic vulnerability in the technical standards for constructions issued by the Ministry of Infrastructure with Ministerial Decree of 17 January 2018 (NTC – Technical Standards for Construction's structures). Summarizing, it was detected that only 48.5% of school buildings have a static certification, 56% has a static certificate of suitability and only 10.1% are built according to earthquake safety criteria. Concluding this brief overview, 50.13% of school buildings on the Italian territory are located in earthquake were affected areas with a medium-high hazard level (Figure 1).

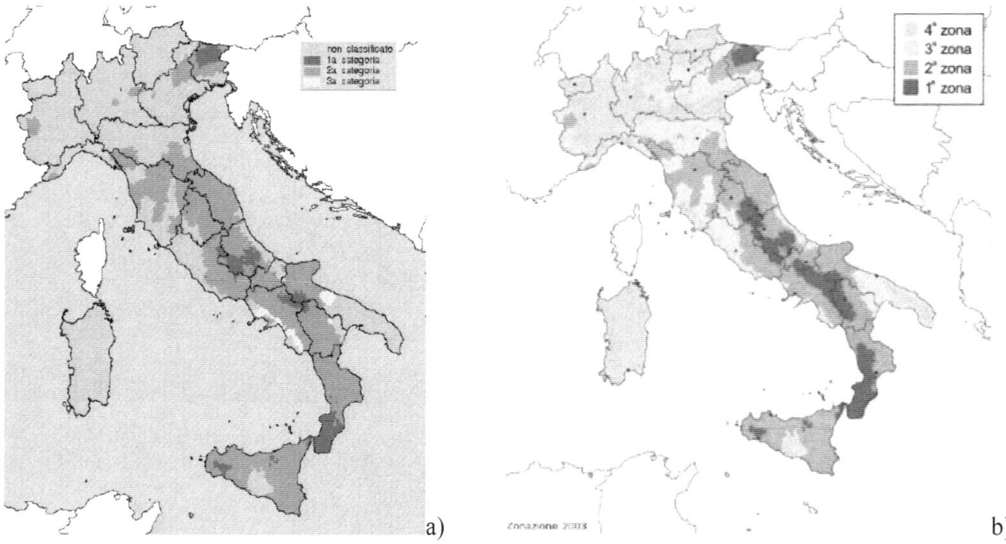

Figure 1. Seismic hazard classification of the Italian territory after the 1980 event in Irpinia region. 45% of the whole territory, 70% of the central-south territory. (a) current Italian earthquake map (OPCM 2003) (b)

3 STANDARDS ITALIAN DEVELOPMENTS

The guidelines for the assessment of seismic risk in Italy has been updated over the years on the basis of catastrophic events that have changed, from time to time, the seismic hazard classification of the territory. In 1980, the Irpinia earthquake, with about 3,000 dead and 10,000 injured, was the first tragic event in recent history and is identified as the moment when an effective policy of seismic risk prevention was firstly introduced, proposing the seismic classification of the earthquake affected areas.

Following the 1984 classification, 45% of Italy's territory is classified as a seismic risk area; about 70% of this is located in central and southern Italy. In addition to this classification, new policies are implemented to prevent and manage emergencies and disaster relief through special actions, such as: improving knowledge through monitoring networks, reducing the vulnerability of older buildings through recovery and redevelopment of the building stock, improving tools for construction interventions. After the collapse of the primary school "Francesco Iovine" in San Giuliano di Puglia (October 31, 2002) that resulted in 28 fatalities, extraordinary measures were taken and the "Plan for the safety of school buildings" and a new earthquake map were issued.

4 METHODOLOGY: WORKFLOW

In order to develop a methodology for using BIM informative model as tool for risk seismic analysis, some details should be defined.

First of all, given the vastness of the school heritage, the most common building technologies will have to be defined, also in relation to the year of construction, and, consequently, the structural characteristics, the mechanical characteristics of structural materials, the distribution of masses, etc. will have to be examined in depth. The seismic hazard, the intended use of the environment and whether any extension, modification, adaptation or improvement work has been carried out or is possible in the future shall be checked. All this information should be contained into BIM model of school existing buildings.

Subsequently a workflow will be created that systematizes two different approaches for the evaluation of the seismic vulnerability, according to a coordination defined by the information BIM modelling (1° step) and to sensors and monitoring data (2° step).

A multi-stage process of increasing complexity will be defined, in which two methodological approaches to seismic analysis will be tested: the first, simplified for "simple" building technologies (e.g., reinforced concrete chassis schools), where information will be collected through visual screening during inspections and through the study of documentation. The second will be further investigated through non-destructive testing and survey. Both procedures will lead to the definition of a new BIM model and will compare the results obtained.

This methodology will lead to use of BIM informative modelling for seismic risk analysis with the final objective of providing a useful tool to asset managers of buildings for the implementation of a screening on a given building stock, in order to define the priorities of intervention.

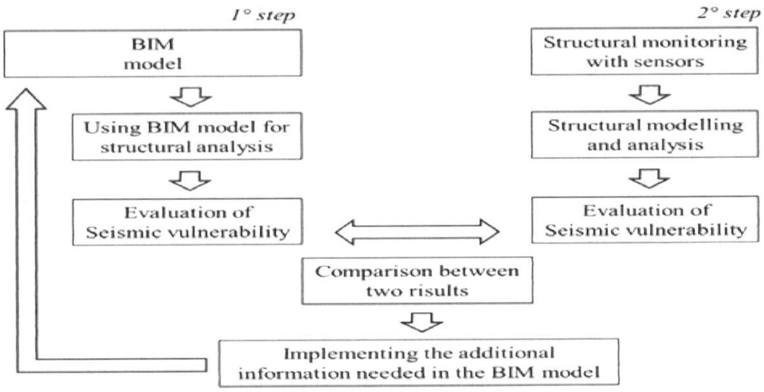

Figure 2. Workflow that shows three different paths of the research.

The experimentation will be examining in depth on selected case studies. Finally, the modalities of interoperability between the different systems and software used will be defined and a method to support the public administrations to define the intervention priorities as well.

4.1 BIM For Structural Analysis

The BIM models can be easily used for structural analysis because, starting from the export of the .ifc model from the architectural BIM software, they contain information on the geometry of the structural elements (beams, pillars, etc.), on the constraints and are updated according to the extension or renovation work carried out over the years. From the BIM model you can directly export the structural 3D model, which can be used to carry out the first structural analyses. (Alirezaei *et al.* 2016). In order to deepen the analyses, it is possible to insert further data such as, for example, the characteristics of the materials, the distribution of the masses, in order to proceed with a general evaluation of the stiffness and the seismic responses of the structure.

The BIM model preparation phase is the most delicate part, because it is necessary to identify the construction technology, the year of construction, the geometric characteristics of the structural elements and mechanical characteristics of the materials (in this case they were not available), the distribution of the masses, etc.

This research will be carried on using the Revit model of an existing secondary school, where the structure of the building and available information on materials are presented.

4.2 Structural Monitoring

Monitoring of structures consists in a step of critical assessment for the structural conditions and employs sensors for measuring physical variables (signals). Signals are sampled (digitalization phase) and processed through amplifiers and filters. For the aim of the present research two types of investigations on the real structures are envisaged. The first one is aimed to evaluate the structural scheme and the components through the existing documentations (design tables and reports). During this first phase the material characteristics have been also assessed: for the concrete material that characterizes the large part of the school asset in Italy the sclerometer will be used.

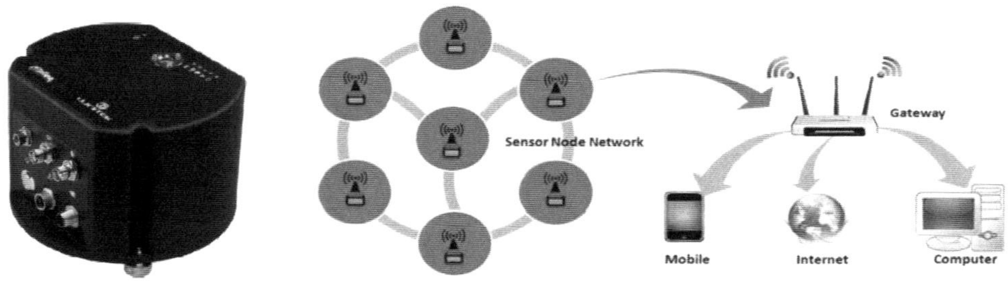

Figure 3. Tools used for the 2° step of the research (wireless sensor network).

In the second step, a wireless sensor network (Figure 3) will be used for collecting the dynamic characteristics of the structures. Indeed, the accelerations collected at different position on the floors can be processed to compute and extract the frequency characteristics of the structures and the vibration shapes.

Both phases will be useful for implementing the numerical model of the structure by using the assessed resisting scheme, materials constitutive models and so on. Then, the validation of the model will be performed through the independent dynamic analyses performed through the wireless sensor network. In particular, the natural vibration frequencies and the mode shapes will be used during the validation procedure.

4.3 Structural Modelling and Validation

The structural model will be developed using the finite element technology starting from the structural scheme and the dimensions defined in the previously determined BIM model. The characterization of the materials will allow the definition of the constitutive model and the appropriate characteristics that will consider in this preliminary stage the linear domain.

The model will then be validated using experimental tests and data collected by the wireless sensor network. In particular, the comparison between the fundamental frequencies and the vibration modes will help to consolidate the procedure. In Figure 4 the numerical model of a building frame is reported.

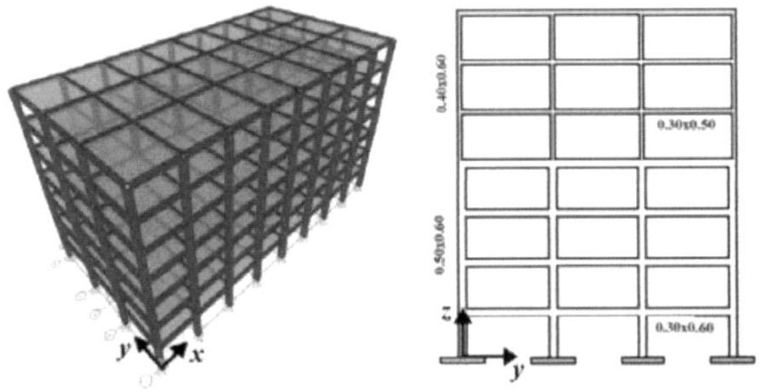

Figure 4. Numerical model of a building frame.

4.4 Seismic Vulnerability

Seismic vulnerability "represents the propensity of a structure to suffer a certain level of damage facing a given seismic event". The current Italian regulations require the assessment of seismic vulnerability for all public buildings and, currently, promote funding for the redevelopment of public buildings, with particular reference to schools. The parameters that can influence seismic vulnerability are the structural type, age of construction, number of floors and maintenance status of the building. In particular, the following vulnerability index can be computed as:

$$\varsigma_E = \frac{F^*}{F_{NTC}} \tag{1}$$

Where F* is the maximum bearable seismic action and FNTC is the seismic action that should be used for designing the same structure accordingly with the current Italian standard DM NTC 2018. The value at the numerator can be computed through different methodologies at increasing complexity. For regular structures, static nonlinear analyses (pushover) could be selected and the capacity curve in Figure 5 can be obtained.

The transition from the linear range could be selected as the maximum bearable seismic action. However, it is worth noting how the standard does not give any other prescription about modelling and it can be interpreted as an implicit recognition of the uncertainties of the problem and the singularity of each structure.

The evaluation of the seismic capacity for the structure should allow estimating a safety margin moving the problem from the seismic hazard characterization at the site to the structural fragility. Such perspective could be also useful for the national authority for tracing the general condition of the existing building asset, built often more than fifty years ago, in times when earthquake engineering was less developed. Conventional symbols should be adopted and used consistently.

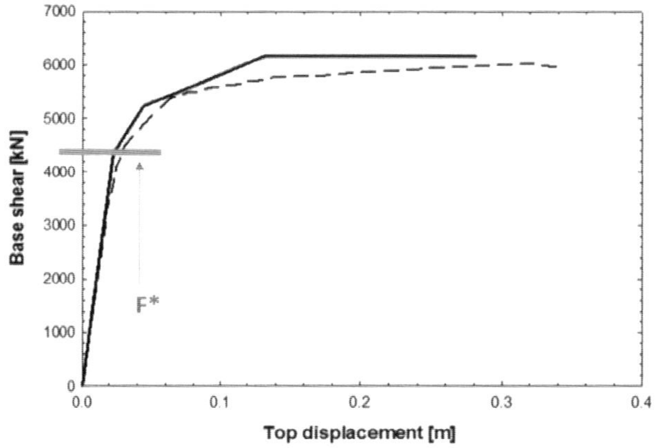

Figure 5. Capacity curve.

There is no constraint on use of units. If they are used as mathematical symbols, standard English letters like x are to appear as x (italicized) in the text.

5 CONCLUSION

The methodology presented defines a path that allows a faster and more efficient evaluation of the coefficient of seismic vulnerability of the building assets managed with the BIM. This allows managers and owners to have a valuable tool for evaluating intervention priorities, based on the deficiencies and seismic risk highlighted. The work carried out has led to the definition of the necessary information detail for a first and second analysis of the models of existing buildings.

References

Alirezaei, M., Noori, M., Tatari O., Macki, K. R., and Elgamal, A., BIM-Based Damage Estimation of Buildings under Earthquake Loading Condition, *Proceedia Engineering*, 145, 1051-1058, 2016.

Becerik-Gerber, B., Jazizadeh, F., Li, N., Calis, G., Application Areas and Data Requirements for BIM-enabled Facilities Management, *Journal of Construction Engineering and Management*, 138(3), 431-442, 2012.

Dong, B., O'Neill, Z., and Zhengwei, L., A BIM-Enabled Information Infrastructure for Building Energy Fault Detection and Diagnostics, *Automation in Construction,* 44: 197-211, 2014.

Eastman, C., Teicholz, P., Sacks, R., and Liston, K., *BIM Handbook: A Guide to Building Information Modeling for Owners, Managers, Designers, Engineers and Contractors* – 2nd Edition, Wiley, Hoboken, New Jersey, 2011.

FEMA Report, *Next-Generation Performance-Based Seismic Design Guidelines: Program Plan for New and Existing Buildings*, 2006.

Tagliabue, L. C., and Villa, V., *Il BIM Per Le Scuole Analisi Del Patrimonio Scolastico*, HOEPLI, Ed., Milano, 2017.

Volk, R., Stengel, J., and Schultmann, F., Building Information Modeling (BIM) for Existing Buildings – Literature Review and Future Needs, *Automation in Construction,* 38, 109–127, 2014.

Welch, D. P., Sullivan, T. J., and Filiatrault, A., Potential of Building Information Modelling for Seismic Risk Mitigation in Buildings, *Bulletin of the New Zealand Society for Earthquake Engineering.,* 47(4), 253–263, 2014.

Streamlining Information Transfer between Construction and Structural Engineering
Edited by Shiau, J., Vimonsatit, V., Yazdani, S., and Singh, A.

ISBN: 978-0-9960437-7-9

DESIGN OF FLEXIBLE PAVEMENT USING MARBLE DUST IN SUBGRADE SOIL

SHUBHAM MOUDGIL and RAJESH PATHAK

Dept of Civil Engineering, Thapar Institute of Engineering and Technology, Patiala, India

Harnessing of industrial waste for improvement of soils is economical and efficient. It helps in enhancing the soil properties and overcoming the disposal problem. Therefore, it is essential to understand the properties of these wastes in order to understand their performance level. In this study, waste marble dust was added to stabilize the soil which was collected from Rajpura Punjab. The various index properties liquid limit, plastic limit was studied by adding Marble dust 10%,15%,20% and 25% by weight of soil. Similarly, California Bearing Ratio (CBR), Unconfined Compressive Strength (UCS) was calculated at maximum dry density obtained from compaction test. It was observed that based on tests that 20% of waste marble dust was optimum for strengthening of the parent soil. Resilient Modulus was calculated afterwards for the subgrade using models specified by Heukelom and Klomp, Thompson and Robnett, Transport and Road Research Laboratory, Erdem Coleri. Based on the results obtained, pavement thickness was determined for designing flexible pavement section as per IRC 37-2012 using IITPAVE. Design was done considering the horizontal tensile strain and vertical compressive strain at the bottom of bituminous base and at top of subgrade, responsible for fatigue and rutting of pavements.

Keywords: CBR, UCS, Resilient modulus, IITPAVE, Crust thickness.

1. INTRODUCTION

Road network is expanding at a reasonable rate worldwide. Construction cost can be substantially decreased by using locally available materials which comprise of construction of granular layers or subgrade/embankment of pavement. Utilization of waste products which are generated by industries is a major issue. Application of these waste materials for engineering purposes can cut down the disposal problems and environmental issues caused by their disposal. Clay is a fine-grained soil with high moisture content generally shows low CBR values as compared to another type of soils. This low CBR value of clayey soil results into high pavement thickness and subsequently into higher construction cost. Marble reserves in India are estimated to be approximately 1200 million tons (Indian Bureau of Mines 2016). Around 70% of marble is wasted in the process of mining, polishing etc. Also, marble slurry generated after cutting gives rise to disposal problems. Therefore, this large amount of waste generated can be used for various engineering purposes, depending upon its properties.

A pavement section mainly consists of bituminous layers resting over the base and/or sub-base which are together compacted over subgrade. There can be rutting in the granular base and subbase layers if the soil subgrade does not have appropriate strength. This rutting in the bottom layers can cause fatigue cracking in the bituminous layers of the pavement section.

Resilient Modulus is the estimation or measurement of the elasticity of the material at a particular stress and temperature. Dione *et al.* (2014) concluded that California Bearing Ratio (CBR) is the most widely used parameter for determining resilient modulus (M_R) of subgrade soil. Over the past few years, various empirical relations have been developed by researchers showing that resilient modulus is not solely dependent on the CBR value. These empirical relations showed that index properties of soil can be used for the prediction of resilient modulus.

2. EXPERIMENTAL PROGRAMME

2.1 Materials

Soil used for the experimental study was obtained from Shmabu Kalan, Punjab (India). Waste marble dust was procured from local a marble cutting store. The various engineering properties of clay (IS: 2720-Part IV:1985) and waste marble dust are shown in Table 1 and 2 respectively.

Table 1. Geotechnical properties of the collected soil sample.

S.No.	Properties	Values
1	Particle Size < 75 μ, %	88.3
2	Liquid Limit (w_L), %	28.7
3	Plastic Limit (w_P), %	16.67
4	Plasticity Index (I_p), %	12.03
5	Flow Index (I_f)	23.7
6	Toughness Index (I_T)	0.507
7	Optimum Moisture Content, %	15.29
8	Maximum Dry Density ($\Upsilon_{d\,max}$), kN/m^3	18.76
9	Unconfined Compressive Strength, kPa	62.34
10	California Bearing Ratio (CBR), %	2.96
11	Soil Classification	CL

Table 2. Physical properties of waste marble dust.

S.No.	Properties	Value
1	Particle Size < 75 μ, %	25.75
2	Specific Gravity	2.62
3	Colour	Grey

2.2 Laboratory Investigation

Marble dust was added at 0%, 10%, 15%, 20% and 25% by weight of soil and are designated as CM0, CM10, CM15, CM20, CM25. Liquid limit and plastic limit values calculated from Atterberg's test are important in determining the plasticity index which determines the range at which soil exhibits plastic properties. The value of liquid limit for each sample was calculated using Casagrande tool as per procedure specified in IS:2720 (Part V):1985.

Light compaction test was performed for each of the samples for determining optimum moisture content and maximum dry density, as per procedure specified in IS:2720 (Part VII):1980. The optimum water content obtained was used to prepare samples of California Bearing Ratio (CBR) test and Unconfined Compressive Strength (UCS) test. The CBR tests for clay and marble dust mixtures were conducted in soaked conditions and the values were determined in accordance with IS:2720 (Part XVI):1987and UCS values were determined for all mixes as per IS:2720 (Part X):1991.

3 RESULTS AND DISCUSSION

There was a decrease in the liquid limit and increase in the plastic limit values upon addition of marble dust. Consequently, the value of the plasticity index also decreased upon addition of marble dust to the parent soil. Values obtained at various percentages of soil and marble dust mixtures are shown in Figure 1.

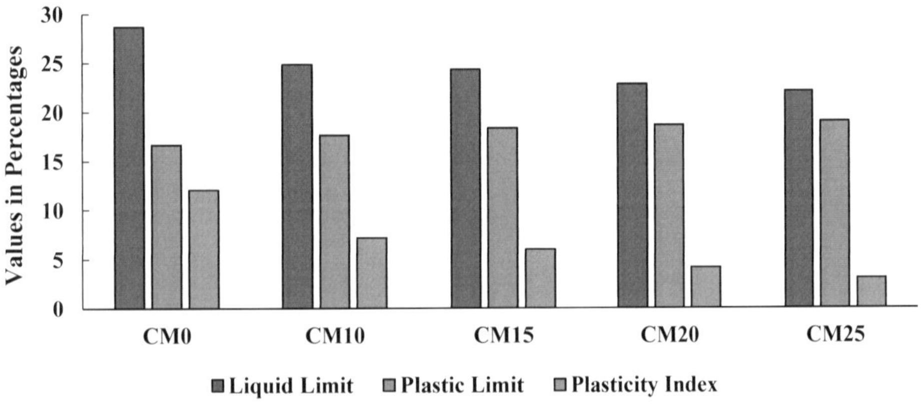

Figure 1. Variation of liquid limit, plastic limit, plasticity index with the addition of marble dust.

Optimum moisture content (OMC) values decreased and maximum dry density (MDD) values increased, up to 20% addition of marble dust in the parent soil. At 25%, the value of OMC increases and MDD decreases. The variation of optimum moisture content (OMC) and maximum dry density (MDD) for different mixes are shown in Table 3.

Table 3. Effect of marble dust on compaction characteristics.

Soil ID	Optimum Moisture Content, %	Maximum Dry Density, kN/m^3
CM0	15.29	18.76
CM10	13.72	18.9
CM15	13.48	19.16
CM20	13.1	19.46
CM25	13.98	19.05

California Bearing Ratio (CBR) and Unconfined Compressive Strength (UCS) values were determined at OMC (from Table 3) at various mixes. Both CBR and UCS values increase up to marble dust content of 20% and then a slight reduction in the value when marble dust was present 25% as shown in Table 4.

Table 4. Effect of Marble dust on CBR and UCS values.

Soil ID	California Bearing Ratio (CBR), %	Unconfined Compressive Strength (UCS), kPa
CM0	2.96	62.34
CM10	3.89	105.25
CM15	4.67	111.34
CM20	6.08	136.75
CM25	6.03	131.77

4 DESIGN OF FLEXIBLE PAVEMENT

4.1 Design Considerations

Pavement design was done for a traffic of 50 msa. The values of Poisson's ratio were considered as 0.35 for all of the layers i.e. bituminous layers, granular layers and subgrade. Tyre pressure and single wheel load were taken as 0.56 MPa and 20,000 N respectively. Dual wheel configuration was used for analysis of fatigue and rutting values.

4.2 Determination of Resilient Modulus (M_R)

For bituminous layers, i.e. DBM and BC, the temperature was assumed to be 35°C and grade of bitumen was taken as VG40. Based on this, resilient modulus value for bituminous layers was taken as 3000 MPa (from Table 7.1, IRC 37:2012). Determination of resilient modulus for subgrade was done as per correlations suggested by Heukelom and Klomp (1962), Thompson and Robnett (1979), Transport and Road Research Laboratory (TRRL), Coleri (2007). Resilient modulus values calculated using these empirical equations are mentioned in Table 5.

Table 5. Subgrade resilient modulus (M_R) values.

Soil ID	Resilient Modulus, MPa			
	Heukelom and Klomp	Thompson and Robnett	TRRL	Erdem Coleri
CM0	29.6	25.06	35.24	65.74
CM10	38.9	38.23	41.98	76.52
CM15	46.7	40.099	47.19	78.82
CM20	60.8	47.87	55.87	82.30
CM25	60.3	46.36	55.58	73.98

Depending on the resilient modulus (M_R) of subgrade, the resilient modulus values of granular layers were computed as per the equation 7.1 of IRC37:2012.

4.3 Determination of Limiting Strains

The value of maximum tensile strain at the bottom of the bituminous layer (ε_t) was computed using equation 6.1 of IRC37:2012. The calculated value of the maximum permissible tensile strain was 155.27 $\mu\varepsilon$. Also, the value of permissible vertical strain in the subgrade was determined using equation 6.4 of IRC37:2012 and the value came out to be 371.7 $\mu\varepsilon$.

4.4 Determination of Actual Strains and Pavement Thickness

Actual strain values were calculated using IITPAVE. Data was inputted as specified in the above sections and pavement thickness was determined using the methods mentioned above. The values of actual strain calculated for various M_R values determined using Coleri (2007), Heukelom and Klomp (1962), Transport and Road Research Laboratory (TRRL), Thompson and Robnett (1979) are tabulated from Table 6 where (εt), $\mu\varepsilon$ is horizontal tensile strain and (εz), $\mu\varepsilon$ is vertical compressive strain.

Total crust thickness calculated using all of the methods is shown in Figure 2 and granular layer thickness is shown in Figure 3. The thickness of bituminous layers was 160 mm for Coleri Method. For Heukelom Klomp method, the bituminous layer thickness was 205 mm for CM0 and 190mm for rest of the cases. Also, for TRRL method, the bituminous layer thickness was

195 mm for CM0 and 190mm for rest of the cases. Thompson and Robnett method had bituminous layer thickness as 215mm for CM0 and 190mm for the remaining cases.

Table 6. Actual strain values by various models.

Soil ID	Erdem Coleri Model		Heukelom and Klomp Model		TRRL Model		Thompson and Robnett Model	
	(ε_t), με	(ε_z), με	(ε_t), με	(ε_t), με	(ε_t), με	(ε_z), με	(ε_t), με	(ε_z), με
CM0	153.4	170.7	153.9	153.7	181.3	198.2	153.8	220.8
CM10	153.2	213.8	153.7	153.6	199.1	178.7	153.7	174.0
CM15	152.7	218.8	153.1	153.0	226.8	228.3	153.6	186.4
CM20	152.1	227.2	152.4	152.5	263.7	281.6	153.4	232.9
CM25	152.6	200.5	152.5	152.8	264.7	294.8	153.5	225.6

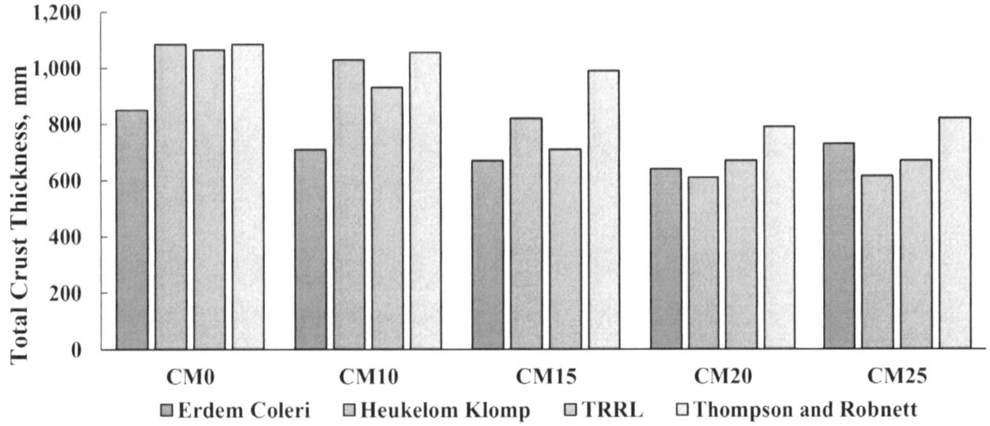

Figure 2. Total crust thickness calculated for clay and marble dust mix calculated using various models.

Figure 3 shows variation of granular layer thickness obtained for different clay and marble dust mixtures using resilient modulus (M_R) values calculated by methods specified above.

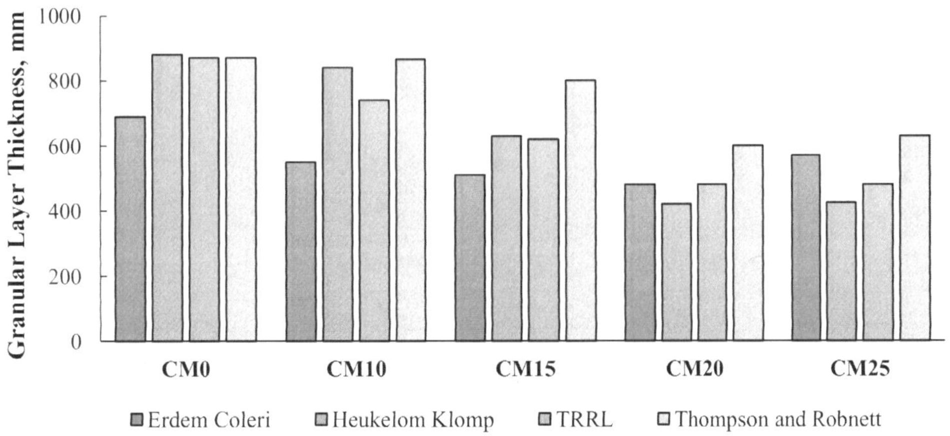

Figure 3. The thickness of granular layers calculated for clay and marble dust mix using various models.

5 CONCLUSION

Addition of 20% waste marble dust showed a maximum increase in CBR and unconfined compressive strength of stabilized soil. Pavement designs also showed 50% decrease in the pavement thickness on the addition of marble dust. So, waste marble dust can be used effectively along with parent soil for subgrade or embankment construction. Also, among various methods, the resilient modulus (M_R) calculated using Coleri (2007) Model yielded the highest value.

References

Coleri, E., *Relation Between Resilient Modulus and Soil Index Properties of Unbound Materials* (master's thesis), CiteSeerX. (10.1.1.632.9765), 2007.

Dione, A., Fall, M., Berthaud, Y., Benboudjama, F., and Michou, A., Implementation of Resilient Modulus–CBR Relationship in Mechanistic Pavement Design, *Sciences Appliquées et de l'Ingénieur*, 1(2), 65-71, 2014.

Heukelom, W., and Klomp, A. J. G., *Dynamic Testing as a Means of Controlling Pavement During and After Construction,* Proceedings of the 1st International Conference on the Structural Design of Asphalt Pavement, University of Michigan: Ann Arbor, Mich, 1962.

Indian Bureau of Mines, Indian Minerals Yearbook (Part- III: Mineral Reviews), 54th Edition, *Marble.* Retrieved from www.ibm.gov.in, on November 2016.

IS: 2720 (Part IV)-1985, *Method of Test of Soils: Grain Size Analysis*, New Delhi, India: Bureau of Indian Standards, 1985.

IS: 2720 (Part V)-1985, *Method of Test of Soils: Determination of Liquid Limit and Plastic Limit*, New Delhi, India: BIS, 1985.

IS: 2720 (Part VII)-1980, *Method of Test of Soils: Determination of Water Content-Dry Density Relation Using Light Compaction*, New Delhi, India: BIS, 1980.

IS: 2720 (Part XVI)-1987, *Method of Test of Soils: Laboratory Determination of CBR*, New Delhi, India: BIS, 1987.

IS: 2720 (Part X)-1991, *Method of Test of Soils: Determination of Unconfined Compressive Strength*, New Delhi, India: BIS, 1991.

IRC: 37-2012, *Guidelines for the Design of Flexible Pavements*, New Delhi, India: Indian Road Congress, 2012.

Thompson, M.R. and Robnett, Q.L., Resilient Properties of Subgrade Soils, *Journal of Transportation Engineering*, 105(1), 71-89, 1979.

Streamlining Information Transfer between Construction and Structural Engineering
Edited by Shiau, J., Vimonsatit, V., Yazdani, S., and Singh, A.

ISBN: 978-0-9960437-7-9

CALIBRATION OF RUTTING AND ROUGHNESS DISTRESS MODELS OF HDM-4 FOR DEVELOPING PAVEMENT MAINTENANCE MANAGEMENT

ADITYA SINGH and TANUJ CHOPRA

Dept of Civil Engineering, Thapar Institute of Engineering and Technology, Patiala, India

The Highway Development and Management model (HDM-4) is a tool developed by the World Bank to aid highway administrators and engineers in the process of decision making for preparing of road investment programme and determining the road network maintenance strategies. HDM-4 essentially models the interaction between the traffic volume, environment and pavement composition to predict the different kinds of distress that develop in pavements over time. Since distress is caused due to different conditions and progresses at different rates, therefore it is necessary to calibrate the HDM-4 model as per the local conditions. The aim of the study is to calibrate the HDM-4 pavement deterioration model in terms of rutting and roughness for the urban road network of Patiala (Punjab, India). In our study, we select 15 road sections and group them based on varying traffic and pavement age. The pavement condition data, which was measured starting from 2012 to the end of 2014, is fed as the input to the HDM-4 distress models. The calibration process is performed using statistical analysis between the observed and predicted value of the distress by keeping minimum Root Mean Square Error (RMSE) and maximum R-square (R^2). The determined calibration factors are validated and further used for developing pavement deterioration models which prove to be helpful in building a Pavement Maintenance and Management system for Patiala.

Keywords: Urban roads, India, Highway administration, Patiala, PMMS.

1 INTRODUCTION

The Road network in India is the second largest comprising of National Highways, State Highways, Major District Roads, other District and Urban Roads (Mandhani 2017). The traffic on the urban road in India in terms of volume and axle load is increasing at an alarming rate with an annual growth of the order of about 10% (MoRT&H 2004). This increase in vehicle population and road usage has put an immense strain on the present urban road leading to a wide-spread deterioration of the pavements. The situation has motivated the development of Pavement Maintenance Management Systems (PMMS) for the roads in our country, most of which belong to the national highway network. It is important to note here that the utilization and maintenance strategies of the urban road networks are totally different from that of the highway networks. The number of trips in the urban road network is small while diversions and intensity of the traffic are very large and non-uniform (Chopra *et al.* 2018), which is very unlike the highway road network. Therefore, the pavement management systems developed for the highway roads cannot be directly used for urban roads. To motivate the use of maintenance strategies for urban road networks as well, in our present study we develop a PMMS using Highway Development and Maintenance (HDM-4) pavement deterioration model for pavement sections in the urban areas of the city of Patiala, Punjab (India). The main objective of a model developed on HDM-4 is to detect pavement deterioration patterns to predict the deterioration of pavement with time under traffic loads (Bannour *et al.* 2017). HDM-4 is made to be used for a broad range of environmental conditions but before that HDM-4 needs to be calibrated as per the local condition to produce better results. If used without calibration, an HDM-4 model will predict the pavement performance, which does not accurately meet with the observed pavement specifications. Therefore, calibration is necessary to predict more indicative output in the environment other than the regions, in which it was developed (Bennet and Paterson 2000). The calibration of HDM-4 model as per local condition has been successfully executed in developed countries. However, in developing countries like India attempts are still being made for successful implementation of calibration of HDM-4 as per varying climate and traffic conditions. Therefore, our present study focuses on the calibration of HDM-4 pavement deterioration model in terms of rutting and roughness for 15 road sections in city Patiala, India for development of PMMS.

Table 1. Details of 15 selected pavement section of the urban road network.

Section ID	Name of Road	Length in km	Width of Carriageway(metre)	Drainage condition	Classification
UR01	ThaparUniv. - Bhadson Road	0.80	6.80	Fair	Other Road
UR02	ThaparUniv. - Bhupindra Road	1.05	7.30	Fair	Sub-Arterial
UR03	Thapar Uni. - Gurdwara Sahib	2.50	7.50	Good	Other Road
UR04	Passey Road - Civil Line	1.00	7.20	Very Poor	Other Road
UR05	Gurudwara Sahib Chowk - Sirhind Road	2.25	7.00	Good	Sub-Arterial
UR06	Leela Bhawan Chowk – Cantonment	0.70	11.50	Good	Sub-Arterial
UR07	Gurdwara Sahib Chowk - Bus Stand Road	0.90	7.50	Poor	Sub-Arterial
UR08	Thikriwala Chowk - Sangrur Road	1.00	7.50	Good	Sub-Arterial
UR09	Thikriwala Chowk - Badungar Road	0.80	11.80	Fair	Other Road
UR10	Bus Stand Chowk - Gurbax Colony	2.10	6.0	Poor	Other Road
UR11	Fountain Chowk - Leela Bhawan	0.70	12.5	Good	Sub-Arterial
UR12	Fountain Chowk - Lower Mall	2.25	7.5	Fair	Other Road
UR13	ThaparUniv. -Gurudwara Sahib	2.25	7.30	Fair	Sub-Arterial
UR14	Leela Bhawan Chowk- 22 No bridge	2.10	7.50	Good	Sub-Arterial
UR15	Leela Bhavan-Guruduwara Sahib	1.46	10.0	Fair	Sub-Arterial

2 URBAN ROAD NETWORK

The urban road network in Patiala comprises of 52 road sections. In our study, 15 road sections were selected for developing the pavement distress model. Each of the 15 selected road sections has been assigned Section IDs such as UR01, UR02 etc. (shown in Table 1). Each pavement section also includes the road length, carriageway width, drainage conditions and classification of the road system. The selected pavement sections have been further classified into groups of 4 based upon the varying pavement age and commercial traffic, as shown in Table 2.

Table 2. Homogenous section group for selected pavement sections.

Homogenous Group	Section Name	Pavement Age	Commercial Traffic
GROUP 1	UR08, UR10, UR15	0-6 Years	Less than 7%
GROUP 2	UR02, UR04, UR06, UR12	6-12 Years	More than 7%
GROUP 3	UR01, UR03, UR05, UR11	0-6 Years	Less than 7%
GROUP 4	UR07, UR09, UR13, UR14	6-12 Years	More than 7%

3 METHODOLOGY

Extensive field work was conducted for 3 successive years beginning from 2012 to the end of the year 2014 for collecting the requisite data about the road sections. The initial work included identification of roads, a collection of pavement data, which included the inventory data such as the geometry of the road, structure evaluation (to determine the load carrying ability of the structure), functional evaluation (to check pavement condition and roughness etc.), climate, temperature and the thickness of the pavement. The traffic volume data was collected from Municipal Corporation Patiala. The next phase involved calibrating the model for the specific local conditions. The inventory data of the pavement was used as input to the HDM-4 pavement deterioration model. The calibration of the model, which has been explained in detail in Section 3.1, was achieved by using regression.

3.1 Calibration of HDM-4

For calibrating the HDM-4 distress model, the software program was executed for all the selected road sections with input data as the traffic volume data and the pavement characteristics. For the first run in HDM-4, calibration factor was determined by obtaining minimum RMSE and max R^2 calculated by two equations:

$$RMSE = \sum_{i=1}^{n} \sqrt{(Ob - Pd)^2/N} \tag{1}$$

$$R^2 = 1 - [\sum((Ob - Pd)^2/(Ob - Oavg)^2)] \tag{2}$$

Where RMSE = root mean square error, R^2 =coefficient of determination, Ob=measured (observed) value of distress, Pd=predicted value of distress by HDM-4 model, Oavg= average observed value of distress, N= no of observations

After obtaining the calibration factor for the first stage, the model was run for the second stage for further improvement by incrementing the calibration factor by 0.01. The data collected for the year 2012 was used as input to the HDM-4 model. The distress estimations made by the model for year 2013 were compared with the actual values of distress measured in the field for the year 2013. A correlation was observed between the two values (measured and estimated) of distress based on the goodness of fit, which was determined by obtaining minimum RMSE and maximum R^2. The calibration factors obtained were different for different groups due to varying traffic volume data and pavement characteristics. The method of calibration for roughness distress model for Group 4 is shown as an example in Table 3. The final calibrated values for each mode of the distress model for different groups are shown in Table 4.

Table 3. Calibration of roughness distress model for group 4.

	UR 07		*UR09*		*UR13*		*UR14*			
Calibration Factor	Obsv	Pred	Obsv	Pred	Obsv	Pred	Obsv	Pred	RMSE	R-Squared
1.00	4.80	4.57	2.20	2.16	5.70	4.74	2.30	2.32	0.4940	0.9743
1.50	4.80	4.67	2.20	2.32	5.70	4.87	2.30	2.49	0.4348	0.9771
1.90	4.80	4.74	2.20	2.46	5.70	4.97	2.30	2.63	0.4221	0.9800
1.95	4.80	4.75	2.20	2.47	5.70	4.99	2.30	2.64	0.4168	0.9808
1.96	4.80	4.75	2.20	2.47	5.70	4.99	2.30	2.64	0.4168	0.9808
1.97	4.80	4.75	2.20	2.48	5.70	5.00	2.30	2.65	0.4163	0.9811
1.98	4.80	4.76	2.20	2.48	5.70	5.00	2.30	2.65	0.4168	0.9808
1.99	4.80	4.76	2.20	2.49	5.70	5.00	2.30	2.66	0.4199	0.9809
2.00	4.80	4.76	2.20	2.49	5.70	5.00	2.30	2.66	0.4199	0.9809
2.01	4.80	4.76	2.20	2.49	5.70	5.01	2.30	2.66	0.4157	0.9817
2.02	4.80	4.76	2.20	2.50	5.70	5.01	2.30	2.67	0.4197	0.9817
2.03	4.80	4.76	2.20	2.51	5.70	5.01	2.30	2.67	0.4215	0.9810

Table 4. Final calibrated value for all groups.

Distress Model	Group 1	Group 2	Group 3	Group 4
Rutting Progression	2.64	2.64	2.74	2.83
Roughness Progression	2.63	2.75	2.42	2.01

4 RESULTS AND DISCUSSIONS

The HDM-4 model requires validation before it can be it for local conditions. Therefore, the results obtained after calibration by HDM-4 distress were further validated for the year 2014 for different group sections to check the dependability and adequacy of the HDM-4 distress model. Validation was done for all groups consisting of homogenous section using regression analysis and percentage variability. The validation of Group 4 consisting of 4 sections is shown as an example in the following sub-sections.

4.1 Rutting Distress Model for Group 4

The measured value of rutting progression is compared with the value of rutting determined by the calibrated HDM-4 model, for the year 2014, for all selected pavement sections of Group 4 (shown in Table 3). The two values are plotted against each other to determine the correlation between them, as shown in Figure 1. The variability in the percentage between the two values (observed and predicted) of distress lies between 4 to 24 %, which is quite reasonable, and the coefficient of relation determined between the two values is 0.7435, which is also quite reasonable and indicates the efficacy and reliability of the calibrated HDM-4 rutting distress model.

4.2 Roughness Distress Model for Group 4

The measured value of roughness progression is compared with the value of roughness as determined by calibrated HDM-4 model for the year 2014 for all selected pavement sections of Group 4, as shown in Table 5. The two values are plotted against each other to determine the correlation between them, which is shown in Figure 2. The variability in the percentage between the two values of distress lies between 0 to 28 %, which is quite reasonable, and coefficient of relation determined between the two values is 0.9857, which is also quite reasonable and tells the efficacy and reliability of the calibrated HDM-4 roughness distress model.

Table 5. Rutting and roughness model validation for group 4.

Section	Observed	Predicted	Variability	R^2
UR07	06	6.24	4	0.743
UR09	11	10.25	6.81	
UR13	09	9.64	7.11	
UR15	08	6.12	23.50	

Section	Observed	Predicted	Variability	R^2
UR07	5.3	5.26	4	0.985
UR09	2.3	2.94	6.81	
UR13	6.1	5.45	7.11	
UR15	2.6	3.05	23.50	

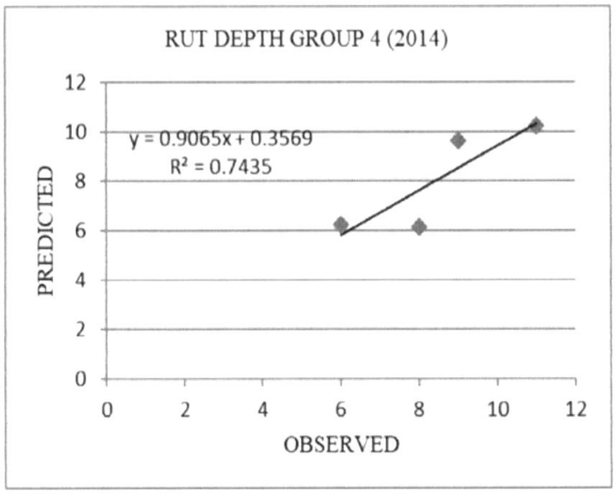

Figure 1. The graph for rutting model validation.

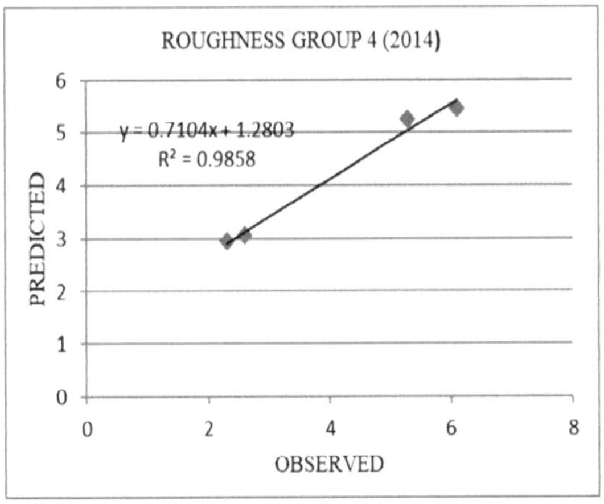

Figure 2. The graph for roughness model validation.

4.3 Chi-Square Test

The Chi-Square test was done for checking the goodness of fit correlation between the measured (observed) distress and the value determined by HDM-4 model. The $X^2_{calculated}$ (from chi-square formula) for distress model for all the different groups were compared with the $X^2_{critical}$ with a level of significance equal to 5% and (N-1) of degrees of freedom. The Chi-Square test was done for all different groups consisting of homogeneous sections and group 4 having homogeneous section is shown as an example in Table 6. The value of Chi(X) is determined using Eq. (3).

$$X^2 = \Sigma(Od - Pd)^2/Pd \quad (3)$$

where Od=observed value of distress Pd=Predicted value of distress by HDM-4 model.

As the value of $X^2_{calculated}$ is less than $X^2_{critical}$ therefore null hypothesis is accepted and a significant relation is established between the two values of the distress model having no statistical difference.

Table 6. Chi-Square Result for Validation of Group 4.

Distress Model	Linear Regression	$X^2_{calculated}$	$X^2_{critical}$	DOF	R^2
Rutting Progression	y = 0.906*x*+0.356	1.537	7.815	3	0.743
Roughness Progression	y = 0.710*x*+1.280	0.283	7.815	3	0.985

5 CONCLUSIONS

- The aim of the study was to calibrate the two-distress model, i.e., rutting and roughness distress model for developing pavement deterioration model for urban road network in India.
- The HDM-4 distress model obtained a good correlation between observed and predicted value of distress after calibration, which tells that a calibrated HDM-4 model can serve as a decent predictive model for pavement distress of urban road networks as well.
- The percentage of variability for rutting distress model lies from 8 to 24% and for roughness distress model from 1 to 7 % is quite reasonable considering different climatic conditions, pavement age, and traffic.
- The calibration factor for rutting distress model ranges from 2.63 to 2.84 and 2.01 to 2.75 for roughness distress model, which are quite reasonable for flexible pavement in urban city road network of Patiala, Punjab, India.
- The calibration factors obtained can be applied to the urban road network of the city having similar traffic and the environmental conditions like city Patiala, Punjab, India.

References

Bannour, A., El Omari, M., Lakhal, E. K., Afechkar, M., Benamar, A., and Joubert, P., Optimization of the Maintenance Strategies of Roads in Morocco: Calibration Study of the Degradations Models of the Highway Development and Management (HDM-4) for Flexible Pavements, *International Journal of Pavement Engineering*, 1-10, 2017.

Bennet, C. R. and Paterson W. D. O., A Guide to Calibration and Adaptation, *HDM-4 Documentation Series*, 5, The World Road Association (PIARC), Paris, France, 2000.

Chopra, T., Parida, M., Kwatra, N., and Chopra, P., Development of Pavement Distress Deterioration Prediction Models for Urban Road Network Using Genetic Programming. *Advances in Civil Engineering*, 2018.

Mandhani, J., *Development of Pavement Maintenance Management System (PMMS) of Patiala. Patiala:* Master Thesis, Department of Civil Engineering, Thapar University, Patiala, 2017.

MoRT&H, Guidelines for Maintenance of Primary, Secondary and Urban Roads, *Ministry of Road Transport & Highways, Government of India*, New Delhi, 2004.

Streamlining Information Transfer between Construction and Structural Engineering
Edited by Shiau, J., Vimonsatit, V., Yazdani, S., and Singh, A.

ISBN: 978-0-9960437-7-9

UTILIZING PLASTIC FOR PRETREATING RECYCLED CONSTRUCTION AGGREGATE TO ELIMINATE BINDER ABSORPTION IN ASPHALT MIXTURE

FARZANEH TAHMOORIAN[1] and JOHN YEAMAN[2]

[1]*School of Engineering and Technology, Central Queensland University, Australia*
[2] *Faculty of Science, Health, Education and Engineering, University of Sunshine Coast, Australia*

The growing quantities of waste materials, lack of natural resources and shortage of landfill spaces represent the importance of finding innovative ways of reusing and recycling waste materials. Due to the large quantities of construction and demolition waste (CDW), recycling and utilization of Recycled Construction Aggregates (RCA) obtained from CDW in construction projects, including asphalt pavement construction, can be the most promising solution to this problem. Asphalt mixtures containing RCA have the problem of high bitumen absorption. Using plastic waste in RCA-contained asphalt mixtures reduces not only bitumen absorption but also the adverse environmental impacts associated with plastic waste disposal due to the non-biodegradability of plastic waste. In addition, the demand reduction for virgin aggregates is another advantage resulting in subsequent economic advantages. This paper characterizes the effects of different types of plastic on the bitumen absorption and properties of asphalt mixtures containing RCA through laboratory investigation. Different types of plastic including High-Density Polyethylene (HDPE) and Low-Density Polyethylene (LDPE) were investigated in this research. The test results indicate that the plastic waste can be a viable material for improving the problem of high bitumen absorption of asphalt mixtures containing RCA.

Keywords: Pavement, Waste, Bitumen, RCA, Non-biodegradability

1 INTRODUCTION

Waste materials are increasingly generated with the rapid growth in the economy and continuously increased consumption. The growing quantities of waste materials, lack of natural resources and shortage of landfill spaces represent the importance of finding innovative ways of reusing and recycling waste materials.

At the same time, as the world's population grows, the demand for public infrastructure and hence the construction materials increase. Today, the demand for aggregates is much greater than the amount of virgin aggregates that could possibly be supplied. Referring to the Australian Quarrying Industry, the average annual consumption of aggregates across Australia is about seven tonnes per person and will grow in the future with the population growth and infrastructure development. The biggest waste generator in Australia is the construction and demolition sector who is responsible for of generation of about 40% of all Australian waste material (CCAA

2008a). If this entire amount were treated as waste, at least 30 major landfill facilities were needed to operate all year round (Australia's Sustainable Aggregate Industry 2013). Thankfully however, of the 160 million tonnes of annual aggregate demand, less than 10% is being covered by sustainable aggregate products that contain recycled materials.

In Australia, RCA obtained from construction and demolition wastes has been used in construction projects as coarse and fine aggregates especially in recent years. Choosing RCA can be considered as a smart option because it reduces the need to excavate more virgin aggregates while extracting value from the waste materials, which generally dispose. Accordingly, in utilizing the waste materials in construction, RCA can be considered as one potential material due to large amount of construction and demolition wastes. To this point, over the last two decades, several studies (e.g. Guimarães and Ribeiro 2005, Paranavithana and Mohajerani 2006, Wong *et al.* 2007, Silva 2009, Mills-Beale and You 2010, Marinho 2011, Chen *et al.* 2011, Pérez *et al.* 2012, Zhu *et al.* 2012, Zulkati *et al.* 2013) have been performed on the various applications of RCA. Utilization of RCA in asphalt mixtures is a sustainable technology due to important role and high portion of aggregates in asphalt mixtures. In addition, for some parameters such as Flakiness Index and Particle Shape, which are two dominant characteristics having significant impact on asphalt mixture strength and stability; RCA displays smaller values in comparison with virgin aggregates such as basalt (Tahmoorian *et al.* 2017). This can be one of the strong points of RCA as flakiness index and particle shape are the two important properties for proper compaction, deformation resistance, and workability of asphalt mixture (Masad *et al.* 2007). However, the major drawback of RCA is its high water absorption compared to conventional aggregates, which subsequently results in high bitumen absorption of asphalt mixtures containing RCA.

On the other hand, recognizing the importance of managing particular waste streams and building the capacity for the implementation of projects regarding the conversion of waste into value-added materials, many researchers (e.g. Justo and Veeravarana 2002, Punith and Veeraragavan 2004, Brozyna and Kowalski 2016, Chandh and Akhila 2016) are investigating more environmentally friendly options for repurposing of plastic wastes. One viable economic solution for preventing vast quantities of plastic wastes from being landfilled or incinerated is using them in asphalt pavements through reliable methods. Utilization of plastic wastes combine the advantages of producing better asphalt pavement as well as plastic waste management and reducing the cost for asphalt industries.

Considering the high bitumen absorption of RCA, by combining plastic, the RCA can be coated and hence the high bitumen absorption of asphalt mixtures containing RCA can be compensated. Using plastic waste in RCA contained asphalt mixtures reduces not only bitumen absorption but also environmental issues associated with plastic waste disposal and the demand for virgin aggregates which will subsequently result in cost savings and economic advantages, as plastic wastes are durable and non-biodegradable wastes that may persist for hundreds or even a thousand years.

However, developing a suitable mix design containing RCA and plastic is essential before deploying this technology as an alternative option for asphalt production. The purpose of the present work is to study the benefit of plastic addition on the bitumen absorption of asphalt mixtures containing RCA in order to optimize the RCA contained asphalt mix design. In that sense, as discussed in the following sections, several tests were conducted on individual aggregates in order to obtain comprehensive information and data of their properties and to compare these properties with the requirements specified in the standards as well as with the properties of the virgin aggregate. Based on the test results on the individual aggregates, necessary tests were conducted on different combinations of these aggregates in asphalt mixtures to investigate their performance in asphalt mixture and to characterize the effects of plastic on the bitumen absorption and volumetric properties of asphalt mixtures containing different percentages of RCA.

2 EXPERIMENTAL WORK

2.1 Materials

In the present study, RCA and basalt have been used as aggregates and the original bitumen studied in this research corresponds to C320, which is the most common binder for wearing courses subjected to heavy loading and/or in hot climates. The crushed virgin basalt aggregate was obtained from a local supplier in Sydney. In addition, RCA passing through 20 mm and retained on 4.75 mm I.S sieve is used throughout the experiments which was collected from a local recycling centre called Revesby Recycling Centre (Revesby, NSW, Australia), a licensed waste facility and transfer station which accepts all construction and demolition wastes from both the residential and commercial waste streams. Two different types of plastic wastes from milk bottles and plastic bags were used for coating RCA in this research.

2.2 Laboratory Tests on Coarse Aggregates

Since the information on aggregate fundamental properties is paramount in designing a durable and sustainable asphalt mixture, the properties of different coarse aggregates used in this research (i.e. RCA and basalt) are investigated through conducting different tests to assess their suitability as coarse aggregate in asphalt. The results of these tests on three samples for each aggregate type are summarized in Table 1.

Table 1. Summary of the test results for the evaluation of coarse aggregate properties.

Test	Test Method	Aggregate		Typical Limit Based on Australian Standards
		RCA	Basalt	
Flakiness Index Test	AS 1141.15	6.9	19.0	25% (max)
Particle Shape Test	AS 1141.14	6.2	18.3	35% (max)
Water Absorption	AS 1141.6.1	**6.30**	1.64	2% (max)
Particle Density	AS 1141.6.1	2.370	2.640	-
Particle Density on Dry Basis	AS 1141.6.1	2.212	2.530	-
Particle Density on SSD Basis	AS 1141.6.1	2.351	2.571	-
Aggregate Crushing Value	AS 1141.21	29.21	8.91	35% (max)
Weak Particles	AS 1141.32	0.23	0.23	1% (max)
Wet/Dry Strength Test	AS 1141.22	26.6	8.5	35% (max)
Wet Strength	AS 1141.22	**119.7**	359.2	150 kN (min)
Dry Strength	AS 1141.22	163.1	392.9	-

As can be observed in this table, the results of preliminary tests on coarse aggregates indicate that all properties of RCA, except for water absorption and wet strength, are within the limits specified by the relevant Australian Standards and hence deemed appropriate for use as aggregate in the asphalt mixture. Importantly, RCA displays smaller value for some parameters such as Flakiness Index and Particle Shape in comparison with basalt. As flakiness index and particle shape significantly affect the final performance of asphalt mixtures in terms of compaction, rutting resistance, and workability, better values for these properties can be one of the strong points of RCA. In addition, as can be seen in Table 1, the water absorption of RCA is higher than the corresponding value of basalt and the Australian Standard's limit due to its high porosity and the great amounts of impurities in RCA. The high water absorption of RCA may result in high bitumen absorption in asphalt mixtures and hence plays an important role in asphalt mixture design. Accordingly, the necessity of finding and studying potential materials to compensate for this problem of RCA has led to the idea of utilization of plastic waste in combination with RCA in asphalt mixture design, which is the main goal of this paper.

2.3 Laboratory Tests on Asphalt Mixtures Containing RCA and Plastic Waste

In this study, the specimen of diameter 100 ± 2 mm and height of 65 ± 5 mm were prepared in accordance with AS2891.2.1 (2014) and AS2891.2.2 (2014). For this experimental work, a group of specimens were made without recycled materials (0%) as reference to specimens made with 25% RCA. In addition, in order to investigate the effect of plastic waste on the bitumen absorption of asphalt mixtures containing RCA, two groups of specimens were also prepared with the addition of 1% of LDPE (plastic bags) and HDPE (milk bottles).

Figure 1. Plastic wastes used in this study.

It is generally recognized that the volumetric composition of mixtures greatly influences their performance. Volumetric properties evaluation of asphalt mixtures is the basis of asphalt mix design and largely determines the performance of asphalt mixture. The asphalt mixture volumetric properties including void content, voids in mineral aggregate (VMA) and voids filled with asphalt (VFA) have been identified as important parameters for the durability and performance of asphalt pavements. The volumetric properties of the basalt-RCA asphalt mixtures with and without plastic were determined and compared accordingly with the standards specifications. To determine the optimum bitumen content for the mixture, the procedure indicated by Australian standards, AGPT04B-14, was followed in this research. Specimens at different bitumen content (4.5, 5, 5.5, 6, and 6.5%) were tested for maximum density, bulk density, and subsequently air voids and VMA calculations. The results of these tests and calculations were used to select the optimum bitumen content. According to the results obtained, Figure 2 illustrates the effect of plastic addition on bitumen content and air voids of samples containing RCA.

As can be clearly observed in these figures, air voids of mixtures containing plastic are lower than the mixtures containing RCA without plastic because of reduction in RCA particles porosity as a result of their coating with plastic.

Importantly, the results of volumetric analysis and air voids calculations based on the bulk density test and maximum density test reveal that the optimum bitumen content of asphalt mixtures varies with the type of plastic waste used, so that mixtures containing LDPE require less bitumen, as presented in Figure 2.

In general, the test results reveal that plastic waste is a viable material for improving the problem of high bitumen absorption of asphalt mixtures containing RCA.

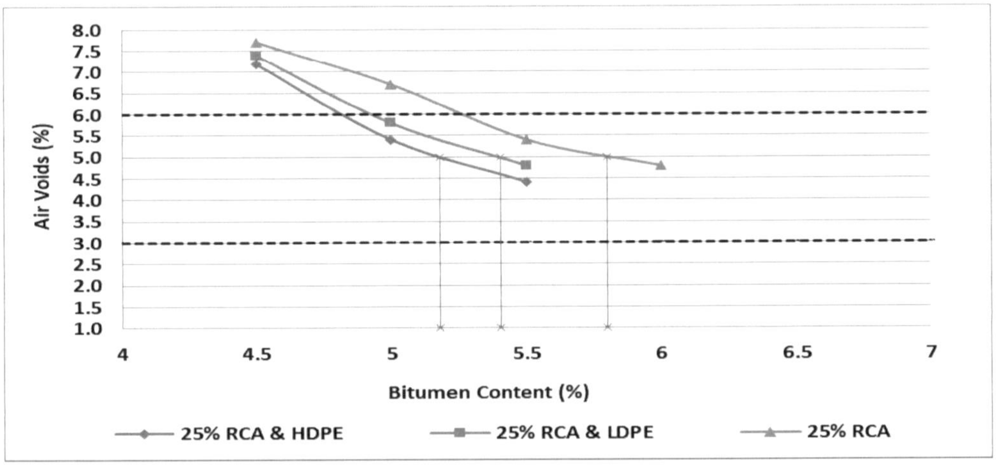

Figure 2. Comparison of optimum bitumen content in different samples.

3 CONCLUSIONS

Considering the rising amount of waste generation in the worldwide, the research project was aimed at evaluating the potential of using plastic waste for compensation of high bitumen absorption of asphalt mixtures containing RCA. The test results on aggregates and asphalt mixtures containing RCA with/without plastic indicate that:

1) The utilization of RCA in asphalt mixtures influences the performance of asphalt mixture and leads to both advantages and disadvantages.
2) RCA has a lower value of flaky and misshapen particles in comparison with basalt. This implies that asphalt mixtures containing a certain amount of RCA can have better workability, deformation resistance, and compaction.
3) RCA exhibits comparatively more absorption and wet/dry strength variation than conventional aggregate, while the results of other tests show that RCA still meets the requirements for aggregate in asphalt mixtures. Cracks and adhering mortar can be significant reasons for the high water absorption of RCA, which needs to be compensated for during mix design.
4) The results of tests on different asphalt mixtures containing different percentages of RCA revealed that a RCA increase will increase optimum bitumen content of the mixtures. Therefore, the selection of the optimum combination of RCA and other aggregates is required to satisfy the relevant requirements.
5) Utilization of plastic waste in combination with RCA was observed to improve the bitumen absorption of these asphalt mixtures.
6) The results of tests on different asphalt mixtures containing RCA and plastic waste indicate that the bitumen absorption depends on the type of plastic waste. In other words, asphalt mixtures containing LDPE have lower optimum bitumen content in comparison with asphalt mixtures containing HDPE and without plastic.

7) The results of volumetric properties of all asphalt mixtures at their optimum bitumen content indicates that asphalt mixtures made by combining 25% RCA and 1% LDPE is the most similar mixture to control samples in terms of optimum bitumen content and volumetric properties requirements.

The natural extension of this work could involve using plastic at different percentages to determine the optimum design content.

Acknowledgments

The authors would like to acknowledge Boral Asphalt Company for providing part of materials required for the research work, and Western Sydney University for providing the testing equipment.

References

Brozyna, D., and Kowalskia, K., Modification of Asphalt Binders by Polyethylene-type Polymers, *Journal of Building Chemistry*, 1, 37-41, 2016.

Chen, M., Lin, J., and Wu, S., Potential of Recycled Fine Aggregates Powder as Filler in Asphalt Mixture, *Construction and Building Materials Journal*, 25, 3909–3914, 2011.

Chandh, K. A., and Akhila, S., A Laboratory Study on Effect of Plastic on Bitumen, *International Journal of Science and Research*, 5(10), 1406-1409, 2016.

Guimarães, R. C. and Ribeiro, H. C., *Utilization of Recycled Aggregates of Solid Waste from Civil Construction in Asphaltic Coatings*, University Unit of Exact Sciences and Technological Civil Engineering. Anápolis, 2005.

Justo, C. E. G., and Veeraragavan, A., Utilization of Waste Plastic Bags in Bituminous Mix for Improved Performance of Roads, *Centre for Transportation Engineering,* Bangalore University, 43-44, 2002.

Marinho, M. N., *Analysis of Mechanical Performance of Hot Mixed Asphaltic Concrete Produced with Recycled Coarse Concrete Aggregate*, MSc Thesis, Federal University of Pernambuco, Brazil, 2011.

Masad, E., Al-Rousan, T., Button J., and Little, D., *Test Methods for Characterizing Aggregate Shape, Texture, and Angularity*, In National Cooperative Highway Research Program Report 555, National Research Council: Washington DC, WA, USA, 2007.

Mills-Beale, J., and You, Z., The Mechanical Properties of Asphalt Mixtures with Recycled Concrete Aggregates, *Construction, and Building Materials Journal*, 24, 230–235, 2010.

Paranavithana, S., and Mohajerani, A., Effects of Recycled Concrete Aggregates on Properties of Asphalt Concrete, *Journal of Resources, Conservation, and Recycling*, 48(1), 1-12, 2006.

Pérez, I., Pasandin, A. R., and Medina, L., *Hot Mix Asphalt Using Construction and Demolition Waste as Coarse Aggregates*, MSc Thesis, 36, 840–846, 2012.

Punith, V. S., and Veeraragavan, A., Behavior of Asphalt Concrete Mixtures with Reclaimed Polyethylene as Additive, *Journal of Materials in Civil Engineering*, 19(6), 500–507, 2007.

Tahmoorian, F., Samali, B., Tam, V. W. Y., and Yeaman, J., Evaluation of Mechanical Properties of Recycled Material for Utilization in Asphalt Mixtures, *Journal of Applied Sciences*, 7, 1-21, 2017.

Silva, C. A. R. S., *Study of Recycled Aggregate from Civil Construction in Bituminous Mixtures for Urban Roads*, MSc Thesis, Federal University of Ouro Preto, Brazil, 2009.

Wong, Y. D., Sun, D. D., and Lai, D., Value-Added Utilization of Recycled Concrete in Hot-Mix Asphalt, *Waste Management*, 27, 294–301, 2007.

Zhu, J., Wu, S., Zhong, J., and Wang, D., Investigation of Asphalt Mixture Containing Demolition Waste Obtained from Earthquake-Damage Buildings, *Journal of Construction and Building Materials*, 29, 466–475, 2012.

Zulkati, A., Wong, Y. D., and Sun, D. D., Mechanistic Performance of Asphalt-Concrete Mixture Incorporating Coarse Recycled Concrete Aggregate, *Journal of Materials in Civil Engineering*, 10, 1299–1305, 2013.

Streamlining Information Transfer between Construction and Structural Engineering
Edited by Shiau, J., Vimonsatit, V., Yazdani, S., and Singh, A.

ISBN: 978-0-9960437-7-9

MECHANICAL PROPERTIES OF WARM MIXED USING POLYMER RECLAIMED ASPHALT WITH NEW AGGREGATE (PRAP-WARM MIX)

WEERAKASET SUANPAGA

Dept of Civil Engineering, Kasetsart University, Bangkok, Thailand

The purpose of this study is to determine the appropriated mixed proportion of asphalt concrete using Polymer Reclaimed Asphalt Pavement, new aggregate and Advera (PRAP- Warm Mixed) that were mixed at warm temperature. Then the Mechanical Properties of new mixed asphalt concrete samples were tested following the standards of the Department of Highway, Thailand. In experimental design, the range of mixing temperature varied as 140°C, 145°C, 150°C, 155°C, and percentage of Advera added varied at 0.20, 0.25, 0.30 and 0.35 percent by weight of aggregate. The engineering properties of PRAP samples obtained from this study showed that the optimum mixing ratio was 0.20 percent at 140°C has a Stability value of 2530 lbs, a flow of 12 in a particle size of VMA 16.2, a strength index, 85 percent of the test results are in line with the requirements of the Department of Highways. Then using 70 percent of PRAP-warm mix with 30 percent of new material at 140 °C and add 0.20 percent of Advera that is the best-mixed proportion. The normal temperature of PMA mixing is higher than 170°C, in this case using PRAP-Warm mix can reduce temperature lesser than 150°C then this mixing process can save fuel consumption. Thus this study is useful for environmental perspectives.

Keywords: Hot mix asphalt, Warm mix asphalt, HMA, WMA, PMA, Pavement.

1 INTRODUCTION GENERAL

Hot Mix Asphalt (HMA) construction requires higher mixing temperatures of 160°C, which consume a lot of fuel to produce asphalt concrete. If we can reduce mixing temperatures lower than 160°C then this mixing process can save fuel consumption. Nowadays, not only a few researchers (Jansen and Smirnovs 2010, You *et al.* 2011, NAPA 2010, and Weerakaset *et al.* 2014) have studied the mechanical properties of warm mix asphalt (WMA), but also some researchers have studied the mechanical properties of warm mixed asphaltic concrete using polymer reclaimed asphalt, and new aggregate and Advera mix together at lower than 160 °C. This new asphaltic concrete is called "PRAP- Warm Mix".

2 OBJECTIVES

To determine the appropriated proportion of asphalt concrete using Polymer Reclaimed Asphalt Pavement, new aggregate and Advera (PRAP- Warm Mix) those were mixed at Warm temperature by considering from physical properties of PRAP- Warm Mix's samples.

3 MATERIALS

The materials of PRAP- Warm Mixed were crush rock size 3/4", 3/8", limestone from Chonburi, Thailand, Polymer Modified Asphaltic (PMA), reclaimed PMA concrete, and a mineral additive (ADVERA). The gradation and mix proportion of Limestone is shown in Table 1, and the gradation curve of a mixed bin of limestone is shown in Figure 1.

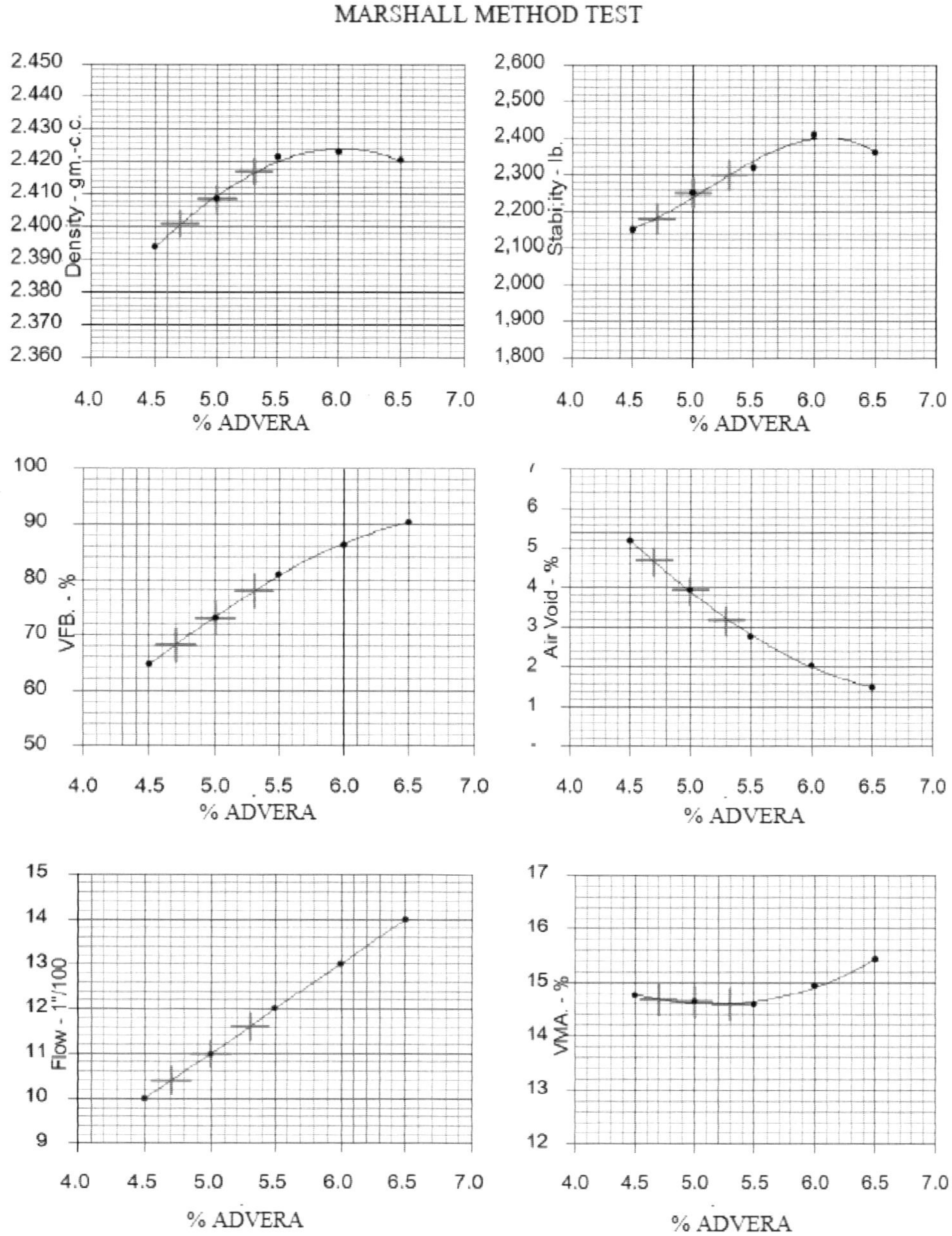

Figure 1. Marshall Method of engineering properties of PRAP- Warm Mix: density, stability, void filled in bitumen (VFB), percentage of air void, flow and voids in mineral aggregate (VMA).

Table 1. Gradation and mix proportion of limestone.

Sieve Size (Inch)	Hot Bin							
	% Passing						Tolerance	
	Rap	Bin 1	Bin 2	Bin 3	Bin 4	mixed	Desired	Limit
3/4"	100	100	100	100	100	100	100	100
1/2"	82.7	100	100	100	2.4	80.6	80-100	76-86
3/8"	63.4	100	100	73.7	0.5	68.2	-	63-73
#4	42.2	100	41.7	18.6		52.5	44-74	48-58
#8	29.3	86.2	0.60	0.50		38.0	28-58	33-43
#16	21.4	58.2				25.6	-	22-30
#30	14.2	37.2				16.4	-	12-20
#50	9.6	25.4				11.2	5-21	7-15
#100	5.6	18.6				8.2	-	6-11
#200	2.8	13.2				5.8	2-10	5-7
Mix proportion	70	13.2	6.0	6.6	4.2	Total 100 (by mass)		

4 METHODS

The specifications standard of testing fine and coarse aggregate were tested both groups referred to American Society for Testing and Materials (ASTM) and the Department of Highway (DOH) standard of testing material.

The specifications of fine materials were tested as follows.

- Gradation testing standard referred to ASTM C136-84A or AASHTO T37-77 or ASTM C136-84A
- Specific Gravity testing standard referred to ASTM C128-84 or AASHTO T84-77 or DOH.209/2517
- Soundness testing standard referred to ASTM C88-83 or AASHTO T104-77 or DOH.213/2531
- The sand Equivalent testing standard referred to ASTM D2419-74 or AASHTO T176-73 or DOH .203/2515

The specifications of course materials were tested as follows.

- Gradation testing standard referred to ASTM C127-84 or AASHTO T85-77 or DOH.207/2517
- Specific Gravity testing standard referred to ASTM C128-84 or AASHTO T84-77 or DOH.209/2517
- Durability testing standard referred to ASTM C88-83 or AASHTO T104-77 or DOH.213/2531
- The flakiness Index testing standard referred to DOH.210/2518
- The elongation Index testing standard referred to DOH. 211/2518
- The sand Equivalent testing standard referred to ASTM D2419-74 or AASHTO T176-73 or DOH .203/2515
- Soundness testing standard referred to ASTM C131-81 or AASHTO T96-77 or DOH. 202/2515

It is to note that some researchers studied fatigue analysis (Bonnaure *et al. 1982*, Monismith *et al.* 1970, Ruth and Olson 1977, and Sousa *et al.* 1998) but for DOH's Thailand this testing standard is not set for main requirement.

Job Mix Formula of all specimens was applied in this study. Adhesive mixtures were polymer modified asphalt (PMA) mixed with "ADVERA" to the ratios of 0.20%, 0.25%, 0.30% and 0.35% by weight of aggregate. Mixing and compaction temperatures were controlled at 140°C, 145°C, 150°C, and 155°C. In this study after found the optimum ADVERA ratio and optimum mixing temperature, this study determined under laboratory conditions the fundamental properties of samples, and evaluated their possible use for WMA, particularly the potential use of ADVERA. These fundamental laboratory conditions included:

- Strength Index
- Temperature Sensitivity
- Maximum Specific Gravity
- Bulk Specific Gravity
- Voids in Asphalt Mixture
- Binder Workability

Table 2 shows the control criteria of engineering properties of each surface course, those course samples should design as follow the DOH Thailand's standard specification.

Table 2. DOH Thailand's standard specification of HMA, WMA and PMA.

Items	Type of Surface Courses				
	Wearing Course 9.5 mm.	Wearing Course 12.5 mm.	Binder Course	Base Course	Shoulder
Blows	75	75	75	75	75
Stability N	8006	8006	8006	7117	7117
(lb)	(1800)	(1800)	(1800)	(1600)	(1600)
Flow 0.25 mm (0.01 in.)	8-16	8-16	8-16	8-16	8-16
Percent air voids	3-5	3-5	3-5	3-5	3-5
Percent voids in material Aggregate (VMA, min)	15	14	13	12	14
Stability/Flow, min N/0.25 mm	712	712	712	645	645
(lb/0.01 in.)	(160)	(160)	(160)	(145)	(145)
Percent strength index	75	75	75	75	75

5 RESULTS AND DISCUSSION

In total, 75 WMA specimens were examined using a general standard specification test according to standards set by DOH Thailand.

The physical engineering characteristics of each item were as follows: stability, flow voids in mineral aggregate (VMA), density, the percentage of air voids, and voids filled in bitumen (VFB) are shown in Figure 2. The physical engineering characteristics of each item were as below.

5.1 Stability and Flow of PRAP- Warm Mix

Normally, the maximum stability's values of all specimens were determined by controlling the settlement rate of specimens (2 inches per minute) with temperature (60 °C). The flows values were determined during the same period. The stability's values of all specimens are 2,530-2,950 lbs. also, the flow values of all specimens are 11-12 (1/0.01 in). The stability per flow value (ASF) is 211-384 (lb/0.01 in.) which higher than 160 (lb/0.01 in.). The results show that all stability and flow values give a higher value than DOH Thailand's standard. Also, the results of using ADVERA as 0.2% of aggregate mixed at 140°C indicated that this is the optimum mix proportion.

5.2 Percentage of Air Void (V) and Void in Mineral (VMA) of PRAP- Warm Mix

The average of the percentage of air void is 3.6-5.8 and the average of the percentage of void in the mineral is 14.2-16.2 these results give a higher value than DOH Thailand's standard.

5.3 Strength Index of PRAP- Warm Mix

The strength indexes of WMA's specimens were tested by following the Ontario vacuum immersion Marshall Test. The strength index values are 84.0-93.0 lbs. These results give a higher value than DOH Thailand's standard (75 lbs).

Table 3 shows the engineering properties of PRAP- Warm Mix specimens using 0.2 - 0.35 % mix at temperature 140°C and Table 4 shows the engineering properties of PRAP- Warm Mix specimens using ADVERA 0.2 %. In conclusion, the engineering properties of PRAP- Warm Mix samples obtained from this study showed that: The optimum mixing ratio was 0.20 percent at 140°C which has a Stability value of 2530 lbs, a flow of 12 in a particle size of VMA 16.2, a strength index, 85 percent. Using 70 percent of PRAP- Warm Mix with 30 percent of new material at 140 °C and add 0.20 percent of Advera that is the best-mixed proportion.

Table 3. Engineering properties of PRAP- Warm Mix specimens using ADVERA 0.2 - 0.35 % mix at temperature 140°C.

Test at mix temperature 140°C	%ADVERA				DOH standard
	0.20	0.25	0.30	0.35	
Average Stability (lb)	2530	2840	2760	3700	1600-1800
Average Flows (1/100 in)	12	11	11	11	11-13
Stability/Flow (1/100 lb/in)	211	258	251	336	>210
Average Density	2.330	2.348	2.359	2.364	2.368-2.385
Percent Air Voids	5.8	5.1	4.7	4.5	3-5
Percent Voids in Mineral Aggregate (VMA)	16.2	15.5	15.1	14.9	>12
Voids Filled with Bitumen (VFB)	63.9	67.1	69.2	70.2	69-79
Strength index	85	85	87	87	>75

Table 4. Engineering properties of PRAP- Warm Mix specimens using ADVERA 0.2 %.

Test	Mix Temperature			
	140°C	145°C	150°C	155°C
Average Stability (lb)	7760	7820	8310	9290
Average Flows (1/100 in)	14	14	13	12
Stability/Flow (1/100 lb/in)	554.2	558.5	639.2	774.1
Average Density	2.282	2.287	2.295	2.312
Percent Air Voids	7.8	7.6	7.3	6.6
Percent Voids in Mineral Aggregate (VMA)	17.9	17.7	17.4	16.8
Voids Filled with Bitumen (VFB)	56.5	57.2	58.4	61.0
Strength index	85	84	84	87

6 CONCLUSION

The results show that the PRAP- Warm Mix process reduces the mixing temperature from 170°C to 140 °C that means this mixing process can save fuel consumption. Also using reclaims PMA. Thus this study is useful for environmental perspectives.

Acknowledgments

We acknowledge Mr.Tanach Sripatin, Department of Highway, Thailand for supporting job mixed testing and PQ Chemicals (Thailand) Co. Ltd. for providing ADVERA.

References

Bonnaure, F., Huibers, A., and Boonders, A., *A Laboratory Investigation of The Influence of Rest Periods on The Fatigue Characteristics Of Bituminous Mixes*, Proceedings of The Association of Asphalt Paving Technologists, St. Paul, MN., 51, 105-121, 1982.

Jansen, J., and Smirnovs, J., *Warm Mix Asphalt Investigation,* Master of Science Thesis, Department of Civil Engineering, Technical University of Denmark, Denmark, 2010.

Monismith, C., Epps, J., Kasianchuk, D., and Mclean, D., Asphalt Mixture Behavior in Repeated Flexure, *Transportation Engineering Report Number 70-5*, University of California, Berkeley, CA., 1970.

NAPA, Warm-Mix Asphalt Best Practice, *Quality Improvement Publication 125*, National Asphalt Pavement Association, Lanham, MD, 2010.

Ruth, E., and Olson, K., Creep Effects on Fatigue Testing of Asphalt Concrete, *Proceedings of AAPT*, Association of Asphalt Paving Technologists, St. Paul, MN., 46, 177-196, 1977.

Sousa, J., Pais, J., Prates, M., Barras, R., Langlois, P., and Leclerc, A., Effect of Aggregate Gradation on Fatigue Life of Asphalt Concrete Mixes, *France Portution Research Record 1630*. 62-70, Paris, 1998.

Weerakaset, S., Watcharin, W., Somsak, C., and Thaweesak, P., *Mechanical Properties of Warm Mix Asphalt Concrete Application Using ADVERA*, Proceedings of Sustainable Solutions in Structural Engineering and Construction, ASEA-SEC02, Bangkok, Thailand, 419-423, 2014.

You, Z., Goh, S., and Dai, Q., Laboratory Evaluation of Warm Mix Asphalt, *Report No. RC-1556*, Michigan Technological University, Houghton, MI, 2011.

Streamlining Information Transfer between Construction and Structural Engineering
Edited by Shiau, J., Vimonsatit, V., Yazdani, S., and Singh, A.

ISBN: 978-0-9960437-7-9

INVESTIGATION OF TEMPERATURE RISING MECHANISM IN LONG TUNNEL BY NUMERICAL SIMULATION

HAEYOUNG KIM, HITOSHI YAMADA, HIROSHI KATSUCHI, and SOICHIRO NAKAMURA

Dept of Civil Engineering, Yokohama National University, Yokohama, Japan

Abnormally high temperature inside the long, undergrounded tunnel is a problem in the urban area. As one of the countermeasures, a ventilation fan has been operated. However, the insufficient temperature reduction effect and high cost are an issue and improvement is required. In this study, in order to improve the operating plan of ventilation, the flow field and temperature distribution characteristics are clarified by computational thermal fluid analysis, and the operation of the optimum ventilation was decided. The dominant factors for the temperature rising were identified as traffic volume, lane axis wind speed, cross flow ventilation, and heat flux between tunnel body and air in tunnel. In the analysis, we focused on these four factors, and applied these factors obtained from long-term on-site measurement to the boundary condition and the initial condition. In addition, the amount of heat from vehicle traffic was calculated based on the measurement and the past report results. The analytical model is 1000 *m* partial tunnel section where the temperature rising was intense. The validation of numerical model was verified from the comparison between the analysis results and the measured values. It was confirmed that the effect of increasing lane axis wind speed as a countermeasure was not significant, and the ventilation amount of crosswind is recommended as 60 m^3/s.

Keywords: Heat flux, Ventilation, Computational thermal fluid analysis.

1 INTRODUCTION

The Yamate Tunnel is the longest road tunnel in Japan, which has 18,200 meters in length. There are inner tunnel and outer tunnel which have two lanes each. Two kinds of ventilation method as shown in Figure 1 are adopted according to the location. One is the cross flow method. This is a method of providing air supply and exhaust ducts in parallel with the roadway and simultaneously ventilating at various places in the tunnel. The other method is a longitudinal flow method, using the car flow and exhausting it at a stroke in a certain place. This long underground tunnel in the urban area has become a problem of high temperature inside the tunnel. In August, the average monthly temperature rises above 40 degrees regardless of day and night, and the maximum temperature becomes 48. 3 degrees as shown in Figure 2. The motorcycle driver and the inspection worker due to the high temperature inside the tunnel are in trouble in terms of the comfort and safety and this should be solved. As one of the countermeasures, ventilation fans have been operated. However, the insufficient temperature reduction effect and high cost are an issue and improvement is required. Therefore, in this study, in order to quantify temperature reduction effect of ventilation operation and to reduce cost, field measurement and 3 dimensional computational thermal fluid analysis were conducted.

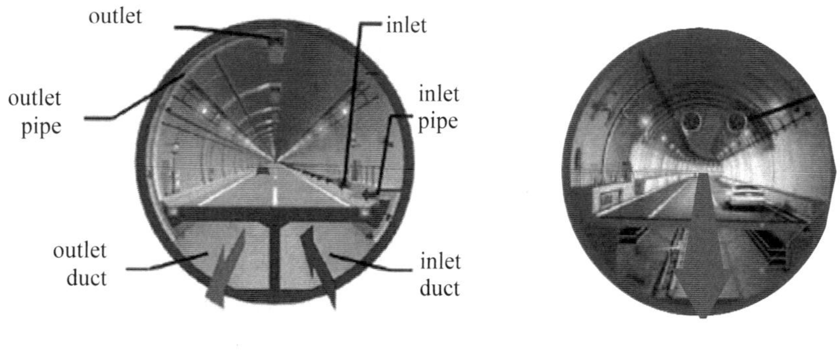

(a) Cross flow method (b) Longitudinal flow method

Figure 1. Ventilation method (red arrow: exhaust, blue arrow: air supply).

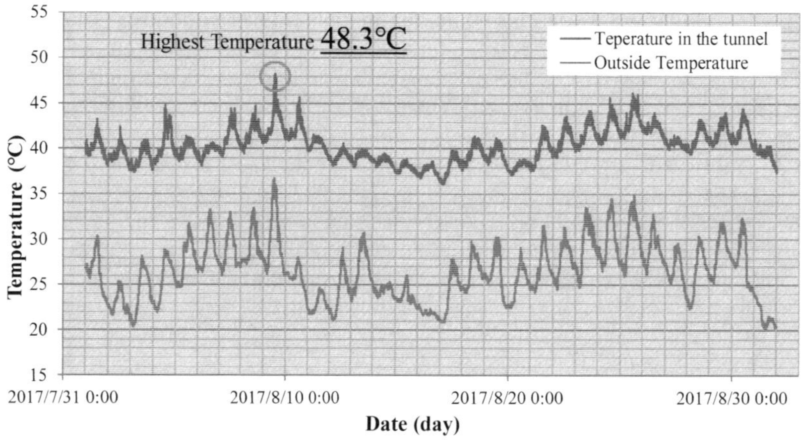

Figure 2. Temperature inside tunnel on August 2017.

2 FIELD MEASUREMENT

Many kinds of observation data such as temperatures, number of cars, wind speed and heat flux between tunnel body and air in tunnel were obtained by Metropolitan Expressway Co. Ltd. and this research team. These data were sorted out and analyzed for figuring out the current situation and setting boundary condition of thermal fluid analysis. From the observation data, it was confirmed that the temperature rise on the downstream side of the tunnel is remarkable throughout the year for both sides of the Yamate tunnel as shown in Figure 3. In addition, temperature distribution inside tunnel taken by the thermal camera is shown in Figure 4. From this temperature distribution, the planar heat source was applied to the asphalt in thermal fluid analysis model.

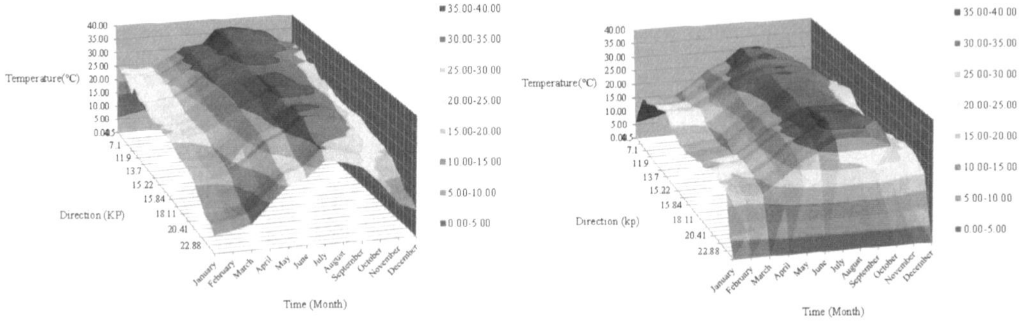

(a) Inner tunnel (b) Outer tunnel

Figure 3. Annual temperature distribution in Tunnel.

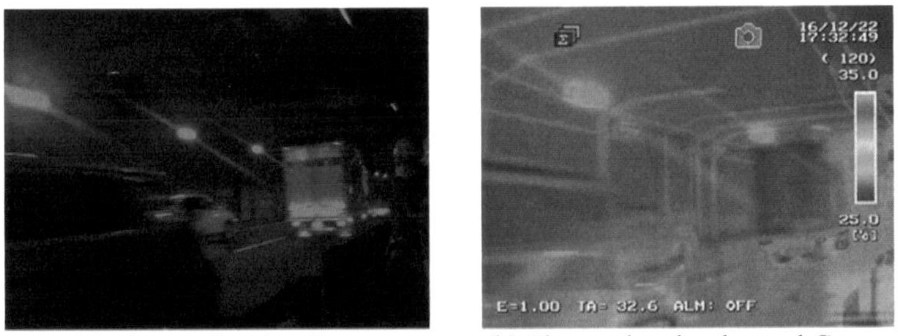

(a) Photo taken by Camera (b) Photo taken by thermal Camera

Figure 4. Photos of inside of tunnel.

3 COMPUTATIONAL METHOD

3. 1 Numerical Set-up

A three-dimensional steady state analysis was conducted and a modified k-ε model was used for the turbulence model. The dominant factors for the temperature rising were identified as traffic volume, lane axis wind speed, cross flow ventilation, and heat flux between tunnel body and air in tunnel. In addition, planar heat source of the pavement part obtained from the long-term on-site measurement were applied to the boundary conditions. The details of numerical set-up and model are shown in Table 1 and Figure 5 respectively.

Table 1. Details of numerical model set-up.

Simulation software	StarCCM+
Turbulence Model	Revised k-ε model
Model length	1000 *m*
Boundary condition	
Tunnel body-air heat transfer	Convection
Pavement part	Planar heat source: Temperature: 28 °C
Inlet duct	Velocity inlet: 4. 0 *m/s*, 26°C
Outlet duct	Velocity inlet: 3. 75 *m/s*, 29°C
Inlet	Velocity inlet: 2. 75 *m/s*, 25. 6°C
Outlet	Zero-pressure condition
Total exhaust heat quantity of vehicles	Input in air part

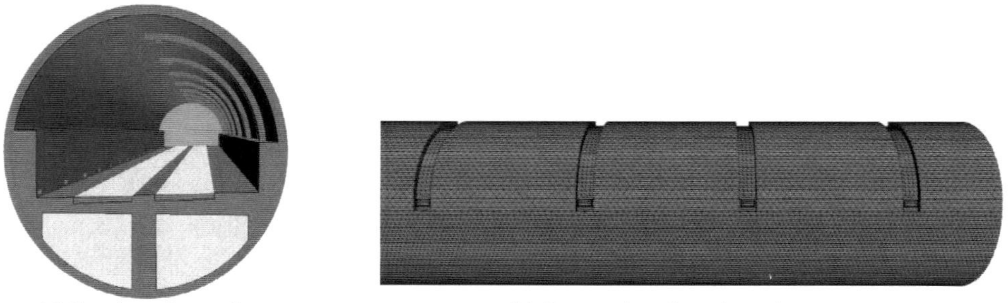

(a) Section plane view (b) Span-wise direction view

Figure 5. Numerical model.

The amount of exhaust heat generated by vehicle traffic are calculated based on the past report by Japanese Ministry of the Environment (2004) and numbers of vehicles. The analysis range was 1000 meters from the tunnel where the temperature rise was intense. The validation of numerical model was verified from the comparison between the analysis result and the actual measurement value as shown in Figure 6.

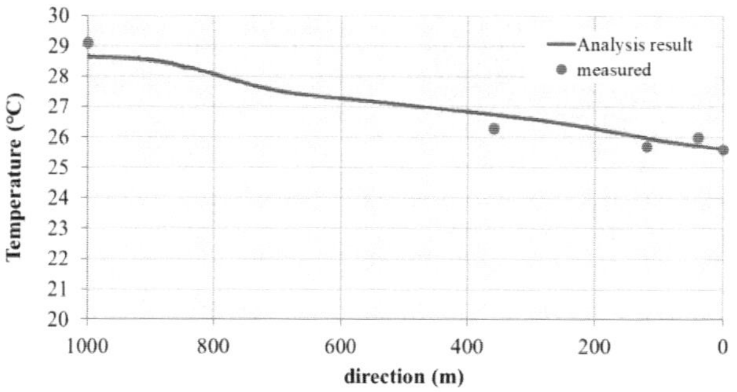

Figure 6. Temperature comparison between numerical result and measurement data.

4 NUMERICAL RESULTS

Table 2. Numerical cases.

No.	Lane direction wind speed [*m/s*]	Air volume of cross flow ventilation [m^3/s]	No.	Lane direction wind speed [*m/s*]	Air volume of cross flow ventilation [m^3/s]
Case11	0	0	Case12	0	30
Case21	0. 75	0	Case22	0. 75	30
Case31	2. 75	0	Case32	2. 75	30
Case41	4. 75	0	Case42	4. 75	30
Case13	0	60	Case14	0	120
Case23	0. 75	60	Case24	0. 75	120
Case33	2. 75	60	Case34	2. 75	120
Case43	4. 75	60	Case44	4. 75	120

We conducted a parametric study in which the cross flow ventilation air flow rate and longitudinal ventilation air flow rate (lane axial wind speed) were varied as shown in Table 2. The effectiveness of the longitudinal ventilation was confirmed from the analysis result in Figure 7(a) that temperature of cross-section increases by 9.13 degrees in case11 of w/o wind. However, Figure 7(b) shows that the effect of increasing lane axis wind speed as a countermeasure with cross flow ventilation was not significant. In addition, Figure 8 shows the temperature distribution at the height where people sense the temperature and Figure 9 shows the temperature distribution of cross section according to the air volume of cross flow ventilation. Table 3 summaries the average cross-sectional temperature according to air flow rate of cross-flow ventilation and the air temperature at the height where people sense the temperature. In case of air flow rates of the cross flow ventilation changes 60 m^3/s to 120 m^3/s, there is no change in the temperature at the height of people sensing position.

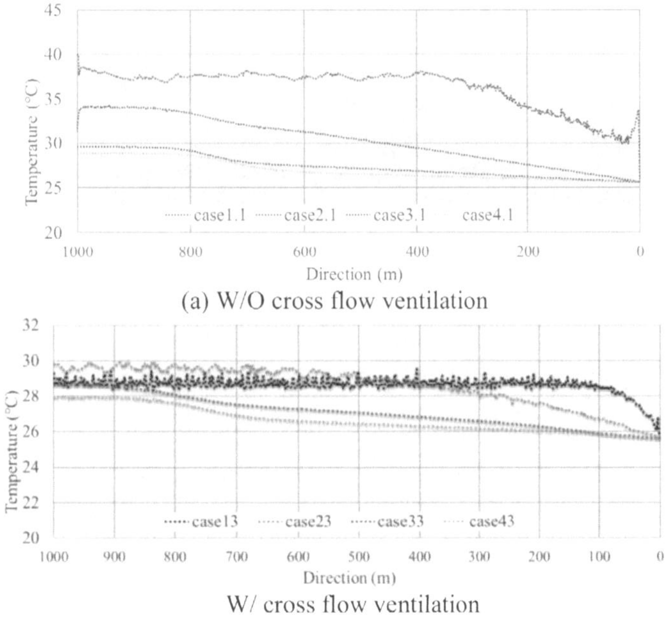

(a) W/O cross flow ventilation

W/ cross flow ventilation

Figure 7. Average temperature of cross section.

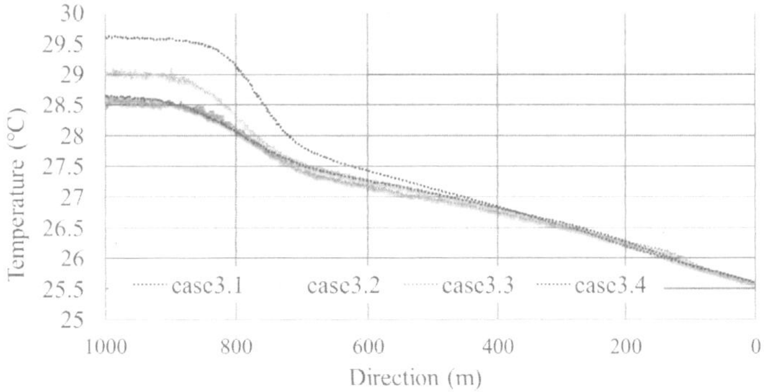

Figure 8. Temperature at the height where the user senses the temperature.

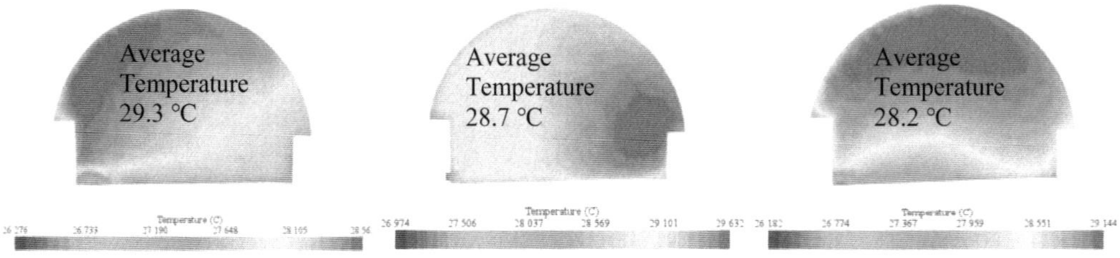

(a) 0 m^3/s (b) 60 m^3/s (c) 120 m^3/s

Figure 9. Temperature distribution of cross-section according to the air volume of cross flow ventilation.

5 CONCLUSIONS

Computational thermal fluid analysis and field measurement was conducted to investigate the mechanism of temperature rising in long urban tunnel. The dominant factors for the temperature rising were identified as traffic volume, lane axis wind speed, cross flow ventilation, and heat flux between tunnel body and air in tunnel. The effectiveness of the longitudinal ventilation without cross flow ventilation was confirmed from the analysis result while the effect of increasing lane axis wind speed (longitudinal ventilation) with cross flow ventilation was not significant and the optimum ventilation amount of crosswind is recommended as 60 m^3/s.

References

Japanese Ministry of the Environment, *Reports on Measurement Against Heat Islands by Urban Population Exhaust*, 74, 2004.

Streamlining Information Transfer between Construction and Structural Engineering
Edited by Shiau, J., Vimonsatit, V., Yazdani, S., and Singh, A.

ISBN: 978-0-9960437-7-9

COMPARATIVE ASSESSMENT OF TRANSPORTATION SUSTAINABILITY RATING SYSTEMS

OLUWATOBI OLUWALAIYE and MEHMET E. OZBEK

Dept of Construction Management, Colorado State University, Fort Collins, USA

The construction industry contributes a significant amount of greenhouse gas emissions to the environment. This highlights the need for this industry to implement sustainable practices to mitigate its negative effects. Sustainability in the construction industry involves considering project development in terms of economic, social, and environmental aspects (triple bottom line). In the transportation sector of the construction industry, several transportation sustainability rating systems (TSRS) have been developed to measure and promote sustainability. However, studies show that these TSRS have not been developed within a consistent "sustainability scope". This raises the question of the consistency of these TSRS in measuring sustainability of transportation projects. This study assesses three prominent TSRS to determine how each measures sustainability with respect to the triple bottom line. The TSRS that are in the scope of this study are: (i) Envision – a third-party rating system, (ii) INVEST– a self-assessed rating and, (iii) GreenLITES – an in-house developed self-assessed rating system for the New York State DOT. Results show that while these three rating systems provide different levels of credits/points with respect to the triple bottom line, they all place the highest focus on the environmental aspect.

Keywords: Envision, Invest, GreenLITES, TSRS.

1 INTRODUCTION AND PURPOSE

The construction industry contributes a significant amount of greenhouse gas emissions (Yudelson 2008). As such, the industry needs to implement sustainable development practices to mitigate its economic, social, and environmental impacts (Simpson 2013). Green buildings are said to reduce energy consumption by 30%, save water by 30 – 50%, diminish carbon emissions by 35% and provide a construction waste reduction of 50 – 90% (Yudelson 2008). However, implementing sustainable development practices in the construction industry should not only be limited to buildings, but also be applied in other sectors of the industry such as the transportation sector – the sector of the industry that is involved in the construction of transportation infrastructure.

Sustainability rating systems are tools that help to guide the assessment of construction projects against a collection of sustainability best practices that cover the three aspects of sustainability: economic, social, and environmental (Griffiths *et al.* 2015). There are several sustainability rating systems developed to measure project performance in both the building and transportation sectors of the construction industry. Transportation Sustainability Rating Systems (TSRS), developed to measure transportation project features and actions against a sustainability standard, can be used to quantify the project's progress against that standard (Veeravigrom *et al.* 2015).

In recent years, there has been an increase in the development of transportation sustainability rating systems (Veeravigrom *et al.* 2015). However, studies show that these TSRS have not been developed with a consistent "sustainability scope" ("the breadth of sustainability addressed and the prioritization within") (Griffiths *et al.* 2015, Veeravigrom *et al.* 2015). Thus, a project, run through more than one of these rating systems may be high performing on one rating system and low performing on another. This brings about the question of the consistency of these rating systems in measuring sustainability. There is, therefore, the need to assess multiple TSRS to determine how each measures sustainability with respect to the triple bottom line. The purpose of this study is to assess three prominent TSRS to determine how each measures sustainability with respect to the triple bottom line. The TSRS that are in the scope of this study are: (i) Envision – a third-party rating system, (ii) INVEST – a self-assessed rating system and, (iii) GreenLITES – an in-house developed self-assessed rating system for the New York State Department of Transportation (NYSDOT). These rating systems were chosen because of their prominence and wide-spread use in the United States as well as the distinct characteristic each rating system provides as stated above.

2 REVIEW OF TSRS STUDIED IN THIS RESEARCH

2.1 Envision

The Envision rating system is a holistic rating system used to rate the sustainability of infrastructure such as bridges, roads, railways, pipelines, water treatment systems, dams, airports, landfills, levees and other civil infrastructure (ISI 2017a). It is a third-party rating system that provides a framework used to rate the economic, community and environmental benefits of infrastructure projects regardless of the size of the project (ISI 2017b). A third-party rating system is one which requires the presence of a sustainability professional accredited by the rating system on a project team to carry out ratings for a project; qualifying the project for verification by a professional employed by the rating system and finally certification.

The Envision rating system has a total of 60 performance objectives, otherwise known as credits/criteria (ISI 2015), that address the impacts of the triple bottom line on sustainability in the design, construction and operation phases of an infrastructure project (ISI, 2017b). These 60 credits that have a maximum achievable point of 809 are classified under five categories: Quality of Life, Leadership, Resource Allocation, Natural World and Climate, and Risk. These five categories are further divided into 14 subcategories (ISI 2015).

Rating a project using the Envision rating system involves the use of three tools: the self-assessment checklist, the online scoring module, and the verification/awards program. These tools can be used independently or in combination based on the project type or the project phase in which it is being applied (Shivakumar *et al.* 2014). The Envision certification/award has four categories: Platinum (50% of total applicable points), Gold (40% of total applicable points), Silver (30% of total applicable points) and Bronze (20% of total applicable points) (ISI 2017c, Shivakumar *et al.* 2014).

2.2 INVEST

Infrastructure Voluntary Evaluation Sustainability Tool (INVEST) is a self-assessed web-based transportation sustainability rating system (FHWA 2017a). A self-assessed rating system is one which does not require an INVEST recognized sustainability professional on the project; neither does it require third-party verification by INVEST. INVEST consists of sustainability best practices known as criteria and it covers the entire lifecycle of transportation projects made up of planning, design, construction and the operations and maintenance phases (FHWA 2017a). INVEST can be used to evaluate the sustainability best practices implemented by a project currently under construction (Brodie *et al.* 2013). It also provides guidance on sustainability best practices that can be applied to a transportation project as well as help the project team set realistic sustainability goals. It can be used to measure the sustainability of completed transportation projects. (Brodie *et al.* 2013).

INVEST is composed of four modules: System Planning for States (SPS), System Planning for Regions (SPR), Project Development (PD) and Operations and Management (OM) (FHWA 2017a, Simpson 2013). The Systems Planning (SPS and SPR) and the Operation and Maintenance modules are designed to assess agencies' programs while the Project Development module assesses projects from the planning phase through the construction phase (FHWA 2017a). The Project Development module has a total of 33 criteria and maximum achievable points of 171 (FHWA 2017b). INVEST has four levels of achievement: Platinum (60% of total applicable points), Gold (50% of total applicable points), Silver (40% of total applicable points) and Bronze (30% of total applicable points) (FHWA 2017c).

2.3 GreenLITES

The GreenLITES Project Design Program began in September, 2008 as a self-assessed program that recognizes transportation projects in which sustainable practices are integrated extensively (McVoy *et al.* 2010). This program is a mandatory tool for use on all NYSDOT projects and is applied to the plans, specifications and estimates of these projects in the design phase (NYSDOT 2017b, Simpson 2013). The GreenLITES program utilizes a scorecard which the project team reviews with the aim of selecting sustainable practices that can be integrated into the project (McVoy *et al.* 2010, Simpson 2013). This program has maximum achievable points of 278 and contains 175 credits (criteria) that are divided into five categories: Sustainable Sites, Water Quality, Materials and Resources, Energy and Atmosphere and Innovation or Unlisted. There are four levels of achievement that can be attained under the GreenLITES Project Design Program (McVoy *et al.* 2010). They are GreenLITES Certified (15 – 29 points), GreenLITES Silver (30 – 40 points), GreenLITES Gold (45 – 59 points) and GreenLITES Evergreen (60 points and above) (McVoy *et al.* 2010, NYSDOT 2017a).

3 COMPARATIVE ASSESSMENT OF TSRS STUDIED IN THIS RESEARCH

All three rating systems apply to the Planning, Design, Construction as well as Operations and Maintenance phases of construction. Tables 1 – 5 display similarities and differences the three rating systems have as discussed under the following five categories that are common to all three rating systems respectively: the environment category, the quality of life category, the materials category, the energy category and the water quality category. The tables also show the percentage of each rating system allocated to the different categories. The points allotted to each category mentioned in the tables below are gotten from the points allotted to these categories in the rating systems' scorecards. These categories do not represent the entirety of the rating systems as there are other categories which are not common to all three rating systems. In the tables, the notation "◊" indicates that the item is covered in another place, whereas "×" indicates that the item is not included in the rating system at all.

Table 1: Points achievable by Credits related to the Environmental Category in each Rating System.

Rating System	Site Vegetation	Siting	Biodiversity	Ecological Connectivity	Habitat Restoration	Improve pedestrian and Bike Access	Environmental Training	Noise Mitigation/ Abatement	Emission Reduction	Monitoring and Maintenance	Reduction of Light Pollution	Percentage allotted in system
Envision	◊	99	55	◊	◊	◊	◊	◊	◊	◊	◊	**25%**
INVEST	6	◊	◊	4	7	6	1	7	2	◊	3	**21%**
GreenLITES	14	13	◊	◊	19	◊	×	◊	◊	◊	3	**29%**

Table 2: Points achievable by Credits related to the Quality of Life/Social Category in each Rating System.

Rating System	Improve Community Quality of Life	Stimulate Sustainable Growth and Development	Enhance Public Health and Safety	Improve Site Accessibility Safety and Wayfinding	Preserve Historic and Cultural Resources	Preserve Views and Local Character	Provide for Stakeholder involvement	Improve Infrastructure Integration	Extend Useful Life	Educational Outreach	Enhance Public Space	Percentage allotted in system
Envision	25	16	16	15	16	14	◊	◊	◊	◊	13	**22%**
INVEST	◊	◊	10	×	3	◊	◊	◊	◊	2	3	**1%**
GreenLITES	◊	◊	◊	×	◊	◊	◊	×	◊	◊	◊	**0%**

Table 3: Points achievable by Credits related to the Material Category in each Rating System.

Rating System	Reduce Net Embodied Energy	Support Sustainable Procurement Practices/Hazardous Material Minimization	Bio-Engineering Techniques	Use Recycled Materials/Reuse and Repurpose Materials	Use Regional Materials	Divert Waste from Landfill	Reduce Excavated Material Taken Offsite	Provide for Deconstruction and Recycling	Percentage allotted in system
Envision	18	9	×	14	10	11	6	12	**10%**
INVEST	3	◊	×	22	×	◊	5	◊	**18%**
GreenLITES	×	6	8	48	4	◊	◊	◊	**24%**

Table 4: Points achievable by Credits related to the Energy Category in each Rating System.

Rating System	Reduce Energy/Electrical Consumption	Traffic Flow Improvement	Use Renewable Energy	Reduce Petroleum Consumption	Commission and Monitor Energy Systems	Stray Light Reduction	Reduction of Pavement Materials Emissions	Infrastructure Resilience	Percentage allotted in system
Envision	18	×	20	◊	11	◊	◊	◊	**6%**
INVEST	8	◊	×	×	×	◊	3	12	**13%**
GreenLITES	10	29	×	15	×	3	◊	◊	**37%**

Table 5: Points achievable by Credits related to the Water Category in each Rating System.

Rating System	Manage Stormwater	Reduce Pesticides and Fertilizer Impacts	Prevent Surface and Groundwater Contamination	Protect Freshwater Availability	Reduce Potable Water Consumption	Monitor Water Systems	Best Management Practices (Permeable Pavement)	Percentage allotted in system
Envision	21	9	18	21	21	11	◊	**12%**
INVEST	6	×	×	×	×	×	5	**6%**
GreenLITES	10	×	×	×	×	◊	10	**7%**

4 DISCUSSION

The Envision rating system is an infrastructure sustainability rating system that applies not only to road transportation projects but also to other infrastructure projects such as pipelines, airports, landfills, levees, railways, water treatment systems, dams and other civil works components (ISI 2015). This rating system, being a third-party verified system, requires that an Envision certified sustainability professional be on the project team for projects that seek Envision certification and recognition (ISI, 2015). With its five categories and 60 credits, the Envision rating system can be said to give the following distribution of its ratings to the economic, social and environmental aspects of sustainability respectively: 15%, 22% and 63%. The Envision rating system has been applied to at least 38 projects since its inception to date (ISI 2017d).

The INVEST rating system has seven project scorecards intended to cater to diverse types of project depending on the project type and its location (FHWA 2017b). Six of these scorecards have pre-determined credits that will apply to projects that fall under these scorecards. However, one of these scorecards, the Custom Scorecard, allows a project that does not fit into the other scorecards to develop criteria (credits) that best suit it (FHWA 2017b). Therefore, a project's final score is determined based on criteria that are relevant to the project (Simpson 2013). The INVEST rating system allocates the following distribution of its ratings to the economic, social and environmental aspects of sustainability respectively: 12%, 16% and 72%. The INVEST rating system has been used to evaluate at least 1844 projects to date (FHWA 2018).

Of the three systems, the GreenLITES rating system places the greatest focus on the environmental aspect of sustainability with a percentage allocation of 86%. The economic and social aspects of sustainability have ratings of 10% and 4% respectively.

The Envision, INVEST and GreenLITES rating systems also have unique aspects in the credits covered by each of them. A unique aspect of the Envision rating system is the category; Quality of Life. This category, along with its credits, specifically focus on the social aspect of the triple bottom line. The other rating systems being studied place some focus on the social aspect of sustainability. The INVEST rating system has its credits individually outlined without being sectioned into categories, however, in the GreenLITES rating system, credits related to the social aspects are embedded in categories that focus on the environmental or economical aspects of sustainability. Another unique aspect of the Envision rating system is the Leadership category which places focus, solely, on stakeholder involvement on a project. This category is important as it highlights the importance and benefits of applying integrated project delivery to a project. While the INVEST and GreenLITES rating systems touch on this aspect, the importance and benefits of involving all stakeholders on a project is not highlighted as much. A unique aspect of the GreenLITES rating system is the focus it places on resources (Material and Energy). 61% of its score are allocated to the resources. The Envision rating system focuses 23% of its score on resources while INVEST rating system focuses 31% of its score on resources.

5 CONCLUSIONS

The comparative analysis of three transportation sustainability rating systems shows that all three rating systems recognize and support sustainability best practices that cut across the triple bottom line. However, even though they all place the highest focus on the environmental aspect, priority and ratings given to each aspect of the triple bottom line by each rating system differs. Another difference these rating systems have is their application. While INVEST can be applied by anyone on a project team, Envision must be applied by a certified Envision sustainability professional who is a part of the project team and GreenLITES can be applied by members of a department of transportation that are part of the project team.

Further research should be carried out to determine how different transportation projects perform when these rating systems are applied to each project. This will help to see in greater detail how these rating systems measure sustainability of transportation projects.

References

Brodie, S., Ingles, A., Colville, Z., Amekudzi, A., Peters, R. W., and Sisiopikou, V., Review of Sustainability Rating Systems for Transportation and Neighborhood-Level Developments, *The Second Conference on Green Streets, Highways, and Development,* 337-354, Texas, USA, 2013.

FHWA, About INVEST, retrieved from https://www.sustainablehighways.org/100/about.html, 2017a.

FHWA, About the Project Development Module. Retrieved from https://www.sustainablehighways.org/900/about-the-project-development-module.html, 2017b.

FHWA, Who Can Use the Project Development Module. Retrieved from https://www.sustainablehighways.org/905/who-can-use-the-project-development-module.html, 2017c.

FHWA, Who is using Invest?. Retrieved from https://www.sustainablehighways.org/1399/who-is-using-Invest.html, 2018.

Griffiths, K., Boyle, C., and Henning, T., Infrastructure Sustainability Rating Tools–How They Have Developed and What We Might Expect to See in The Future, *IPWEA 2015: Sustainable Communities Sharing Knowledge*, Rotorua, New Zealand, 2015.

ISI, Envision Guidance Manual. Retrieved from https://sustainableinfrastructure.org/portal/files/GuidanceManual.pdf, 2015.

ISI, Envision. Retrieved from https://sustainableinfrastructure.org/envision/, 2017a.

ISI, Envision - How it Works, retrieved from https://sustainableinfrastructure.org/envision/how-it-works/, 2017b.

ISI, Envision Fact Sheet, retrieved from https://sustainableinfrastructure.org/brochure/fact-sheet/, 2017c.

ISI, Project Awards. Retrieved from http://sustainableinfrastructure.org/envision/project-awards/#3/48.22/-95.89, 2017d.

McVoy, G. R., Nelson, D. A., Krekeler, P., Kolb, E., and Gritsavage, J. S., *Moving towards Sustainability: New York State Department of Transportation's GreenLITES Story,* 2010.

NYSDOT, GreenLITES. Retrieved from https://www.dot.ny.gov/programs/greenlites, 2017a.

NYSDOT, Project Design Program. Retrieved from https://www.dot.ny.gov/programs/greenlites/project-design-cert, 2017b.

Shivakumar, S., Pedersen, T., Wilkins, S., and Schuster, S., Envision: A Measure of Infrastructure Sustainability, *Pipelines 2014: From Underground to the Forefront of Innovation and Sustainability,* 2249-2256, 2014.

Simpson, S. P., *A Framework for Assessing Transportation Sustainability Rating Systems for Implementation in US State Departments of Transportation.* Colorado State University, 2013.

Veeravigrom, M., Muench, S. T., and Kosonen, H., A Global Framework for Sustainable Roadway Rating Systems, *Transportation Research Board 94th Annual Meeting*, Washington D.C., USA, 2015.

Yudelson, J. a. M. T. A., Book Reviews-The Green Building Revolution, *Journal of Real Estate Literature,* 16(2), 253, 2008.

Streamlining Information Transfer between Construction and Structural Engineering
Edited by Shiau, J., Vimonsatit, V., Yazdani, S., and Singh, A.

ISBN: 978-0-9960437-7-9

STUDY ON THE INFLUENCE OF CURING TEMPERATURE ON MECHANICAL PROPERTIES OF MODIFIED FLY ASH MORTAR

HIROTAKA MATSUO[1], KOJI TAKASU[2], HIDEHIRO KOYAMADA[2], and HIROKI SUYAMA[2]

[1]*Graduate School of Engineering, University of Kitakyushu, Kitakyushu, Japan*
[2]*Faculty of Engineering, University of Kitakyushu, Kitakyushu, Japan*

Using fly ash as an admixture for concrete can contribute to environmental load reduction and concrete quality improvement. However, as the quality of fly ash fluctuates depending on the ash source, quality stabilization is required. It was proved that concrete with fly ash of Japanese Industrial Standardized class II has different strength properties depending on curing temperature, but it is not obvious whether concrete with modified fly ash by flotation method has similar properties. In this study, the influence was examined on the mechanical properties when changing the curing temperature of mortar using fly ash modified by the flotation technique. The sealing curing was set to 5°C, 20°C, 40°C and 60°C. Also, after 7 days, 5°C, 40°C and 60°C, is changed to 20°C and compression strength and static elasticity coefficient were measured. The value of compressive strength and static elastic modulus showed that mortar using modified fly ash had the same characteristics as mortar with ordinary fly ash. Because it was represented by one strength compressive estimation curve regardless of the curing temperature, it became clear that compressive strength can be evaluated by roughly using accumulated temperature as an indicator.

Keywords: Flotation method, Curing method, Compressive strength, Cumulative temperature, Static modulus of elasticity.

1 INTRODUCTION

Fly ash (FA), which is an industrial by-product of coal-fired power plants, can be used as an admixture for concrete, helping to reduce environmental load and improve concrete quality. However, because the quality depends on the ash source, quality stabilization of FA is required. Therefore, we used the flotation method, which has been used to remove the unburned carbon in FA (Takasu *et al.* 2014), to improve the quality of FA by modifying the ignition loss so that the FA conforms to Japanese Industrial Standards (JIS) class I. The strength properties of concrete containing JIS class II FA depend on curing temperature (Watanabe and Suda 2007). However, it is unclear what effect FA modified by the flotation method has on concrete properties. In this study, we examined the effect of FA modified by the flotation method on the mechanical properties of concrete cured at various temperatures.

2 SUMMARY OF THE EXPERIMENT

We made modified FA with the unburned carbon removed by the flotation method. In this method, kerosene was used as a collector for unburned carbon and pine oil was used as a foaming agent. FA containing a large amount of unburned carbon was eliminated with air bubbles. The FA before modification did not meet the JIS. However, the modified FA had a density, ignition loss, and specific surface area conforming to JIS class I.

Table 1 shows the materials and Table 2 shows the mix proportions used in this study. Ordinary Portland cement (OPC) and sea sand as the fine aggregate were used. The mix proportions in Table 2 exclude the coarse aggregate. In all mix proportions, the water/cement ratio was 50% and the FA replaced the fine aggregate. The unmodified FA had a density of 50 g/m3, whereas the modified FA had densities of 50, 100, and 150 kg/m^3.

Table 1. The materials.

Material	Sign	Type	Physical properties
Cement	C	Ordinary Portland cement	Density: 3.16g/cm^3
Admixture	FA	Fly ash(unmodified)	Density: 2.10g/cm^3
			Specific surface area: 3890cm^2/g
			Loss on Ig: 11.95%
		Fly ash(modified)	Density: 2.10g/cm^3
			Specific surface area: 3610cm^2/g
			Loss on Ig: 1.23%
Water	W	Tap water	—
Fine aggregate	S	Sea sand	Density in sarface-dry:2.74g/cm^3
			Water absorption rate:1.26%
			Solid content rate:62.4%

Table 2. The mix proportions.

Sign	W/C (%)	Unit Weight(kg/m^3)				
		W	C	FA	S	G
FA50	**50.0**	**180**	**360**	**50**	**744**	~~**949**~~
FA100	**50.0**	**180**	**360**	**100**	**718**	~~**915**~~
FA150	**50.0**	**180**	**360**	**150**	**692**	~~**882**~~

Table 3 shows the curing method. Three curing conditions were used: underwater curing, air curing, and sealing curing. The underwater curing and air curing were performed at 20°C. Sealing curing was carried out in a chamber in which the temperature was set to 5, 20, 40, or 60°C. Some of the specimens cured at 5, 40, and 60°C were then cured at 20°C after 7 days. The specimens were prepared by mixing cement, tap water, and FA for 30 s in a mortar mixer, and then sea sand was added and mixed for 4 to 5 min. The amount of air-entraining agent added was adjusted, and the target air amount was set to 6.7 ± 1.0%. We used a plastic cylindrical mold of ϕ 50 × 100 mm to produce specimens for the compressive strength test. We immediately stored the underwater curing and air curing test specimens in a curing room at a temperature of 20 ± 1°C. The specimens were removed from the mold after 1 day and transferred to the curing conditions. The sealing cured specimens were cured in a chamber at each temperature.

Table 3. Curing condition.

Sign	Curing temprtuer (°C)		Curing method
	Before 7days	After7days	
5s-20s	5	20	Seal
40s-20s	40	20	
60s-20s	60	20	
5s	5		
20s	20		
40s	40		
60s	60		
W	20		Under water
A	20		Air

The compressive strength was measured in accordance with JIS A 1108 "Method of Test for Compressive Strength to Concrete" and the static modulus elasticity was measured in accordance with JIS A 1149 "Method of Test for Static Modulus Elasticity". The compressive strength was tested after 7, 28, and 91 days of aging.

3 RESULTS AND CONSIDERATION

3.1 Compressive Strength

Figure 1 shows the compressive strength for each replacement rate. For underwater curing, all the specimens showed increased compressive strength. The compressive strengths of the other curing conditions for the modified FA 10% and the modified FA 15% specimens were higher than those of the underwater curing specimens. Focusing only on the sealing curing, the compressive strength at 7 days of age tended to be lower as the curing temperature was lower in the specimen. Focusing on compressive strength increase by age, 60s hardly increased. These results confirmed that continuing high-temperature curing did not increase the compressive strength. The compressive strength increased with the FA replacement rate. This result confirmed that the modified FA increased the compressive strength. The specimen containing unmodified FA 5% was compared with that containing modified FA 5%. For A, 40s and 60s of, no increase in compressive strength for unmodified FA was confirmed at 91 days compared with 28 days. The compressive strength of 60s modified FA specimens did not increase, whereas that of A, 40s, increased. These results indicated that modified FA increased the compressive strength more than unmodified FA. The compressive strengths of the samples cured under these conditions at 91 days were similar. For all replacement rates, the compressive strength at 91 days increased as the curing temperature decreased. These results confirmed that the difference in the initial curing temperature strongly affected the subsequent strength increase. We compared the unmodified and modified FA 5% 40s-20s and 60s-20s specimens. The modified 5% FA specimens showed a large increase in compressive strength from 7 to 28 days, whereas the modified FA 5% showed a large increase in compressive strength from 28 to 91 days. In addition, the increase in compressive strength increased with the rate of FA replacement.

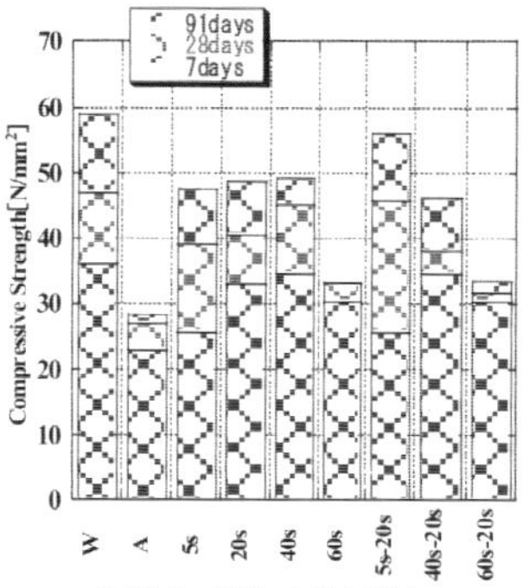

(a)Modified FA5%.

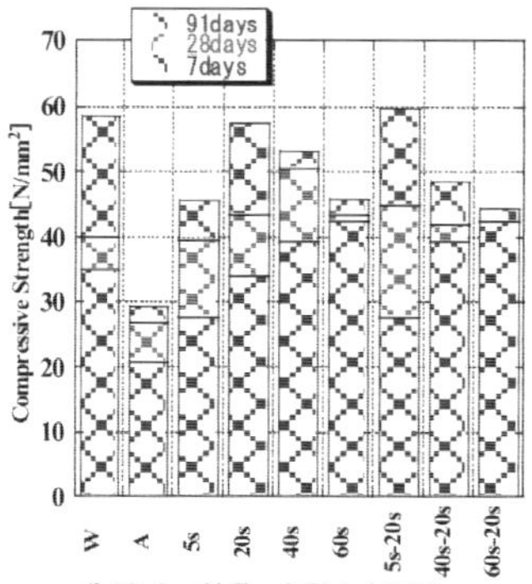

(b)Modified FA10%.

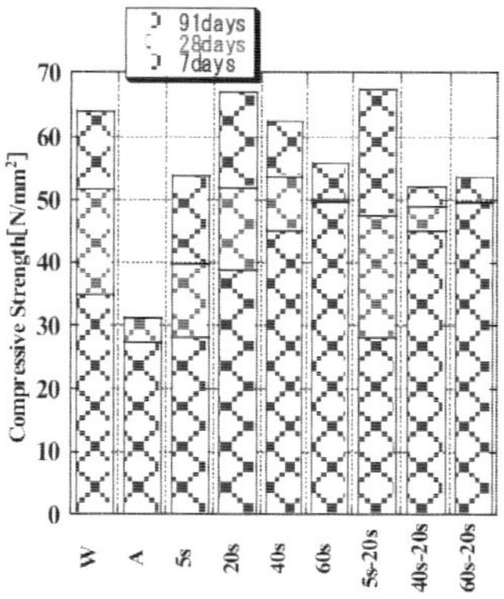

(c)Modified FA15%.

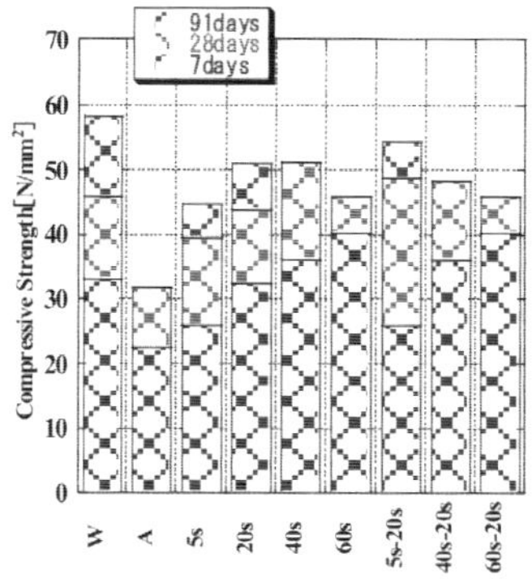

(d)Unmodified FA5%.

Figure 1. The compressive strength for each replacement rate.

Figure 2. shows the relationship between compressive strength and cumulative temperature. The cumulative temperature is used for the strength prediction equation for normal strength concrete. The cumulative temperature is expressed by equation (1).

$$\sum_{\theta}^{t}(\theta + 10) \times \Delta t \tag{1}$$

M: Cumulative strength (℃・day), Δt: Time (day), θ: Δt temperature during time(℃)

However, the relationship between cumulative temperature and compressive strength depends on cement type, fineness of powder, and mix proportions. Therefore, we predicted compressive strength using exponential function (2), which is a growth curve.

$$F = a \times exp\left(-\frac{b}{M}\right) \tag{2}$$

F: Compressive strength (N/mm^2) a,b: Experiment coefficient,

M: Cumulative strength (℃・day)

The experimental values in previous work tended to vary from the strength prediction curve for small amounts of cement (Mizobuchi *et al.* 2012). However, in this work, it was possible to secure at least some minimum amount of cement because it was part of the fine aggregate. Therefore, the experimental values generally followed the compressive strength prediction curve. In addition, the experimental coefficient, a, increased as the FA replacement rate increased, and the value of the strength prediction equation also increased. This result demonstrated that the strength increase with the FA replacement rate could be expressed by the strength prediction equation using the cumulative temperature even when modified FA was used. Therefore, one strength prediction curve was used for all curing temperatures.

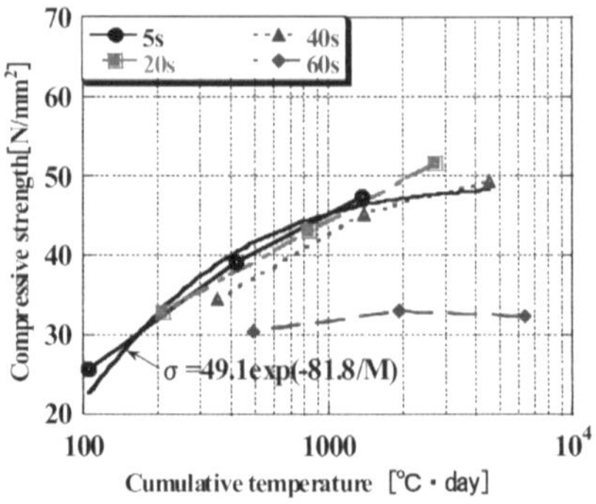

(a) Modified FA5%.

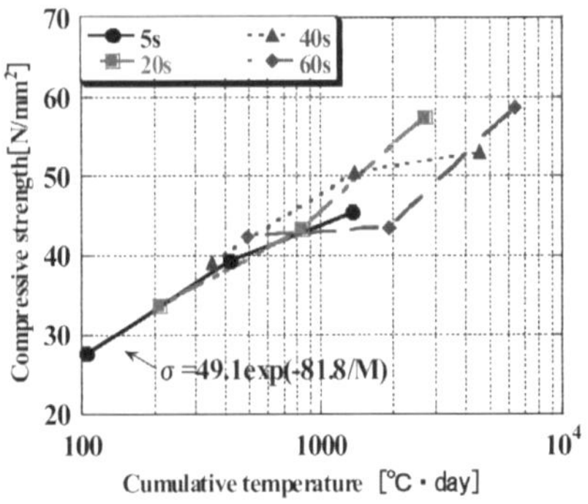

(b) Modified FA10%.

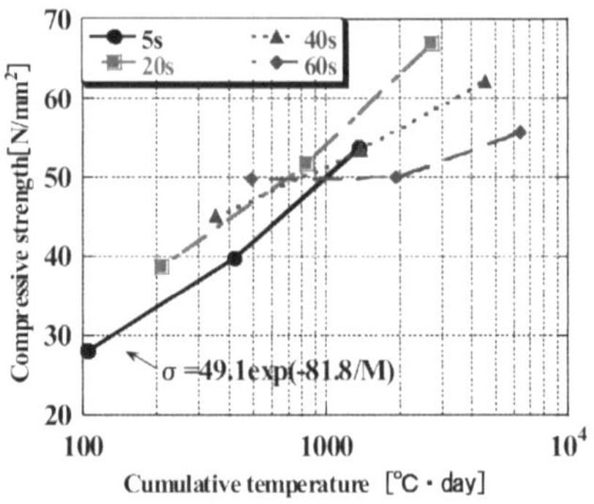

(c) Modified FA15%.

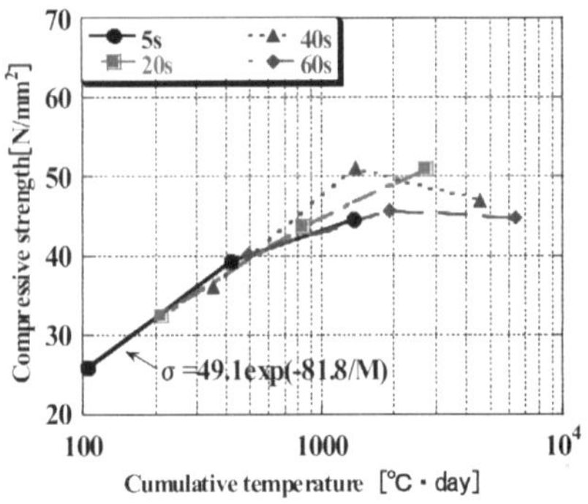

(d) Unmodified FA5%.

Figure 2. The relationship between compressive strength and cumulative temperature.

3.2 Static Modulus of Elasticity

Figure 3 shows the relationship between the static modulus of elasticity and compressive strength. The experimental values were distributed below those calculated with the Architectural Institute of Japan standard for structural calculation of reinforced concrete structures equation (The new RC equation) because the experiment was conducted with mortar. The experimental values followed the new RC equation; therefore, the new RC equation was suitable for describing the distribution of values, even for different curing temperatures. The distribution of the experimental values of the modified FA 15% specimens was wide because of the increase in compressive strength due to the increase in the FA replacement rate. These results showed that the static modulus of elasticity could be expressed in relation to compressive strength even for different curing temperatures.

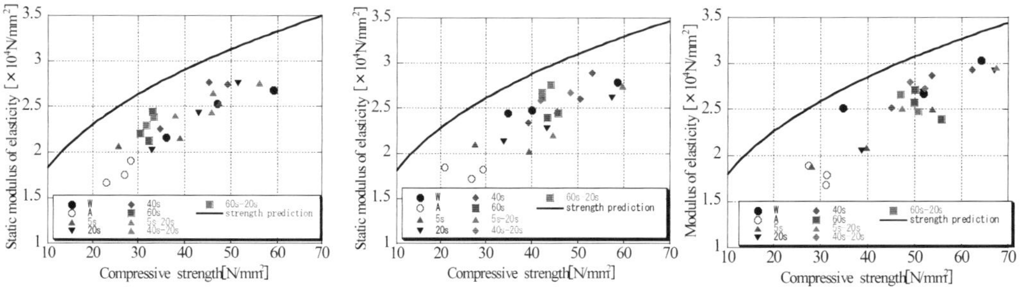

(a) Modified FA5%. (b) Modified FA10%. (c) Modified FA15%

Figure 3. The relationship between the static modulus of elasticity and compressive strength.

4 CONCLUSION

The findings obtained in this research are shown below.

(1) In mortar using modified FA, compressive strength at initial hardening increases with higher curing temperature, but it can not be expected to increase long-term compressive strength. The lower the initial curing temperature, the greater the improvement in long - term compressive strength

(2) In mortar using modified FA, compressive strength also increases as FA replacement rate increases in any curing condition. Moreover, in the specimens given the temperature change, the strength increase rate increased as the replacement rate of FA increased.

(3) In addition, the experimental coefficient, a, increased as the FA replacement rate increased, and the value of the strength prediction equation also increased. This result demonstrated that the strength increase with the FA replacement rate could be expressed by the strength prediction equation using the cumulative temperature even when modified FA was used.

(4) In mortar using modified FA, static modulus of elasticity could be expressed in relation to compressive strength even for different curing temperatures.

Acknowledgments

The authors acknowledge the assistance in this work provided by Eiji Mikura and Yuto Murakami. This research was financially supported by Environmental Restoration and Conservation Agency ERCA) No.3-1703 "Development of advanced recycling technology for fly ash to enable cement-free concrete."

References

Japanese Standards Association, Japanese Industrial Standards Handbook Ready-mixed Concrete, July 31, 2018.

Mizobuchi, A., Kobayashi, T., Chikamatsu, R., and Ichise, K., Effects of Curing Condition for the Strength Properties of Low Carbon Concrete (in Japanese), *Cement Science and Concrete Technology,* 66, 332-337, 2012.

Takasu, K., Suyama, H., and Koyamada, H., Experimental Study on Removal of Unmurnt Carbon in Fly Ash by Flotation Method and Properties of Concrete with the Fly Ash Slurry (in Japanese), *Journal of Structural and Construction Engineering (Transaction of AIJ),* 79(697), 331-340, Mar, 2014.

Watanabe, S., and Suda, N., Effect of Curing Temperature on Strength and Shrinkage of High-Strength Concrete with Fly Ash (in Japanese), *Concrete Engineering Annual Memoirs,* 29(1), 285-290, 2007.

Streamlining Information Transfer between Construction and Structural Engineering
Edited by Shiau, J., Vimonsatit, V., Yazdani, S., and Singh, A.

ISBN: 978-0-9960437-7-9

MECHANICAL PROPERTIES AND SHRINKAGE PROPERTIES OF THE CONCRETE WITH RECYCLED FINE AGGREGATE AND ADMIXTURE

YUJI NAKATA[1], KOJI TAKASU[2], HIDEHIRO KOYAMADA[2], and HIROKI SUYAMA[2]

[1]*Graduate School of Engineering, The University of Kitakyusyu, Kitakyusyu, Japan*
[2]*Faculty of Engineering, The University of Kitakyusyu, Kitakyusyu, Japan*

In Japan, it is forecasted that massive amounts of concrete waste material will be generated in the future as a result of demolition of many buildings, and expansion of the use of recycled aggregate is expected. In this study, it was verified the effect when relatively large amount of admixture is mixed, a combination of recycled fine aggregate of different quality and various admixtures, combination of each admixture in order to realize high strength and high durability by using recycled aggregate. The increase in the drying shrinkage ratio due to the deterioration of the recycled fine aggregate quality was larger than the fluctuation due to the admixture mixing ratio and the drying shrinkage ratio was distributed by forming a group for each quality of recycled fine aggregate. In the relationship between the pore volume and the compressive strength, when evaluated with pore volume of 2 μm or less in both cases, a good linear relationship could be confirmed. The relationship between the pore volume and the drying shrinkage rate was similar. Therefore, it was suggested that compressive strength and drying shrinkage ratio of mortar contained composite recycled fine aggregate and admixture could be predicted by evaluating with the pore volume of 2 μm or less.

Keywords: Demolition wastes, Fly ash, Ground granulated blast furnace slag, Compressive strength, Drying shrinkage, Pore volume.

1 INTRODUCTION

In Japan, it is forecasted that massive amounts of concrete waste material will be generated in the future as a result of demolition of many buildings, and expansion of the use of recycled aggregate is expected. It is necessary to actively use unused resources in resources recycling society. The construction industry in Japan is working on the recycling of construction waste in this situation. In 2000, there were 85 million tons of construction waste from wood, dirt, asphalt concrete mass, concrete mass and others; 72 million tons (81%) was recycled. The concrete mass waste was 49% and 96% of that was recycled (JCI 2005). However, considering the fact of recycling, most of the recycling of concrete masses is used for the road board material. It is necessary to take out the aggregate from concrete masses and use it for concrete again as a recycled aggregate. The research of the recycled aggregate is being studied by researchers, and the recycled aggregate was standardized by JIS now. In this study, the effects were studied when a relatively large amount of admixture is mixed using a combination of recycled fine aggregate of different quality and various admixtures, and combinations of each admixture, in order to realize high strength and high durability by using recycled aggregate.

Table 1. Materials.

Item	Type	Physical properties	Symbol
Cement	Ordinary portland cement	Density 3.16g/cm³	C
Water	Top water		W
Fine aggregate	standard sand	Density 2.57g/cm³	NS
		Water absorption rate 1.26%	
		Solid content rate 62.4%	
		Fineness modulus 2.60	
	M class recycled fine aggregate [JIS] A 5022	Density 2.21g/cm³	MS
		Water absorption rate 6.55%	
		Solid content rate 63.8%	
		Fineness modulus 3.21	
	L class recycled fine aggregate [JIS] A 5023	Density 1.96g/cm³	LS
		Water absorption rate 11.1%	
		Solid content rate 73.7%	
		Fineness modulus 3.21	
Admixture	Ground granulated blast furnace slag	Density 2.96g/cm³	BS
		Specific surface area 3970cm²/g	
	Fly ash（JIS class II）	Density 2.25g/cm³	FA
		loss on Ignition 2.25%	
		Specific surface area 3770cm²/g	
AE agent	AE Water reducing agent	Alkyl ether type anionic surfactant	AE

2 SUMMARY OF THE EXPERIMENT

Table 1 shows the materials. We used ordinary Portland cement, and standard sand, M class recycled fine aggregate, or L class recycled fine aggregate as fine aggregate. Fly ash for concrete class II was used as fly ash. We used ground granulated blast furnace slag (GGBS) for concrete conforming to the 4000 class. Table 2 shows the mixture proportions, and N indicates standard sand, M indicates M class recycled fine aggregate, and L indicates L class recycled fine aggregate depending on the quality of the fine aggregate. Hereafter, we refer to samples made with standard sand as N series, those made with M class recycled fine aggregate as M series, and those made with L class recycled fine aggregate as L series. The replacement of cement with 10%, 20%, and 30% fly ash is indicated by FA10, FA20, and FA30, respectively. The replacement of the cement with 10%, 20%, and 30% GGBS is indicated by BS10, BS20, and BS30, respectively. We prepared a total of 21 samples. The water content per unit volume of concrete was 170 kg/m³ and the cement content was 567 kg/m³ without admixture. The mortar did not contain coarse aggregate. The target mortar flow was 18 ± 2 cm and the target air content was 6.2 ± 1.5%. The compressive strength, drying shrinkage rate, and pore volume were measured. We used cylindrical specimens ϕ50 × 100 mm in size to measure compressive strength. The specimens were demolded after 1 day of aging and cured in water at 20 ± 1 °C. The compressive strength test was conducted after 28 and 91 days of aging. We used 40 × 40 × 160 mm prism specimens 5021 for the drying shrinkage test. We measured the specimens after up to 91 days of aging. The specimens were demolded after 1 day of aging and cured in water at 20 ± 1.0 °C for 7 days. We sealed both ends and measured the original length. The samples were then cured in a thermostatic chamber at 20 ± 1.0 °C and relative humidity of 60 ± 5.0%, and the drying shrinkage rate was measured at a predetermined material age. For the pore volume measurements, after the compressive test at 28 days, the specimen was crushed and the hydration was stopped with acetone. Thereafter, D-dry drying was performed and the pore volume from 6 nm to 90 μm was measured by a mercury intrusion porosimeter.

Table 2. Mix proportions.

Symbol	W/C	W/B	Unit content (kg/m³)					
	(%)	(%)	W	C	FA	BS	S	G
N							700	
M	30.0			567	0		643	(945)*
L							630	
N+FA10							692	
M+FA10	33.3			510	57		623	(934)*
L+FA10						0	580	
N+FA20							683	
M+FA20	37.5			453	113		628	(923)*
L+FA20							615	
N+FA30							675	
M+FA30	42.9	30	170	397	170		621	(912)*
L+FA30							608	
N+BS10							698	
M+BS10	33.3			510		57	628	(943)*
L+BS10							585	
N+BS20							696	
M+BS20	37.5			453	0	113	640	(940)*
L+BS20							627	
N+BS30							694	
M+BS30	42.9			397		170	638	(938)*
L+BS30							625	

※Estimated amount of coarse aggregate

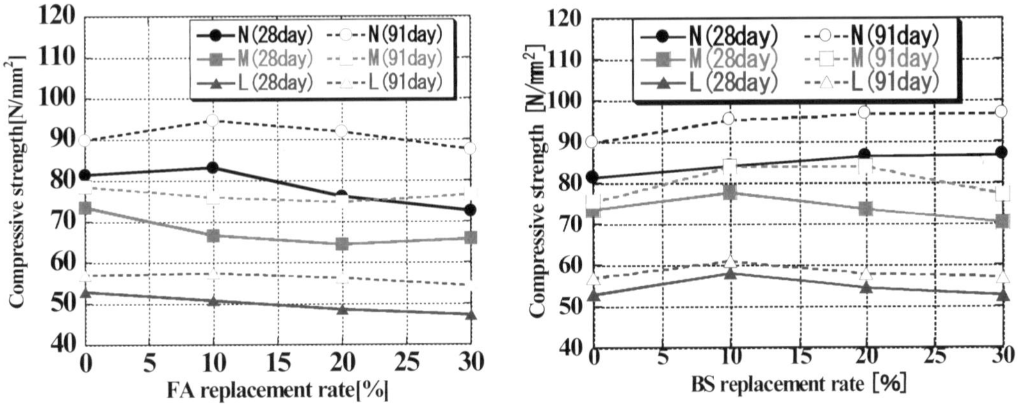

Figure 1. The relationship between the replacement rate of admixture and compressive strength.

3 RESULT AND DISCUSSION

3.1 Compressive Strength

Figure 1 shows the relationship between the replacement rate of admixture and compressive strength after 28 and 91 days of aging. When standard sand was used in the FA series, the strength increased at a replacement rate of 10% after 28 and 91 days of aging. Subsequently, the compressive strength decreased as the replacement rate increased. The rate of decrease was smaller at 91 days than at 28 days. The strengths of the specimens containing fly ash at 28 days were lower than those containing recycled aggregate, whereas at 91 days, the specimens showed the same strength. This increased strength was explained by the pozzolanic reaction with the fly ash. Therefore, when the water/cement ratio was 42.9 or less, even if the specimen contained fly ash, the amount of cement necessary for the pozzolanic reaction was present and the compressive strength increased. For the BS series, the compressive strength also increased with the replacement rate of GGBS at 28 and 91 days. For mortar using recycled aggregate M, the compressive strength increased at the replacement rates of 10% and 20% and decreased at 30%. However, all these specimens had higher compressive strengths than the specimens without fly ash. In the FA and BS series specimens, the compressive strength was reduced by the difference in the quality of the fine aggregate rather than the effect of increasing the replacement rate from 0 to 30%. In particular, the difference in the compressive strengths of M class and L class aggregates was large.

3.2 Drying Shrinkage

Figure 2 shows the drying shrinkage rate for FA series and BS series. When standard sand was used in the FA series, specimen N + FA 10 had a drying shrinkage rate similar to specimen N. However, the shrinkage rate increased as the replacement rate increased. Compared with the specimens containing M class recycled fine aggregate, the shrinkage rate was equivalent to the specimens without fly ash at a replacement rate of 10%. However, in the M + FA 30 specimen, the shrinkage rate was lower than that without fly ash. In the L series, the shrinkage rate was increased L + FA 10 more than without FA. Moreover, the shrinkage rate was smaller L + FA 30 than without FA. In the BS series, compared with the specimens containing standard sand, the shrinkage rate was equal to that without fly ash, regardless of the GGBS replacement rate. However, in the M and L series, the dry shrinkage rate was reduced by GGBS. Therefore, in mortar containing recycled fine aggregate, the drying shrinkage rate was decreased by replacing cement with 30% fly ash or up to 30% GGBS. In both the FA series and the BS series, the shrinkage rate fluctuation due to the deterioration of the quality of the regenerated fine aggregate was larger than that due to the replacement rate of the admixture. Although the shrinkage rate was smaller in the BS series than in the FA series, there was no decrease in the shrinkage ratio compared N indicates standard sand and without admixture.

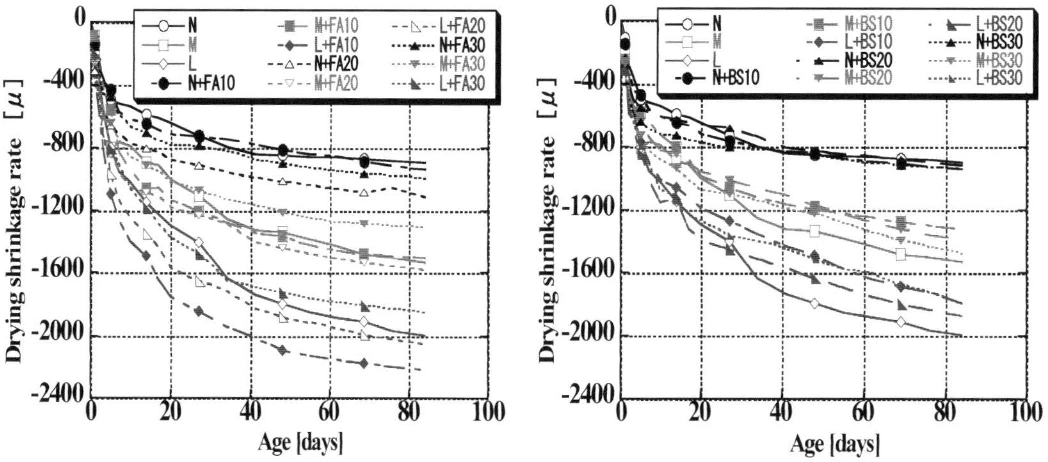

Figure 2. The drying shrinkage rate for FA series and BS series.

3.3 Relation Between Pore Volume, Compressive Strength and Drying Shrinkage

Figure 3 shows the relationship between the pore volume from 50 nm to 2 μm at 28 days aging and the compressive strength. Uchikawa *et al.* (1990) reported that a capillary void volume of 50 nm or more affected the compressive strength in normal concrete. However, in the present experiment, the coefficient of determination between the pore volume from 50 nm to 2 μm and the compressive strength was 0.28 and we did not observe a good correspondence. Figure 4 shows the relationship between the pore volume from 4nm to 2 μm at 28 days and the compressive strength. Hanehara *et al.* (1993) reported that when the water to cement ratio was 40% or less, almost no transition zone was formed, so the correlation between the pore void volume, including the small diameter, and the compressive strength was high. In the present experiment, the compressive strength decreased as pore volumes increased below 2 μm. The relationship was linear, and the coefficient of determination was 0.94. Therefore, the compressive strength of samples with a pore volume from 4nm to 2 μm could be evaluated regardless of the quality of the recycled fine aggregate, the kind of admixture, and the replacement rate.

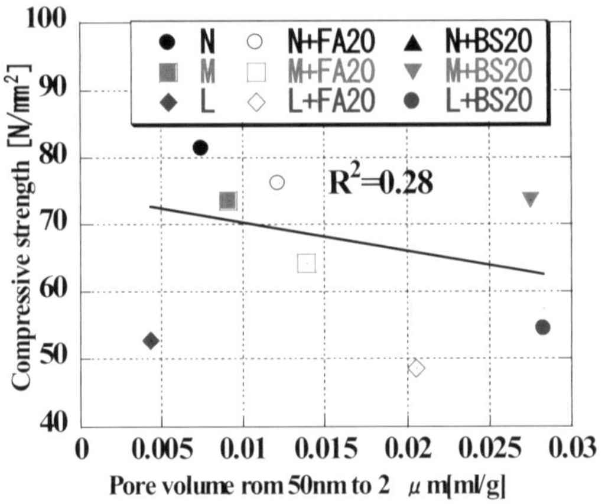

Figure 3. The relationship between the pore volume from 50 nm to 2 μm at 28 days aging and the compressive strength.

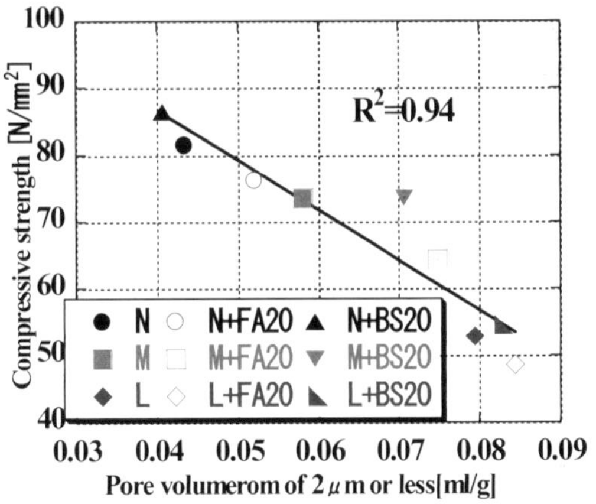

Figure 4. The relationship between the pore volume from 4nm to 2 μm at 28 days and the compressive strength.

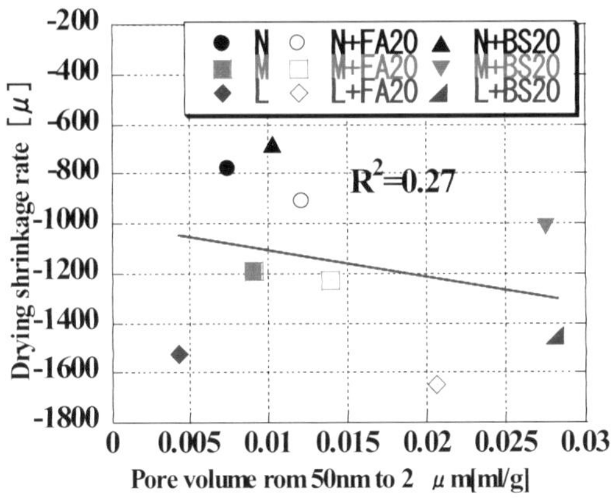

Figure 5. The relationship between the pore volume from 50 nm to 2 μm at 28 days aging and the drying shrinkage rate.

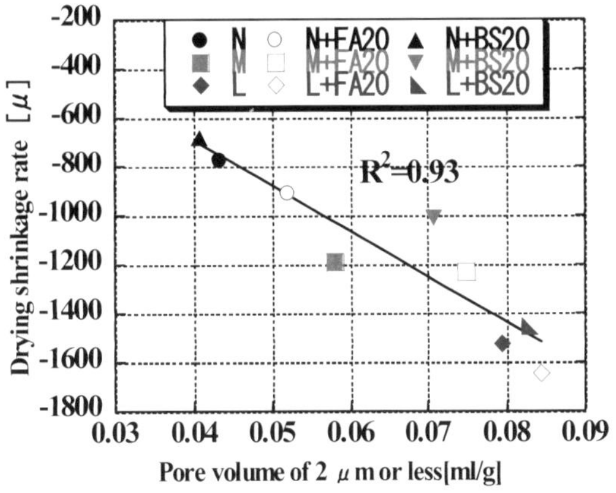

Figure 6. The relationship between the pore volume from 4nm to 2 μm at 28 days and the drying shrinkage rate.

Figure 5 shows the relationship between the pore volume from 50 nm to 2 μm at 28 days aging and the drying shrinkage rate. Similar to the compressive strength, the coefficient of determination between the drying shrinkage rate and the pore volume from 50 nm to 2 μm was 0.27, and we did not observe a good correspondence.

Figure 6 shows the relationship between the pore volume from 4nm to 2 μm at 28 days and the drying shrinkage rate. The shrinkage rate increased linearly with the pore volume and the coefficient of determination was 0.93. Therefore, the drying shrinkage rate of specimens with a pore volume from 4nm to 2 μm could be evaluated for mortar containing recycled fine aggregate and admixture for high compressive strength.

4 CONCLUSION

In the FA and BS series specimens, the compressive strength was reduced by the difference in the quality of the fine aggregate rather than the effect of increasing the replacement rate from 0 to 30%. In particular, the difference in the compressive strengths of M class and L class aggregates was large. The shrinkage rate fluctuation due to the deterioration of the quality of the regenerated fine aggregate was larger than that due to the replacement rate of the admixture. Although the shrinkage rate was smaller in the BS series than in the FA series, there was no decrease in the shrinkage ratio compared N indicates standard sand and without admixture.

In relation to the variation of pore volumes, from 50 nm to 2 μm, the results did not show a good correlation between the compressive strength and the drying shrinkage. The compressive strength and drying shrinkage of samples with a pore volume from 4nm to 2 μm could be evaluated regardless of the quality of the recycled fine aggregate, the kind of admixture, and the replacement rate.

Acknowledgments

The authors acknowledge the assistance in this work provided by Mr. Eiji Mikura and Ema Kawano.

References

Hanehara S., Structure of Concrete and Its Physical Properties, *Easy-to-Understand Cement Science. Japan Cement Association*, 78-104, 1993.

JCI, Symposium on Promotion of Recycled Aggregate for Concrete, *Japan Concrete Institute*, 102-170, 2005.

Uchikawa H., Hanehara S., and Sawaki D., Efect of Hardened Structure of Blended Cement Mortar and Concrete on Their Strength, *Cement Science and Concrete Technology*, 44, 330-335, 1990.

Streamlining Information Transfer between Construction and Structural Engineering
Edited by Shiau, J., Vimonsatit, V., Yazdani, S., and Singh, A.

ISBN: 978-0-9960437-7-9

EXPERIMENTAL STUDY ON COMPRESSIVE YOUNG'S MODULI AND TENSILE YOUNG'S MODULUS OF FLY ASH CONCRETE

YOICHI MIMURA[1], VANISSORN VIMONSATIT[2], YUKI WATANABE[3], ITARU HORIGUCHI[1], and ISAMU YOSHITAKE[4]

[1]*Dept of Civil and Environmental, National Institute of Technology, Kure College, Kure, Japan*
[2]*School of Civil and Mechanical Engineering, Curtin University, Perth, Australia*
[3]*Advanced Course of Project Design Engineering, National Institute of Technology, Kure College, Kure, Japan*
[4]*Dept of Civil and Environmental, Yamaguchi University, Ube, Japan*

Initial cracks due to volume changes at an early age affect the durability of concrete structures, so numerical simulations are often conducted in order to predict cracks. Such prediction requires some mechanical properties of early age concrete. Tensile Young's modulus is directly dependent on the prediction of tensile stress and is one of the important input data for FEM analysis. However, direct tension test for tensile Young's modulus needs a unique apparatus and specimen, and such test is not suitable for evaluating Young's modulus at early ages of concrete. The present study compared tensile Young's modulus with compressive Young's moduli of Fly ash concrete. Compressive Young's moduli used in this study were secant modulus and initial tangent modulus. In addition, linear modulus taken from a regression line of a compressive stress-strain curve in the range of stresses less than the splitting tensile strength was also evaluated. It is found that the secant modulus, which is generally used as Young's modulus in Japan was clearly smaller than the tensile Young's modulus, which means that, tensile stresses evaluated using a secant modulus might be underestimated. On the other hand, linear modulus and initial tangent modulus were almost equal to the tensile Young's modulus. This result indicates that tensile stresses can be evaluated using Young's modulus obtained from a compression test with general apparatus and specimens.

Keywords: Direct tension test, Linear regression, Secant modulus, Splitting tensile strength.

1 INTRODUCTION

It is well known that cracking due to volume change at an early age are undesirable for the durability of concrete structures. Numerical simulations are often performed in order to predict thermal cracking. Tensile properties, such as Young's modulus and creep, are required for an estimation of tensile stress, which causes thermal cracking. Tensile Young's modulus is directly dependent on the prediction of tensile stress and is one of the important input data for FEM analysis. Young's modulus in Japan is generally a secant modulus obtained from a compression test. Mimura *et al.* (2011) report that the tensile Young's modulus based on the direct tension test by using the slender reinforced concrete (RC) specimen is higher than the compressive modulus (secant modulus) at early age. In this study, several Young's moduli obtained from compression test, which can be performed with general equipment, were compared with the tensile Young's modulus obtained from the direct tension test by using the unique apparatus and specimen. Compressive Young's moduli in the present study were secant modulus and initial tangent modulus. In addition, linear modulus, which is taken from a regression line of a compressive stress-strain curve in the range of stresses less than the splitting tensile strength was also evaluated.

2 EXPERIMENTAL PROGRAM

2.1 Materials and Specimens

The concrete materials and mixture proportion in this study are given in Table 1 and Table 2, respectively. The amount of fly ash (FA) used in the present study was 20% by mass of the cement of the normal concrete used for infrastructures in Japan. The targeted slump and the targeted air amount were 8 ± 2.5 cm and 4.5 ± 1.5%. The unburned carbon contained of FA decreases the effect of the air entrainment (AE) water reducing agent, so the AE agent for restraining the influence of unburned carbon was used in FA mixes. In this study, direct tension test for obtaining tensile Young's modulus, splitting tensile test (JIS A 1113 2006) for obtaining tensile strength and compression test for obtaining Young's modulus were carried out.

Table 1. Concrete materials.

Concrete materials	Type	Density
Water (W)	Tap water	1.00 g/cm^3
Cement (C)	Ordinary Portland cement	3.16 g/cm^3
Fly ash (FA)	Class II	2.24 g/cm^3
Fine aggregate (S)	Crashed sand	2.62 g/cm^3
Coarse aggregate (G1)	Crashed stone 20-15 mm	2.67 g/cm^3
Coarse aggregate (G2)	Crashed stone 15-5mm	2.67 g/cm^3
Admixture (Ad1)	AE water reducing agent	1.07 g/cm^3
Admixture (Ad2)	AE agent for fly ash	1.04 g/cm^3

Table 2. Mixture proportion.

	W/B [a]	W	C	FA	S	G1	G2	Ad1	Ad2 [b]
FA concrete	55%	165	240	60	833	493	493	2.40	16.8

[a] Water-binder ratio (B=C+FA), [b] diluted solution (diluted 100 times with water)

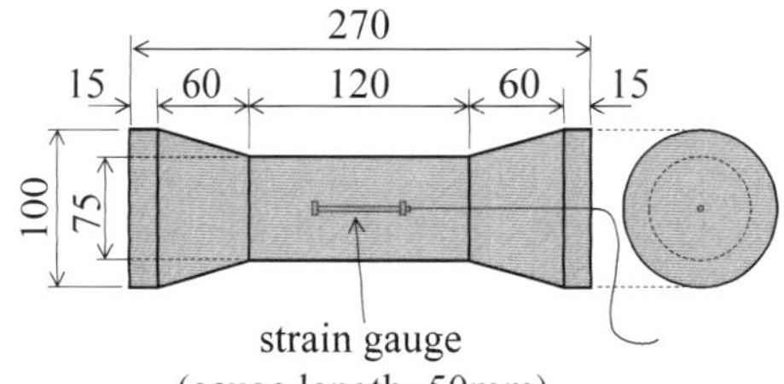

Figure 1. Dog-bone-shaped specimen for direct tension test.

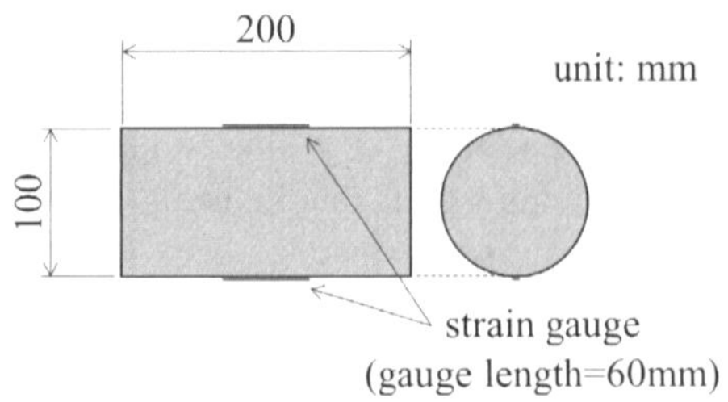

Figure 2. Cylindrical specimen for splitting tensile strength test and compression test.

The dog-bone shaped specimen for the direct tension test is shown in Figure 1. A small mold strain gauge, having a thermometric function and gauge length of 50 mm, was embedded in the center of the cross-section of the specimen. A cylindrical specimen with a diameter of 100 mm and a height of 200 mm as shown in Figure 2 was used for splitting tensile test and compression test. Strain gauge with gauge length of 60 mm was attached at the center of both side faces of each specimen for compression test. After casting concrete, all specimens were carried into the curing room to be cured at a room temperature of 20 ± 1 °C for 24 hours before demolding. After that, the specimens were cured underwater in a tank installed in the curing room until load tests were to be carried out at the age of 7 days. Water temperature in the tank was 16 ± 1 °C.

2.2 Test Method

In this investigation, Young's moduli evaluated by several procedures, based on data obtained from the direct tension test and compression test, were compared. The direct tension test by using a dog-bone shaped specimen and a loading apparatus is shown in Figure 3 was carried out in order to obtain a tensile Young's modulus. The load was manually applied to the tip of the steel arm. The specimen was loaded until its failure. The load and strain were recorded at every step of strain increment of 10×10^{-6}. After each recording, the load was temporarily removed and the strain at unloading was measured. Such loading/unloading cycle was performed twice at each strain level. The reason for this loading procedure is to eliminate the influence of inelastic strain as much as possible for the Young's modulus.

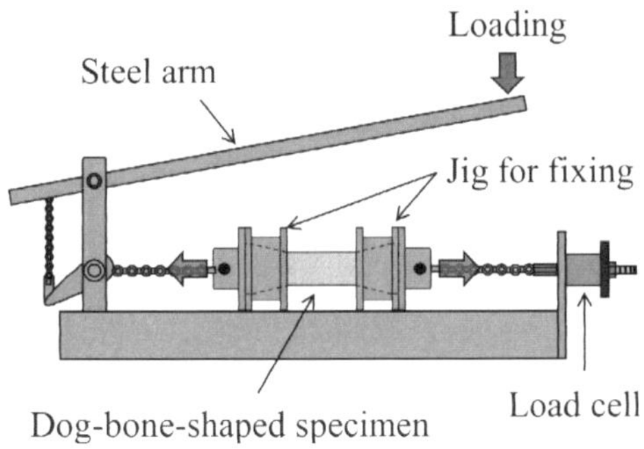

Figure 3. Test setup of the direct tensile test.

Compression test based on JIS A 1149 (2010) with a cylindrical specimen (diameter of 100 mm and height of 200 mm) was also conducted for obtaining a compression Young's modulus, as shown in Figure 4. The load and strain were recorded at every load increment of 10 kN. The specimen was loaded until its failure. The general compressive Young's modulus (hereinafter referred to as secant modulus) was evaluated from the data at the strain of 50 x 10^{-6} and the data at the load of 1/3 of the maximum load. In addition to compression test based on JIS, the compression test with the same procedure of the direct tension test, repeating loading and unloading, was also conducted. In this compression test, the loading ended after the stress corresponding to the splitting tensile strength was recorded. Thereafter, the same specimen for this compression test was used to perform a compression test based on JIS described above. In this investigation, the direct tension test and compression test by using specimens with FA concrete at the age of 7 days; four specimens were used for each test.

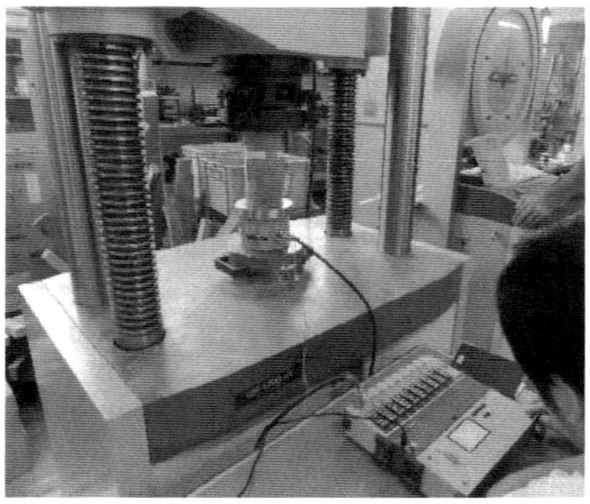

Figure 4. Test setup of the compression test.

3 TEST RESULTS AND DISCUSSION

Figure 5 shows an example of the stress-strain relationship obtained from the direct tension test. The strain shown in Figure 5 does not include the residual strain at unloading. The stress-strain relationship was almost linear up to the maximum point, as shown in Figure 5, so the gradient of the regression line was taken as tensile Young's modulus.

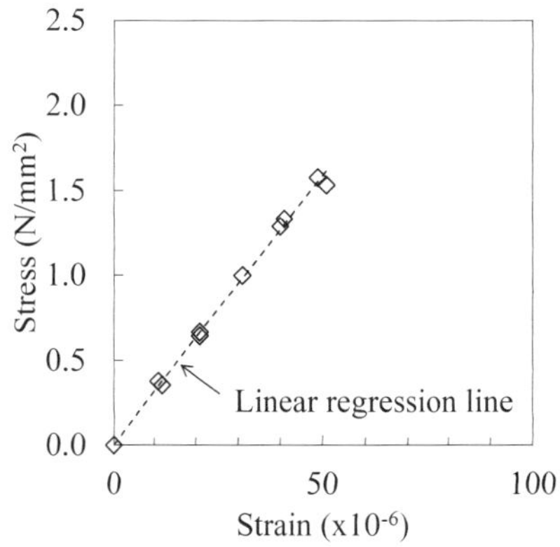

Figure 5. Stress-strain relationship of the direct tensile test.

Figure 6 presents the result of the compression test with the same procedure of the direct tension test. The f_t in Figure 6 means the splitting tensile strength. One of the compressive Young's modulus in this study (hereinafter referred to as linear modulus) was obtained from the slope of this linear regression line. The result of the compression test based on JIS is demonstrated in Figure 7.

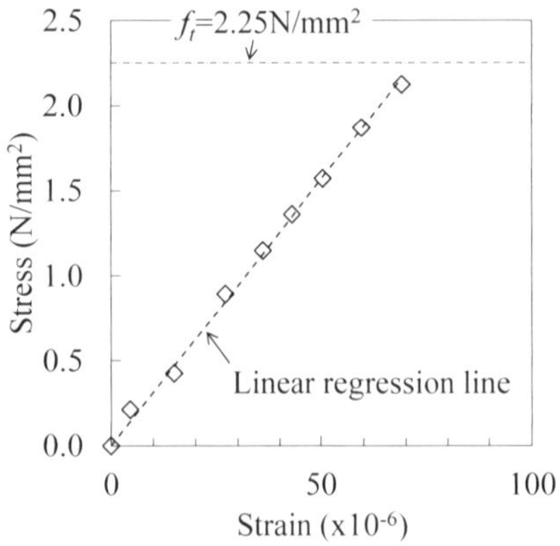

Figure 6. Stress-strain relationship of compression test with the same procedure of the direct tensile test.

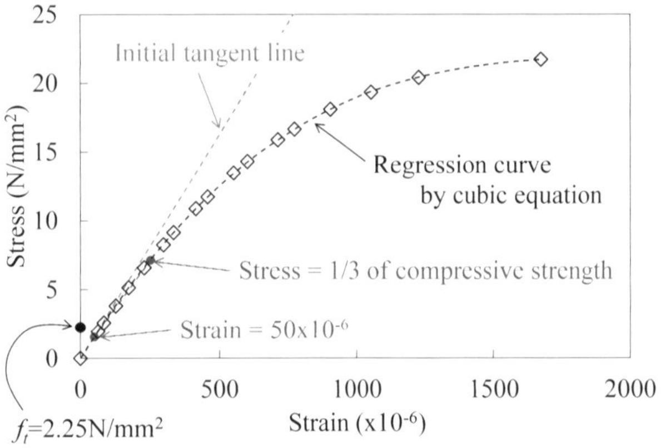

Figure 7. Stress-strain relationship of the compression test based on JIS.

Table 3. Results of Young's modulus

	No.1	**No.2**	**No.3**	**No.4**	**Ave.**
Tensile Young's modulus (kN/mm^2)	31.7	30.3	32.2	33.5	31.9 (1.00)
Secant modulus (kN/mm^2)	27.8	26.5	27.6	29.4	27.8 (0.87)
Initial tangent modulus (kN/mm^2)	32.5	31.4	32.5	34.4	32.7 (1.03)
Linear modulus (kN/mm^2)	29.6	31.7	31.3	33.5	31.5 (0.99)

The result in Figure 7 was regressed by the cubic equation shown in Eq. (1).

$$\sigma = a \cdot \varepsilon^3 + b \cdot \varepsilon^2 + c \cdot \varepsilon \tag{1}$$

where σ is stress (N/mm^2), ε is strain (x 10^{-6}), a, b and c are coefficients.

The experimental data was accurately regressed by the cubic curve in Eq. (1), and R^2 was more than 0.99. The coefficient c in Eq. (1) regresses the stress-strain relationship, which corresponds to the initial tangent modulus. The secant modulus, applied as a general compressive Young's modulus of concrete in Japan, was evaluated from the data at the strain of 50 x 10^{-6} and the data at the stress of 1/3 of the compressive strength, as shown in Figure 7. Table 3 gives the results of the tensile Young's modulus, the secant modulus, the initial tangent modulus, and the linear modulus. The numerical values in brackets () in Table 3 represent the ratio of each Young's modulus to the tensile Young's modulus. The secant modulus was obviously smaller than the tensile Young's modulus, with the average value being 87% of the tensile Young's modulus. This result indicates that tensile stresses evaluated using a secant modulus might be underestimated. On the other hands, the initial tangent modulus and the linear modulus were nearly equal to the tensile Young's modulus. This was because the stress-strain relationship shown in Figure 6 was almost linear in the stress range less than the tensile strength. However, the stress was not proportional to the strain as the stress increased, particularly beyond 1/3 of the compressive strength, as shown in Figure 7. Therefore, the difference between the tensile Young's modulus and the secant modulus can be caused by the difference in the level of the stress and strain used for the evaluation of the Young's modulus. In other words, the Young's modulus evaluated from the compressive stress-strain curve at the stress range less than the splitting tensile strength may be suitable for crack estimation even if the stress field differs. In addition, the loading apparatus and dog-bone shaped specimen used in the direct tensile test shown in Figure 1 and Figure 3 are not typical, so it is difficult to obtain the development evaluation of Young's modulus at early ages based on many tests. The compression tests by using a typical cylindrical specimen can be relatively easily performed at the various ages of concrete.

4 CONCLUSIONS

This paper describes experimental investigations for obtaining the development of Young's modulus at an early age for thermal crack prediction. For comparison with the tensile Young's modulus, several compressive Young' moduli were evaluated as the secant modulus generally used in Japan, the initial tangent modulus and the linear modulus obtained from the gradient of the regression line. The conclusions of this paper are as follows:

(i) The secant modulus was found smaller than the tensile Young's modulus, which was 87% in the present study. Therefore, tensile stresses evaluated using a secant modulus might be underestimated.

(ii) The tensile Young's modulus of FA concrete at the age of 7 days is almost equal to the linear modulus evaluated with the compressive stress-strain curve at the stress range less than the splitting tensile strength.

(iii) The Young's modulus evaluated from the compressive stress-strain curve at the stress range less than the tensile strength may be suitable for crack estimation even if the stress field differs.

Acknowledgments

The authors would like to thank Mr. K. Yamada, Mr. S. Michimoto (former students of National Institute of Technology, Kure College) and the Chugoku Electric Power Co. Ltd.

References

JIS A 1113, *Method of Test for Splitting Tensile Strength of Concrete (in Japanese),* Japanese Industrial Standard (JIS) 2006.

JIS A 1149, *Method of Test for Static Modulus of Elasticity of Concrete (in Japanese),* Japanese Industrial Standard (JIS), 2010.

Mimura, Y., Yoshitake, I., and Zhang, W., Uniaxial Tension Test of Slender Reinforced Early Age Concrete Members, *Materials*, MDPI, 4(8), 1345-1359, August, 2011.

Streamlining Information Transfer between Construction and Structural Engineering
Edited by Shiau, J., Vimonsatit, V., Yazdani, S., and Singh, A.

ISBN: 978-0-9960437-7-9

INFLUENCE OF SAND PERCENTAGE TO FRESH CONCRETE PROPERTIES

TAKUJI GENNAI[1], TAKAHIRO TAMURA[2], and ATSUSHI SHIMABUKURO[2]

[1]*Advanced Course of Environmental and Civil Eng., NIT, Tokuyama College, Shunan, Japan*
[2]*Dept of Civil Eng. and Architecture, NIT, Tokuyama College, Shunan, Japan*

In recent years, the target value of slump in general concrete construction is changed from 8 cm to 12 cm in Japan so that constructors will aim to improve workability and ensure quality. However, increasing the concrete slump may cause material separation. As one of the evaluation methods of material separation of fresh concrete, although TAISEI CORPORATION developed a "T-post slump test," this test result depends on the engineers' experience because this test is the qualitative evaluation, not the quantitative evaluation. Therefore, in order to propose the simple evaluation method of this test, we investigate the relationship between the result of the T-post slump test and bleeding amount of fresh concrete in each sand percentage. As a result, it shows that the material separation, which water exudes on the surface, in the T-post slump test occurs at the sand percentage of 40.0% and 50.0% and the final bleeding amount also obtains 0.39 cm^3/cm^2 and 0.53 cm^3/cm^2.

Keywords: Slump, Material separation, Bleeding, Quantitative evaluation, Qualitative evaluation, Evaluation method.

1 INTRODUCTION

In Japan, since the high aseismic capacity has been required after the Hanshin-Awaji earthquake, the density of the reinforcing bars in the concrete is increasing. With the conventional slump target value of 8 cm, since the fresh concrete placement efficiency into the densified reinforced concrete is lowered, and the quality of the hardened concrete is expected to deteriorate due to the insufficient filling of concrete, we aim to improve these problems by modifying the slump target value of the concrete construction from 8 cm to 12 cm. However, it is generally considered that material separation occurs by enlarging the concrete slump. Here, Figure 1 shows the situation of T-post slump test. This test is one of the simple evaluation methods for separation resistance.

Figure 1. T-post slump test situation.

T-post slump test is the test conducted after slump test. First, when the slump test is conducted, we put the Phenolphthalein solution on the top surface of the fresh concrete stuffed in a slump cone before pulling out a slump cone. Then, we hit the corner of slump board with a hammer until the fresh concrete spreads to 47 cm and 52 cm after measuring the slump value of fresh concrete and confirm the change of the circular shape of the top surface with the Phenolphthalein solution at that time. However, the many experiences are required in this test because it is a qualitative evaluation, not a quantitative evaluation. Therefore, in order to evaluate this test quantitatively, we conduct the T-post slump test and the bleeding test, which focus on the sand percentage, in this study. From the relationship between both tests results, we try the quantitative evaluation of T-post slump test.

2 EXPERIMENTAL OUTLINE

2.1 Experimental Factors

Table 1 shows the mix proportion and the required quantity of each material in the concrete using in this study. The water cement ratio was 0.55 at all mix proportions. The standard mix proportion was 45.5% of the sand percentage and we changed only the sand percentage at each mix proportion to investigate the relationship between the T-post slump test and the bleeding test.

Table 1. Concrete mix.

Composition	W/C	s/a	Unit Weight (kg/m³)						F
	(%)	(%)	W	C	S	G1	G2	(C × %)
Base	55	45.5	185	336	815	402	599	0.25
Change s/a	55	50.0	185	336	896	369	550	0.25
	55	40.0	185	336	716	443	659	0.25

2.2 Employed Material

Table 2 shows the type and density of the employed material. Two kinds of coarse aggregate and one type of fine aggregate were used in this study. Normal Portland cement is used, and Air Entraining (AE) water reducing agent was used as the admixture.

Table 2. Material and density.

Code	Type	Density (g/cm³)
C	Normal Portland cement	3.15
S	Crushed Sand	2.7
G1	Crushed Stone (particle diameter : 5-15mm)	2.77
G2	Crushed Stone (particle diameter : 10-20mm)	2.78
F	AE Water Reducing Agent (normal type)	1.25

3 EXPERIMENTAL RESULT AND CONSIDERATION

3.1 Aggregate Dispersion Distance

In this study, the aggregate dispersion was calculated by Kennedy's thickness of excess paste theory (Kennedy 1940). In this theory, concrete is made as a two-phase material of aggregate and cement paste. The gaps between the aggregates are filled with gap paste and concrete flows due to excess paste as shown in Figure 2. Assuming that the excess paste covers the aggregate, twice the thickness of excess paste is the aggregate dispersion distance.

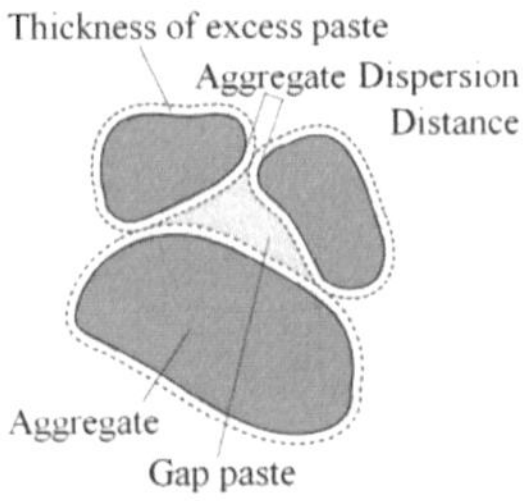

Figure 2. Aggregate dispersion distance.

The larger aggregate dispersion distance shows the higher the fluidity (Powers 1968). The aggregate dispersion distance is calculated by

$$D_{ep} = \left(\sqrt[3]{\frac{C_a}{V_a}} - 1 \right) D_a \tag{1}$$

Where, D_{ep} is the aggregate dispersion distance ($\times 10^3 \mu m$), C_a is the percentage of the absolute volume of aggregate (%), V_a is the aggregate volume ratio (%), and D_a is the mean size of aggregate (mm). C_a used in this study was a mixture value of fine aggregate and coarse aggregate. The mean size of aggregate is calculated by the following Eq. (2) (Tokumitsu 1965 and Teranishi *et al.* 2015).

$$D_a = \frac{1}{\sum_{n=1} \frac{(ln\, D_{n+1} - ln\, D_n)}{D_{n+1} - D_n} V_n} \tag{2}$$

In Eq. (2), n is the order of sieves counted from the smallest nominal size, D_n is the nominal dimension of nth sieve (mm), and V_n is the aggregate volume ratio staying at the nth sieve (%). Here, Figure 3 shows the relationships between the aggregate dispersion distance, the slump value and the sand percentage. In this figure, the horizontal axis is the sand percentage and the vertical axis on the left is the aggregate dispersion distance and the vertical axis on the right is the slump value. In general, when the sand percentage is increased, the aggregate dispersion distance decreases. This result is consistent with the past experimental results (Teranishi *et al.* 2015).

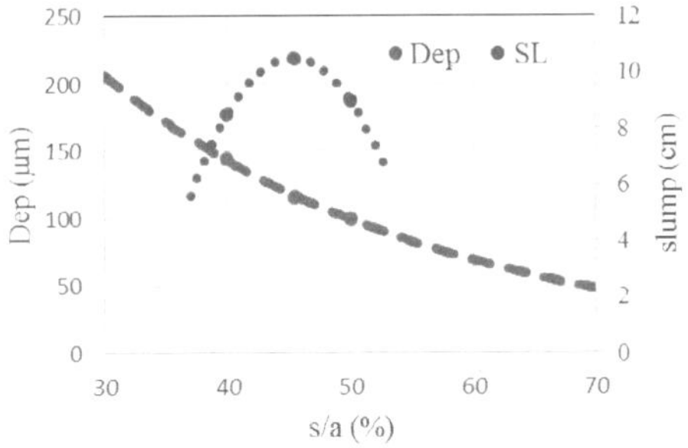

Figure 3. Aggregate dispersion distance and sand percentage.

However, there was no clear relationship between aggregate dispersion distance and slump value.

3.2 T-post Slump Test

For the determination of the material separation resistance in the T-post slump test, it is preferable that the colored circular shape of the top is kept circular at 47cm of the slump flow. When the colored circular shape collapses before 47 cm of the slump flow, it tends to separate. Also, the circular shape has to be lost at 52cm of the slump flow (Liang *et al.* 2012), and if a colored circle remains, it means that the separation resistance is too strong. Here, Figure 4 shows the condition of the slump flow at 47 cm for each sand percentage. Although the colored shape retained circular at 45.5% of the sand percentage, the circular shape collapsed at 40.0% and 50.0%. In addition, the material separation which water exudes on the surface was observed at 40.0% and 50.0%. In general, the viscosity and the material separation resistance of concrete with increasing the sand percentage. However, when the sand percentage is excessively increased and decreasing such as this result, material separation would occur. Also, Figure 5 shows the condition of the slump flow at 52cm of 45.5% of sand percentage. The colored circular shape showed the collapsed shape in this figure, so that separation resistance of the concrete in this study was appropriate at 45.5% of sand percentage in combination with the result at Figure 4.

Therefore, it is considered that the adequate design of the sand percentage is required to satisfy appropriate the material separation resistance for fresh concrete.

Figure 4. Result of T-post slump test.

Figure 5. Slump flow 52cm of 45.5% of sand percentage.

Figure 6 shows the relationship between the center height of the sample after the T-post slump test and the slump flow. The center height was examined at 47 cm and 52 cm of the slump flow. The sand percentage 40.0% and 50.0% changed the height from 8 cm to 6 cm. The sand percentage 45.5% changed from 11 cm to 6.5 cm. Considering the described above consideration, the center height of the fresh concrete, which was 45.5% of the sand percentage, with the appropriate material separation was higher than others at 47 cm of the slump flow. It is considered that the fresh concrete with the appropriate material separation resistance spreads its surround while keeping its center height due to its appropriate viscosity. On the other hand, all center height of 52 cm of the slump flow showed about 6 cm regardless of the sand percentage. From this result, it is effective to measure the center height at 47 cm of the slump flow after the T-post slump test to evaluate the material separation resistance.

Therefore, measuring the center height after the T-post slump test would be the effective method as a quantitative evaluation to investigate the material separation resistance of fresh concrete.

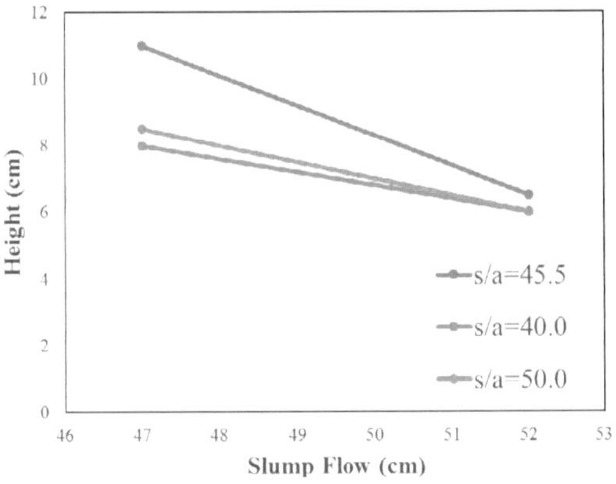

Figure 6. Aggregate dispersion distance and sand percentage.

3.3 Bleeding Test

The bleeding test was conducted by, a metallic cylindrical container with a diameter of 250 mm and a height of 285 mm. The concrete was filled in three layers. Measurement of bleeding was picked up by dropper on the surface. We measured every 10 minutes for the first hour and then measured every 30 minutes until there was no bleeding. Figure 7 shows the relationship between the elapsed time and the bleeding amount.

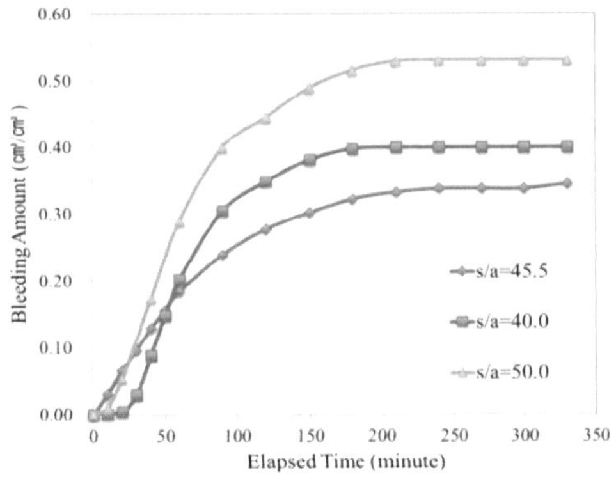

Figure 7. Aggregate dispersion distance and sand percentage.

It is considered from this figure that the bleeding amount occurs with the elapsed time even when the sand percentage is too small or too large. Also, when the sand percentage is at 45.5%, the bleeding amount is the smallest at 0.34 cm^3 / cm^2. From these results and the T-post slump test results, we find that there is a tendency of material separation when the bleeding amount is large.

4 CONCLUSIONS

In this study, we investigated the properties of concrete by the change of sand percentage. We obtained the aggregate dispersion distance and performed the T-post slump test and the bleeding test. From the experimental results, we can be concluded the following.

(1) There was no clear relationship between aggregate dispersion distance and slump.
(2) When the sand percentage is greatly changed from 45.5%, the material separation which water exudes on the surface occurs.
(3) The amount of bleeding of concrete in which material separation occurred in the T-post slump test tends to increase.
(4) The height of the slump flow 52 cm converges to approximately 6 cm.

Acknowledgments

The author would like to thank Takahiro Tamura and Atushi Shimabukuro for him a matchless technical competence.

References

Kennedy, C. T., *The Design of Concrete Mixes*, Proceedings of the American Concrete Institute, 36, 373-400, 1940.
Liang, J., Tsuyoshi M., Sakamoto J.and Kimitaka U., *Development of a Simple Quantitative Method for Evaluating Concrete Separation Resistance,* Journal of Advanced Concrete Technology, 15, 143-150, 2017.
Powers, T. C., *Properties of Fresh Concrete*, John Wiley & Sons, Inc, New York, USA, 1968.
Teranishi, K., Masuda, Y., Hiroshi, J., Kage, T., and Tomoaki, I., Effects of Fine Aggregate and Mix Proportion on Workability of Concrete, *Journal of Structural and Construction Engineering*, 80(707), 9-18, Jan, 2015.
Tokumitsu, Y., About Packing Powder, *Powder Engineering,* 29-34, 1965.

Streamlining Information Transfer between Construction and Structural Engineering
Edited by Shiau, J., Vimonsatit, V., Yazdani, S., and Singh, A.

ISBN: 978-0-9960437-7-9

EVALUATING PERMEABILITY AND MECHANICAL PROPERTIES OF WASTE MARBLE DUST MIX CONCRETE AND BENTONITE MIX CONCRETE

SHAMS UL KHALIQ, IRFAN JAMIL, and HANIF ULLAH

Dept of Civil Engineering, University of Engineering and Technology, Peshawar, Pakistan

In this research attempt is made to examine Waste Marble Dust Mix (WMD) Concrete and Bentonite Mix Concrete. For physical and chemical properties of both concrete mixes, cylinders were examined in the laboratory for permeability testing, scanning electronic microscopy test, slump test, compressive strength test, and split tensile test. Cylinders were prepared in the ratio 1:2:4 with varying marble dust and bentonite content with 5% increment up to 20%. At 28 days with 10% replacement of marble dust and bentonite with cement, the permeability of WMD and Bentonite concrete were decreased by 21% and 34%. Increasing the percentage of WMD and bentonite from 10 to 20%, the MD concrete permeability was drastically increased. After crossing 10% limit it starts working as over burdening additives and allows water to flow through the cylinders, while with bentonite mix concrete by 20% replacement with cement, the permeability was further decreased up to 43%. The compressive and split strength test results of both mixes were totally different from each other. The compressive strength of WMD mix concrete decreased as the percentage of WMD was increased. At 28 days with 10% replacement, the compressive and split tensile strength were reduced by 8.9% and 2.1%. Whereas in bentonite mix, the concrete compressive and split strength were increased by 7.2% and 1.45 and were furthermore increased up to 20% replacement.

Keywords: Concrete strength, Consistency, Impermeable concrete, Environment.

1 INTRODUCTION

In the last few decades, the marble industry has generated a large amount of waste marble dust (WMD). About 50% of marble is wasted due to blasting, poor quality product, unskilled workforce, and largely unrecognized market. This wasted/leftover material, known as waste marble dust, needs permanent dumping because it is harmful for human health by polluting the environment, affecting the agricultural land, water and air due to its alkaline properties. Therefore, the reusing of these extract products from the marble industry needs to be managed permanently in any form, so that their effects may be minimized.

Bentonite clay (Montmorillonite clay) is a geological material having pozzolanic/cementitious properties, mined from earth, formed after volcanic ash in the presence of water with passage of time. Bentonite clay has many uses in the construction industry as a partial replacer to cement because it has pozzolanic properties. Pozzolanic material when reacted with water, forms products that has cementitious properties. Many researchers have made attempts to utilize the WMD and bentonite as a partial replacer to cement.

According to a previous research, in which the WMD and Dolomite (pozzolanic material combination of calcium and magnesium carbonates $CaMg(CO_3)_2$), as a partial replacement of cement, was added up to 20% to study the mechanical properties. According to their conclusion, for to 15% replacement, the flexural strength of concrete mix was improved by 11% whereas the compressive strength was increased by 42% with 10% replacement (Ergun 2011). While in another research, focused on the study of the physical properties (consistency) of waste marble dust mix concrete, it was discovered that as the percentage of marble dust was increased, the slump value of the concrete was increased. So, according to their recommendation, WMD can be used in self-compaction concrete (Alyamac and Ince 2009).

2 METHODOLOGY

This research methodology consists of three phases, which are briefly explained below.

The **first phase** consists of fabrication of concrete specimens according to ASTM C192 (2016) standard. Concrete cylinders in molds having dimensions of 150mm x 300mm were prepared to investigate the mechanical and physical properties of both the concrete mixes.

The **second phase** is lab experimentation: the following tests were performed to investigate the mechanical properties – (i) splitting tensile strength test, (ii) tensile strength at 28 days for 30 cylinders for both WMD and Bentonite mix concrete samples by Universal Testing Machine under ASTM C496 (2017) standard, (iii) compressive strength test for 60 cylinders were tested to find out crushing strength for 7 and 28 days for both WMD and Bentonite mix concrete samples in the Universal Testing Machine under ASTM C39 (2018) standard, (iv), slump test was performed to find the consistency under ASTM C143 (2015) standards, (v) the water permeability test was performed by Blain's apparatus, shown in Figure 1(a) under the ASTM C204 (2018) Standard. After preparing and placing the concrete specimen in the apparatus shown in Figure 1(b), the sample was then fully saturated with water leaving no bubbles on its surface. 30 bar pressure was applied to the specimen for a defined time. The water shall penetrate through the sample with pressure and get collected in a beaker underneath. Then the coefficient of permeability can be derived by Eq. (1).

(a) (b)

Figure 1. (a) Blain's Apparatus (b) top view having cylinder placed in Blain's apparatus.

$$K = (Q \,.\, X) / (A \,.\, h) \quad (1)$$

Where; Q = Volume flow rate (m^3/s), A=Cross-sectional Area (m^2), h= head of water (m), X= Specimen thickness in the direction of thickness (m), K= Permeability coefficient.

To determine the elemental composition of waste marble dust and bentonite, Scanning Electron Microscopy (SEM) apparatus was used, shown in Figure 2(a). SEM is a type of electron microscope that produces pictures of a sample by scanning the area with a confined beam of electrons. The electrons intermingle with the sample of the atoms and provides a surface variation, composition, and topography of the area by producing various signals. The chemical properties obtained from the examination of Electron Microscopy (SEM/ EDS analysis) for marble dust and bentonite fall under under ASTM E2809 (2013) standards.

The **third phase** is the analysis of results of all the lab experimentation mentioned in phase two, and is briefly explained below.

(a)

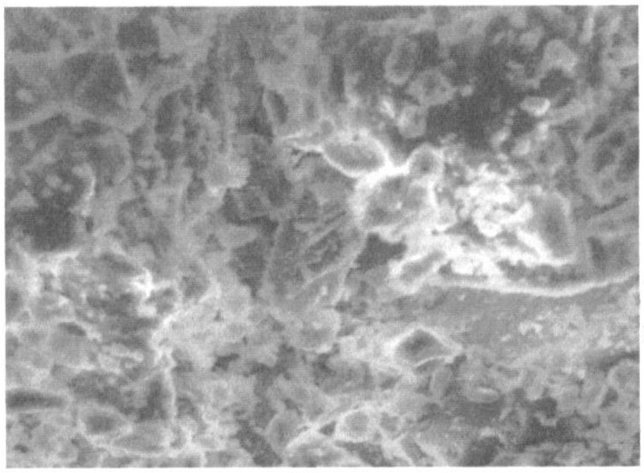

(b)

Figure 2. (a) SEM apparatus (b) Chemical composition of WMD.

2.1 Compressive Strength Test Results

The results shown in Figure 3(a) elaborates that as the percentage of Waste Marble Dust was increased, the compressive strength of the concrete kept decreasing. At 28 days with 20% replacement of cement with waste marble dust, strength gradually decreased from 3361 psi to 1979 psi.

While at 28 days, the Bentonite mix concrete strength was increased. The strength increased from 3361 psi of control sample to 3606 psi at 15% replacement of cement with bentonite, while at 20 % replacement the strength dropped as compared to other replacement, but still was 0.5% as that of control sample as shown in Figure 3(b).

2.2 Splitting Tensile Strength Results

The results show that as the percentage of Waste Marble Dust was increased, the tensile strength of the concrete kept decreasing. At 28 days with 20% replacement of cement with waste marble dust, strength gradually decreased from 310.1 psi to 232.8 psi. Results are shown in Figure 4(a).

But, at 28 days the Bentonite mix concrete strength was increased from 310.1 psi of control sample to 315.2 psi at 15% replacement of cement with bentonite. But at 20% replacement, the Split strength decreased by 0.47% to that of 15% replacement of Bentonite, as shown in Figure 4(b).

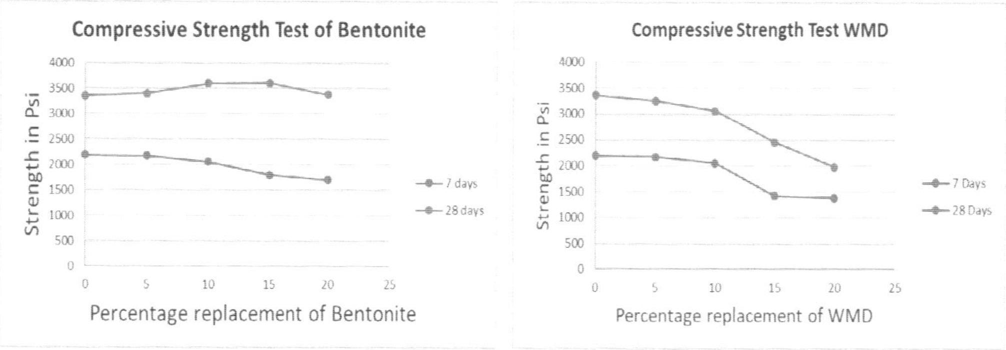

(a) (b)

Figure 3. (a) WMD Compressive Strength Test graphs of 7 and 28 days, (b) Bentonite Compressive strength test graphs of 7 and 28 days.

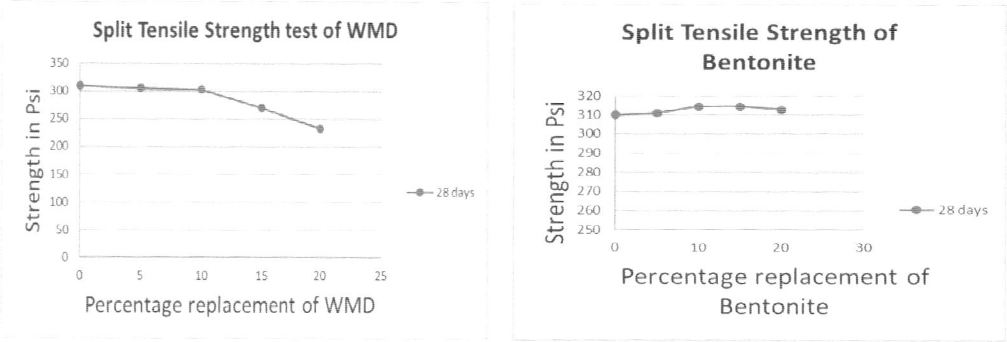

(a) (b)

Figure 4. (a) WMD 28 days split tensile strength graph, (b) Bentonite 28 days split tensile strength graph.

2.3 Permeability Test Results

The permeability results show that as we keep increasing the percentage of marble dust from 0 to 10%, the permeability of concrete mix was improved/decreased by 21%. The reason is that marble dust acts as a filler and dust makes the concrete compacted, and stops the ingress of water through the sample. But after crossing 10% and reaching 20%, the permeability was increased (not improved). After 10% replacement of marble dust it starts acting as an over burdening material. Results are shown in Figure 5(a).

While in the bentonite mix concrete up to 20% replacement, the permeability was gradually improved/decreased by 43% as compared to the control sample as shown in Figure 5(b).

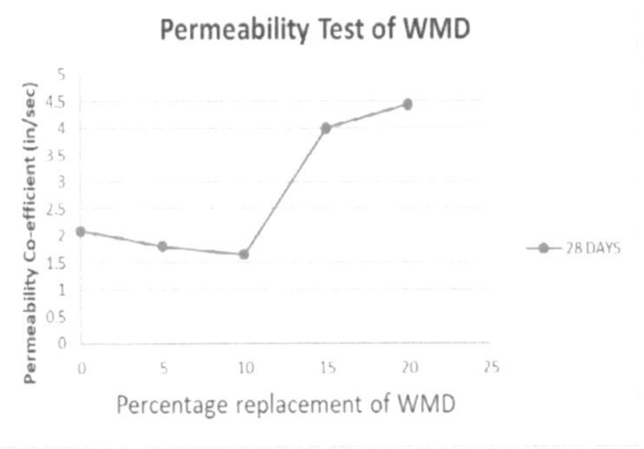

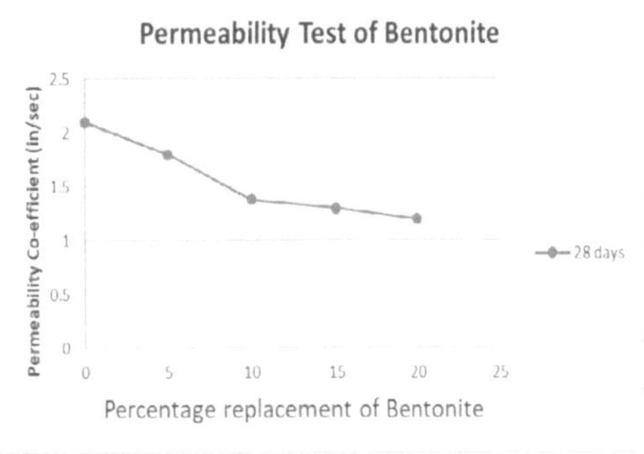

(a) (b)

Figure 5. (a) WMD 28 days permeability test graph, (b) Bentonite 28 days permeability test graph.

2.4 Scanning Electron Microscopy

The chemical composition of the both the waste marble dust and bentonite are examined via SEM test. The increase in the strength of Bentonite mix concrete is due to the high percentage of calcium oxide (74.9%). The chemical composition of both the waste marble dust bentonite samples are shown below in Tables 1 and 2.

Table 1. SEM analysis of bentonite.

Composition	Content (wt. %)
SiO_2	14.13
Al_2O_3	4.9
Fe_2O_3	2.32
MgO	3.11
CaO	74.9
Na_2O, TiO_2, K_2O	Where less than 1% each.

Table 2. SEM analysis of WMD.

Composition	Content (wt. %)
CaO	26.1
SiO_2	10.5
MgO	12.1
Loss in ignition	43.5
K_2O, Na_2O, Fe_2O_3, Al_2O_3	Where generally less than 1% each

2.5 Slump Tests

The slump test results show that as the percentage of marble dust was increased in the mix, the slump value decreases. Because marble dust doesn't absorb water actually, it entraps it, and as a result, the concrete workability is decreased as shown in Figure 6 (a).

While in the bentonite mix concrete, the slump value was continuously increasing as the percentage of bentonite was increased, the reason is that bentonite particles have the tendency to absorb water thus resulting increasing the workability of concrete as shown in Figure 6 (b).

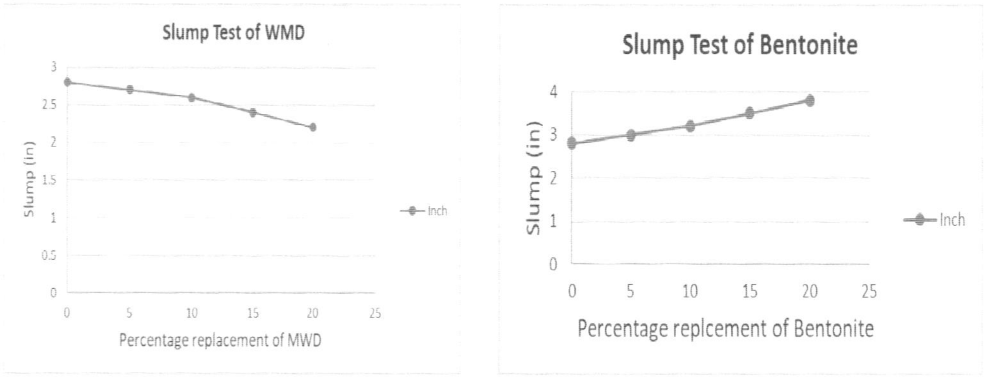

(a) (b)

Figure 6. Slump test of WMD and bentonite samples.

3 CONCLUSIONS

Foregoing experimentations show that for initial optimum replacement of cement: the compressive strength and tensile strength decreases with Waste Marble Dust (WMD) and increases with Bentonite; the permeability decreases/improves with WMD and decreases/improves with Bentonite; the slump decreases with WMD and increases with Bentonite.

References

Alyamac, K. E. and Ince, R., A Preliminary Concrete Mix Design for SCC with Marble Powders, *Construction and Building Materials,* 23, 12011210, 2009.

ASTM E2809, *Standard Guide for Using Scanning Electron Microscopy/X-Ray Spectrometry in Forensic Paint Examinations,* West Conshohocken, PA, USA, 2013.

ASTM C192, *Standard Practice for Making and Curing Concrete Test Specimens in the Laboratory, West* Conshohocken, PA, USA, 2016.

ASTM C143, *Standard Test Method for Slump of Hydraulic-Cement Concrete,* West Conshohocken, PA, USA, 2015,

ASTM C204, *Standard Test Methods for Fineness of Hydraulic Cement by Air-Permeability Apparatus,* West Conshohocken, PA, USA, 2018.

ASTM C39, *Standard Test Method for Compressive Strength of Cylindrical Concrete Specimens, West* Conshohocken, PA, USA, 2018.

ASTM C496, *Standard Test Method for Splitting Tensile Strength of Cylindrical Concrete Specimens,* West Conshohocken, PA, USA, 2017.

Ergun, A., Effects of The Usage of Diatomite and Waste Marble Powder as Partial Replacement of Cement on the Mechanical Properties of Concrete, *Construction and Building Materials*, *25*(2), 806-812, 2011.

Streamlining Information Transfer between Construction and Structural Engineering
Edited by Shiau, J., Vimonsatit, V., Yazdani, S., and Singh, A.

ISBN: 978-0-9960437-7-9

IMPRESSION EVALUATION OF CONCRETE SURFACE USING BYPRODUCTS

DAIKI NAKAYAMA[1], HIROKI SUYAMA[2], KOJI TAKASU[2], and HIDEHIRO KOYAMADA[2]

[1]*Graduate School of Engineering, The University of Kitakyushu, Kitakyushu, Japan*
[2]*Faculty of Engineering, The University of Kitakyushu, Kitakyushu, Japan*

Cement concrete requires strength, durability, and workability. Furthermore, the appearance of concrete must be good. Recently, the concrete industry has a tendency to actively utilize byproducts, all over the world, but consumers dislike the color of concrete using byproducts. This is an obstacle to the active use of byproducts. However, it is not clearly understood how consumers make the impression to color of concrete using byproducts. Therefore, we conducted a questionnaire survey using concrete using byproducts, and analyzed the answers with factor analysis and covariance analysis. As a result, the following was clarified: (1) three factors, such as beauty, massiveness and familiarity, were extracted as a factor of exposed concrete appearance, (2) the beauty of concrete was enhanced by the whiteness of the surface and was worsened by pits and surface yellowing, (3) the massiveness of concrete was worsened by the beauty and surface whiteness, and (4) even though the subjects knew that recycled byproducts were used in concrete, the evaluation of beauty did not change.

Keywords: Psychological impact, Path analysis, Covariate structure analysis.

1 INTRODUCTION

Exposed concrete is widely utilized for not only its structural strength, durability and workability, but also for its decorativeness as a finishing material. In recent years, industrial byproducts, most notably fly ash and ground granulated blast furnace slag, have been actively employed as admixtures for such concrete. This said, concrete surfaces made with such byproducts tend to be shunned because of the coloration to which they are prone. And yet, no effort has been made to assess how such coloration impressionistically impacts viewers (Akiyama 1992).

Lingzhi *et al.* (2007) undertook research on the impressionistic impact of exposed concrete, where test subjects assessed them by applying factor analysis. We prepared concrete samples of varying colorations by content of byproducts. Then, we surveyed test subjects to learn their impressions, applied factor analysis to extract main factors behind the impressions, and covariate structure analysis to find the impact of sample porosity, color blurring, brightness and coloration.

2 EXPERIMENT OVERVIEW

2.1 Sample Overview

2.1.1 *Materials investigated*

Materials utilized to produce exposed concrete samples for these impressionistic assessments are: normal Portland cement; tap water; sea sand; crushed stone; five classes (C through G) of fly ash;

and ground granulated blast furnace slag. Table 1 shows the brightness L* and colorations a* and b* of each of these materials, based on JIS Z 8729/CIE1976. We also added a high-performance AE (air entraining) water reducing agent to assure a suitable degree of fluidity upon pouring.

Table 1. Materials.

Concrete samples	L*	a*	b*	Density	Blaine value
Normal Portland cement	47.87	-0.48	9.79	3.16	-
fly ash C	52.71	0.99	11.71	2.21	3568.45
fly ash D	37.72	0.45	5.28	2.24	3449.11
fly ash E	52.36	0.48	12.49	2.25	3703.19
fly ash F	37.90	0.23	2.15	2.14	3802.02
fly ash G	47.23	0.35	2.83	2.28	3727.21
Ground granulated blast furnace slag	45.81	-0.17	7.60	2.31	3522.22

2.1.2 *Mixing and pouring*

We prepared nine types of exposed concrete samples for assessment. Samples A and I were made of normal concrete having a water content of 175 kg/m^3 and a water-cement ratio of 50%. Sample B is of high-strength concrete having a water-cement ratio of 40%. Samples C through G (5 samples) are comprised of an ordinary concrete base to which is added (substituted/mixed) a fly ash aggregate of, respectively, class C to G at a proportion of 120kg/m^3. And, Sample H is a concrete prepared by a 45% substitution of cement with ground granulated blast furnace slag.

Test sample dimensions were 450mm x 450mm x 30mm. The samples were prepared by pouring cement into frames fabricated with coated concrete panels, and allowed to cure for seven days. After removal, samples were cured in air for seven more days. They were subjected to impressionistic assessment only after confirming that their surface brightness had stabilized.

2.1.3 *Surface condition of the samples*

Tabulated in Table 2 are their porosity fraction, brightness L*, colorations a* and b*, and standard deviation of brightness L*. Porosity fraction was determined by first photographing sample surfaces and then measuring the proportion of pixels indicative of a pore. Because some samples exhibited much more porosity than others, we applied the Weber–Fechner approach and, in our analysis, used logarithmic values. For the measurement of brightness L* and coloration a*, we utilized a Minolta CR-300 Chroma Meter (colorimeter), with results taken as the average value of 37 measurements of each sample surface. We found the fluctuation in red-green coloration a* to be smaller than the fluctuation in either brightness L* or in yellow-blue coloration b*. All ten observers within a preliminary survey could discern a yellowing of the samples along the yellow-blue balance but none could detect any change in the red-green balance. We thus decided to exclude red-green coloration a* as an item for analysis under this study.

2.2 Survey Overview

2.2.1 *Query items*

We next conducted a survey to assess the psychological/impressionistic impact of each of the nine types of exposed concrete samples. Here, we turn to a set of 18 adjectival pairs utilized in previous research. We assessed the results by means of the semantic differential method, grading each over seven levels ranging from -3 to +3. Query items are listed in their entirety in Table 3.

Table 2. Sample surfaces.

Concrete samples	Porosity fraction	L*	a*	b*	Standard deviation of brightness L*
A	1.794	67.54	-0.77	3.56	2.02
B	0.003	70.11	-0.69	4.48	0.84
C	0.152	70.44	-0.39	5.38	1.10
D	0.079	62.48	-0.66	1.20	0.84
E	0.068	72.28	-0.64	3.56	0.85
F	0.230	62.35	-0.81	1.44	1.02
G	0.206	66.87	-0.82	1.46	4.63
H	0.742	74.56	-0.58	3.47	1.43
I	0.273	68.39	-0.43	4.75	2.20

Table 3. Query items.

No.	Queryitems are listed in their entirety
A01	Coarse⇔Smooth
A02	Sober ⇔Flashy
A03	Heavy⇔Light
A04	Bright⇔Dark
A05	Loose⇔Moist
A06	Clean⇔Dirty
A07	Plain⇔ Special
A08	Rich⇔Poor
A09	Quiet⇔Active
A10	Hot⇔Cool
A11	Shiny⇔Dull
A12	Like⇔Hate
A13	Concrete⇔Not-concrete
A14	Supernatural⇔Artificial
A15	Hard⇔Soft
A16	Beautiful⇔Ugly
A17	Mannish⇔Woman
A18	Airy⇔ Substantial

2.2.2 *Implementing institution, survey participant affiliations*

We surveyed 200 participants (i.e., N = 200) over 7-13 November 2015. Participant affiliations are tabulated within Figure 1. We divided the participants into two equal groups, one for which we presented a detailed written explanation of survey questions (N = 100), and one for which we presented no explanation (N = 100). The verbiage on the questionnaire sheet for the first (detailed explanation) group is as below:

Samples A, I: Ordinary concrete: typically used for buildings and other structures.

Sample B: High-strength concrete: reinforced concrete with additional cement.

Samples C through G: Fly-ash concrete, containing fly ash in addition to cement (fly ash is a byproduct of combustion at coal-fired electric power plants).

Sample H: Blast furnace slag concrete, containing ground granulated blast furnace slag in addition to cement (blast furnace slag is a byproduct of the iron-making process at steelworks).

2.2.3 *Method of sample presentation*

The survey was conducted on a wooden deck. The samples were illuminated with only natural outdoor light (a MODEL LX-105 digital light meter (CUSTOM Corporation) indicated a surface illuminance of 960 to 3,118x). Figure 1 illustrates the manner in which the samples were shown to the participants. They were presented as 400mm discs placed at a distance of 1,500mm. Per the groundbreaking proxemics research of Edward T. Hall, we know that humans can see most acutely over a visual field of 12°. We designed our arrangement accordingly (specifically, we placed the samples at what would be a 10° downward angle for a person with an eye level of 1,531mm, typical for the 20- to 24-year-old Japanese men and women recruited for the test. We placed the samples at 100mm intervals on a dark cloth background so as to prevent anything else from entering the participants' field of vision.

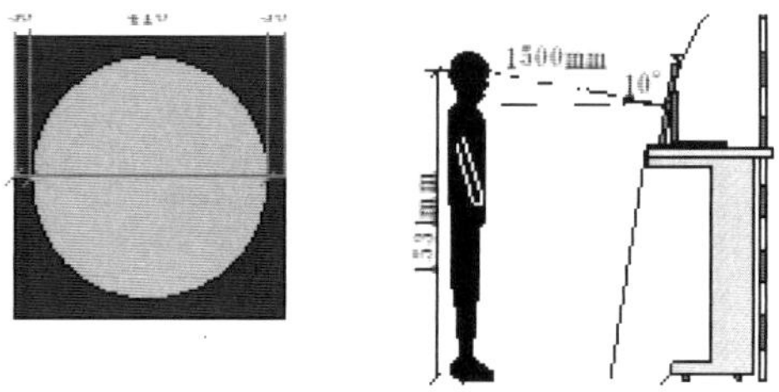

Figure 1. The manner in which the samples were shown to the participants.

2.2.4 *Analysis method*

We utilized IBM SPSS Statistics 22.0 and IBM SPSS Amos 23 (a structural equation modeling (SEM) program) to analyze the survey results.

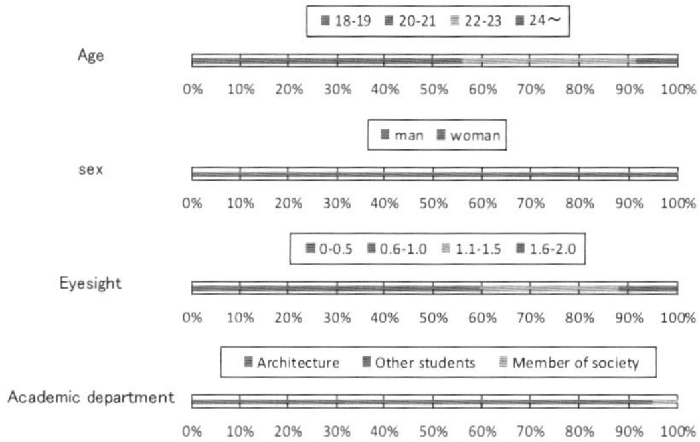

Figure 2. Survey participant affiliations.

3 SURVEY OVERVIEW

3.1 Factors Comprising the Impressionistic Impact of Concrete

With results obtained with the 18 adjectival pairs, we conducted factor analysis by means of the maximum likelihood method. Changes in eigenvalue were: 5.80, 1.85, 1.56, 1.30, 1.00.... We hypothesized three factor structures (upon consideration of the number of queries) and once again conducted a factor analysis, this time with the maximum likelihood method / promax rotation. We excluded four pairs not having a sufficient factor loading (0.35 minimum) and, by this, arrived at the factor pattern of Table 4.

The first factor comprises six pairs and, at a contribution rate of 39.2%, accounts for a large portion of the total. We consider this factor – "appearance"– to be the most important and note high factor loadings in such appearance-related pairs as beautiful/ugly, like/hate and clean/dirty.

The second factor is comprised of five pairs and shows high factor loadings in such texture-related pairs as heavy/light and airy/substantial. We refer to this factor as "presence."

The third factor is comprised of three pairs and shows high factor loadings in perceptions generally considered contrasting to typical concrete, including plain/special and subdued/active. We refer to this factor as "familiarity." Correlations are shown in Table 5.

3.2 Impact of Concrete Surface Condition on Viewer Impressions

To examine the effect of porosity, color blurring, brightness L* and coloration b*on each of the above three factors, we carried out a path analysis under covariate structure analysis.

Table 4. Pattern matrix.

No.	Adjectival pairs	FactorI	FactorII	FactorIII
A16	Beautiful⇔Ugly	0.884	0.036	0.002
A12	Like⇔Hate	0.842	0.194	-0.024
A06	Clean⇔Dirty	0.769	-0.097	0.151
A08	Rich⇔Poor	0.610	0.065	-0.282
A01	Coarse⇔Smooth	-0.550	0.214	-0.173
A11	Shiny⇔Dull	0.516	-0.139	-0.060
A03	Heavy⇔Light	0.125	0.878	0.020
A18	Airy⇔ Substantial	-0.007	-0.749	0.108
A17	Mannish⇔Woman	-0.136	0.672	-0.104
A04	Bright⇔Dark	0.167	-0.632	-0.041
A15	Hard⇔Soft	0.122	0.523	0.045
A07	Plain⇔ Special	-0153	-0.150	0.541
A09	Quiet⇔Active	0.301	0.083	0.526
A02	Sober ⇔Flashy	-0.222	0.347	0.508
Eigenvalue		5.49	1.64	1.44
Contribution rate		39.2%	50.9%	61.2%
Factor's name		Appearance	Presence	Familiarity

Table 5. Factor correlation matrix.

Factor	I	II	III
Appearance	1.000	-.646	.090
Presence		1.000	.010
Familiarity			1.000

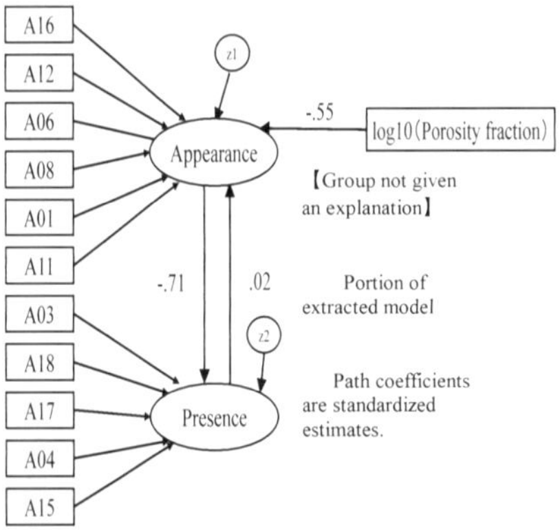

Figure 3. Measurement model for "appearance" and "presence".

We consider the cause and effect relation among the three factors (i.e., appearance, presence and familiarity). Under assumptions that (a) appearance and presence have some influence on each other and (b) porosity has an influence on only appearance, we analyzed questionnaire results from the group to which we did not present a preliminary explanation (Figure 3). We found that the path from presence to appearance does not have significance ($p = .853$). By this, we learn that although appearance does have a negatively impact on presence, presence does not have a significant impact on appearance.

We hypothesized the cause-and-effect relationships among the three factors to be only that appearance has an impact on weightiness /presence. We also assume that color blurring on the surface sample (standard deviation of L*), brightness L* and coloration b* have an impact on the three factors. We analyze results from the group to which no explanation was presented. We find that the path from color blurring to presence is significant ($p = .018$), as are the paths from color blurring to familiarity ($p =.393$), from brightness L* to familiarity ($p =.370$), and from coloration b* to familiarity ($p = .374$). We thus eliminate the paths found to be not significant (leaving us with paths for which $p < .001$) and, this time adding questionnaire data from the group to which an explanation was presented. This brings us to Figure 4, our final model, from which we have that $\chi 2 = 1078.845$, df = 166, $p< 0.01$, GFI=.918, AGFI=.881, and RMSEA=.055. From this figure, we find that concrete whiteness (magnitude of brightness L*) acts to enhance appearance, whereas concrete yellowness (magnitude of coloration b*) detracts from it. Also, none of the concrete surface related variables shows a significant path to familiarity.

3.3 Bias to Impressionistic Assessments by Awareness of Byproduct Presence

We also utilized this model to evaluate how an awareness of byproduct utilization within a concrete biases viewers' impressionistic assessments of that concrete. Here, we conducted path analysis for both groups (those given an explanation, and those not) under a hypothesis that the addition of a byproduct does have an effect on concrete appearance (Figure 5). We find that with regards to the group given no explanation, the path from "use of byproduct" to "appearance" is not significant ($p = .062$). This is to be expected, because the respondents did not know that some byproduct was added to the cement during its preparation. This said, there was also no significant path from "use of byproduct" to "appearance" even with regards to the group given an explanation ($p = .459$). To put it another way, viewer assessments of concrete appearance are almost entirely the same whether or not those viewers know that some byproduct was added to the concrete during its preparation.

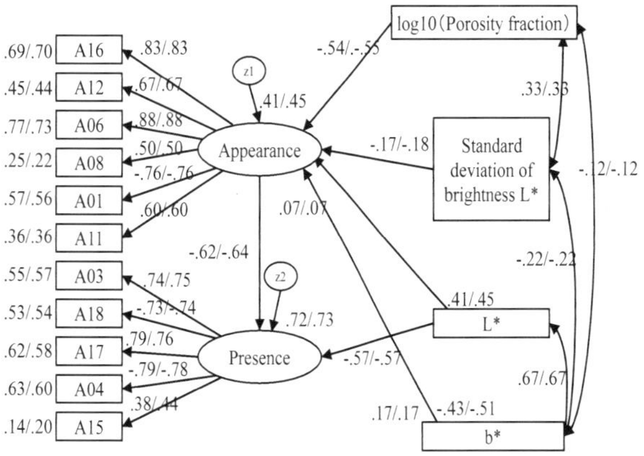

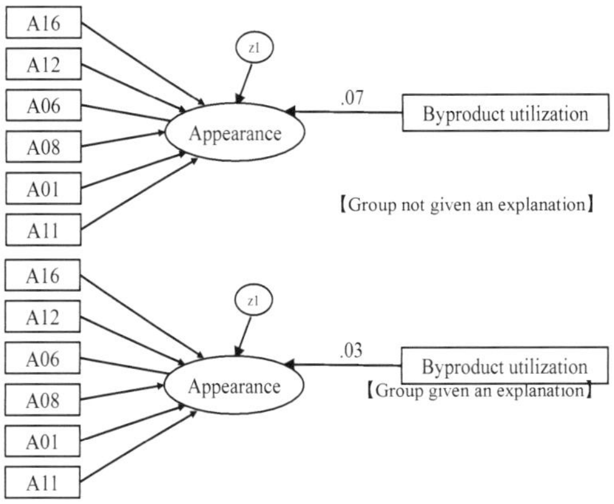

Figure 4. Formation model for exposed concrete appearance and presence.

Figure 5. Measurement model for "use of byproduct" and "appearance".

4 CONCLUSIONS

This paper described how we first prepared samples of exposed concrete of differing colorations by varying the content of industrial byproducts within them. Then, with these samples, we surveyed test subjects to learn their impressions of them. We next subjected the results to factor analysis to extract the main factors behind such impressions and, from there, to covariate structure analysis.

As factors to impressionistic assessments of exposed concrete, we were able to extract "appearance," "presence" and "familiarity." Although the appearance of exposed does have a negative impact on presence, presence does not have a significant impact on appearance. Exposed concrete whiteness acts to enhance appearance, whereas yellowness detracts from it.

Viewer assessments of the appearance of exposed concrete are almost entirely the same, whether or not those viewers know that a byproduct was added to concrete during its preparation.

Acknowledgments

This work was supported by JSPS KAKENHI Grant Number 18K13872.

References

Akiyama, T., *et al.,* Is there a Future for Exposed Concrete?, *Architectural Institute of Japan*, 588, 24-25, 1992.

Lingzhi, L., *et al.*, Influence of Color of Cement and Fine Aggregate on the Color of Exposed Concrete, 612, 17-22, 2007.

Streamlining Information Transfer between Construction and Structural Engineering
Edited by Shiau, J., Vimonsatit, V., Yazdani, S., and Singh, A.

ISBN: 978-0-9960437-7-9

RELIABILITY BASED MULTI-OBJECTIVE OPTIMIZATION OF CONCRETE MIX PARAMETERS USING NSGA II

MAHESH SHARMA[1], RACHNA AGGARWAL[2], RAJENDRA SHARMA[3], and MANEEK KUMAR[4]

[1] *School of Mathematics, Thapar Institute of Eng. & Technology, Patiala, India*
[2]*Dept of Mathematics, M.L.N. College, Yamunanagar, Haryana, India*
[3]*Dept of Computer Science, Thapar Institute of Eng. & Technology, Patiala, India*
[4]*Dept of Civil Engineering, Thapar Institute of Eng. & Technology, Patiala, India*

This article presents a method to optimize concrete mix proportions with respect to different goals of economy and reliability or, equivalently, probability of failure. This method is based on a quadratic generalized ridge regression model to predict compressive strength of concrete for 28 days curing period and a linear regression model to predict cost of concrete. NSGA II is used to obtain reliable Pareto-optimal fronts with non-dominated solutions for different compressive strength requirements. Pareto-optimal fronts evolved by varying compressive strength requirements and probability of failure are analyzed. It is found that there is a nominal rise in cost as probability of failure decreases up to a certain limit for a given compressive strength requirement. However, there is a sharp rise in cost of concrete below that limit.

Keywords: Compressive strength, Regression model, Probability of failure.

1 INTRODUCTION

Reliability based design techniques are getting more acceptance for practical optimization in civil engineering, due to uncertainties involved in design processes (Dimou and Koumousis 2009, Behnam and Eamon 2013, Torii *et al.* 2012, Tao and Tam 2013, Barone and Frangpol 2014). These uncertainties owe themselves to inherent randomness, limited information, imperfect knowledge, human errors, structural idealizations in formulating the mathematical model of the structure to predict its response or behavior and the limitations of numerical techniques. In Reliability Based Design Optimization (RBDO) technique, some constraints are replaced by probabilistic constraints allowing these constraints to be violated within the given limits of failure. RBDO sacrifices true optimum in order to find a reliable optimum corresponding to given probabilities of failure. However, studying the impact of probability of failure on optimal solutions can be of great practical importance as it gives some idea regarding how the design variables change so as to make corresponding solution more and more reliable. This can be realized by making optimization problem multi-objective by considering minimization of probability of failure of one or more constraints as additional objectives and obtaining a set of non-dominated solutions, *i.e.*, a Pareto-optimal front.

In this study, focus is on optimizing concrete mix designs as concrete plays a major role in the performance of a structure. Much work has been reported in literature to find optimal mixture composition of concrete satisfying specific performance. Yeh (1999, 2003, 2007, 2009) developed computer-aided design system to find optimum concrete mix compositions based on artificial neural networks and non linear optimization techniques. Karihaloo and Kornbak (2001) employed nonlinear mathematical programming technique in design of fibre reinforced concrete mixes which have both high ductility and high tensile strength. Lim *et al.* (2004) proposed a method to design high performance concrete mix based on genetic algorithms. Lee *et al.* (2009) proposed a design methodology based on artificial neural network, convex hull and genetic algorithms for optimal mixture proportioning of concrete composition. Jayaram *et al.* (2009) developed elitist genetic algorithm models for the optimization of high volume fly ash concrete.

The aim of this study is to find reliable concrete mix compositions, for normal strength concretes, meeting a specific compressive strength requirement and to study the impact of reliability level on optimal solution in multi-objective environment. However, a multi-objective problem demands to find as many Pareto-optimal solutions as possible. Over last two decades, a number of multi-objective Evolution Approaches [EA] have been suggested which are capable of finding multiple Pareto-optimal solutions in one single simulation run (Deb 2010). Elitist Non-Dominated Sorting Algorithm (NSGA-II) proposed by Deb *et al.* (2002) is one of such approaches that has low computational requirements, elitist approach, parameter less niching approach and simple constraint handling strategy. This technique is used to obtain reliable Pareto-optimal front in this article.

2 EXPERIMENTAL DATASET

The Compressive strength data explored in this study was generated by Kumar (2002) by conducting experiments under controlled laboratory conditions. The concrete mixes were proportioned using four basic ingredients, namely, water, cement, coarse aggregate and fine aggregate. A set of 49 normal strength concrete mixes were prepared by varying water-cement ratio, cement contents and aggregates fractions. Water-cement content ratio was kept between 0.42 and 0.55. For each mix, 15 cubes of 150 mm size were cast and were tested at 28 days of curing period. Thus, a sufficiently large data bank was generated and the same has been used in the present work for analyzing compressive strength of concrete. Also, unit cost of each material is determined by taking into account the price rates in India. Based on the prices, cost of 1 m^3 of concrete is calculated for each mixture and is measured in Indian rupees.

3 MATHEMATICAL MODELS FOR CONCRETE MIX PARAMETERS

There are four design variables, namely, water content (w), fine aggregate content (fa), coarse aggregate content (ca) and cement content (c). All the design variables are measured in kg/m^3. There are two response variables, namely, cost of concrete ($cost$) and 28 days compressive strength ($st28$). Compressive strength of concrete is measured in MPa and cost of concrete is measured in Indian rupees.

In this study, no cost is associated with the water content and as such, cost of concrete is a linear function of fine aggregate content, coarse aggregate content and cement content. Linear regression model developed for cost of concrete is shown in Eq. (1).

$$cost = 236.461 + 0.629fa + 0.333ca + 4.892c \tag{1}$$

However, compressive strength of concrete may not be linearly related to the constituent materials of concrete. In the present study, compressive strength of concrete has been modeled as a quadratic function of w/c ratio, fa/c ratio, ca/c ratio and cement content c. These predictor variables have been selected on the basis of correlation analysis. Table 1 shows correlation matrix of w, fa, ca, c, w/c ratio, fa/c ratio, ca/c ratio and $st28$.

Table 1. Correlation matrix.

Parameter	w (kg/m^3)	fa (kg/m^3)	ca (kg/m^3)	c (kg/m^3)	w/c	fa/c	ca/c	$st28$ (MPa)
w (kg/m^3)	1.000	0.805	-0.305	0.541	0.171	0.261	-0.492	0.000
fa (kg/m^3)		1.000	0.102	0.026	0.617	0.751	0.046	-0.462
ca (kg/m^3)			1.000	-0.375	0.194	0.320	0.870	-0.214
c (kg/m^3)				1.000	-0.734	-0.637	-0.776	0.821
w/c					1.000	0.960	0.517	-0.968
fa/c						1.000	0.546	-0.900
ca/c							1.000	-0.581

All the four selected predictor variables have strong correlation with $st28$ in comparison to remaining variables. Further, it can also be observed from the table that data used for modeling suffers from multi-collinearity as correlation between each pair of predictor variables is numerically greater than 0.5. To tackle this problem, Generalized Ridge Regression (GRR) technique proposed by Hoerl and Kennard (1970) has been employed for model development. The quadratic GRR model developed for compressive strength of concrete is given in Eq. (2):

$$\begin{aligned} st28 = {} & -0.034569 - 0.110144\,w/c - 35.27014\,fa/c + 0.83211\,ca/c + 0.38787c - 0.11413(w/c)^2 \\ & - 0.39153(fa/c)^2 + 0.34431(ca/c)^2 - 0.00041c^2 - 0.20540(w/c * fa/c) \\ & - 35.14388(w/c * ca/c) + 11.14388(fa/c * ca/c) \end{aligned} \tag{2}$$

4 MULTI-OBJECTIVE PROBLEM FORMULATION

Multi-objective concrete mix optimization problem is formulated in Eq. (3):

$$\left.\begin{aligned} & \underset{\mu_w,\mu_{fa},\mu_{ca},\mu_c}{Minimize} \quad cost \\ & \underset{\mu_w,\mu_{fa},\mu_{ca},\mu_c}{Minimize} \quad Pof\_st28 \\ & where\ Pof\_st28 = 1 - Prob(st28 \geq f_c) \\ & Subject\ to{:}\quad Pof\_st28 \leq P_f \\ & \qquad 0.42 \leq w/c \leq 0.55 \\ & \qquad w_l \leq w \leq w_u \\ & \qquad fa_l \leq fa \leq fa_u \\ & \qquad ca_l \leq ca \leq ca_u \\ & \qquad c_l \leq c \leq c_u \end{aligned}\right\} \tag{3}$$

where $\mu_w, \mu_{fa}, \mu_{ca}$ and μ_c are, respectively, the mean values of water content, fine aggregate content, coarse aggregate content and cement content; Pof_st28 denotes probability of failure of compressive strength constraint. f_c denotes the target 28 days compressive strength of concrete; P_f is the upper bound on probability of failure of compressive strength constraint. Water-cement content ratio w/c is kept between 0.42 and 0.55. w_l, fa_l, ca_l, c_l, respectively, are lower bounds for water, fine aggregate, coarse aggregate and cement content. w_u, fa_u, ca_u, c_u, respectively, are upper bounds for water, fine aggregate, coarse aggregate and cement content.

In this article, all the four design variables are considered as random design variables. There are no deterministic variables and random parameters. It is assumed that all the four design variables follow normal distribution with their respective means and standard deviations listed in Table 2. The lower and upper bounds for the design variables are also taken from Table 2. The constraint on w/c is taken as deterministic constraint.

Table 2. Descriptive statistics.

Variable	Minimum (kg/m^3)	Maximum (kg/m^3)	Mean (kg/m^3)	Standard deviation (kg/m^3)
w	180.00	230.00	202.44	12.69
fa	416.93	642.18	535.64	57.29
ca	798.48	1252.05	1064.85	133.42
c	350.00	475.00	424.49	37.32
$st28$	31.66	54.49	45.80	5.42

Since the two objectives given in (3) are of conflicting nature, therefore, the problem is to find a set of non-dominated solutions forming Pareto-optimal front depicting the tradeoff between the objectives efficiently. Here, NSGA II has been used to locate Pareto-optimal fronts.

5 RESULTS AND DISCUSSION

The optimization is run for target compressive strength of 25 MPa, 30 MPa, 35 MPa and 40 MPa, which are normal strength concretes used for general applications. The results are obtained for P_f as 0.05. Probability of failure is calculated using mean value approximation method. VisualDoc7.2 developed by Vanderplaats Research & Development, Inc. has been used to carry out reliability analyses and optimizations.

5.1 Selection of NSGA II Parameters

The maximum number of generations is varied up to 100 in steps of 25. The Pareto-optimal fronts obtained for different values of number of generations for f_c = 25 MPa, P_f = 0.05 are demonstrated in Figure 1. It is seen that difference between the graphs is not much beyond generation size of 75.

Similar observations are made in remaining cases also. Therefore, the maximum generation size is logically set as 75. Also, Population size is taken as 16, 18, 20, 22, 24 in each case under study. The Pareto-optimal fronts obtained for different population sizes for f_c = 25 MPa, P_f = 0.05 are shown in Figure 2. It can be seen from the figure that best Pareto-optimal front is obtained for population size of 20. Similar results are obtained in all the remaining cases. So, the optimal value of population size is taken as 20.

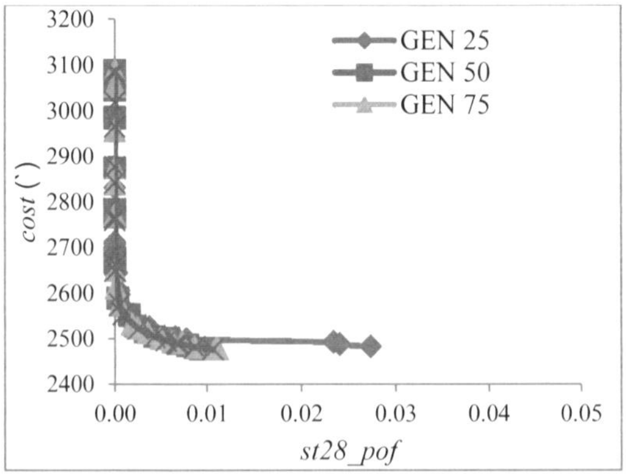

Figure 1. Pareto-optimal fronts obtained for 28 days compressive strength of 25 MPa and P_f of 0.05 for different number of generations.

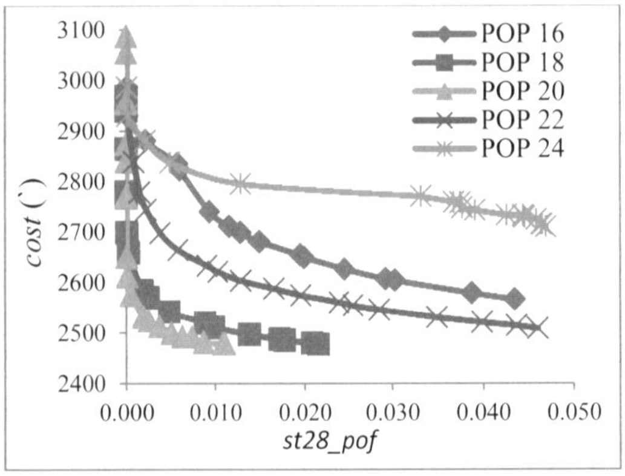

Figure 2. Pareto-optimal fronts obtained for 28 days compressive strength of 25 MPa and P_f of 0.05 for different population size.

5.2 Optimization Results

Figure 3 depicts the locations of Pareto-optimal fronts in all the four cases under study. It can be noted from Figure 3 that Pareto-optimal front for f_c of 25 MPa does not cover entire range of allowed probability of failure. This can be attributed to the fact that 25 MPa is a very small value of compressive strength and can be attained with a very high level of reliability for range of parameters under study.

Pareto-optimal fronts depict that the two objectives of cost and probability of failure are not linearly related to each other. It can be noted that Pareto-optimal fronts are divided into two parts. For very low values of P_f (10^{-5}or below for f_c = 25 MPa, 10^{-4}or below for f_c = 30 MPa, 35 MPa and 10^{-3} or below for f_c = 40 MPa), the rate of rise in optimal cost is more rapid. But, as the probability of failure increases, this rate becomes slow. This means lowering probability of failure below a certain level may not be cost effective and the designer should assess the required reliability level properly to get reliable economic design.

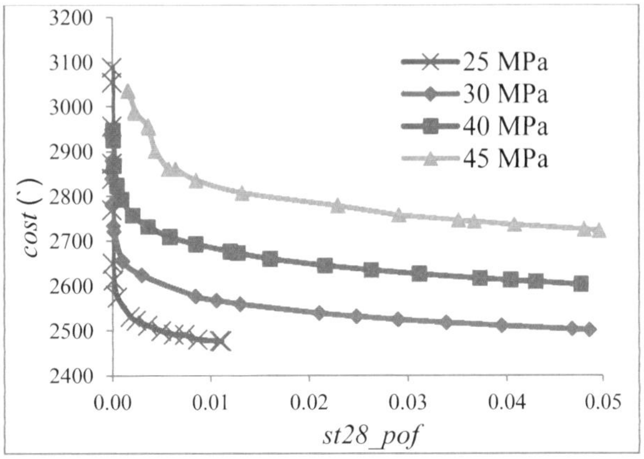

Figure 3. Pareto-optimal fronts for $st28\_pof \leq 0.05$.

6 CONCLUSIONS

A multi-objective optimization problem, where cost of concrete and probability of failure of compressive strength constraint are simultaneously minimized, is formulated. Pareto-optimal solutions are obtained using NSGA II. Pareto-optimal fronts evolved by varying compressive strength requirements are analyzed. It is found that rise in cost with decreasing pof_st28 is nominal up to a certain limit of P_f for a given compressive strength requirement. Below that limit, there is a sharp rise in cost for a small fall in P_f. It is also observed that cement content plays a major role in optimization of concrete mix parameters.

References

Barone, G. and Frangpol, D. M., Life-Cycle Maintenance of Deteriorating Structures by Multi-Objective Optimization Involving Reliability, Risk, Availability, Hazard and Cost, *Structural Safety*, 48, 40-50, 2014.

Behnam, B. and Eamon, C., Reliability-Based Design Optimization of Concrete Flexural Members Reinforced with Ductile FRP Bars, *Construction and Building Materials*, 47, 942-950, 2013.

Deb, K., Multi-Objective Optimization Using Evolutionary Algorithms, John Wiley & Sons, U.K., 2010.

Deb, K., Partap, A., Aggarwal, S., and Meyarivan, T., A Fast and Elitist Multi-Objective Genetic Algorithm, NSGA II, *IEEE Transactions on Evolutionary Computation*, 6(2), 182-197, 2002.

Dimou, C. and Koumousis, V., Reliability-Based Optimal Design of Truss Structures Using Particle Swarm Optimization, *Journal of Computing in Civil Engineering*, 23(2), 100–109, 2009.

Hoerl, A. E., and Kennard, R. W., Ridge Regression, Biased Estimation for Nonorthogonal Problems, *Technometrics*, 12, 55-67, 1970.

Jayaram, M. A., Natraja, M. C., and Ravikumar, C. N., Elitist Genetic Algorithm Models, Optimization of High Performance Concrete Mixes, *Materials and Manufacturing Processes*, 24, 225-229, 2009.

Karihaloo, B. L., and Kornbak, D. L., Optimization Techniques for the Design of High Performance Fibre-Reinforced Concrete, *Structural and Multidisciplinary Optimization*, 21 (1), 32-39, 2001.

Kumar, M., *Reliability Based Design of Structural Elements,* PhD Dissertation, Thapar University, Patiala, India, 2002.

Lee, B. Y., Kim, J. H., and Kim, J.K., Optimum Concrete Mixture Proportion Based On a Database Considering Regional Characteristics, *Journal of Computing in Civil Engineering*, 23, 258-265, 2009.

Lim, C. H., Yoon, Y. S., and Kim, J. H., Genetic Algorithm in Mix Proportioning of High–Performance Concrete, *Cement and Concrete Research*, 34, 409-420, 2004.

Tao, R. and Tam, C. M., System Reliability Theory Based Multiple-Objective Optimization Model for Construction Projects, *Automation in Construction*, 31, 54-64, 2013.

Torii, A. J., Lopez, R. H., and Biondini, F., An Approach to Reliability-Based Shape and Topology Optimization of Truss Structures, *Engineering Optimization*, 44(1), 37-53, 2012.

Yeh, .I C., Design of High Performance Concrete Mixture Using Neural Networks and Non Linear Programming, *Journal of Computing in Civil Engineering*, 13 (1), 36-42, 1999.

Yeh, I. C., A Mix Proportioning Methodology for Fly Ash and Slag Concrete Using Artificial Neural Networks, *Chung Hua Journal of Science and Engineering*, 1 (1), 77-84, 2003.

Yeh, I. C., Computer Aided Design of Optimum Concrete Mixtures, *Cement and Concrete Composites*, 29(3), 193-202, 2007.

Yeh, I. C., Optimization of Concrete Mix Proportioning Using a Flattened Simplex-Centroid Mixture Design and Neural Networks, *Engineering with Computers*. 25 (2), 179-190, 2009.

Streamlining Information Transfer between Construction and Structural Engineering
Edited by Shiau, J., Vimonsatit, V., Yazdani, S., and Singh, A.

ISBN: 978-0-9960437-7-9

COMPILATION AND INNOVATION OF TECHNICAL SPECIFICATION OF ELECTROCHEMICAL TECHNOLOGY FOR DURABILITY OF CONCRETE STRUCTURES

WEILIANG JIN[1], P. Z. WANG[1], J. XIA[1], J. H. MAO[2], W. J. FAN[2], and X. SONG[1]

[1]*Zhe Jiang University, Hang Zhou, China*
[2]*Ningbo Institute of Technology, Zhejiang University, Ning Bo, China*

The Chinese *Technical Specification of Electrochemical Technology for Durability of Concrete Structures* was compiled on the basis of the overseas and domestic development, practice and research of durability electrochemical technology for concrete structures. The Specification stipulated the general provisions, scope of application, main parameters, quality control and inspection etc of electrochemical technology. Main innovation of Technical Specification appears as: 1. An extensive introduction highlighted the operation and requirement of electrochemical technology for durability of concrete structures, including cathodic protection, electrochemical desalination, electrochemical deposition, bi-directional electromigration etc. 2. The mechanism and application of a new-type electrochemical technology, bi-directional electromigration, was described and its advantages were illustrated in case study. Accompanied by the implementation and application of the technical specification, the durability enhancement and control of concrete structures in China will have another level.

Keywords: Bidirectional electromigration, Organic corrosion inhibitors, Hydrogen embrittlement, Cathodic polarization curve.

1 INTRODUCTION

With the extensive application of concrete structure, the durability of concrete structure has attracted more and more attention. Durability of concrete can be influenced by several factors such as the quality of cement, steel and other materials, the design and construction. There are also environmental factors, including carbonation, corrosion (chloride salt erosion), freeze-thaw, alkali aggregate reaction and so on while the corrosion (chloride salt erosion) is considered to be the main factor (Jin 2012, Ihekwaba 1996, Jin and Zhao 2014). It is generally accepted by scholars at home and abroad that electrochemical technology is the only non-destructive repair technology to stop steel corrosion in concrete structures (Sohanghpurwala *et al.* 2002). Since the 1970s, electrochemical techniques for durability of concrete have emerged and developed such as cathodic protection, electrochemical re-alkalization, electrochemical chlorine removal, electrochemical deposition, bi-directional electromigration etc (Bi *et al.* 2007, Jin *et al.* 2016). However, there are still many problems to be solved, such as defining application condition of these electrochemical technologies in the types of concrete structures as well as concrete deterioration stages, specifying the range of electrochemical parameters and the implementation process of these technologies, proposing a set of acceptance indicators appropriate for the domestic present status. Therefore, Zhejiang University united with domestic universities, research institutes, design, construction and management departments, to form a force to focus on the compilation of the technical specifications.

This paper will focus on the compiling process of the specification, expounding the main innovations as well as the case study. The implementation and application of the technical specification will play an important role in promoting and controlling the durability of concrete structures in China.

2 COMPILING PROCESS

2.1 Preliminary Work

Investigation, experiment as well as monographic studies focusing on the electrochemical technology has provided the theoretical basis to the compilation. During the compilation of the specification, the National Natural Science Foundation of China, the International Cooperation Fund of the Ministry of Science and Technology of China and Zhejiang Province Natural Science Foundation etc have given a great help.

2.2 Chapters and Sections

The specification set up a systemic frame of electrochemical technology by defining the range of application, introducing specific methods of implementation and recommending acceptance indicators. The specification consists of nine chapters in all, including three chapters to overview named general provisions, terms and symbols, basic requirements and six chapters to introduce the electrochemical technologies named impressed current cathodic protection (ICCP), sacrificial anode cathodic protection (SACP), electrochemical realkalization, chloride extraction, electrochemical deposition, bi-directional electromigration. Every technology introducing chapter consists of five sections including general requirements, design, installation and commissioning, quality control and inspection, operation and maintenance.

2.3 Review and Approval

During the compilation process, two expert consultation meetings were held, relevant experts were invited, and their latest research results were also absorbed to form a consultation paper. Through an extensive consult on the *National Standardization Information Network for Construction of Engineering* and 32 copies of consultation paper sending to relevant units and experts, the compiling group received 213 suggestions and with a painstaking sorting and studying, the draft paper came into being. Afterward, an expert review committee was convened to examine the draft paper and 90 suggestions were proposed to form the version to be approved.

3 MAIN INNOVATION

3.1 Bi-directional Electromigration

Bi-directional electromigration (BIEM) is a new-type electrochemical technology focusing on both chloride and corrosion inhibitor migration, distinct from the chloride extraction technology. Shown as Figure 1, the electric field was formed with a metal mesh in the solution as anode and the rebar in the concrete as cathode, causing the corrosion inhibitor cation migrates to the rebar while the chloride ion migrate to the metal mesh (Wu H. 2018).

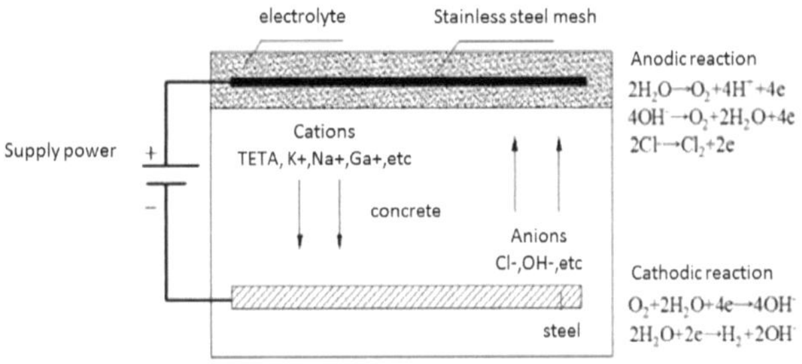

Figure 1. BIEM interactions diagram.

The corrosion inhibitor apply to bi-directional electromigration must have follow performance: 1. Effectively preventing or delaying the occurrence of corrosion of steel bars, can reduce the corrosion rate of corroded steel bars or stop the corrosion of steel bars. 2. Soluble in water and will provide positively charged ions while the solution is alkaline. 3. Can't react with the concrete composition when it migrates into the concrete. 4. Showing a great chemical stability in concrete and can persist for a period of time. 5. Economic and environmentally-friendly, conducive to extensive application in engineering.

After preliminary selection, several organic corrosion inhibitors were tested with potentiodynamic polarization and electrochemical impedance spectroscopy (EIS) to find out their impact on steel corrosion in chlorinated simulated concrete pore solution. At the same time, the electromigration ability of these organic corrosion inhibitors in concrete, the effect on chloride ions and reinforcement corrosion were studied by using organic element analyzer, rapid chloride test (RCT), EIS, weak polarization method and scanning electron microscopy (SEM). Finally, triethylenetetramine (TETA) behaved the best comprehensive performance and was selected as the recommended electromigration corrosion inhibitor (Zhang 2012).

During the bi-directional electromigration, provided that the protection current density is too high, hydrogen evolution reaction will occur at the cathode. When hydrogen enters into the stressed steel bar, it will induce crack, reduce the plasticity of the steel bar and lead to brittle failure of the steel bar, named hydrogen embrittlement of reinforcement (He and Wang 1995, Liu *et al.* 2008). Hydrogen embrittlement of prestressed reinforcement is the main technical bottleneck in the application of electrochemical technology, and how to evaluate and control the hydrogen embrittlement of prestressed reinforcement is the premise and foundation for the further application of electrochemical technology. The compiling group has studied the chloride removal effect, corrosion inhibitor migration effect and hydrogen embrittlement risk of ordinary concrete structure and prestressed concrete structure under different current densities. The research results show that with a current density less than $1000mA/m^2$, it is not possible to ensure the removal of chloride while provided that the current density exceeds $3000mA/m^2$ (the prestressed structure $2000mA/m^2$), the plasticity of steel bar will decrease to less than 80% of the initial plasticity with a high risk of hydrogen embrittlement. The effective range of bi-directional electromigration current density in ordinary concrete structure is $1000\text{-}3000mA/m^2$ (the prestressed structure $1000\text{-}2000mA/m^2$) (Chen 2016, Wu X. 2018). At the same time, it is suggested that the current density implemented on the prestressed concrete structures should be lower than the hydrogen evolution current density during the bidirectional electromigration, which can be obtained by analyzing the cathodic polarization curve.

It was found that the surface strength of the specimens decreased linearly with the prolongation of the electrification time. When the current density is $3A/m^2$ and the electrification time reaches 15 days, the ratio of the corrosion inhibitor and chlorine has exceeded 1, presenting a great effect on corrosion resistance. When the current density is $3A/m^2$ and the electrification time is longer than 30 days, the loss of concrete surface strength and steel bar bonding force is close to 50% of the initial plasticity (Huang 2014). Therefore, the specification suggests that the normal duration of bi-directional electromigration is 15-30d.

3.2 Electrochemical Deposition

The purpose of electrochemical deposition is to repair cracks and fill pores in reinforced concrete structures, as shown in Figure 2. The area of reinforced concrete structure to be repaired is immersed with electrolyte, and the positive electrode of the power supply is connected with the metal plat while the negative electrode of the power supply is connected with the internal steel bar. Under the action of electric field, cations migrate to the cathode reinforcement, anions migrate to the external anode electrode, and electrochemical deposition occurs, so the deposited products fill the pores of reinforced concrete structure and finally repair the cracks.

Current density is an important parameter of electrochemical deposition (Otsuki and Ryu 2001): the smaller the current density, the longer the time needed for crack healing; and the deeper the filling depth of sediment, the better the healing effect. The compiling group studied the effects of different current densities on electrochemical deposition and took the effects of crack repair and desalination into consideration, setting $1.0A/m^2$ as the most appropriate current density (Chinese code 2012, Japanese code 2001). Referring to the existing specification parameters and experimental research results, the suggested range of current density is $500mA/m^2$ to $1000mA/m^2$ and the suggested electrolyte are $MgCl_2$, $ZnSO_4$ and Mg $(NO_3)_2$.

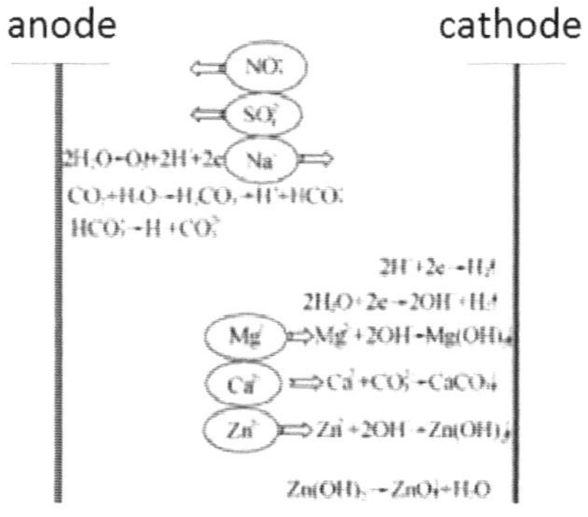

Figure 2. Electrochemical deposition interactions diagram.

3.3 Impressed Current Cathodic Protection Structure Strengthening (ICCP-SS)

Structural reinforcement and cathodic protection are often needed in engineering practice to address both the symptoms and root-cause, while since they belong to structural engineering and electrochemistry respectively, causing each one has its own material system, technical requirements and design methods. The compiling group studied and compounded these two technologies and put forward the normative provisions suitable for the popularization of ICCP-SS. The main innovations include: The mechanical strength of carbon fiber mesh cloth is required according to the current national standards. The shear strength of adhesive (polymer mortar, epoxy resin adhesive) is tested (Dai *et al.* 2016), and the ultimate shear capacity is proposed to prevent interlaminar failure (Xu 2015). The resistivity of the adhesive is required to ensure the normal operation of the cathodic protection. The bearing capacity amplification of the repaired members is limited to prevent the brittle fracture (Chen 2017, Chu 2000).

4 CASE STUDY

A bridge is Chinese first sea-crossing bridge built on the same plane by highway and railway. The width of the bridge is 28.2 meters, of which 7.2 meters are single track and others are lanes and sidewalks on each side. It was opened to traffic on April 27, 2001 and has been in operation for 15 years with annual traffic volume of approximate four million 620 thousand times and daily average of 12600 times. Site investigation shows that the pier exists rust induct cracking, the main reasons are: The pier is in the splashing zone, high concentration chloride and thin thickness of protective layer of certain part result in significant rust induct cracking. Since the cracks provide a smooth path for chloride ions to reach the surface of steel bars, the durability of concrete will deteriorate significantly, and the corrosion rate of steel bars will be multiple increased.

In order to ensure the life expectancy of the bridge, the durability safety hazards of concrete must be reduced to a minimum. The project adopts bi-directional electromigration technology to rehabilitate the pier. As shown in Figure 3, the trend map of residual chloride content before rehabilitation and after rehabilitation in different parts of the pier is drawn.

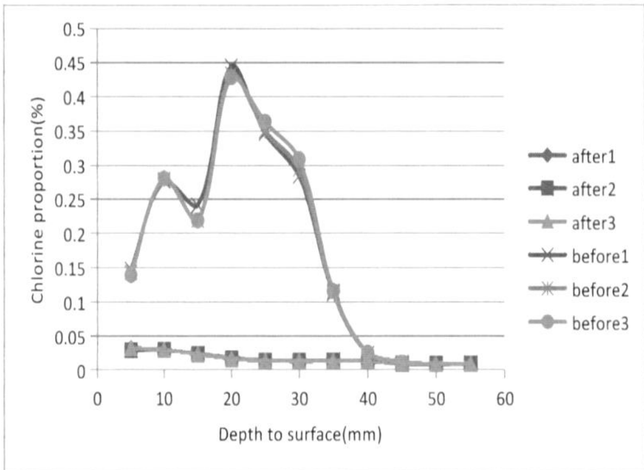

Figure 3. Distribution of chloride before and after rehabilitation.

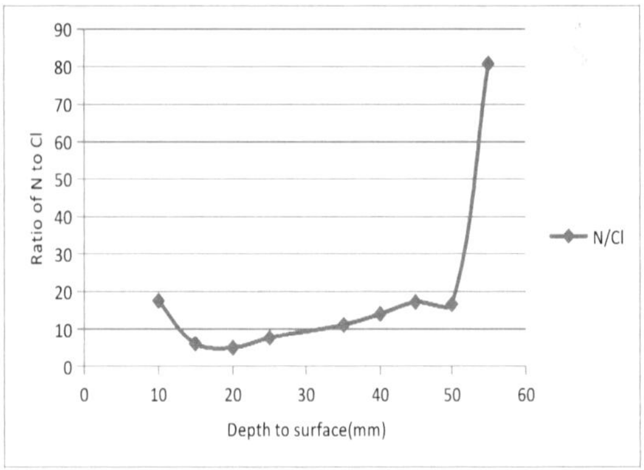

Figure 4. Evaluation of rehabilitation effect.

The chloride ion concentration before rehabilitation, from the surface of concrete to the depth of 35mm, was higher than the standard threshold, while the chloride ion concentration after rehabilitation has decreased obviously even to 80% of initial concentration and every part's chloride concentration is within the security range. The ratio of Nitrogen concentration to chlorine ion is an index to evaluate the effect of bi-directional electromigration and the higher ratio, the greater the effect. The N/Cl- distribution of different parts of the pier is drawn according to the collected data as shown in Figure 4. The ratio is much higher than 1.0 at the stress reinforcement of the pier, indicating that the overall effect of bi-directional electromigration is satisfactory, especially concrete surround the reinforcement.

5 CONCLUSIONS

It is necessary to establish a set of electrochemical technology for improving durability of concrete since the electrochemical methods for improving durability of concrete structures are developing and new technologies and methods are emerging constantly. According to the accumulated engineering experience and the new frontier technologies at home and abroad, the compiling group of Chinese *Technical Specification of Electrochemical Technology for*

Durability of Concrete Structures has perfected the traditional electrochemical technology, so as to popularize the new technologies and test their effectiveness and innovation by engineering practice.

The bidirectional electromigration technology has undergone the theoretical exploration, the selection of corrosion inhibitors, the optimization of electrification parameters, the evaluation of rehabilitation effect, and has already achieved the practical application in major projects. Bi-directional electromigration technology has been proved to have the advantages of both chloride extraction and corrosion inhibitor migration, and has a broad engineering application prospects.

Several mature and having engineering background technologies have been incorporated into this specification, while other new electrochemical technologies will be considered for adoption in the revision process of the specification after subsequent research and mature application.

References

Bi, J. B., Ding, Z., and Xu, J. Z., Electrochemical Technology for Protection and Renovation of Concrete Structures, *China Building Materials Science and Technology*, 03, 1-5, 2007.

Chen, J. Y., Effect of Electrochemical Repair Technology on Reinforced Bars in Simulated Concrete Solution, *Zhejiang University*, 2016.

Chen, X. L., Flexural and Shear Properties of FRP Composite Reinforced Concrete Beams, *Shenzhen University*, 2017.

Chinese code, Ministry of Communications and Communications of the People's Republic of China, *Technical Specification for Electrochemical Anticorrosion of Reinforced Concrete Structures for Harbour Projects*, July 19, 2012.

Chu, W. Y., *Fracture and Environmental Fracture*, Science Press, 2000.

Dai, X. Y., Jin, Z. Q., and Chen, Y. F., Research of Carbon Fiber Reinforced Cement-based Anode Electro Conductibility and Toughness, *Bulletin of the Chinese Ceramic Society*, 12, 4144-4148, 2016.

He, S. K. and Wang, G. S., Hydrogen Embrittlement of High-Strength Steel for Aviation, *Aviation Science and Technology*, 1, 9-12, 1995.

Huang, N., Repair Effect and Comprehensive Effect of Bi-Directional Electromigration on Chloride-Eroded Reinforced Concrete Structures, *Zhejiang University*, 2014.

Ihekwaba, N. M., Hope, B. B., and Hansson, C. M., Structural Shape Effect on Rehabilitation of Vertical Concrete Structures by ECE Technique, *Cement and Concrete Research*, 26(1),165, 1996.

Japanese code, Concrete Committee of Japan Civil Society. Design and Construction Guidelines for Electrochemical Anticorrosion Methods, *Series of Japanese Civil Society Concrete*, 107, 2001.

Jin, W. L., *Corroded Concrete Structures*, Beijing: Science Press, 2012.

Jin, W. L. and Zhao, Y. X., *Durability of Concrete Structures*, Beijing: Science Press. 2014.

Jin, W. L., Chen, J. Y., and Mao, J. H., The Effect of Electrochemical Rehabilitation on Service Performance of Reinforced Concrete Structures, *Engineering Mechanics*, 02, 1-10, 2016.

Liu, Y, Du, R. G., and Lin, C. J., Progress in Electrochemical Treatment Applied to Reinforced Concrete, *Corrosion Science and Protection Technology*, 20(2), 125-129, 2008.

Otsuki, N. and Ryu, J.S., Use of Electrodeposition for Repair of Concrete with Shrinkage Cracks, *Journal of Materials in Civil Engineering*, 13(2), 136-142, 2001.

Sohanghpurwala, A. A., Scannell, W. T., and Hartt, W. H., Repair and Rehabilitation of Bridge Components Containing Epoxy-Coated Reinforcement, *Transportation Research Board,* 2002.

Wu, H. T., Performance Improvement and Control Optimization of Bi-Directional Electro-migration Rehabilitation on Reinforced Concrete, *Zhejiang University*, 2018.

Wu, X. X., Experimental Study on the Effect of Electrochemical Rehabilitation on Hydrogen Embrittlement of Prestressed Tendons in Concrete Structures, *Zhejiang University*, 2018.

Xu, W. B., Finite Element Analysis of FRP Shear Strengthened RC Beam Based on Plastic Damage Theory of Concrete, *Shenzhen University*, 2015.

Zhang, S. Y., A Study of Corrosion Inhibitors for Bidirectional Electromigration Rehabilitation. *Zhejiang University*, 2012.

Streamlining Information Transfer between Construction and Structural Engineering
Edited by Shiau, J., Vimonsatit, V., Yazdani, S., and Singh, A.

ISBN: 978-0-9960437-7-9

SUSTAINABILITY IMPLICATIONS OF THE USE OF HYBRID POLYVINYL ALCOHOL FIBRE-REINFORCED FERROCEMENT AND FLY ASH

VANISSORN VIMONSATIT and WAHIDUL K. BISWAS

School of Civil and Mechanical Engineering, Curtin University, Perth, Australia

Extensive research has been conducted on the use of fly ash as a partial replacement of cement in order to promote the sustainable use of cement. Most of these research has focused on the investigation of the cementitious properties of the blended cement and the engineering properties of the end products, such as fly ash concrete. The sustainability benefit of using fly ash is often qualitatively perceived without any quantitative assessment. A recent study on the performance of hybrid polyvinyl alcohol fibre-reinforced ferrocement (HPVAF) shows that adding moderate amounts of fly ash in the mixes could maintain the ultimate flexure and tensile strength of HPVAF. The increased service life/durability and the use of FA up to a 25% replacement for cement in HPVAF not only conserve virgin resources for producing energy-intensive construction materials but also avoid associated environmental impacts due to the manufacturing of these materials. This certainly offers socio-economic benefits in terms of cost saving, enhance affordability and guaranteed material supply for the people both in current and future generations. Life cycle sustainability assessment (LCSA) was conducted to determine these triple bottom line benefits associated with the use of HPVAF and FA in building construction.

Keywords: Ferrocement composite, Fly ash, Environmental impact, Life cycle cost.

1 INTRODUCTION

The construction industry is the main contributor to the greenhouse gas emission worldwide. Concrete which is the main construction material has a high carbon cost due to the production of cementitious materials. Another reason for concern is the depletion of resources such as sand and coarse aggregates which are heavily used in producing concrete. A recent study shows that the world is facing a global sand crisis because of the skyrocketing demand and sand mining is harmful to humans and the environment (Torres *et al.* 2017). A more efficient use of resources is therefore needed which requires innovative ideas and technologies in making concrete less resource-intensive. One of the improvement strategies is to use industry wastes or by-products a partial replacement of cement and/or aggregates (Kulasuriya *et al.* 2014). The main challenge is to maintain the engineering properties while using a cement replacement and/or recycled aggregates. Another strategy is on construction innovations, for example, to improve construction efficiency such as modular, prestressed and precast constructions, and to develop energy-efficient alternatives of conventional structural sections such as steel/concrete composites or other innovative composites (Banthia *et al.* 2012). Most of these research focused on the engineering aspects of the development to achieve desirable mechanical properties and durability requirements. The use of HPVAF in concrete could potentially reduce the amount of energy and carbon-intensive construction materials such as coarse aggregates, and also it enables the use of industrial by-product as a replacement of cementitious materials. Thus, cost as well as environmental saving opportunities will be resulted to achieve a sustainable built environment. The sustainability aspects in a sense of the triple bottom line benefits (i.e. environmental, social and economic) are often perceived achieve but not quantitatively measured. This paper presents a life cycle assessment of a new reinforced concrete composite which was recently proposed to improve eco-efficiency performance of the construction sector by providing the same output (i.e. concrete with a required level of compressive strength) with less cost and environmental impacts. In this new composite, fly ash was used as a cement replacement and engineered ferrocement was used as a permanent form that would lead to the reduction in construction cost and time-saving.

2 HPVAF COMPOSITE

A further development of HPVAF led to a combined use of hybrid polyvinyl alcohol (PVA) as fibre reinforcements in ferrocement panels (Abushawashi and Vimonsatit 2015), which was named hybrid PVA ferrocement, or HPVAF. The proposed HPVAF composite is made of an HPVAF panel, which is engineered with two types of PVA fibres, short (8 mm) and long (30 mm), at the tension part of the composite. HPVAF panels can be precast off-site to improve the productibility and quality control of the HPVAF fabrication. Thus, the HPVAF panels can be used as a permanent formwork in the construction of the HPVAF composite. The upper part of the HPVAF composite can be cast on-site and can be engineered further to achieve its weight efficiency, thus achieving a further reduction in materials use. Figure 1 shows the HPVAF composite and construction sequence. The main components of a typical HPVAF composite section are HPVAF panel, conventional concrete (CC) and steel reinforcements, with lightweight infill as optional.

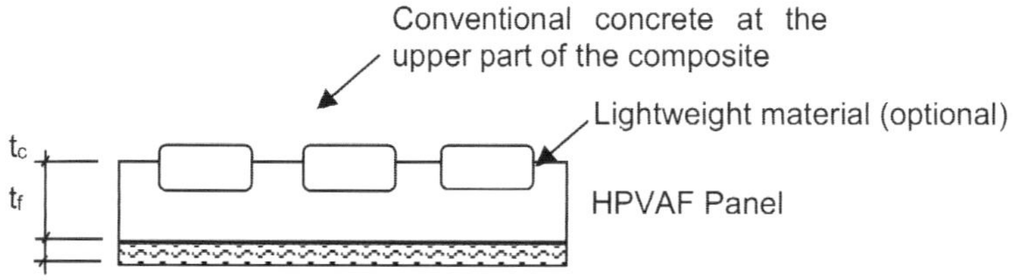

Figure 1. Section of HPVAF Composite.

The thickness, t_f, of the HPVAF panel is varied, from 20 mm to 40 mm. The thickness, t_c, of the upper concrete section can be designed to suit the design requirements for bending and shear using a conventional approach. A lightweight material can be used as an infill in the ineffective region of the upper concrete (Wahyuni *et al.* 2012) to achieve up to a 30% weight reduction when compared to a conventional reinforced concrete section with the same total depth ($t_f + t_c$). Using HPVAF panels as the permanent form of the composite section provides numerous construction benefits, especially for a large span construction. The permanent form eliminates the need for installing a temporary formwork and falsework system, thus reduces construction cost and time.

2.1 HPVAF Panel VS Conventional Concrete

Various mix proportions of HPVAF mortar have been tested for determining the compressive strength (Abushawashi 2015) using Ordinary Portland cement (OPC) and fly ash (FA) as the main binders of the mortar. The mortar mixture for the HPVAF panels used in the present study contained 400 kg/m^3 cement, with a ratio of 0.75:0.025:0.40:1:0.05 by weight of cement, FA, water, sand, and silica fume, respectively. The compressive strength of this mix was 46 MPa. The upper part of the HPVAF composite could be made of conventional concrete, with or without fibre reinforcement, depending on the design requirement. The cement:sand:aggregate ratio of the conventional concrete slabs was 1.0:2.1:3.2, and the water:cement ratio was 0.45. This mix was designed to have the compressive strength of 32 MPa at 28 days.

2.2 Mixtures of HPVAF Composites

Considering an HPVAF composite section which is made of a 40mm-thick HPVAF panel and a 115mm-thick of the upper concrete portion (Abushawashi and Vimonsatit 2014). Thus, the percentage of the HPVAF panel and of the CC are 26% and 74%, respectively. Table 1 shows the mixtures of the conventional concrete CC and the HPVAF and CC composites.

Table 1. Life Cycle Inventory of CC and HPVAF composites.

Energy and materials	**CC1**	**CC2**	**CC3**	**HPVAF +CC1**	**HPVAF +CC2**	**HPVAF +CC3**	**HPVAF +CC3+LW**
Mixture proportions							
OPC (kg/m^3)	400	350	300	374	337	300	264
Fly ash (kg/m^3)	0	50	100	26	63	100	100
Coarse agg. (kg/m^3)	1280	1280	1280	947	947	947	795
Sand (kg/m^3)	840	840	840	726	726	726	626
Water (kg/m^3)	180	180	180	175	175	175	147
Silica fume (kg/m^3)	0	0	0	3.9	3.9	3.9	3.9
PVA fibres (kg/m^3)	0	0	0	5	5	5	5
Wire mesh (kg/m^3)	0	0	0	13.7	13.7	13.7	13.7
Tensile steel bars (kg/m^3)	55	55	55	0	0	0	0
Lightweight (kg/m^3)	0	0	0	0	0	0	28
Total (kg/m^3)	2755	2755	2755	2271	2271	2271	1983
Transportation (tkm)	61	69	77	55	61	67	61
Manufacturing (kWh)	91	91	91	76	76	76	66

Note: LW = lightweight infill made up of 16% of the CC volume in the present case.

All materials were sourced locally in Western Australia. An estimate of the energy consumption at the concrete batching plant is 0.034kWh/kg of concrete (Nath *et al.* 2018). Two types of wire mesh were used, hexagonal woven wire mesh (chicken mesh) with a diameter of 1.4 mm and wire spacing of 40 mm, and flexible galvanized welded square mesh of 1.24-mm diameter and 25x25-mm holes. An estimate of the mesh production using a 2.2kW motor is 130 m^2 and 210 m^2 per machine hour for wire mesh and welded square mesh, respectively. The energy required for hot-dip galvanized steel is 2.5 GJ/tonne. The energy consumption during the construction (Biswas 2014) includes the off-site precast of HPVAF panels, transport operation from the casting site to the construction site, concrete ready mix, and concrete pouring. It should be noted that the consumption of other associated construction activities such as the installation of formwork and falsework are not included in this study. The HPVAF composites do not require the installation of formwork and falsework which means a significant reduction in the energy consumption compared to the CC construction. The environmental advantage of the avoidance of formwork and falsework savings that are resulted from the use of HPVAF in concrete is beyond the scope of this study and hence, will be considered in the future work.

A life cycle assessment (LCA) was carried out to estimate global warming, embodied energy consumption, solid waste, water consumption and land use of seven concrete mixtures following ISO14040-44 guideline (ISO 2006), which consists of four steps, namely: goal and scope, life cycle inventory (LCI), impact assessment and interpretation. We had considered the estimation of first two environmental indicators in this research as the construction industries contribute about 23% of Australia's annual GHG emissions and 20% of the total energy consumption (Nath *et al.* 2018). In addition, water is a scarce resource in WA and this construction sector produces a large amount of wastes (i.e. 19 million ton per year, (DoSEWPCQ 2011)). Finally, the land use changes due to quarrying and manufacturing of construction materials cause the loss of biodiversity (Biswas and Cooling 2013). The functional unit of this LCA study is 1 m^3 of concrete, which was used for conducting an inventory analysis to estimate energy and materials used during concrete life cycle (Table 1). Input and output data from the LCI in Table 1 were incorporated into SimaPro 8.4 LCA software (Pré Consultants 2017) and then relevant materials were linked to local Australian emission databases. In the absence of local databases, new libraries were created (i.e. wire mesh, lightweight). Australian emission databases were used for cement, fly ash, sand, electricity, transportation and water (Life Cycle Strategies Pty Ltd. 2015). The unit of tkm or tonne-kilometers was used to calculate the environmental impacts from the transportation of construction materials. Since the emission databases for coarse aggregate are unavailable, a new database was developed. The emissions associated with the combustion of diesel are available in the Australian databases. The foreign emission databases, such as Eco-invent and USLCI (US Life Cycle Inventory) that are available in the software, was used to calculate the impacts of silica fume and galvanized steel, respectively. It should be notable that PVA fibres only account for 0.2% of the total weight. Therefore, the exclusion of this chemical is not expected cause notable impact on the LCA results. Also its emission factor is not currently available. Finally, Australian indictor method has been used to calculate five aforementioned impacts (Life Cycle Strategies Pty Ltd. 2015).

3 RESULTS AND DISCUSSIONS

Table 2 presents five different environmental impacts resulting from mining, transportation and manufacturing of seven concrete classes. Global warming, land use, solid waste and embodied energy consumption are reduced from 573 kg CO_2 e-, 16.3 m^2, 1.3 kg and 3GJ to 488 kg CO_2 e-, 14.4 m^2, 1.4 kg and 2.8GJ due to the replacement of CC1 with CC3. The use of industrial by-products as a partial replacement for energy-intensive OPC has mainly reduced all these impacts. For example, Figure 2a confirms that OPC accounted for a significant portion of GHG emissions (i.e. 68%). These environmental impacts can further be reduced if HPVAF is added to these concrete classes. This is because of the use of HPVAF that increased the tensile strength of the concrete and thus completely eliminates the use of carbon-intensive tensile steel. Whilst overall GHG emissions are reduced due to use of HPVAF, Figures 2a and b show that OPC still remains as the hotspot.

CC3+HPVAF+LW was found to offer the higher environmental savings than other concrete classes (Figure 3) due to the reduction in 34% of OPC, 28% of coarse aggregates, 18% of water and complete replacement of tensile steel. Land use (i.e. 37- 48%) and water use (i.e. 37 – 46%) impacts decreased significantly when HPVAF is added to the conventional concrete. Firstly, the use of HPVAF composite completely reduces the use of steel which means that the land required for mining and processing of steel is reduced. Secondly, water consumption is reduced by 18% due to a reduction in the mixing of concrete ingredients (i.e. coarse aggregates, sand, and cementitious materials). Solid wastes are reduced quite considerably due to the avoidance of land clearing and inputs reduction for ingredient production. This means that inter-generational social equity aspect of sustainability can be enhanced by conserving land and non-renewable material resources for the future generations. The other indirect environmental impacts that could be avoided in due using HPVAF composite is the loss of biodiversity associated with quarrying and mining activities for concrete ingredients production. Reduction in expensive materials such as tensile steel and OPC is expected to reduce to the overall cost of HPVAF based concrete classes.

Table 2. Environmental impacts of seven concrete classes.

Impact category	CC1	CC2	CC3	HPVAF +CC1	HPVAF + CC2	HPVAF +CC3	CC3 +HPVAF+LW
Global Warming (kg CO_2e-)	573.6	525.2	478.4	488.3	453.8	419.2	374.0
Land use (m^3)	16.3	15.3	14.4	11.0	10.3	9.6	8.5
Water Use (m^3)	3.2	3.2	3.2	2.1	2.0	2.0	1.7
Solid waste (kg)	1.4	1.3	1.3	1.2	1.1	1.1	1.0
Cumulative energy demand (MJ)	3007.0	2880.4	2779.3	2512.2	2437.5	2362.7	2134.7

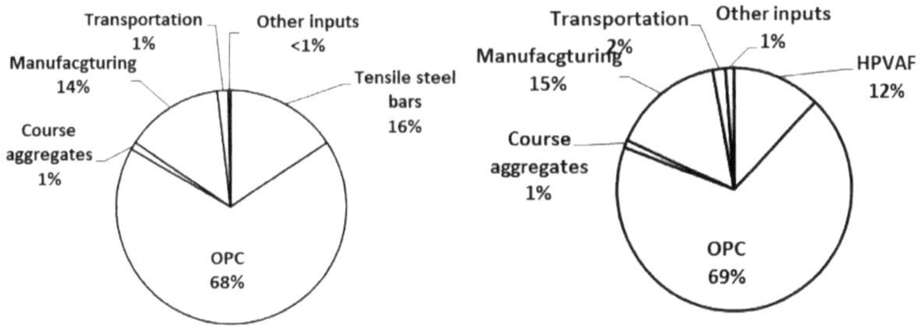

a) b)

Figure 2. Breakdown of GHG emissions in terms of inputs a) CC1, b) HPVAF+CC3+LW.

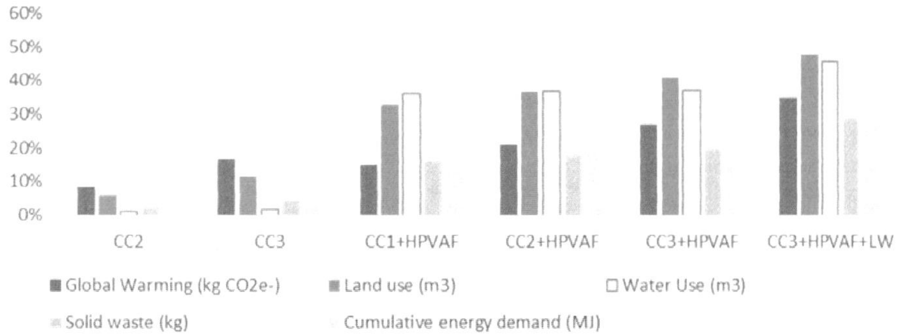

Figure 3. Environmental impacts saving associated with the use of HPVAF and industrial by-products.

4. CONCLUSIONS

There are environmental, social and economic implications associated with the use of HPVAF in concrete. The amount of conventional concrete that consists of energy and carbon-intensive construction materials can substantially be reduced due to the use of HPVAF in concrete. This ferrocement also maximizes the use of industrial by-products such as fly ash and silica fume without compromising the structural performance of concrete. The incorporation of these by-products with HPVAF could further increase sustainability benefits. A detailed economic analysis and the estimation of the service life of concrete classes of this study will be carried out to further highlight the sustainability benefits of these HPVAF concrete classes.

References

Abushawashi, N. F., *Development of Hybrid Fibre Ferrocement Panels as Permanent Formwork of Concrete Slabs with Lightweight Blocks Infill,* PhD Thesis, Curtin University, Australia, 2015.

Abushawashi, N. F., and Vimonsatit, V., Flexural Behaviour of Hybrid Polyvinyl Alcohol Fibre-Reinforced Ferrocement Containing Different Fibre Types and Volume Fractions, *Australian Journal of Structural Engineering*, 16(4), 252-261, 2015.

Abushawashi, N., and Vimonsatit, V., *Use of Ferrocement Panel as Reinforced Concrete Slabs with Lightweight Blocks Infill*, Second Australasia and Southeast Asian Conference, Bangkok, Thailand, 245-250, Nov 3-5, 2014.

Banthia, Nemkumar, Bindiganavile, V., Jones, J., and Novak, J., Fiber-Reinforced Concrete in Precast Concrete Applications: Research Leads to Innovative Products, *PCI Journal,* 57(3), 33–46, 2012.

Biswas, W. K., Carbon Footprint and Embodied Energy Consumption Assessment of Building Construction Works in Western Australia, *International Journal of Sustainable Built Environment,* 3, 179–186, 2014.

Biswas, W. K., and Cooling D., Sustainability Assessment of Red Sand as a Substitute for Virgin Sand and Crushed Limestone, *Journal of Industrial Ecology*, 17(5), 756-762, 2013.

DoSEWPCQ, Management of Construction and Demolition Waste in Australia, *DoSEWPCQ (Department of Sustainability, Environment, Water, Population and Communities, Queensland)*, Queensland, 2011.

ISO14040, ISO 14040:2006 - *Environmental Management -- Life Cycle Assessment -- Principles and Framework,* International Organization for Standardization, Geneva, 2006.

Kulasuriya, C., Vimonsatit, V., Dias, WPS, and De Silva, P., Design and development of Alkali Pozzolan Cement (APC), *Construction and Building Materials,* 68, 426-433, 2014.

Life Cycle Strategies Pty Ltd., Australasian Unit Process LCI Library and Methods. Version 2015_02_06. 2015. Retrieved from http://www.lifecycles.com.au/#!australasian-database/cbm5 on June 30, 2015.

Nath, P., Prabir, K. S., and Biswas, W. K., Effect of Fly Ash on The Service Life, Carbon Footprint and Embodiedenergy of High Strength Concrete in The Marine Environment, *Energy and Buildings,* 158, 1694–1702, 2018.

Pré Consultants, SimaPro database. Pré Consultants, The Netherlands, 2017.

Torres, A., Brandt, J., Lear, K., Liu, J., A Looming Tragedy of The Sand Commons, 357(6355), 970-971, DOI: 10.1126/science.aao0503, 2017. Also featured in http://theconversation.com/the-world-is-facing-a-global-sand-crisis-83557.

Wahyuni, A. S., Vimonsatit, V., and Nikraz, H., Shear Behaviour of Lightweight Sandwich Reinforced Concrete Slabs, *Advances in Structural Engineering*, 15(10), 1705-1715, 2012.

Streamlining Information Transfer between Construction and Structural Engineering
Edited by Shiau, J., Vimonsatit, V., Yazdani, S., and Singh, A.

ISBN: 978-0-9960437-7-9

A STUDY ON INTENSITY DISTRIBUTION OF CONCRETE IN STRUCTURE WITH DIFFERENT BUILDING AGES

RYOTA UMEBAYASHI[1], HIDEHIRO KOYAMADA[2], KOJI TAKASU[2], and HIROKI SUYAMA[2]

[1]*Graduate School of Engineering, The University of Kitakyushu, Kitakyushu, Japan*
[2]*Faculty of Engineering, The University of Kitakyushu, Kitakyushu, Japan*

Recently, seismic diagnosis and earthquake resistant reinforcement works of buildings built before the new earthquake resistant standards are conducted in various places. However, there is no uniformity in sampling methods, and it seems that there are occasional extreme variations in the determination of representative intensity of the building. In addition, the physical properties of structure concrete which factor of the above-mentioned variations are considered to reflect the horizontal and vertical distribution at the time of concrete placing. Therefore, this study, the core of the existing buildings to be disassembled was collected, and the distribution of physical properties of structure concrete was examined. All target buildings were reinforced concrete constructions, and cores were collected from pillars and waist walls. In addition, there are parts that were built at the time without the ready-mixed concrete plant (from 1960) and pump truck (from 1964). Experimental results of horizontal and vertical distributions of compressive intensity at the pillars and waist walls revealed that there was no greatest decrease in strength and there was not much difference between the maximum strength of the pillar and 300 mm from the vicinity of the pillar and about 300 mm from F. L. Therefore, from this study, it seems that the position of 300 mm from the pillar and F. L. is suitable for the measurement position which develops the representative intensity of the building.

Keywords: Strength, Carbonation depth, Demolition.

1 INTRODUCTION

In this study, the distribution of compressive strength and carbonation depth of concrete was collected from multiple demolition buildings, and the analysis was done considering the influence of the time background and the construction conditions.

Recently, seismic diagnosis and earthquake resistant reinforcement works of buildings built before the new earthquake resistant standards are conducted in various places. As a result, there are many cases of dismantling in various buildings. However, there is no uniformity in sampling methods, and it seems that there are occasional extreme variations in the determination of representative intensity of the building.

This is considered to be a cause of variation by the part of structure concrete. Also, the physical properties of structure concrete which factor of the above-mentioned variations are considered to reflect the horizontal and vertical distribution at the time of concrete placing. The vertical distribution has been studied in the past and the reason was clarified, the report on the horizontal distribution of the wall, which is most suitable to collect the core from a real building is limited.

Therefore, this study, the core of the existing building which was dismantled target was collected, and distribution of the physical properties of the concrete structure as a main issue, that was examined with the change of the building age and the building technology.

Moreover, it is difficult to collect the core usually to target the demolition building. Therefore, it is a feature of this research that can be taken from a pillar.

2 RESEARCH OVERVIEW

2.1 Buildings Surveyed

The specifications of the target buildings that are investigated in the present structural assessment are listed in Table 1. The structural components of all the buildings are made from reinforced concrete. The five building types are labeled A to E. The most common building-type A comprised three subgroups, which included four-, six-, and six-story buildings. The four-story buildings were built in 1952, 1953, 1954, 1955, 1958, 1960, whereas the first type of six-story buildings were built in 1966, 1967, and 1968; the remaining buildings were built in 1970 and 1979. Overall, eleven buildings of different age groups were studied. The buildings that were constructed at different times were built using different construction equipment and materials. Some of the buildings were built before the raw concrete factories were introduced in 1960, whereas some were built without pump trucks, which were introduced in 1965.

Table 1. The present structural assessment of the target buildings.

building \ details	building age	use	story
A-1～A-6	1952,1953,1954,1955,1958,1960	university	four-story
A-7～A-9	1966,1967,1968		six-story
A-10～A-11	1970,1979		six-story
B	1943	primary school	two-story
C	1933	primary school	two-story
D	1947	middle school	two-story
E	1971	cultural center	six-story

2.2 Survey Components

This study discusses five main aspects that are related to the structural analysis of these buildings, which are as follows:

1) Dimensions of the test specimen
2) A method for measuring the carbonation depth in the sample core
3) Effects of the distribution of carbonation depth and finishing
4) Horizontal and vertical distributions of the compressive strength and density in the lower half of the walls
5) A vertical distribution of the compressive strength in column and its relation to the wall strength

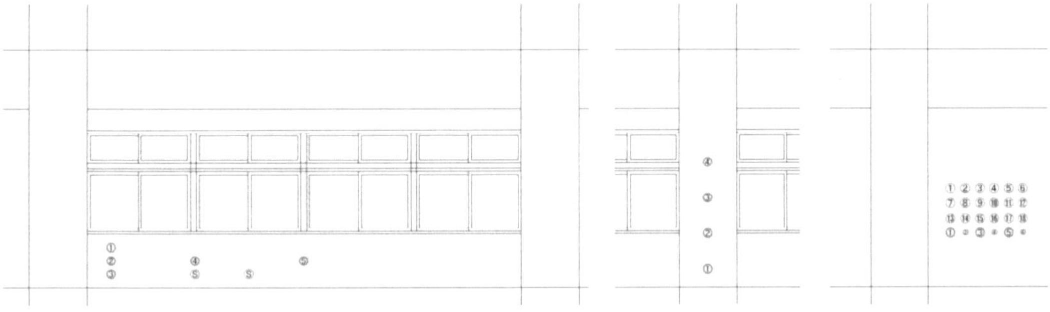

Figure 1. Core sampling position (from left, wainscot, pillar, no window wall).

2.3 Core Collection Position

Core positions are shown in Figure 1. Cores were collected from a spandrel, a column, and a windowless wall in each building. At the spandrel, points ①–③ were approximately 300 mm from the column. The heights of these positions from the floor line (FL) were ① 450, ② 300, and ③ 150 mm. ⑤ is located at the center of the span, whereas ④ is located midway between ① –③ and ⑤. Both ④ and ⑤ are horizontal with ②. Ⓢ is close to ④ and ⑤. On the column, ① is 300 mm from the FL, ② is 900 mm from the FL, ③ is 1500 mm from the FL, and ④ is 2100 mm from the FL. The measurements were recorded for ④ only if the wall was tall enough. At the windowless wall, ①–⑱ exhibit the locations of the cores that are used for storage, whereas ①–⑥ exhibit the locations of the cores used to measure the effect of the dimensions. ①, ⑦, ⑬, and ① are approximately 300 mm from the column. The height from the FL for all of these cores was approximately 1000–1500 mm. Although the standard core diameter was ϕ 75 mm, these cores were collected at ϕ 50mm because of the arrangement of the reinforcing bars in the concrete. More than 300 cores were collected in total, and approximately 150 samples were tested.

3 EXPERIMENTAL RESULTS AND CONSIDERATIONS

3.1 Dimensions of the Test Specimen

To examine the influence of the core diameter on the test results, I examined the correlation between the average measurements that were recorded using three cores of φ 50mm and φ 75mm that were obtained from five locations in the same windowless wall. The results are shown in Figure 2. The average values of the core test body of φ 50mm and φ 75mm were observed to differ if they were obtained from walls that were built at different times. However, this test confirmed that the core diameter did not exhibit much effect on the relevant measurements.

3.2 A Method for Measuring the Carbonation Depth in the Sample Core

To ensure adherence with the JIS3 standards, structural analysis must proceed using samples from five or more locations in the structure. However, it is difficult to evenly distribute the choice of locations for these five points across the different structural components of a building.

Thus, cores were taken from more than five points in this study, and the carbonation depth measurements were performed using a selection of measurement points. Further, the carbonation depth was measured from all the cores that were taken from the spandrel ②. The results are shown in Table 2. From the table, some differences can be observed between the variations and the number of measurements; however, the difference between the mean values is generally less than 0.30 mm. Based on this result, the measurement of the carbonation depth can be conducted using the average of the maximum value and based on eight other measurements.

Table 2. A study of the number of carbonation measurements.

Number of measurements	8-point (in) Average (mm)	Stdev (mm)	16-point (in) Average (mm)	Stdev (mm)	Difference (mm)	8-point (out) Average (mm)	Stdev (mm)	16-point (out) Average (mm)	Stdev (mm)	Difference (mm)
A-1 South	21.64	5.22	21.66	4.88	0.02	13.53	8.94	13.65	8.44	0.12
A-1 North	32.91	6.76	32.63	6.38	0.28	46.86	4.32	46.88	3.85	0.02
A-2 South	23.85	5.30	23.46	5.13	0.40			-		
A-2 North	18.31	8.62	17.61	8.96	0.70	0.53	1.15	0.42	0.90	0.11
A-3 South	0.97	0.70	0.54	0.69	0.43	0.97	0.22	1.21	0.24	0.24
A-3 North	1.36	0.91	1.26	0.77	0.10	1.76	1.67	1.91	1.44	0.15
A-4 South	37.75	24.77	36.87	23.35	0.88	1.65	0.89	1.66	0.79	0.00
A-4 North	46.27	3.58	46.32	3.31	0.05	1.27	0.68	1.09	0.69	0.18
A-5 South	3.40	0.95	3.45	0.96	0.05	-				
A-5 North	2.31	0.41	2.28	0.42	0.04	1.26	1.42	1.17	1.35	0.09
A-6 South	16.25	2.20	16.12	2.08	0.13	3.96	1.75	3.72	1.78	0.24
A-7 South	21.39	2.23	20.78	2.03	0.62	0.48	0.42	0.53	0.40	0.05
A-8 South	1.62	0.27	1.72	0.34	0.10	2.49	2.05	1.96	1.63	0.53
A-8 North	2.62	2.11	2.39	1.98	0.23	1.05	0.68	1.08	0.66	0.03
A-10 South	14.20	5.05	14.25	4.59	0.05			-		
A-11 South	13.62	2.40	13.42	2.07	0.20			-		
B	9.98	1.42	9.84	1.40	0.14	1.50	0.49	1.41	0.41	0.09
C	1.08	0.62	0.83	0.54	0.25	0.54	0.36	0.66	0.38	0.11
D	All sections are neutering									
E	43.75	6.20	43.85	5.98	0.10	2.69	0.65	2.12	0.60	0.57

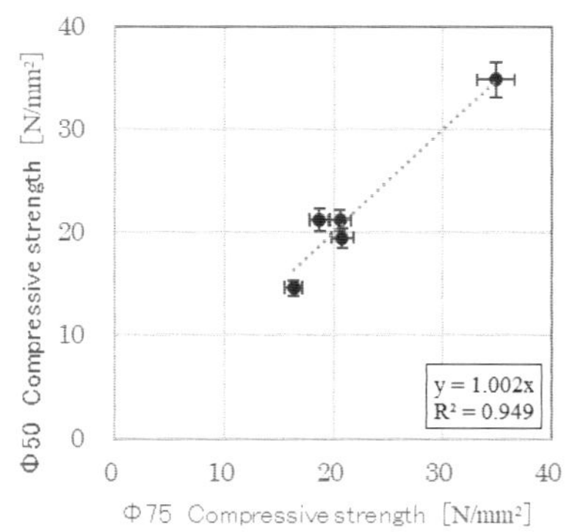

Figure 2. Effect of diameter difference.

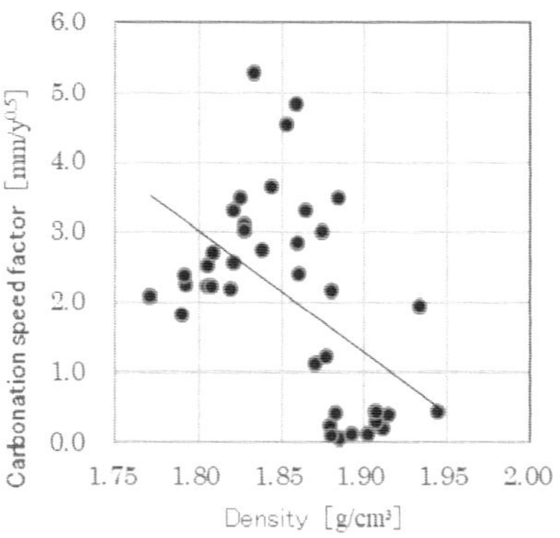

Figure 3. Relationship between finish density and carbonation speed factor.

3.3 Effects of the Distribution of Carbonation Depth and Finishing

As reported previously (Yanagibashi 2010), the carbonation depth of concrete is affected by four main factors. The carbonation speed is affected by ① the environmental conditions, ② the finishing material that is used on the concrete surface, ③ the quality of concrete, and ④ aging. This study focuses on the relation between the finishing material and the carbonation speed. The results are shown in Figure 3. The two values are not strongly correlated; however, a slight inverse correlation is apparent. If the finishing material is dense, carbonation proceeds somewhat slowly.

3.4 Horizontal and Vertical Distributions of the Compressive Strength and Density in the Lower Half of the Walls

Figures 4 and 5 depict the compressive strength distribution within the spandrel in the horizontal (②, ④, ⑤ in Figure 1) and vertical (①, ②, ③ in Figure 1) directions. Further, the strength is observed to vary in the horizontal direction in all the specimens. The strength is considered to be the highest at sampling position ② and begins to decrease as it approaches ⑤. Some variations were observed in the vertical direction with age; however, the compressive strength was generally observed to be the highest at ②, ①, and ③, whereas the strength at ③ tended to be lower. From this result, position ② is the optimal location from which a core sample that will most accurately represent the wall strength can be obtained.

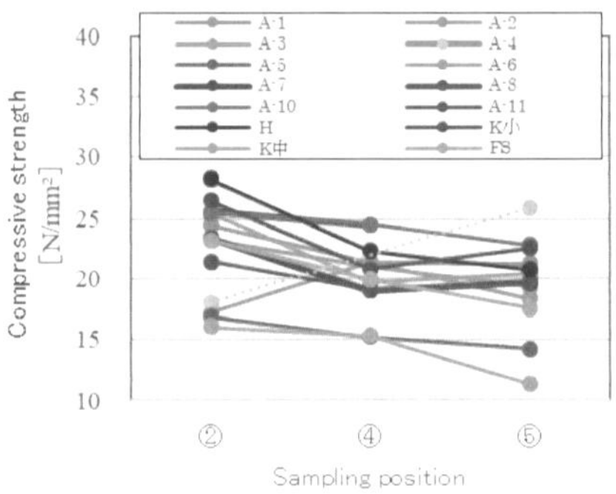

Figure 4. Horizontal distribution of compressive compressive strength (in spandrel).

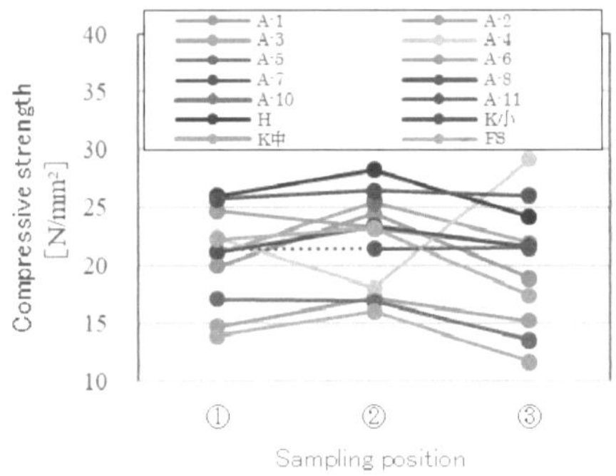

Figure 5. Vertical distribution of strength (in spandrel).

3.5 A Vertical Distribution of the Compressive Strength in Column and its Relation to the Wall Strength

The strength distribution in the vertical direction of the column was concentrated at the bottom of the pillars, which was similar to that observed in previous reports (JIS A 1107 2012). The concrete strength gradually decreased toward the top of the column, and the strength tended to increase at the boundaries of some components. This increased strength seems to be a result of compaction as the concrete was poured. Here, we investigated the core strength of the spandrel section ② at the same height as that of ①, which exhibited the highest strength in the column. It is shown in Figure 6. From this, it can be observed that the strength of the column is 1.6 times higher than the strength of the spandrel. However, the difference between the strengths of the spandrel and column is not statistically significant, and the two values exhibit a weak positive correlation. This trend indicates that the measurements of the difference in strength between the FL and the column are suitable for estimating the overall strength of the structure.

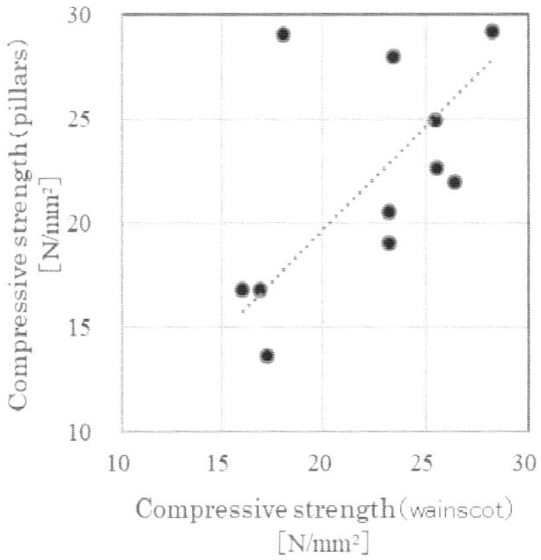

Figure 6. Relationship between compressive strength in column ① and spandrel ②.

4 CONCLUSIONS

This study investigated the distribution of physical properties by collecting cores from reinforced concrete buildings built in various ages. Summarizes the results obtained are shown below.

1) The compressive strength and density were confirmed by the horizontal and vertical distributions from the core sampling position.
2) The collected concrete core of approximately 300 mm from the F.L. of the wainscot and near the pillar (about 300 mm in this study) is seems to be suitable for the measurement position, which expresses the strength that represents the building.

Acknowledgements

The authors acknowledge the assistance in this work provided by Mr. Eiji Mikura and Ms. Saori Itanami.

References

JIS A 1107, *Method of Sampling and Testing for Compressive Strength of Drilled Cores of Concrete,* JIS Handbook, 2012.

Yanagibashi K., *Materials and Techniques to Support the Ultra-Life "500 Years Concrete",* Japan Testing Center for Construction Materials, 2010.

Streamlining Information Transfer between Construction and Structural Engineering
Edited by Shiau, J., Vimonsatit, V., Yazdani, S., and Singh, A.

ISBN: 978-0-9960437-7-9

A STUDY ON THE FLUIDITY OF MORTAR WITH FINE AGGREGATE ADJUSTED GRAIN-SIZE AND POWDERS

SHOHEI KOIZUMI[1], KOJI TAKASU[2], HIDEHIRO KOYAMADA[2], and HIROKI SUYAMA[2]

[1] *Graduate School of Engineering, The University of Kitakyusyu, Kitakyusyu, Japan*
[2] *Faculty of Engineering, The University of Kitakyusyu, Kitakyusyu, Japan*

It has been clarified that the particle shape and particle size distribution of fine aggregate have influenced the fresh properties of concrete and mortar in the past research. But it has not been sufficiently studied how each particle size and particle size distribution of fine aggregate affects the fresh properties of mortar and concrete. In this study, we adjusted the particle size of fine aggregate, and evaluated fresh properties of mortar by indicators. We investigated the fluidity of mortar, which changed depending on the composition of particle size distribution. In addition, the behavior of fresh properties of mortar contained powder substituted as part of cement and fine aggregate adjusted particle size distribution was compared with that when no powder was mixed. Its behavior was discussed. It became clear that, the factor influencing the fluidity varied depending on the composition of particle size distribution. The influence of fine particles on fluidity of mortar varied depending on the content of fine particle and the particles size distribution. When combined with the large particles, the fluidity was improved as compared with the combination with the intermediate particles. In addition, the same tendency as in powder-free mixing was confirmed at powder mixing and increase rate of the relative flow area ratio when fly ash was contained became larger than when the limestone fine powder was contained.

Keywords: Relative flow area ratio, Mini-slump, Particle size distribution, Solid content rate, Water absorption rate, Fine particle, Fly ash, Limestone fine powder.

1 INTRODUCTION

Research on fluidity of mortar and concrete has been done so far, and it has been clarified that the aggregate particle size affects the fresh state of mortar and concrete. Fine particle content of fine aggregate as one of the factors that affects it, can be mentioned. Also, it has been clarified that the particle shape and particle size structure of fine aggregate have influenced the fresh properties of concrete and mortar in the past research. But it has not been sufficiently studied how each particle size and Particle size distribution of fine aggregate affects the fresh properties of mortar and concrete. In addition, in recent years, as one of the movements contributing to low carbon society, it is attracting attention to replace a part of cement with a by-product type powder having a small amount of CO_2 emission.

In this research, we adjust the particle size of fine aggregate, and evaluate fresh properties of mortar using indicators such as water absorption, solid content and fineness modulus, and we investigated the fluidity of mortar, which changes depending on the particle size composition. In addition, the behavior of mortar fresh when combining powder substituted as part of cement and fine aggregate with controlled particle size was compared with that when no powder was mixed, and its behavior was discussed.

2 SUMMARY OF THE EXPERIMENT

Table 1 shows the materials used in the experiment. We investigated the properties of mortar made with aggregates that had five particle size distributions. We used three kinds of oven-dried fine aggregates (two types of crushed sand [CSa, CSb], and one type of sea sand [SS]) in this experiment. The crushed sand met the criteria specified in Japanese Industrial Standards (JIS) A 5005 "Crushed stone and manufactured sand for concrete. We also used two kinds of fly ash and fine limestone powder, and the fly ash was equivalent to JIS II class. We kept the materials for more than 24 h at a constant room temperature of 20 ± 1 °C. Next, table 2 shows mix proportions. As explained below, "N-CSa", "N-CSb", and "N-SS" without admixture were the basic mix proportions, and the amount of admixture was 30% replacement of the cement volume. The volume ratio of water to powder was constant in all mix proportions. The mix proportions were indicated by "FA" for fly ash and "LP" for fine limestone powder followed by a number showing the substitution rate. The type of sand was indicated by "CSa", "CSb", or"SS". "P" in the mix proportions list indicated the powder, and "Wv/Pv" expressed the water- to-powder volume ratio. Next, table 3 shows physical properties and measured value. There were five particle size distribution patterns for the fine aggregates: no particle size adjustment, 2.5–1.2 mm, 1.2-0.6mm, less than 1.2mm, 2.5-1.2mm and less than 0.6 mm. The content ratio of 1.2–0.6 mm to less than 0.6 mm particles in pattern III was defined as the mass ratio calculated from the particle size distribution. In addition, the content ratio of 2.5–1.2 mm to less than 0.6 mm particles in pattern V was defined as a mass ratio of 1:1 because the amount of fine particles of sea sand was increased in pattern V if the particle size distribution ratio was used. The content of 2.5–1.2 mm particles in sea sand was low, and it was excluded from the mix proportions because it was difficult to obtain the amount necessary for casting. Mortar mixing was carried out with a mortar mixer in a mixing room at room temperature of 20 ± 1 °C. The measurements were mortar flow, mini-slump test, air content, solid content rate of fine aggregate, and water absorption rate.

Table 1. Materials used in the experiment.

Item	Type	Physical properties	Symbol
Cement	Ordinary portland cement	Density3.16g/cm^3	C
Water	Tap water	-	W
Fine aggregate	Crushed sand A	Water absorption rate0.75% Density2.71g/cm^3 FM3.1 Solid content rate65.5%	CSa
	Crushed sand B	Water absorption rate1.96% Density2.74g/cm^3 FM2.7 Solid content rate64.7%	CSb
	Sea sand	Water absorption rate0.76% Density2.59g/cm^3 FM2.4 Solid content rate62.3%	SS
Admixture	Fly ash (JIS class Ⅱ)	Density2.43g/cm^3 Loss on ignition2.00% Specific surface area4014cm^2/g	FA
	Limestone fine powder	Density2.77g/cm^3 Specific surface area5221cm^2/g	LP

Table 2. Mix proportions.

Symbol	W/C	Wv/Pv	Unit mass[kg/m^3]				
			W	C	FA	LP	S
N-CSa	50	1.58	280	560	0	0	1256
N-CSb	50	1.58	280	560	0	0	1282
N-SS	50	1.58	280	560	0	0	1223
FA30-CSa	71	1.58	280	392	121	0	1256
FA30-CSb	71	1.58	280	392	121	0	1282
FA30-SS	71	1.58	280	392	121	0	1223
LP30-CSa	71	1.58	280	392	0	147	1256
LP30-CSb	71	1.58	280	392	0	147	1282
LP30-SS	71	1.58	280	392	0	147	1223

A 5201 "Physical testing methods for cement", and the mini-slump test was measured in accordance with JIS A 1171 "Test methods for polymer-modified mortar". Air content was measured with a mortar air meter, and the solid content rate of fine aggregate was measured in accordance with JIS A 1104 "Methods of test for bulk density of aggregates and solid content in aggregates". In the water absorption rate measurements for each particle size composition, the smaller the particle size, the more difficult it was to identify the saturated surface-dry state of the aggregate, and we could not measure it by the JIS water absorption test. Therefore, oven-dried specimens (90 g) for each particle size composition were placed in a Le Chatelier flask containing water and allowed to stand at 20 ± 1 °C for 2 h. The water absorption rate was calculated by the value obtained by dividing the volume decrease by 90 g. The relative flow area ratio (Γ) is used as a self-filling index, and the fluidity was high so the value was big. The formula for calculating the relative flow area ratio is shown below.

$$\Gamma = \frac{(d_1 \cdot d_2 - d_0^2)}{d_0^2} \tag{1}$$

Table 3. Physical properties and measured value.

Symbol			Unit mass[kg/m^3] 2.5-1.2 [mm]	1.2-0.6 [mm]	0.6-0.0 [mm]	Frow value [mm]	Relative flow area ratio(Γ)	Mini-slump[cm]	Air content [%]	FM [%]	Solid content rate[%]	The ratio of paste volume to fine aggregate void volume	Water absorption rate[%]
N	CSa	Ⅰ	527	410	319	178.0	2.17	5.4	4.5	3.1	65.5	315.3	1.48
		Ⅱ	1256	0	0	149.0	1.22	5.1	4.0	3.4	59.7	255.5	1.04
		Ⅲ	0	707	550	151.5	1.30	2.4	4.0	1.9	62.0	279.9	1.04
		Ⅳ	0	1256	0	150.0	1.25	3.0	5.0	2.8	57.6	245.0	1.37
		Ⅴ	628	0	628	168.0	1.82	4.2	1.3	2.3	66.4	344.4	1.26
	CSb	Ⅰ	382	372	528	167.0	1.80	4.5	4.0	2.4	64.7	288.8	0.38
		Ⅱ	1282	0	0	142.0	1.00	3.5	4.0	3.1	53.2	225.6	0.26
		Ⅲ	0	530	752	135.0	0.80	1.8	3.8	1.7	62.5	268.0	0.15
		Ⅳ	0	1282	0	133.0	0.80	1.5	6.5	2.5	52.6	222.8	0.26
		Ⅴ	641	0	641	161.5	1.60	4.8	1.9	1.8	64.7	342.8	0.49
	SS	Ⅰ	54	450	719	167.0	1.79	1.4	7.0	2.7	62.3	302.1	0.67
		Ⅱ	1223	0	0					3.3	52.4	222.4	0.05
		Ⅲ	0	471	752	154.0	1.37	1.9	5.2	1.7	61.2	288.6	0.61
		Ⅳ	0	1223	0	149.0	1.22	1.1	9.0	2.6	56.3	239.6	0.56
		Ⅴ	612	0	612	203.5	3.14	8.3	3.7	2.0	63.0	310.3	0.11

3 RESULT AND CONSIDERATION

3.1 Solid Content Rate of Fine Aggregate and Fluidity

Table 4 shows the ratio of the fine particle of 75 μm or less in the fine aggregate input amount of Pattern I, III, V, Figs.1 shows relationship between solid content rate of fine aggregate and the relative flow area ratio, mini-slump. Figure 1 shows that the relative flow area increased as the solid content rate of the fine aggregate increased. It has been reported that the flow increases as the ratio of the paste volume to the void volume of aggregate increases (Hirata *et al.* 1995). In the present work, because the paste volume was constant in all mix proportions, we assumed that the solid content rate of the fine aggregate was equal to the ratio of the paste volume to the void volume of the aggregate. However, for each particle size configuration, although the solid volume rate of fine aggregate was large, there were some points at which the relative flow area ratio did not increase. Therefore, we assumed that the factors affecting relative flow area ratio changed. We compared the relative flow area ratios for each particle size composition and examined the factors.

Compared with pattern I, for patterns II and IV, the solid content rate of the fine aggregate and the relative flow area ratio decreased for all fine aggregates. This is because the fine aggregates in patterns II and IV are 2.5–1.2 mm particles only or 1.2–0.6 mm particles only, respectively, and the intermeshing between the particles is poor. Therefore, the dispersion distance of fine aggregates in the mortar is small and the fluidity is poor. Consequently, for 2.5–1.2 mm particles or 1.2–0.6 mm particles, the most important factor affecting fluidity is the solid content rate of the fine aggregate.

Compared with the equivalent solid content rate of pattern I, the relative flow area ratio of pattern III decreased greatly for all fine aggregate. In addition, there was insufficient water in the mortar in pattern III immediately after mixing. These results suggested that as the amount of particles less than 1.2 mm in size increased, the water absorption rate increased and the fluidity became poor. However, the water absorption rate of pattern III was smaller than that of pattern I (Table 3). Therefore, although a factor other than the water absorption rate of fine aggregate affected the fluidity strongly, we could not identify this factor and intend to investigate it in future work.

The relative flow area ratio of pattern V was the largest of those with particle size adjustment, and the viscosity of the mortar immediately after mixing was the lowest. We assumed that this was because fine particles 90 μm or less in size (Yamaguchi *et al.* 1994), which function as a powder, were added to the paste, and the amount of paste increased; thus, the dispersion distance of fine aggregate decreased and the fluidity became poor. However, although the fine particle content in pattern III was comparable to those in patterns I and V, the relative flow area ratio of pattern III was smaller than those in patterns I and V, and there was a large difference in relative flow area ratio (Table 4). Therefore, the fluidities of pattern III composed of medium-sized particles (1.2–0.6 mm) and fine particles and pattern IV composed of only medium-sized particles were equivalent. On the other hand, the fluidity of pattern V, composed of fine particles, added to pattern II, with only large particles (2.5–1.2 mm), improved. Therefore, we assumed that influence on the fluidity of fine particles is different by the particle size of the particle constituted with the fine particle. In addition, the relative flow area ratio of the sea sand in pattern V was very large because the solid content rate of fine aggregate was large and contained a lot of 2.5–1.2 mm particles with very low water absorption. Thus, the water absorption rate of the fine aggregate decreased and the paste fluidity increased.

In this experiment, the particle size distribution that gave the highest fluidity was "no particle size adjustment", and the particle size of the crushed sand was adjusted with constant water–cement ratio (W/C), but no increase in fluidity was observed. Therefore, it is necessary to evaluate the fluidity of mortar when using crushed sand with particle size adjusted by changing W/C. In addition, Figure 3 shows that mini-slump increased as the solid content rate of fine aggregate increased, although the relationship was not linear. Therefore, mini-slump was not affected strongly by the solid content rate of fine aggregate. The same behavior was also observed for the solid content rate of the fine aggregate and the air content.

Table 4. The ratio of the fine particle of 75 μm or less in the fine aggregate input amount of Pattern I, III and V.

	Fine particle content[%]		
	I	III	V
CSa	0.07	0.12	0.14
CSb	0.54	0.76	0.65
SS	0.69	0.73	0.59

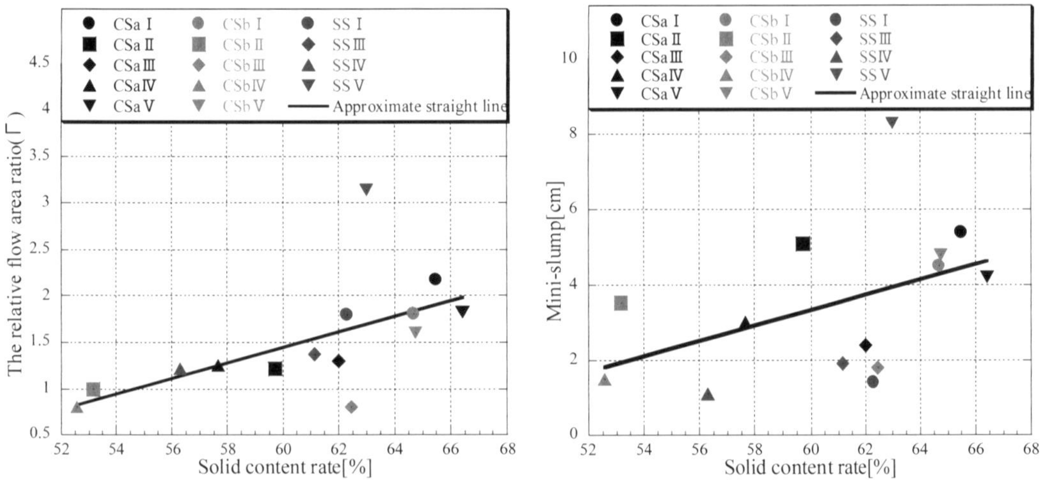

Figure 1. Relationship between solid content rate and the relative flow area ratio, mini-slump. (Left: the relative flow area ratio/Right: mini-slump)

3.2 Solid Content Rate of Fine Aggregate and Fluidity with Powder

Table 5 shows measured value with FA and LP, Figs.2 shows relationship between solid content rate of fine aggregate and the relative flow area ratio when FA and LP are mixed. For powder-free mixing, the relative flow area ratio increased as the solid content rate of the fine aggregate increased. When fly ash was added, the relative flow area ratio increased compared with powder-free mixing, and the rate of increase of the relative flow area ratio became large. This result was explained by the ball bearing effect of the smooth, round fly ash particles increasing the paste fluidity. In addition, when limestone fine powder was added, the rate of increase of the relative flow area ratio did not change much, even compared with powder-free mixing (Figure 1). Therefore, the change in fluidity was caused by the powder characteristics of the fine aggregates with adjusted particle size. In pattern III of CSb with limestone fine powder, there was insufficient water to mix the mortar. We could not explain this result, and we intend to investigate it in future work.

Table 5. Measured value with FA and LP.

Symbol			Frow value[mm]	Relative flow area ratio(Γ)	Mini-slump [cm]	Air content [%]
FA (LP)	CSa	Ⅰ	201.1(187.3)	3.04(2.51)	8.3(4.3)	1.1(3.6)
		Ⅱ	153.2(161.5)	1.35(1.61)	5.1(4.5)	1.8(4.5)
		Ⅲ	162.7(154.5)	1.65(1.39)	3.2(0.6)	2.5(2.8)
		Ⅳ	151.5(141.5)	1.30(1.00)	3.5(1.6)	3.5(4.0)
		Ⅴ	193.9(178.0)	2.76(2.17)	7.6(3.0)	1.0(1.9)
	CSb	Ⅰ	182.0(169.5)	2.80(1.87)	4.4(4.4)	1.2(3.5)
		Ⅱ	130.5(131.0)	0.70(0.72)	1.5(1.8)	3.2(5.4)
		Ⅲ	153.5(-)	1.40(-)	2.3(-)	0.8(-)
		Ⅳ	135.0(133.5)	0.80(0.78)	0.9(0.3)	5.5(8.2)
		Ⅴ	192.0(170.0)	2.70(1.89)	7.1(3.6)	1.2(2.9)
	SS	Ⅰ	187.5(166.5)	2.52(1.77)	4.6(2.6)	3.5(5.8)
		Ⅱ				
		Ⅲ	192.0(164.5)	2.68(1.71)	4.3(2.5)	2.8(5.8)
		Ⅳ	163.5(164.5)	1.67(1.71)	2.8(2.5)	5.1(6.5)
		Ⅴ	227.5(200.5)	4.18(3.02)	11.2(7.5)	1.2(2.7)

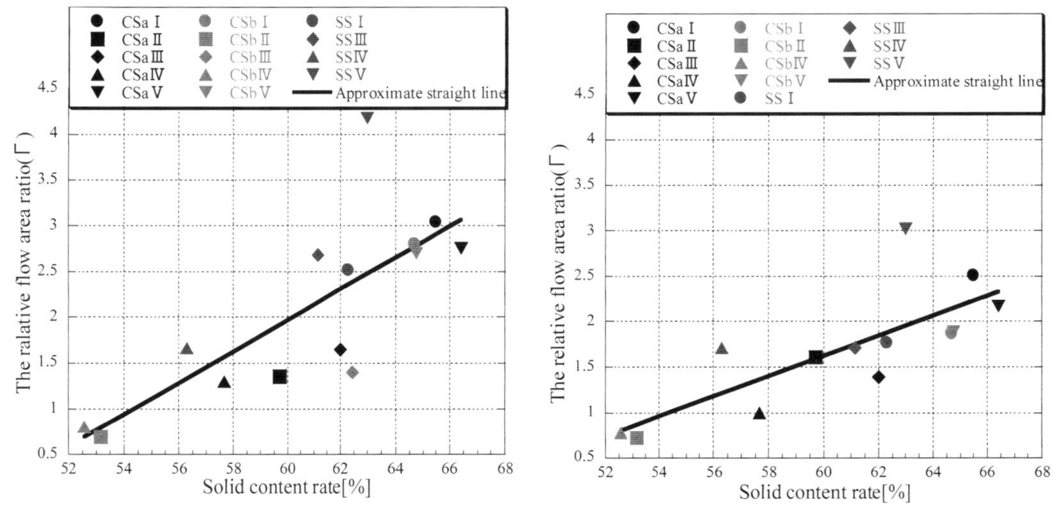

Figure 2. Relationship between solid content rate and the relative flow area ratio when FA, LP are mixed. (Left: FA/Right: LP)

4 CONCLUSION

The findings obtained in this research are shown below.

(1) Even when the particle size was adjusted, there was a tendency for the relative flow area to increase as solid volume rate of fine aggregate became increase. But, although the solid volume rate of fine aggregate was large, there were some points that the relative flow area ratio was not increased. Therefore, we assumed that the factors affecting relative flow area ratio changed by particle size composition.

(2) For 2.5–1.2 mm particles or 1.2–0.6 mm particles, the most important factor affecting fluidity was the solid content rate of the fine aggregate. In addition, for less than 1.2 mm, although a factor other than the water absorption rate of fine aggregate affected the fluidity strongly, we could not identify this factor and intend to investigate it in future work. In addition, we assumed that influence on the fluidity of fine particles is different by the particle size of the particle constituted with the fine particles.

(3) Even when powder was mixed, there was a tendency for the relative flow area to increase as solid volume rate of fine aggregate became increase. When compared with powder-free mixing, fly ash had a smooth particle shape for limestone fine powder, so the increase rate in the relative flow area ratio when fly ash was mixed was larger than when limestone fine powder was mixed.

Acknowledgements

The authors acknowledge the assistance in this work provided by Mr. Eiji Mikura, Mr. Yuki Takaki. This research was financially supported by Environmental Restoration and Conservation Agency (ERCA) No.3-1703 "Development of advanced recycling technology for fly ash to enable cement-free concrete".

References

Hirata, T., Takada, N., Miura, N. and Sogou, S., *The Influence of the Aggregate Grading on Fresh Highly-Workable-Concrete,* Proceedings of the Japan Concrete Institute, 17(1), 81-86, June, 1995.

Yamaguchi, S., Edamatsu, Y., and Okamura, H., *Characterization of Sand in View of Mortar Flow*, Proceedings of the Japan Concrete Institute, 16(1), 83-88, 1994.

Streamlining Information Transfer between Construction and Structural Engineering
Edited by Shiau, J., Vimonsatit, V., Yazdani, S., and Singh, A.

ISBN: 978-0-9960437-7-9

TENSILE CREEP OF FLY ASH CONCRETE AT EARLY AGE CONSIDERING THE YOUNG'S MODULUS DEVELOPMENT

YOICHI MIMURA[1], VANISSORN VIMONSATIT[2], YUKI WATANABE[3], ITARU HORIGUCHI[1], AND ISAMU YOSHITAKE[4]

[1]*Dept of Civil and Environmental, National Institute of Technology, Kure College, Kure, Japan*
[2]*School of Civil and Mechanical Engineering, Curtin University, Perth, Australia*
[3]*Advanced Course of Project Design Engineering, National Institute of Technology, Kure College, Kure, Japan*
[4]*Dept of Civil and Environmental, Yamaguchi University, Ube, Japan*

Tensile properties are important for predicting tensile stress which causes thermal cracking. Fly ash, a by-product from coal-fired power plants, has been recently used to reduce such thermal cracks. However, investigations dealing with tensile properties of fly ash concrete are still limited. This study focuses on the tensile properties of concrete mixed with fly ash at an early age. Fly ash was mixed in general purpose concrete with a cement-replacement ratio of 20% by mass to simulate fly ash concrete used in Japan. To examine tensile Young's modulus and tensile creep, direct tension test was conducted using dog-bone shaped specimens. The tensile creep tests were conducted at the age of 3 days or 7 days, and the loading (30% of splitting tensile strength at the loading age) was sustained for 14 days. Past investigations usually assumed a constant elastic strain during creep test. It should be noted however that elastic strain at early age decreases with the age of concrete as hydration continues. This study takes a consideration of Young's modulus development during creep test to distinguish actual creep and elastic strains. Test results show that creep strain has been underestimated when assuming constant elastic strain.

Keywords: Creep behavior, Thermal cracking, Creep strain, Elastic strain, Splitting tensile strength.

1 INTRODUCTION

Tensile properties, such as Young's modulus and creep, are required for estimating tensile stresses which cause thermal cracking. Creep, which is inelastic strain, is more severe under the influence of thermal stresses due to strains restraint. Fly ash (FA), a by-product from coal-fired power plants, has been recently used to reduce such thermal cracks. FA is generally replaced with cement, so that hydration heat which can cause thermal cracking is decreased.

In this study, the tensile creep tests using the dog-bone shaped specimen were conducted on FA concrete at the age of 3 or 7 days, and the loading was sustained for 14 days. The loading stress was constant during the tensile creep test and was set at 30% of the splitting tensile strength. It should be noted that elastic strain at an early age decreases with the age of concrete. However, the effect of the decrease in elastic strain during creep test has hardly been mentioned in most investigations (Benoît and Michel 1995, Matthieu *et al.* 2012). This study takes a consideration of Young's modulus development during creep test to distinguish actual creep and elastic strains based on superposition principle. In order to obtain the stiffness development during creep test, Young's modulus and elastic strain were examined at various ages. The elastic strains during the tensile creep test were also measured by temporary unloading.

2 EXPERIMENTAL PROGRAM

2.1 Materials and Specimens

The concrete materials and mixture proportions in this study are given in Table 1. Normal concrete with the water to cement ratio of 55% as shown in Table 1 is generally used for infrastructures in Japan. The amount of fly ash used in the present study was 20% by mass of the cement of the normal concrete. The targeted slump and the targeted air amount were 8 ± 2.5 cm and 4.5 ± 1.5%. The unburned carbon contained of FA decrease the effect of the AE water reducing agent, so the AE agent for restraint the influence of unburned carbon was mixed. In this study, tensile creep test for obtaining tensile creep, direct tension test for obtaining tensile Young's modulus, splitting tensile test (JIS A 1113 2006) for obtaining tensile strength, and compression test for obtaining Young's modulus were carried out.

Table 1. Concrete materials and mixture proportions.

Property	**Materials**	**Proportion (kg/m^3)**		**Density (g/cm^3)**
		Normal concrete	**FA concrete**	
Mix ID	---	N	FA	
Water (W)	Tap water	165	165	1.00
Cement (C)	Ordinary Portland cement	300	240	3.16
Fly ash (FA)	Class II	0	60	2.24
Fine aggregate (S)	Crashed sand	844	833	2.62
Coarse aggregate (G1)	Crashed stone 20-15 mm	499	493	2.67
Coarse aggregate (G2)	Crashed stone 15-5mm	499	493	2.67
Admixture (Ad1)	AE water reducing agent	3.00	2.40	1.07
Admixture (Ad2)[a]	AE agent for fly ash	0	16.8	1.04

[a] diluted solution (diluted 100 times with water)

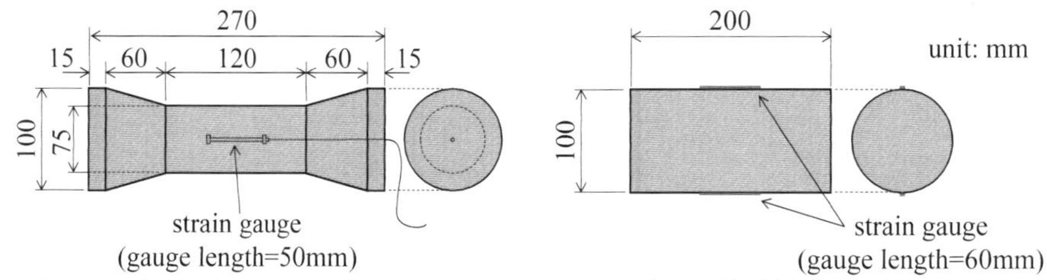

(a) Dog-bone shaped specimen. (b) Cylindrical specimen.

Figure 1. Dog-bone shaped specimen and cylindrical specimen.

The dog-bone shaped specimen for the direct tension test and the tensile creep test is shown in Figure 1(a). A small mold strain gauge having a thermometric function was embedded at the center of the cross-section. A cylindrical specimen as shown in Figure 1(b) was used for splitting tensile test and compression test. Strain gauge was attached at the center of both side faces of each specimen for compression test. After casting, all specimens were carried into a curing room to be cured at a room temperature of 20 ± 1°C for 24 hours before demolding. After that, the specimens were cured underwater in a water tank (16 ± 1°C) installed in the curing room until load tests were to be carried out at the age of 3 days or 7 days.

2.2 Test Method

The test setup of the tensile creep test in the present study is shown in Figure 2. A weight was placed on the tip of the steel arm for loading. The relationship between the weight and the load stress was obtained beforehand, so the load cell was not used in this test. The loading apparatus was installed in the water tank in the curing room (room temperature = 20 ± 1°C). The water temperature was kept at 16 ± 1°C. In this study, the tensile creep test was conducted in water in order to eliminate the influence of drying.

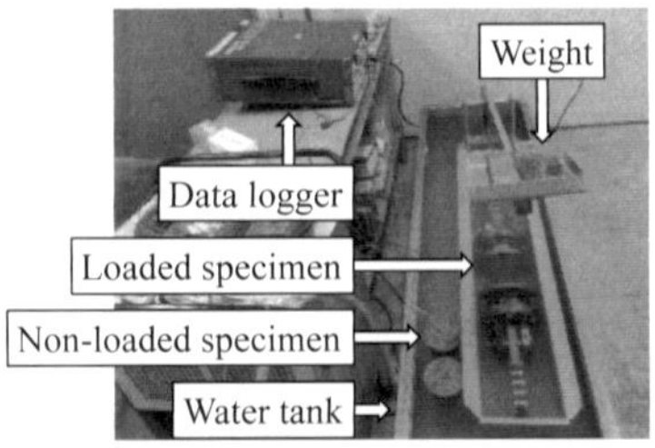

Figure 2. Test setup of the tensile creep test.

The non-loaded specimen was placed in the same tank. The strain of the non-loaded specimen was subtracted from that of the loaded specimen, so strains caused by factors other than loading were removed as much as possible. Measurement was carried out at one-hour intervals by using a data logger. The measurement items were strain and temperature of the dog-bone shaped specimen and water temperature. Table 2 gives the details of the tensile creep test. The loading started at the age of 3 days, or 7 days, as it was found in a separate FEM analysis of a wall structure that thermal stress turned from compression to tension from 3 days to 7 days. Thermal cracks occur at the age of around 2 weeks, so the loading was sustained for 14 days. According to Davis-Granville's law, stress less than 1/3 of strength is proportional to creep strain, so the loading stress in this study was set to 30% of the splitting tensile strength at the loading-start age. The strain at temporary unloading was recorded in order to measure the elastic strain during the creep test, and the Young's modulus was examined at the ages given in Table 2.

Table 2. Details of tensile creep test.

I.D.	Concrete	Loading age	Measurement of Young's modulus	Measurement of elastic strain
N3a	N	3 days	3, 7, 10, 14 and 17 days	At starting and ending
N3b	N	3 days	3, 5, 7, 10, 14 and 17 days	Every day
N3c	N	3 days	3, 5, 7, 10, 14 and 17 days	Every day
N7	N	7 days	7, 9, 11, 14, 17 and 21 days	At starting and ending
FA3a	FA	3 days	3, 7, 10, 14 and 17 days	At starting and ending
FA3b	FA	3 days	3, 5, 7, 10 and 17 days	Every day
FA3c	FA	3 days	3, 5, 7, 10 and 17 days	Every day
FA7	FA	7 days	7, 9, 11, 14, 17 and 21 days	At starting and ending

In determining the Young's modulus from the compression test, the cylindrical specimen was loaded using a universal testing machine until reaching the splitting tensile strength. The load and strain were recorded at every strain increment of 10 x 10^{-6}. After each recording, the load was temporarily removed and the strain at the unloading was measured. The Young's modulus was evaluated from the gradient of the regression line of the stress – strain relationship, which is linear at this loading range.

3 TEST RESULTS AND DISCUSSION

The splitting tensile strength at the loading age and the loading stress are given in Table 3. Each result is the average of the 3 test results. Figure 3(a) shows an example of the strain behavior of the loaded specimens and the non-loaded specimens. The data between the loading strain and non-loading strains are the strains at temporary unloading in order to measure the elastic strain during the tensile creep test. Creep strains as shown in Figure 3(b) were evaluated by subtracting the non-loading strain and the elastic strain from the loading strain, assuming that the elastic strain was constant at the start of the loading.

Table 3. Splitting tensile strength at loading-start age and loading stress.

I.D	**Splitting tensile strength *c***	**Loading stress (30% of f_t)**	**I.D.**	**Splitting tensile strength**	**Loading stress (30% of f_t)**
N3a	2.38 N/mm^2	0.714 N/mm^2	FA3a	1.51 N/mm^2	0.453 N/mm^2
N3b	1.22 N/mm^2	0.366 N/mm^2	FA3b	1.52 N/mm^2	0.456 N/mm^2
N3c	2.09 N/mm^2	0.627 N/mm^2	FA3c	1.52 N/mm^2	0.456 N/mm^2
N7	3.11 N/mm^2	0.933 N/mm^2	FA7	2.74 N/mm^2	0.822 N/mm^2

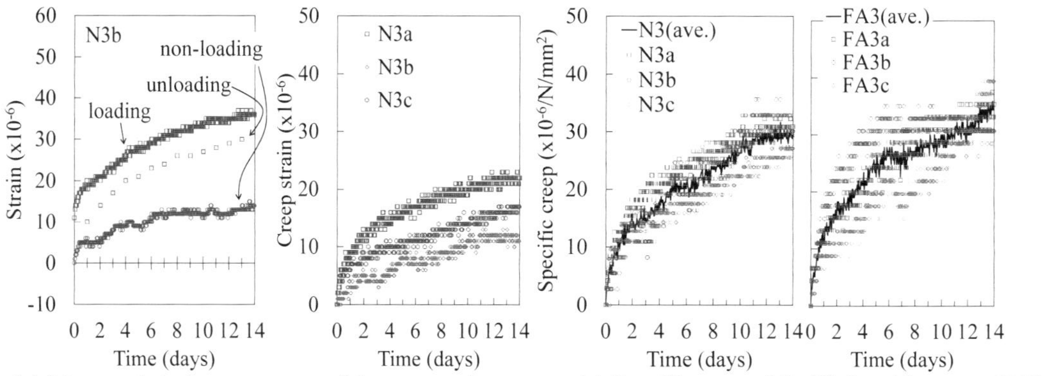

(a) Measured strain (b) creep strain (c) Specific creep (N) (d) Specific creep (FA)

Figure 3. Strain behavior based on constant elastic strain.

The loading stress of N3a was approximately twice that of N3b, so the creep strain of N3a was more than twice that of N3b. The specific creep of normal concrete obtained by dividing the creep strain by the loading stress is shown in Figure 3(c). In Figure 3(c), the average of the 3 results is also demonstrated. The difference in the specific creep of the normal concrete became smaller than the difference in the creep strain. The average specific creep was in the range of ± 6 x 10^{-6}/N/mm^2. The average specific creep of the FA concrete had larger errors than that of the normal concrete and was in the range of ±10 x 10^{-6}/N/mm^2.

It should be noted that the elastic strain at early age decreases as the concrete age increases. The Young's modulus obtained from the tensile creep test is shown in Figure 4. The development of the Young's modulus of FA concrete is more gradual than the development of the Young's modulus of the normal concrete.

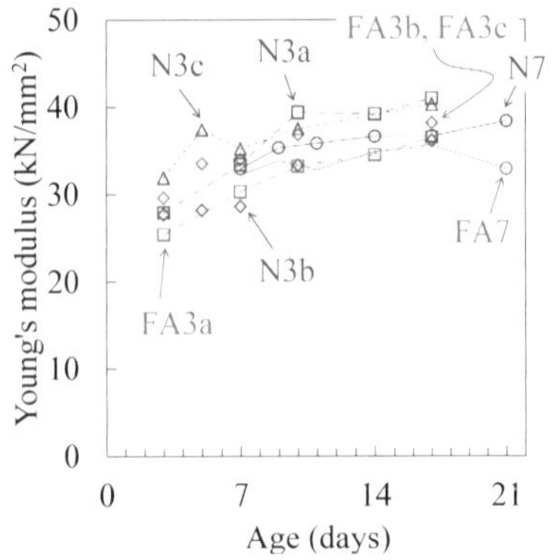

Figure 4. Young's modulus during the tensile creep test

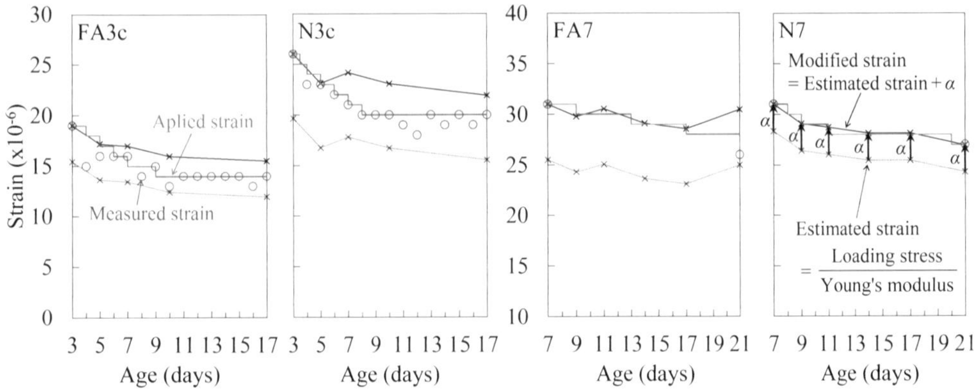

Figure 5. Decreasing elastic strains during the tensile creep test.

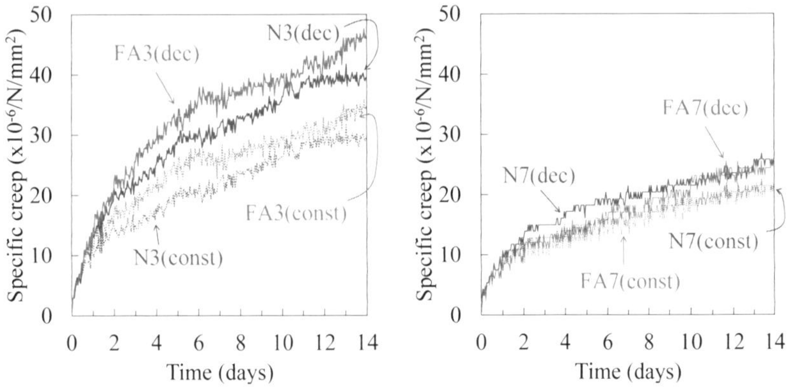

Figure 6. Specific creep based on the decreasing elastic strain.

Two sets of elastic strains were evaluated, one was from the measurement during the tensile creep test, and the other was estimated by dividing the loading stress to the Young' modulus based on the Hooke's Law, as shown in Figure 5. The estimated strains at the start of the loading were not equal to the elastic strains from the measurement. The modified strains were thus calculated by adding the difference (α) between the estimated strain and the measured strain at the start of the loading to the estimated strain. Both the measured strains and the modified strains were decreased with increasing ages. This result indicates that the creep determined from assuming a constant elastic strain has been underestimated. The modified strain of N7 was equivalent to the measured strain at the end of loading (at the age of 21 days). However, the modified strains such as FA3c and N3c had a different behavior with the measured strains during the tensile creep test. The elastic strains, shown as red lines in Figure 5, for evaluating the creep in this study were therefore determined with reference to the measured strains rather than the modified strains. It can be seen that the applied strains of FA3c and N3c decreased until the age of 8 or 9 days, while the elastic strains of FA7 and N7 decreased more gradually.

Figure 6 shows the specific creep based on the decreasing elastic strains. The dotted lines in Figure 6 represent the results when using the constant elastic strain. In the case when the load started at the age of 3 days, the difference in the specific creep caused by considering the development of the Young's modulus expanded with decreasing the elastic strain. In this case, the maximum increase of the specific creep of the normal concrete was 10.1 x $10^{-6}/N/mm^2$, and was 8.0 x $10^{-6}/N/mm^2$ on average. The FA concrete also had the specific creep's maximum increase of 11.7 x $10^{-6}/N/mm^2$, and 8.4 x $10^{-6}/N/mm^2$ on average. Such increase was larger than the difference in the specific creep caused by mixing FA into the normal concrete. In other words, the decrease of the elastic strain contributes more in the evaluation of the specific creep than the use of the FA mixing. On the other hands, the specific creep when starting the load at the age of 7 days was less affected by the FA mixing and the decreasing elastic strain than by starting the load at the age of 3 days.

4 CONCLUSIONS

This paper describes experimental investigations for obtaining development of the Young's modulus and tensile creep behavior at early age. Tensile creep behavior based on decreasing elastic modulus during creep test was examined. The conclusions of this paper are as follows:

(i) When loading started at the age of 3 days, the elastic strain determined by referring to the measured strain or the estimated strain decreases until the age of 8 or 9 days. This result indicates that the assumption of a constant elastic strain causes the underestimation of a specific creep.

(ii) The decrease in the elastic strain contributes more to the specific creep when loading started at the age of 3 days than the use of FA mixing. The specific creep when loading started at the age of 7 days was less affected by the FA mixing and the decreasing elastic strain.

Acknowledgments

The authors would like to thank Mr. K. Yamada, Mr. S. Michimoto (former students of National institute of Technology, Kure College) and the Chugoku Electric Power Co. Ltd.

References

Benoît, B., and Michel P., Tensile Creep at Early Ages of Ordinary, Silica Fume and Fiber Reinforced Concretes, *Cement and Concrete Research*, 25(5), 1075-1085, July, 1995.

JIS A 1113, *Method of Test for Splitting Tensile Strength of Concrete (in Japanese),* Japanese Industrial Standard, 2006.

Matthieu, B., Farid, B., Jean-Michel, T., and Georges, N., Concrete Early Age Basic Creep: Experiments and Test of Rheological Modelling Approaches, *Construction and Building Materials*, 36, 373-380, November, 2012.

Streamlining Information Transfer between Construction and Structural Engineering
Edited by Shiau, J., Vimonsatit, V., Yazdani, S., and Singh, A.

ISBN: 978-0-9960437-7-9

STUDY ON REDUCTION OF UNIT WATER AMOUNT OF CONCRETE WITH WASTE GYPSUM

ATSUSHI SHIMABUKURO and KEIJI KUWAJIMA

Dept of Civil Engineering and Architecture, NIT, Tokuyama College, Shunan, Japan

This study discovered that the combination of waste gypsum and fly ash was capable of being used as fine aggregate in concrete. However, results showed that the unit water amount of the concrete with waste gypsum had to be increased to assure the designated workability. In general, if the unit water amount of concrete increases, hardened concrete will have various issues. Therefore, in this study, we try to reduce the unit water amount of this concrete by the high-performance AE (Air Entraining) water reducing agent containing viscosity improver. In order to investigate the effect of the above AE water reducing agent for the reduction of the unit water amount of the concrete with waste gypsum, the workability of above concrete with this agent from slump test and air containing test was examined. Also, compression, tension, and bending, of the concrete with waste gypsum from each strength test was measured. On the basis of the test results, it showed that we are able to assure the workability of fresh concrete with waste gypsum by the above agent used in this study, even if the unit water amount is reduced. Also, it was confirmed that the compressive strength of concrete with waste gypsum is equivalent to the strength of concrete without waste gypsum.

Keywords: High-performance, Air entraining water reducing agent containing viscosity improver, Trial mixing, Workability, Strength test, Strength ratio.

1 INTRODUCTION

The sea sand is used as a fine aggregate of concrete. To protect marine resources, the mining of sea sand is prohibited and the Japanese government limits the expansion of mining regions in the coastal regions of Setouchi of Japan. In recent years, the use of crushed sand as a fine aggregate in concrete has increased, and crushed sand is widely used to take the place of sea sand. However, the rocks which are the raw material of crushed sand are used as a fine aggregate of concrete as well as a material for asphalt concrete. The supply of rocks will be more and more difficult in future years. Therefore, we use waste gypsum to replace a part of sand as the fine aggregate of concrete. The amount of waste gypsum generated in Japan is next to concrete waste, construction sludge and wood waste. It is expected that the amount of waste gypsum will keep increasing in the future. Hence, it is necessary to find a good method for the practical use of waste gypsum.

In a previous study, we verified that the combination of waste gypsum and fly ash was able to be used as a fine aggregate of concrete (Shimabukuro and Kuwajima 2017). However, it shows that the unit water amount of the concrete with waste gypsum increased in order to secure the designated workability. In general, if the unit water amount of concrete increases, there will be various problems with hardened concrete. Therefore, in this study, we try to reduce the unit water amount of concrete by high-performance air entraining (AE) water reducing agent containing viscosity improver, which is named as AEV in this paper.

In order to investigate the effect of AEV on reducing the unit water amount of concrete containing waste gypsum, we make and test trial mixes for the concrete with waste gypsum and AEV until the target values of the slump and the air content are satisfied. Then, the appropriate mix proportion of each material for the concrete with waste gypsum and AEV is determined, based on the results of trial mixing. In addition, we conduct the strength tests on the concrete specimens that are made with this mix proportion. According to these above tests, the effect of AEV on the concrete with waste gypsum for (i) fresh concrete characteristics, and (ii) strength characteristic of the hardened concrete are evaluated.

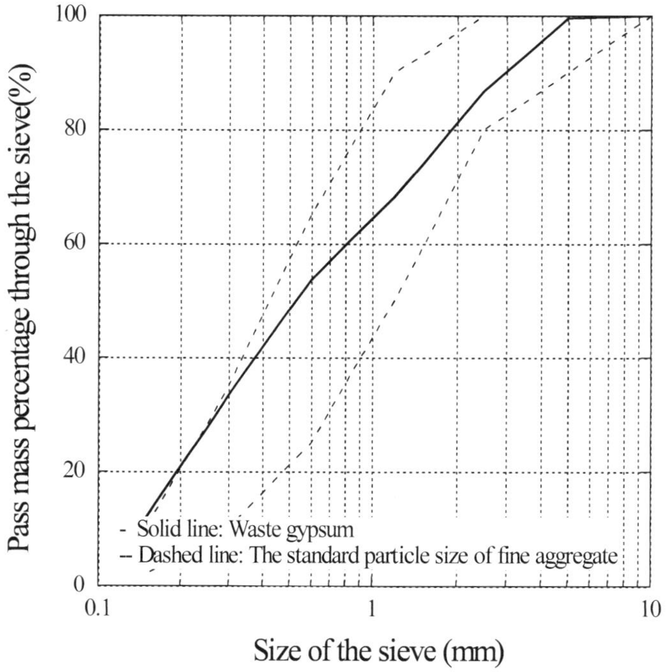

Figure 1. The result of the sieving test.

2 ABOUT WASTE GYPSUM

Gypsum board is widely used as a building material because it has a low price and some useful properties, such as fire resistance, adiabaticity, and sound insulation. The waste gypsum is produced when dismantling buildings having gypsum board. According to the extent of crystallization of water, the gypsum is classified into the gypsum dihydrate, the gypsum hemihydrate and the anhydrous gypsum (Kanto Branch of Japanese Geotechnical Society 2013). The waste gypsum used in this study is the gypsum hemihydrate. The property of the gypsum hemihydrate is to react with the water and to solidify. The results of the sieving test of the waste gypsum used in this study are shown in Figure 1 (Shimabukuro and Kuwajima 2017). From this result, we consider that the waste gypsum can be applied to substitute the sand as a fine aggregate, because the particle size distribution of the waste gypsum is within the standard particle size range of fine aggregate.

3 ABOUT AEV

AEV used in this study is a kind of an air entraining and high-range water reducing agent that integrates viscosity improvement into a high-performance AE water reducing agent. The properties of AEV are described as follows:

- It is classified as a high-performance AE water reducing agent in Japanese Industrial Standards (JIS).
- We can upgrade ordinary concrete with a relatively small cement amount to high flowing concrete.
- We can improve the workability of the fresh concrete and reduce the amount of compaction required.

From these properties, it is considered that workability can be assured by applying AEV to concrete with waste gypsum, even if the unit water amount in concrete is reduced.

4 EXPERIMENTAL OVERVIEW

4.1 Mix Design

We conducted the mix design of the concrete in which a part of the fine aggregate is replaced with waste gypsum. For each design condition of the concrete, the design strength is 18 N/mm^2, the targeted slump value is 8 cm, and the targeted air content is 4.5%. And, the water-cement ratio is 0.64 because we use this value when we set the concrete strength to 18 N/mm^2 in our laboratory. The maximum size of crushed stone used as coarse aggregate is 20 mm in this study. On the other hand, the sea sand with the fineness modulus of 2.70 is used as the fine aggregate. Each density of the coarse aggregate and the fine aggregate is 2.66 g/cm^3 and 2.60 g/cm^3, respectively. Also, the density of waste gypsum used in this study is 2.63 g/cm^3. The replacement ratio of waste gypsum to the fine aggregate is 10% of the volume ratio. AEV is mixed depending on the ratio of the weight to the cement amount.

4.2 Slump Test and Air Content Test

In this study, we made trial mixes to find the required amount of each material in the concrete until the slump and the air content satisfy the targeted value. The slump and the air content are measured by the method standardized by JIS. The tolerances of the slump value and the air content are ±2.5cm and ±1.5%, respectively.

4.3 Curing Method and Curing Period

The curing method in this study uses water curing for a week, four weeks, and 12 weeks for the compressive strength test.

4.4 Strength Test

The compressive strength test is carried out according to the method standardized by JIS. The shape of the specimen is a cylinder with 100 mm diameter and 200 mm height. The compressive load is sustained until the specimen fractures entirely. We calculate the compressive strength from the maximum load. And we carry out the tensile strength test and the bending strength test to confirm their relationship with compressive strength. The tensile strength specimens are the same shape as the compressive strength test specimens, and the bending strength specimens are the prismatic shape with width of 150 mm, height of 150 mm, and length of 530 mm. These strength tests are also performed under the method standardized by JIS.

Table 1. Mix proportion quantity and the results of our previous study.

Mixing Case	W (kg/m^3)	C (kg/m^3)	Wg (kg/m^3)	S (kg/m^3)	G (kg/m^3)	AE (kg/m^3)	Note	Slump (cm)	Air (%)	Evaluation (Satisfying target values)
OC	169	264	0	816	1036	0.911	Our previous study	8	4.1	○
PC	192	300	79	700	985	0.912		7	5	

W: Water, C: Cement, Wg: Waste gypsum, S: Fine aggregate (sea sand), G: Coarse aggregate, AE: AE water reducing agent

Table 2. Mix proportion quantity in this study and the results of the slump tests and air content tests.

Mixing Case	W (kg/m^3)	C (kg/m^3)	Wg (kg/m^3)	S (kg/m^3)	G (kg/m^3)	AEV (kg/m^3)	Addition amount of AEV	Slump (cm)	Air (%)	Evaluation (Satisfying target values)
Trial 1	168	264	83	735	1036	2.64	1%	4.5	-	×
Trial 2	173	275	84	749	989	2.75	1%	4.5	-	×
Trial 3	175	283	83	742	976	5.66	2%	7.5	4.4	○

AEV: Air Entraining water reducing agent containing viscosity improver

5 RESULT AND CONSIDERATION OF TRIAL MIXING

We performed trial mixes on the basis of OC and PC in Table 1, which were shown in our previous study (Shimabukuro and Kuwajima 2017). OC is ordinary concrete without waste gypsum and PC is the concrete with waste gypsum, which replaces 10% of the fine aggregate volume with waste gypsum volume. From this table, we find that the unit water amount increases by containing waste gypsum to satisfy the targeted values of the slump and the air content values. Therefore, we try to reduce the unit water amount by AEV after some trial mixing. The results are shown in Table 2 and the detail of the modified method is as follows.

First, we changed the air entraining water reducing agent to AEV based on the values of OC in Table 1. We set AEV amount as 1% to the cement weight in Trial 1 shown in Table 2. From the results of Trial 1, the slump value was 4.5 cm at Trial 1 even if AEV was contained in the concrete with waste gypsum. Because the target value of the slump was not satisfied, we had to perform the next trial mixing.

Secondly, the unit water amount in Trial 2 was corrected based on the result of Trial 1. However, the targeted slump was not satisfied even in Trial 2.

Based on these results, we modified the unit water amount slightly and changed AEV amount from 1% to 2% in Trial 3. As a result, both the target values of the slump value and the air content were satisfied. Therefore, we consider that Trial 3 was the most appropriate mixing for the concrete with waste gypsum and AEV. Comparing Trial 3 with PC, unit water amount of the concrete with waste gypsum was reduced by containing AEV. Hence, AEV is considered to be an effective agent for reducing the unit water amount of the concrete with waste gypsum.

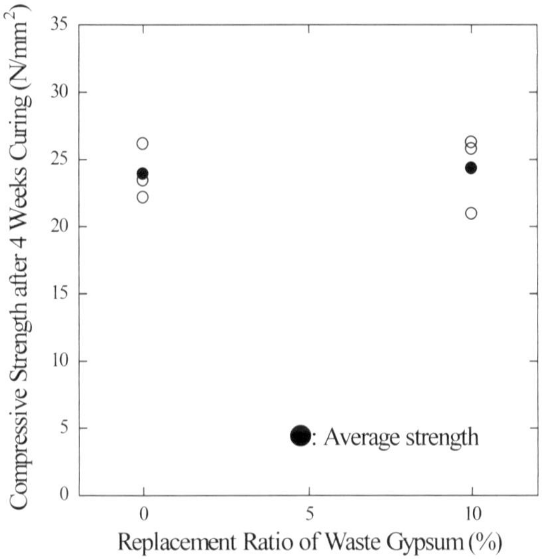

Figure 2. Results of compressive strength test.

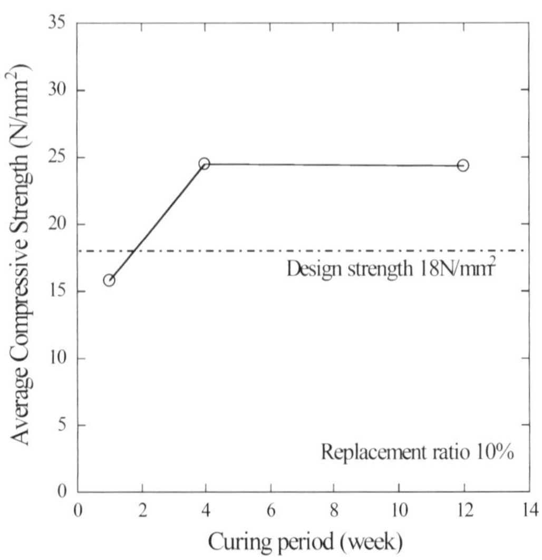

Figure 3. The average compressive strength of the concrete with waste gypsum and curing period relation.

Table 3. Each strength ratio when the compressive strength ratio is 1.0.

Strength ratio	Replacement ratio 10%
Compressive strength ratio	1.0
Tensile strength ratio	1/11
Bending strength ratio	1/6

6 RESULT AND CONSIDERATION OF STRENGTH TESTS

6.1 Compressive Strength Characteristic of Concrete with Waste Gypsum and AEV

The strength tests using the mixing result of Trial 3 were carried out to investigate and confirm the strength characteristic of the concrete with waste gypsum and AEV.

Figure 2 shows the result of the compressive test after curing for four weeks. According to this figure, the compressive strength of the concrete with waste gypsum is equivalent to that of the ordinary concrete without waste gypsum. Figure 3 shows the relationship between the compressive strength and the curing period. The values of Figure 3 are the average values of 3 specimens. From this figure, the compressive strength of the specimen with curing for a week does not reach the design strength which is 18 N/mm^2, and the strength of the specimen with curing for four weeks satisfies the design strength. Besides, the compressive strength of the specimen with curing for 12 weeks is almost equivalent to the strength of four weeks and does not also decrease from four weeks of strength.

From these, it is concluded that the concrete with waste gypsum and AEV can be used as same as the ordinary concrete.

6.2 Relationship between Compressive Strength and Each Strength for Concrete with Waste Gypsum and AEV

Table 3 shows the ratios of each strength to the compressive strength of the concrete with waste gypsum and AEV. From this table, for each strength ratio of ordinary concrete, the tensile strength ratio is 1/10 to 1/13 and the bending strength ratio is 1/5 to 1/7 (Kobayashi 2006). The values of the concrete with waste gypsum and AEV show almost the same tendency of the ordinary concrete. Therefore, in conjunction with the above consideration, it is considered that the concrete with waste gypsum and AEV can be applied as construction material, regarding the compressive strength characteristic.

7 CONCLUSIONS

We examined the workability of the fresh concrete with waste gypsum and AEV and measured various strengths of this hardened concrete. As a result, the knowledge obtained in this study is shown as follows:

(1) From the trial mixing results, it is possible to reduce the unit water amount of the concrete with waste gypsum by containing AEV. Therefore, AEV is considered to be an effective agent for reducing the unit water amount of the concrete with waste gypsum.

(2) From the strength tests results, the compressive strength of the concrete with waste gypsum and AEV was equivalent to that of ordinary concrete, and the tendency of each strength ratio to the compressive strength showed the same tendency as that of the ordinary concrete. Hence, it is considered that the concrete with waste gypsum and AEV can be used as a construction material like ordinary concrete, in relation to the compressive strength characteristic.

References

Kanto Branch of Japanese Geotechnical Society, *Recycling of Waste Gypsum Board Mainly for Ground Improvement Materials*, Kanto Branch of Japanese Geotechnical Society, Japan, 2013.

Kobayashi, I., *Latest Concrete Engineering 5th Edition*, Morikita Publishing Co. Ltd, Japan, 2006.

Shimabukuro, A. and Kuwajima, K., *The Basic Study of the Concrete Using Waste Materials*, Proceedings of the 15th East Asia-Pacific Conference on Structural Engineering and Construction, 1414-1419, 2017.

Streamlining Information Transfer between Construction and Structural Engineering
Edited by Shiau, J., Vimonsatit, V., Yazdani, S., and Singh, A.

ISBN: 978-0-9960437-7-9

UNBURNT CARBON REMOVED FROM FLY ASH BY FLOTATION METHOD

RINTARO TAKEHARA[1], HIROKI SUYAMA[2], KOJI TAKASU[2], and HIDEHIRO KOYAMADA[2]

[1]*Graduate School of Engineering, The University of Kitakyushu, Kitakyushu, Japan*
[2]*Faculty of Engineering, The University of Kitakyushu, Kitakyushu, Japan*

Because fly ash has pozzolanic activity, it has been used for a long time as an admixture for concrete. However, unburnt carbon in the fly ash adversely affects the fresh properties of concrete. For this reason, when fly ash having a high unburnt carbon content is used as an admixture for concrete, it is desired to reduce the amount of unburnt carbon. So far, at our university, we have developed a technique to remove unburnt carbon by flotation method. It is necessary to add a scavenger and a foaming agent when using this equipment. However, the appropriate amount is unknown. Therefore, we investigated a method to easily estimate the amount of chemicals suitable for flotation treatment from the physical properties of raw fly ash. It is obvious that efficient treatment can be performed when the addition amount of scavenger and the addition amount of foaming agent is set to 10: 1, and that the addition amount of the agent can be accurately estimated by using the methylene blue adsorption amount of fly ash.

Keywords: Foaming agent, Scavenger, Ignition loss, Flocculent.

1 INTRODUCTION

In recent years, much progress has been made in Japan on the recycling of industrial byproducts. One such move has been toward the use of fly ash, a byproduct of coal-fired power generation, as concrete admixture. In our research described herein, we develop a technique for removing unburned carbon from fly ash by means of froth separation, a technology commonly employed in the mining industry to separate minerals from gangue (waste rock) (Takasu *et al.* 2007). With this technique, we begin by pumping a fly ash containing slurry into a cylindrical water tank. Then, by adding a scavenger agent and a flocculant to the slurry, we induce micro bubbling and, in this case, create a froth that contains a comparatively high proportion of unburned carbon. Finally, we recover the tailing ash, which thus contains a relatively low proportion of unburned carbon. The tailing ash obtained with this technique is essentially a fly ash for which ignition loss can be kept to 3.0% or lower, a characteristic that facilitates its application to concrete admixtures.

The authors and colleagues previously analyzed the costs entailed by fly ash improvement/separation by means of froth flotation. A breakdown of treatment costs reveals that materials, especially chemical additives (scavenger agent and flocculant), account for a high proportion of the total (Figure 1). Such costs can be substantially reduced by lowering the amount of such additives. An issue, though, is that with an insufficient amount of additives, it becomes difficult to obtain a tailing ash having an ignition loss of 3.0% or less; while, with an excessive amount of additives, it becomes difficult to keep treatment costs acceptably low. In our research, we seek to develop a method of readily predicting the properties of fly ash to result from additions of suitable amounts of chemical additives to the froth flotation process.

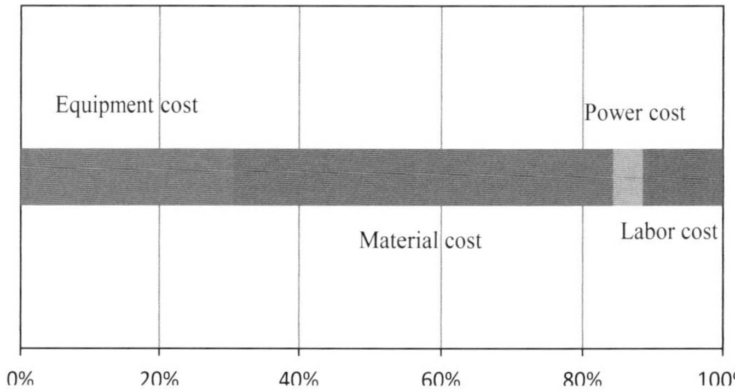

Figure 1. Breakdown cost of flotation treatment.

2 RELATIONSHIP BETWEEN ADDITIONS OF SCAVENGER AGENT AND ADDITIONS OF FLOCCULENT

2.1 Experiment Overview

The effectiveness of an addition of a scavenger agent can vary depending on a simultaneous addition of a flocculant, and vice versa. In this research, we keep our flocculant additions at a constant amount will varying our additions of scavenger agent. And, by this, we investigate the impact of differing amounts of scavenger agent on the ignition loss of tailing ash recovered from the froth flotation process.

Table 1. Physical properties of raw material fly ash.

Fly ash	Production area	Ignition Loss	BET specific surface area (m^2/g)	MB adsorption amount (mg/g)
L	I	11.07%	7.90	4.14
S	I	12.81%	16.95	4.20
T	I	11.95%	16.48	4.25
M	II	4.43%	2.81	0.53
V	II	5.21%	5.07	1.64
N	III	8.48%	6.69	1.71
W	III	7.22%	8.74	1.78
X	III	9.16%	7.47	1.73
Q	IV	12.29%	17.22	2.32
U	IV	12.30%	11.11	2.46
P	V	7.58%	8.38	1.65
R	VI	4.52%	2.16	0.24

Presented in Table 1 are the properties of fly ash used within our research. For the work outlined in this chapter, we utilize Material S. Ignition loss with this material (i.e. fly ash) was measured to be 12.81%. We prepared three differing levels of flocculant additions to improve the properties of this fly ash: 0.1% (weight percent vs. fly ash), 0.2%, and 0.3%. Initially, we added 3.0% (weight percent) of scavenger agent and, from there, steadily added more scavenger agent until we could observe a clear trend in the ignition loss of the resulting tailing ash. Preprocessing for froth flotation consisted of a preliminary stirring stage, for which a 0.5L juice mixer was utilized to stir/mix a slurry of fly ash, tap water and scavenger agent for 3 minutes. The stirred sample was then put into a 16L container together with flocculant and, by means of a froth flotation tank, froth floated for 30 minutes.

2.2 Experimental Results and Discussion

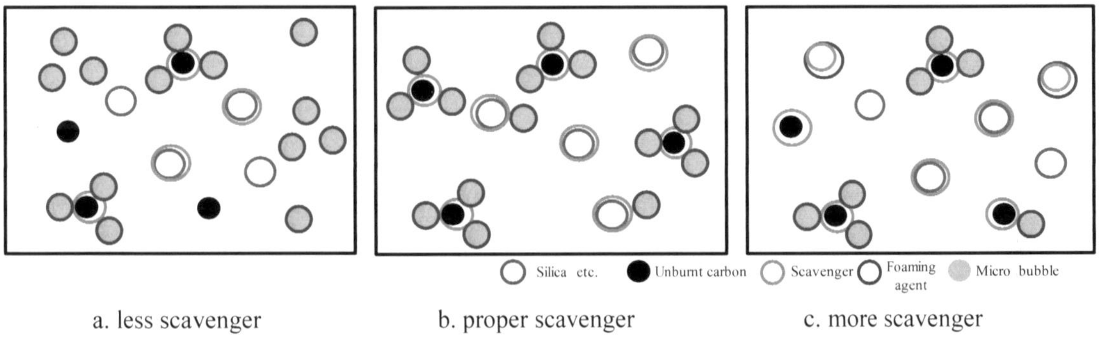

Figure 2. Amount of scavenger vs ignition loss of tail ash.

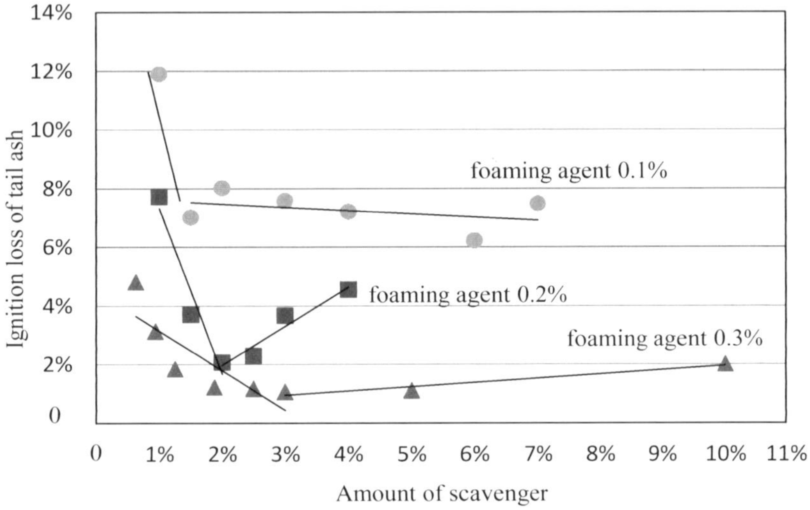

Figure 3. Mechanism to remove of unburned carbon.

Experimental results are shown in Figure 2. We see that tailing ash ignition loss tends to decrease with higher flocculant additions. With a constant 0.1% flocculant addition, the ignition loss decreases with increasing scavenger agent additions up to 1.5%. With further scavenger agent additions beyond the range, however, ignition loss remains essentially unchanged. With a constant 0.2% flocculant addition, ignition loss decreases with increasing scavenger additions up to 2.0%, at which point ignition loss reaches a minimum. Beyond that point, however, ignition loss increases with increasing scavenger additions. With a constant 0.3% flocculant addition, tailing ash ignition loss reaches a minimum at a 3.0% scavenger addition. From these results, it appears that adding scavenger agent and flocculant in a 10:1 proportion provides an efficient/effective way to reduce ignition loss.

Figure 3 illustrates our discussion of the manner in which the ignition loss of tailing ash tends to vary with differing additions of scavenging agent. Figure 3a shows a case in which there is a shortage of scavenging agent. Because of this paucity, not all unburned carbon surfaces can be coated by the agent. Uncoated carbon, thus unable to merged with/get removed by micro bubbles, remains in place. Conversely Figure 3c shows a case in which there as an excess of scavenging agent. Here, excess agent emulsifies and, in the process, consumes flocculant, thereby impeding the formation of microbubbles and, by extension, hindering process performance.

3 IMPACT OF CHEMICAL ADDITIVE AMOUNT ON TAILING ASH IGNITION LOSS

3.1 Experimental Overview

In this chapter, we examine the influence of differing amounts of chemical additions on tailing ash ignition loss by varying the amount of scavenger agent/ flocculant added to differing types of fly ash samples. Following on the results of the previous chapter, we set the ratio of scavenger agent addition (mass) to flocculant addition (mass) to 10:1. Of the differing types of fly ash covered within our research, we find that Material G has a comparatively low ignition loss. Materials D, E, H and I, on the other hand, have an ignition loss in excess of 5.0% and would thus be expected to have a deleterious impact on the fluidity and air entrainment characteristics of mixed concrete. The other materials have an ignition loss in excess of 8.0% and are thereby unsuitable for use as not only a concrete admixture, but even as an aggregate substitute. The procedure for froth flotation here is the same as that of the previous chapter.

3.2 Experimental Results and Discussion

Figure 4 shows relations between the amounts of scavenger agent/ flocculant additions and tailing ash ignition loss. For each of the fly ash materials, ignition loss tends to decrease with increasing amounts of these chemical additives. This said, depending on the type of fly ash, the effect on ignition loss can vary substantially, even at essentially the same levels of scavenger agent/flocculant additions. Setting control (target) values for tailing ash ignition loss to 2.0%, 1.5% and 1.0%, we determine (by linear interpolation) the amounts of scavenger agent additions necessary to reach those values (we hereafter refer to these amounts as the amount of scavenger agent necessary to attain, respectively, 2% ignition loss, 1.5% ignition loss, or 1.0% ignition loss). Interpolation results are shown in Figure 5. Likewise, Figure 6 shows relationships between these necessary amounts of scavenger agent and fly ash properties.

Within Figure 6, we observed a positive correlation between Brunauer-Emmett-Teller (BET) surface area of the fly ash and the amount of scavenger agent necessary to attain 2% ignition loss, although we also note that the cross-correlation is comparatively weak. In contrast, the correlation between methylene blue (MB) absorption amount and the amount of scavenger agent necessary to attain 2% ignition loss is both positive and comparatively strong. These results suggest that the apparent relation between BET surface area and scavenger agent necessary for 2% ignition loss is spurious (due to lurking variables). Here, the methylene blue adsorption amount, we believe, serves as a more suitable parameter for estimating the amount of chemical additives necessary for froth flotation. As a factor behind this result (i.e., as a factor for why the methylene blue absorption amount is highly correlated with the amount of scavenger agent necessary to attain 2% ignition loss), we note that methylene blue has a molecular weight of approximately 320, which is close to the molecular weight of scavenger agents (170 to 254). We thus infer that methylene blue adheres to fly ash particles during testing in much the same way as scavenger agent adheres to fly ash particles during froth flotation, a similarity that acts to raise the correlation between them. SEM photographs of Material P and Material Q are presented in Figure 7. Unburned carbon on the surface of Material P is comparatively smooth, while that on the surface of Material Q is deeply corrugated.

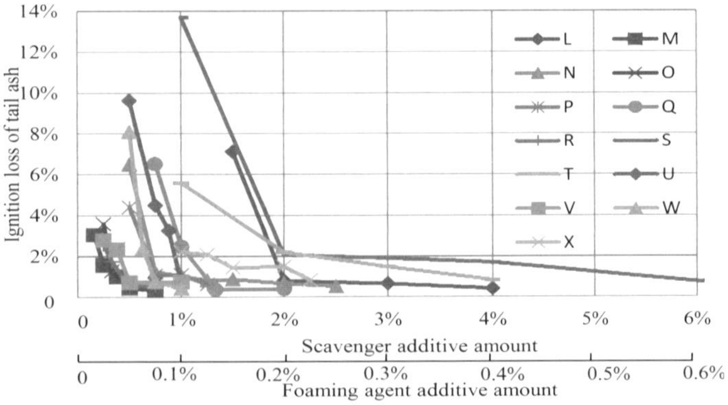

Figure 4. Amount of drug vs ignition loss of tail ash.

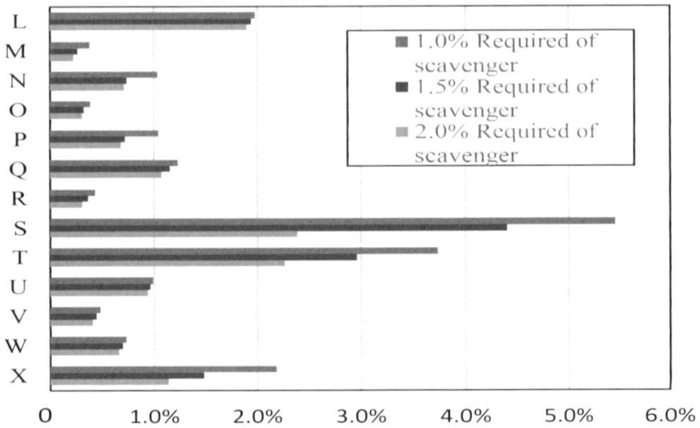

Figure 5. Physical properties of raw material fly ash vs required amount of scavenger.

Referring to Figure 6, we see that values for Material P remain close to the best-fit line, while those for Material Q are widely scattered. This, we infer, reflects differences in the way that scavenging agent adheres to unburned carbon. If the surface of the unburned carbon is highly corrugated, it requires that much more scavenging agent to cover. Thus, with a given (set) amount of scavenging agent, a corrugated surface would not be covered to the same degree of uniformity as a smooth surface; and, as a result, a coated corrugated particle would not adhere to a microbubble with the same degree of stability as a coated smooth particle.

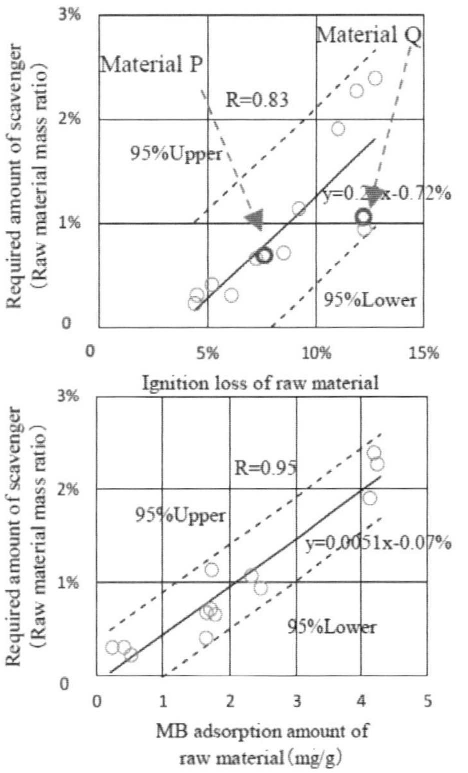

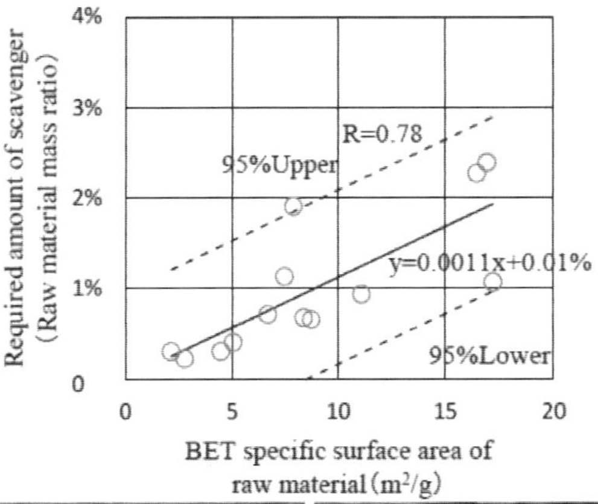

Figure 6. Physical properties of raw material fly ash vs. required amount of scavenger.

Figure 7. SEM Image of raw material fly ash.

4 CONCLUSION

- When using approximately equal amounts of scavenger agent, the ignition loss of the resultant tailing ash is lower with higher additions of flocculant.
- Adding scavenger agent and flocculant in a 10:1 proportion is thought to offer an efficient/effective way to reduce ignition loss.
- Even with essentially the same amount of chemical additions, the effect on ignition loss can vary substantially depending on the specific type of fly ash material.
- Measurements of methylene blue adsorption amount are thought to serve as an accurate indicator of the amount of chemical additives necessary for froth flotation.

Acknowledgements

The authors acknowledge the assistance in this work provided by Mr. Eiji Mikura and Ms.Ikumi Otsuki.

References

Takasu, K., *et al.*, Characteristics of Concrete Using Unburned Carbon and Its Fly Ash Slurry in Fly Ash by Japanese Floor Ash, 7914 No. 697 of the Japanese Institute of Architectural Engineers, March 2014.

Streamlining Information Transfer between Construction and Structural Engineering
Edited by Shiau, J., Vimonsatit, V., Yazdani, S., and Singh, A.

ISBN: 978-0-9960437-7-9

EXPLANATION OF CONCRETE FILLING DEFECT ON MORTAR LOSS MODEL

HIROAKI OKU[1], HIROKI SUYAMA[2], KOJI TAKASU[2], and HIDEHIRO KOYAMADA[2]

[1]*Graduate School Engineering, The University of Kitakyushu, Kitakyushu, Japan*
[2]*Faculty of Engineering, The University of Kitakyushu, Kitakyushu, Japan*

Problems with filling can lead to rebar corrosion and impair the durability and all other functional aspects of reinforced concrete structures. However, the prevention of concrete filling defects is left to the skill, judgment, and experience of on-site operators. In this research, we develop a methodology for quantitatively estimating the rate of occurrence of concrete filling defects. By this, we hope to promote the efficient and effective prevention of such defects at construction sites. Therefore, as a model for defective filling, we propose a mortar loss model in which the mortar component of concrete free falling into the frame adheres to the reinforcing bars in the frame, and based on that, we conducted an experiment. Using actual measurement data, we quantitatively estimated the risk of defective filling.

Keywords: Concrete work, Filling defect, Honeycomb, Mortar loss model.

1 INTRODUCTION

Modern concrete is a building material that is semi-permanent in durability. This said, this durability can be significantly impaired by problems during construction. One such problem is poor filling. Problems with filling can lead to rebar corrosion and impair the durability and all other functional aspects of reinforced concrete structures. Likewise, the poor filling is well known as a factor behind the peeling, separation and eventual fallout of concrete fragments, a phenomenon that puts passersby at risk of serious injury.

Concrete filling defects occur with alarming frequency and remain a serious problem. A research committee of the Japan Concrete Institute found that filling defects existed within 39% of medium- to high-rise reinforced concrete (RC) multifamily dwellings, with only 3% of potential occurrences successfully prevented beforehand (JCI 2008). Over the years, much research has been conducted on the prevention of concrete cracking; yet, in contrast, there have been few academic studies on the defective filling. It is true that construction companies generally have in-house guidelines to prevent such defects. This said, there is no industry-wide set of uniform views or methodology. Usually, the prevention of concrete filling defects is left to the skill, judgment, and experience of on-site operators.

In this research, we develop a methodology for quantitatively estimating the rate of occurrence of concrete filling defects. By this, we hope to promote the efficient and effective prevention of such defects at construction sites. Working with a "mortar loss model" (described within the paper), the authors measure "mortar loss," which here means the amount of mortar found to adhere (stick) to rebar mesh within a test frame. They then apply their measurements to a wall construction case study and, within it, estimate the risk of defective concrete filling.

2 PROPOSITION OF A MORTAR LOSS MODEL

Cited as a major factor behind defective concrete filling is the intricate, excessively dense rebar mesh characteristic of many construction projects. The conventional model for rebar/concrete separation (Figure 1) has that aggregate tends to get caught in gaps between reinforcing bars, blocking any further flow of concrete beyond that point. Note, however, that the conventional model fails to explain why filling defects occur with particular frequency when pouring walls. Here, we point out that when pouring a concrete wall, the tremie pipe often cannot be inserted all the way to the bottom of the frame and so the concrete must be poured down from above. It is this characteristic, we think, that accounts for this high frequency of filling defects.

In this research, we propose a new model for defective filling. Under our "mortar loss model" (Figure 2), concrete pours freely down into the frame, whereupon a portion of the mortar component adheres to reinforcing bars within the frame. Here, an arrival of concrete containing excess aggregate at the bottom of the frame is deemed to be indicative of defective filling.

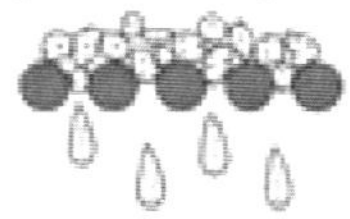

Figure 1. The conventional model.

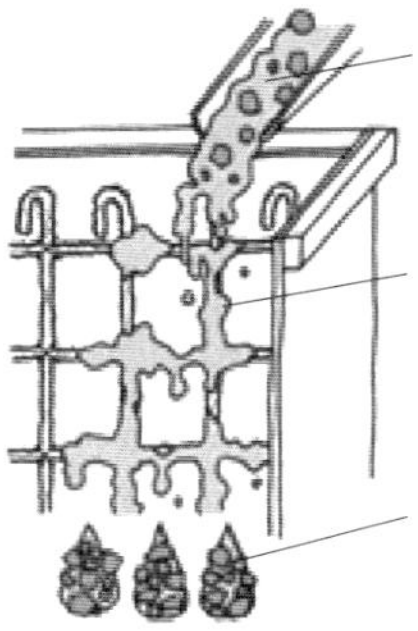

Figure 2. Mortar lost model atoncrete placing.

3 MEASUREMENT OF LOSS MORTAR

3.1 Experimental Method

3.1.1 Concrete samples: formulation and mixing

Table 1 tabulates the materials used to concrete test samples: ordinary Portland cement (OPC), tap water (W), sea sand (S) and crushed stone (G). Mix proportions of concrete test samples are shown in Table 2. We prepared for four mix proportions and assigned target slump and target slump flow values to each. All were prepared by first dry mixing ordinary Portland cement and sea sand, injecting water, mixing to uniformity and then adding crushed stone. The air content of the mixed concrete was 4.5 ± 1.5%. In consideration of the time dependence of slumping behavior, we subjected the concrete samples to testing within two hours after water injection.

Table 1. Using materials.

OPC: density 3.16g/cm^3
S: density(surface-dry) 2.59g/cm^3, F.M.2.4
G: density(surface-dry) 2.70g/cm^3, F.M.6.9

Table 2. Mixture proportions; kg/m^3.

	Type	W/OPC	W	OPC	S	G
I	Slump 8cm	50%	161	322	750	1,087
II	Slump 18cm	50%	175	350	764	1,010
III	MFC	50%	175	350	823	949
IV	SCC	35%	170	486	871	796

3.1.2 Measurement method

We poured our concrete samples into test frames (200mm wide, 200mm thick, 200mm high) and measured the amount (mass) of mortar that adheres to rebar within them ("loss mortar"). The frames had no bottom plate. Thus, all concrete other than loss mortar simply flowed out through the bottom. Test parameters are, type of concrete formulation; the amount of concrete flowing through the frame; and diameter/arrangement of rebar within each frame. For rebar, we used a deformed reinforcing bar of diameter 13, 16, 25, or 32 mm. The rebar was arranged within each frame as one horizontal bar; as one vertical bar; or as two horizontal bars/ two vertical bars (four points of intersection). On the other hand, we collected experimental data about the influence of rebar position and sparseness on loss mortar. Frames arranged nine horizontal bars at intervals of 100~200mm or one horizontal bar 200~1,800mm long were used for the experiments.

3.2 Experimental Results and Discussion

Figures 3 and 4 show relations between (1) the product of (a) the square root of flow volume and (b) total rebar periphery, and (2) loss mortar mass. We see that the amount of loss mortar is proportional to this product. Here, we propose Equations (1) and (2) as means to calculate loss mortar mass due to horizontal and vertical bars.

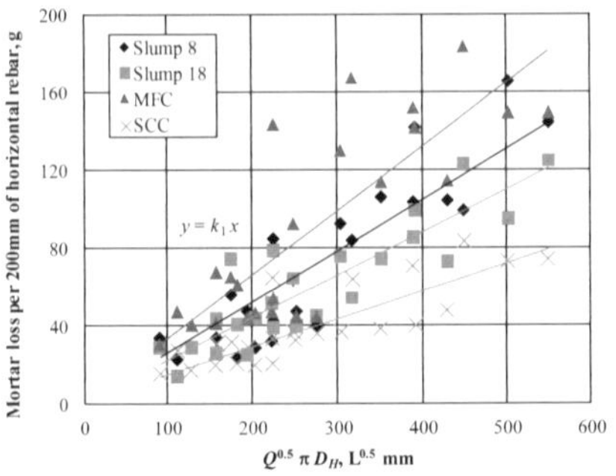

	Slump 8	Slump 18	MFC	SCC
R^2	0.95	0.95	0.92	0.93
k_1	0.26	0.22	0.33	0.14

Figure 3. The change in loss mortar mass with horizontal bar.

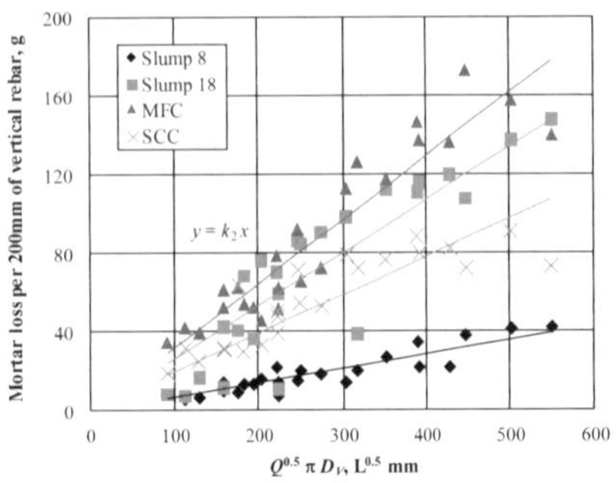

	Slump 8	Slump 18	MFC	SCC
R^2	0.96	0.94	0.98	0.95
k_2	0.071	0.27	0.32	0.19

Figure 4. The change in loss mortar mass with vertical bar.

$$M_{H200} = k_1\sqrt{Q}\pi D_H \quad (1)$$

$$M_{V200} = k_2\sqrt{Q}\pi D_V \quad (2)$$

M_{H200}: Mortar loss per 200mm of horizontal rebar, g
M_{V200}: Mortar loss of 200mm of vertical rebar, g
Q: Concrete flow volume, L
D_H: Diameter of horizontal rebar, mm
D_V: Diameter of vertical rebar, mm
k_1, k_2: Constant, g/$L^{0.5}$mm

The change in loss mortar mass with horizontal/vertical bar intersection is shown in Figure 5. We note that loss mortar mass is proportional to the product of (a) the square root of the sum of mortar loss at horizontal and vertical rebar and (b) the square of rebar diameter. Here, we propose Equation (3) as a function with which to calculate the amount of loss mortar as a result of such intersections.

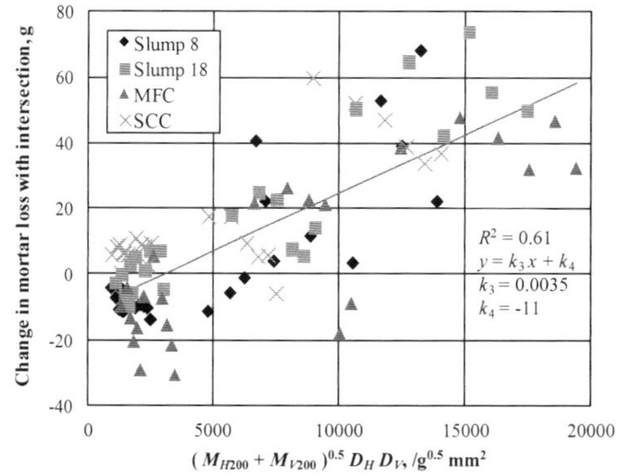

Figure 5. The change in loss mortar mass with horizontal/vertical bar intersection.

$$M_I = k_3\sqrt{M_{H200} + M_{V200}}D_H D_V + k_4 \tag{3}$$

M_I: Change in mortar loss with horizontal/vertical bar intersection, g
k_3: Constant, g$^{0.5}$/mm^2
k_4: Constant, g

Figure 6. shows the influence of horizontal bar position and sparseness on loss mortar. The most loss mortar adheres to the second bar from the top. On the third and under the bar, the amount of loss mortar decreases with sequence and converges on a constant value. The constant value increases with increasing interval of the horizontal bar.

Figure 7 shows the influence of vertical bar position on loss mortar. The most loss mortar adheres to the vertical bar in the range from the top down to 200mm. The vertical bar in the others range has loss mortar of about 0.4 times as compared with the range from the top down to 200mm.

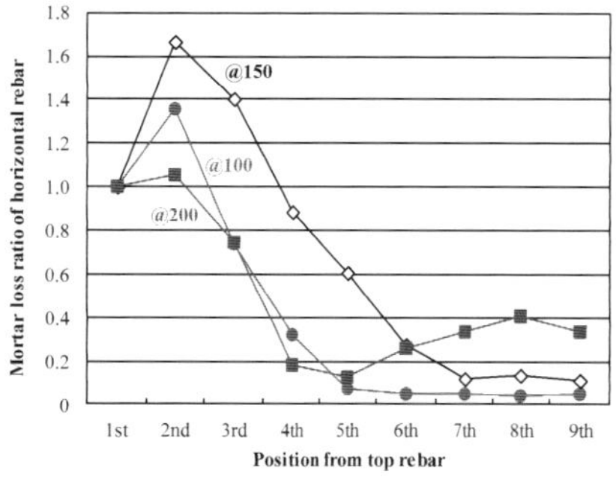

Figure 6. The influence of horizontal bar position and sparseness on loss mortar.

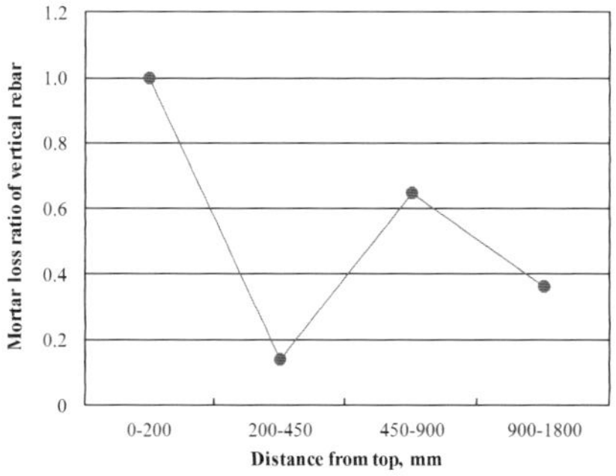

Figure 7. The influence of vertical bar position on loss mortar.

4 ESTIMATION OF DEFECTIVE FILLING RISK

We next utilize our mortar loss model to estimate the risk of defective filling when pouring concrete to form a wall of a reinforced-concrete structure. The rebar mesh for this test wall was determined in reference to a wall formwork presented by the Architectural Institute of Japan as a standard for structural calculations (AIJ 2010). Concrete materials and their formulation are as described earlier in this paper. The dimensions of the wall are taken as a floor height of 3,600mm and a thickness of 200mm. The mortar component of the concrete is assumed to adhere to rebar over a horizontal distance of 200mm. The angle of repose of concrete containing excess aggregate at the bottom of the frame is assumed to be 45°: value of crushed stone (Glover 1995). Here, we utilized Equations (4) and (3') to calculate the total amount of loss mortar to result as concrete is poured into rebar-meshed frames.

$$M = \sum \frac{M_{H200} L_H}{200} \gamma_{Hi} + \sum \frac{M_{V200} L_V}{200} \gamma_{Vi} + \sum M'_I \tag{4}$$

$$M'_I = k_3 \sqrt{M_{H200}\gamma_{Hi} + M_{V200}\gamma_{Vi} D_H D_V} + k_4 \tag{3'}$$

M: Total amount of mortar loss, g
L_H: Horizontal rebar length, mm
L_V: Vertical rebar length, mm
γ_{Hi}, γ_{Vi}: Correction factor
M'_I: Corrected change in mortar loss with horizontal/vertical bar intersection, g

γ_{Hi} is assumed to be the value in Table 3. γ_{Vi} is assumed to be 1.00 in the range from the top down to 200mm, 0.38 in the range under 200mm. k_1~k_4 are assumed to be the value in Figures 3~5.

Table 3. Correction factor.

Interval	Position from top rebar				
	1st	2nd	3rd	4th	5th, Under
100mm	1.00	1.35	0.74	0.32	0.05
200mm	1.00	1.05	0.74	0.27	0.27

The unit coarse aggregate bulk volume within the concrete arriving at the bottom of the frame is calculated with Equation (5) as a function of the amount of loss mortar thus obtained.

$$G_{BV} = \frac{G_V + \frac{\Delta M}{\Delta Q}\frac{1}{\rho}}{G_{lim}} * 10^{-3} \tag{5}$$

G_{BV}: Unit coarse aggregate bulk volume, m^3/m^3
G_V: Absolute volume of coarse aggregate in sound concrete, L/m^3
ρ: Density of mortar component, kg/L
G_{lim}: Solid content of coarse aggregate, m^3/m^3

For the purposes of this paper, we take a risk of defective filling to arise when the mortar fraction is no longer able to fill in gaps/voids produced as aggregate fills on its own. More specifically, we judge this risk to occur when the unit coarse aggregate bulk volume is over $0.76 m^3/m^3$.

Calculated results are shown in Figure 8. Under rebar interval of 100mm and concrete slump 18cm, the unit coarse aggregate bulk volume is over 0.76 from the bottom of the frame up to a distance of about 130mm into the frame (a situation suggestive of a risk of forming a defective filled layer). The rebar arrangement, which is highly intricate, is predicted to present a comparatively high risk of defective filling (and from a higher level from the bottom of the frame). This is because of the increase of mortar loss at the vertical bar with narrowing of the interval. On the other hand, SCC is predicted to present a comparatively low risk of defective filling. This is because of a little absolute volume of coarse aggregate in SCC.

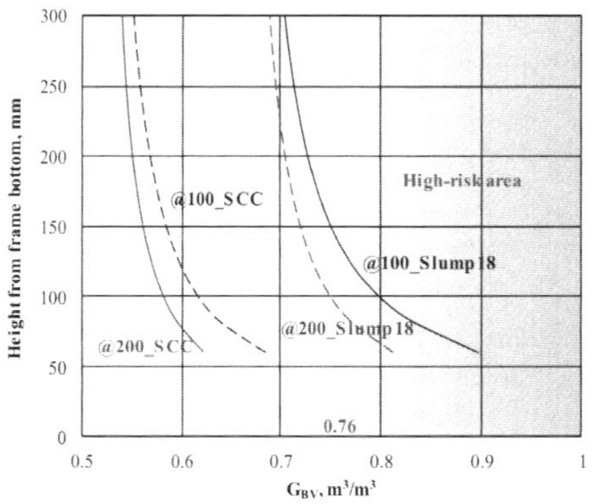

Figure 8. Calculated results.

5 CONCLUSIONS

Our findings are as follows:

(1) As a model for defective filling, we propose a mortar loss model in which the mortar component of concrete falling freely down into a frame is considered to adhere to rebar inside the frame. Concrete arriving at the bottom of the frame thus contains an excess amount of aggregate, a situation considered prone to the defective filling.

(2) The amount of loss mortar adhering to horizontal and vertical bars is taken to be proportional to the product of (a) the square root of concrete flow volume and (b) rebar periphery. The change in the amount of loss mortar resulting from rebar intersections can be approximated as a linear function of the product of (a) the square root of the sum of mortar loss at horizontal and vertical rebar and (b) the square of rebar diameter.

(3) The most loss mortar on of horizontal bar adheres to the second bar from the top. On the third and under the bar, the amount of loss mortar decreases with sequence and converges on a constant value. The constant value increases with increasing interval of the horizontal bar. On the other hand, the most loss mortar of vertical bar adheres to the range from the top down to 200mm.

(4) Using actual measurement data, we quantitatively estimated the risk of defective filling. Here, an intricate rebar arrangement is more prone to defective filling, and SCC is less prone to the defective filling.

Acknowledgments

This work was supported by JSPS KAKENHI Grant Number 16K18191.

References

AIJ, *Architectural Institute of Japan Standard for Structural Calculation of Reinforced Concrete Structures*, 2010.

Glover, T. J., *Pocket Ref.*, Sequoia Publishing, 1995.

Japan Concrete Institute, Study on the Occurrence Probability of Risk Factors in Concrete Construction, *Concrete Construction Board Meeting Report*, ISBN 978-4-931451-87-2-C3050, 2008.

Streamlining Information Transfer between Construction and Structural Engineering
Edited by Shiau, J., Vimonsatit, V., Yazdani, S., and Singh, A.

ISBN: 978-0-9960437-7-9

DURABILITY AND POZZOLANIC REACTION OF FLY ASH CONCRETE EXPOSED TO VARIOUS OUTSIDE ENVIRONMENTS

YOSHITAKA ISHIKAWA

Electric Power Development Co., Ltd., Chigasaki Research Institute, Chigasaki, Japan

In this study, outside exposure tests of fly ash concrete were started in 2009 at various places in Japan in order to assess the effect of the differences of the exposure environment on the variation of long term durability of fly ash concrete and the degree of pozzolanic reaction of fly ash. This paper shows the results of durability monitoring over seven years from the start of exposure testing and the degree of pozzolanic reaction after seven years. The durability of fly ash concrete is improved due to pore structure complicated by the pozzolanic reaction. The degree of pozzolanic reaction of fly ash in concrete exposed under any conditions ends up being the same when enough time has elapsed, such as seven years, regardless of differences in exposure conditions.

Keywords: Compressive strength, Volume resistivity, EPMA, Pore size distribution, Tortuosity, Chloride ion permeation resistance.

1 INTRODUCTION

Fly ash (FA) is a byproduct of coal-fired power generation. When mixed with concrete, fly ash contributes to improved long-term strength, reduced water content per unit volume, prevention of temperature cracking, improved durability, and so on. The mechanism that improves the durability of FA concrete is considered to be the densification of the hardened cement structure that proceeds with the pozzolanic reaction introduced by the FA (Japan Fly Ash Association 2015). Generally, concrete durability is assessed by accelerated testing of various kinds. However, the relationship between the degradation observed under accelerated test conditions and the degradation observed in concrete structures in their real service environments is not always clear. Real concrete structures are exposed to many variations in climatic conditions, temperature conditions and other conditions, making it difficult to evaluate their durability in a uniform manner. To evaluate the durability of FA concrete in real environments, we have been conducting outdoor exposure testing of FA concrete at three locations in Japan since 2009. We used wall-shaped large specimens for exposure testing in order to better simulate conditions experienced by real structures. Each year, we have been extracting concrete core samples from these test specimens in our year-to-year concrete durability monitoring activities. This paper covers FA concrete monitoring results over seven years from the start of exposure, providing data on the aging degradation of durability-related physical properties such as strength and volume resistivity in different environments. Aiming to find out how pozzolanic activity may vary according to environment, we selected three test locations differing greatly in terms of environmental conditions, and analyzed data from concrete core samples in three test locations for seven years from the start of exposure, paying attention to hydrate and pore size distribution in concrete structure.

2 OVERVIEW OF EXPOSURE TESTING

2.1 Tested Concrete Types and Composition

Table 1 lists the materials composing the tested concrete specimens. Table 2 shows the composition in terms of mixing ratios. The FA used is FA of Type II as defined in Japan Industrial Standard. The exposure testing addressed FA concrete of two types that differ in the FA substitution ratio: 20 % substitution (FA20) and 30 % substitution (FA30). For comparison, testing was performed also on normal concrete made of Portland cement (OPC). As to concrete composition, we chose the widely used mixing ratios of W/P = 55 %, s/a = 43 %, slump = 12 cm, and air mixture = 4.5 %.

2.2 Exposure Specimens and Exposure Sites

As shown in Figure 1 each exposure test specimen is given a wall-like geometry: 770 mm (W) x 540 mm (H) x 300 mm (D). For the evaluation of year-after-year degradation, a single concrete core sample (74 mm in diameter) was extracted from each specimen every year. The specimen was kept in the formwork for seven days after casting. After the formwork was removed, each specimen was left alone indoors for four weeks to age the material and then placed outside in test location to start outdoor exposure testing. The specimens were exposed to environments at three locations that differ greatly in terms of environmental conditions as shown in Table 3. All specimens are undergoing exposure testing since 2009.

2.3 Durability Monitoring through Concrete Core Sampling

We have been extracting a single concrete core sample from each exposure test specimens every year. The following tests and measurements performed on concrete core samples every year.

(1) Compressive strength

The specifications of the samples used in compressive strength testing were as follows: cylindrical samples of 100 mm in diameter and 200 mm in height for test performed at the material age of 28 days; cylindrical samples of 74 mm in diameter and 100 mm in height cut out from concrete core samples for tests performed annually from one to seven years from the start of exposure.

(2) Volume resistivity

Volume resistivity of concrete is highly related with the chloride ion permeation resistance. From this viewpoint, we used the volume resistivity as an indicator of durability in this study. The volume resistivity of each core sample was measured according to JSCE K 562-2008 (2008).

(3) Pore size distribution

The structural density of the concrete is identified as a factor contributing to durability. In order to evaluate this structural density, we performed pore size distribution measurement using mercury press-in method. Each sample for pore size distribution measurement was prepared by using a cutter blade to cut an approximately 5 mm square block out of the mortar portion of the concrete core sample at a depth of 150 to 160 mm from the surface, and each extracted sample was used after freeze drying.

Table 1. Materials used.

Specification
Ordinary Portland Cement Density: 3.16 g/cm^3, Blaine specific surface area: 3,310cm^2/g
JIS type II, Density: 2.25 g/cm^3, Blaine specific surface area: 3,800 cm^2/g
Sea sand, Density: 2.57 g/cm^3, Water absorption: 1.30%
Crushed stone (G_{max}=20mm), Density: 2.85 g/cm^3, Water absorption: 1.58%
AE water reducing agent AE agent

Table 2. Mix proportion of Concrete.

Type of Concrete	W/P [%]	s/a [%]	F/(C+F) [%]	Unit content [kg/m^3] W	C	F	S	G
OPC	55	43	0	159	289	0	779	1,144
FA20			20	151	220	55	785	1,154
FA30			30	146	186	80	790	1,161

Table 3. Exposure site.

Site (Symbol)	Exposure environment
Hokkaido (Site H)	Subarctic zone, Inland area
Kanagawa pref. (Site K)	Temperate zone, Inland area 2km far from seashore
Okinawa pref. (Site O)	Subtropical zone, Marine environment

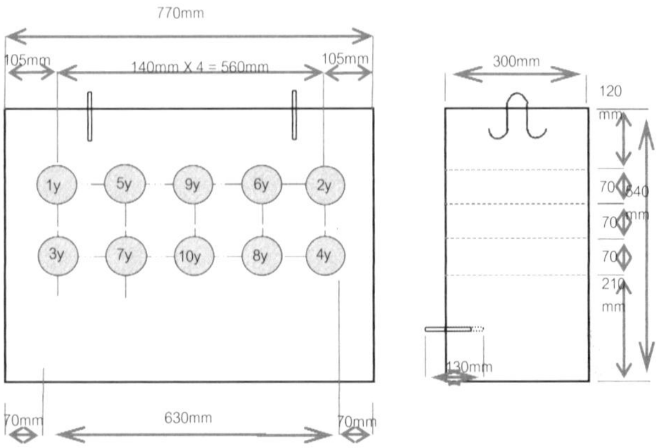

Figure 1. Outline of Exposed specimen.

2.4 Evaluation of Pozzolanic Activity

As targets, we chose specimens exposed to greatly different environments at three different locations: Site H in a cold zone, Site K in a temperate zone, and Site O in a sub-tropical zone. We evaluated pozzolanic activity in FA concrete specimens exposed for seven years. We evaluated pozzolanic activity using the following approach:

Presence or absence of $Ca(OH)_2$: We determined the presence or absence of $Ca(OH)_2$, which is required for sustaining pozzolanic activity, by means of TG analysis and XRD analysis. The sample for analysis was taken from the mortar portion of the core samples and preconditioned like a sample for pore size distribution measurement. This time, however, the material was analyzed after being crushed using a disk mill.

3 TEST RESULTS AND DISCUSSIONS

3.1 Year-to-Year Changes in Compressive Strength

Figure 2 shows year-to-year changes in the compressive strength (at three locations that greatly differ in environmental conditions). At every exposure test site, the difference in compressive strength between OPC concrete and FA concrete, which was evident before the start of exposure, almost disappeared by the third year of exposure. The results of measurements up to seven years from the start of exposure demonstrate some variation in strength over time because measurements were taken every year on a single core sample only. Nevertheless, at every site and irrespective of the FA substitution ratio, the strength was found to be similar between OPC concrete and FA concrete even after seven years had elapsed from the start of exposure.

3.2 Year-to-Year Changes in Volume Resistivity

Figure 3 shows year-to-year changes in the volume resistivity ρ. The curve in the chart shows regression by $\rho(t) = abt/(1+bt)$, where a and b are constants while t represents the exposure time. The greatness of volume resistivity value was in the order of FA30 > FA20 > OPC. The volume resistivity measured on OPC was low and remained almost constant. We found that the volume resistivity of FA concrete continued to increase in the first three to four years, then the rate of increase slowed down until the value almost stabilized at a certain level.

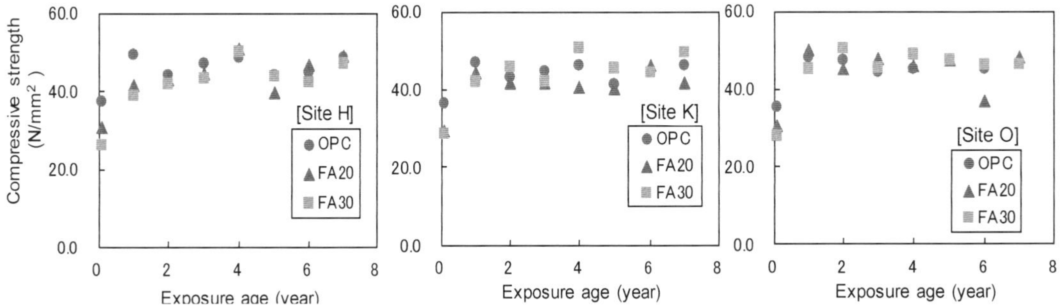

Figure 2. Changes in compressive strength with exposure age.

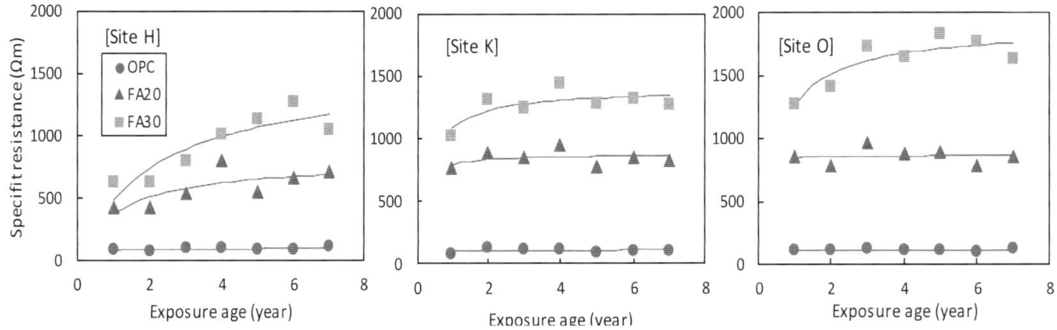

Figure 3. Changes in volume resistivity with exposure age.

3.3 Relationship between Pore Structure and Volume Resistivity

Figure 4 shows the relationship between the total pore specific surface area, derived from the pore size distribution measurement, and the volume resistivity. As shown in the chart, the increase in the total pore specific surface area was found to be in a linear relationship with the increase in the volume resistivity. The larger the volume resistivity of FA concrete, the larger is the total pore specific surface area. The greatness of volume resistivity suggests complexity of the pore structure, which may be referred to as "tortuosity". From the relationship between diffusion coefficient D, volume resistivity ρ, and tortuosity[1], we determined the ratio of FA to OPC in terms of tortuosity in pore structure (See Table 4. Note that the volume resistivity was derived from a regression curve). Based on this, it is surmised that the substitution of OPC by FA reduces the

[1] $\tau = D \propto \rho^{-1}, D \propto \tau^{-1}$

diffusion coefficient of the FA concrete by a factor of about 1/8 to 1/7 in the case of FA20, or by a factor of about 1/15 to 1/11 in the case of FA30.

Table 4. Tortuosity ratio of FA to OPC.

Symbol	Exposure site H	A	K	F #1	F #2	O
FA20	7.22	7.97	7.69	8.34	7.09	7.55
FA30	12.30	11.24	12.00	15.07	13.81	15.46

Table 5. Value of Ca/Si at the peak of frequency distribution curve.

Symbol	Exposure site H	K	O	Average
OPC	1.95	2.00	1.90	1.95
FA20	1.70	1.70	1.75	1.72
FA30	1.65	1.60	1.65	1.63

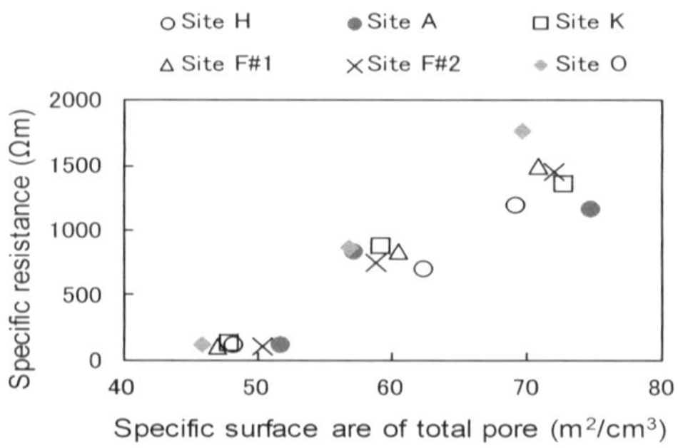

Figure 4. The relationship between the Specific surface area of total pore and volume resistivity.

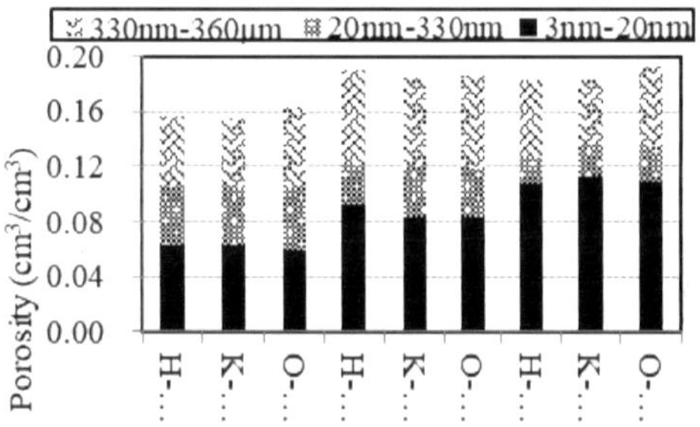

Figure 5. Pore size distribution.

3.4 Changes in the Pore Size Distribution with the Progress of Pozzolanic Reaction

Figure 5 shows pore size distribution determined by mercury press-in method. Pore diameters were divided into the following three ranges: 3 to 20 nm, 20 to 330 nm, and 330 nm to 360 μm. We can see from the chart that, as the FA substitution ratio increases, the total volume of pores with a diameter in the range between 3 to 20 nm increases while the total volume of pores with a diameter in the range between 20 to 330 μm decreases. The pores increased with a diameter in the range between 3 to 20 nm and the decrease of pores with a diameter in the range between 20 to 330 μm resulting from substitution by FA. It is assumed to be caused by changes in pores that take place with the progress of the pozzolanic reaction introduced by the FA. As to variation across sites, we found little variation in the volumes of pores of the above mentioned ranges that are affected by the pozzolanic reaction.

As a conclusion from the above, the study of the pore size distribution did not demonstrate any significant variation across sites in the progress of pozzolanic activity introduced by FA or in changes to pores resulting from it. This observation is consistent with the absence of variation across sites in the results of compressive strength tests performed on specimens after seven years from the start of exposure (Figure 2). In this study, we tested concrete after seven years from the start of exposure, and the test results suggest that, after the passage of this much of time, there would be little variation in the progress of pozzolanic activity across different sites distributed over the subarctic, temperate, and subtropical zones of Japan.

3.5 Presence or Absence of $Ca(OH)_2$

Figure 6 shows DTG analysis results of the mortar portion to seven years from the start of exposure. The chart shows for OPC a reduction of the weight of $Ca(OH)_2$ in the 400 to 500 °C temperature range. For FA concrete, however, the chart does not show any significant reduction of the weight of $Ca(OH)_2$. Figure 7 shows XRD analysis results. For OPC, the XRD analysis results indicate a clear peak of $Ca(OH)_2$ around 18.1° and also around 34.1°. For FA concrete, however, irrespective of exposure sites, the XRD analysis did not indicate any significant presence of $Ca(OH)_2$. As a conclusion from the above results, in all FA concrete specimens, almost all $Ca(OH)_2$ in the hardened substance had been consumed by the pozzolanic reaction, preventing any further continuation of pozzolanic activity. That is to say, even though the FA concrete specimens were exposed to different environmental conditions, the progress of the pozzolanic reaction in them over a long period of time had consumed almost all $Ca(OH)_2$ in them before seven years had elapsed, preventing any further continuation of pozzolanic activity.

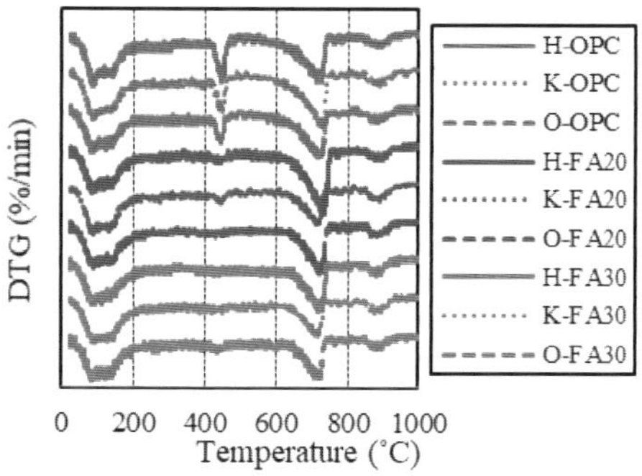

Figure 6. Results of DTG analysis.

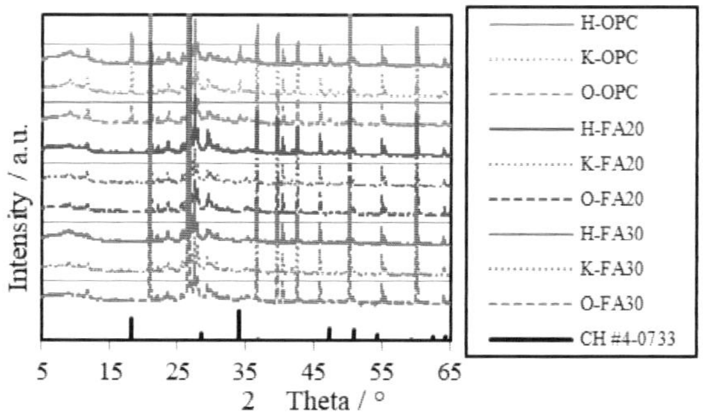

Figure 7. Results of XRD analysis.

4 SUMMARY

The following lists the findings in this study:

(1) The strength of FA concrete becomes approximately similar to that of OPC concrete by the third year from the start of exposure. The FA concrete compressive strength was found to vary across exposure sites up to two years had elapsed from the start of exposure, but no such variation was found in tests performed after three or more years from the start of exposure.
(2) The volume resistivity of FA concrete continued to increase for the first three to four years, then the rate of increase slowed down until the value was mostly stabilized at a certain level.
(3) It is surmised that the substitution of OPC by FA reduces the diffusion coefficient of FA concrete by a factor of about 1/7 to 1/8 in the case of FA20, and by a factor of about 1/11 to 1/15 in the case of FA30. We believe that this happened as pore structures became more complex as a result of the pozzolanic reaction introduced by the FA.
(4) We believe that the absence of variation across exposure sites is due to the fact that, at every exposure site, the calcium hydroxide in specimens was almost totally consumed by the time seven years had elapsed, preventing pozzolanic activity introduced by FA from continuing further.

References

Japan Fly Ash Association, *Coal Ash Handbook*, Japan, 2015.
JSCE K 562-2008, *Procedure for the Measurement of Volume Resistivity of Profile Repair Compound by the Four Electrode Method*, Japan Society of Civil Engineers, 2008.

Streamlining Information Transfer between Construction and Structural Engineering
Edited by Shiau, J., Vimonsatit, V., Yazdani, S., and Singh, A.

ISBN: 978-0-9960437-7-9

A REVIEW OF DEM-BASED SIMULATION OF FRESH CONCRETE FLOW

LIWEI HENG, YAN ZHUGE, MD. MIZANUR RAHMAN, and XING MA

School of Natural and Built Environments, University of South Australia, Adelaide, Australia

Recently, there has been growing research interest in simulation on the flow behaviors of fresh concrete, because these properties have significant effects on construction quality of casting and forming process, and also the properties of hardened concrete. The behaviors of fresh concrete are usually described by workability, consistency, flowability, mobility, pumpability, shootability, compactibility, and finishability. While various numerical simulations for concrete flow are available, District Element Method (DEM) has been applied widely due to its unique advantages of both macro and micro analysis i.e. the flow of the concrete mixture as a whole and particulate behavior of individual constituents in a mixture. This paper first presents a brief overview on the rheological models of fresh concrete, followed by a detailed review on simulating the fresh concrete flow using the DEM method. Moreover, recommendations for future work and directions will be provided.

Keywords: District element method, Rheology, Single-Phase Element Model, Separate Single-Phase Element Mode.

1 INTRODUCTION

Concrete plays a significant role in civil engineering materials all around the world. There has been a tremendous amount of research on hardened concrete, but only recently fresh concrete is drawing researchers' attention since the properties of fresh concrete have massive impacts on construction quality of casting and forming process, and the properties of hardened concrete as well (Jiao *et al.* 2017). Tattersall (1976) attempted to study flow behaviors of fresh concrete using a rheological method. The rheology is an effective tool to characterize the behaviors of fresh concrete, which are normally described as workability, consistency, flowability, mobility, pumpability, shootability, compactibility and finishability (Wallevik and Wallevik 2011). In this paper, various computational models of fresh concrete rheology will be overviewed first, followed by a detailed review on DEM simulation.

According to Roussel *et al.* (2007), there are mainly three kinds of computational modelling of concrete flow, which are *single fluid simulation*, *suspension flow simulation*, and *simulating of discrete particle flow*. Single fluid simulation is the most commonly used method in modelling fresh concrete flow. Mori and Tanigawa (1992) introduced Viscoplastic Finite Element Method (VFEM) and Viscoplastic Divided Element Method (VDEM) to model fresh concrete flow, assuming that fresh concrete was a homogeneous single fluid. Computational Fluid Dynamics (CFD) was also applied to simulate single fluid through the numerical solution of the governing partial differential equations or other mathematical equations of motion (Wesseling 2000).

Simulation of suspension flow models the particles suspended in a fluid. Visco-plastic suspension element method (VSEM) was derived from VFEM (Mori 1992) and used to predict the rheological properties of the concrete suspension from the rheological properties of the matrix and the coarse aggregate volume fraction.

Distinct element methods include standard discrete (distinct) method (DEM) and Dissipative Particle Dynamics method (DPD). DEM will be reviewed further in this paper. DPD has been presented by Martys (2005) in the field of cementitious material recently. In DPD, the chosen inter-particle interactions allow for much larger time steps. This makes it possible to study physical behavior on different levels of time scales.

2 DEM SIMULATION ON FRESH CONCRETE AND RHEOLOGICAL MODELS

DEM was initially developed by Cundall (1971) to analyze rock mechanics problems and then applied to soil problems and was introduced to analyse fresh concrete flow as a numerical method using the discontinuous body model (Petersson 2001).

The flow of fresh concrete should be defined as two parameters. One of them is the yield stress, since it shows an initial resistance to the flow, and the other is plastic viscosity, which governs the flow after it was initiated. Thus, fresh concrete and other cementitious suspensions are normally considered as a Bingham model material, a plastic material shows little or even no deformation up to a certain level of stress. Bingham model describes the rheological properties of fresh concrete with two parameters, yield stress and plastic viscosity.

The constitutive equation can be expressed as in Eq. (1) and Eq. (2):

$$\tau = G\gamma \quad \text{for } \tau < \tau_0 \tag{1}$$

$$\tau = \tau_0 + \mu\dot{\gamma} \quad \text{for } \tau \geq \tau_0 \tag{2}$$

where τ is shear stress; τ_0 is yield stress, which gives the maximum value for the static behavior of the material; γ is the shear deformation; $\dot{\gamma}$ is the shear rate; μ is a constant plastic viscosity, G is the spring constant.

When $\tau < \tau_0$, the mixture is in a solid state. When $\tau \geq \tau_c$, the mixture is yield to a liquid state and start to flow. Once the external force is reduced, and the condition is back to $\tau < \tau_0$, the concrete will stop flowing and reach a new solid state. Therefore, the yield stress is the most important parameter, which affects the shape and diameter of the flow spread.

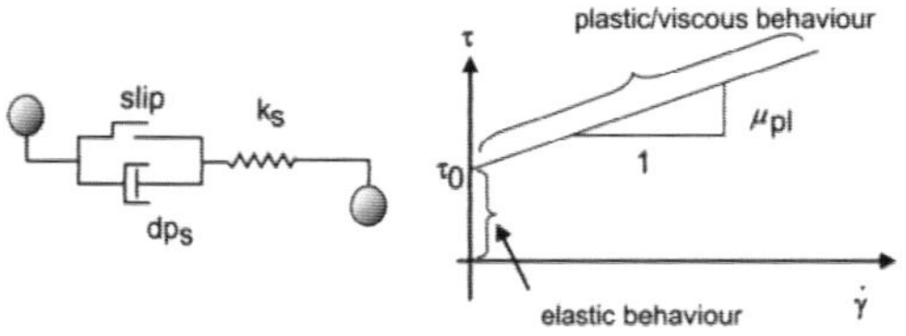

Figure 1. A Bingham material model (Mechtcherine *et al.* 2014).

One may note, there are many other yield-stress fluid methods commonly used in addition to the Bingham fluid model, for example, the Casson and the Herschel-Bulkley fluids (Mitsoulis 2007). However, the details of these models are not included in this paper.

3 DEM MODELS OF FRESH CONCRETE BEHAVIOR

The key point of simulating fresh concrete with DEM is to develop a reasonable DEM model. The available DEM for flow analysis of a concrete mixture can be divided into three main subgroups: single-phase element model, two-phase element model and multi-phase element model (Hoornahad and Koenders 2014, Quoc and Uomoto 2008). Because fresh concrete is made up of coarse aggregates and mortar with characteristics of a fluid, coarse aggregate can be described as discrete elements, and the assumption of mortar becomes the only difference in each model.

3.1 Single-Phase Element Model

According to Nabeta (1989), who was the first one to introduce DEM into fresh concrete simulation, the fresh concrete was considered as a single-phase material.

When developing a fresh concrete DEM model, Nabeta (1989) and Maki (1998) simplified the concrete mixture as follows: coarse aggregates are enclosed completely by a layer of mortar, but the mixture was considered as a single-phase material. In this case, contacts and collisions only take place between mortar cells (Figure 2). This model is quite similar to the traditional DEM model applied on fracture mechanics and granular systems for rock and soil (Quoc and Uomoto 2008).

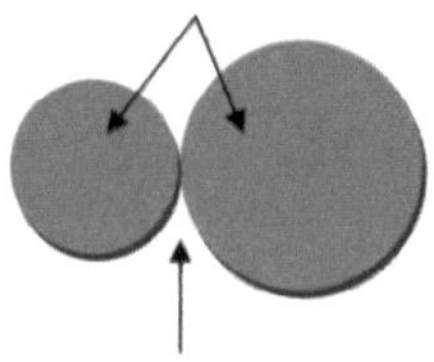

Figure 2. Single-phase model with gravel completely enclosed by mortar (Quoc and Uomoto 2008).

In this model, the motion of mortar can only take place with the movement of gravel but cannot move without gravel, and the interactions between coarse aggregates were ignored. Therefore, large deviations between the numerical and experimental results were usually observed. Although the single-phase model has significant drawbacks, it is a milestone in fresh concrete modeling, offering a possibility for the later research.

3.2 Two-Phase Element Model

In 1996, Chu developed a modified DEM model based on the work of Nabeta (1989). The mixture was divided into a mortar phase and a coarse aggregate phase, represented by an assembly of two-phase elements consisting of inner hard core coarse aggregates covered with a layer of the mortar which was very low in stiffness (Hoornahad and Koenders 2014). Although it was simplified that the coarse aggregates enclosed by mortar ring, which was similar to single-phase model, the aggregates could contact and collide with either mortar or aggregates based on the extent of overlapping. As shown in Figure 3, if the overlapping exceeds the thickness of mortar ring, contacts and collisions will take place between aggregates.

This two-phase assumption makes a step closer to the real situation, but it still considers mortar as a whole mixture with only mechanical behavior, but without specific geometrical characteristics (Quoc and Uomoto 2008). Besides, there are simplifications about the homogeneous distribution of mortar. This model was applied for simulation flow behavior and shootability of shotcrete (Puri *et al.* 2002).

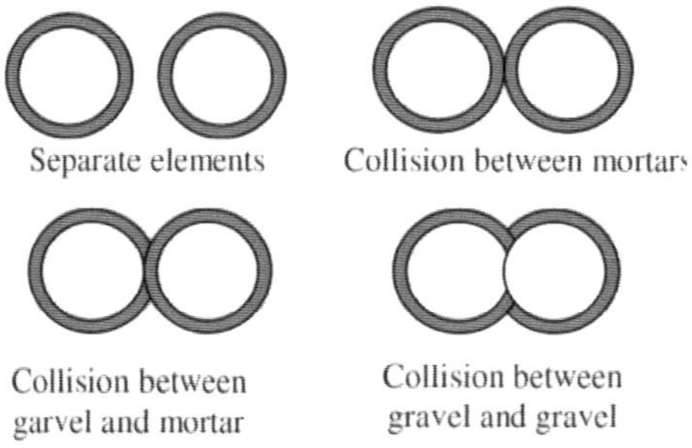

Figure 3. Illustration of different types of contact in two-phase modelling(Quoc and Uomoto 2008).

Two-phase element model is the most widely used DEM model for fresh concrete flow. For example, Hoornahad (2014) used a two-phase model to simulate the self-compacting mixture behaviors. Li *et al.* (2016) applied the two-phase model on the prediction of time-dependent flow behaviors of fresh concrete, and achieved a reasonable comparison between experimental and numerical result (see Figure 4).

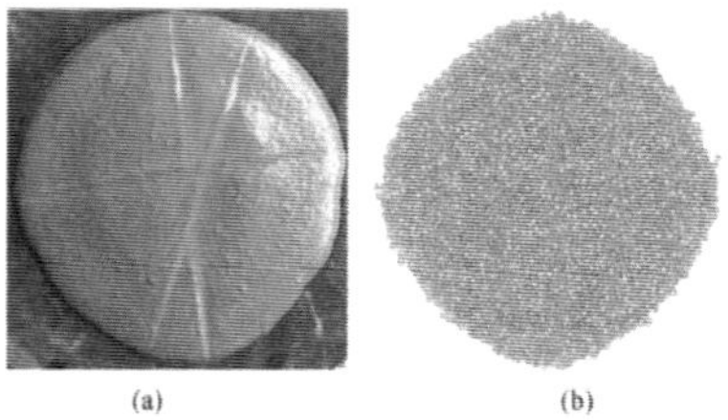

Figure 4. The final shape of the mortar on the flow table: (a) experiment (mean flow value: 210 mm), and (b) numerical simulation (mean flow value: 212mm) (Li *et al.* 2016).

3.3 Multi-Phase/ Separate Single-Phase Element Mode

In this model, the concrete mixture is divided into a mortar phase and a coarse aggregate phase, represented by a combination of separate-single phase elements including mortar and coarse aggregate (Hoornahad and Koenders 2014).

Noor (2000) adopted a new model in which aggregate and mortar elements had been modelled using separate elements as shown in Figure 5.

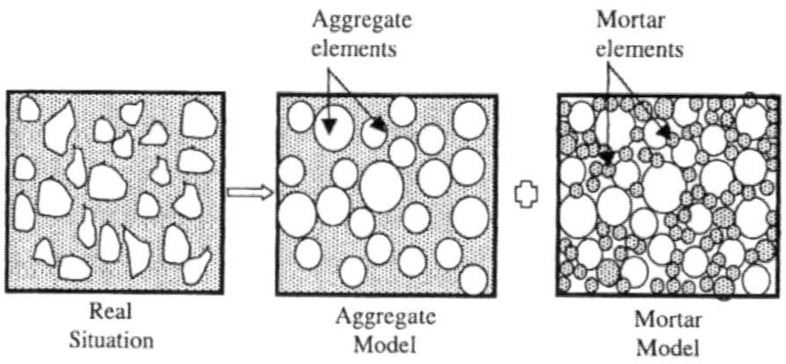

Figure 5. Model with separate single-phase particles for mortar and gravel (Noor 2000).

Since there were a vast number of fine aggregates, which would make the numerical simulation complicated and become time-consumed, the DEM parameters of mortar and aggregate elements were set and verified separately, for example, contact stiffness, bond strength for both shear and tangential direction, and friction factor between two elements at each contact point (Noor 2000). Noor (2000) gave a new method to choose appropriate simulation parameters and successfully reduced the element number. The numerical results using this new method was compared with experimental results and showed a good match. Therefore, the separate single-model can be applied to analyze the flow behaviors of fresh concrete and predict some common phenomenon such as the coarse aggregate blocking.

Taking all the advantages and disadvantages of existing DEM models into consideration, Quoc (2002) presented a new DEM model, in which the advantage of setting parameters of fine mortar and coarse aggregate element separately was adopted, while mortar element was divided into two parts, known as the spacing mortar and bonding mortar (as shown in Figure 6). Spacing mortar element was used to keep coarse aggregate element at a certain distance from each other, and bonding mortar was used to fill the space between space particles and coarse gravel. As the key point to the model, the parameters of these two mortar elements were required to be set up separately. Quoc (2002) introduced neural network to study the relationship between the rheological properties parameters and the numerical parameters of fresh concrete.

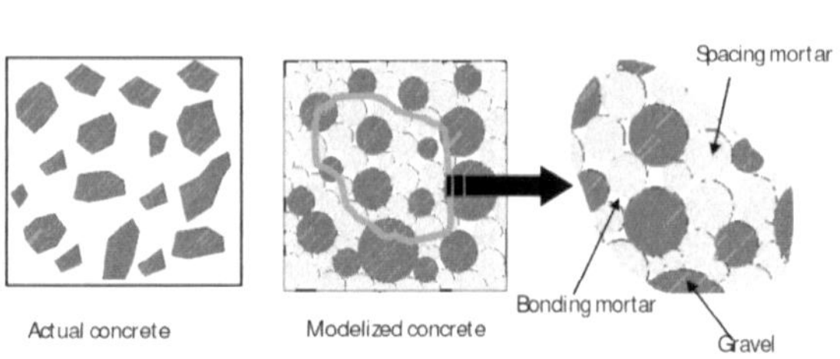

Figure 6. Modified model with two phases in separate particles (Quoc and Uomoto 2008).

4 CONCLUSIONS

In this paper, several numerical modelling methods of fresh concrete flow have been overviewed, including DEM. The paper focused on the rheological models and DEM models used in fresh concrete simulation.

The assumption of the rheology model is based on Bingham Model, however, De Larrard *et al.* (1998) pointed out that the relationship between torque and rotation speed was not exactly linear and even a negative yield stress countered with the Bingham model in their rheological tests. Thus, more research is required in the future to consider other rheology models such as Newtonian model (Mitsoulis 2007), Herschel- Bulkley (H-B) model (De Larrard *et al.* 1998), and modified Bingham model, etc.

Although DEM model has been successfully used to simulate the fresh concrete flow by many researchers, there are a lot more research could be done in the future. For example, the current model used only elastic component (spring), plastic component (slip/slider) and viscous component (dashpot) as particles' contact model. In reality, the interactions between mortar and coarse aggregates are much more complicated and the accuracy of the simple combination of the above contact model does not seem satisfactory. In the future research, other models such as liquid bridge model, hard-core soft-shell model will be studied and discussed.

Besides, there are some drawbacks of DEM modelling on fresh concrete, for example, it was not clear of how to determine DEM parameters used to simulate the fresh concrete flow. The parameters were fitted to obtain the best comparison between numerical predictions and experimental results.

In this paper, only a small section of DEM modelling is reviewed. There are several procedures of DEM modelling needs to be studied, for instance, constitutive relationships and parameter estimation.

References

Cundall, P. A., *A Computer Model for Simulating Progressive, Large Scale Movement in Blocky Rock Systems*, Proceedings of the Symposium of the International Society for Rock Mechanics, , 129-136, France, II-8, 1971.

De Larrard, F., Ferraris, C.F., and Sedran, T., Fresh Concrete: A Herschel-Bulkley Material, *Materials and Structures*, 31(7), 494-498, 1998.

Hoornahad, H., and Koenders, E. A. B., Simulating Macroscopic Behavior of Self-Compacting Mixtures with DEM, *Cement and Concrete Composites,* 54, 80-88, 2014.

Jiao, D., Shi, C., Yuan, Q., An, X., Liu, Y., and Li, H., Effect of Constituents on Rheological Properties of Fresh Concrete-A Review, *Cement and Concrete Composites,* 83,146-159, 2017.

Li, Z., Cao, G., and Tan, Y., Prediction of Time-Dependent Flow Behaviors of Fresh Concrete, *Construction and Building Materials,*125, 510-519, 2016.

Maki, T., *Fundamental Study on Mechanism of Shotcrete-Application of 2D-DEM for Estimation of Rebound, The Sixth East Asia Pasific Conference on Structure Engineering and Construction*. 1998.

Martys, N. S., Study of a Dissipative Particle Dynamics-Based Approach for Modeling Suspensions, *Journal of Rheology,*49(2), 401–424, 2005.

Mechtcherine, V., Gram, A., Krenzer, K., Schwabe, J. H., Shyshko, S., and Roussel, N., Simulation of Fresh Concrete Flow using Discrete Element Method (DEM): theory and Applications, *Materials and Structures,*47(4), 615-630, 2014.

Mitsoulis, E., Flows of Viscoplastic Materials: Models and Computations, *Rheology Reviews,* 135-178, 2007.

Mori, H. T., Yasuo. Simulation Methods for Fluidity of Fresh Concrete, *Memoirs of the Faculty of Engineering,* 44, 71-134, 1992.

Nabeta, K., *Flow Simulation of Fresh Concrete by Distinct Element Method*, Proceedings of JCI 16(1), 479-484, 1989.

Noor, M. A.; Three-Dimensional Discrete Element Simulation of Flowable Concrete, 2000.

Petersson, Ö., and Hakami, H., *Simulation of Self-Compacting Concrete- Laborator Experiments and Numerical Modeling of Slump Flow and L-Box Tests*, Proceedings of 2nd International Symposium on Self-Compacting Concrete, Japa*n,* 79–88, 2001.

Puri, U. C., Puri, T., and Uomoto, T., Characterization of Distinct Element Modeling Parameters for Fresh Concrete and Its Application in Shotcrete Simulations, *Journal of Materials in Civil Engineering* 14(2), 137-144, 2002.

Quoc, P. H. D., 3D Simulation Using Distinct Element Method for Prediction of Shotcrete Shootability, *The University of Tokyo*, 2002.

Quoc, P. H. D., and Uomoto, T., *Review on Simulation of Fresh Concrete Using Distinct Element Method, The 3rd ACF International Conference, Vietnam,* 585591, *Nov*. 2008.

Roussel, N., Geiker, M. R., Dufour, F., Thrane, L. N., and Szabo, P., Computational Modeling of Concrete Flow: General Overview, *Cement and Concrete Research,* 37(9), 1298-1307, 2007.

Tattersall, G. H., *The Workability of Concrete,* Cement and Concrete Association; 1976.

Wallevik, O. H., and Wallevik, J. E., Rheology as a Tool in Concrete Science: The Use of Rheographs and Workability Boxes, *Cement and Concrete Research*, 41(12),1279-1288, 2011.

Wesseling, P., *Principles of Computational Fluid Dynamics,* 2000.

Streamlining Information Transfer between Construction and Structural Engineering
Edited by Shiau, J., Vimonsatit, V., Yazdani, S., and Singh, A.

ISBN: 978-0-9960437-7-9

ON THE REDUCTION OF PRESTRESSING FORCE NEAR SUPPORTS IN PARTIALLY PRESTRESSED CONCRETE FLEXURAL MEMBERS

NAZAR OUKAILI

Dept of Civil Engineering, University of Baghdad, Baghdad, Iraq

Straight tendons in pretensioned members can cause high-tensile stresses in the concrete extreme fibers at end sections because of the absence of the bending stresses due to self-weight and superimposed loads and the dominance of the moment due to prestressing force alone. Accordingly, the concrete tensile stresses at the ends of a member prestressed with straight tendons may limit the service load capacity of the member. It is therefore important to establish limiting zone in the concrete section within which the prestressing force can be applied without causing tension in the extreme concrete fibers. Two practical methods are available to reduce the stresses at the end sections due to the prestressing force. The first method based on changing the eccentricity of some tendons by raising them towards the end zone. The second method is based on bond prevention by encasing some of the tendons in plastic sheathing, effectively moving the point of application of prestressing force inward toward midspan for part of tendons. The present study focuses on a proposed third method to reduce the effect of the prestressing force near end supports by using straight strands with limited initial prestressing value in compression zone. New equations were suggested for the cracking moment and the prestressing force which consider the prestressed tendons in compression zone.

Keywords: Pretension, Strands, Stress, Top fibers, Compression zone, Cracking.

1 INTRODUCTION

It should be recognized that almost prestressed concrete members implicit significant amount of nonprestressed bonded reinforcement in addition to the prestressed steel. The nonprestressed reinforcement may be included in the flexural member to facilitate its fabrication, or due to strength and serviceability consideration. Such concrete members, which reinforced with combination of prestressed and nonprestressed reinforcement, are called partially prestressed concrete members. They occupy the whole spectrum of reinforcing range between fully reinforced and fully prestressed concrete members (Libby 1990), (Lin and Burns 1982), (Tadros *et al.* 1985).

Prestressing can be used to best advantage by varying the position of the prestressing force. Tendons whether bars, wires, strands, or made up cables may be used either straight or curved. Straight steel tendons are still the most commonly used tendons in pretensioned concrete members. Continuously curved tendons are used primarily in posttensioning applications. Cast-in ducts are positioned in the concrete unit to a continuous curve chosen to suit the varying bending moment distribution along the members.

From the theoretical viewpoint, it should be recognized that the profile of the prestressed steel, in accordance with (Figure 1. a, b, c, d) is considered to be the most appropriate placement of the longitudinal steel in the flexural structural member because of this distribution is compatible with the principal tensile stresses trajectory. At the same time it is considered as the most complicated and costly from the practical viewpoint (Libby 1990), (Lin and Burns 1982).

Two practical methods are available to reduce the stresses at the end sections due to the applied prestressing force. These common methods are: (1) Changing the eccentricity of some tendons by raising them towards the support zone (see Figure 1. a, b, c, d). This leads to reduce the moment value due to prestressing force. In sequence, the stresses at the ends can be reduced, and the service load capacity of the member can be increased, and (2) Sheathing portion of the tendons by plastic tube or a heavy paper or cloth tape having a waterproof adhesive at the immediate ends of the member. This eliminates the prestress transfer of part of the tendons at some distance from the support section. Here portion of the tendons prevented from bonding to the concrete at the immediate ends of the member, and, in so doing, preventing the unbounded tendons from prestressing the concrete at the ends.

If the path of the tendons is to be other than straight, vertical forces must be applied to the tendons to change their path. The devices used to apply the vertical forces and deflect the tendons from the straight path are called tendon deflectors or tendon-deflecting mechanisms. These mechanisms consist of devices that support the tendons in high position and devices that hold the tendons in the lower position. These devices are referred to as hold-up devices and hold-down devices, respectively (Libby 1990).

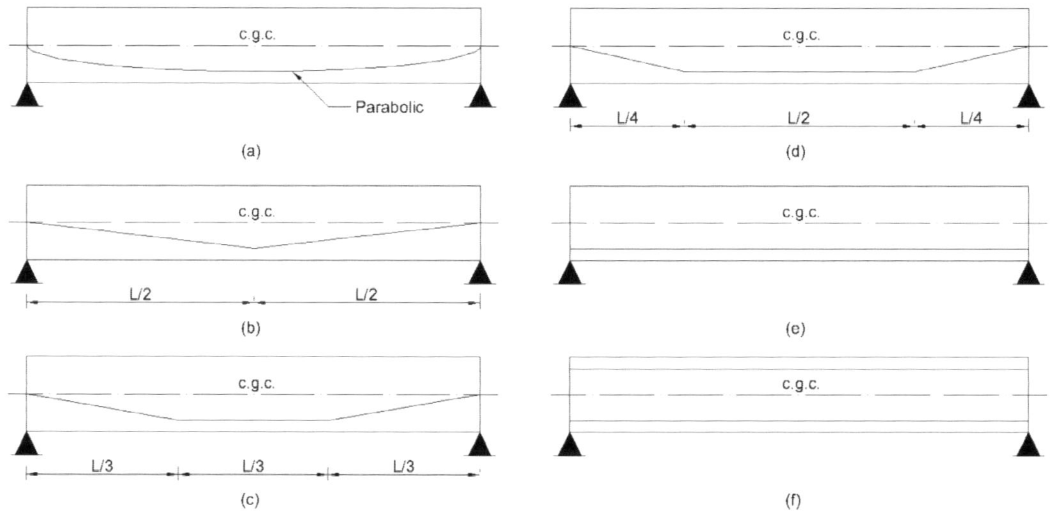

Figure 1. Profile of prestressed steel in flexural concrete members.

Significant disadvantages of the use of deflected tendons include the large capital investment in the stressing bed, the cost of the deflecting mechanisms, the friction losses that occur along the deflected tendons, the secondary stresses in the tendons where they pass over small-diameter pins and rollers, and the extra labor required to deflect the tendons (Libby 1990).

It is believed also that bond prevention can be used to advantage with complete safety if the tendons that will remain unbounded. The main disadvantage of this method is that AASHTO Standards (AASHTO 2014) restricted the number of partially debonded tendons which should not be exceeding 25% of the total number of tendons. Also, the number of debonded tendons in any horizontal row shall not exceed 40% of the number of tendons in that row. Additionally, if the bond is prevented at the end of a strand tendon, and the design allows tension in the pre-compressed tensile zone, the length required for flexural bond stresses to develop the strength of the tendon, (ℓ_d), is taken to be twice as great as that for a tendon that is bonded to the end of the member, see section 25.4.8.1 in ACI building code (ACI 318 2014).

These restrictions are considered to be the main disadvantages of the sheathing method because of the following: (1) Many practical cases may be encountered in which it is required to debond number of tendons more than the mentioned above percentages, and (2) Members that are of such a nature that high moments may occur near their ends, such as short simple spans and short cantilevers, require special consideration with respect to transfer and development lengths.

2 REDUCTION OF PRESTRESSING FORCE NEAR SUPPORT SECTIONS

It is obvious that the use of the straight profile (Figure 1.e) for the longitudinal prestressing steel in tension zone will be more economic than using other profiles if an approach is used to achieve a particular control for limiting or eliminating the tensile stresses which produced at the pretensioned compression zone of the section due to the effective prestressing force.

The proposed approach in this study is based on limiting or eliminating the tensile stresses at the extreme pretensioned compression zone fibers of the section by using straight longitudinal prestressing steel in compression zone with cross sectional area, A'_{ps}, not more than 15% of the area of the prestressing steel in tension zone, A_{ps} (Figure 1.f). Additionally, the value of the prestress introduced in the prestressing steel of the compression zone, f'_{pe}, shall be determined in such a manner that: (1) at the moment of transfer of the prestressing force, the tensile stress at the extreme concrete fibers shall not exceed 75% of the concrete modulus of rupture, f_r, and (2) at the moment of the flexural failure of the member these tensile stresses, f'_{pe}, must be nullified or progressed as compressive stresses. The reason for this is attributed to that "the prestressed steel in compression zone adversely affects the moment capacity of the structural member in the case if the generated prestress value not depleted entirely".

3 MODIFIED PRESTRESSING FORCE

ACI building code (ACI 318 2014), AASHTO Standards (AASHTO 2014) and PCI Design Handbook (2010) determine the effective prestressing force, P_e, according to the following equation:

$$P_e = A_{ps}\, f_{pe} \tag{1}$$

In partially prestressed concrete members and due to the existing bond between the nonprestressed steel and the surrounding concrete, compressive stresses are generated in nonprestressed steel. These stresses are created due to the shrinkage and creep of concrete after the exposure of the concrete member to the compressive prestressing force (Tadros *et al.* 1985). Accordingly, Eq. (1) becomes invalid to assess the effective prestressing force.

The proposed in the present study equation for determining the effective prestressing force, which considers the above mentioned factors, takes the following form:

$$P_e = A_{ps}\, f_{pe} + A'_{ps}\, f'_{pe} - A_s\, f_s - A'_s\, f'_s - \int f_c\, dA \tag{2}$$

where A_{ps} - cross-sectional area of the longitudinal prestressing steel in tension zone, mm^2; f_{pe} - effective prestress, after all losses, in the longitudinal prestressing steel in tension zone, N; A'_{ps} - cross-sectional area of the longitudinal prestressed steel in compression zone, mm^2; f'_{pe} - effective prestress, after all losses, in the longitudinal prestressed steel in compression zone, MPa; A_s – cross-sectional area of the longitudinal nonprestressed steel in tension zone, mm^2; f_s - compressive stress in the longitudinal nonprestressed steel in tension zone due to shrinkage and creep of concrete, MPa; A'_s - cross-sectional area of the longitudinal nonprestressed steel in compression zone, mm^2; f'_s - compressive stress in the longitudinal nonprestressed steel in compression zone due to shrinkage and creep of concrete, MPa; f_c – stress in concrete discrete parts, MPa; and dA – cross-sectional area of the concrete discrete part, mm^2.

The eccentricity of the effective prestressing force relative to the center gravity of the concrete section can be determined by equality the moment from the prestressing force and the moment from its components.

$$e_{ce} = \frac{A_{ps}\, f_{pe}\, y_{ps} - A'_{ps}\, f'_{pe}\, y'_{ps} - A_s\, f_s\, y_s + A'_s\, f'_s\, y'_s \mp \int f_c\, y_c\, dA}{P_e} \tag{3}$$

where y_{ps}, y'_{ps} - distance from the center gravity of the concrete section and the point of existence of the resultant of prestressing forces in tension and compression zones, respectively, mm; y_s, y'_s - distance from the center gravity of the concrete section and the point of existence of the resultant forces of the nonprestressed steel in tension and compression zones, respectively, mm; and y_c - distance between the center gravity of the concrete section and the center of the concrete discrete part under consideration, mm.

4 CRACKING MOMENT AT SERVICEABILITY STAGE

It is worth to mention out that if the point of application of the resultant prestressing force in a structural concrete member is restricted to a field that does not exceed r_c^2/c_b above the centroidal axis and r_c^2/c_t below the centroidal axis, tensile stresses will not occur in the concrete fibers of the section. This field in which the prestressing force can be applied without tensile stresses is called the elastic kern zone.

It is generally recommended to evaluate the magnitude of the cracking moment M_{cr} of the fully or partially prestressed concrete members in order to determine the reserve strength and overload limits that the designed section has. As already noted, for flexural prestressed concrete members is of great importance to consider the nonlinear strains of concrete in compression zone. Adoption of a triangular stress block in the compression concrete often leads to significant inaccuracies in the calculation. This is clearly shown by experimental works with concrete and reinforced concrete compression members subjected to bending moment. When the longitudinal compression force located on the boundary of elastic kern of the section and even within it, but near the boundary, it is indicated that the longitudinal compression force causes cracks on the extreme concrete fibers of the section, which according to the formulas of the strength of materials must have had either zero or compressive stresses (Zalesov *et al.* 1988).

At serviceability stage, the moment producing first cracks at bottom fibers in partially prestressed concrete girders is usually calculated by the elastic theory, assuming that cracking starts when the tensile stress in the extreme concrete fibers reach their modulus of rupture f_r. Most available test data seem to indicate that the elastic theory is sufficiently accurate up to the point of cracking, and the method is currently used. The ACI 318 code value for modulus of rupture, f_r, is $0.62\sqrt{f'_c}$ where both f_r and f'_c are in MPa.

It should be pointed out that the concrete modulus of rupture, f_r, is only a measure of the beginning of opening hair cracks which are often invisible to the naked eye. A tensile stresses which are greater than f_r required to produce visible cracks (Lin and Burns 1982).

To produce zero stress at the bottom concrete fibers, the center of the pressure (C-line) must be raised from the prestressing steel level to the upper kern point.

The distance from the center gravity of the section to the upper elastic kern point, k_t, can determined by the following equation

$$k_t = \frac{I_c}{A_c\, c_b} = \frac{r_c^2}{c_b} \tag{4}$$

Hence the elastic moment, M_{s1}, which required to move the compressive resultant from the level of the steel centroid to the upper kern point can be evaluated by multiplying the prestressing force by its lever arm, thus

$$M_{s1} = P_e\,(e_{ce} + k_t) = P_e\left(e_{ce} + \frac{I_c}{A_c\, c_b}\right) = P_e\,(e_{ce} + \frac{r_c^2}{c_b}) \tag{5}$$

Consequently, the additional moment, M_{s2}, required to develop the first flexural crack at the extreme tension fibers due to overload, such as the bottom fibers at mid-span of a simply supported beam, is

$$M_{s2} = \frac{f_r\, I_c}{c_b} = \frac{0.62\sqrt{f_c'}\, I_c}{c_b} \tag{6}$$

Accordingly, the total moment at cracking of the extreme bottom fibers at serviceability stage, M_{crs}, is given by

$$M_{crs} = M_{s2} + M_{s1} = \frac{0.62\sqrt{f_c'}\, I_c}{c_b} + P_e\left(e_{ce} + \frac{I_c}{A_c\, c_b}\right) \tag{7}$$

where A_c - gross cross-sectional area, mm^2; I_c - moment of inertia of gross concrete section about centroidal axis, neglecting reinforcement, mm^4; c_b - distance from centroidal axis of the gross section, neglecting reinforcement, to the extreme fiber in tension, mm; e_{ce} - eccentricity of the effective prestressing force relative to center gravity of the concrete section (c.g.c), mm; r_c – radius of gyration of cross-section of a flexural member, mm; and f_c' - specified 28-days compressive strength of concrete, MPa.

5 CRACKING MOMENT AT TRANSFER STAGE

Following the elastic theory, the moment producing first cracks at top fibers in partially prestressed concrete members can be determined at transfer stage. To produce zero stress at the top concrete fibers of the simply supported beam, the center of the pressure (C-line) must be shifted down from the level of the steel centroid to the lower kern point.

The distance from the center gravity of the section to the lower elastic kern point, k_b, calculated from the following equation:

$$k_b = \frac{I_c}{A_c\, c_t} = \frac{r_c^2}{c_t} \tag{8}$$

The elastic moment, M_{t1}, which is required to shift the C-line to the lower kern point is determined as follows

$$M_{t1} = P_i\,(k_b - e_{ci}) = P_i\left(\frac{I_c}{A_c\,c_t} - e_{ci}\right) = P_i\left(\frac{r_c^2}{c_t} - e_{ci}\right) \tag{9}$$

The additional moment, M_{t2}, required to develop the first flexural crack at the extreme tension fibers such as the top fibers at mid-span of a simply supported beam, is

$$M_{t2} = \frac{f_r\,I_c}{c_t} = \frac{0.62\sqrt{f'_{ci}}\,I_c}{c_t} \tag{10}$$

where c_t - distance from centroidal axis of the gross section, neglecting reinforcement, to the extreme fiber in compression, mm; e_{ci} - eccentricity of the initial prestressing force relative to center gravity of the concrete section, (c.g.c), mm; P_i – initial prestressing force, N; and f'_{ci} - specified compressive strength of concrete at the time of initial prestress, MPa.

Accordingly, the total moment at cracking of the extreme top fibers at transfer stage, M_{crt}, is

$$M_{crt} = M_{t2} + M_{t1} = \frac{0.62\sqrt{f'_{ci}}\,I_c}{c_t} - P_i\left(e_{ci} - \frac{I_c}{A_c\,c_t}\right) \tag{11}$$

6 CONCLUSION

An approach is proposed that based on limiting or eliminating the tensile stresses at the extreme pretensioned compression zone fibers of the section. That can be achieved by using straight longitudinal prestressing steel in compression zone with cross-sectional area not more than 15% of the area of the prestressing steel in tension zone and limited value of initial prestress.

Two equations were proposed to determine the cracking moment in flexural prestressed concrete members at transfer stage of prestressing force and at serviceability stage. These equations are considering the presence of prestressing steel in compression zone of the section.

References

AASHTO, *AASHTO LRFD Bridge Design Specifications*, 7th ed., American Association of State Highway and Transportation Officials, Washington, DC: USA 2014.

ACI 318, *Building Code Requirements for Structural Concrete (ACI 318M-14) and commentary (ACI 318R-14)*, American Concrete Institute, Farmington Hills, MI, USA, 2014.

Libby, J., *Modern Prestressed Concrete*, 4th ed., Van Nostrand Reinhold, New York, USA, 1990.

Lin, T., and Burns, N., *Design of Prestressed Concrete Structures,* 3rd ed., John Wiley & Sons, New York, USA, 1982.

PCI Design Handbook, *Precast and Prestressed Concrete*, MNL-120., 7th ed. PCI Industry Handbook Committee, Chicago, IL, USA, 2010.

Tadros, M., Ghali, A., and Meyer, A., Prestress Loss and Deflection of Precast Concrete Members, *PCI Journal*, 30(1), 1-29, 1985.

Zalesov, A., Kodish, E., Lemish, L., and Nikitin, I., *Strength, Cracking Control, and Deformation of Reinforced Concrete Structures*, Stroiizdat, Moscow, 1988.

Streamlining Information Transfer between Construction and Structural Engineering
Edited by Shiau, J., Vimonsatit, V., Yazdani, S., and Singh, A.

ISBN: 978-0-9960437-7-9

BOND STRESS-SLIP OF STRAND IN PRESTRESSED CONCRETE SLEEPERS

HOSSEIN ASKARINEJAD [1], WEIWO XIE[1], SON LE[1], and T. G. SUNTHARAVADIVEL[2]

[1]*Dept of Engineering and Architectural Studies, Ara Institute of Canterbury (Christchurch Polytechnic), Christchurch, New Zealand*

[2]*School of Engineering & Technology, Central Queensland University, Rockhampton, Australia*

Continuous vertical force on sleepers gradually degrades the bonding between the concrete and the prestressing strands. In particular, the vertical wheel loads can be significantly amplified due to irregularities on the wheel or on the track. This can eventually result in cracking or splitting of concrete along the strands leading to the structural failure of concrete sleepers. The bond between strand and concrete is an important factor which can directly or indirectly influence the cracking and splitting failure of the sleeper. This paper reports a series of pullout tests undertaken on specimens with 85 MPa concrete (nominal strength) and 7-wire prestressing strands replicating typical prestressed concrete sleepers produced in New Zealand. The results are discussed and compared in terms of the loads causing the first strand slip and also the bond stress – slip relationship. The results show that longer water curing time does not necessarily increase the bond between strand and concrete in prestressed concrete sleepers. This behaviour can be attributed to the effect of shrinkage and the trade-off between strength and shrinkage influenced by water curing. The obtained results can contribute to determining optimum sleeper curing time to achieve required strength and bonding and is valuable for concrete sleeper manufacturers where curing process is associated with considerable cost and logistical issues.

Keywords: Sleeper failure, Prestressing strands, Pullout test, Strand slip.

1 INTRODUCTION

Railway prestressed concrete sleepers are located between the rail and ballast in transverse direction to the rail. The sleepers transfer the loads from the rail to the track substructure and also restrain the rail in the lateral, longitudinal and vertical directions. Concrete sleepers experience high magnitudes of shear force and bending moment under repeated vertical loads from the passing trains which can result in structural failures such as different cracking, rail seat crushing and strand slippage in the sleeper. In particular, the vertical loads are significantly increased at the sleepers next to the rail welds or rail joints or in vicinity of other rail and wheel surface irregularities due to high wheel-rail forces. The bond of steel bars or strands with concrete is an important factor, which can affect the performance and durability of concrete sleepers. For deformed bars and strands embedded in concrete, the bond is described as the transfer of force from the reinforcement to the surrounding concrete through adhesion, friction and mechanical interlock.

In this paper, the loading and failure mechanisms of prestressed concrete sleepers are briefly reviewed, and the importance of strand bond is discussed. Next, a series of the pullout tests used to investigate bond behaviour of prestressing strand are presented. The general testing process is based on ASTM A1081/A1081M (2012) and AS/NZS 4672.1 (2007). This study is a part of an ongoing research on the failure of concrete sleepers and track support damages and its effect on the loads and progressive deterioration of track.

2 LOADING AND FAILURE OF PRESTRESSED CONCRETE SLEEPERS

A view of a prestressed concrete sleeper in rail track is shown in Figure 1. Sleepers are subject to a high cycle of vertical loads due to the running trains. The vertical wheel load can vary from 25kN to 150kN depending on the train type and its properties (Esveld 2001). These vertical loads generate a bending moment in sleeper where the bending scenario is dependent upon the sleeper-ballast contact pressure distribution. The pressure distribution beneath the sleeper and in turn the sleeper bending scenario is highly variable depending on the track substructure condition (Sadeghi 2008); therefore, causing variability in the location of peak bending stresses along the sleeper in different rail tracks.

Figure 1. Typical prestressed concrete sleeper in railway track.

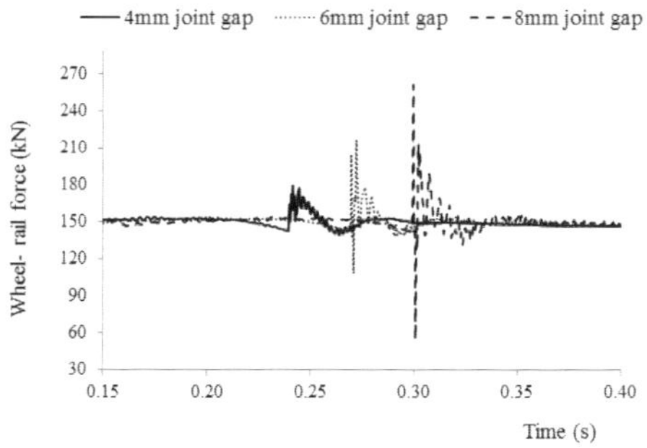

Figure 2. Typical wheel-rail force amplification at a rail joints (Askarinejad and Dhanasekar 2016).

Additionally, the vertical wheel loads may be significantly amplified due to irregularities on the wheel or on the track. This includes wheel flats, rail surface corrugation and roughness, rail welds, rail joints and track geometry irregularities. The dynamic wheel load amplifications have been studied by many researchers. For examples, Grossoni *et al.* (2013) and Askarinejad and Dhanasekar (2016) are among the recent ones presenting simulation and experimental results for wheel-rail forces at rail joints. Figure 2 shows a sample wheel-rail force scenario at a rail joint. It can be seen that the vertical force can reach nearly double the static wheel load at the rail joints. This value can be significantly higher when the rail joint is located on a track geometry dip. Moreover, some studies have reported much higher vertical forces generated due to wheel flats; for instance, a wheel-rail force of 600 kN due to 20-25 mm wheel flats (Kaewunruen and Remennikov 2009).

Considering the repetitive and high magnitudes of forces, the adjacent concrete sleepers can get severely damaged. Various types of damage or failure have been reported in the literature such as rail seat deterioration, cracking in different locations along the sleeper as well as splitting and strand slip within the concrete (Thun *et al.* 2008, Lutch *et al.* 2009, Kaewunruen and Remennikov 2013, Zakeri and Rezvani 2012, Taherinezhad *et al.* 2013). Figure 3 shows two sample sleeper failure types.

a) Longitudinal crack (Zakeri and Rezvani 2012 b) Splitting failure (Kaewunruen and Remennikov, 2013)

Figure 3. Longitudinal crack and splitting failure along the strands in prestressed concrete sleeper.

The bond between strand and concrete is an important factor which can directly or indirectly influence the cracking and splitting failure of the sleeper. Insufficient strand bond can adversely affect the overall flexural performance of the sleeper resulting in various flexural cracks. These cracks may also gradually aggravate under repeated loads leading to the splitting along the strands. The concrete-steel bar bond in various applications have been investigated by many researchers. For example, Hadi (2008) conducted a series of pull out tests using specimens fabricated with steel bars ranging from 12mm to 36 mm. He also looked into the effect of concrete cover around the bar using 240 mm and 300mm diameter concrete cylinders. Lim *et al.* (2013), studied the bond behavior and transfer length of prestressing strands using experimental and finite element analyses. Dang *et al.* (2014) investigated the correlation of strand surface quality and bond strength. They concluded that the lubricants and residue chemicals that accuse the variation may reduce bond between concrete and steel bar. Additionally, some researchers have focussed on the bond stress and strand slip behaviour in special or newly emerged concrete materials such as geopolymer concrete and self-consolidation concrete. Kim and Park (2015) studied the bond properties of reinforcements embedded in Geopolymer Concrete. Looney *et al.* (2012) reported an experimental study on bond strength of reinforcing steel in self-consolidating concrete.

This paper reports a series of pullout tests undertaken on specimens with 85 MPa concrete (nominal strength) and 7-wire prestressing strands replicating typical prestressed concrete sleepers produced in New Zealand. The results provide good information on strand stress-slip magnitudes in a typical prestressed concrete sleeper, which will be useful in understanding strand slippage failure behaviour. Additionally, particular attention is paid to the effects of sleeper curing condition on the strand bond as curing is a major logistical aspect of prestressed concrete production process.

3 BOND SLIP TEST

The test specimens were produced with dimensions of 150x150x450 mm using 85MPa concrete (nominal strength) and 7-wire prestressing strands. Considering suitable dimensions and embedment length are critical to obtain the expected bond stress-slip behaviour. For example, if the embedded length of the bar is not suitable, the strand may yield and fail before slipping through the concrete. The bond stress acts parallel to the bar along the concrete interface (Figure 4). Therefore, assuming uniform stress distribution, the bond stress (u) can be calculated as the average stress between the reinforcing bar and the surrounding concrete along the embedded length of the bar as below:

$$u = \frac{F}{\pi L d_b} \tag{1}$$

Where, F is the pullout force, L is the strand embedded length in the concrete, and d_b is the strand diameter. The specimens were fabricated in a prestressed concrete plant in Christchurch, NZ. The strands were 7-wire low relaxation strand with a diameter of 8mm and cross section area of 38.82 mm^2. The ultimate tensile strength and yield strength of the strand are 2060MPa and 1967MPa respectively and elongation ratio is 3.79%. The concrete was placed into the moulds in a number of layers where each layer of concrete was vibrated and compacted thoroughly with the penetrating vibrator. After 24 hours, the moulds were removed and the specimens were split into four groups (three specimens in each group). The first group (Group A) was kept in room temperature. The second group (Group B) was kept in room temperature for 21 days and then placed in a water tank for 7 days. The third group (Group C) was kept in room temperature for 14 days and then in water tank for 14 days. The fourth group (Group D) was cured in a water tank for 28 days. The tests were undertaken using an Avery Universal Testing Machine (UTM) at Ara's Engineering Laboratory. The concrete blocks were attached under the cross-bar where the strand was clamped on the top of the machine. A dial gauge was clamped at the bottom end of the strand to measure the strand slip.

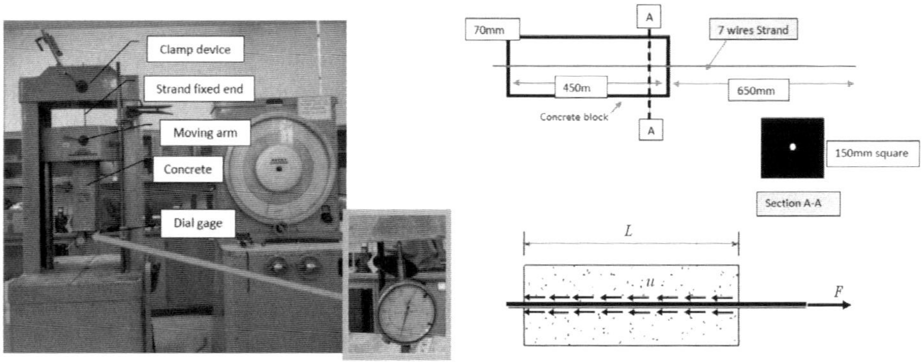

Figure 4. Test setup and the specimens' size and dimensions.

Figure 4 demonstrates one of the test specimens, which has been set up ready for testing. As shown in this figure, the test specimens were suspended via the strand. Moving UTM table upwards would force the test specimens against the underside of the stationary cross-bar. Once the specimens rested against the cross-bar, a pullout force would be applied to the strand. The load was applied with the loading rate of 0.1kN/s.

4 TEST RESULTS AND DISCUSSION

The pullout forces and the corresponding gauge readings were recorded. The loading stopped when the strand slip reached about 1mm. The same procedure was repeated for each specimen. Observation of the specimens after the tests shows the damage at the strand-concrete interface due to the strand slip as expected. Figures 5 presents the magnitudes of pullout force, which caused the first slip in the strands.

It can be seen in Figure 5 that the first slip occurred at higher loads in the Groups B and D specimens compared to other groups. The pullout force causing the first slip in the strands in these two groups are measured between 45 kN and 55 kN. Figure 6 shows the bond stress vs strand slip comparing the bond stress-slip relationship for the groups of specimens. All three curves show a nonlinear pattern up to about 0.2mm slip where low rate of slip is observed for high values of bond stress. When the slip exceeds 0.2mm, the strand slip increases nearly linearly as the bond stress rises.

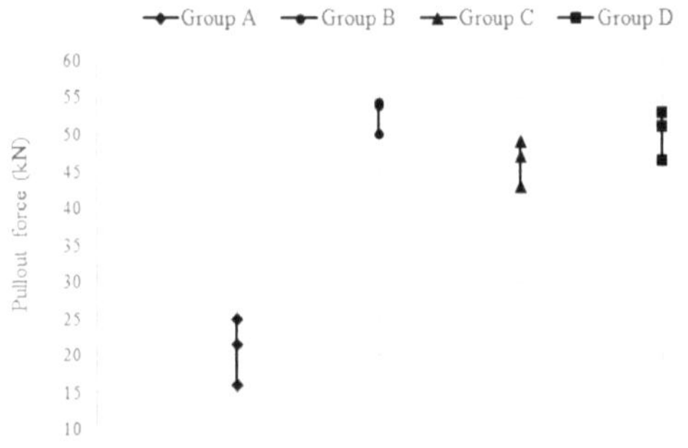

Figure 5. Pullout force at the first slip.

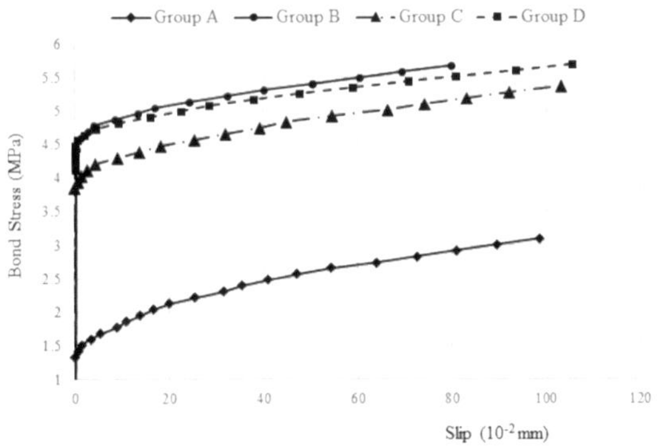

Figure 6. Bond stress vs strand slip.

It is interesting to see that Group B samples (with 7 days water curing) show consistently higher bond stress with magnitudes very close to Group D (with 28 days curing). This shows that longer water curing time does not necessarily increase the bond between strand and concrete in prestressed sleepers. This behavior can be attributed to the effect of shrinkage.

It should be noted that moist curing could have two significant effects on concrete. The moist curing would harden the cement paste and thus increases the concrete strength as well as the concrete-steel bond; additionally it would decrease the shrinkage. Even though shrinkage can cause problems such as dimensional changes, it can be beneficial to increase the concrete-steel bond. Concrete shrinkage would cause the concrete to grip the strands more tightly, increasing the friction and hence increase the bond between concrete and strand. This means there is an optimum curing time to achieve the required compressive strength and strand bond; such that whichever is dominate (concrete strength effect or shrinkage effect) would determine if the longer moist curing would increase or decrease the bond between concrete and steel. This result can contribute to determining optimum sleeper curing time to achieve required strength and bonding and is valuable for concrete sleeper manufacturers where curing process is associated with considerable cost and logistical issues.

In the next stage, further research is in progress to link the measured bond stress-slip with the vertical wheel loads applied to the sleeper in track and the strand slippage failure mechanism. In addition, it should be noted that a damaged sleeper can cause a progressive rise in wheel-rail forces and subsequent damages to the surrounding sleepers and the underlying substructure.

5 CONCLUSIONS

As part of an ongoing research, an attempt has been made to investigate strand bond behavior of the prestressed concrete sleepers. Strand bond is an important factor, which can control the performance of concrete sleepers in track where high magnitudes of shear force and bending moment are generated under wheel loads. Lab experiments were undertaken to determine the relationship between bond stress and strand slip in specimens with high strength concrete and 7-wire prestressing strands replicating typical prestressed concrete sleepers produced in New Zealand. This was done by measuring the strand movement (slip) under pullout force. The measured bond stress-slip showed a nonlinear pattern up to about 0.2 mm slip and then the strand slip increased almost linearly as the bond stress increased. The tests provided interesting results where the specimens that water cured for 7 days showed bond stress as high as the ones cured for 28 days, which was attributed to the effects of concrete shrinkage.

Acknowledgments

Roger Richardson provided advice during the experimental design and tests. His assistance was invaluable to this study.

References

Askarinejad, H., and Dhanasekar, M., A Multi-Body Dynamic Model for Localised Track Responses in Vicinity of Rail Discontinuities, *Journal of Structural Stability & Dynamics*, 16(9), 1550058, 2016.

ASTM A1081/A1081M, Standard Test Method for Evaluating Bond of Seven-Wire Steel Prestressing Strand, West Conshohocken, U.S.: ASTM International, 2012.

AS/NZS 4672.1, Steel Prestressing Materials-General Requirements, Australian/New Zealand Standard, 2007.

Dang, C. N., Murray, C. D., Floyd, R. W., Hale, W. M., and Martí-Vargas, J. R., Correlation of Strand Surface Quality to Transfer Length, *ACI Structural Journal*, 111(5), 1245-1252, 2014.

Esveld, C., *Modern Railway Track*, 2nd ed, Delft University of Technology, 2001.

Grossoni, I., Iwnicki, S., Bezin, Y., and Gong, C., *Dynamic Response of Vehicle–Track Coupling System with an Insulated Rail Joint*, in 11th Int. Conf. Vibr. Prob., Lisbon, Portugal, September 9–12, 2013.

Hadi, M. N., Bond of High Strength Concrete with High Strength Reinforcing Steel, *The Open Civil Engineering Journal*, 2, 143-147, 2008.

Kaewunruen, S., and Remennikov, A. M., Progressive Failure of Prestressed Concrete Sleepers under Multiple High-intensity Impact Loads, *Engineering Structures,* 31(10), 2460–2473, 2009.

Kaewunruen, S., and Remennikov, A. M., On the Residual Energy Toughness of Prestressed Concrete Sleepers in Railway Track Structures Subjected to Repeated Impact Loads, *Special Issue: Electronic Journal of Structural Engineering*, 13(1), 41-61, 2013.

Kim, J. S., and Park, J. H., An Experimental Investigation of Bond Properties of Reinforcements Embedded in Geopolymer Concrete, *WASET International Journal of Civil, Environmental, Structural, Construction and Architectural Engineering*, 9(2), 96-99, 2015.

Lim, S. N., Choi, Y. C., Oh, B. H., Kim, J. S., Shin, S., and Lee, M. K., *Bond Strength and Transfer Length of Prestressing Strand in Pretensioned Concrete Structure*, VIII International Conference on Fracture Mechanics of Concrete and Concrete Structures, Spain, 2013.

Looney, T. J., Arezoumandi, M., Volz, J. S., Myers, J. J., An Experimental Study on Bond Strength of Reinforcing Steel in Self-Consolidation Concrete, *International Journal of Concrete Structures and Materials,* 6(3), 187-197, 2012.

Lutch, R. H, Harris D. K., and Ahlborn T. M., *Pre-stressed Concrete Ties in North America*, Proceeding of AREMA Annual Conference, 1–39, Chicago, USA, 2009.

Sadeghi, J., Experimental Evaluation of Accuracy of Current Practices in Analysis and Design of Railway Track Sleepers, *Canadian Journal of Civil Engineering*, 35(9), 881-893, 2008.

Taherinezhad, J., Sofi, M., Mendis, P.A., and Ngo, T., A Review of Behaviour of Prestressed Concrete Sleepers, *Special Issue*: *Electronic Journal of Structural Engineering*, 13(1), 1-16, 2013.

Thun, H., Utsi H., and Elfgren L., Load Carrying Capacity of Cracked Concrete Railway Sleepers, *Structural Concrete,* 9 (3), 153–161, 2008.

Zakeri, J. A., and Rezvani, H., F., Failures of Railway Concrete Sleepers during Service Life, *International Journal of Construction Engineering and Management*, 1(1), 1-5, 2012.

Streamlining Information Transfer between Construction and Structural Engineering
Edited by Shiau, J., Vimonsatit, V., Yazdani, S., and Singh, A.

ISBN: 978-0-9960437-7-9

RESPONSE OF COMPOSITE STEEL PLATE GIRDER BRIDGES SUBJECTED TO DEBRIS FIRES

AMILA DISSANAYAKE, SRIKANTH VENKATESAN, and SUJEEVA SETUNGE
School of Engineering, RMIT University, Melbourne, Australia

Bridge fires have been a serious concern for asset owners for decades. Considerable research work has been published on the assessment of bridges subjected to hydrocarbon fires and Wildland Urban Interfaces (WUI) fires like bushfires. However, the impact from fires that may have been generated due to onsite-accumulated debris, has been largely overlooked in the past. Prolonged duration of debris fires might well cause significant damage to the bridges, unless properly accounted during the design stage. The current study investigates the influence from debris-generated fires on steel girder bridges, using a reasonably validated advanced numerical modelling framework. Fire development was modelled first as a static, and then a travelling fire, using fire dynamic simulation (FDS) to capture the Adiabatic Surface Temperature (AST) development of the structure. The resulted AST was coupled with thermo-mechanical analysis using sequentially coupled thermo-mechanical analysis procedure in ABAQUS. The numerical model was used to estimate the temperature and displacement development of the bridge. The outcome from the study will facilitate asset managers for conducting necessary risk assessments incorporating the influence of onsite flammable debris for bridges.

Keywords: Wildland urban interface (WUI), Fire dynamic simulation (FDS), Adiabatic surface temperature (AST), Thermo-mechanical analysis.

1 INTRODUCTION

Occurrence of bridge fires is considered rare incidents. However, New York Department of Transportation survey in 2008 (Garlock *et al.* 2012) reported that bridge fire failure was nearly 3 times more than seismic induced failure. Giuliani *et al.* (2012) has also reported that the fire induced bridge failures in USA accounts for about 3% of the total cases surveyed. At the same time, the rapid expansion of urbanization has led to the increased vulnerability of infrastructures to WUI fires. The impact from fires that may have been generated due to onsite-accumulated debris, has largely been overlooked in the past. Prolonged duration of debris fires compared to bushfires or grassland fires might well cause significant damage to the bridges. Given the scant research on wildland urban interface (WUI) fires, this research presents an investigation of a possible bridge response to a debris driven WUI fire.

2 CASE STUDY BRIDGE USED FOR ANALYSIS

A 25m long and 13m wide composite steel plate girder bridge with 5 numbers of girders with 2.5m spacing is modeled to represent a typical bridge structure. Steel plate girder is 1.2m deep and attached to a 175mm thick reinforced concrete slab. Five steel plate girders of 1.2m deep support a 13m wide reinforced concrete slab a depth of 175mm. Refer to Figure 1 for information.

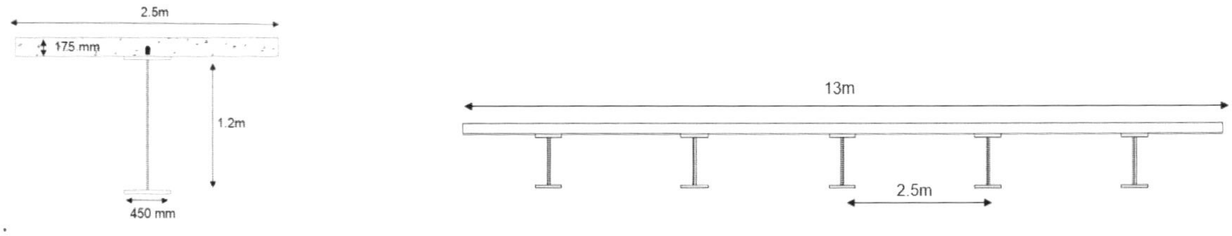

Figure 1. Case study bridge configurations.

3 DEVELOPMENT OF THE COMPUTATIONAL FLUID DYNAMIC (CFD) MODEL

Heating curves used in simple fire models are originally designed for building structures, where fire is naturally constrained in an enclosed space, whereas bridge fires are open environment fires. This difference makes the simple fire models inadequate for capturing the realistic bridge fire behaviours. A more realistic modelling of fire in a bridge can be obtained using computational fluid dynamics modelling. Fire Dynamics Simulator (FDS) (McGrattan *et al.* 2013a) software can be used to develop fire models to predict the fire engineering related variables such as temperatures, heat fluxes or gas pressures in fire events. The simulation results in a three-dimensional, time-dependent, prediction of fire behavior. FDS software has been developed at National Institute of Standards and Technology (NIST) and has been extensively validated experimentally (McGrattan *et al.* 2013b).

Building the FDS model requires defining a control volume with its boundary conditions relevant to the analysis. The geometry representing the case study structure should be included in the control volume with a suitable mesh or discretization of the control volume. The control volume used in this study includes the bridge as well as its two abutments. It measures 31m X 23m X 15m along the x, y and z-directions respectively. The control volume cells have approximate dimensions of 0.2 m.

Results of a vegetative based heavy fuel bed burning experiment reported by Cheney (1990) was used as the representative debris pile in the simulation process (Figure 2). During the experiment, a brigalow (Acacia harpophylla) fuel bed that was composed of 410 kg of fuels less than 25mm and 1145kg of fuels 25-75 mm in diameter with an average fuel load of 100kg/m^2 was burnt on a weighted platform. Figure 2 is the reaction intensity curve obtained for the burning fuel bed. It was reported that the flames spread rate is around 1m/min across the fuel bed.

Fire footprint was modelled as a 5x5m^2 placed directly below the centre of the bridge with the 3.5m effective clearance to the steel girders. The effect of wind was not considered in the FDS model.

Adiabatic surface temperature is a practical tool to express the thermal exposure of a surface to fire. Utilizing the AST helps to reduce the data flow in to the structural models by eliminating the dependency of the surface temperature on the net heat flux. AST temperature development depends on the incident flux and geometry of the structural members and the boundary conditions imposed on the control volume.

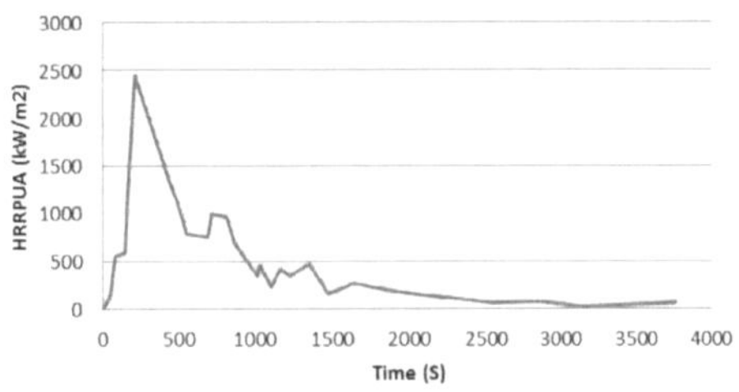

Figure 2. Experimental HRRPUA curve of a debris pile burning (Cheney 1990).

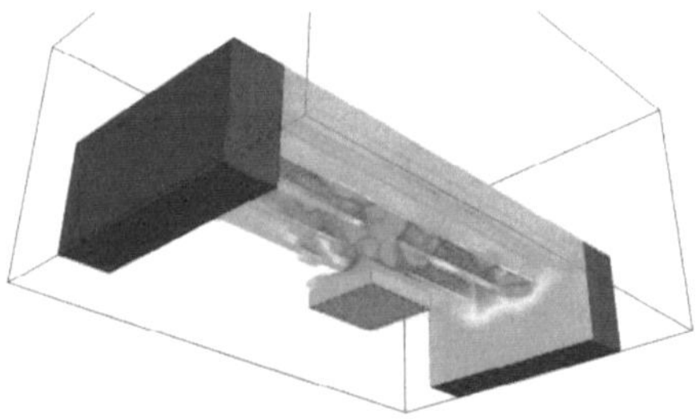

Figure 3. Adiabatic Surface temperature development of the structure.

The FDS model monitored the Adiabatic surface temperature development along the middle girder and the concrete surfaces on either side. AST Sensors measured the 52 cross-sections along the girder length with 9 sensors at each section. Figure 3 shows the colour contour of the AST development of the structure.

4 FINITE ELEMENT MODEL FOR THERMO- MECHANICAL ANALYSIS

Bridge girder is analysed using a sequentially coupled thermo-mechanical analysis with ABAQUS finite element software. It should be noted that FDS solves the CFD problem without considering the deformation of the structure. The thermal analysis provides the transient nodal temperature developments of the structure. Temperature distribution of the system determined with a standard heat transfer analysis that was used in the subsequent thermo-mechanical analysis.

Temperature dependent mechanical properties such as the stress-strain relationship, elastic modulus, and thermal properties such as specific heat, conductivity, and expansion were defined according to the Eurocode for concrete and steel elements respectively. The structural steel mechanical property was defined as nominal yield stress of 359 MPa, Young's modulus of 210 GPa and a Poison's ratio of 0.3. concrete compressive strength was taken as 40 MPa to build elevated temperature stress-strain curves.

Adiabatic surface temperatures obtained with FDS model were applied as thermal boundary conditions to the heat transfer analysis of the plate girder structure. Unexposed surfaces were defined to be at ambient temperature. In the thermal analysis a convective heat transfer coefficient of 25, 50W/m^2 K and an emissivity coefficient, of 0.7 were used for the exposed surfaces and a convective heat transfer coefficient of 9W/m^{2}K (EN 1991)was used for unexposed surfaces.

The bridge girder studied in this paper is a composite steel plate girder bridge. The steel girders and the concrete slab are connected with shear connectors to worked together to sustain the loads acting on the bridge. Pin and roller support conditions are used to represent the mechanical boundary conditions of the girder. Lateral movement of the girder is restrained at the support points.

Self-weight of the steel girder and the concrete slab are included automatically by the Abaqus software. In addition to that, a dead load corresponding to the weight of the wearing surface has been also considered. The analysis excludes the effects of live load because of the unlikeliness of having traffic on the deck during a fire situation.

5 PARAMETRIC STUDY AND RESULTS

Fire source footprint was taken to be 5x5m^2, which is placed below the center of the middle girder of the bridge structure. Fire was modeled first as a static fire where the whole surface of fire footprint starts to burn at once to follow the corresponding HRRPUA curve. In a separate analysis fire foot print was divided in to a series of 0.5x5m^2 width rectangular strips perpendicular to the bridge girder. Then, each of those strips were assigned to follow the same HRRPUA curve but with a time lag for the initial ignition time that is calculated based on the experimentally observed progress rate of the flame front across the fuel bed of 1m/minute. In the text this analysis is called as the progressive fire or travelling fire scenario.

5.1 Temperature Development

A large variation of the temperatures along the bridge length was observed in the analysis results. Temperature close to the midspan of the girder was considerably higher than to the support areas as it is expected. Assuming a constant fire load along the bridge span is therefore unrealistic.

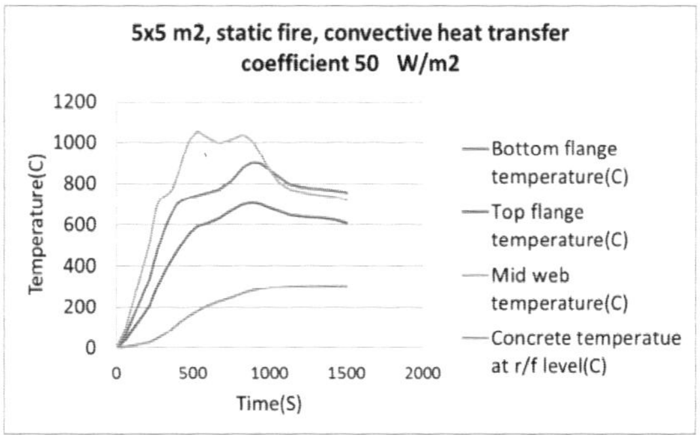

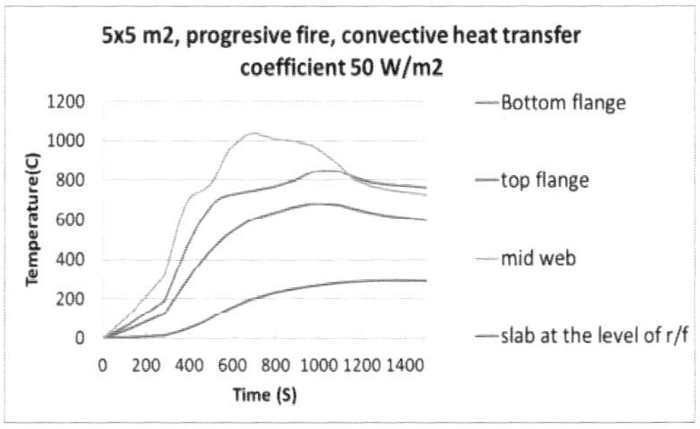

Figure 4. Temperature development of the mid-section of the structure.

Figure 4 shows the time history plots of temperatures for girder midspan cross section. This shows the temperature profiles at the top and bottom flanges, mid web and in the concrete slab at the level of reinforcement. The temperature of steel girder has been observed to increase to a maximum and then decayed gradually with the fire is gradually extinguished. Temperature development of the concrete slab remains below 400°C for the entire time. Top flange temperature development of the structure is well below the temperature development of the web or the bottom flange. This is mainly due to the presence of the concrete slab thermally attached to the top flange of the structure that acts as a heat sink during the fire exposure. Mid web temperature of the cross section goes above 1000ºC. Maximum temperature development of the static fire case is slightly higher than when the fire is modeled as a travelling fire.

5.2 Effect of Heat Transfer Coefficient

Heat transfer from fire zone to boundaries of the girder is mainly through convection and radiation, while conduction is the main mechanism for heat transmission within the girder assembly. Selecting a suitable convective heat transfer coefficient to represent the characteristics of a fire is an important step. According to the Eurocode (EN 1991-1-2 2002), 25, 50, 35 $W/m^2.K$ values are recommended for standard fire, hydrocarbon fire and simplified fire models respectively. Based on this recommendation, CFD based fire response evaluation programs can selects the appropriate values to represent the convective part of the heat transfer mechanism. However, there is lack of guidelines to represent a vegetative based fire. In the current study two convective heat transfer coefficient values 25 and 50 $W/m^2.K$, were utilized to understand the effect along with emissivity coefficient 0.7.

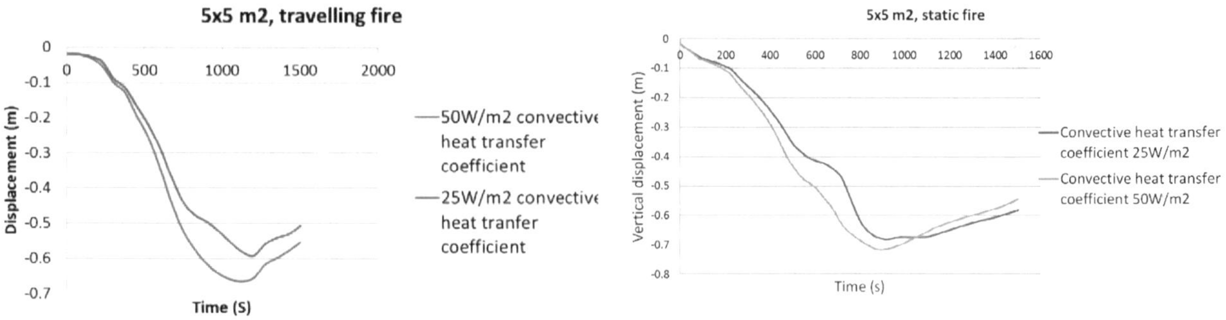

Figure 5. Midspan displacement development of the structure due to different fire conditions.

Both travelling fire and the static fire cases were analyzed using two convective heat transfer coefficients. Figure 5 shows the results related to the midspan displacement development of the girder. Displacement development is slower, and the maximum displacement is lower when a low coefficient of convective heat transfer coefficient is used in the analysis. Effect is more prominent.

5.3 Static vs Travelling Fire Effect

Fire footprint was modelled as a $5x5m^2$ squire placed directly below the centre of the bridge with the 3.5m effective clearance to the steel girders. The effect of wind was not considered in the FDS model.

As it is mentioned earlier, the debris pile burning was modelled first as a static fire and extended to a more realistic progressive fire. Midspan displacement development was monitored in both cases convective heat transfer coefficient was taken as $50W/m^2.K$. Displacement development is slower when the fire is modelled as a progressive fire (Figure 6). This is mainly due to constrained fuel consumption of a progressive fire. However, the maximum displacement development is almost similar in engineering point of view.

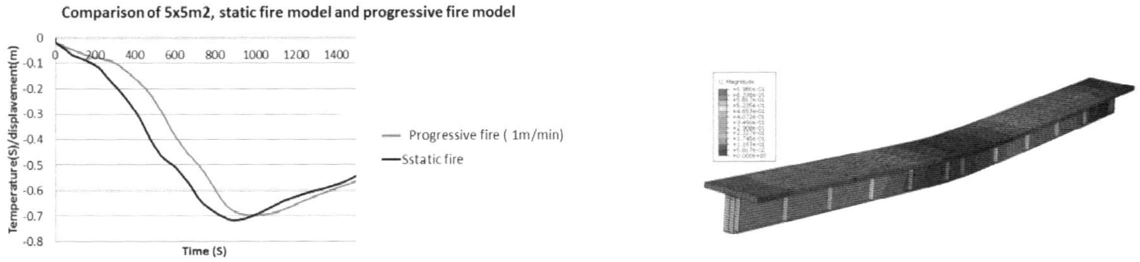

Figure 6. Effect of travelling and static fire on the bridge response.

6 CONCLUSIONS

A systematic approach to investigate the response of a bridge subjected to a debris driven WUI fire event has been presented in this paper. Sequentially coupled thermal-stress analysis has been carried out to evaluate the behaviour of the bridge girders subjected to a vegetative based fire. Fire was modelled in both ways to understand the difference between the modelling the fire as a static vs travelling. Travelling fire modelling resulted a slower displacement development of the girder, however the maximum displacement response is comparable in both cases. Effect of convective heat transfer coefficient on the midspan vertical displacement is prominent when the fire is modelled as a travelling fire than a static fire. Maximum displacement exceeded the allowable serviceability criteria of the structure in all the cases. This case study clearly indicates the risk associated with debris collected around a bridge structure in an accidental fire event.

References

Cheney, N. P., Quantifying bushfires, *Mathematical and Computer Modelling*, 13, 9-15, 1990.

EN 1991-1-2, Eurocode 1: Actions on Structures—Part 1-2: General Actions—Actions on Structures Exposed to Fire', *British Standards*, 2002.

Garlock, M., Ignacio, P.-Z., Venkatesh, K., and Li, G., Fire Hazard in Bridges: Review, Assessment and Repair Strategies, *Engineering Structures*, 35, 89-98, 2012.

Giuliani, L., Crosti, C., and Gentili, F., *Vulnerability of Bridges to Fire*. Proceeding of 6th International Conference On Bridge Maintenance, Safety And Management (IABMAS 2012), Stresa, Lake Maggiore, Italy, 8-12, 2012.

McGrattan, K. B., McDermott, R. J., Weinschenk, C. G., and Forney, G. P., *Fire Dynamics Simulator*, technical reference guide (No. Special Publication (NIST SP)-1018), 2013a.

McGrattan, K., Hostikka, S., McDermott, R., Floyd, J., Weinschenk, C., and Overholt, K., *Fire Dynamics Simulator*, user's guide. NIST special publication, 1019, 20, 2013b.

Streamlining Information Transfer between Construction and Structural Engineering
Edited by Shiau, J., Vimonsatit, V., Yazdani, S., and Singh, A.

ISBN: 978-0-9960437-7-9

REMAINING STRENGTH EVALUATION OF FERRY BRIDGE CORRODED LOCALLY

FUMIYA TAKAMI and KATASHI FUJII

Dept of Civil and Environmental Engineering, Graduate School of Engineering, Hiroshima University, Hiroshima, Japan

This paper presents the results of a nonlinear finite element analysis to determine the remaining ultimate strength of a ferry bridge with local corrosion at the girder flange. The ultimate strength decreases with the corrosion thinning of the flange, when the corrosion damage locates at the center of the span. On the other hand, when the corrosion damage exists near the support, the ultimate strength did not decrease, which shows the repair of the flange thinning near the support is not so effective in strength rehabilitation of the bridge. The reason of these facts is; the collapse of the bridge is due to bending moment in the former case, but in the latter, it is due to not bending moment but shear force, therefore, flange thinning does not concern with the shear strength. Thus, rational maintenance will be possible after clarifying the collapse mode and the remaining strength of the bridge.

Keywords: Plate girder, Ultimate remaining strength, Nonlinear FEM analysis, Safety factor of ultimate load, Flange local corrosion.

1 INTRODUCTION

In Japan, there has been a growing consciousness of importance of maintenance because serious damages have been appearing in a lot of deteriorated infrastructures constructed from 1960s to 1980s. The proportion of bridges exceeding 50 years since construction is currently about 20%, but it will increase to about 40% in 10 years and to about 70% in 20 years, consequently, to prolong the life of bridge is becoming an urgent and grave task, when the future serious economic condition is taken into account.

Although soundness of deteriorated bridge has been usually being judged currently from the corrosion-damaged grade, the safety measures decided by that judgement may not necessarily be appropriate in terms of strength and cost, from the point of whole bridge. Therefore, it will be more appropriate in the future to maintain based on the remaining strength of the bridge.

According to the above concept, the remaining strength of a whole bridge has been investigated by Tai *et al.* (2015), Yamaguchi *et al.* (2011) and Connor *et al.* (2005) who evaluated the ultimate remaining strength of general road bridges corroded by FEM. However, in their researches, the effect of the corrosion location and grade on ultimate remaining strength has not been clarified yet. By conducting nonlinear finite element analysis considered with geometrical and material nonlinearities, the ultimate remaining strength of ferry bridge as well as that of general road bridge built with steel girders has been investigated by Takami and Fujii (2014, 2017). Then, it was revealed that the ultimate remaining strength of the ferry bridge is far lower than that of general road bridge, because the ferry bridge is a simple structure consisted of main girder and steel floor deck and has few secondary members. Adding to this, ferry bridge is in the environment that is easier to corrode than general road bridge, because it is constructed at sea shore in most cases.

Therefore, it seems more important to grasp ultimate behavior and remaining strength of ferry bridge than road bridge. In this paper, conducting nonlinear finite element analyses, ultimate behavior and remaining strength are investigated when the local corrosion progresses at the tensile flange of the main girder, varying the location and the grade of the corrosion.

2 ANALYSIS MODEL OVERVIEW

The bridge to be analyzed is a movable ferry bridge completed in 1977 at Fukuyama Port which locates in the southeastern area of Hiroshima prefecture. The dimensions of the bridge are shown in Figure 1. The bridge length is 20.6 m, the span is 20.0 m, the effective width of the steel deck is 4.0 m, the girder height is 1.3 m, and the checker plate with a thickness of 12 mm is used for the floor slab. The main girder web and the cross beam are joined by welding and the steel floor deck is attached by welding on the cross beam. Vertical ribs are attached to the steel floor slab by welding at approximately 340 mm intervals. As shown in Figure 1, in section A and C, each cross-section changes because the flange has a gradient.

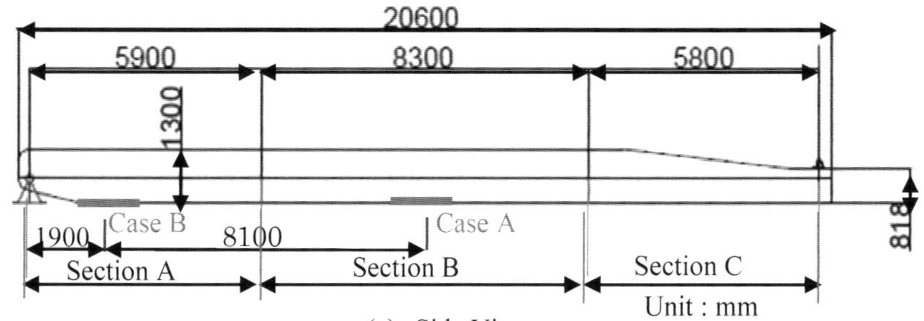

(a) Side View

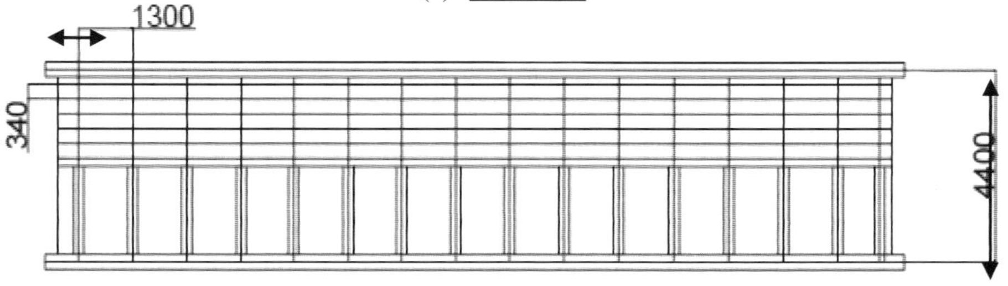

(b) Plan View Unit : mm

Figure 1. Dimensions of analyzed ferry bridge.

Nonlinear FEM analysis is conducted on this bridge using ABAQUS. All steel members are modeled by using 4-node iso-parametoric shell element, and the supports at both ends are modeled by using rigid element. The material properties of steel member are assumed as perfectly elasto-plastic material and are shown in Table 1. The support on the ground side is a pin support allows only the rotation around the axis of the direction perpendicular to the bridge axis, and the support on the sea side is a suspension support restrains only the vertical displacement.

Table 1. Materials properties of steel member.

Young's Modules [N/mm^2]	Yield Stress [N/mm^2]	Poisson's Ratio [-]	Unit Weight [kN/m^3]
210000	253.2	0.3	78.5

The dead load L_d is subjected to firstly, and then the live load L_l is subjected and increased up to the maximum load, consequently, the total load L_{all} is given by Eq. (1) in which α in Eq. (1) is the load increment parameter. Since α can be regarded as a kind of safety factor on live load, we call α safety factor hereafter. According to Japanese Specifications for Highway Bridges (JSHB), two kind of distributed loads (live load) P_1 and P_2 are subjected on the floor deck in order that the location of corrosion damage becomes most dangerous, as shown in Figure 2. Then, the location of live load changes depending on the location of local corrosion.

$$L_{all} = L_d + a \cdot L_l \tag{1}$$

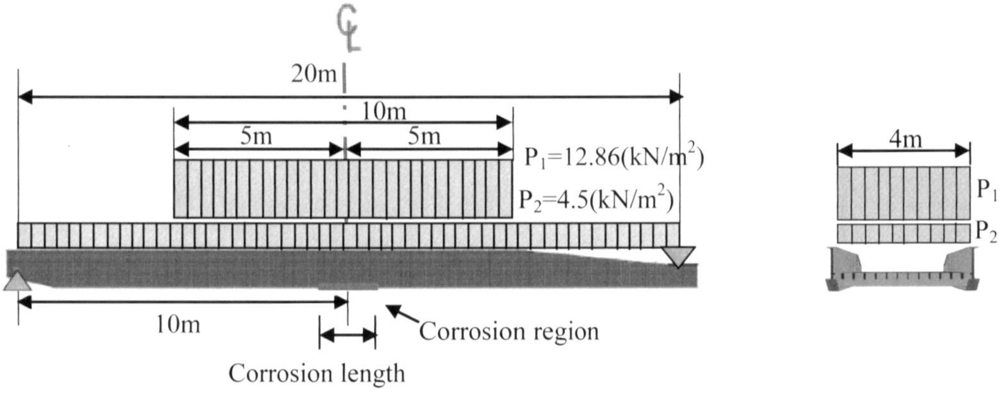

(a) Case A (corrosion at the center of the span)

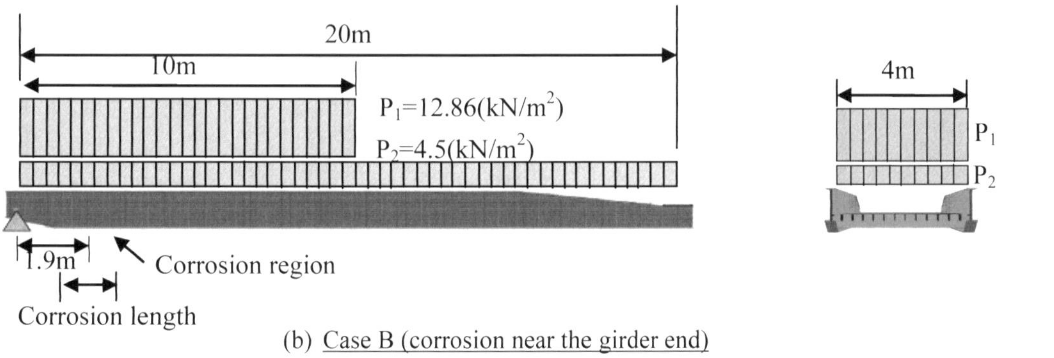

(b) Case B (corrosion near the girder end)

Figure 2. Location of live load (distributed L-load in JSBH).

In this paper, we assume that the location of local corrosion is at the center of the span and near the support shown by the red line in Figure 1 and 2, and let them be Case A and Case B, respectively. For the corrosion damage, the degree of corrosion is expressed by the thinning rate R shown in Eq. (2), and the thinning rate is changed from 0% to 100%. Moreover, the length of the corrosion is changed to 200 mm, 1000 mm, 2000 mm, respectively. Also, we assume the thickness of the flange is uniformly reduced from both surfaces, then the eccentricity of neutral plane does not change after corrosion.

$$R = (Reduced\ thickness\ of\ steel) / (Original\ thickness\ of\ steel) \times 100[\%] \quad (2)$$

3 ANALYSIS RESULT AND DISCUSSION

Figure 3 shows the relationship between the safety factor α_u at the maximum load and the thinning rate R. In the figure, "200" of "Case A-200", for example, means the length of corroded region of the local corrosion. It can be noticed from Figure 3 (a) that the safety factor α_u is 1.9 at the initial state without corrosion, which is significantly lower than general road bridges (usually α_u=4~5), and also that the safety factor decreases gradually with thinning progression. The figure shows the safety factor reduces to 1.45 when the lower flange at the center disappears. On the other hand, when the local corrosion locates near the support, it can be noticed that the safety factor does not change α_u=2.11 irrespective the thinning rate and also the length of corrosion region, as shown in Figure 3 (b).

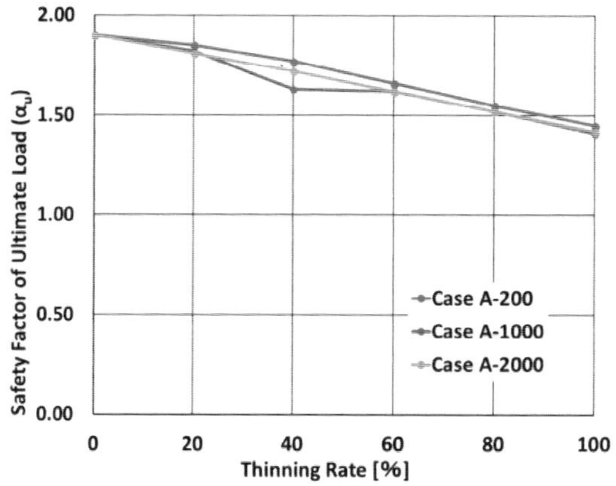

(a) Case A (corrosion at the center of the span)

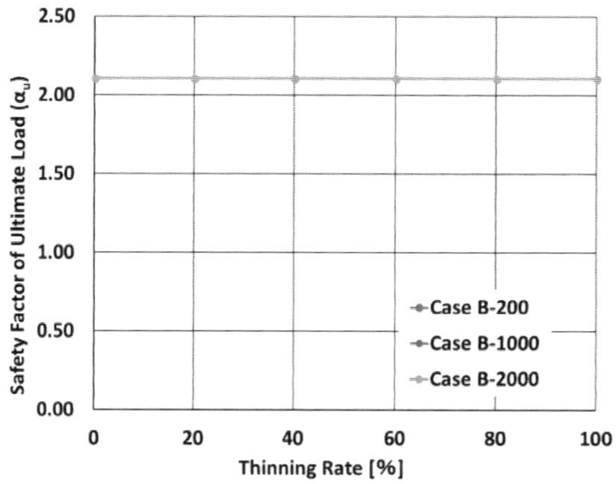

(b) Case B (corrosion near the girder end)

Figure 3. The relationship between safety factor and thinning rate.

The collapse state and the contour of von Mises stresses at the maximum load are shown in Figure 4. The collapse is caused by flange yielding in both cases. However, it is noticed that the location of flange yielding is different in both cases. Namely, at the center of the bridge in Case A, which is the same as the location of the local corrosion, but for Case B, the yielding occurs at the butt joint, where cross-sectional dimensions change, different from the corroded location. This is the reason that the safety factor does not concern with the corrosion thinning. Also, Case B shows the remaining strength does not enhance even though the corrosion is repaired. Thus, clarifying remaining strength seems to be important for the economical and efficient maintenance.

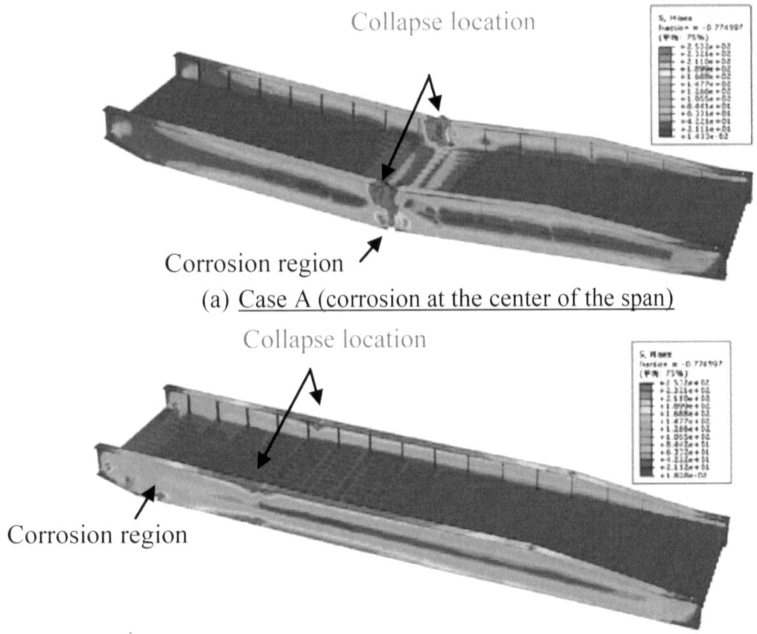

(b) Case B (corrosion near the girder end)

Figure 4. Collapse mode and von Mises stresses at maximum load.

4 CONCLUSIONS

In this study, remaining strength of entire bridge was analyzed for a ferry bridge corroded locally, by nonlinear FEM. Local corrosion damage was given in the lower flange only of main girder, changing the location and the grade of corrosion. The following terms can be listed from analytical results.

(1) Although the corrosion damage makes the bridge strength decrease in almost case, the possibilities to be unrelated to ultimate remaining strength is shown as a case. If the location of local corrosion is different from the trigger point of the collapse, the ultimate strength may not decrease due to corrosion. Consequently, it is important for adequate bridge maintenance that ultimate behavior and collapse mode after corrosion are analyzed and clarified from the view point of whole bridge.

(2) The safety factor for live load was 1.9 at initial state, though the ferry bridge analyzed in this paper is a typical type. This value is significantly smaller than general road bridge which has at least 4 commonly. Adding to this, considering flying salt environment of ferry bridge, more carefulness will be necessary in maintenance of ferry bridge.

References

Connor, R. J., Dexter, R., and Hussam, Md., Inspection and Management of Bridges with Fracture-Critical Details, National Cooperative Highway Research Program, *NCHRP Synthesis,* 354, 2005.

Tai, M., Shimozato, T., Tamaki, Y., Arizumi, Y., and Yabuki, T., Analytical Investigation on Collapse Mechanism of Steel Girder Bridge Due to Severe Corrosion Damage and Damage Recovery Evaluation at Bridge End Span, *Journal of Structural Engineering*, Japan Society of Civil Engineers, 61A, 416-428, March, 2015.

Takami, F., and Fujii, K., Capacity Recovery Case and Capacity Recovery Design Method of Corroded Steel Structures, *Steel Structures Series 23,* Japan Society of Civil Engineers, August, 2014.

Takami, F., and Fujii, K., Analytical Study on Ultimate Strength of Ferry Bridge Focusing on The Corrosion Location and Grade on Lower Flange of Main Girder, Research Presentation in Chugoku Branch, *Japan Society of Civil Engineers,* May, 2017.

Yamaguchi, S., Fujii, K., Fujii, M., Yamamoto, T., and Uenoya, M., Mechanical Behavior of a Composite Box Girder Bridge Used for 40 Years, *Journal of Structural Engineering*, Japan Society of Civil Engineers, 57A, 1074-1086, March, 2011.

Streamlining Information Transfer between Construction and Structural Engineering
Edited by Shiau, J., Vimonsatit, V., Yazdani, S., and Singh, A.

ISBN: 978-0-9960437-7-9

INFLUENCE OF INITIAL DEFLECTIONS ON COUPLED BUCKLING STRENGTH OF UNSTIFFENED SQUARE BOX SECTION COLUMNS

XIANG CHEN, SHOZO NAKAMURA, TOSHIHIRO OKUMATSU, and TAKAFUMI NISHIKAWA

Dept of Civil and Environmental Engineering, Nagasaki University, Nagasaki, Japan

Although it is well known that initial deflections have a significant influence on coupled buckling strength, their quantitative influence is not clear. In this paper, a series of nonlinear finite element analyses to reveal the quantitative influence of initial deflections on coupled buckling strength are conducted with MSC.Marc. There are nine kinds of combinations about the size of local and overall initial deflections including 1/75, 1/150, 1/450 of the plate width and 1/500, 1/1000, 1/3000 of the column length. In addition, the width-thickness ratio and slenderness ratio parameters are also varied in the parametrical study. Both ranges of slenderness and the width-thickness ratio parameters are from 0.1 to 2.1. As a result, the initial deflection influence coefficient is proposed to describe the quantitative influence of local and overall initial deflections on coupled buckling strength. The numerical results reveal that the coefficient decreases with the increase of initial deflections. The sensitivity of the strength to initial deflections depends on both width-thickness and slenderness ratio parameters. The maximum decrease of 21.4% on load-bearing capacity caused by local initial deflection appears when width-thickness and slenderness ratio parameters equal to 1.2 and 2.1, respectively. Column strength is quite sensitive to overall initial deflection at the slenderness ratio parameter equal to 1.4. The maximum decrease in load-bearing capacity reaches 18.2%.

Keywords: Axial compression, Local and overall initial deflection, Amplitude of initial deflection, Quantitative influence, Initial deflection influence coefficient, Load-bearing capacity.

1 INTRODUCTION

Thin-walled compression members are subject to interactive buckling instability due to high slenderness and width-thickness ratio parameters. And it is well known that initial deflections have a significant influence on coupled buckling strength. Initial deflections consist of local initial deflection on every plate and overall initial deflection on the column. High amplitude of initial deflections will lead to lower load-bearing capacity in most cases.

Many researchers had measured the amplitude of initial deflections of box section columns. In Usami and Fukumoto's (1982) tests, amplitudes of overall initial deflection on 19 specimens were measured. Their maximum is $L/1590$ and average is $L/3850$, where L is the column length. In Kuhlmann and Detzel's (2004) tests, amplitudes of overall initial deflection on six specimens were measured. Their maximum is $L/3560$ and average is $L/2500$. However, these values are much smaller than those allowed in some specifications such as $L/1000$ of American (AISC 2016), Japanese (JRA 2017) and Chinese (GB 50017 2003) specifications.

Although many tests on coupled buckling strength have been conducted, there are not many studies to evaluate the quantitative influence of initial deflections. In this paper, unstiffened square cross-section columns under axial compression are chosen as a target. Nonlinear finite element models are developed to analyze the coupled buckling strength of columns. Initial deflection influence coefficient is proposed to describe the quantitative influence of local and overall initial deflections on the strength.

2 VALIDATION OF FE MODEL

2.1 Existing Experiment on Unstiffened Box Section Columns

The tests of six columns were carried out in the Laboratory of Otto-Graf Institute, MPA Stuttgart at the University of Stuttgart (Kuhlmann and Detzel 2004). Cross section is presented in Figure 1. The sections consist of 6mm thick plates with a width of 230mm for the webs and 282mm for the flanges by fillet welds. Three kinds of length, 2200mm, 3700mm and 4900mm are considered. The steel grade of the plates is S355 with measured material properties of 390MPa for the yield stress (f_y) and of 187GPa for Young's modulus (E). Two models were prepared for each column, leading to a total of six tests. The columns were tested under the pin-ended condition with the buckling axis being set parallel to the flange plates as shown in Figure 2.

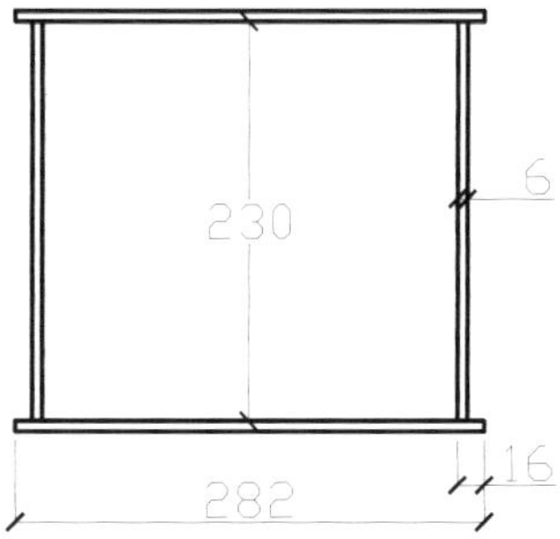

Figure 1. Dimension of cross section.

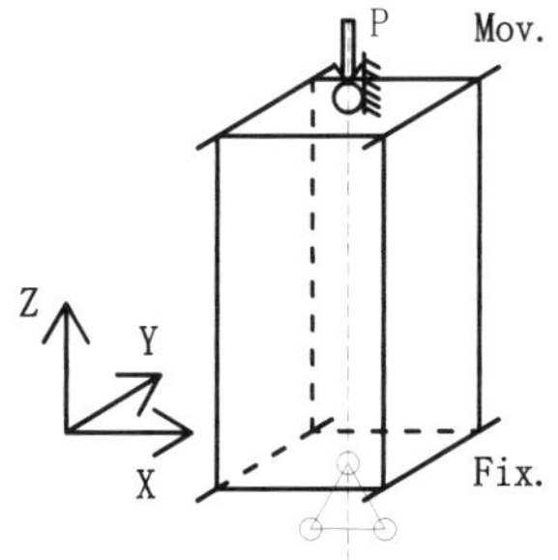

Figure 2. Boundary condition.

2.2 FE Model

The general-purpose FE software MSC.Marc was applied to the numerical analysis on coupled buckling strength of unstiffened box section columns. Thick shell element (No.75) is used in the model. Considering the results of preliminary exanimation on the influence of mesh size and the convenience of residual stress setting, plates are divided into twenty elements in the width direction with a size of 14.1mm on the flange and of 11.5mm on the web. The von Mises yield criterion is adopted in the analysis. Bilinear model with strain hardening coefficient of $E/100$ is used to describe the material property of S355. Measured material properties are used for the yield stress and Young's modulus. The Poisson ratio is set to 0.3.

Residual stress is assumed to exist in every plate with compressive residual stress in the middle area and tensile residual stress at both sides of the plate as shown in Figure 3 so as to be in self-equilibrium condition. Assumed residual compressive stress is $0.25f_y$, which is the measured maximum of residual stress in the past tests.

Images of both overall and local initial deflection shapes are shown in Figure 4. Half-sinusoidal wave shape expressed by Eq. (1) is assumed for the overall deflection. On the other and, the double trigonometric function as shown in Eq. (2) is assumed for the local deflection. The amplitude of initial deflections is set to the measured value as shown in Table 1.

$$w_g = A\sin(\frac{\pi z}{L}) \tag{1}$$

$$w_l = w_{l0} \sin\frac{m\pi z}{L}\cos\frac{\pi x}{b} \text{ Or } w_l = w_{l0} \sin\frac{m\pi z}{L}\cos\frac{\pi y}{b} \tag{2}$$

In the formula, A is the amplitude of overall initial deflection, w_{l0} is the amplitude of local initial deflection, m is the number of half sinusoidal wave giving the minimum buckling strength determined by the aspect ratio of the plate, and b is the plate width.

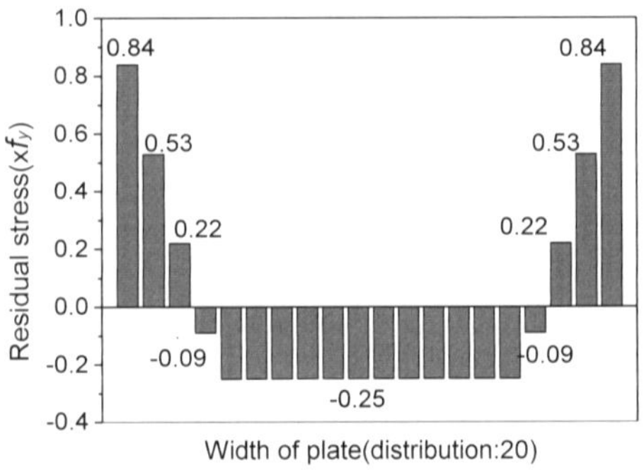

Figure 3. Residual stress distribution in a plate.

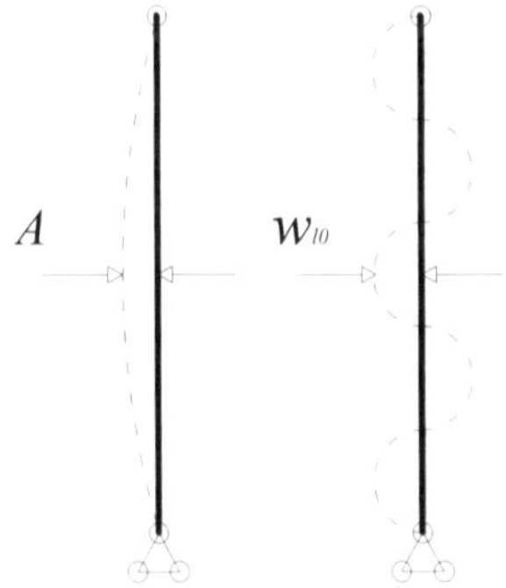

Figure 4. Initial deflection shape.

Table 1. Measured initial deflections.

Initial deflections	S1a	S1b	S2a	S2b	S3a	S3b
Overall initial deflection	$L/2800$	$L/2500$	$L/4000$	$L/4000$	$L/4350$	$L/3700$
Local initial deflection			$b/250$			

2.3 Validation Results

The comparison of ultimate strength between experiments and FE analyses is shown in Table 2. It can be seen that FEA results are slightly lower than test results. This underestimation is considered to be due to the use of the measured maximum initial deflection in the FE models. The average and standard deviation of errors are 3.2% and 1.73%, respectively. Consequently, it can be concluded that the FE models have sufficient accuracy.

Table 2. Comparison of ultimate strength between FE analysis and test.

Specimen	S1a	S1b	S2a	S2b	S3a	S3b	AVG(%)	STDEV (%)
Test results F_T(kN)	2053	2139	2065	2062	2114	2044	-	
FE results F_F (kN)	2036.1	2035.3	2016.6	2016.6	1987.3	1982.8		
$(F_F - F_T)/F_T$ (%)	-0.82	-4.85	-2.34	-2.20	-5.99	-2.99	-3.2	1.73

3 PARAMETRICAL STUDY

3.1 FE Models of Unstiffened Square Box Section Column

Using the verified models, an extensive numerical study on unstiffened square box section column is carried out. Further numerical models have been defined to cover a range of moderate slenderness and width-thickness ratio parameters shown in Table 3, in which coupled buckling is expected to occur. In addition, the amplitude of local and overall initial deflections is also varied. Table 4 shows nine combinations of the amplitude of initial deflections.

The Combi5 of which the local initial deflection is $b/150$ and the overall initial deflection is $L/1000$ is set as the standard since it is the combination of allowable initial deflections in Japanese design code. Initial deflection influence coefficient (ID influence coefficient) is proposed as the ratio of load-bearing capacity of a combination to that of Combi5 to describe the quantitative influence of initial deflections.

Table 3. Range of main parameters of the numerical specimen.

Parameter	Range of variable
Slenderness ratio parameter λ	0.1-2.1
Width-thickness ratio parameter R	0.1-2.1
Length of the column	1524-32004mm
Thickness of plate	5.28-110.79mm

Table 4. Combinations of amplitude of initial deflections.

	Combi1	Combi2	Combi3	Combi4	Combi5	Combi6	Combi7	Combi8	Combi9
Local	$b/75$	$b/75$	$b/75$	$b/150$	$b/150$	$b/150$	$b/450$	$b/450$	$b/450$
Overall	$L/500$	$L/1000$	$L/3000$	$L/500$	$L/1000$	$L/3000$	$L/500$	$L/1000$	$L/3000$

3.2 Influence of Local Initial Deflection

Results of the models whose R and λ are respectively 0.9 and 1.4 are set as examples to demonstrate the relationship between ID influence coefficient and local initial deflection as shown in Figure 5(a). It can be seen that ID influence coefficient decreases approximately linearly as the amplitude of local initial deflection increases. And the decreasing slope is nearly the same for different amplitudes of the overall initial deflection. Since the decrease of ID initial deflection can be regarded as the decrease of load-bearing capacity, the decrease of load-bearing capacity is 19.2% when the amplitude of local initial deflection increases from $b/450$ to $b/75$.

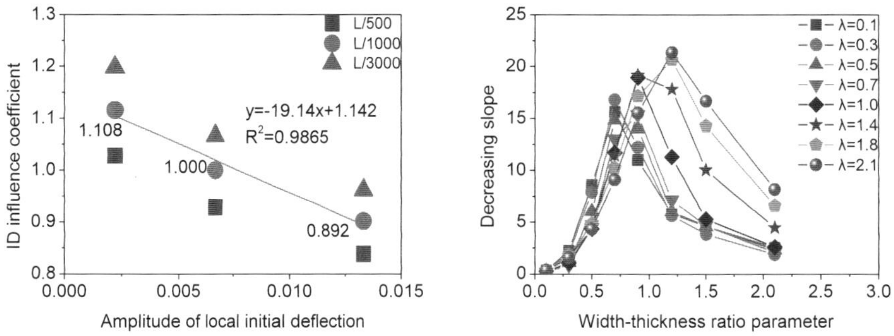

(a) ID influence coefficient (b) Deceasing slope of ID influence coefficient

Figure 5. Influence of local initial deflection.

The decreasing slope of ID influence coefficient depends on R and λ. Figure 5(b) and Table 5 show the relationship between the absolute value of the decreasing slope and R-value in the case of the overall initial deflection of $L/1000$. When R is small (i.e., $R \leq 0.5$) or large (i.e., $R > 1.5$), the influence of local initial deflection seems to be slight. In the width-thickness ratio parameter range of 0.7-1.5, column strength becomes comparatively sensitive to local initial deflection.

When λ is small (i.e., $\lambda \leq 0.7$), the influence reaches the maximum with the decreasing slope about 16.79 at R equal to 0.7. It corresponds to the 17.2% decrease of load-bearing capacity. When λ is large (i.e., $\lambda \geq 1.0$), the strength of columns with higher R-value becomes sensitive to the local initial deflection. With R and λ equal to 1.2 and 2.1, respectively, the influence of local initial deflection reaches the maximum with the decreasing slope of 21.37 corresponding to the 21.4% decrease of load-bearing capacity.

Table 5. Decreasing slope caused by local initial deflection (for overall initial deflection of $L/1000$).

R	λ=0.1	λ=0.3	λ=0.5	λ=0.7	λ=1.0	λ=1.4	λ=1.8	λ=2.1
0.1	0.32	0.29	0.17	0.18	0.27	0.34	0.37	0.37
0.3	2.25	1.95	1.16	0.88	1.06	1.54	1.77	1.57
0.5	8.60	7.90	6.05	4.43	4.35	4.96	4.83	4.33
0.7	15.63	16.79	14.85	13.04	11.71	11.48	10.24	9.11
0.9	10.99	12.22	14.02	15.37	18.92	19.14	17.13	15.53
1.2	5.81	5.64	6.03	7.14	11.30	17.80	20.63	21.37
1.5	4.59	3.83	4.60	4.64	5.29	10.05	14.29	16.67
2.1	2.52	1.88	2.34	2.08	2.62	4.46	6.60	8.17

3.3 Influence of Overall Initial Deflection

The influence of amplitude of the overall initial deflection on coupled buckling strength is shown in Figure 6(a). ID influence coefficient decreases approximately linearly as the amplitude of the overall initial deflection increases. And the decrement is nearly the same for the different amplitude of local initial deflection. As the amplitude of the overall initial deflection increases from $L/3000$ to $L/500$, the load-bearing capacity decreases by 15.1%.

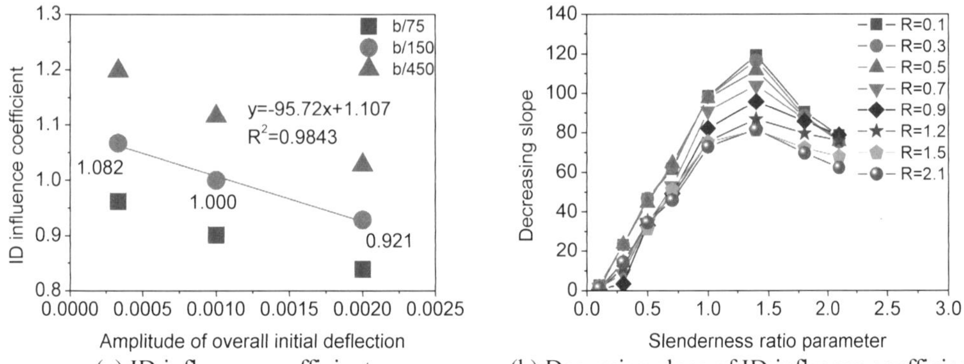

(a) ID influence coefficient (b) Deceasing slope of ID influence coefficient

Figure 6. Influence of overall initial deflection.

Figure 6 (b) and Table 6 show the relationship between the decreasing slope and slenderness ratio parameter in the case of the local initial deflection of $b/150$. As the λ-value increases, the decreasing slope increases first, and then decreases after the peak value. In the range of slenderness ratio parameter from 0.7 to 2.1, column strength becomes comparatively sensitive to overall initial deflection. With λ equal to 1.4, the maximum decreasing slope appears with a range from 81.05 to 119.09.

In addition, the influence of overall initial deflection on coupled buckling strength depends on R. With the decrease of R, the influence of overall initial deflection increases. A probable reason is that column strength is mainly dominated by overall buckling when R is small. With R and λ equal to 0.1 and 1.4, respectively, the influence reaches the maximum with the decreasing slope of 119.09 corresponding to the 18.2% decrease of load-bearing capacity.

Table 6. Decreasing slope caused by overall initial deflection (for local initial deflection of $b/150$).

λ	R=0.1	R=0.3	R=0.5	R=0.7	R=0.9	R=1.2	R=1.5	R=2.1
0.1	2.59	2.38	0.99	1.73	0.63	0.76	0.47	1.23
0.3	23.34	23.34	23.54	8.54	3.47	11.36	14.54	14.34
0.5	46.39	46.78	44.56	34.02	33.12	35.48	31.43	34.64
0.7	61.71	62.71	64.51	52.84	49.27	48.11	51.05	45.89
1.0	98.57	98.05	97.47	91.04	82.10	75.44	75.04	72.80
1.4	119.09	116.62	111.51	104.04	95.72	86.77	81.05	81.51
1.8	90.42	88.37	87.30	86.64	85.67	79.46	72.24	69.60
2.1	78.32	75.27	75.61	77.06	78.66	75.84	67.76	62.25

4 CONCLUSIONS

Main conclusions of this study can be summarized as follows.

(i) Coupled buckling strength linearly decreases as the amplitude of local/overall initial deflection increases. The decreasing slope is nearly the same for different amplitudes of overall/local initial deflection.

(ii) As the width-thickness and slenderness ratio parameters increase, the decreasing slope increases first, then decreases after the peak value.

(iii) With the width-thickness and slenderness ratio parameters equal to 1.2 and 2.1, respectively, the influence of local initial deflection reaches the maximum with the decreasing slope of 21.37 corresponding to the 21.4% decrease of load-bearing capacity.

(iv) With the width-thickness and slenderness ratio parameters equal to 0.1 and 1.4, respectively, the influence of the overall initial deflection reaches the maximum with decreasing slope of 119.09 corresponding to the 18.2% decrease of load-bearing capacity.

In the future, formulae to evaluate the quantitative influence of local and overall initial deflections will be proposed based on abundant FEA results.

References

AISC, *Specification for Structural Steel Buildings*, American Institute of Steel Construction, USA, 2016.

GB 50017, *Code for Design of Steel Structures*, China, 2003.

JRA, *Specifications for Highway Bridges*, Part II Steel Bridges and Members, Japan Road Association, Japan, 2017.

Kuhlmann, U., and Detzel, A., *Competitive Steel and Composite Bridges by Improved Steel Plated Structures*, WP 3.3 Test report, ComBri/USTUTT-44. Internal Report; December 2004.

Usami, T., Fukumoto, Y., Local and Overall Buckling of Welded Box Columns, *Journal of the Structural Division*, ASCE, 108(ST3), 525-542, 1982.

Streamlining Information Transfer between Construction and Structural Engineering
Edited by Shiau, J., Vimonsatit, V., Yazdani, S., and Singh, A.

ISBN: 978-0-9960437-7-9

NOISE REDUCTION DUE TO VIBRATION DAMPING ESTIMATED BY MEASUREMENTS OF PARTICLE VELOCITY

MASAAKI HIROE[1] and NOZOMU TANIGUCHI[2]

[1]*Kobayasi Institute of Physical Research, Kokubunji, Japan*
[2]*Maebashi Institute of Technology, Maebashi, Japan*

We proposed a new method to evaluate noise reduction due to damping treatment by using the particle velocity of sound measured directly in the vicinity of a vibrating object. Moreover, the proposed method was verified based on the comparison between vibration and particle velocity reductions. In Japan, a procedure to revamp old railway bridges, which had no specific fissure damages or serious corrosions, by applying to composite remodeling process have been widely used. During the process of such composite remodeling, we have often used magnetic rubber damper or rubber-latex mortar to reduce remarkable large power of structure-borne sound radiated from steel bridge of railway. This proposed method enables us to estimate precisely noise reduction of these countermeasures, because we grasp the behavior of sound radiated from the vibrating object in detail. In this paper, we present the results of laboratory and field experiments on the effects of noise reduction by using two magnetic rubber dampers, type-A and type-B. In the field measurement, the type-B more significantly reduced the sound radiated from vibrating beam. However, the reduction of radiated sound and that of vibration did not coincide. The reason was clarified from comparison between vibration and particle velocity, which occur due to vibrating a steel plate of simulated beam of bridge in laboratory experiment. Thus, the proposed method enables us to evaluate noise reduction precisely.

Keywords: Steel bridge, Magnetic rubber damper, Environmental countermeasures, Field measurement, Laboratory experiment.

1 INTRODUCTION

Steel railway bridges made up of members of both longitudinal and cross girders have centuries-old history, and many steel bridges have still existed in various places even nowadays. In recent years, steel bridges have also been shown to have long life-cycle costs. In addition, many of the steel members are both high quality and high reliability of factory products, so they have many advantages such as short construction period than concrete bridge. However, on the other hand, steel members with a thin structure are caused to vibrate easily, so there is a disadvantage that the steel bridge has a significantly larger radiated sound due to vibration than a concrete bridge. The residents in major urban areas have high level of consciousness to the surrounding environment, so there are tendencies that the number of complaints at areas along the steel bridge is larger than that at areas along the concrete bridge. Therefore, despite the various advantages such as long life-cycle cost and shortening of construction work, there are some cases where construction of a steel bridge is avoided in consideration of the sound environment along railway.

In order to deal with the urgent issue of reducing radiated sound due to vibration of steel members, magnetic-vibration damping (Masanori and Mifune 1998) and rubber-latex mortar (Lin *et al.* 2017) have been devised as countermeasures applicable to the existing steel railway bridge. Either countermeasures increase the internal loss of the steel member by adding the material, and result in reducing the energy of the vibration which causes structure-born sound. In many cases which such countermeasure is applied, there are often difference of reduction effect between the vibration and radiated sound (Taniguchi *et al.* 2011). The reduction effect of structure-born sound may be not precisely estimated from that of vibration as for countermeasure of steel railway bridge. Therefore, in this study, we propose a method to evaluate the reduction effect based on the particle velocity of the sound directly measured near the vibrating body. We verified the validity of the proposed method by comparing reduction among vibration velocity, particle velocity and intensity of sound due to the vibration (Hiroe and Taniguchi 2017).

2 METHOD OF MEASUREMENTS AND EVALUATION

2.1 Measurement Method

In this study, two kinds of magnetic-vibration damping (type-A and type-B) fixed with the magnetic force were used as shown in Figure 1. The measurement for the steel girder bridge was performed at the Kasukawa railway bridge of the Jomo Electric Railway Co., Ltd. as shown in left side of Figure 2. We implemented a countermeasure to five unit spaces bounded by adjacent intermediated stiffeners, and five pieces of magnetic-vibration dampers which were size of 0.45 m * 0.30 m were attached per one unit space. At the center of five unit spaces, we set 5 sensors of vibration acceleration $\boldsymbol{a}$ on inside of the plate girder and 1 sensor of sound intensity $\boldsymbol{I}$, that consists of 2 sensors of sound pressure $\boldsymbol{p}$, near outer side of the plate girder as shown in right side of Figure 2.

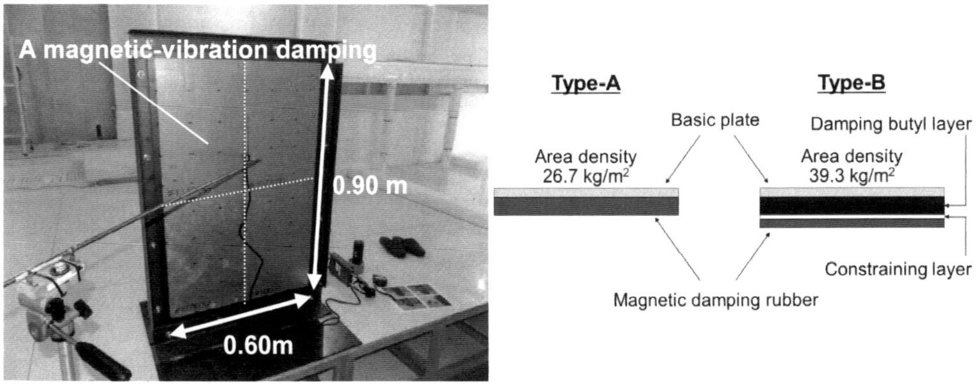

Figure 1. A photograph and structures of magnetic-vibration damping.

Figure 3 shows an overview of a steel plate and the arrangement of both 3 exciting points (x-mark) and 70 measurement points (black circle) on/near the plate in a primary experiment. The steel plate used in the primary experiment was size of 1.055 m * 0.740 m, and attached to 4 pieces of magnetic-vibration dampers. As shown in Figure 3, the particle velocity $\boldsymbol{u}$ (m/s) in the vicinity of site for attachment of magnetic-vibration dampers was measured, and the vibration acceleration $\boldsymbol{a}$ (m/s^2) was measured on the opposite side of the particle velocity.

We used an impact hammer PH-51 with a hard tip (Figure 3), sensors of vibration acceleration PV-85/86 (Figures 2 and 3) and an intensity microphone SI-34 (Figure 2) for the field measurement at steel girder bridge. For the primary experiment using the steel plat, PH-51, PV-86 and a sensor of particle velocity KIT-PA-SP0 (Figure 3) were used. All output signals caused by excitation force of the impact hammer were recorded using a data recorder LX-110 for the field measurement (sampling frequency f_s = 24.0 kHz) or DA-20 for the primary experiment (f_s = 25.6 kHz). We secured the effective data which exceed 10 dB of signal to noise substantially of more than 5 for each combination of excitation point and measurement point.

Figure 2. An overview of deck-type plate girder bridge and the arrangement of point excitations (x-mark) and measurement points on/near the girder.

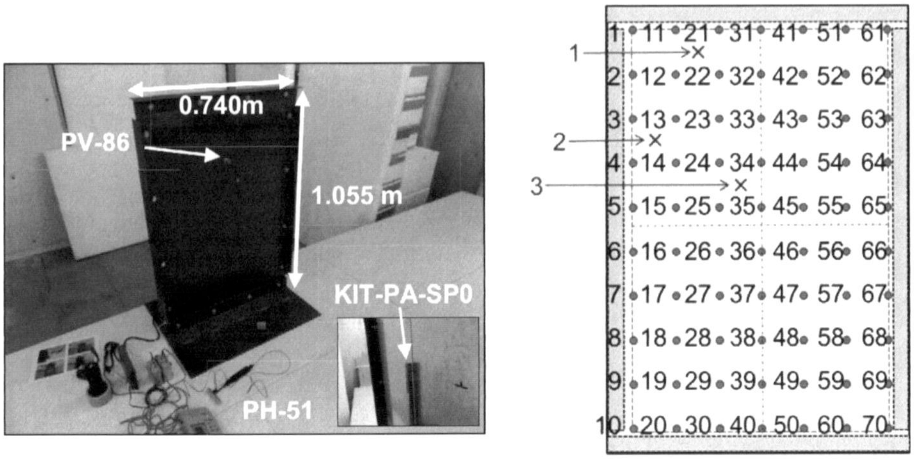

Figure 3. A photograph of the steel plate and the arrangement of 3 point excitations (x-mark) and 70 measurement points (black circle) on/near the plate.

2.2 Evaluation Procedure

Firstly, the electrical signals recorded were converted to those of excitation force $\boldsymbol{F}$ (N), sound pressure $\boldsymbol{p}$ (Pa), vibration acceleration $\boldsymbol{a}$ (m/s^2) and particle velocity $\boldsymbol{u}$ (m/s). Then, we converted the signal of vibration acceleration $\boldsymbol{a}$ to that of the vibration velocity $\boldsymbol{v}$ (m/s) by pass through the filter in the inverse ratio of frequency. The signal of sound intensity $\boldsymbol{I}$ (Watt/m^2) was calculated according to the formula from two signals of the sound pressure $\boldsymbol{p}$ by directly using digital integrators (Tachibana and Yano 1988).

Secondary, we calculated the time-integrated energy value of each variable $\boldsymbol{Q}$ which indicated $\boldsymbol{F}^2$, $\boldsymbol{v}^2$, $\boldsymbol{u}^2$ and $\boldsymbol{I}$ as shown in the following equation (1).

$$P_{ijk} = \int_{-\infty}^{+\infty} Q_{ijk}(t)dt \tag{1}$$

where, subscripts of $\boldsymbol{i}$, $\boldsymbol{j}$ and $\boldsymbol{k}$ mean excitation position (N_i=3), measurement position (N_j=5 in the field measurement / N_j=70 in the primary experiment) and iteration count (N_k=5). After normalizing by excitation force $\boldsymbol{F}$ according to the equation (2), the averaged value $\boldsymbol{P_{ij/F}}$ of each variable $\boldsymbol{P_{ijk/F}}$ was calculated according to the equation (3). Further, the area-averaged value $\boldsymbol{P_{i/F}}$ of each variable $\boldsymbol{P_{ij/F}}$ was calculated according to the equation (4).

$$P_{ijk/F} = P_{ijk}/F_{ijk} \tag{2}$$

$$P_{ij/F} = \left(\sum_{k=1}^{N_k} P_{ijk/F}\right)/N_k \tag{3}$$

$$P_{i/F} = \left(\sum_{j=1}^{N_j} P_{ij/F}\right)/N_j \tag{4}$$

Finally, in accordance with the equation (5), the level ratio of the standardized exposure value for vibrating body between with damper (type-A or type-B) attachment $\boldsymbol{P_{i/F}}^{(A,B)}$ and without damper attachment $\boldsymbol{P_{i/F}}^{0}$ was calculated as reduction effect $\boldsymbol{\Delta P_{i/F}}^{(A,B)}$. Finally, we calculated the difference of reduction $\boldsymbol{\Delta P_{i/F}}^{B/A}$ between type-B $\boldsymbol{\Delta P_{i/F}}^{(B)}$ and type-A $\boldsymbol{\Delta P_{i}}^{(A)}$ in according to the equation (6).

$$\Delta P_{i/F}^{(A,B)} = -10 \times log_{10}\left(P_{i/F}^{(A,B)}/P_{i/F}^{0}\right) \tag{5}$$

$$\Delta P_{i/F}^{B/A} = \Delta P_{i/F}^{(B)} - \Delta P_{i/F}^{(A)} \tag{6}$$

3 RESULTS AND DISCUSSIONS

3.1 The Result in the Field Measurement at a Deck-Type Steel Girder Bridge

Figure 4 shows a comparison of reduction effects $\boldsymbol{\Delta v}^{B/A}{}_{/F}$ between (a) the field measurement and (b) the primary experiment. The left side figure (a) shows two kinds of reduction difference, $\boldsymbol{\Delta v}^{B/A}{}_{/F}$ for vibration velocity and $\boldsymbol{\Delta I}^{B/A}{}_{/F}$ for sound intensity in the field measurement at the deck-type steel girder bridge. The right shows two kinds of difference reduction, $\boldsymbol{\Delta v}^{B/A}{}_{3/F}$ for vibration velocity and $\boldsymbol{\Delta u}^{B/A}{}_{3/F}$ for particle velocity in the primary experiment using the steel plate for No.3 of excitation position.

As shown in Figure 4(a), there is a clear difference between the sound intensity $\boldsymbol{\Delta I}^{B/A}{}_{/F}$ and the vibration velocity $\boldsymbol{\Delta v}^{B/A}{}_{/F}$ in the reduction effect at the frequency range from 630 Hz to 2 kHz. We notice that there is discrepancy between the vibration of girder plate and the radiated sound due to the vibration with magnetic-vibration damper attachment at the frequency range from 630 Hz to 2 kHz. On the other hands, as shown in Figure 4(b), the result of the primary experiment shows that there is a clear difference between the particle velocity $\boldsymbol{\Delta u}^{B/A}{}_{3/F}$ (μ-Flown) and the vibration velocity $\boldsymbol{\Delta v}^{B/A}{}_{3/F}$ (PV) in the reduction effect at the frequency range from 2 kHz to 4 kHz. Similarly to the result in the field measurement, as far as it relates to the particle velocity of sound radiated from vibrating body ($\boldsymbol{\Delta u}^{B/A}{}_{3/F}$), the reduction of type-B is larger than that of type-A in all frequency range, despite the fact that the damper of type-A has large reduction in the vibration than the damper of type-B near the frequency of around 2.5 kHz ($\boldsymbol{\Delta v}^{B/A}{}_{3/F}$).

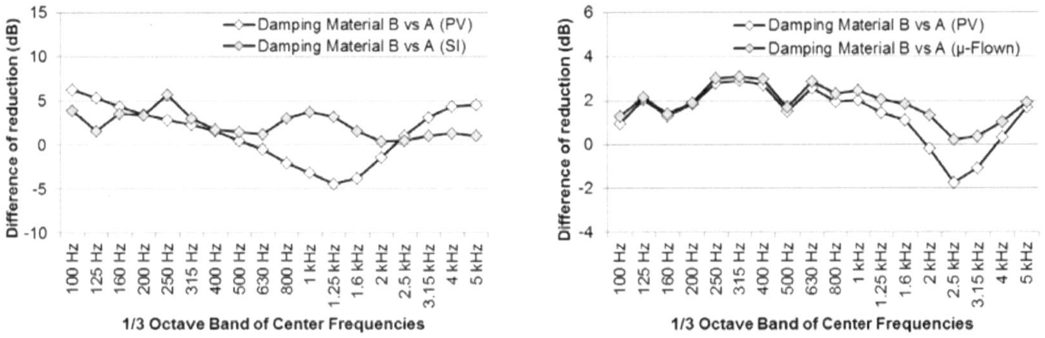

(a) The field measurement (b) The primary experiment

Figure 4. The reduction effects in (a) the field measurement: vibration velocity PV and sound intensity SI in the frequency characteristic at the steel girder bridge, and (b) the primary experiment: PV and particle velocity μ-Flown for excitation position No.3 at the steel plate.

3.2 The Result in the Primary Experiment Using a Steel Plate

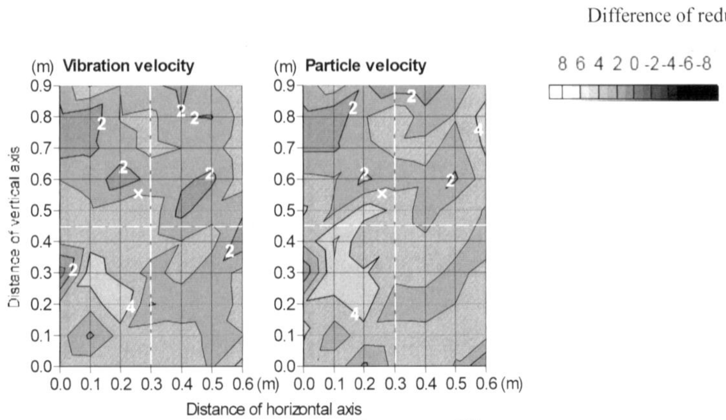

(a) Frequency 315 Hz: $\Delta v^{B/A}{}_{3/F}$ vs $\Delta u^{B/A}{}_{3/F}$

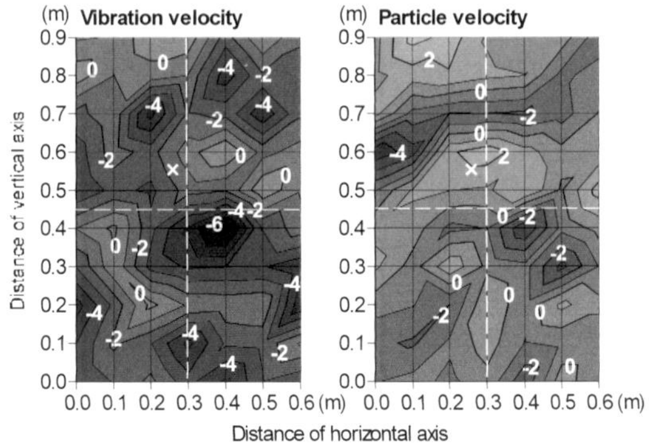

(b) Frequency 2 kHz: $\Delta v^{B/A}{}_{3/F}$ vs $\Delta u^{B/A}{}_{3/F}$

Figure 5. Comparisons of reduction effect between vibration velocity $\Delta v^{B/A}{}_{3/F}$ and particle velocity $\Delta u^{B/A}{}_{3/F}$ for excitation position 3 in the primary experiment.

Figure 5 shows contour maps of difference reduction compared type-B with type-A for both particle velocity $\boldsymbol{\Delta u}^{B/A}{}_{3/F}$ and vibration velocity $\boldsymbol{\Delta v}^{B/A}{}_{3/F}$. Figure 5(a) is the contour maps in the frequency band of 315 Hz, and Figure 5(b) shows those in the frequency band of 2.5 kHz.

These contour maps make clear that the consistency or discrepancy of reduction effect between particle velocity $\boldsymbol{\Delta u}^{B/A}{}_{3/F}$ and vibration velocity $\boldsymbol{\Delta v}^{B/A}{}_{3/F}$ appear over the entire surface of the steel plate. In regard to the reduction effect caused by the magnetic-vibration damper attachment, the behavior of particle velocity in the vicinity of the vibrating plate seems to differ with that of its vibration velocity above the middle frequency of around 500 Hz. As shown in Figure 5(a), because the wavelength of bending wave on the steel plate becomes larger than the size of a piece of magnetic-vibration damper below the frequency of 500 Hz (in the frequency band of 315 Hz), each piece of the magnetic-vibration damper express similar behavior to the vibrating steel plate. On the other hands, at the frequency range far higher than 500 Hz (in the frequency band of 2.5 kHz), there are a few peaks and troughs for reduction effect of vibration velocity in each piece of the magnetic-vibration damper. As shown in Figure 5(b), at such frequency range, the behavior of steel plate seems to differ from that of the magnetic-vibration damper. Therefore, our proposed method to evaluate based on the particle velocity near a vibrating body is superior to the method based on vibration velocity of the body in determining noise reduction of a countermeasure precisely.

4 CONCLUSIONS

In this study, we propose a method to evaluate reduction effect based on the particle velocity of sound measured directly near the vibrating body. Through both the field measurement at the deck-type steel girder bridge and the primary experiment using the steel plate, we verified the validity of the proposed method. In regard to two magnetic-vibration dampers, the behavior of particle velocity seems to differ with that of vibration velocity above the middle frequency of around 500 Hz. Thus, it is necessary to evaluate reduction effect based on the particle velocity rather than vibration velocity in order to determine noise reduction of a countermeasure precisely.

Acknowledgments

This research was funded by JSPS Grant-in-Aid for Research (15K 06189: Research Associate, Maebashi Inst. Tech. Associate Professor Taniguchi). Jomo Electric Railway Co., Ltd. got a lot of cooperation in the measurement of steel girder bridge. In addition, we received great support from Toyo Technical Co., Ltd. in particle velocity measurement. I am deeply grateful to their lot of great help.

References

Hiroe, M., and Taniguchi, N., Investigation on Estimation of Noise-Reduction Effect Due to Vibration Damping of Steel Railway Bridge by Using Particle Velocity Measurement (Japanese), *Proceedings of Constructional Steel,* 25, November 2017.

Lin, W., Taniguchi, N., and Yoda, T., Novel Method for Retrofitting Superstructures and Piers in Aged Steel Railway Bridges, *Journal of Bridge Engineering*, 22(11), 05017009, 2017.

Masanori, H., and Mifune, N., Study on the Vibration Damping Properties of the Damping Material Applying Rubber Compounding Magnetic Powder (DRM), *Journal of the Acoustical Society of Japan (E)*, 19(1), 13-21, 1998.

Tachibana, H., and Yano, H., Sound Intensity Measurement for Impulsive Sounds. *The Journal of the Acoustical Society of America,* 84(S1), S33-S33, 1988.

Taniguchi, N., Hansaka, M., Koide, N., Ohgaki, K., Ohkubo, F., Satake, S., and Sugio, Y., Examination about Measures of Noise and Vibration for Steel Railway Bridges Using Rubber-Latex Mortar (Japanese), *Proceedings of Bridge Vibration Colloquium 2011*, 195-200, 2011.

Streamlining Information Transfer between Construction and Structural Engineering
Edited by Shiau, J., Vimonsatit, V., Yazdani, S., and Singh, A.

ISBN: 978-0-9960437-7-9

DEFORMATION PERFORMANCE IMPROVEMENT IN TIMBER COLUMN–GROUND SILL JOINT REINFORCED USING ARAMID FIBER SHEET

NORIKO TAKIYAMA, AKARI YAMAGUCHI, XINYAN CHEN, and SHO KOIKE

Dept of Architecture and Building Eng., Tokyo Metropolitan University, Tokyo, Japan

We report the progress of an experimental study conducted to understand the seismic performance of a timber column–ground sill joint reinforced using an aramid fiber sheet, and to improve the deformation property of the reinforced joint. An aramid fiber sheet is a new material that weaves high-performance aramid fibers in one or two directions. In previous research, certain problems were found: (a) even when applying a similar reinforcement, the sheet did not necessarily demonstrate the same failure mode, and (b) when the sheet was peeled off in stretches, the joint is destroyed through brittleness. In this study, based on the preceding research, we proposed a new sheet-pasting method for an improvement in the deformation property, and conducted a bending test under cyclic loading for some column–ground sill joint specimens to verify the seismic performance and failure behavior. It was found that, by splitting the sheet, the problems of the preceding research are avoidable.

Keywords: New material, Cyclic loading, Restoring force, Failure mode, Deformation property, Maximum force.

1 INTRODUCTION

Many timber structures collapse during earthquakes because of joint damage. To prevent this, the Japanese Building Standard Law was revised in 2000 to include joint specifications for timber structures (Ministry of Land, Infrastructure, Transport and Tourism 2016). However, the seismic resistance of timber frames with steel joints can be compromised through a partial loss of the member sections such as the bolt holes (Sawai 2012).

High-performance aramid fiber sheets are a new class of composite material made up of weaved polyamide fibers. Recently, a method for reinforcing timber joints using aramid fiber sheets was proposed, through which existing timber structures can be reinforced without a large-scale demolition because the aramid fiber sheets can be easily glued to the joints using an adhesive agent.

In a previous study, to evaluate the seismic performance and failure behavior of joints reinforced with aramid fiber sheets, we conducted bending tests under cyclic loading for some column–ground sill specimens with T-shaped joints (Yamaguchi *et al.* 2016, Yamaguchi *et al.* 2017), and found a few problems: (a) even when applying a similar reinforcement, the sheet did not necessarily demonstrate the same failure mode, and (b) when the sheet was peeled off in stretches, the joint is destroyed through brittleness.

In this study, we conducted additional bending tests under cyclic loading to solve the problems discovered in our previous research. We propose a new sheet-adhesive approach for an improvement in the deformation property, and conducted a bending test under cyclic loading for eight T-shaped column–ground sill joint specimens to verify the seismic performance and failure behavior. Herein, we attempted using a single sheet split into a certain number of parts before adhering them onto the timber.

2 IMPROVED METHOD AND COLUMN–GROUND SILL JOINT SPECIMENS

2.1 Results of the Previous Study

The major findings from our previous study are as follows:

(i) Not covering a resin around the joint surface prevented the aramid fiber sheet from easily breaking, as shown in Figure 1(a), because compression does not occur in the aramid fiber reinforced plastic layer, as indicated in Figure 1(c).

(ii) The deformation performance of the restoring force was higher when the sheet is peeled off sequentially from the outermost edge, as shown in Figure 1(c), as compared to when it is entirety peeled off instantly, as shown in Figure 1(b).

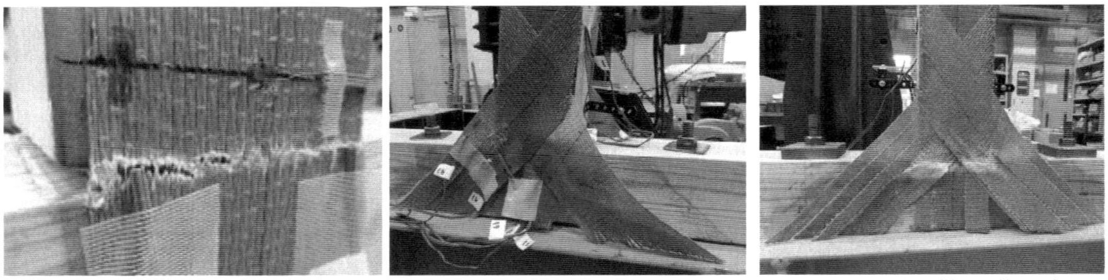

(a) Linear fracture of boundary (b) Sheet peeled off entirely (c) Split in the fiber direction

Figure 1. Main damage indicated in a previous study.

2.2 Improved Sheet Adhering Method and Test Specimens

For an improvement in the deformation performance of the bending property for a timber column–ground sill joint, we propose splitting the aramid fiber sheet into pieces by cutting the horizontal threads that connect the fiber ribs in advance. Thus, we can control the failure mode in which a splitting failure mode continuously occurs, as shown in Figure 1(c).

T-shaped column–ground sill specimens were designed. The columns and ground sills (105 mm × 105 mm) were made from cedar wood, and the joints were fastened using V-shaped steel members. After a load was applied to the specimen, V-shaped steel members were removed and reinforced with aramid fiber sheets, and the load was reapplied to the specimen.

The sheets were attached to both sides of the joints of the specimens using a vertical and crossing sheet. To study the effects of the improved method, we prepared some specimens with split sheets. The aramid fiber sheet used in this study has 30 ribs. Thus, we tried to split the sheet, as shown in Figure 2. For a vertical sheet, in the case of V1 shown in Figure 2(a), 30 ribs are divided into five groups of six ribs each. Then, V2 is divided into 9-5-3-5-9 ribs, and V3 is divided into 3-6-12-6-3 ribs. For the crossing sheet, C1 is divided into 6-6-6-6-6 ribs, C2 is divided into 9-8-6-5-3 ribs from the outside, and C3 is divided into 3-5-6-8-9 ribs, also from the outside.

The types of specimens are shown in Table 1. For the sheet-attachment method, vertical sheets V1–V3 and horizontal sheets C1–C3 are combined. Moreover, to avoid affecting the compression in the aramid fiber reinforced plastic layer, resin was not covered around the joint surface. The specifications of the materials used to fabricate the specimens are listed in Table 2.

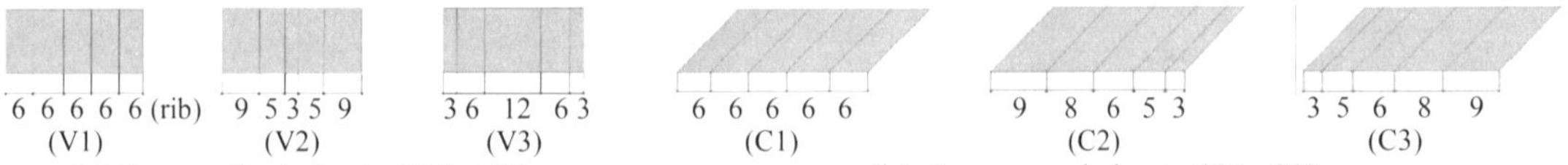

Figure 2. Number of rib partitions.

Table 1. Joint specimens and sheet adhering style.

		Crossed Sheet			
		Non	(C1)	(C2)	(C3)
Vertical Sheet	(V1)	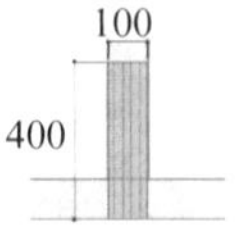 Specimen V01	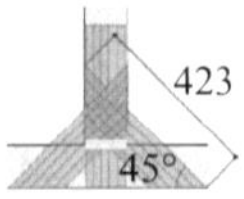 Specimen VC4		
	(V2)	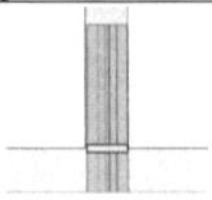 Specimen V02		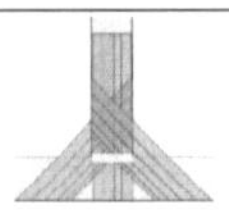 Specimen VC5	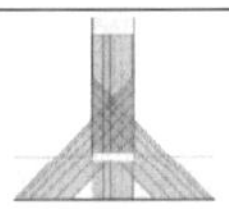 Specimen VC7
	(V3)	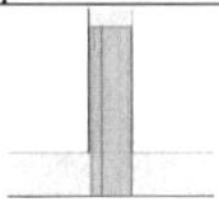 Specimen V03		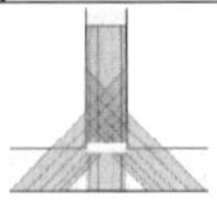 Specimen VC6	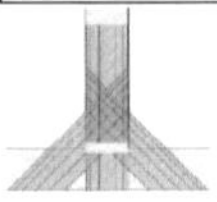 Specimen VC8

Table 2. Materials used to fabricate specimens applied during the tests (Fibex 2016, Asahi Bond 2016).

Material	Item	Notes	Item	Notes
Aramid Fiber Sheet	Description	FiBRA Sheet, AK-40	Manufacturer	Fibex Co., Ltd
	Tensile strength	2060 N/mm^2	Young's Modulus	118 kN/mm^2
	Design thickness	0.193 mm	Width	100 mm
Resin	Description	Asahi bond, 701	Manufacturer	Asahi Bond, Inc.
	Tensile strength	38.3 N/mm^2	Bending strength	66.0 N/mm^2

3 CYCLIC LOADING TEST

3.1 Loading System

The loading system is illustrated in Figure 3. The vertical displacement at the top of the column was measured using a displacement transducer, and the rotation angle R was determined trigonometrically from the measured vertical displacement and the known column length. Cyclic loading was applied to the specimen such that the amplitude of R gradually increased. For the analysis, the left side of Figure 3 was assumed to be in the positive direction; here, the front was designated as the S side, and the back was designated as the N side.

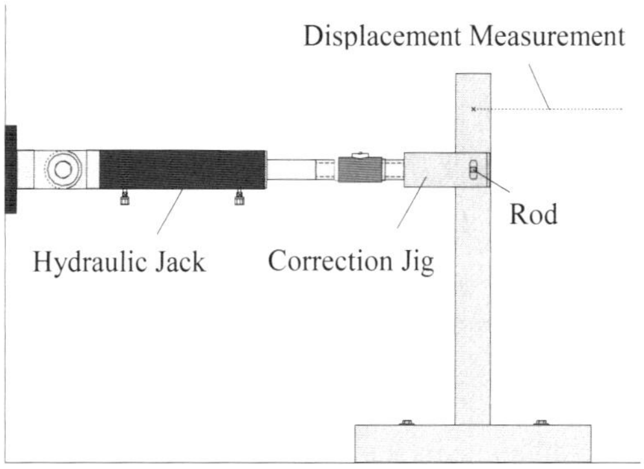

(a) Schematic diagram

(b) Loading state

Figure 3. Loading system.

3.2 Failure Mode and Restoring Force

The main failure mode is shown in Figure 4, and representative restoring force of specimens V02 and VC7 are shown in Figure 5. The main values are indicated in Table 3, including the maximum horizontal force, rotation angle, and failure mode. A comparison of the restoring force of specimens V02 and VC7 with the previous results is provided in Figure 6.

(a) Linear fracture of boundary (failure mode I)

(b) Fracture or peeling from the outside (failure mode II)

Figure 4. Main damage.

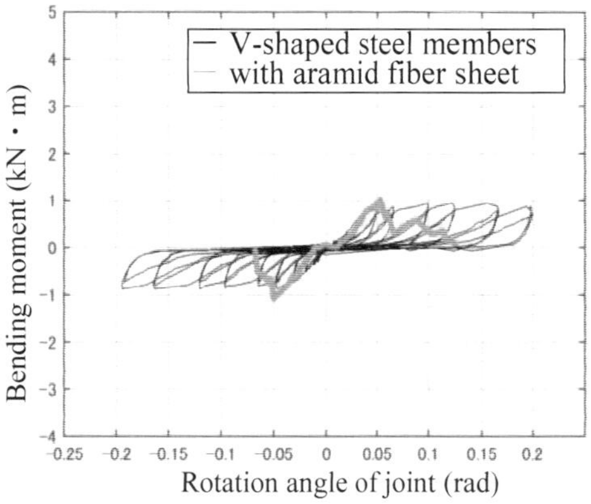

(a) Specimen V02

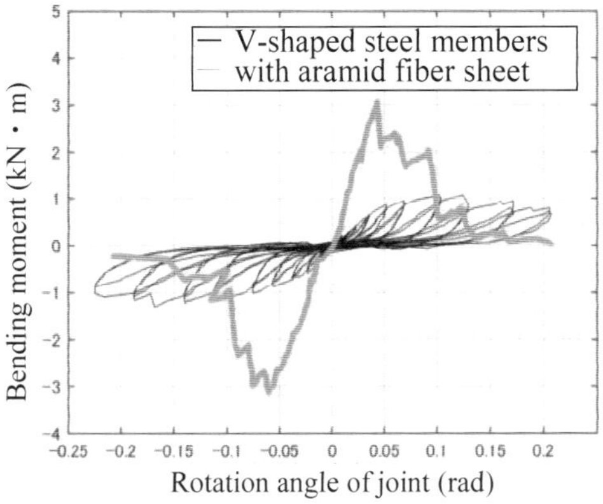

(b) Specimen VC7

Figure 5. Restoring force.

Failure mode I is a linear fracture of the boundary. This occurred in specimens V01 and V03, which do not have a no-resin area. Failure mode II is a fracture or peeling from the outside. This occurred on the other specimens, which have a no-resin area. The restoring force of the specimens with vertical and crossing aramid fiber sheets is higher than that with V-shaped steel members, as indicated in Figure 5(b), for example. The restoring force of the specimens reinforced using a split sheet achieves a higher deformation performance than when using the existing method, as indicated in Figure 6.

Table 3. Results of the bending test.

Specimen	Maximum horizontal force (kN)	Rotation angle when Max. (rad)	Failure mode
V01	0.5	1/30	I
V02	1.6	1/15	II
V03	0.8	1/30	I
VC4	4.3	1/15	II
VC5	5.2	1/15	II
VC6	5.2	1/10	II
VC7	4.3	1/15	II
VC8	3.8	1/15	II

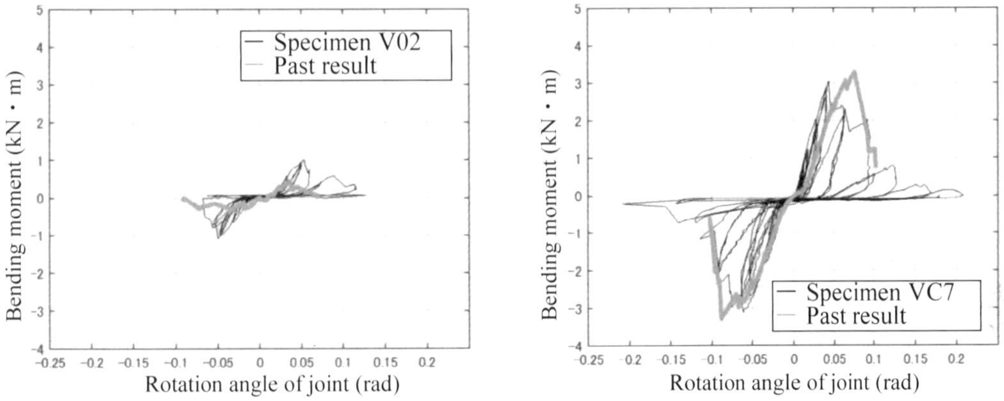

(a) With vertical sheet (b) With vertical and crossing sheet

Figure 6. Comparison of restoring forces.

4 DISCUSSIONS

The major findings from this study are discussed in the following section:

(i) For reinforcement using both a vertical sheet and a vertical and crossing sheet, a column–ground sill joint using a split sheet has a greater deformation property than one using an existing sheet under a bending load. A sudden drop in force from the maximum value does not occur, as indicated in the results from the improved sheet adhesion method.

(ii) The maximum force is not affected by the improved sheet adhesion method. The maximum force of a joint reinforced using a split sheet is as strong as that using an existing sheet.

5 CONCLUSIONS

In this paper, we proposed a new method for the attachment of aramid fiber sheets to timber column–ground sill joints in order to develop deformation performance. As a part of this study, we also conducted additional bending tests under cyclic loading.

The major findings from this study are summarized as follows:

(i) For an improvement in the deformation performance of the bending property for a timber column–ground sill joint, we proposed splitting an aramid fiber sheet in advance, allowing the splitting failure mode to continuously occur.

(ii) A joint reinforced by a split sheet has a higher deformation property than one reinforced by an existing sheet. However, as the results indicate, the maximum force is not affected by the improved sheet adhesion method.

Acknowledgments

This research was supported by the Obayashi Foundation and Grants-in-Aid for Scientific Research (C) (No. 17K06649). We are grateful to Fibex Co., Ltd., for providing the aramid fiber sheets and to Asahi Bond, Inc., for providing the resin used in this study. We would like to thank the graduate students of Tokyo Metropolitan University for their assistance with the cyclic loading tests.

References

Asahi Bond, Inc. (in Japanese). Retrieved from http://www.asahibond.co.jp/ on June 1, 2016.

Fibex Co., Ltd. (in Japanese). Retrieved from http://www.fibex.co.jp/index.html on June 1, 2016.

Ministry of Land, Infrastructure, Transport and Tourism. Retrieved from http://www.mlit.go.jp/index.html on June 1, 2016.

Sawai, S., *Timber Structure Seismic Construction Perfect Guide* (in Japanese), X-Knowledge Co., Ltd., Japan, 2012.

Yamaguchi, A., Chen, X., and Takiyama, N., *Performance Confirmation Test for Timber Column-Ground Sill Joints Reinforced with Aramid Fiber Sheets*, Proceeding of the ASEA-AEC-3, Kuching, Sarawak, Malaysia, November 2016.

Yamaguchi, A., Chen, X., and Takiyama, N., *Analysis of Timber Column-Ground Sill Joints Reinforced with Improved Aramid Fiber Sheets,* Proceeding of The Ninth International Structural Engineering and Construction Conference, Valencia, Spain, July 2017.

Streamlining Information Transfer between Construction and Structural Engineering
Edited by Shiau, J., Vimonsatit, V., Yazdani, S., and Singh, A.

ISBN: 978-0-9960437-7-9

SEISMIC DAMAGE PREDICTION FOR JAPANESE-STYLE WOODEN HOUSE IN THE 2016 KUMAMOTO EARTHQUAKE

TOMIYA TAKATANI[1] and HAYATO NISHIKAWA[2]

[1]*Dept of Civil Engineering & Architecture, National Institute of Technology, Maizuru, Japan*
[2]*Dept of Civil Engineering & Architecture, Fukui Institute of Technology, Fukui, Japan*

Thousands of wooden houses were destroyed by the 2016 Kumamoto Earthquake. A seismic damage prediction function for wooden houses taking into consideration the consecutive strong earthquake motions, the amplification effect of ground surface layer, and the rupture propagation effect of seismic fault was proposed in this paper. Relationship between three ground characteristics above mentioned and the seismic damage for wooden house in the 2016 Kumamoto earthquake was analytically investigated by 3-D collapsing process analysis. The maximum drift angle was evaluated in this collapsing analysis of two-story wooden house.

Keywords: Seismic performance, Collapsing process analysis, Drift angle, Empirical Green's function.

1 INTRODUCTION

A tremendous seismic damage of collapse to wooden houses was caused by the 2016 Kumamoto Earthquake occurred on both April 14 and 16 (Building Research Institute 2016). In particular, Mashiki town located at near the hypocenter of these earthquakes has twice earthquake ground motions with the Japan Meteorological Agency (hereafter referred as JMA) seismic intensity of "7" level successively. As main factors to the destruction of wooden houses in the 2016 Kumamoto Earthquake, strong earthquake motions due to fore-shock and main shock, the amplification effect of ground surface layer, and the pulse wave due to seismic fault rupture propagation have been indicated by many researchers and engineers. Also, Nozu (2017) reported that the effect of the rupture propagation of seismic fault in the main shock in the 2016 Kumamoto Earthquake affects seismic damage of bridge in the north-east area in its epicenter region (Building Research Institute 2016, Goto *et al.* 2017). The purpose of this paper is to propose a seismic damage prediction function for two-story wooden house by taking into consideration the consecutive strong earthquake motions, the amplification effect of ground surface layer, and the rupture propagation effect of seismic fault.

2 GENERATION OF INPUT EARTHQUAKE GROUND MOTION WAVE

In this paper, several earthquake motion waves are evaluated by an empirical Green's function method using seismic fault model considering a hypocenter location of the main shock and its asperity in the 2016 Kumamoto Earthquake. These earthquake motion waves have three seismic motion characteristics in this earthquake mentioned above. Difference between the rupture propagation effects on earthquake motion waves at 50 location points around a seismic fault is investigated. Earthquake ground motion waves observed at KiK-net Mashiki in the earthquake (M_{JMA}=4.2) occurred at 00:50 on April 15, 2016 are employed in the empirical Green's function method. In order to investigate the site amplification characteristics, earthquake ground motion wave at each location is calculated using the site amplification characteristics with three different frequencies. In the calculation of earthquake motion wave on the ground surface, the non-linearization of ground due to a strong seismic motion can be considered by the multi non-linearity effect.

Using earthquake motion waves calculated by the empirical Green's function method, seismic collapsing behavior of wooden house is numerically investigated by 3-D collapsing process analysis of "wallstat", which was developed by Nakagawa *et al.* (2010). To investigate the effect of consecutive earthquake motion wave on seismic performance of wooden house, earthquake motion waves in the foreshock, the main shock, and both the fore and main shocks are employed in this 3-D seismic collapsing process analysis of wooden house (Takatani and Nishikawa 2017). Based on the maximum drift angle of wooden house obtained from collapsing analytical results, seismic damage state of wooden house is classified, and also seismic damage prediction map is made.

2.1 Empirical Green's Function Method

Empirical Green's Function Method can calculate an acceleration wave at the estimation point by some small to medium size earthquake motion waves occurring on a seismic fault of the target earthquake. In this paper, the empirical Green's function proposed by Nozu (2017) was used to numerically calculate some acceleration waves. Propagation path characteristics were referred to a research paper by Kato (2001). 61 location points in the analytical area in Kumamoto prefecture and the asperity locations of A1, A2, and A3 are employed in this paper. The fore-shock (M_{JMA}=6.5) in 2016 Kumamoto Earthquake occurred on 21:26 on April 14, 2016.

2.2 Hypocenter Model

Hypocenter model employed in the empirical Green's function method is referred to a characterized hypocenter model of both the foreshock and main shock in the 2016 Kumamoto Earthquake proposed by Nozu (2017). In this paper, characterized hypocenter model is not described due to the limited space.

Figure 1 shows an outline image around a hypocenter of the fore-shock mentioned above. A mark of ★means an epicenter of the fore-shock, ★means an epicenter of the main shock, □ means an epicenter of small to medium size earthquakes in making an evaluation of phase characteristics, ▲ means analytical location points. Earthquake ground motion waves at 61 location points nearby a seismic fault, which is supposed the foreshock and the main shock in the 2016 Kumamoto Earthquake, are created by the empirical Green's function method. In order to investigate the effect of the site amplification characteristics, these earthquake ground motion waves at 61 location points are evaluated by the site amplification characteristics with three different frequency characteristics. In this paper, earthquake motion waves at KMM008, KMM011 and KMM018 in the strong motion seismograph networks (K-NET) are employed as three different frequency characteristics.

Figure 2 indicates a referred characterized hypocenter of the foreshock in the 2016 Kumamoto Earthquake. "Asperity 1" and "Asperity 2" are set in two location points of the seismic fault, where the maximum slip velocity value is estimated as a reference of earthquake motion wave inversion result. Table 1 shows some parameters regarding each asperity. Failure at each asperity is assumed to expand to the radical direction from a rupture start point of each asperity.

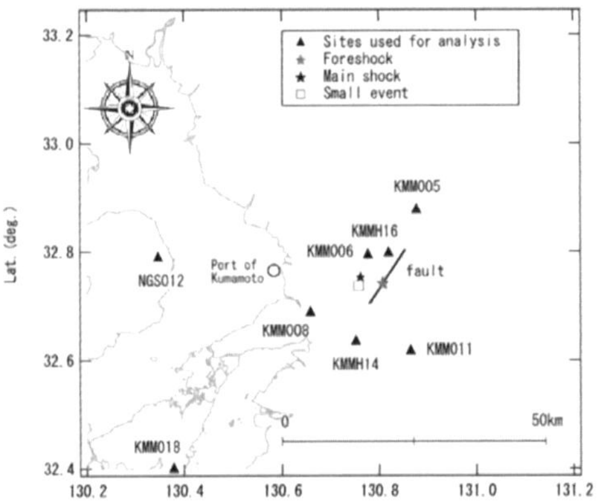

Figure 1. Information around a hypocenter of the foreshock in 2016 Kumamoto Earthquake.

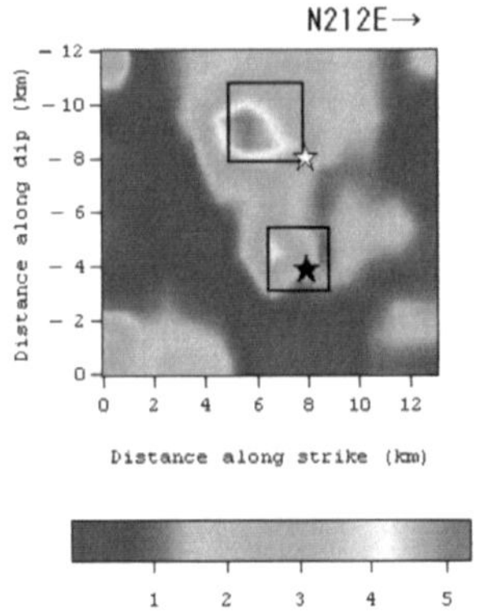

Figure 2. Characterized hypocenter of the foreshock in the 2016 Kumamoto Earthquake.

Table 1. Parameters of characterized hypocenter model of foreshock in the 2016 Kumamoto Earthquake.

	Asperity 1	Asperity 2
East Longitude at Start Point (deg.)	130.808	130.809
North Latitude at Start Point (deg.)	32.742	32.742
Depth of Start Point (km)	11	7
Length (km) * Width (km)	2.5 * 2.5	3 * 3
Seismic Moment, Mo (N·m)	1.50E+24	1.30E+24
Relative Rupture Start Time (s)	0	2.7
Rupture Propagation Velocity (km/s)	2.8	2.8
Rise Time (s)	0.33	0.4
Division Number	5 * 5 * 5	5 * 5 * 5

2.3 Site Amplification Characteristics

Site amplification characteristics are to display the surface soil layered ground amplification in the frequency domain, which is located on the earthquake engineering base rock (Approximately shear wave velocity 3km/s). In order to investigate the seismic intensity and the difference on seismic damage of structure, three different site amplification characteristics, KMM08, KMM011 and KMM018, are used in this paper. First natural frequencies at KMM08, KMM011 and KMM018 are 0.85Hz, 3.1Hz and 9,4Hz, respectively. Based on the ground classification of the Building Standards Act in Ja-pan, KMM011 and KMM018 correspond to the second kind of ground, and KMM008 corresponds to the third kind of ground.

Figure 3 shows the maximum velocity, PGV (cm/s), distribution map and the maxi-mum displacement, PGD (cm), distribution map, which are obtained by the empirical Green's function method using three site amplification characteristics mentioned above. It is found that PGV and PGD values trend to increase toward the north-east direction. Three seismic motions have a tendency to become larger in order of KMM008, KMM011 and KMM018.

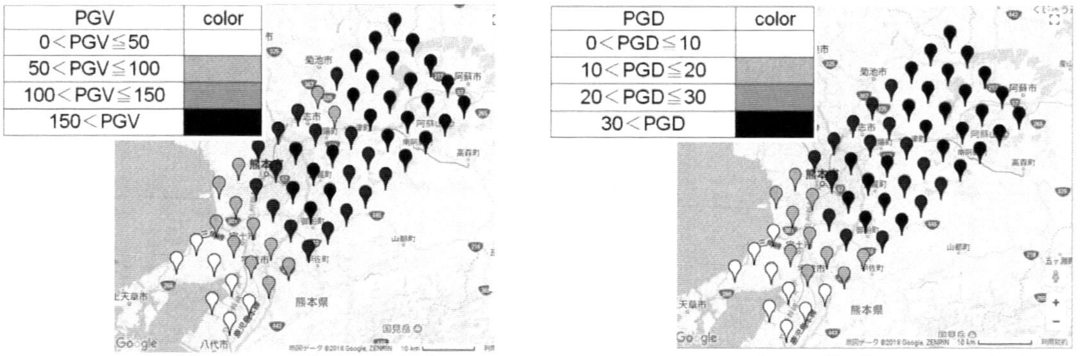

Maximum velocity (PGV) Maximum displacement (PGD)

Figure 3. Maximum velocity and displacement distribution map (KMM008).

3 SEISMIC DAMAGE PREDICTION OF WOODEN HOUSE

3.1 Evaluation of Maximum Drift Angle of Wooden House

The maximum drift angle is defined with the ratio of a horizontal displacement of each story layer to the height of each layer, and the maximum value in each direction is chosen from the maximum drift angles at four corners of wooden house. Figure 4 shows the maximum drift angle distribution maps of Rx and Ry when EW and NS components of earthquake motion wave considering the site amplification characteristics of KMM008 were employed to X direction. Rx means the maximum drift angle chosen in four drift angles in X direction, and Ry means the maximum drift angle chosen in four drift angles in Y direction.

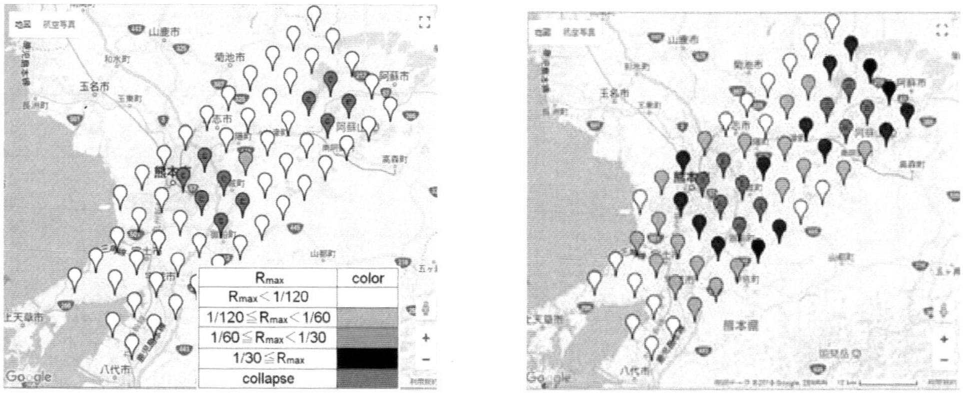

(a) Rx (b) Ry

Figure 4. Maximum drift angle distribution maps (KMM008, X-EW component).

3.2 Maximum Drift Angle of Wooden House

Based on the seismic collapsing process simulation results of wooden house described in the previous chapter, a prediction equation of the maximum drift angle of two-story wooden house can be written by

$$\log R_{max} = c_1 + c_2 \log D + c_3\left(X^2 + S\right) + c_4 X + c_5 \log f_g + c_6 f_g \tag{1}$$

where, D is directivity coefficient, X is the minimum distance (km) of seismic fault, f_g is the first natural frequency (Hz) of the site amplification characteristics, c_1 to c_6 are recurrence coefficients, and S is the constant coefficient regarding the peak of seismic motion around a hypocenter. Eq. (1) can be obtained by the recurrence analysis changing S value. Table 2 shows each recurrence coefficients used in Eq. (1).

Seismic damage prediction of wooden house around Kumamoto region may be possible by Eq. (1) proposed in this paper. This seismic damage prediction of wooden house was obtained from the recurrence analysis based on some earthquake ground motion waves in the 2016 Kumamoto Earthquake, and also is used in estimating seismic damage prediction of two-story wooden house with high seismic performance and satisfying the seismic standard in the Building Stand ards Act in Japan amended in 2010. Therefore, a recurrence analysis considering the site amplification characteristics in other region may be needed in order to evaluate seismic damage prediction of wooden house in other regions. Moreover, the seismic damage prediction of old wooden house built before 2010 can be easily obtained from the recurrence analysis using seismic collapsing process results of this kind of wooden house.

Table 2. Recurrence coefficients in Eq. (1) (Input motion wave used in X direction).

	R_{max}	c_1	c_2	c_3	c_4	c_5	c_6
NS component	X	-1.769	-3.147	-0.287	-0.17	2.147	-0.375
	Y	-1.119	-0.292	-0.221	-0.009	1.892	-0.338
EW component	X	-1.246	-0.511	-0.146	-0.018	3.273	-0.492
	Y	-1.378	-0.069	-0.025	-0.027	0.608	-0.207

3.3 Analytical Results

A seismic damage prediction equation of the maximum drift angle of wooden house proposed in the previous session is investigated in this session. Comparison the maximum drift angle, R_{max} (Observation value), obtained from 3-D seismic collapsing process analysis with the maximum drift angle, R_{max} (Prediction value), calculated from Eq. (1) is conducted. R_{max} in X direction is described Rx, and R_{max} in Y direction is described Ry. Coefficient of determination, R^2, of Rx is 0.93, and that of Ry is 0.88 as shown in Figure 5. Any coefficient of determination is very close to 1.0. This implies that the prediction equation of the maximum drift angle, R_{max}, written in Eq. (1) has an accurate precision to seismic damage of wooden house and can utilize seismic damage of wooden house in Kumamoto region in future.

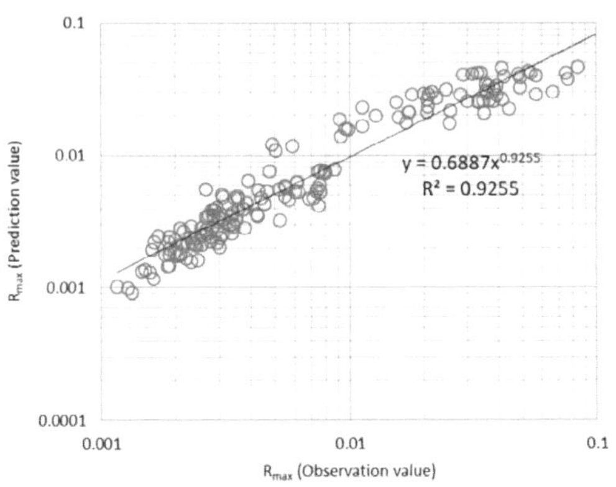

(a)Rx

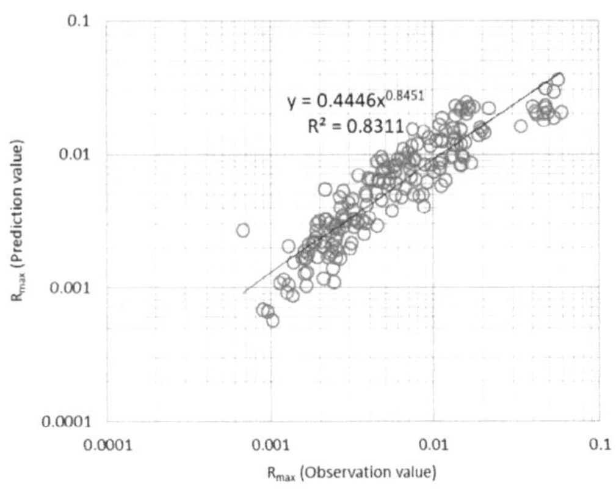

(a)Rx

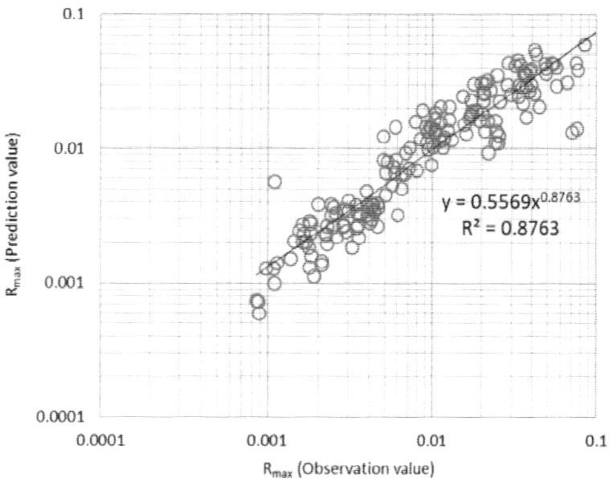

(b)Ry
(EW component in X direction)

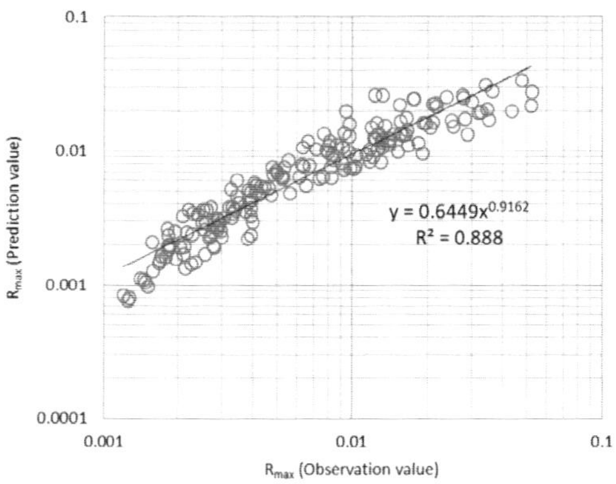

(b)Ry
(NS component in X direction)

Figure 5. Comparison between R_{max} (Prediction value) and R_{max} (Observation value).

4 CONCLUSIONS

In this paper, a seismic damage prediction function for two-story wooden house by taking into consideration the consecutive strong earthquake motions, the amplification effect of ground surface layer, and the rupture propagation effect of seismic fault was proposed.

The summary of this paper is as follows,

1) The effect of consecutive strong earthquake motions with both the fore-shock and main shock in the 2016 Kumamoto earthquake on the seismic performance of wooden house was analytically revealed. There may be a possibility of a collapse of the wooden house with low seismic performance against consecutive strong earthquake motions.
2) Seismic damage prediction of wooden house around Kumamoto region may be possible by the maximum drift angle prediction equation. This seismic damage prediction of wooden house can be obtained from the recurrence analysis based on some earthquake ground motion waves in the 2016 Kumamoto Earthquake.

References

Building Research Institute, 2016. Kumamoto Earthquake of April 16, 2016, April 16, 2016. Retrieved from http://iisee.kenken.go.jp/quakes2/20160416kumamoto.html on October 2018.

Goto, H., Hata, K., and Yoshimi, M., Relationship between Site Amplification Characteristics and Seismic Damage Concentrated Region in the 2016 Kumamoto Earthquake, *Journal of Science*, 87(2), 186-191 (in Japanese), 2017.

Kato, K., Evaluation of Hypocenter, Propagation Path and Ground Amplification Characteristics of the 1997 Kagoshima North-West Earthquake Based on Strong Motion Seismograph Networks K-NET, *Journal of Architectural Institute of Japan*, 66 (543), 61-68 (in Japanese), 2001.

Nakagawa, T., Ohta, M., Tsuchimoto, T., and Kawal, N., Collapsing Process Simulations of Timber Structures Under Dynamic Loading III: Numerical Simulations of the Real Size Wooden Houses, *Journal of Wood Science*, 56(4), 284-292, 2010.

Nozu, A., *On Seismic Motion Wave, Report of the First Anniversary of the 2016 Kumamoto Earthquake*, Japan Society of Civil Engineers (in Japanese), 2017.

Takatani, T. and Nishikawa, H., *Seismic Performance of Japanese-Style Two-Story Wooden House Against the 2016 Kumamoto Earthquake*, Proceedings of the 2017 World Congress on Advances in Structural Engineering and Mechanics (ASEM2017), T5F-5, Ilsan (Seoul), Korea, 2017.

Streamlining Information Transfer between Construction and Structural Engineering
Edited by Shiau, J., Vimonsatit, V., Yazdani, S., and Singh, A.

ISBN: 978-0-9960437-7-9

BEHAVIOR OF NONSTRUCTURAL COMPONENT SUPPORTED BY HANGER BOLTS UNDER SEISMIC EXCITATIONS BY SHAKING TABLE TEST

HAEYOUNG KIM[1], KUNIO MIZUTANI[2], and SYOJIRO MOTOYUI[3]

[1]*Dept of Civil Engineering, Yokohama National University, Yokohama, Japan*
[2]*Dept of Architectural Engineering, Tokyo Polytechnic University, Atsugi, Japan*
[3]*Dept of Architectural Engineering, Tokyo Institute of Technology, Yokohama, Japan*

During the Great East Japan Earthquake of March 2011, nonstructural components, such as pipe systems, ducts, cable racks and ceilings were severely damaged while main structural members in the building were not damaged seriously. Pipes, cable racks, apparatus and ducts' hanger bolts were ruptured causing the equipment to fall down. Because of these damages, buildings cannot be used for a long period of time and one person was killed by pipe's falling in Japan. In this study, the behaviors of nonstructural components are investigated by conducting shaking table tests to verify the cause of damage. More specifically, damage to hanger bolts is investigated by simulating its rupturing mechanism through shaking table test. To simulate the real installation condition of nonstructural components, apparatus-duct-pipe system supported by hanger bolts is selected as specimen. Roof floor response wave at the actual 5-story steel building under the Great East Japan Earthquake and sweep wave are used for the input waves. The maximum response acceleration was about 4 G in X direction under response wave 75% and the damage occurred at the metal fitting which is the connection part between braces and hanger bolt. And without installing braces, the upper hanger bolts at the fixed supporting part were ruptured easily since the natural frequency of the specimen closed to those of target building during excitations and the response became huge.

Keywords: Great East Japan earthquake, Apparatus-duct-pipe system, Rupturing mechanism.

1 INTRODUCTION

Under the Great East Japan Earthquake of March 2011, the damage on the nonstructural components was seen in many buildings as shown in Figure 1 summarized from Mizutani and Kim (2012, 2013). By the Japan Building Mechanical and Electrical Engineers Association (JBMEE) and the Society of Heating, Air-Conditioning and Sanitary Engineers of Japan (SHASE), the questionnaire from members and company of these associations was performed about the nonstructural components damage caused by this earthquake disaster. We analyzed the data from questionnaire. The rate of nonstructural components damage factor is shown in Figure 2(a). The damage report by shaking is as the highest as 87%, and the damage report according the damage caused by ground subsidence and liquefaction to 6% and tsunami has become only 1%. In addition, the damage case, which is hit by tsunami to the whole building, is not included in this questionnaire. The damaged equipment is classified into the air conditioning equipment, plumbing equipment, electric equipment, fire protection equipment, and elevators, and the rate is shown in Figure 2(b). For the characteristics of damage, the rupture of the hanger bolt, which supported the facilities or duct or plumbing from a slab floor, many were seen. Therefore, in this study, the damage to hanger bolts is investigated by simulating its rupturing mechanism through shaking table test.

(a) apparatus (b) ceiling (c) sprinkler pipe

Figure 1. Damages on the nonstructural components.

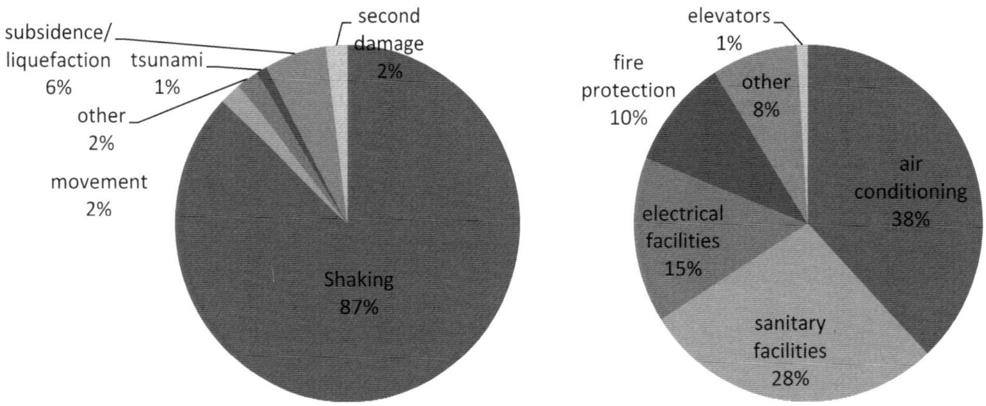

(a) Damage factor (b) Damage situation

Figure 2. Damage situation.

2 SHAKING TABLE TEST

2.1 Experiment Outline

To simulate the real installation condition of nonstructural components, apparatus-duct-pipe system supported by hanger bolts is selected as specimen as shown in Figure 3. The shaking table and the installed apparatus-duct-pipe system are shown in Figure 4 (a) and (b), respectively. The braces installed in each side of the apparatus's hanger bolt with X –shape shown in Figure 4 (c). Table 1 shows the outline of each specimen.

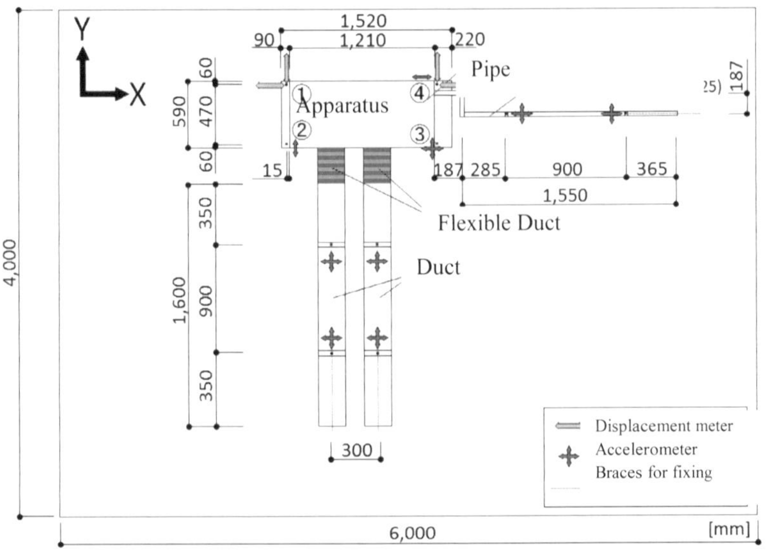

Figure 3. Plan view of specimen (apparatus-duct-pipe system).

(a) shaking table (b) installation of specimen (c) braces installed in apparatus

Figure 4. View of shaking table.

Table 1. Outline of each specimen.

Specimen	Hanging length	Weight	No. of hanger bolt	Size of hanger bolt
Apparatus	600 mm	57.5 kg	4	W3/8
Duct	360 mm	4.5 kg	2	W3/8
Pipe	400 mm	1.5 kg	2	W3/8

2.2 Experiment Method

The target building was selected as the low-rise steel frame building which was seen many damages of ceiling and apparatus during the Great East Japan earthquake. It is the fifth stories steel frame building located in Hitachi City, Ibaraki Prefecture. The response of the roof floor of the target building as shown in Figure 5, obtained by the earthquake response analysis is used as the input wave. Figure 6 shows the response spectrum of the input wave and it contains two peak frequencies which are the 1st and 2nd frequencies of the target building. Those are 1.1 Hz and 3.57 Hz. Input waves are shown in Table 2. After exciting every response wave, the sweep wave is used for investigating the natural frequency of specimen. Table 3 shows the changes of the natural frequency of specimen. It was changed from 2.807 Hz to 2.624 Hz because of weakened stiffness of hanger bolts after exciting the response waves.

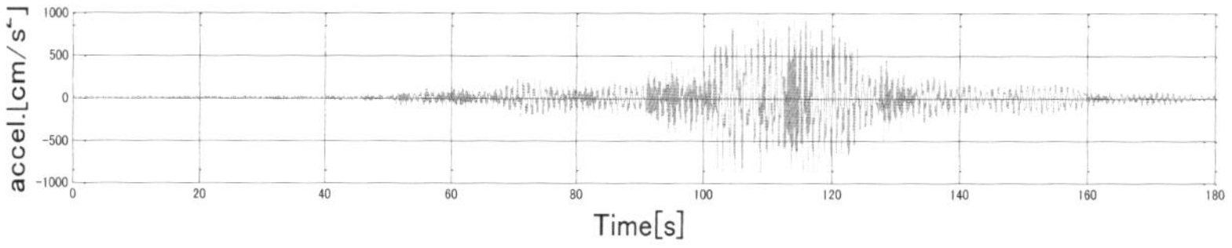

Figure 5. Response wave.

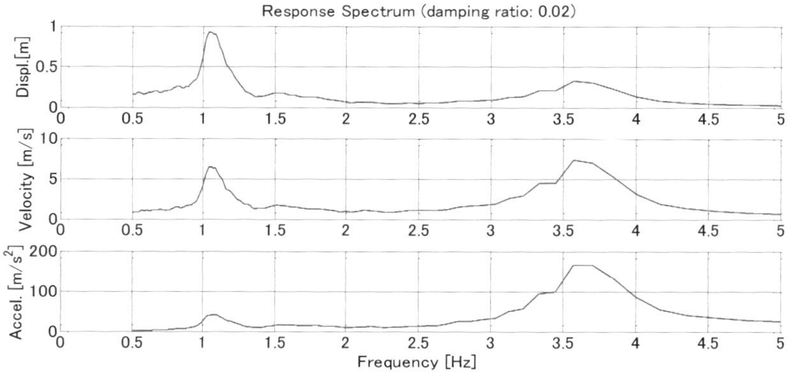

Figure 6. Response spectrum of the response wave.

Table 2. Input waves.

No.	Name	Input wave	Remarks
1	sweep1	sweep wave 1	0.2Hz~2.5Hz, 30 gal
2	RW25	response Wave 25%	Max. 0.3 G
3	sweep2	sweep wave 2	0.2Hz~2.5Hz, 30 gal
4	RW50	response Wave 50%	Max. 0.6 G
5	sweep3	sweep wave 3	0.2Hz~2.5Hz, 30 gal
6	RW75	response Wave 75%	Max, 1.2 G
7	sweep4	sweep wave 4	0.2Hz~2.5Hz, 30 gal

Table 3. Changes of the natural frequency of the apparatus.

Input wave	Natural frequency	Natural period	Remarks
sweep1	2.807 Hz	0.356 s	initial condition
sweep2	2.807 Hz	0.356 s	after exciting the RW25
sweep3	2.750 Hz	0.364 s	after exciting the RW50
sweep4	2.624 Hz	0.381 s	after exciting the RW75

3 EXPERIMENT RESULTS

3.1 The Responses under the Response Waves

The maximum responses of specimen are shown in Table 4. There is a gap between the apparatus and the hanger bolt and it is about 15 mm as shown in Figure 7. If the response displacement is larger than this gap, the apparatus impacts the hanger bolt. Under response wave 25%, the response displacement was about 20 mm and the impacting occurred 25 times. The maximum response acceleration was about 1.5 G and the maximum restoring force on hanger bolts was about 1.2 kN. After exciting RW25, there were no released bolts and the natural frequency was same as one in the initial condition. Under response wave 50%, the impacting between the apparatus and the hanger bolt was occurred 70 times. The maximum response acceleration was about 3 G and the maximum restoring force on hanger bolts was about 2 kN. After exciting RW50, some released bolts were seen in the connection part of brace and hanger bolt. And the natural frequency became low from 2.807 Hz to 2.750 Hz. Next, under response wave 75%, the maximum response displacement was about 27 mm and the impacting between the apparatus and hanger bolts was occurred about 116 times. The maximum response acceleration was about 4 G in X direction and about 1 G in Y direction. The relationship between restoring force and relative response displacement at each hanger bolt is shown in Figure 8. The impacting at the hanger bolt 1 was occurred severely and the restoring force reached 1.5 kN at the hanger bolt 1. The apparatus is connected with the pipe and ducts. It caused the apparatus vibrated in torsional motion and the hanger bolt 1 was loaded severely. As shown in Figure 8, the stiffness of the apparatus became very high after impacting and it caused the hanger bolt was loaded to the horizontal direction excessively. After exciting RW75, there were released bolts in the connection part of brace and the metal fittings were stretched. The natural frequency became 2.624 Hz.

Table 4. Maximum responses under RW25 and RW50.

Input	Input level	Max. accel.	Max. restoring force	Numbers of impacting
RW25	0.3 G	1.5 G	1.2 kN	25
RW50	0.6 G	3 G	2 kN	70
RW75	1.2 G	4 G	3 kN	116

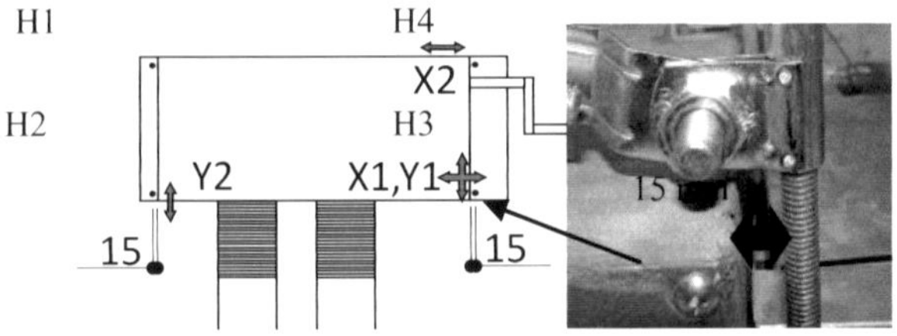

Figure 7. Gap between apparatus and hanger bolt.

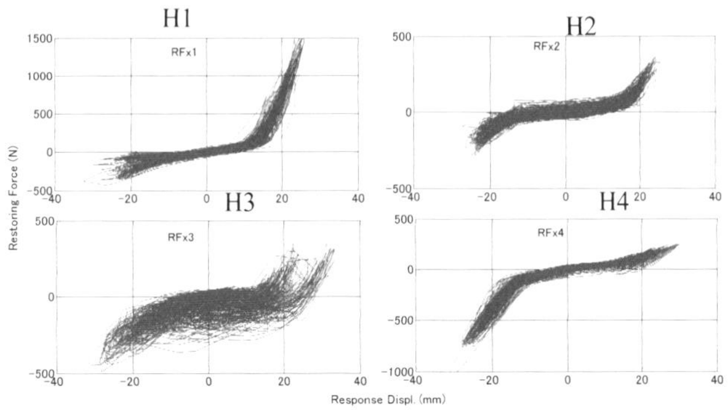

Figure 8. Relation between restoring force and displacement at each hanger bolt under RW75.

3.2 The Responses under the Sine Wave

After exciting the response wave 75%, the bolts and nuts were released on the connection parts of braces and hanger bolts and metal fittings were stretched. We considered the braces didn't work out any more. Therefore, we took the all braces off. And the natural frequency of the apparatus became 1.75 Hz. To investigate the failure of specimen, the sine wave, which contains 1.75 Hz was used as input wave. As a result, the hanger bolt 1 at the apparatus was ruptured on the upper fixed part after only 10 seconds. Figure 9 shows the response acceleration. The pipe was put out of joint and the flexible ducts were ruptured. This situation of failure was seen many in the apparatus-pipe-duct system under the Great East Japan Earthquake.

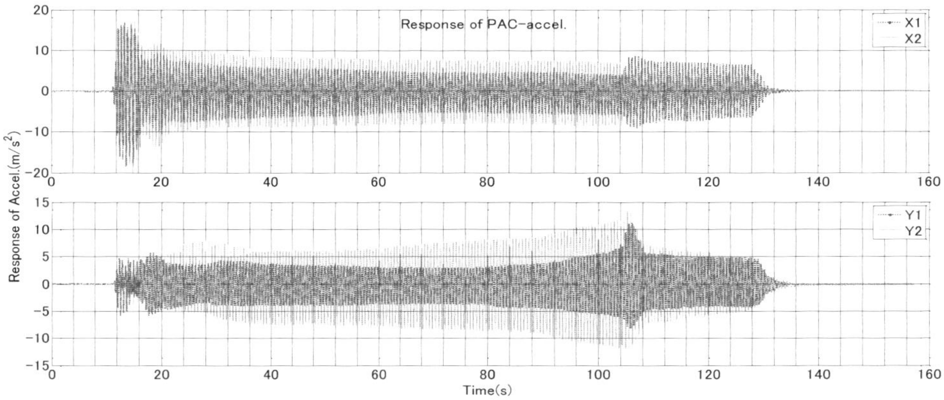

Figure 9. Response acceleration of apparatus w/o braces under sine wave.

4 CONCLUSIONS

In this study, the behaviors of nonstructural components are investigated by conducting shaking table tests to verify the cause of damage. To simulate the real installation condition of nonstructural components, apparatus-duct-pipe system supported by hanger bolts is selected as specimen. We found that the damage occurred at the metal fitting which is the connection part between braces and hanger bolt. And without installing braces, the upper hanger bolts at the fixed supporting part were ruptured easily because the natural frequency of the specimen closed to those of the target building during excitations, and the response became huge.

Acknowledgments

This study is supported by Japanese Grants-in-Aid for Scientific Research (24246094).

References

Mizutani, K. and Kim, H., *The Damage of the Building Equipment under The 2011 Tohoku Pacific Earthquake*, 9th International Conference on Urban Earthquake Engineering, 6-8, March, 2012.

Mizutani, K. and Kim, H., *Behavior of Building Equipment Under Seismic Excitations*, 10th International Conference on Urban Earthquake Engineering, 1-2, March, 2013.

Streamlining Information Transfer between Construction and Structural Engineering
Edited by Shiau, J., Vimonsatit, V., Yazdani, S., and Singh, A.

ISBN: 978-0-9960437-7-9

EFFECTS OF THE INITIAL DEFLECTION SHAPE ON THE MOMENT AMPLIFICATION FACTOR OF BEAM-COLUMNS

HARUNA UTSUNOMIYA[1], MASAYUKI HARAGUCHI[2], MASAE KIDO[1], and KEIGO TSUDA[1]

[1]*Dept of Architecture, The University of Kitakyushu, Kitakyushu, Japan*
[2]*Hojo Structure Research Institute Co, Ltd., Osaka, Japan*

In the design of slender steel beam-columns, the moment amplification factor is used to estimate the maximum moment along with the longitudinal direction. While formulas for evaluating the factor have been presented on the basis of elastic or elastic-plastic analysis, the initial deflection of the column is not considered. The effect that the initial deflection on the strength and behavior of the column has been shown only when the initial deflection shape is half sine wave. This paper discusses the effect of the initial deflection shape on the value of the moment amplification factor by performing the analytical work. The analytical model is the hinged-end beam-column subjected to constant axial compressive force and end moments. First of all, the equilibrium differential equation which governs the problem is solved and the formula for calculating the bending moment is presented. In the parametric study, magnitude of initial deflection, initial deflection shape, axial load ratio, slenderness ratio and end moment ratio are selected as the parameters. In this paper, we discuss the effects of the amount of the initial deflection and the initial deflection shape.

Keywords: Steel beam-columns, End moment ratio, Axial load ratio, Slenderness ratio, Strength formula, Analytical solution.

1 INTRODUCTION

In the design formulas of the steel beam-columns of AIJ Recommendation for Limit State Design of Steel Structures (hereinafter referred to as LSD guideline) (AIJ 2010), when the maximum value of the bending moment occurs between the point of supports, the bending moment is evaluated by using the moment amplification factor a_m. In LSD guideline, the moment amplification factor ${}_d a_m$ is adopted as the factor at the maximum strength in the inelastic state (Sakamoto *et al.* 1968). On the other hand, the moment amplification factor ${}_e a_m$ can be analytically calculated for a linear elastic body steel beam-columns. Moment amplification factor a_m, ${}_d a_m$ and ${}_e a_m$ have been obtained by analysis in the absence of the initial deflection. Regarding the effect of the initial deflection on the strength and behavior, we examined only when the initial deflection shape was sine half wave (Utsunomiya *et al.* 2016). This paper discusses the effect of initial deflection expressed by a half sine wave only or the sum of a half sine wave and one-cycle sine wave and the dimensionless factors on the value of moment amplification factor by performing the analytical work.

2 ANALYSIS

2.1 Problem Setting

As shown in Figure 1, the analytical model is the hinged-end beam-column subjected to constant axial compressive force N and end moments M_1 and $M_2(=\kappa M_1, |\kappa|\leq 1)$. The value of κ is the end moment ratio. The end moment M_1 is numerically larger than M_2. The value of κ is positive when the member is bent in reverse curvature. Boundary conditions are pin supports at both ends. In this paper, the moment amplification factor a_m ($=M_{max}/M_1$:M_{max} is the maximum bending moment in the entire region of the member) is analytically calculated when the initial deflection y_0 is given.

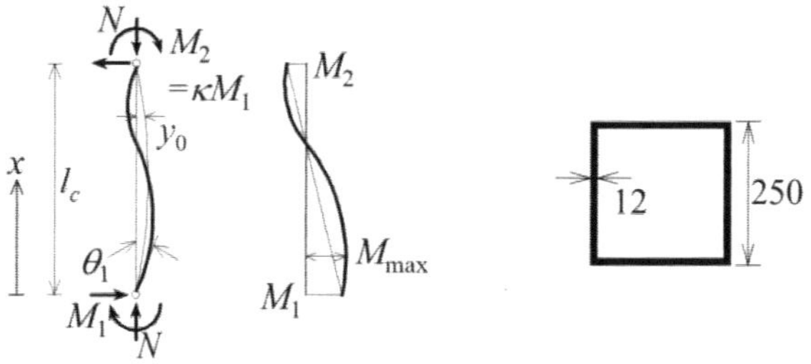

Figure 1. Analytical model.

2.2 Equilibrium Differential Equations and the Solutions

The moment considering the initial deflection y_0 is given in Eq. (1) (Utsunomiya *et al.* 2016).

$$\frac{M(\xi)}{M_1}=\cos\left(\pi\sqrt{n_y\lambda_c^2}\,\xi\right)-\frac{\kappa+\cos\left(\pi\sqrt{n_y\lambda_c^2}\right)}{\sin\left(\pi\sqrt{n_y\lambda_c^2}\right)}\sin\left(\pi\sqrt{n_y\lambda_c^2}\,\xi\right)+\sum_{j=1}^{n}\frac{Nc_j}{M_1}\frac{j^2}{j^2-n_y\lambda_c^2}\sin(j\pi\xi) \tag{1}$$

In the above equation, ξ is $\xi\equiv x/l_c$ (x is the coordinate in the member length direction with the origin at the column base), the axial load ratio n_y and the normalized slenderness ratio λ_c in the Eq. (1) are defined by Eqs. (2) and (3), respectively.

$$n_y=\frac{N}{N_y}\quad\left(N_y=A\sigma_y\right) \tag{2}$$

$$\lambda_c=\sqrt{\frac{N_y}{N_e}}\quad\left(N_e=\frac{\pi^2 EI}{l_c^2}\right) \tag{3}$$

Where N_y is yield axial force, A is cross-sectional area, σ_y is yield stress, E is the Young's modulus (=205,000N/mm^2), I is the moment of inertia.

Initial deflection c_j is the coefficient that defines the initial deflection y_0 expressed in Eq. (4).

$$y_0=\sum_{j=1}^{n}c_j\sin(j\pi\xi) \tag{4}$$

M_{max}/M_1 obtained by Eq. (1) gives the moment amplification factor a_m. From Eq. (1), a_m is related to the end moment ratio κ, $n_y\lambda_c^2$, and Nc_j/M_1. The value of Nc_j/M_1 is divided into the n_y, c_j, and η.

The dimensionless amount η is defined in Eq. (5) equation, where M_y is a yield moment.

$$\eta = \frac{M_1}{M_y} \tag{5}$$

2.3 Moment Amplification Factor (Elastic and Inelastic)

Equation (6) shows the moment amplification factor ${}_e a_m$ obtained by the elastic analysis in the case there is no initial deflection, and Eq. (7) is the moment amplification factor ${}_d a_m$ at the maximum strength in the inelastic state prescribed in LSD guideline.

$$\begin{aligned} {}_e a_m &= 1 && (\kappa \geq -\cos(\pi \sqrt{N/N_e})) \\ &= \frac{\sqrt{1+\kappa^2+2\kappa\cos\left(\pi\sqrt{N/N_e}\right)}}{\sin\left(\pi\sqrt{N/N_e}\right)} && (\kappa < -\cos(\pi\sqrt{N/N_e})) \end{aligned} \tag{6}$$

$$\begin{aligned} {}_d a_m &= 1 && \left(n_y \lambda_c^2 \leq 0.25\left(1+\kappa\right)\right) \\ &= \frac{1-0.5\left(1+\kappa\right)\sqrt{N/N_e}}{1-N/N_e} && \left(n_y \lambda_c^2 > 0.25\left(1+\kappa\right)\right) \end{aligned} \tag{7}$$

2.4 Analytical Parameters

We targeted two initial deflection shapes, 1) a half sine wave only, 2) the sum of a half sine wave and one-cycle sine wave.

We selected the value of initial deflection shape n, the coefficient of the initial deflection c_1 and c_2, end moment ratio κ, axial load ratio n_y, and η values in Eq. (5) as the analytical parameters, and they vary as follows.

1) n=1 (a half sine wave only)
n=2 (the sum of a half sine wave and one-cycle sine wave)
2) $c_1=l_c/1500$, $c_2=l_c/15000$, $l_c/7500$, $l_c/3000$
$c_1=l_c/1000$, $c_2=l_c/10000$, $l_c/5000$, $l_c/2000$
$c_1=l_c/500$, $c_2=l_c/5000$, $l_c/2500$, $l_c/1000$
3) □□=-1.0, -0.5, 0, 0.5, 1.0
4) n_y=0.1, 0.3, 0.5, 0.7
5) □ =1, 2/3

The cross section used in the analysis is a square steel tube, the width of the cross section is 250 mm, and the plate thickness is 12 mm (see Figure 1). The yield stress σ_y is 325 N/mm^2.

3 RESULT AND DISCUSSION

3.1 $a_m/{}_e a_m$-$n_y\lambda_c^2$ Relations

3.1.1 *Influence of initial deflection*

Figure 2 shows the influence of the initial deflection c_1, c_2 on the moment amplification factor a_m when the axial load ratio n_y=0.3, η=1. In the figure, $c_1=l_c/500$, $l_c/1000$, and $l_c/1500$ are indicated by solid lines, ● mark, and ○ mark, respectively.

From Figure 2, it can be seen that the moment amplification factor a_m increases as the initial deflections c_1 and c_2 are larger.

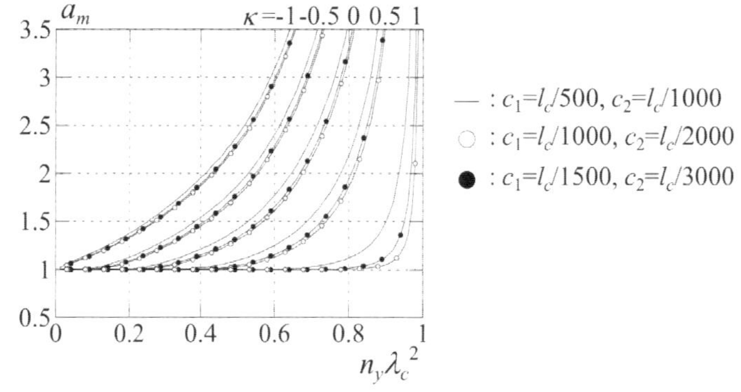

Figure 2. a_m-$n_y\lambda_c^2$ Relations (Influence of initial deflection c_1, n_y=0.3, η =1).

3.1.2 *Influence of initial deflection shape*

Figure 3 shows the influence of c_2 on the moment amplification factor a_m when the axial load ratio n_y=0.3, η=1. Fig. 3(a), (b) and (c) show the case of the initial deflections $c_1=l_c/500$, $l_c/1000$, $l_c/1500$ respectively. Regarding the initial deflection c_2, there is a relationship of $c_2=c_1/2$. In the figure, the initial deflection of a half sine wave only with n=1 is shown by the dotted line, and the initial deflection of the sum of a half sine wave and one-cycle sine wave is shown by the solid line.

According to Fig. 3, it can be seen that in the case of the initial deflection made of the half sine wave and one-cycle sine wave, the moment amplification factor a_m is larger than the case where the initial deflection is only the half sine wave.

3.2 Influence of Initial Deflection Shape

Figure 4 shows the relationship between a_{m2}/a_{m1} and $n_y\lambda_c^2$. a_{m2}/a_{m1} is the value obtained by dividing the moment amplification factor a_{m2} in the case of n=2 by the moment amplification factor a_{m1} in the case of n=1. Figs. 4 (a)-(c) show the cases where the initial deflection c_1 is $l_c/500$

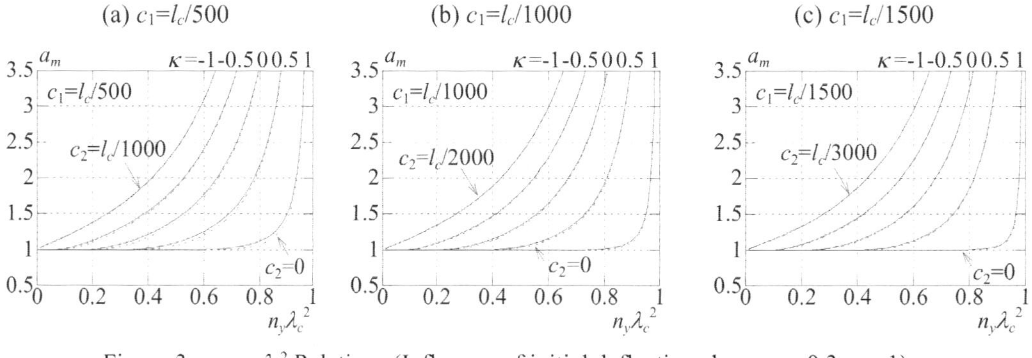

Figure 3. a_m-$n_y\lambda_c^2$ Relations (Influence of initial deflection shape, n_y=0.3, η =1).

and c_2 is $l_c/1000$, $l_c/2500$ and $l_c/5000$ respectively. Figs. 4 (d)-(f) show the case of initial deflection c_1 is $l_c/1000$ and c_2 is $l_c/2000$, $l_c/5000$ and $l_c/10000$ respectively. Figs. 4 (g)-(i) show the case of initial deflection c_1 is $l_c/1500$ and c_2 is $l_c/3000$, $l_c/7500$ and $l_c/15000$ respectively. In the figure, ○, dotted line, ●, Dashed line, and ■ marks in the figure indicate cases where end moment ratio κ is -1, -0.5, 0, 0.5, 1, respectively.

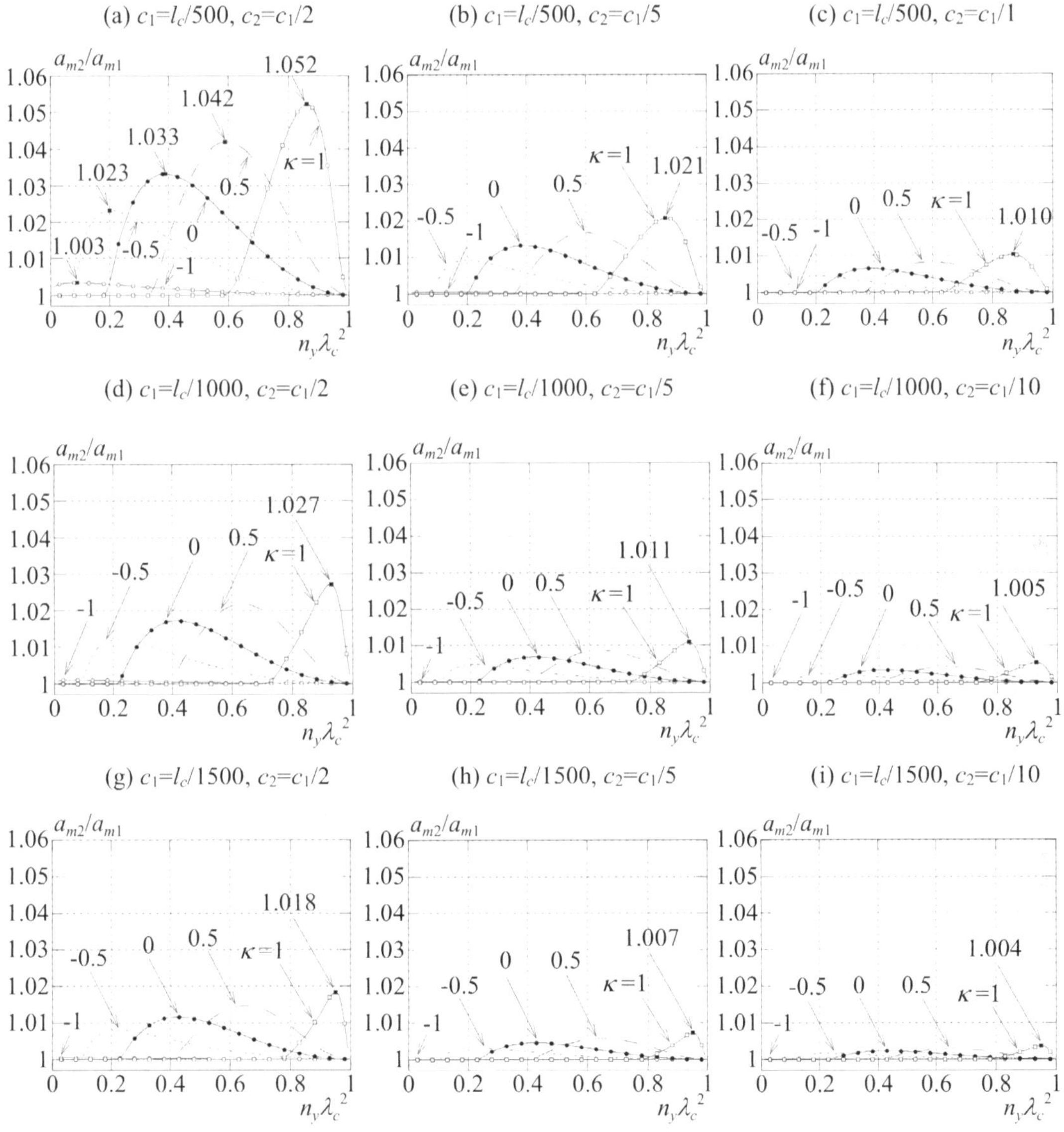

Figure 4. Influence of initial deflection shape n.

According to Figure 4, it is observed that the maximum value of a_{m2}/a_{m1} increases as the end moment ratio κ increases. Figs. 4 (a), (b), and (c), when the value of the initial deflection c_1 is constant, the maximum values of a_{m2}/a_{m1} are 1.052, 1.021, and 1.010, respectively, and the larger the initial deflection c_2 is, the maximum value of a_{m2}/a_{m1} is larger. Figures 4 (a), (d), and (g), when the value of the initial deflection c_2 is equal to $c_1/2$, the maximum value of a_{m2}/a_{m1} is 1.052, 1.027 and 1.018 respectively, the larger the initial deflection c_1 is, the larger the maximum value of a_{m2}/a_{m1}.

Also, when $c_1=l_c/500$ and $c_2=l_c/1000$ in Fig. 4 (a) where the effects of the initial deflection is the largest, the maximum values of a_{m2}/a_{m1} in the cases of the end moment ratio κ=-1, -0.5, 0, 0.5 and 1 are 1.003, 1.023, 1.033, 1.042 and 1.052, respectively, at most about 5%.

From the above, the effects of the initial deflection shape on the moment amplification factor increases as the value of c_2 increases. Even when c_2 is large as $c_2=c_1/2$ and the influence of one-cycle sine wave is at most about 5% larger than the case of an only half sine wave.

4 CONCLUSIONS

The conclusions derived from this study are as follows:

1) As the initial deflection c_1 and c_2 increase, the value of the moment amplification factor $\underline{a_m}$ increases when $\underline{n_y}\lambda_c^2$ is equal.
2) In the case of initial deflection is the sum of a half sine wave and one-cycle sine wave, the moment amplification factor $\underline{a_m}$ is larger than the case where the initial deflection is only the half sine wave. The effects of the initial deflection shape composed of the sum of a half sine wave and one-cycle sine wave on the moment amplification factor becomes larger as the value of c_2 becomes larger, but even if c_2 is large as $c_2=c_1/2$, a_{m2}/a_{m1} is about 1.052 and 5% at the maximum, and the effects of one-cycle sine wave is not large (see Fig. 4).

Acknowledgments

This study was supported by the Grant-in-Aid for Scientific Research of Japan Society for the Promotion of Science.

References

AIJ, *Recommendation for Limit State Design of Steel Structures* (in Japanese), Architectural Institute of Japan, 2010.

Sakamoto, J., Miyamura, A., and Watanabe, M., The Plastic Interaction Formulae of The Steel Beam-columns (in Japanese), *Transactions of the AIJ*, No.149, 9-15, 1968.

Utsunomiya, H., Haraguchi, M., Kido, M., and Tsuda, K., Moment Amplification Factor of Beam-Columns with Initial Deflection (in Japanese), *Journal of Constructional Steel*, 24, 822-829, 2016.

Streamlining Information Transfer between Construction and Structural Engineering
Edited by Shiau, J., Vimonsatit, V., Yazdani, S., and Singh, A.

ISBN: 978-0-9960437-7-9

HYSTERESIS CHARATERISTICS OF SQUARE CFT COLUMNS SUBJECTED TO AXIAL FORCE AND BENDING MOMENT

KOUTA JYOUZAKI and MASAE KIDO

Dept of Architecture, The University of Kitakyushu, Kitakyushu, Japan

Hysteresis characteristics of concrete filled steel tubular columns have been proposed. However, it has not been clarified if the skeleton curve could be applied for slender columns, and there are few studies about the unloading stiffness of the hysteresis loop. The purpose of this study is to compare the skeleton curves of square CFT columns, which have been proposed by the moment-rotation angle relationships and the experimental data by the previous research. The relationship between the initial stiffness obtained by the theoretical calculation and the unloading stiffness obtained by the experimental study are shown. Also the effects of the axial force ratio and the width-thickness ratio on the above mentioned relationships are discussed. The loading condition of the experimental study is monotonic and cyclic loading. It is found that the skeleton curve evaluates conservatively the bending moment-rotation angle relationships obtained by the test in both loading conditions, except for the width-thickness ratio 43 specimen with cyclic loading. The ratio of unloading stiffness ${}_{ex}K_r$ obtained from the experimental data and calculated theoretical stiffness K was shown. The effect of the axial force ratio on the ${}_{ex}K_r/K$ was observed when the rotation chord angle was more than 1% and the width-thickness ratio increased as ${}_{ex}K_r/K$ decreased.

Keywords: Theoretical stiffness, Skeleton curve, Width-thickness ratio, Unloading stiffness, Monotonic loading, Cyclic loading.

1 INTRODUCTION

The factors which effects the hysteresis characteristics of concrete filled steel tube columns (hereinafter CFT) are the width-thickness ratio, axial force ratio, the effective length-depth ratio, strength of steel tube and concrete, etc. Furthermore, for the reasonable performance design, it is necessary to formulate the hysteresis characteristics that can consider the damage condition and the parameters mentioned above, which affect the structural performance.

There are several studies on the skeleton curve and hysteresis rule of CFT columns. Among them, the model of hysteresis characteristics with wide application range proposed by the U.S.-Japan research program (Mori *et al.* 1998) is shown in the Recommendations for design and Construction of Concrete Filled Steel Tubular Structures (hereafter CFT Recommendations) (AIJ 2008). The skeleton curve is a normal trilinear type model. According to the CFT Recommendations, the skeleton curve and the hysteresis rule correspond well to the experimental data, however, the strength evaluates conservatively the experimental strength and the hysteresis energy absorption is overestimated when the rotation angle is more than 1%. There are few studies about the unloading stiffness of CFT columns and clarifying the unloading stiffness is important for evaluating appropriately the hysteresis energy absorption when the rotation angle of CFT columns is large.

In this study, the skeleton curves are compared with the experimental data of the square CFT columns subjected to the constant axial force and lateral force. Furthermore, the relation between the initial stiffness based on the theory and the unloading stiffness are shown and the effect of the width-thickness ratio and axial force ratio is shown.

2 INFLUENCE OF AXIAL FORCE RATIO, WIDTH-THICKNESS RATIO, CONCRETE STRENGTH ON SKELETON CURVE

2.1 Outline of the Proposed Skeleton Curve

This section outlines the skeleton curve of the CFT column shown in AIJ CFT Recommendations. Figure 1 shows a schematic diagram of a skeleton curve. The first break point is the yield strength M_Y, and the second break point is the ultimate strength M_U according to the CFT Recommendations. The theoretical initial stiffness is calculated by assuming that the total cross-section is elastic. The stiffness reduction rate α_Y is 0.7. In addition, the limit rotation angle R_U is defined as the rotation angle when the maximum strength is reduced by 95%, and it is expressed by the Eq. (1). Where, in the Eq. (1), γ_r is 1 when L_k/D is less than or equal to 10 and is 0.8 when L_k/D is larger than 10. In Eq. (1), β is defined by Eq. (2).

$$R_U(\%) = \frac{\gamma_r \times 100}{0.15 + 3.79\dfrac{N}{N_0}} \cdot \frac{t}{D} \cdot \beta \tag{1}$$

$$\beta = 1.0 - \frac{{}_c\sigma_B - 40.3}{566} \leq 1.0 \tag{2}$$

Where, N is the axial force, N_0 is the compressive strength, D is the section width, t is the plate thickness and ${}_c\sigma_B$ is the concrete strength.

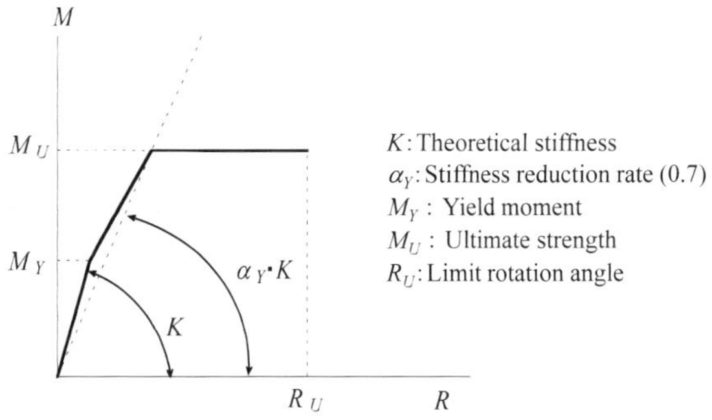

Figure 1. Skeleton curve of the CFT column defined in CFT Recommendation.

2.2 Effect of the Parameters

The skeleton curve shown in the previous section was calculated by selecting the following parameters for square cross sections.

- Axial force ratio n (=N/N_0): 0, 0.1, 0.2, 0.3, 0.4, 0.5, 0.8
- Width-thickness ratio D/t: maximum value of FA and FC rank

- Concrete strength ${}_{c}\sigma_B$ (N/mm²): 24, 60

In addition, the depth D of the cross section is and 500 mm, the effective length-depth ratio L_k/D (L_k: effective length) is 12 and the steel tube F value is 325 N/mm². The width-thickness ratio's rank is defined in the Technical requirements for structural safety (MLIT 2015).

Figure 2 shows the skeleton curves when the concrete strength is 24 and 60N/mm² and the width-thickness ratio is the FA and FC rank. Some limit rotation angle R_U exceed 5% and it is expressed by arrows in the figure.

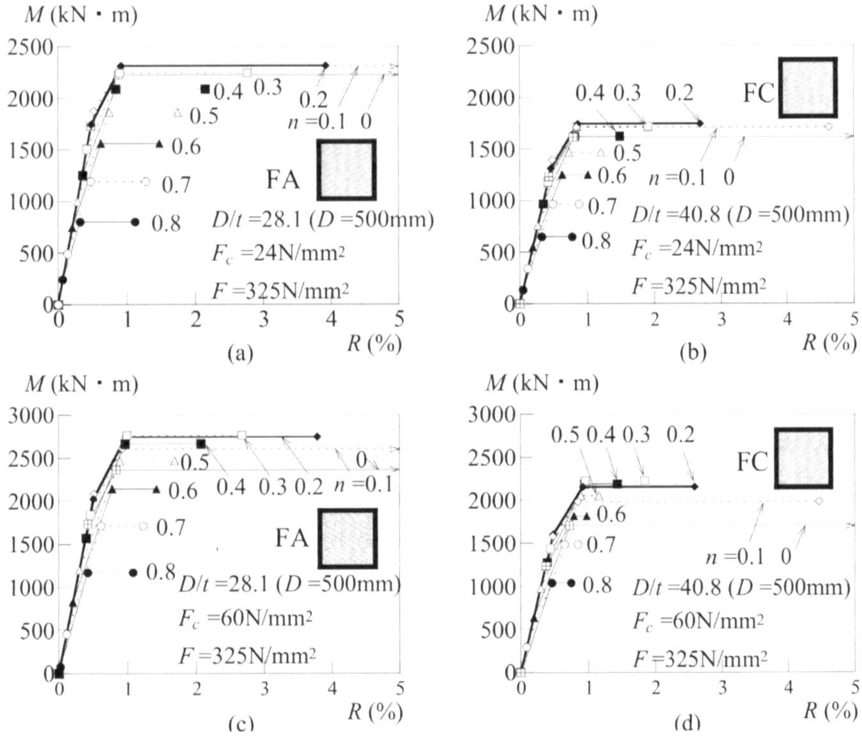

Figure 2. Influence of axial force ratio, width-thickness ratio and concrete strength on skeleton curve.

3 COMPARISON OF PAST EXPERIMENT RESULTS

3.1 Outline of the Experimental Study

Results of the previous experimental study are used. As shown in Figure 3, the loading condition is a cantilevered column which is one end fixed and other end free and subjected to the constant axial force N and lateral force Q.

In Table 1, No.1~No.3 series specimen's L_k/D is 10 (Matsui *et al.* 1998) and No.4 series specimen's L_k/D is 20 (Jyouzaki *et al.* 2018). The effective length L_k is equal to $2L$. The experimental parameters of No.1 ~ No.3 series are the width-thickness ratio D/t of the steel tube, the axial force ratio n, and the loading program. The cyclic loading program is shown in Figure 4 and the rotation angle of the column R ($=\delta/L$, δ: lateral displacement) is increased by 0.25% every four cycles. The experimental parameter of No.4 series is the axial force ratio n, and the loading program is monotonic loading.

The steel tube used in the experiment is a square steel tube of STKR 400 (Japanese Industrial Standard) and BCR295. The measured dimensions of the steel tube are shown in Table 1, the results of the tensile test of the steel tube and the concrete compressive strength are shown in Table 2.

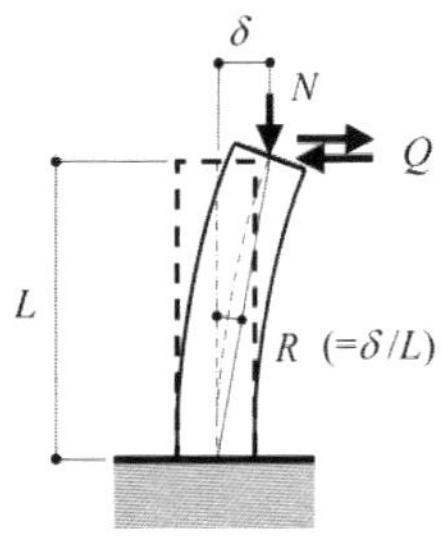

Figure 3. Loading condition.

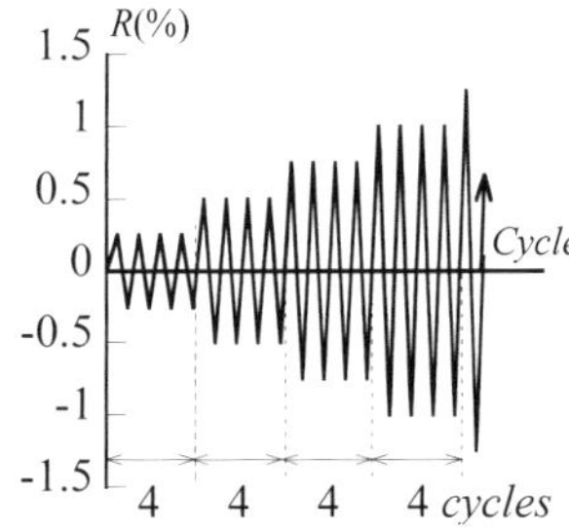

Figure 4. Loading program.

Table 1. Measured dimension of steel tube.

Series No.	Diameter ratio	L_k/D	Section depth D (cm)	Thickness t(mm)	D/t	Sectional area $_sA$ (cm^2)
1	22	10	10.0	4.40	22.7	16.3
2	31	10		3.21	31.1	12.2
3	43	10		2.24	44.6	8.6
4	25	20	15.0	5.97	25.1	33.3

Table 2. Tension test results and concrete strength.

Series No.	Diameter ratio	Steel tube	Yield stress $_s\sigma_y$ (N/mm^2)	Tensile strength σ_u (N/mm^2)	Concrete strength $_c\sigma_B$ (N/mm^2)	
					R*	M*
1	22	STKR	407	463	34.1	36.3
2	31	STKR	449	505	36.0	36.3
3	43	STKR	415	465	36.2	36
4	25	BCR	378	457	-	55.1~61.6

Note: *R: cyclic loading, M: monotonic loading

3.2 Comparison of Experiment Results

3.2.1 *In case of monotonic loading*

Figure 5 shows the bending moment M-rotation angle R relationship obtained by the experimental study and the calculated skeleton curves. The bending moment is calculated as $M = Q{\cdot}L + N{\cdot}\delta$ (δ=RL). In the figure, the square marks indicate the point where the strain at the outer edge reaches the yield strain by the tensile test for the first time (steel tube yield), the arrows indicate the point where the local buckling occured and the circle marks indicate the maximum bending strength. When the L_k/D=20 and n=0.45 and 0.6, the limit rotation angle R_u calculated by Eq. (1) are smaller than the second break point's rotation angle.

According to the figure, the ultimate strength of the skeleton curve evaluates conservatively the experimental value except for the case of D/t=43 and n=0.5. In case of the D/t=43, n=0.3 and L_k/D=20, n=0.45 and n=0.6, the ultimate strength of the skeleton curve is almost same as the experimental maximum load. As for the lateral stiffness, in case of L_k/D=20, the skeleton curve's stiffness is larger than that of the experiment. This is because that these specimens are slender and the first breaking point strength is larger than the actual one and the reduction of the stiffness after the concrete cracking are not considered.

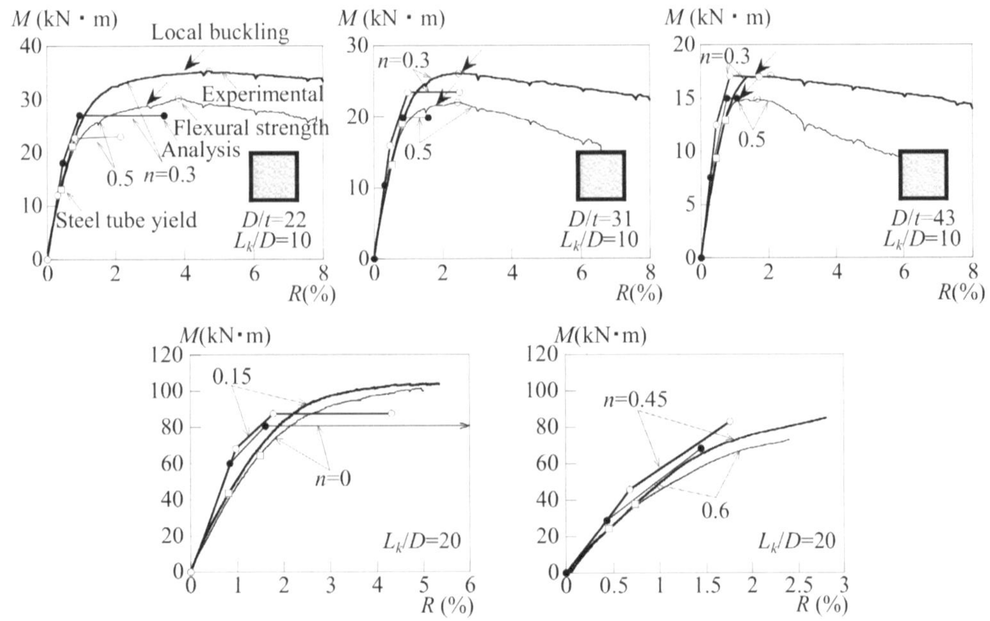

Figure 5. Comparison between experimental value and calculated value (Monotonic loading).

3.2.2 *In case of cyclic loading*

Figure 6 shows the comparison between the *M-R* relationship and the skeleton curve in the case of the axial force ratio n=0.3. The dotted line in the figure shows the result of monotonic loading. When the width-thickness ratio is 43, the experimental strength is smaller than the calculated ultimate strength.

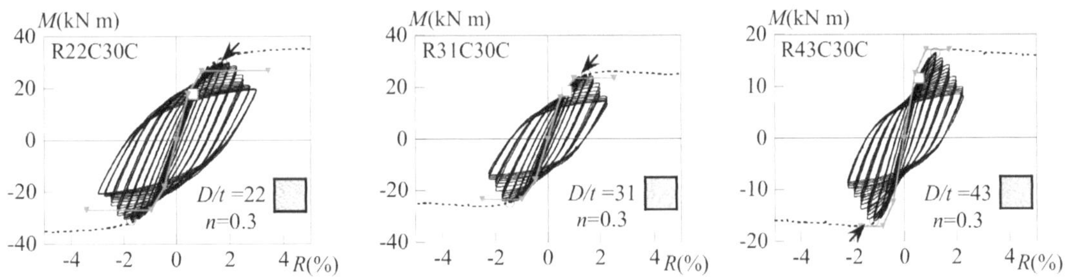

Figure 6. Comparison between experimental value and calculated value (Cyclic loading).

4 INITIAL STIFFNESS AND UNLOADING STIFFNESS

Unloading stiffness ${}_{ex}K_r$ was obtained from experimental data. Where, the values of the experienced rotation angle and the residual rotation angle obtained by Kido and Tsuda (2005) were used.

Figure 7 shows the ratio of the unloading stiffness and the theoretical stiffness K. The axial force ratio is 0.3 and 0.5. The ratio ${}_{ex}K_r/K$ decreases as the rotation angle increases. The effects of the axial force ratio is observed after the rotation angle exceeding 1%, and the value of ${}_{ex}K_r/K$ in case of n=0.5 is smaller than those of n=0.3. The value of ${}_{ex}K_r/K$ is smaller in case of D/t=43 than those of D/t=22 and 31.

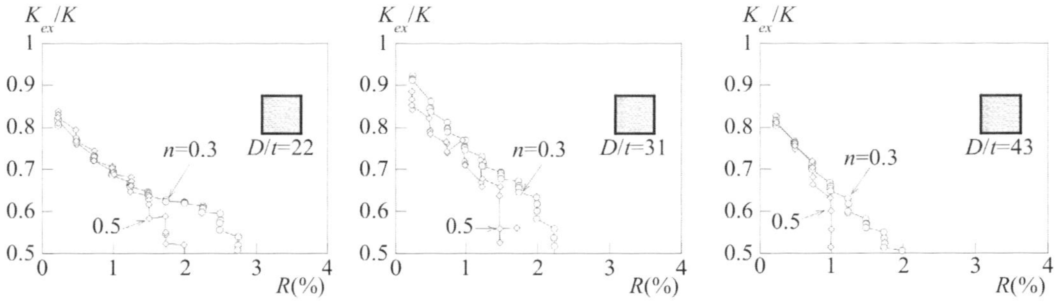

Figure 7. Comparison between experimental value and calculated value (Cyclic loading).

5 CONCLUSIONS

The conclusions derived from this study are as follows:

- The skeleton curves shown in AIJ CFT Recommendation were compared with the moment-rotation angle relationship obtained by the experimental study. The ultimate strength of the skeleton curves evaluate well except for the D/t=43 and n=0.5 specimen in monotonic loading and D/t=43 and n=0.3 specimen in cyclic loading. The lateral stiffness of the skeleton curves disagree with the test results in case of specimens with L_k/D=20.
- The relationship between the theoretical stiffness and the unloading stiffness obtained by the experimental study were shown. The effects of the axial force ratio is observed after the rotation angle exceeding 1%.

References

AIJ, *Recommendations for Design and Construction of Concrete Filled Steel Tubular Structures (in Japanese),* Architectural Institute of Japan, Japan, 2008.

Jyouzaki, K., Kido, M., Fujioka, D., and Tsuda, K., *Study on The Fatigue Properties of Slender Square CFT Columns under Cyclic Lateral Loading with Constant Deflection Part 3 in The Case Of L_k/D=20 Specimens Subjected to High Axial Force (in Japanese),* AIJ Kyushu Chapter Architectural Research Meeting, No. 57, Japan, March, 2018.

Kido, M. and Tsuda, K., Relation between Experimental Lateral Drift Angle and Residual One of Concrete Filled Steel Square Tubular Column (in Japanese), *Steel Structure Annual*, Vol. 13, pp.503-508, November, 2005.

Matsui, C., Tsuda, K., Yamaji, Y., and Fujinaga, T., Structural Performance and Axial Force Limit of Concrete Filled Steel Square Tubular Columns (in Japanese), *Journal Structural and Construction Engineering*, AIJ, No. 504, 103-110, Japan, February, 1998.

MLIT, *The 2015 Commentary on Structure Related Technical Standards for Buildings (in Japanese)*, Building Guidance Division, Housing Bureau, the Ministry of Land, Infrastructure and Transport (supervisor), June, 2015.

Mori, O., Noguchi, T., Fujimoto, T., and Morino, S., *US-Japan Cooperative Structural Research Project on Composite and Hybrid Structures (CFT-33) Hysteresis Characteristics Model of Concrete Filled Steel Tubular Columns (in Japanese)*, AIJ, Japan, September, 1998.

Streamlining Information Transfer between Construction and Structural Engineering
Edited by Shiau, J., Vimonsatit, V., Yazdani, S., and Singh, A.

ISBN: 978-0-9960437-7-9

EXPERIMENTAL STUDY OF INNOVATIVE COMPOSITE BUCKLING-RESTRAINED FUSE FOR CONCENTRICALLY BRACED FRAMES UNDER CYCLIC LOAD

MASOUD MOHAMMADI[1, 2], MOHAMMAD ALI KAFI[1], ALI KHEYRODDIN[1], and HAMID REZA RONAGH[2]

[1]*Dept of Civil Engineering, Semnan University, Semnan, Iran*
[2]*Centre for Infrastructure Engineering, Western Sydney University, Sydney, Australia*

Concentrically Braced Frames (CBFs) are among the most commonly used lateral resisting systems utilized in the construction of steel structures due to their rigidity, low lateral displacement and ease of implementation. However, the lack of ductility due to the buckling that occurs in the bracing elements before yielding is their main disadvantage. This study presents an innovative Composite Buckling Restrained Fuse (CBRF) to be used as a bracing segment in concentrically braced frames that improves the ductility and eliminates premature buckling. The proposed CBRF with relatively small dimensions is a hysteretic damper consisting of thin steel plate core and extra tensile elements embedded in a composite encasement. Two CBRF samples are designed and tested experimentally. The results indicate that the proposed structural fuse has a ductile behaviour with high energy absorption and sufficient strength along with a reasonably stable hysteretic response under cyclic load.

Keywords: Steel structure, Structural fuse, Ductility.

1 INTRODUCTION

Concentrically Braced Frames (CBFs) have become prevalent as a lateral load resisting system over the last three decades. Large lateral stiffness, low lateral displacement associated with ease of implementation at a low-cost are the advantages of the CBFs which encourages the increasing use of this system in construction. However, damage of the whole brace as a result of buckling have been frequently reported under severe lateral loads. In many instances such as Northridge earthquake in 1994, buckling of the whole brace before yielding was the main problem reported of this lateral system which limited the ductility and affected the energy dissipation capacity of the frame (Bruneau *et al.* 2011) and collapse propagation (Mohajeri Nav *et al.* 2017, Abbasnia *et al.* 2016). Many studies since have been focused on increasing the ductility of CBFs. In most of these studies, researchers tried to make a modification in connections of the brace or placed a ductile member as a hysteretic damper into the brace to increase the deformation capacity of the system. Among these, sliding friction mechanism dampers and connections (Mualla and Belev 2002, Rodgers *et al.* 2017) and yielding parts such as T-shaped dampers (TahamouliRoudsari *et al.* 2018) and ring element fuses (Andalib *et al.* 2018, Deihim and Kafi 2017) can be named. From a different perspective, some of the researchers were looking at methods to prevent premature buckling of the braces. Confining braces by mortar encase and steel cases were some of the techniques gradually implemented till the modern Buckling Restrained Braces (BRBs) were evolved (Xie 2005).

The current study aims to present an innovative Composite Buckling Restrained Fuse (CBRF) to be used as sacrificial brace segment. CBRF with relatively small dimensions is a hysteretic damper with different tension and compression capacity. Extra tensile elements have been used innovatively so that no reduction of the tensile capacity occurs unlike what happens when an ordinary fuse is used. In this paper, the experimental results on the proposed CBRF are described, and the results are presented.

2 GENERAL DESCRIPTION OF CBRF AND INNOVATIVE TENSILE ELEMENT

Structural fuses are the sacrificial yielding elements embedded in different parts of a structure such as braces, beams or columns, depending upon their bearing load type, to absorb the energy of the loads exerted to the system. CBRF proposed in this paper is a kind of replaceable axial structural fuse with a short length. This fuse comprised tensile elements and thin steel core plate restrained in a composite encase. This reduced length BRB is placed at each end of the bracing members, with different tension and compressive capacity, to localise the structural failures and dissipate the energy of a severe load. Figure 1, illustrates CBRF components and its placement in a diagonal bracing frame.

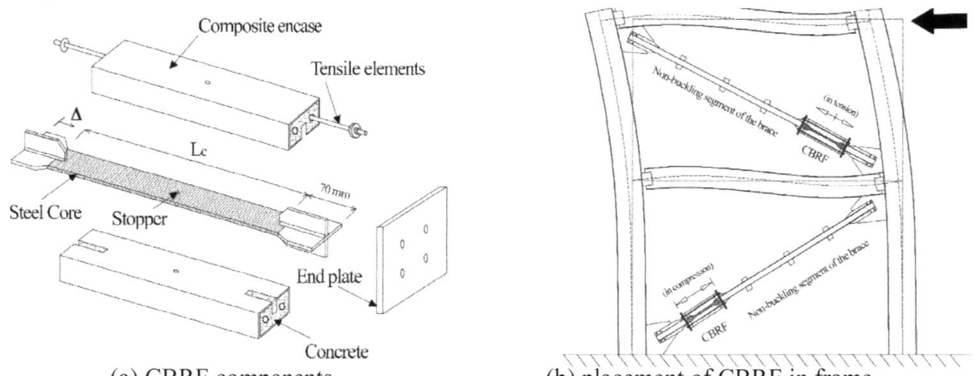

(a) CBRF components (b) placement of CBRF in frame

Figure 1. CBRF assemblage and placement.

Assume that the frame, shown in Figure 1(b) without any fuse, is subjected to an severe lateral load. Depending on the force, one of the braces might be under tension while the other may buckle under compression. Utilizing an ordinary fuse in each bracing member causes these elements to fail prior to any other segments allowing the dissipation of energy. These sacrificial elements which have almost the same tensile and compressive capacity are designed based on the critical compressive load of the bracing members to prevent the premature buckling by considering a reduction factor. Since the bracing member has different tensile and compressive capacity, this leads that limitation happens in both tension and compression capacity of the bracing member and reduces the total lateral bearing capacity of the frame. In the proposed CBRF the tensile capacity is designable. Extra tensile elements, steel bars, have been used innovatively so that no limitation of the tensile capacity of the whole brace occurs.

3 DESIGNING AND DETAILING OF CBRF

The cross-sectional area of the steel core, A_C, is designed based on the method mentioned in AISC 341 (AISC 2016) by considering a reduction factor, ϕ, which depends on the compressive strength of fuse, β, the average strain hardening of the steel core, ω, and the expected yield stress, R_y, for the core plate section, Eq. (1).

$$\phi \leq \frac{1}{\beta\omega R_y} \tag{1}$$

Moreover, Tensile elements can be designed based on the desired maximum tensile force that would occur in the whole bracing member upon the anticipated story drift. Determining the optimal length of CBRF is another essential parameter that plays an important role in absorbing the energy and plastic behavior of the fuse. Due to the short length of CBRF, at the same displacement, Δ, the core average strain, $\varepsilon_c=\Delta/L_c$, would be more than the similar longer ones. From the practical metallurgical point of view, according to the protocol of loading and number of inelastic cycles, the minimum length of the core plate, L_c, can be evaluated based on preventing the fracture due to low-cyclic fatigue phenomena (Mirtaheri *et al.* 2011), Eq (2).

$$L_c \geq (\frac{1}{\varepsilon'_c}) \times \lambda^{|c|} \tag{2}$$

Where, ε'_c is the real ultimate strain for steel material, c is the fatigue ductility exponent and for λ we have:

$$\lambda = \sum_{1}^{m}(2n_i \times \Delta_i^{\left|\frac{1}{c}\right|}) \tag{3}$$

Where m is the total number of in-elastic cycles of loading protocol and n_i is the number of cycles with the same amplitude cycle of Δ_i.

4 EXPERIMENTAL STUDY

In order to investigate the performance of CBRF and the effect of extra tensile elements, an experimental program was conducted. Two specimens were designed and tested subjected to cyclic loads. The specimen 1, was a Reduced Length Buckling Restrained Braces, RL-BRB, without the tensile bar unlike specimen 2 which contained the tensile bar.

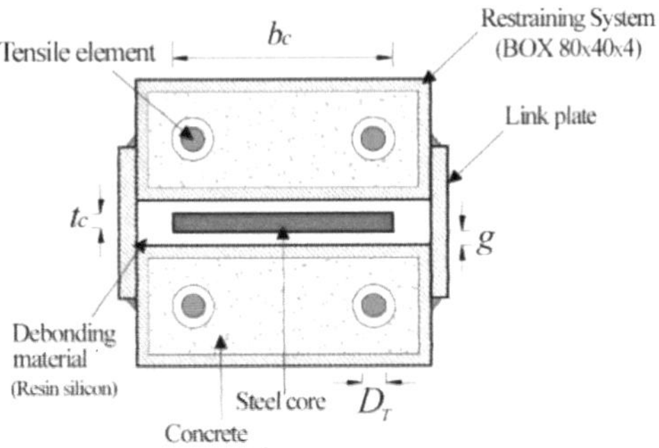

Figure 2. CBRF cross-section.

Steel with the yield stress of 290 MPA and ultimate stress of 400 MPA at the strain of 0.25% was used for the core steel plate. Considering the critical load of 120kN for a diagonal bracing member and employing the safety factor of 0.5, the cross-sectional area of the core, A_c, was obtained to be 2.1 cm^2. The composite encase was made of two steel box filled with normal concrete connected by two steel link plates. 2mm gap size was considered for the free space between the core and the composite encase, shown in Figure 2. Length of the core plate, L_c, was calculated as 30cm based on the Eq. (2) as discussed in Section 4.1. Specimen parameters are summarized in Table 1. The t_c and b_c are the thickness and the width of core plate and g is the gap size.

Table 1. Specimen parameters.

No.	**Specimens**	**Core Plate**			**Restraining system**	**Tensile bars**
		L_c (mm)	**t_c (mm)**	**b_c (mm)**	**g (mm)**	**D_T (mm)**
1	RL-BRB	300	5	42	2	-
2	$B_{co}L_3D_8$	300	5	42	2	8

4.1 Loading Protocol

ATC24 loading protocol was utilized for evaluating the cyclic performance of the specimens. This loading protocol includes cycles which are a multiplier of the yield deformation, Δ_y, of the segment that was calculated based on the material properties (Krawinkler 1992). The slow and manageable rate of deterioration in this loading protocol allows prediction of the load-deformation response with more confidence without any missing cycles. At the beginning of the loading protocol, nine cycles of the loading were considered with the elastic amplitude lower than Δ_y to measure the initial axial stiffness of the specimen. The subsequent inelastic phase of the loading was a set of variable deformation amplitude gradually increasing by multipliers of Δ_y at 2, 3, 4, 6, 8, 11, 14, 18, 22, 27, 32, 38 times, which follow a geometric progression. The optimized length of the fuse was calculated based on the Eq. (2) as $L_c \geq 25.1$ cm. In order to avoid the anticipated fracture due to low-cyclic fatigue phenomena, core length was taken $L_c = 30$ cm for the specimens.

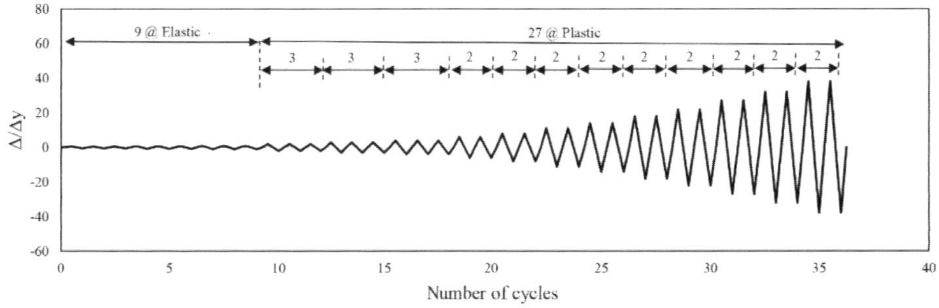

Figure 3. Loading protocol (Krawinkler 1992).

4.2 Test Setup

The uniaxial test setup composed of a hydraulic jack capable of exerting cyclically up to a maximum of 2000kN compressive load and a maximum of 1000kN tensile load while accommodating a maximum stroke of ±100 mm. It was also equipped with a 1000kN load-cell. As shown in Figure 4, the setup consisted of two reaction blocks which were connected to the strong floor. One end of the specimens was attached to the reaction block, whereas the other end was connected to the hydraulic actuator. Linear guideways (wagons) were used in the setup to prevent the lateral displacement of the load-cell during the test and exert the load axially to the specimen without rotation. Two displacement-control linear LVDTs with high accuracy were mounted on the load support to monitor the exact displacement. Moreover, two other LVDTs were placed on the composite encase of the specimen. Figure 4 depicts the assembled CBRF specimen in the setup.

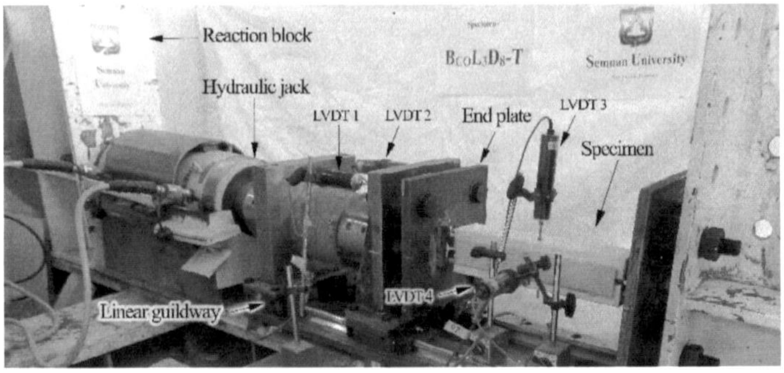

Figure 4. Assembled CBRF specimen in the setup.

5 EXPERIMENTS AND RESULTS

The hysteresis curves of the two specimens that were mentioned in Table 1 are presented in Figure 5, where the failure point of each specimen is marked with a red triangle and the envelope curves are depicted by the dashed line.

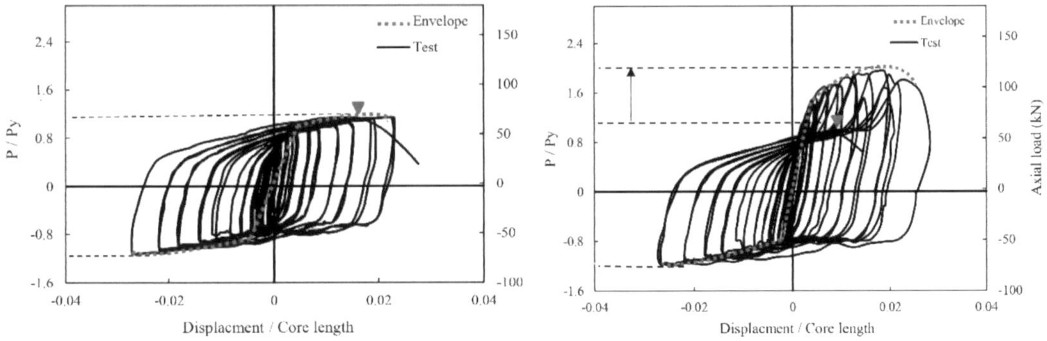

(a) Specimen 1, RL-BRB (b) Specimen 2, $B_{co}L_3D_8$-T

Figure 5. The hysteretic response of specimens.

The hysteresis results of two specimens were investigated closely. The RL-BRB, specimen 1, was the control sample to compare the performance of using extra tensile elements in the fuse. As shown in Figure 5(a), the hysteretic loops of the specimen 1 are steady and stable without pinching. The fuse has almost the same bearing capacity in the compression and tension and the ductility is improved. Passing through the critical load of the core and occurrence of buckling in the first mode, the composite encase prevents the early global buckling of the central core and leads the core to reach higher modes as it displaces axially. The sawtooth parts of the curves show the resulting degradation due to the progress of the bucking modes in compression phase. The detected differences between the compression and the tension load bearing ratio are not impressive in the specimen 1. The maximum axial bearing ratios of the RL-BRB are P_{max} / P_y=1.14 in tension and 1.13 in compression. The elastic stiffness of RL-BRB obtained from the initial elastic cycles is about 113 kN/mm. The maximum inelastic axial average core strain after the end of 32 cycles at the failure point is 3.4%, which is interpreted by the appropriate behaviour of the core in ductility and energy dissipation. As shown in Figure 5(b), the hysteresis curve of the CBRF demonstrate the steady stable wide loops similar to Specimen which is a desirable improvement on the tensile strength and energy dissipation for the CBRF. The maximum axial compressive bearing ratio of the CBRF is about 1.15 which is almost the same as that in Specimen 1. In addition, The maximum tensile bearing ratio is 1.94 (axial load of 120kN) due to the utilisation of the extra tensile bars. The elastic stiffness of the CBRF obtained from the initial elastic cycles as 110 kN/mm and the maximum inelastic axial average core strain after the end of 32 cycles at the failure point is about 3.4%, which are almost similar to those of Specimen 1.

6 CONCLUSION

This study presents an innovative structural fuse, CBRF, to be used as a segment in the concentric braces. This kind of replaceable hysteresis damper, with different capacity in tension and compression, has the ability to compensate the limitation of the tensile capacity unlike what occurs when an ordinary fuse is used in a bracing member. Two specimens were tested experimentally. The results indicate that CBRF offers favourable improvement in the tensile capacity and energy dissipation along with reasonably stable hysteretic response under cyclic loads.

References

Abbasnia, R., Mohajeri Nav, F.,Usefi, N., and Rashidian, O., A New Method for Progressive Collapse Analysis of RC Frames, *Structural Engineering and Mechanics,* 60(1), 31-50, 2016.

AISC. *ANSI/AISC 341-16, Seismic Provisions for Structural Steel Buildings*, United States of America, Chicago: American Institute Of Steel Construction, 2016.

Andalib, Z., Kafi, M. A., Bazzaz, M., and Momenzadeh, S., Numerical Evaluation of Ductility and Energy Absorption of Steel Rings Constructed from Plates, *Engineering Structures,* 169, 94-106, 2018.

Bruneau, M., Uang, C. M., and Sabelli, R., *Ductile Design of Steel Structures, 2nd Edition*, McGraw-Hill, 2011.

Deihim, M. and Kafi, M. A., A Parametric Study into the New Design of a Steel Energy-Absorbing Connection, *Engineering Structures,* 145, 22-33, 2017.

Krawinkler, H., *ATC24, Guidelines for Cyclic Seismic Testing of Components of Steel Structures*, Applied Technology Council, Redwood City, California, 1992.

Mirtaheri, M., Gheidi, A., Zandi, A. P., Alanjari, P., and Samani, H. R., Experimental Optimization Studies on Steel Core Lengths in Buckling Restrained Braces, *Journal of Constructional Steel Research,* 67(8), 1244-1253, 2011.

Mohajeri Nav, F., Usefi, N., and Abbasnia, R., Analytical Investigation of Reinforced Concrete Frames Under Middle Column Removal Scenario, *Advances in Structural Engineering,* 2017.

Mualla, I. H. and Belev, B., Performance of Steel Frames with w New Friction Damper Device Under Earthquake Excitation, *Engineering Structures,* 24(3), 365-371, 2002.

Rodgers, G. W., Chase, J. G., Causse, R., Chanchi, J., and MacRae, G. A., Performance and Degradation of Sliding Steel Friction Connections: Impact of Velocity, Corrosion Coating and Shim Material, *Engineering Structures,* 141, 292-302, 2017.

TahamouliRoudsari, M., Eslamimanesh, M. B., Entezari, A. R., Noori, O., and Torkaman, M., Experimental Assessment of Retrofitting RC Moment Resisting Frames with ADAS and TADAS Yielding Dampers. *Structures,* 14, 75-87, 2018.

Xie, Q., State of the Art Of Buckling-Restrained Braces in Asia, *Journal of Constructional Steel Research,* 61(6), 727-748, 2005.

Streamlining Information Transfer between Construction and Structural Engineering
Edited by Shiau, J., Vimonsatit, V., Yazdani, S., and Singh, A.

ISBN: 978-0-9960437-7-9

YIELD STRESS BASED ON NATURAL STRAIN THEORY UNDER CYCLIC TENSION-COMPRESSION LOAD AFTER LARGE SIMPLE SHEAR

YASUYUKI KATO[1] and HIROKI UCHIDA[2]

[1]*Dept of Mechanical Engineering, College of Science and Technology, Nihon University, Tokyo, Japan*
[2]*Graduate Student of College of Science and Technology, Nihon University, Tokyo, Japan*

Many researches on the yield behavior under cyclic loads have been conducted for many years. However, almost these studies have been done within the range of relatively small size deformation. Hence, the detailed studies for yield phenomenon under cyclic loads after applying a large pre-deformation have not been sufficiently elucidated. Therefore, in the series of our studies, using test pieces of high purity tough pitch copper, the yield behavior generated under cyclic loads after applying the pre-deformation of large uniaxial tension or large simple shear have been investigated. However, as for the case that the type of pre-deformation differs from the type of deformation in the cyclic load, detailed studies have not been conducted yet. Hence, in the present study, the yield behavior generated under cyclic loads for tension and compression after applying a large pre-deformation of simple shear is examined based on Natural Strain theory. Furthermore, experimental results in the present study are compared with results of previous studies concerning cyclic loads for tension and compression after applying the large uniaxial tension. Consequently, it is revealed that the decreasing tendency of yield stress at the compression side appears more strongly as compared with the experimental result by previous study that the pre-deformity is a uniaxial tension.

Keywords: Finite deformation, Elasto-plastic analysis, Pre-deformation, Anisotropy, Deformation history, Slope of tangent, Proof stress, Ductile materials.

1 INTRODUCTION

The Natural Strain theory used in this study is obtained by integrating an infinitesimal strain rate on an identical line element over the whole process of the deformation path. Consequently, since the shearing strain component is represented by pure angular strain, the rigid body rotation can be clearly removed from the shearing strain component. Moreover, since the additive law of strain on an identical line element can be satisfied, the strain rate can be decomposed into the elastic component and the plastic component. Moreover, the elastic component can also be decomposed into deviatoric part and volumetric part in the same manner as the conventional infinitesimal deformation theory. Therefore, the Natural Strain is an effective strain representation, which can be systematically treated from infinitesimal deformation to the large deformation.

In general, when the structure undergoes cyclic loads due to earthquake etc., a plastic failure arises in the structure even if the number of cycles is small. Therefore, it is an important issue in the structural engineering to find out the actual mechanisms of low-cycle fatigue failure and the yield behavior under cyclic loads. However, although many of researches have been done within the range of an infinitesimal deformation, the detailed studies for yield behavior under cyclic loads after applying a large pre-deformation have not been fully elucidated. Therefore, in a series of studies (Kato 2017, Kato 2018), the yield behavior generated under cyclic loads after applying the pre-deformation of large uniaxial tension or large simple shear have been investigated. In those studies, as the method for determining yield stress in each cycle, the slope of tangent in the principal deviatoric stress -principal deviatoric strain curve has been adopted instead of the estimation method by conventional proof stress. However, concerning the case that the type of pre-deformation differs from the type of deformation in the cyclic load, detailed studies have not been conducted yet. Hence, in this research, the yield behavior generated under cyclic loads for tension and compression after applying a large simple shear is examined.

2 PRE-DEFORMATION AND CYCLIC LOAD

Figure 1 (a) and (b) shows the cyclic load of tension and compression after large pre-deformation. As shown in Figure 1(a), different size of large pre-deformation is firstly applied to the test piece. During this process, the anisotropy has been formed in the material. As a consequence, the value of yield stress at the pre-deformation side is different from the opposite side for Bauschinger effect, see S_{1YB} and S_{1YC} in Figure 1(b). Figure 1(c) and (d) represent the schematic diagram of the anisotropy of yield surface in a stress space, which occurs immediately after pre-deformation. In these figures, Figure 1(c) shows the yield surface concerning the pre-deformation of uniaxial tension, and the principal axis of anisotropy is formed in the tension and the compression side.

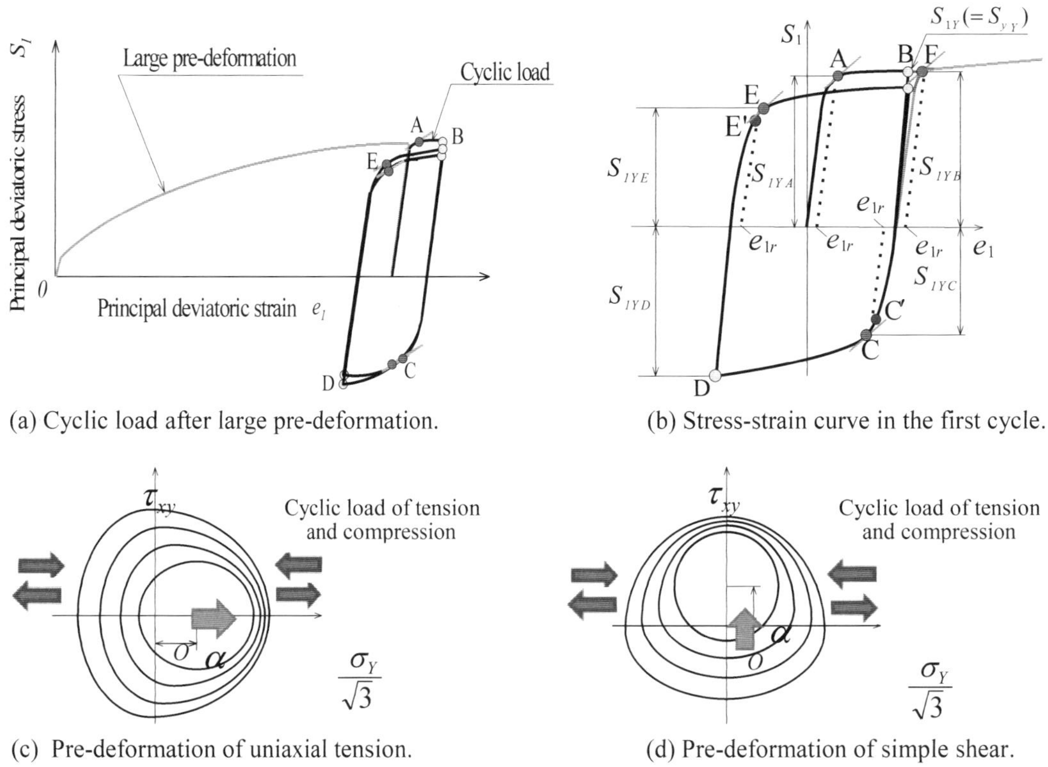

(a) Cyclic load after large pre-deformation.

(b) Stress-strain curve in the first cycle.

(c) Pre-deformation of uniaxial tension.

(d) Pre-deformation of simple shear.

Figure 1. Relation between the anisotoropy in yield surface after large pre-deformation and the cyclic load.

Hence, the direction of the principal axis of anisotropy is the same as the direction of cyclic load for tension and compression. On the other hand, Figure 1(d) shows the yield surface concerning the pre-deformation of simple shear in this research, and the principal axis of anisotropy is formed in the shear deformation side. Hence, under this condition, the direction of the principal axis of anisotropy is different from the direction of cyclic load for tension and compression.

3 ESTIMATION METHOD OF YIELD STRESS UNDER CYCLIC LOAD

Figure 1(b) shows the first cycle for a deviatoric stress and deviatoric strain diagram when the tensile load and the compressive load are subjected alternately to the test pieces after applying a pre-deformation of a large simple shear. The reloading process when the tension applies again after unloading process of tension from a point B is represented by a green curve. Since the value of stress at a point B is preliminary determined, the yield stress during the reloading process is already determined. And, in this figure, it is represented by a point F. In the neighborhood of a yield point F, the deviatoric stress and deviatoric strain curve becomes a shallow curve, and in a series of our studies (Kato and Kazama 2013, Kato 2014), it is formulated as follows.

$$S_1 = a\left(1 - exp(be_1)\right) + ce_1 + d \tag{1}$$

However, a, b, c and d in Eq. (1) are coefficients determined by using the *Levenberg-Marquardt Method*, which is one of the non-linear least-squares methods. Furthermore, the slope of tangent is also obtained by differentiating Eq. (1), and it is represented by Eq. (2).

$$\frac{dS_1}{de_1} = -a\,b\,\exp\left(be_1\right) + c \tag{2}$$

Hence, the slope of the tangent at yield point F can be derived by using Eq. (2). Therefore, the slope of tangent at yielding can be specified in advance and it is indicated by the straight blue line.

As for the yield stress in a cycle, firstly, the yield stress obtained after applying the tensile strain with specified constant strain amplitude is represented by a point B just before unloading of tension. Secondly, the yield stress in compression side is obtained by representing the shallow curve by using the Eq. (1). And, the stress, which has the same value as the slope at a point F, i.e., the stress at point C in this figure, is assumed to be a yield stress. Thirdly, the yield stress obtained after applying the compressive strain with specified constant strain amplitude is represented by a point D. Finally, the yield stress in tension side, which is derived by applying a tensile load again after unloading of compression, is obtained by representing the shallow curve by using Eq. (1) once again. Similarly, the stress, which has the same value as the slope of tangent measured at the point F, i.e., the stress at point E in this figure, is assumed to be a yield stress.

On the other hand, the yield stress by proof stress is determined by using the value of the residual strain e_{1r} that is obtained by unloading from a point F. Hence, as shown in point C' and E', the proof stress is estimated smaller as compared with the estimated yield stress in this study.

4 EXPERIMENTAL METHOD

In the experiments, the cylindrical specimens made from high purity tough pitch copper are used, i.e., purity 99.99 % and outer diameter 22 [mm], inner diameter 16 [mm], gauge length 30 [mm].

Next, as for the experimental condition, this experiment is composed of two stages. Namely, experiments applying the pre-deformation of the large uniaxial tension to the specimens are firstly conducted. Subsequently, the cyclic loading tests for tension and compression are conducted.

(1) Experimental condition of pre-deformation of large simple shear. Three different types of shear deformation, namely, values of slip k = 0.37, 0.53, 0.98 [-], i.e., values of the principal stretch λ = 1.2, 1.3, 1.6 [-], are applied to the test pieces.

(2) Experimental condition of cyclic loading tests for tension and compression. After attaching the strain gauges to the test pieces, the experiments of the cyclic loading tests for tension and compression are conducted with the constant strain amplitude, namely, $\triangle e_I = \pm 0.006$ [-]. In these experiments the numbers of cycles n are all ten times. Then, the values of yield stress at points B, C, D and E in Figure 1 are determined, and the changes of yield stress with an increase of the number of cycle n are investigated. Moreover, as for the values of the principal stretch λ =1.2 and λ =1.3 [-], the low-strain amplitude, i.e., $\triangle e_I = \pm 0.003$ [-], is also examined in this research.

5 EXPERIMENTAL RESULTS

Figures 2 and 3 show some examples of experimental results obtained under cyclic loading tests for tension and compression after applying the pre-deformation of large simple shear with respect to the constant strain amplitude (the strain amplitude $\triangle e_I = \pm 0.006$ [-]). In these figures, (a) is the deviatoric stress and deviatoric strain diagrams and (b) shows the variation of yield stress with an increase of the number of cycles. In Figure 2 (b) and 3(b), black plots show the stress at B and D (i.e., $S_{I\mathrm{YB}}$ and $S_{I\mathrm{YD}}$ in Figure 1(b)) and red plots show the stress at C and E (i.e., $S_{1\mathrm{YC}}$ and $S_{1\mathrm{YE}}$ in Figure 1 (b)), respectively. As obviously from these figures, the yield stress at C and D on the compression side and yield stress at E and B on the tension side have a decreasing tendency as number of cycles increase, and as is clear from a comparison of Figures 2 and 3, these tendencies are remarkable as the value of pre-deformation large. On the other hand, results by conventional proof stress, namely, yield stresses at point C' and E' in Figure 1(b), are represented by blue curves in these figures. Hence, the estimated yield stresses at point C and E, namely, red curves, is larger as compared with the results by proof stress.

Figure 4 represents the experimental result in the case of low-strain amplitude ($\triangle e_I = \pm 0.003$ [-]). It is seen that the decreasing tendency of yield stress is small as compared with the results of Figure 3, where the value of pre-deformation is the same.

On the other hand, Figure 5 shows the the previous result in the case of pre-deformation of uniaxial tension (principal stretch λ = 1.3 [-] and strain amplitude $\triangle e_I = \pm 0.006$ [-]). From the comparison of the result of Figure 3, it is seen that the decreasing tendency of the yield stress at point D on the compression side is smaller than the yield stress at point B on the tension side. Moreover, it can be confirmed form this figure that the shape of stress-strain curve obtained during the process of tension after compression, which is surrounded red ellipse in Figure 5, is different. This is due to a difference in the direction that the principal axis of anisotropy of yield surface is formed (see Figure 1(c) and (d)).

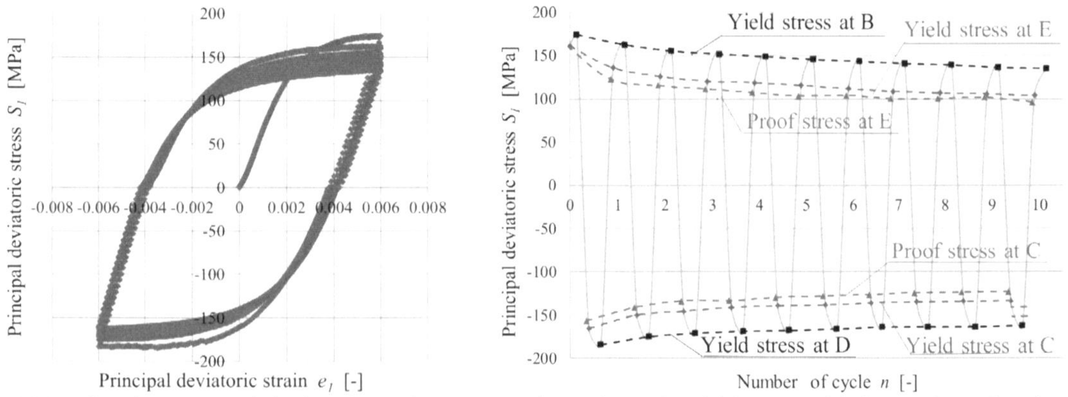

(a) Deviatoric stress and deviatoric strain. (b) Variation in yield stress with the number of cycles.

Figure 2. Yield stress in cyclic load for tension and compression after simple shear (λ=1.6, Δe_I = 0.006).

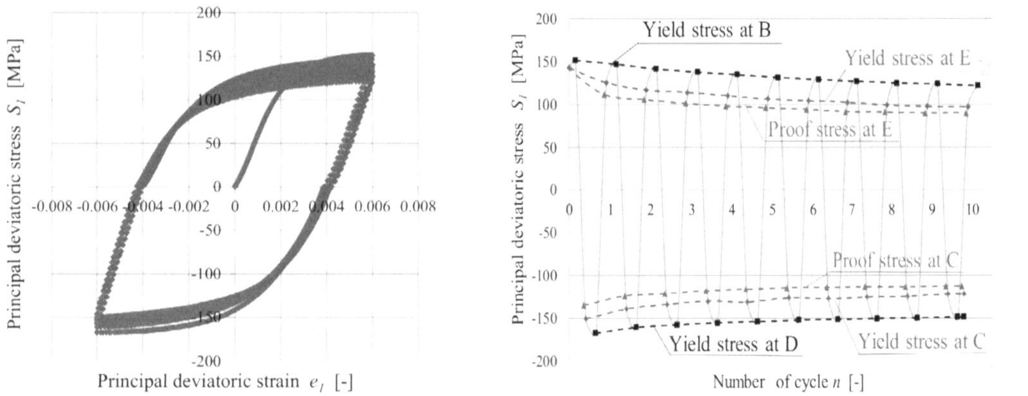

(a) Deviatoric stress and deviatoric strain. (b) Variation in yield stress with the number of cycles.

Figure 3. Yield stress in cyclic load for tension and compression after simple shear (λ=1.3, Δe_I = 0.006).

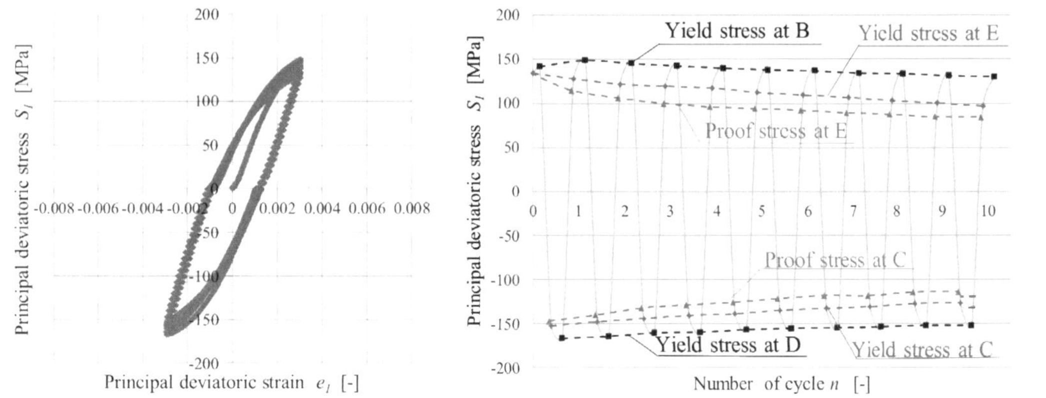

(a) Deviatoric stress and deviatoric strain. (b) Variation in yield stress with the number of cycles.

Figure 4. Yield stress in cyclic load for tension and compression after simple shear (λ=1.3, Δe_I = 0.003).

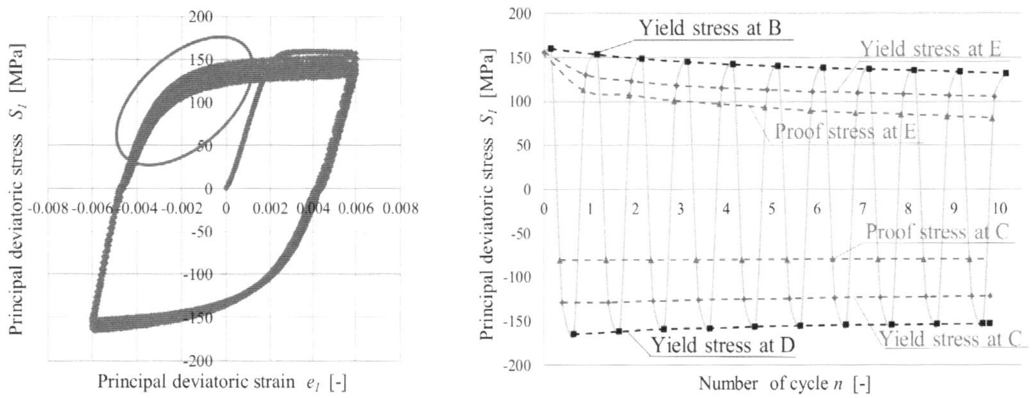

(a) Deviatoric stress and deviatoric strain. (b) Variation in yield stress with the number of cycles.

Figure 5. Yield stress in cyclic load for tension and compression after uniaxial tension (λ=1.3, Δe_I = 0.006).

6 CONCLUSIONS

In this study, using the test pieces, which was already applied the large simple shear, the yield stress under cyclic loads for tension and compression was estimated. As a result, the following conclusions are obtained.

(1) The yield stresses have a decreasing tendency with the number of cycle increase. However, the decreasing tendency reduces as the number of cycles increases, and they approach to the constant value.

(2) As a value of the pre-deformation of simple shear is larger, the decrease tendency of the yield stress at the tension side and compression side during the early stage of cycle load is more emphasized.

(3) The yield stress by proof stress is small as compared with the yield stress estimated in this study at both the tension side and the compression side.

(4) The decreasing tendency of the yield stress on the compression side in this experiment appears slightly strong as compared with results in the case of pre-deformation of uniaxial tension.

References

Kato, Y., and Kazama, A., Elasto-plastic Analysis using Natural Strain: Shape of Yield Surface Generated After Pre-deformation of Large Simple Shear, *Advanced Materials Research*, 875-877, 518-523, Feb 2013.

Kato, Y., *Natural Strained Yield Surface Shape Estimated under Pre-deformation of Tension and Torsion,* Sustainable Solutions in Structural Engineering and Construction, Chantawarangul, K., Suanpaga, W., Yazdani, S., Vimonsatit,V., Singh, A. (eds.), 153-158, ISEC Press, 2014.

Kato,Y., *Yield Stress Based on Natural Strain Theory under Cyclic Tensile-Compressive Loading after Large Uniaxial Tension*, Resilient Structures and Sustainable Construction, Pellicer, E., Adam, J. M., Yepes, V., and Singh, A. (eds.), ISEC Press, 2017.

Kato,Y., *Yield Stress Based on Natural Strain Theory Under Cyclic Shear Loading After Large Simple Shear*, Responsible Design and Delivery of the Constructed Project, Abdul-Malak, M., Khoury, H., Singh, A., and Yazdani, S. (eds.), ISEC Press, 2018.

Streamlining Information Transfer between Construction and Structural Engineering
Edited by Shiau, J., Vimonsatit, V., Yazdani, S., and Singh, A.

ISBN: 978-0-9960437-7-9

FINITE ELEMENT ANALYSIS OF HYBRID COLD-FORMED STEEL SHEAR WALL PANELS

NIMA USEFI, HAMID RONAGH, and MASOUD MOHAMMADI

Centre for Infrastructure Engineering, Western Sydney University, Sydney, Australia

Improving the lateral performance of cold formed steel systems by employing different sheathing materials and bracing systems has been of particular interest to researchers in these days. However, due to their relatively low seismic capacity, the need for higher shear resistance of these structural systems is still felt. Therefore, this study aims to propose three hybrid shear wall panels in order to provide better performance and higher resistance in seismic zones. The numerical method is verified based on the experimental results in the literature and then is used for evaluating the hybrid systems. The proposed walls are analyzed under lateral and vertical loads and then their performances are compared to each other. The results showed better performance and higher shear capacity for hybrid wall system compared to the ordinary cold formed steel shear wall. In addition, the strength to weight ratio approved the economic application of hybrid panels for high seismic regions.

Keywords: Hybrid system, Numerical method, Lateral performance, Hot rolled steel.

1 INTRODUCTION

In recent years, researchers have proposed various systems in order to increase the lateral performance of the cold-formed steel (CFS) frames (Sharafi *et al.* 2018). Application of sheathing boards such as plywood (Serrette and Nolan 2009), oriented strand board (OSB) (Nava and Serrette 2015), steel sheets (Javaheri-Tafti *et al.* 2014) and gypsum wall board (GWB) (Moghimi and Ronagh 2009b) can be mentioned as different methods of increasing lateral capacity. Different configurations of bracing systems through lateral load bearing spans including K brace (Zeynalian *et al.* 2012), Knee brace (Zeynalian and Ronagh 2011, 2012), and strap brace systems (Moghimi and Ronagh 2009a), are the other methods which can be used for increasing the lateral resistance of walls. Hybrid systems (Mortazavi *et al.* 2018, Usefi and Ronagh 2018) including CFS and hot-rolled steel to accommodate the advantages of both structural systems, are also new, interesting fields for enhancing the lateral capacity of CFS structures. In this study, three different hybrid wall panels (HWP) system are proposed and their lateral performances are compared with an ordinary CFS wall. A finite element (FE) method with ABAQUS package (Hibbitt *et al.* 2001) is utilized here for simulation of both hybrid and ordinary shear walls. First, the shear walls are analyzed under lateral load only, and then the effect of vertical load on these systems is also investigated.

2 VERIFICATION OF FINITE ELEMENT METHOD

Specimens C1 and C3 from an experimental work by Balh (2010) were used for verifying the FE method. More details about material properties and specimen dimensions can be found in the literature (Balh 2010). A fix restraint was used to simulate the fixity conditions such as the location of hold downs. The top track was assumed to have no displacement and rotation out of the wall plane. Fasteners with Cartesian criteria were utilized for modeling of screw connections. Based on the mesh convergence study for identifying an appropriate mesh density, the 15 mm with S4R shell element was utilized for modeling the members. Further details of modeling techniques for verifying FE method can be found in Usefi *et al.* (2018). The shear walls were simulated, and the results of shear resistance-lateral displacement of the walls were compared with the experimental data. As it is shown in Figure 1, the results of FE method have relatively good agreement with the experimental data, and therefore can be used for simulation of different wall panels. The differences between experimental and numerical results could be due to simplifications in FE method and experimental errors (Mohajeri Nav *et al.* 2018, Usefi *et al.* 2016).

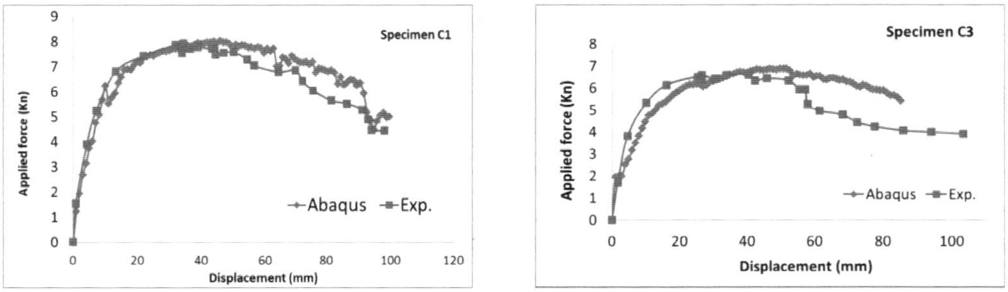

Figure 1. Comparison of experimental and FE results for specimens C1 and C3.

3 LATERAL CAPACITY OF HYBRID SHEAR WALLS

Based on the verified FE approach, a numerical simulation in ABAQUS platform was performed in order to compare the lateral capacity of different hybrid shear walls against an ordinary CFS shear wall. Three different hybrid systems and one ordinary strap braced CFS shear wall were employed for this comparison. The details of the walls and the materials used for simulation of members are given in Table 1 and Table 2, respectively.

Table 1. Details of the wall specimens.

Type of wall	Span (mm)	Stud dimension (mm)	Track dimension (mm)	Hot Rolled (mm)	Strap brace (mm)
CFS wall	2400×2400	89-36-0.75 Couples C section for chord Stud	92-36-1.15	----	90×1
Hybrid System 1		89-36-0.75		Square hollow section 89×89×3.5	150×2
Hybrid System 2					----
Hybrid System 3					

Table 2. Material properties of walls.

CFS members		Hot rolled Members		CFS strap	
Yield stress (MPa)	Ultimate stress (MPa)	Yield stress (MPa)	Plastic strain (MPa)	Yield stress (MPa)	Plastic strain (MPa)
364	400	400	480	360	420

For the ordinary CFS shear wall, coupled c sections were used as the chord studs. The configuration of walls is shown in Figure 2; the white color represents the hot rolled elements (square hollow section), the green color shows the CFS elements, and the red color indicates the strap brace in hybrid wall. A monotonic load, with displacement control method (target of 60 mm lateral displacement), was applied to the top of the walls in order to compare the load-displacement of the three different walls. First, the lateral performance of the walls was studied based on only applying lateral load. Then, in addition to lateral load, a vertical load of 32 kN was also applied to the top track of each wall, and the behavior of walls under both vertical and lateral load was assessed.

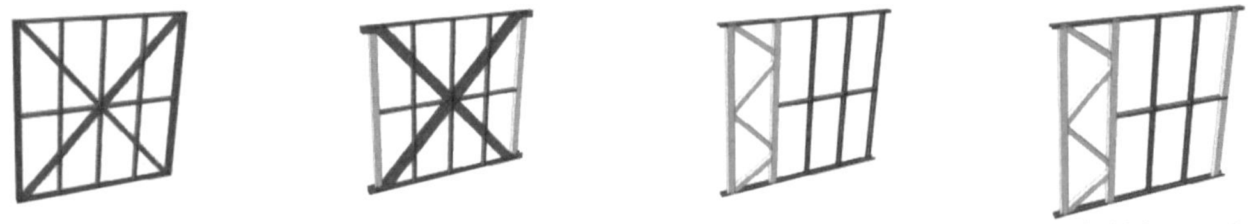

Ordinary CFS wall   Hybrid System 1   Hybrid System 2   Hybrid System 3

Figure 2. Configuration of walls.

3.1 Comparison of Walls Under Lateral Load Only

Lateral shear resistance of all four walls is shown in Figure 3. According to this figure, the application of hot rolled steel in hybrid system can significantly increase the lateral capacity of the CFS shear walls. The shear resistance of all three hybrid systems is around four times greater than the ordinary CFS shear wall. Moreover, the elastic stiffness of hybrid systems is roughly twice the stiffness of an ordinary CFS shear wall, which is a benefit for a hybrid system to be used in high seismic regions. All walls experienced a plastic deformation after reaching the 8 mm lateral displacement. The shear resistance for hybrid systems increased slightly after first yielding point, while for the ordinary CFS shear wall the resistance decreased after reaching to yield point. All three hybrid panels have approximately similar resistance and stiffness; however, the overall performance of hybrid system 2 and 3 is better compared to hybrid system 1.

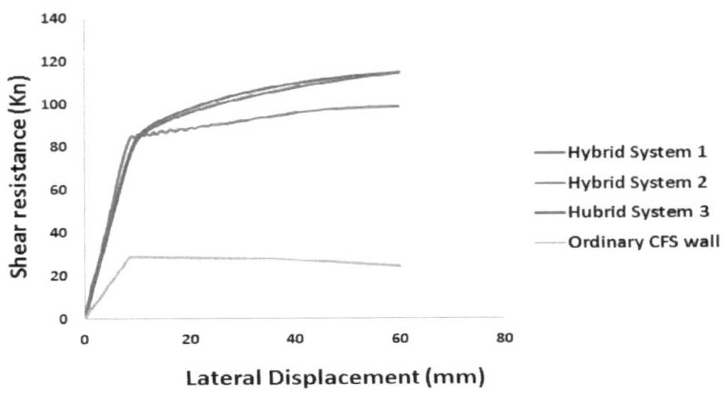

Figure 3. Shear resistance vs lateral displacement of the walls (vertical load ignored).

The results of von-mises stress are also presented in Figure 4. As it is indicated in this figure, the lateral load has been distributed in elements with higher stiffness such as hot rolled, bracing and chord stud elements. In hybrid system 2, only the left hot rolled frame resists the lateral load, while for the hybrid system 1 the overall load is distributed in hot rolled columns and bracing elements. Hybrid system 3 is in better condition compared to other walls since the two sides of the wall is composed of square hollow sections.

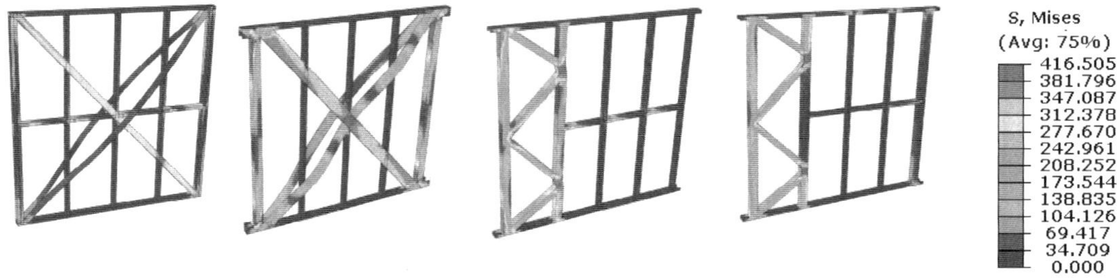

Ordinary CFS wall | Hybrid System 1 | Hybrid System 2 | Hybrid System 3

Figure 4. Comparison of von mises stress for shear walls (3D).

3.2 Comparison of Walls Under Lateral and Vertical Loads

A shear wall shall resist all dead, live, gravity and lateral loads in a real condition. Hence, in another attempt, the abovementioned walls were subjected to both vertical and lateral loading. The load displacement curves of the walls under lateral and vertical loading are presented in Figure 5.

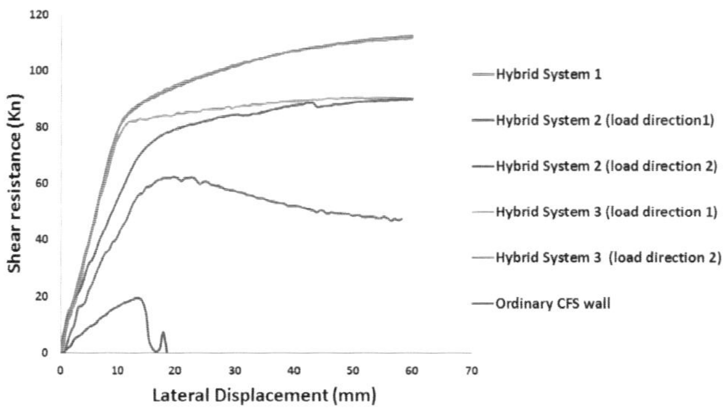

Figure 5. Shear resistance vs lateral displacement of the walls under lateral and vertical loads.

According to this figure, the capacity of all the shear walls has been decreased due to the applying vertical load on the top track. The performance of the ordinary CFS wall under lateral load is completely different from when both lateral and vertical loads are applied. Based on the results, the ordinary CFS wall under vertical and lateral loads failed at lateral displacement of 18 mm and the analysis of the wall was terminated. This indicates that this type of the ordinary CFS shear wall is not suitable for resisting both lateral and vertical loading. For hybrid system 2, the stiffness and shear strength decreased considerably when two loads were applied, while for the hybrid system 1 there is a slightly reduction for these values. As hybrid systems 2 and 3 are not symmetric, lateral load was applied in two directions and the results were compared. In hybrid system 2, when the loading is in direction of CFS frame, the capacity of wall is less than when the direction of load is along the hot rolled part. In hybrid system 3, there is no significant difference between two types of loading, which shows the high capability of this wall to be stable under lateral and vertical load.

Figure 6 shows the deformation and von-mises stress for all walls at the end of analysis under lateral and vertical loading. Hybrid systems 2 and 3 experienced a local failure in middle stud due to the presence of vertical load; however, hybrid system 3 was stable to the end of analysis with no major buckling. The ordinary CFS wall collapsed at the beginning of the analysis and could not undergo both vertical and lateral loads at the same time.

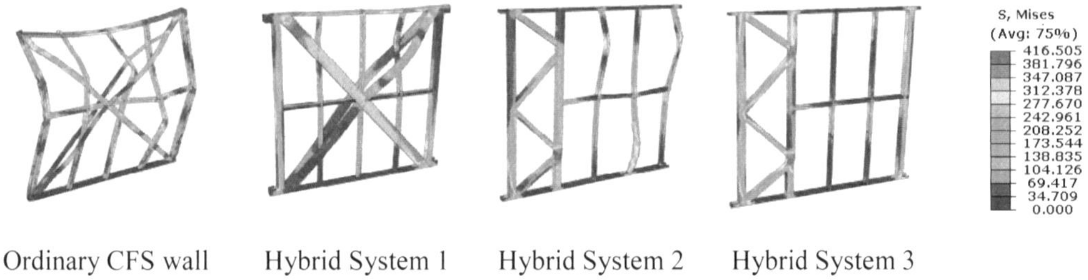

Figure 6. Comparison of von mises stress for shear walls under lateral and vertical load.

3.3 Strength to Weight Ratio of the Walls (Strength/Weight)

Since the weights of the hybrid systems and ordinary CFS shear walls are different, a criterion is required in order to reasonably compare the performance of the walls. Therefore, in this section, the strength to weight ratio of the walls was calculated and the results were compared with each other. The maximum elastic capacity of the walls was considered for the strength value. Table 3 shows the details of weight, maximum strength, and strength to weight ratio of the walls for both types of loadings. According to the strength/weight ratio of the walls, all hybrid systems have higher strength to weight ratio when both lateral and vertical loads are applied. On the other hands, the ordinary CFS wall has a very low strength to weight ratio, which is not reasonable nor suitable to be used in high seismic regions.

Table 3. Details of weight, maximum strength, and strength to weight ratio of the walls.

Type of wall	Weight (kg)	Maximum strength (lateral load) (kN)	Maximum strength (vertical & lateral load) (kN)	Strength to weight ratio (lateral load) (kN/kg)	Strength to weight ratio (lateral & vertical load) (kN/kg)
Hybrid System 1	91	83.3	82	0.91	0.9
Hybrid System 2	86	84	75	0.97	0.87
Hybrid System 3	102	85	90	0.83	0.88
Ordinary CFS wall	33	28	19	0.83	0.56

4 CONCLUSIONS

Lateral performance of three different hybrid cold-formed/ hot rolled steel shear walls was numerically investigated in this research. It was indicated that the application of square hollow section in the ordinary CFS walls can increase the maximum lateral capacity of the system as well as the stiffness. The results also showed that hybrid system 3 was in better condition compared to other walls when both lateral and vertical loads are applied to the wall. Experimental research on this field is still required to approve the advantageous of hybrid CFS shear walls.

References

Balh, N., Development of Seismic Design Provisions for Steel Sheathed Shear Walls, Masters Abstracts International, 2010.

Hibbitt, Karlsson, and Sorensen, *ABAQUS/Explicit: User's Manual,* v. 1, Hibbitt, Karlsson and Sorenson Incorporated, 2001.

Javaheri-Tafti, M. R., Ronagh, H. R., Behnamfar, F., and Memarzadeh, P., An Experimental Investigation on the Seismic Behavior of Cold-Formed Steel Walls Sheathed by Thin Steel Plates, *Thin-Walled Structures,* 80, 66-79, 2014.

Moghimi, H., and Ronagh, H. R., Better Connection Details for Strap-Braced CFS Stud Walls in Seismic Regions, *Thin-Walled Structures,* 47, 122-135, 2009a.

Moghimi, H., and Ronagh, H. R., Performance of light-Gauge Cold-Formed Steel Strap-Braced Stud Walls Subjected to Cyclic Loading, *Engineering Structures,* 31, 69-83, 2009b.

Mohajeri Nav, F., Usefi, N., and Abbasnia, R., Analytical Investigation of Reinforced Concrete Frames under Middle Column Removal Scenario, *Advances in Structural Engineering,* 21, 1388-1401, 2018.

Mortazavi, M., Sharafi, P., Ronagh, H., Samali, B., and Kildashti, K., Lateral Behaviour of Hybrid Cold-Formed and Hot-Rolled Steel Wall Systems: Experimental Investigation, *Journal of Constructional Steel Research,* 147, 422-432, 2018.

Nava, M., and Serrette, R., Strength Reduction Factors for High-Aspect-Ratio OSB Cold-Formed Steel Frame Shear Walls, *Journal of Structural Engineering,* 141, 04014148, 2015.

Serrette, R., and Nolan, D. P., Reversed Cyclic Performance of Shear Walls with Wood Panels Attached to Cold-Formed Steel with Pins, *Journal of Structural Engineering,* 135, 959-967, 2009.

Sharafi, P., Mortazavi, M., Usefi, N., Kildashti, K., Ronagh, H., and Samali, B., Lateral Force Resisting Systems in Lightweight Steel Frames: Recent Research Advances, *Thin-Walled Structures,* 130, 231-253, 2018.

Usefi, N., Nav, F. M., and Abbasnia, R., Finite Element Analysis of RC Elements in Progressive Collapse Scenario, *Gradevinar,* 68, 1009-1022, 2016.

Usefi, N., and Ronagh, H., *Numerical Evaluation of a New Cold Formed Steel Shear Wall Panel,* The 25th Australasian Conference on Mechanics of Structures and Materials (ACMSM25*),* Brisbane, Queensland, Australia, 2018.

Usefi, N., Ronagh, H. R., Kildashti, K., and Samali, B., *Macro/Micro Analysis of Cold-Formed Steel Members Using ABAQUS and OPENSEES,* Volume of Abstracts: 13th International Conference on Steel, Space and Composite Structures (SS18), The University of Western Australia, Perth, Australia, Jan 31-Feb 2, 2018.

Zeynalian, M., and Ronagh, H. R., A Numerical Study on Seismic Characteristics of Knee-Braced Cold Formed Steel Shear Walls, *Thin-Walled Structures,* 49, 1517-1525, 2011.

Zeynalian, M., and Ronagh, H. R., An Experimental Investigation on The Lateral Behavior of Knee-Braced Cold-Formed Steel Shear Walls, *Thin-Walled Structures,* 51, 64-75, 2012.

Zeynalian, M., Ronagh, H. R., and Hatami, S., Seismic Characteristics of K-Braced Cold-Formed Steel Shear Walls, *Journal of Constructional Steel Research,* 77, 23-31, 2012.

Streamlining Information Transfer between Construction and Structural Engineering
Edited by Shiau, J., Vimonsatit, V., Yazdani, S., and Singh, A.

ISBN: 978-0-9960437-7-9

RETROFITTING OF RC BEAMS USING ULTRA HIGH PERFORMANCE FIBER REINFORCED CONCRETE

VARUN GARG, PREM PAL BANSAL, and RAJU SHARMA

Dept of Civil Engineering, Thapar Institute of Engineering and Technology, Patiala, India

Shear deficient structures have been strengthened or repaired by using various methods, e.g., external pre-stressing, shotcreting, steel plate bonding, polymer impregnation, fiber reinforced polymers (FRPs). But there were problems with each of the above techniques as the use of FRP gives sudden failure due to de-bonding, steel plates got corroded after some time and several losses in the case of external pre-stressing. To overcome all the above problems faced in different retrofitting techniques, a new cementitious material with ultra high performance characteristics like very high compressive strength (above 110 MPa), high tensile and flexural strength due to the addition of fibers, very less permeable and high workability i.e. UHP-FRC came into existence. In the present study, shear deficient RC beams initially stressed to a prefixed percentage, i.e. 60% of the ultimate failure load are retrofitted using UHP-FRC. Retrofitting is done using side face retrofitting. From the study it is seen that the ultimate load carrying capacity of RC beams retrofitted with UHP-FRC is significantly increased with side face retrofitting. Moreover, this technique also increased the ductility and energy absorption of RC beams which resulted in the usefulness of UHP-FRC as a retrofitted material.

Keywords: Shear deficient beam, Initially stressed RC beams, De-bonding, Under reinforced concrete beam, Over reinforced concrete beam.

1 INTRODUCTION

There are two primary modes of failure in reinforced concrete (RC) beams, flexural failure and shear failure. Flexural failure of a beam is ductile in nature, whereas shear failure is brittle in nature and does not allow substantial redistribution of loads. Thus, shear failure occurs without any prior warning and is often catastrophic. Shear failure of RC structures may be due to many factors, e.g., insufficient shear reinforcement, reduction of the steel area due to corrosion and spalling of concrete caused by aggressive environmental conditions, increased service load due to change-in-usage of the structure, and any detailing, design, and/or construction error. Various materials and techniques such as shotcreting, steel plate, polymers, FRP and pre-stressed FRP, anchor bolt being used to retrofit the shear deficient structure. But the limitation of each material forced the researchers to develop the new retrofit material. The last four decade's efforts have now established a new cementitious material with ultra high performance characteristics i.e. UHP-FRC.

UHP-FRC is the outcome of a demand that began in the 1930s to seek for approaches to increase the mechanical properties of concrete, in particular its compressive strength. UHP-FRC can be successfully used when durable and strong but light members are required, or in cases where a considerable liberty with regard to form is desirable. Aïtcin (1998) reported that the world's first engineering structure designed with UHPC was the Sherbrooke footbridge in Sherbrooke, Quebec, built in 1997. Brühwiler and Denarié (2008) evaluate the behavior of 30 mm thick UHP-FRC overlay for the rehabilitation of short span road bridge with heavy traffic. The protective function of the UHP-FRC overlay was verified by air permeability tests which confirmed the extremely low permeability; approximately 30 times lower than normal concrete. Tayeh *et al.* (2012) experimentally evaluated the relationship between concrete substrate roughness and the bonding performance of UHP-FRC used as a repair material. The results showed the significant influence of substrate surface preparation method on bonding strength between UHP-FRC and the concrete substrate. The composite with sand blasted surface behaved closely as a monolithic structure.

In the present study shear deficient RC beams initially stressed to a prefixed percentage, i.e. 60% of the ultimate failure load are retrofitted using UHP-FRC. The side face retrofitting is used to retrofit the shear deficient beam. From the study, it is seen that the ultimate load carrying capacity of RC beams retrofitted with UHP-FRC is significantly increased with side face retrofitting. Moreover, this technique also increased the ductility, energy absorption and initial stiffness of RC beams which resulted in the usefulness of UHP-FRC as a retrofitted material.

2 MATERIAL CHARACTERIZATION

The ordinary Portland 43 grade cement is used in this study. The specific gravity, initial and final setting time of cement is 3.04, 40 min and 320 min respectively. The standard consistency is 33% and 28 day compressive strength is 44.5 MPa. The specific gravity and fineness modulus of 20 mm aggregates are 2.61 and 6.87 respectively. The specific gravity and fineness modulus (F.M) of 10 mm aggregates were 2.68 and 6.37 respectively. The specific gravity, F.M and water absorption of sand are 2.46, 2.56 and 0.83% respectively. Silica fume is obtained from KGR Agro Fusions (P) Ltd., Ludhiana, Punjab. To develop UHPC mixes, polycarboxylic ether based MASTERGLENIUM SKY 8866 Superplasticizer was obtained from BASF solutions, Chandigarh. The crimped steel fiber and hooked steel fiber (SHAKTIMAN ® Steel Fibers) are used which were obtained from Stewols India (P) Ltd., Nagpur. The metakaolin and epoxy were also used in the experimental study. HYSD steel of grade Fe-500 D (Tata Tiscon) of 16 mm, 12 mm and 8 mm diameters were used as reinforcement. The yield strength and ultimate strength of the reinforcement are 540 N/mm^2 and 600 N/mm^2 respectively. In the current study potable water available in the material testing laboratory was used for making concrete.

2.1 Concrete Mix Design

The M25 concrete mix was designed by using the properties of cement, fine aggregate and crushed stone coarse aggregates as per are: IS 10262 (2009) guidelines. The mix proportions are given in Table 1. The cubes were prepared with this mix proportions and were tested at the age of 7 days and 28 days. The average compressive strength of cubes at 7 days and 28 days was 20.9 MPa and 34.40 MPa respectively. The cement, fine aggregate, coarse aggregate and water in M 25 grade concrete are 417.9 kg, 544.38 kg, 1190 kg and 192 kg respectively. UHP-FRC was used for retrofitting of the beams. For the development of UHP-FRC, various trials were done, out of which three best trials were taken by varying amount of superplasticizer, metakaolin and use of different sand particle sizes to densify the matrix to achieve the target compressive strength of 120 MPa. These are shown in Table 1.

Table 1. Mix Trials for the Development of UHP-FRC.

	Trial Mix 1	**Trial Mix 2**	**Trial Mix 3**
Items	**Quantity(Kg/m^3)**	**Quantity(Kg/m^3)**	**Quantity(Kg/m^3)**
Cement	875	875	1000
Sand (<300μm)	-	-	270
Sand (<600μm)	270	270	-
Sand (<300μm-	-	-	750
Sand(1.18-2.26mm)	1055	1055	-
Silica fume	300	300	150
Metakaoline	-	-	150
Superplasticizer	45	35	45
Fibers (hooked)	1.5% by vol. of concrete	1.5% by vol. of concrete	1.5% by vol. of concrete
Fibers (crimmed)	0.5% by vol. of concrete	0.5% by vol. of concrete	0.5% by vol. of concrete
Water	200	200	230
7 Day Strength(MPa)	48	52	75.8
28 Day Strength(MPa)	75	79	118

3 EXPERIMENTAL PROGRAM

3.1 Casting of Beams

RCC beams of size 700 x 150 x 150 mm are designed and cast using M25 grade concrete and Fe 500D grade steel. The RCC beam is designed using the limit state method and both under-reinforced and over-reinforced sections are designed. The beam is designed having 2 steel bars of 8 mm diameter at compression face and 2 bars of 16 mm diameter at tension face for over-reinforced section; 2 bars of 12 mm diameter at tension face for under-reinforced section and the stirrups used are of 8 mm diameter and at the spacing of 85 mm for under-reinforced section and at 75 mm for the over-reinforced section.

3.2 Testing Arrangement

The beams were tested after 28 days from the date of casting. All the specimens were tested as simply supported beams subject to a two-point loading with loads applied at a distance of L/3 from each end; where L is the effective span i.e. 600 mm. The shear span to effective depth ratio (a/d) was equal to 1.6. The testing of beams was carried out in the Universal testing machine (UTM) of 1000 kN load capacity.

3.3 Process of Retrofitting

Firstly, all beams are initially stressed to 60% of its ultimate load carrying capacity and then retrofitted with UHP-FRC using epoxy as a bonding agent between old concrete and UHP-FRC. Retrofitting of beams is done in three stages- chipping off side cover to expose stirrups, the epoxy layer applied and then finally retrofitting with the new UHP-FRC layer. Inside face retrofitting, all the three stages are shown in the Figure 1. The section is retrofitted till it is restored to its original dimensions of 150x150 mm with UHP-FRC.

Figure 1. a) Side face chipped off beam (Stage 1); b) Epoxy layer applied after chipping off (Stage 2); c) Beam retrofitted with UHP-FRC (Stage 3).

4 RESULTS AND DISCUSSION

The beams are tested to failure and deflection at the center of the span is recorded using dial gauges. The results of the beams CB1 and CB2 are then averaged to obtain a single set of results (designated as beam CB-U/R) for under-reinforced section control beam and (designated as beam CB-O/R) for over-reinforced section control beam. The load-deflection curves for both beams (CB-U/R and CB-O/R) are presented in Figure 2 and Figure 3. The first flexural crack in beams CB-U/R and CB-O/R was observed at a load of 32 kN and 48 kN respectively, and it occurred between the left point load and the center of the beam respectively.

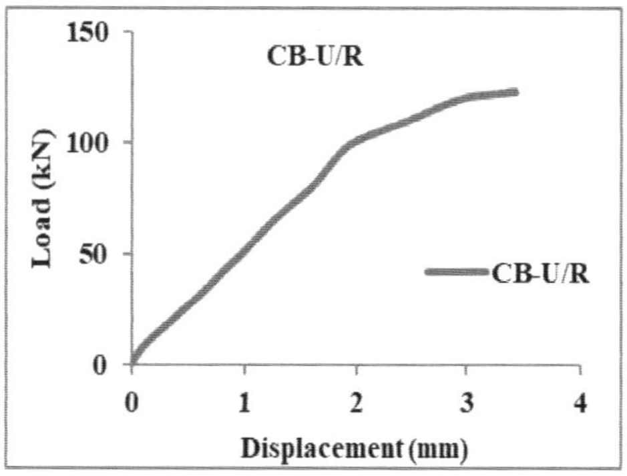

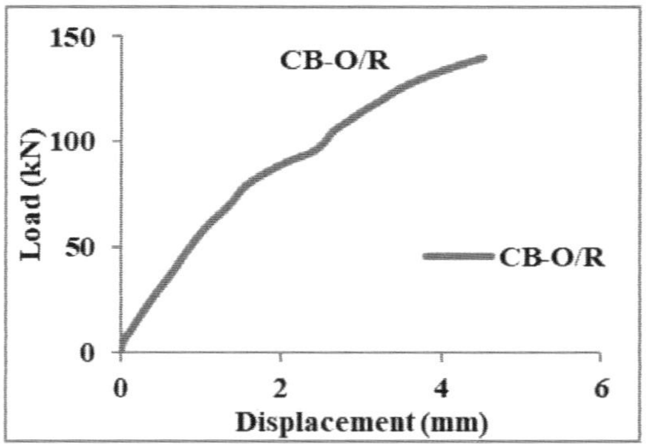

Figure 2. Load v/s mid-span deflection of CB-O/R. Figure 3. Load v/s mid-span deflection of CB-U/R.

The shear cracks started appearing on the beam in shear span at a load of 70 kN and 84 kN for CB-U/R and CB-O/R beams respectively. On increasing the load further, shear cracks started to widen and propagate upward toward the point at which loads are applied to the beam and finally beam failed in shear with a maximum deflection at the center 3.42 mm and 4.52 mm for beams CB-U/R and CB-O/R respectively.

4.1 Effect of Retrofitting of Strength of Beam

The beams were initially stressed to 60 % of their corresponding ultimate load i.e.73.8 kN for U/R section and 84 kN for O/R section beams and then retrofitted using UHP-FRC by side face retrofitting techniques. The retrofitted beams were then tested to failure and result of each beam was recorded. It has been observed that after retrofitting the first crack load, yield load, initial stiffness, and maximum load carrying capacity increased. After retrofitting the maximum load carrying capacity was increased to 147.9 kN for SF-U/R from 123 kN (CB-U/R) in side face retrofitting technique whereas for O/R side face retrofitted beam the load carrying capacity increased to 160.75 kN for SF-O/R from 140 kN (CB-O/R). In addition to that, the first crack load of CB-U/R shifted from 32 kN to 36.75 kN respectively, whereas, the yield load shifted from 115 kN to 121.6 kN. The first crack load of CB-O/R shifted from 140 kN to 160.75 kN whereas the yield load shifted from 140 kN to 160.75 kN.

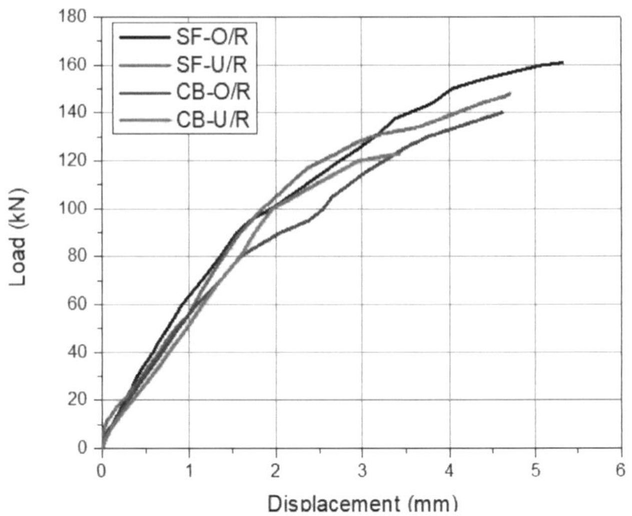

Figure 4. Load vs mid-span deflection for all control and retrofitted specimen.

Retrofitting of beams (SF-U/R) and (SF-O/R) resulted in 20.24% and 14.82% improvement in load carrying capacity respectively. It has been observed that the mid-span deflection of the retrofitted beams increased. The possible reasons for the above results are (a) UHP-FRC has a higher compressive strength as compared to control concrete (118 MPa v/s 25 MPa). In addition, the higher modulus of elasticity of UHP-FRC confined the initial damaged beam from the side faces (b) The hybrid fibers able to retard and delay the crack propagation so that the load carrying capacity of side face retrofitted initially stressed beam is increased.

4.2 Effect of Retrofitting on Ductility and Energy Dissipation for Control U/R, Control O/R and Side Face Retrofitted Specimen

To measure ductility, it is generally accepted to use ductility factor (μ) which represents the ratio of deflection (Δ) at the ultimate load to corresponding property when the reinforcement starts yielding (Afefy and Mahmoud 2014). It is observed that the ductility increased when the beams are retrofitted by Side Face for U/R section beams. For over-reinforced section, beams failed in shear before the steel yields so the ultimate load and the yield load are same. It can also be noted that the ductility values are 1 for O/R section beams, whereas, the ductility of control section increased from 1.266 to 1.708. The energy absorption of CB-U/R is 266.14 jouls and increases 77% after side face retrofitting. Likewise, CB-O/R the energy absorption is 401.96 jouls and 44.23% increment is observed after side face retrofitting. Initial stiffness is calculated from the initial slope of the load-deflection curve. The initial stiffness of CB-U/R and CB-O/R is 42.53 kN/mm and 51.5 kN/mm respectively. The 21.09% and 15.84% increment in initial stiffness of SF-U/R and SF-O/R is observed.

5 CONCLUSIONS

Based on the results of an experimental study carried out the following conclusions can be drawn;

- The yield load, first crack load, and load carrying capacity increases when the initially stressed (60%) beams are retrofitted with UHP-FRC.
- Retrofitting of initially damaged beams by replacing side face concrete cover with UHP-FRC in the shear span increases the ultimate load carrying capacity and yield load of under-reinforced (U/R) section beams by 20.24% and 5.74% respectively.
- The deflection at the failure load increased when beams are retrofitted with UHP-FRC in the side face retrofitting configurations.
- Maximum ductility and energy absorption capacity have been observed for side face retrofitting. The corresponding value of ductility factor (μ) is 1.70 and increment in energy absorption is 77.05% for U/R section beam and 44.23% for O/R section beams. The initial stiffness of the retrofitted beams also increased as compared to control beams.

References

Aïtcin, P. C., *High Performance Concrete*, 1st edn, E & FN Spon, 1998.

Afefy, H. M. E. and Mahmoud, M. H., Structural Performance of RC Slabs Provided by Pre-cast ECC Strips in Tension Cover Zone, *Construction and Building Materials*, Elsevier, 65, 103-113, 2014.

Brühwiler, E. and Denarié, E., *Rehabilitation of Concrete Structures Using Ultra-High Performance Fiber Reinforced Concrete*. Second International Symposium on Ultra High Performance Concrete, Kassel, Germany, 2008.

IS 10262, *Indian Standard Concrete Mix Proportioning- Guidelines,* Bureau of Indian Standards (BIS): July 2009.

Tayeh, B. A., Bakar, B. H. A. Johari, M. A. M., and Voo, Y. L., Mechanical and Permeability Properties of The Interface Between Normal Concrete Substrate and Ultra-High Performance Fibre Concrete Overlay, *Construction Building Material,* Elsevier, 36, 538-548, 2012.

Streamlining Information Transfer between Construction and Structural Engineering
Edited by Shiau, J., Vimonsatit, V., Yazdani, S., and Singh, A.

ISBN: 978-0-9960437-7-9

FLEXURAL STRENGTH OF SQUARE CFT BEAM-COLUMNS WHEN DAMAGED WITH CONCRETE CRACKS

DAIJIRO FUJIOKA[1], MASAE KIDO[1], MAO LIU[2], and HIROKI IKEDA[1]

[1]*Dept of Architecture, University of Kitakyushu, Kitakyushu, Japan*
[2]*Dept of Civil Engineering and Architecture, National Institute of Technology, Tokuyama College, Shunan, Japan*

The lateral stiffness of CFT beam-columns is considered to decrease because of the damage by the concrete crack or yielding of the materials however the relationship between two phenomena is not clear. It is necessary to clarify the strength when the CFT columns damaged by the concrete crack for evaluating the lateral stiffness of CFT beam-columns properly. The purposes of this study are to calculate the flexural crack strength of a square CFT section and to compare them with the yield strength. The analysis is carried out as follows; 1) assuming the stress-strain relationships of materials and the strain and stress distribution of the cross-section, 2) setting the limit of the tensile strain of concrete and 3) calculating the relation between the axial force and the bending moment when the strain of concrete reaches the limit tensile strain. Parameters are width-thickness ratio, the yield strength of a steel tube and compressive strength of concrete. As a result of the analysis, we showed that the axial force ratio is 0.2 ~0.3 when the flexural crack strength and the yield strength are the same. It is concluded that the lateral stiffness decreases because of the crack of concrete – and not from the yielding of materials -- when the axial force ratio is smaller than 0.2~0.3.

Keywords: Axial force ratio, Bending analysis, Lateral stiffness, Restoring force characteristics, Steel-concrete composite structures, Tensile strength, Yield strength.

1 PURPOSE

The skeleton curve of concrete-filled steel tube (hereafter CFT) members are shown in AIJ CFT Recommendations (AIJ 2008). The first break point is the yield strength, and the second break point is the ultimate bending moment. Until the first break point, the initial lateral stiffness is calculated on the condition that all cross sections are effective and elastic in the CFT Recommendations. However, the effect of the concrete cracks is not considered and there is a possibility that the deformation of the CFT beam-column is evaluated conservatively.

Regarding a decrease of the lateral stiffness, the following can be considered. When the CFT columns are subjected to the axial force and the flexural moment, and the axial force ratio is relatively small, firstly the concrete cracks occur, and then the lateral stiffness decreases due to concrete cracking. As the bending moment further increases, a part of the cross section yields and the lateral stiffness further decreases. Where "a part of the cross section yields" means a steel tube yields or the stress of concrete reaches two-thirds of the compressive strength. On the other hand, when the axial force ratio is relatively large before the cracks occur, a part of the cross section yields, and the lateral stiffness decreases. As described above, it is presumed that the factors resulted in the decrease of lateral stiffness depends on the axial force ratio. However, it is not clear that the strength when the CFT columns are damaged by the concrete cracks, and moreover the relationship between two strengths, when the CFT beam-columns are damaged by the concrete cracks and yield. For the proper evaluation of the lateral stiffness of the CFT beam-columns, it is necessary to clarify the relationship between the axial force and flexural moment of the CFT columns damaged by the concrete cracks, and the relationship between the crack moment and the yield moment. In the previous research, it has been shown that the parameters such as the material strength and width-thickness ratio affect the yield strength of the CFT columns (Liu *et al.* 2011).

The purposes of this study are to calculate the flexural moment when the square CFT beam-column is damaged by the concrete cracks and to show the crack flexural moment on the *M-N* relationships. The effects of the material strength and width-thickness ratio on the crack flexural moment are clarified. The correspondence of the crack flexural strength to the yield strength is shown. The range of the axial force ratio when the two strengths are the same is clarified.

2 METHOD

2.1 Analysis Model and Analysis Condition

Figure 1 shows the CFT cross-section used in the analysis. The cross-section consists of the rectangular steel tube whose cross section (width) is ${}_{s}D$ and the plate thickness is t, and it is subjected to the axial force N and the flexural moment M. Figure 2 shows the strain distribution and the stress distribution of the concrete portion when the tensile edge stress of the concrete portion has reached the tensile strength ${}_{c}\sigma_{cr}$ of concrete. ${}_{c}D$ is the depth/width of the concrete part.

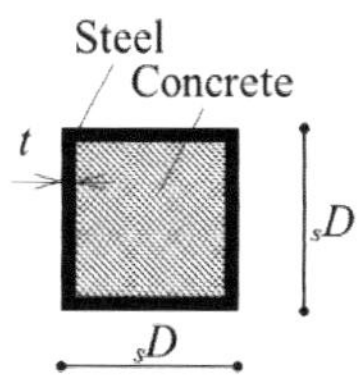

Figure 1. Beam-column cross-section.

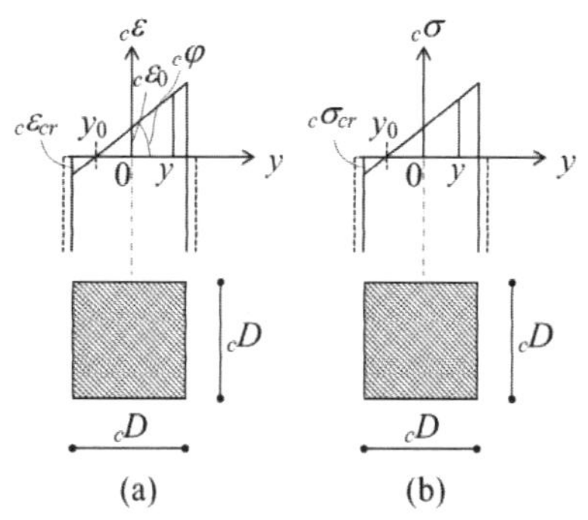

Figure 2. The strain distribution and stress distribution of the concrete part.

In this study, we use the tensile strength ${}_c\sigma_{cr}$ recommended by JSCE (2017).

$$ {}_c\sigma_{cr} = 0.23 \cdot {}_c\sigma_B^{\frac{2}{3}} \tag{1} $$

The tensile strength of concrete is ${}_c\sigma_{cr}$ = 3.53N/mm^2, when concrete compressive strength ${}_c\sigma_B$ = 60N/mm^2. The strain at the tensile strength ${}_c\varepsilon_{cr}$ is about 0.01% although it depends on Young's modulus. In addition, the stress-strain relationship of concrete was assumed to be linear in both compression and tension for simplifying the analysis. Figure 3 shows the linear stress-strain relationship when the compressive strength of the concrete ${}_c\sigma_B$ = 60 and 100N/mm^2. The Young's modulus ${}_cE$ of concrete was calculated by using the equation shown in AIJ RC Standard (AIJ 2010). The value of Young's modulus ${}_cE$ is also shown in Figure 3.

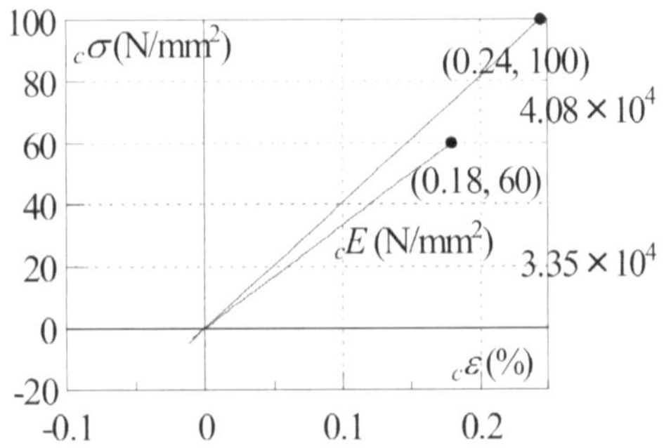

Figure 3. The stress-strain relationship of concrete.

2.2 Flexural Moment and Axial Force

2.2.1 *Concrete portion*

When the coordinate axis is taken as shown in Figure 2(a) the strain at the position of y expressed by the Eq. (2) according to the Navier's hypothesis.

$$ {}_c\varepsilon(y) = {}_c\varepsilon_0 + y\,{}_c\varphi \tag{2} $$

From the definition of the section force, the axial force ${}_cN$ and the flexural moment ${}_cM$ of the concrete part can be calculated. Furthermore, using the following various quantities and dimensionless quantities, the dimensionless axial force, and the dimensionless flexural moment ${}_cn$, ${}_cm$ are given by the Eq. (3) respectively. Where ${}_c\varepsilon_0$ is the strain of the concrete at the axis of the member, ${}_c\varepsilon_m$ is the strain at the compressive strength, ${}_c\varphi$ is the curvature and y_0 is the distance from the axis of the member to the neutral axis.

$$ {}_c\tilde{\varepsilon}_0 = {}_c\varepsilon_0 / {}_c\varepsilon_m,\ {}_c\tilde{\varphi} = {}_c\varphi\,{}_cD / {}_c\varepsilon_m,\ {}_cN_0 = {}_cD^2 \cdot {}_c\sigma_B, $$
$$ {}_cM_0 = {}_cD^3 \cdot {}_c\sigma_B,\ {}_cn = {}_cN / {}_cN_0,\ {}_cm = {}_cM / {}_cM_0 $$
$$ {}_cn = \frac{{}_cE \cdot {}_c\varepsilon_m \cdot {}_c\tilde{\varepsilon}_0}{{}_c\sigma_B},\quad {}_cm = \frac{{}_cE \cdot {}_c\varepsilon_m \cdot {}_c\tilde{\varphi}}{12\,{}_c\sigma_B} \tag{3} $$

2.2.2 *Steel tube portion*

The stress-strain relationship of the elastic-perfectly plastic model was used for the steel tube portion. The dimensionless axial force and the dimensionless flexural moment ${}_sn$ and ${}_sm$ of the steel tube portion are given by the Eq. (4).

$$\left.\begin{aligned} {}_sn &= \frac{{}_sN}{{}_sN_y} = \frac{{}_c\varepsilon_m}{{}_s\varepsilon_y}\,{}_c\tilde{\varepsilon}_0 \\ {}_sm &= \frac{{}_sM}{{}_sM_y} = \frac{1}{2}\frac{{}_sD}{{}_cD}\frac{{}_c\varepsilon_m}{{}_s\varepsilon_y}\,{}_c\tilde{\varphi} \end{aligned}\right\} \tag{4}$$

Where, ${}_sN_y = {}_sA\sigma_y$ (${}_sA$ is the sectional area of the steel tube), ${}_sM_y = Z\sigma_y$ (Z is the section modulus of the steel tube) and ${}_s\varepsilon_y$ is the yielding strain of the steel tube.

2.3 Crack Condition

Assuming that the flexural crack occurs in the CFT column when the strain at the tensile edge of the concrete reaches the strain ${}_c\varepsilon_{cr}$ at the tensile strength of the concrete, the following equation is derived by the Eq. (2).

$${}_c\varepsilon\left(-\frac{{}_cD}{2}\right) = {}_c\varepsilon_0 - \frac{{}_cD}{2}\,{}_c\varphi = {}_c\varepsilon_{cr} \tag{5}$$

Therefore, the crack condition is given by the following expression with dimensionless quantities.

$${}_c\tilde{\varepsilon}_0 - \frac{{}_c\tilde{\varphi}}{2} = \frac{{}_c\varepsilon_{cr}}{{}_c\varepsilon_m},\quad {}_c\tilde{\varphi} = \frac{1}{{}_c\tilde{\varepsilon}_0/{}_c\tilde{\varphi} - 1/2}\cdot\frac{{}_c\varepsilon_{cr}}{{}_c\varepsilon_m} \tag{6}$$

2.4 Flexural Crack Strength of CFT Column

The flexural crack strength of CFT column M and N are obtained by the following procedure; 1) calculating the value of ${}_c\tilde{\varepsilon}_0$ and ${}_c\tilde{\varphi}$ by the Eq. (6) with varying the value of ${}_c\tilde{\varepsilon}_0/{}_c\tilde{\varphi}$ from -0.5 to 0.5, which is the condition that the neutral axis position is in the concrete portion, 2) calculating ${}_cn$, ${}_cm$, ${}_sn$, and ${}_sm$ by substituting ${}_c\tilde{\varepsilon}_0$ and ${}_c\tilde{\varphi}$ into Eqs. (3) and (4). The dimensionless crack flexural moment m and the axial force n of CFT cross section is calculated by the following equation. ${}_cA$ is the sectional area of the concrete portion.

$$\left.\begin{aligned} n &= \frac{N}{N_0} = \frac{{}_cn\cdot{}_cN_0 + {}_sn\cdot{}_sN_y}{{}_sA\cdot\sigma_y + {}_cA\cdot{}_c\sigma_B} \\ m &= \frac{M}{{}_sM_y} = \frac{{}_cm\cdot{}_cM_0 + {}_sm\cdot{}_sM_y}{{}_sM_y} \end{aligned}\right\} \tag{7}$$

3 RESULTS AND DISCUSSION

3.1 Analytical Parameters

Analysis parameters are as follows:

- Width to thickness ratio of square steel tube ${}_sD/t = 20, 40$
- Yield strength of square steel tube $\sigma_y = 325, 440$ N/mm^2
- Concrete compression strength ${}_c\sigma_B = 60, 100$ N/mm^2

3.2 Results of Calculation

Figure 4 shows the crack strength on the *m-n* relationship. Also, the yield strength obtained by the method presented in the past research (Liu *et al.* 2011) is indicated by thin lines. The yield strength is determined by the lower of the strengths the stress of the concrete reaches $2/3{}_c\sigma_B$ or the steel tube yielding. When calculating the yield strength, the stress-strain relationship of the concrete is assumed by a quadratic function and the strain ${}_c\varepsilon_m$ at the compressive strength is 0.2% and 0.25%.

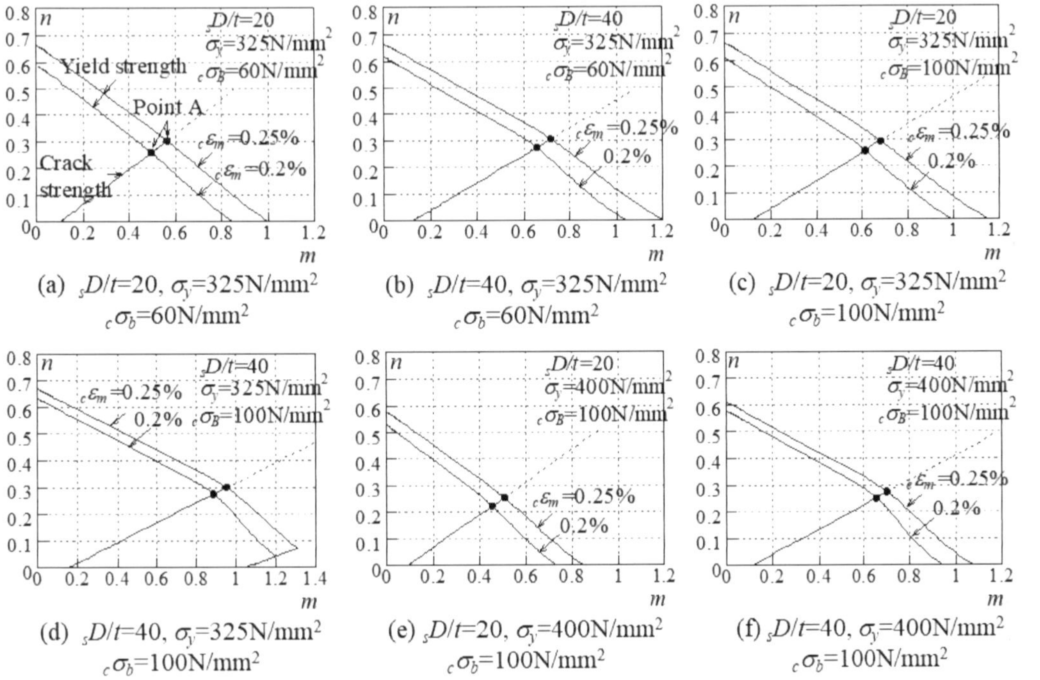

(a) ${}_sD/t$=20, σ_y=325N/mm^2 ${}_c\sigma_b$=60N/mm^2 (b) ${}_sD/t$=40, σ_y=325N/mm^2 ${}_c\sigma_b$=60N/mm^2 (c) ${}_sD/t$=20, σ_y=325N/mm^2 ${}_c\sigma_b$=100N/mm^2

(d) ${}_sD/t$=40, σ_y=325N/mm^2 ${}_c\sigma_b$=100N/mm^2 (e) ${}_sD/t$=20, σ_y=400N/mm^2 ${}_c\sigma_b$=100N/mm^2 (f) ${}_sD/t$=40, σ_y=400N/mm^2 ${}_c\sigma_b$=100N/mm^2

Figure 4. The bending moment-axial force relationship.

3.3 Effect of the Parameters on the Crack Flexural Strength

The effect of parameters on the crack strength is discussed. With regards to the effects of the width-thickness ratio ${}_sD/t$ and comparing Figures 4 (a) with (b), (c) with (d), (e) with (f) respectively, it is observed that the value of the crack flexural moment *m* at the same axial force ratio increases as ${}_sD/t$ increases. With regards to the effect of the compressive strength of concrete ${}_c\sigma_B$ and comparing Figures 4 (a) with (c), (b) with (d) respectively, as ${}_c\sigma_B$ increases, it is observed that the value of *m* at the same axial force ratio increases. As for the effect of the steel tube yield strength σ_y and comparing Figures 4 (c) with (e), (d) with (f) respectively, as σ_y increases, it is observed that the value of *m* at the same axial force ratio decreases.

3.4 Relationship between the Crack Flexural Strength and Yield Strength

According to the relation between the yield moment and the crack moment in Figure 4, the following can be considered. For example, in Figure 4(a), the crack moment is less than the yield moment at n = 0.1 and the yielding moment is reached after cracking occurs. On the other hand, when n = 0.4, the crack moment is larger than the yielding moment, and cracks would not occur before yielding. In other words, in the case where the axial force is relatively small, it is considered that the lateral stiffness decreases due to the occurrence of cracks.

The point A in Figure 4 shows the point when the crack moment is equal to the yielding moment. The axial force ratio to be point A is approximately 0.2 to 0.3. Therefore, when the axial force ratio is less than or equal to this axial force ratio, cracks occur and the lateral stiffness would start decreasing. This result suggests that the concrete cracking should be considered when we evaluate the lateral stiffness of the CFT beam-columns.

4 CONCLUSIONS

The flexural crack strength of the CFT column was calculated by taking the width-thickness ratio of the steel tube, the yield strength of the steel tube and the compressive strength of the concrete as parameters. The flexural crack strength was shown on the *m-n* interaction diagram and the influence of each parameter was clarified. In addition, we compared the flexural crack strength with the yield strength calculated based on the method by the past research. The conclusions derived from this study are as follows:

- As the width-thickness ratio of the steel tube and the compressive strength of concrete increase, the value of the flexural crack moment at the same axial force ratio increases. As the yield strength of the steel tube increases, the value of the flexural crack moment at the same axial force ratio decreases.
- The axial force ratio when the flexural crack strength is equal to the yield strength is approximately in the range of 0.2 to 0.3.

It is considered that the lateral stiffness decreases due to the occurrence of the flexural cracks when the axial force ratio is less than 0.2~0.3. We would now like to go on to calculate by expanding the range of the parameters and present the skeleton curve of the CFT beam columns, which can consider the effect of the concrete cracking by using these results.

Acknowledgments

This study was supported by the Grant-in-Aid for Scientific Research of Japan Society for the Promotion of Science.

References

AIJ, *AIJ Standard for Structural Calculation of Reinforced Concrete Structure* (in Japanese), Architectural Institute of Japan, Japan, 2010.

AIJ, *Recommendations for Design and Construction of Concrete Filled Steel Tubular Structures* (in Japanese), Architectural Institute of Japan, Japan, 2008.

JSCE, *Standard Specifications for Concrete Structures 2017* (in Japanese), Japan Society of Civil Engineers (JSCE), Japan, 2017.

Liu, Q. S., Kido, M., and Tsuda, K., *Study on Relationship Between Superposed Strength and Yield Strength of the Square CFT Columns*, Modern Methods and Advances in Structural Engineering and Construction proceeding of ISEC-6, Zurich, June 21-26, 2011.

Streamlining Information Transfer between Construction and Structural Engineering
Edited by Shiau, J., Vimonsatit, V., Yazdani, S., and Singh, A.

ISBN: 978-0-9960437-7-9

BEHAVIOR OF REINFORCED CONCRETE SLABS UNDER ACCIDENTAL IMPACTS

SHAMSOON FAREED

Dept of Civil Engineering, NED University of Engineering and Technology, Karachi, Pakistan

Loads resulting from activities such as rock fall, heavy drop weights (for e.g. equipment's, heavy machines during installation), missile and aircraft interaction with slabs may results in loading intensity which have higher magnitude as compared to static loading. Based on the velocity of the impacting object at the time of contact, these activities may result in impact loading. Therefore, slabs designed should provide resistance to these accidental loading during their entire operational life. In this study, a dynamic non-linear finite element analyses were conducted to investigate the behavior of the reinforced concrete slabs subjected to high-mass low-velocity impacts. For this purpose, initially an already published impact test results were used to validate the numerical predictions. Following validation, a study was conducted to investigate the influence of the impact velocity on the behavior of the reinforced concrete slab. Based on the numerical investigation, it was found that the velocity of the impacting object has a significant influence on the behavior exhibited by slab under impact loading. Furthermore, it was also found that the behavior of slab under impact is both local and global. Local behavior is associated with the damage caused at the contact area of the slab and the impactor, whereas global behavior refers to the overall deformation of the slab when stress waves move away from the impact zone and travel towards the supports.

Keywords: Dynamic analysis, Finite element, Non-linear analysis, Loading rates.

1 INTRODUCTION

Reinforced concrete slabs based on their operational use and location may be subjected to the accidental loading stemming from rock fall, heavy drop weights, missile and aircraft collisions. These loading activities depending on the magnitude of the velocity at the time of collision with the slab may result in impact loading. The behavior of the reinforced concrete slabs subjected to impact loading may result in the three common types of failure (i) perforations, (ii) penetrations and (iii) scabbing (Chen and May 2009).

The behavior of Reinforced Concrete (RC) slabs subjected to impact loads was investigated by Kojima (1991). For this purpose, 12 RC slabs having side lengths of 1200 mm each with five different thicknesses of 60 mm, 90 mm, 120 mm,180 mm and 240 mm were investigated. The impact load was applied by firing a 2 kg missile with three different velocities (i) 100 m/s (ii) 150 m/s and (iii) 200 m/s. It was observed that when slab was impacted with identical impact energy, the size of the spalling area increases with increase in the thickness of the slab. It was also observed that the soft nose missile shows significant deformation due to its contact with RC slab, however no deformation was observed for the case of hard missile. Furthermore, the damage caused to the reinforcing bars, the scabbing and penetration produced due to the soft nose missile collision with slab was less as compared to the hard nose missile collision.

The behavior of reinforced concrete slab under high-mass low-velocity impacts was investigated by Chen and May (2009). For this purpose, six reinforced concrete slabs having two different cross-sectional dimensions were used. In type 1 slabs with cross section of 760 mm x 760 mm and thickness of 76 mm, whereas in type 2 slabs having cross section of 2300 mm x 2300 mm and thickness of 150 mm were used. It was observed that contact force in slabs 1-6 increases until it reaches its maximum value. However, the behavior of Type I (slab 1-4) and Type II (slabs 5 and 6) differ significantly after reaching maximum load. The Type I slabs show a plateau before contact is over, whereas, in Type II (slabs 5 and 6), a number of contacts were formed between slab and impactor before complete separation occurred and impactor rebound.

In this study, initially an already published impact test result of Chen and May (2009) were used to validate the numerical predictions. Following the validation, a study was conducted to study the influence of the impact velocity on the behavior of the reinforced concrete slab.

2 NUMERICAL INVESTIGATION

In order to study the behavior of reinforced concrete slabs subjected to impact loading a numerical investigation is conducted herein. Abaqus (2013) was used to carry out dynamic nonlinear finite element analysis (DNLFEA). Initially the predictions obtained were validated using the test data of Chen and May (2009). Furthermore, an investigation was also carried out to study the influence of the impact velocity on the behavior of the reinforced concrete slabs under impact loads.

2.1 Description of The Problem

Chen and May (2009) investigated the behavior of slabs under high mass low velocity impact. Drop weight impact tests were carried out on 6 reinforced concrete slabs. The emphasis of numerical investigation presented herein focuses on studying the behavior of slab having length of 760 mm on both sides and a thickness of 76 mm. Impact (drop-weight) test was carried out by allowing a 98.7 kg to drop on the surface of slab with an impact velocity of 6.5 m/s and impact energy of 2.1 kJ. Figure 1 shows the geometry of impactor and test setup used in the experimental study.

(a) (b)

Figure 1. (a) Cross-sectional dimensions of impactor (b) impact test setup (Chen and May 2009).

2.2 Development of The Finite Element Model

The dynamic NLFEA were carried out using Abaqus (2013). Figure 2 represents the FE model used in this investigation. A finer mesh was used at the contact area of the slab and impactor in order to capture the variation of stresses and strains more accurately, however, coarser mesh having a uniform size of 60 x 60 was used for the slab away from the impact region as the variation of the stresses and strains are not significant.

Figure 2. Finite element model used to investigate the behavior of the reinforced concrete slab.

3 VALIDATION OF THE NON-LINEAR DYNAMIC FINITE ELEMENT ANALYSIS

Figure 3 shows the comparison of the contact force time histories obtained from the impact tests and the DNLFEA. The comparison shows very good agreement between the numerical predictions and the corresponding impact test results for the case of the maximum contact force. Furthermore, the contact force time history curves corelate very well from the start of the impact event i.e. When the impactor come into contact with the slab to the maximum contact force (see Figure 3). However, diversions between the numerical predictions and the impact test results were found after maximum contact force has been attained, as in experimentally recorded contact force a plateau of contact duration has higher magnitude and contact duration as compared to that exhibited numerically. The diversion was also found between the total impact duration exhibited experimentally and numerically.

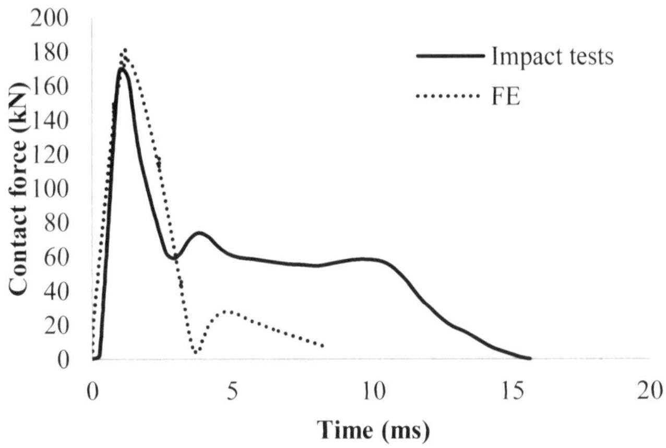

Figure 3. Comparison of the contact force time histories exhibited experimentally and numerically.

Figure 4 shows the damage caused to the slab due to the impact load exhibited experimentally and numerically. As can be seen from Figure 4 that good agreement was observed between the experimentally and numerically exhibited failure caused to the slab at its top and bottom surface. It was also observed that the damaged caused to the slab at its top surface due to the impact load is localized at the impact area, however, the damage significantly reduces away from the contact area of the slab and impactor. Furthermore, it was also observed that the intensity of damage caused to the slab due to the impact load was higher at the bottom face as compared to its top face. As can be seen from Figure 4 that concrete cover teared off and reinforcement was exposed at the bottom face of the slab due to the impact load in the contact area of the slab and the impactor.

(a)

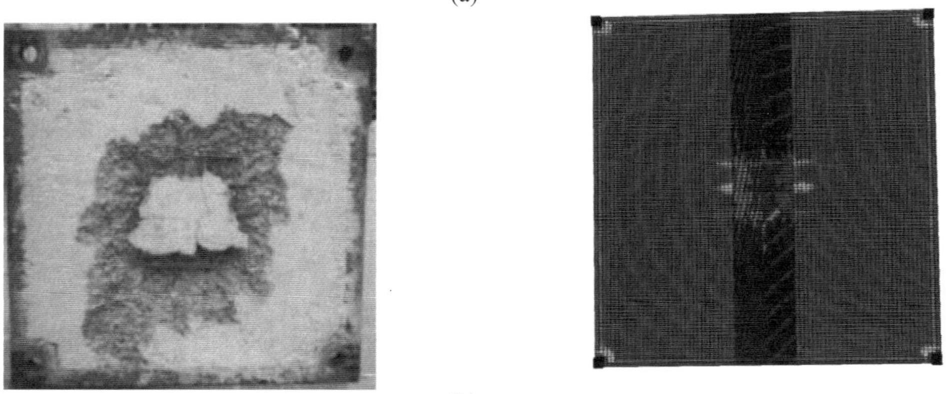

(b)

Figure 4. Comparison of the damage observed at the (a) top and (b) bottom surface of slab exhibited experimentally and numerically.

4 INFLUENCE OF VELOCITY

In order to investigate the effect of velocity with which the impactor strikes the slab on the impact behavior of the RC slab a study was also conducted. For this purpose, three different impact velocities having magnitude equal to the experimental (Chen and May 2009) (case study 1), half of the experimental (case study 2) and twice of the experimental (case study 3) were used to study numerically the influence of the impact velocity on the behavior of the slab.

Figure 5 and Figure 6 shows the comparison of the contact force time histories and damage exhibited numerically for the case of slabs impacted with velocity equals to (a) experimental (case study 1), (b) half of the experimental (case study 2) and twice of the experimental (case study 3) respectively. It was observed that with the increase in the impact velocity the magnitude of the maximum contact force also increases and impact duration decreases (see Figure 5). It was also observed that with the increase in the impact velocity the damage caused to the slab increases at the contact area of the impactor and the slab. Although damage increases with the impact velocity but its influence is negligible away from the contact area (see Figure 6).

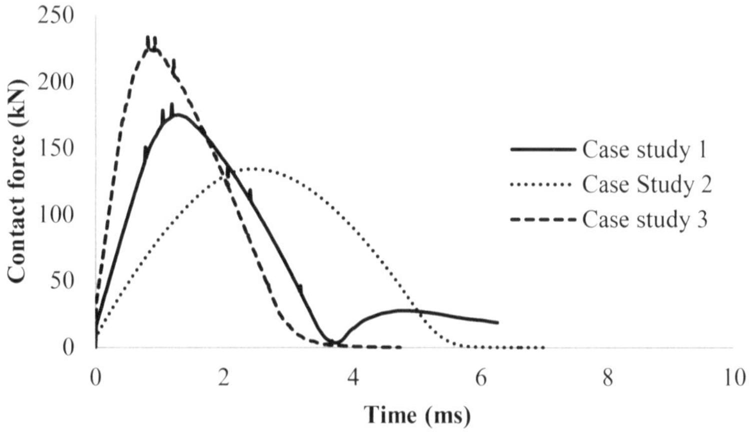

Figure 5. Comparison of the contact force time histories exhibited numerically for the case of slabs impacted with velocity equals to (a) experimental (case study 1), (b) half of the experimental (case study 2) and twice of the experimental (case study 3).

(a) (b) (c)

Figure 6. Comparison of the damage exhibited numerically for the case of slabs impacted with velocity equals to (a) experimental (case study 1), (b) half of the experimental (case study 2) and twice of the experimental (case study 3).

5 CONCLUSIONS

Based on the numerical investigation, it was found that the velocity of the impacting object has a significant influence on the behavior exhibited by slab under impact loading. Furthermore, it was also found that the behavior of slab under impact is both local and global. Local behavior is associated with the damage caused at the contact area of the slab and the impactor, whereas global behavior refers to the overall deformation of the slab when stress waves travels away from the contact area and move towards the supports. It was also found that with the increase in the impact velocity the damage caused to the concrete in the impact region also increases.

References

Abaqus, 6.13-1, User's Manual, ABAQUS Inc., 2013.

Chen, Y., and May, I. M., Reinforced Concrete Members Under Drop-Weight Impacts, *Proceedings of the Institution of Civil Engineers-Structures and Buildings*, 162(1), 45-56, 2009.

Kojima, I., An Experimental Study on Local Behavior of Reinforced Concrete Slabs to Missile Impact, *Nuclear Engineering and Design 2*, 130, 121-132, 1991.

Streamlining Information Transfer between Construction and Structural Engineering
Edited by Shiau, J., Vimonsatit, V., Yazdani, S., and Singh, A.

ISBN: 978-0-9960437-7-9

ULTIMATE SLIP BEHAVIOR OF PERFOBOND STRIP UNDER THE LATERAL CONFINEMENT PARALLEL TO THE PBL PLATES

RYO ISUDA[1], SHINTARO TANI[2], and KATASHI FUJII[1]

[1]*Dept of Civil and Environmental Engineering, Graduate School of Engineering, Hiroshima University, Hiroshima, Japan*
[2]*Kyokuto Kowa Corporation, Hiroshima, Japan*

Perfobond strip (called PBL, hereafter) proposed by Leonhardt in Germany is widely used as the validated shear connector in steel-concrete composite structures. Fujii et al. proposed a fracture mechanism of PBL as follows: Slip force (shear force) acting on a PBL causes the expanding force which splits the concrete around the PBL. If restraint force, which is caused by cover concrete, rebar etc., resisting the expanding force exist, the slip strength is enhanced. Consequently, the fracture of PBL will happen when the expanding force exceeds the restraint force. Also, based on the above concept, they have given the evaluation formula of PBL slip strength and verified it by comparing previous experimental results. However, bonding force and frictional force are not taken into account in the formula. Through the loading tests of the composite girder, we found that the bonding and the frictional force, which appear along the boundary surface between concrete deck and steel flange, work as resistance force to not only the slip strength but also the expanding force. In this paper, the effects of the bonding force and the frictional force on the ultimate slip behavior of PBL are investigated experimentally, and an evaluation formula of PBL slip strength is improved by taking their effects into account.

Keywords: Slip strength, Lateral force, Bonding force, Friction force.

1 INTRODUCTION

In Japan, perfobond strip invented by Leonhardt *et al.* (1987) is widely used in many steel-concrete composite structures as an effective shear connector. Fujii *et al.* (2014) have investigated and shown the fracture mechanism of PBL as follows: Slip force (shear force) acting on PBL causes "Expanding force", which splits the concrete around the PBL and which brings the fracture of PBL. On the other hand, if some restraint forces resisting the expanding force exist, the slip strength will rise because the restraint force delays the crack-outbreak of the concrete at PBL hole. The expanding force becomes large with the increase of the slip force; consequently, the fracture of PBL will occur when the expanding force exceeds the resultant of the restraint forces, which are caused by cover concrete, a rebar set through the hole, etc. Based on the above assumption, they have given an evaluation formula of the ultimate slip strength of PBL, and verified it by comparing previous experimental results. However, bonding force and frictional force are not taken.

By the way, through the loading tests of the composite girder, it was suggested that the bonding and the frictional forces, which appear along the boundary surface between concrete deck and steel flange, work as resistance force to not only the slip strength but also the expanding force. In this paper, the effects of the bonding force and the frictional force on the ultimate slip strength of PBL are investigated experimentally, and the evaluation formula of PBL slip strength is improved by considering their effects. In the followings, two types of push out test are conducted for two purposes: to clarify the bonding force and the friction coefficient between concrete and steel plate: to get the ultimate slip strength of PBL with effects of bonding and frictional forces. The mix proportion and the material properties of concrete are shown in Table 1 and Table 2, respectively.

Table 1. Mix condition of the concrete.

Compressive sterength (N/mm^2)			Tensile strength (N/mm^2)			Elastic modulus (kN/mm^2)		
Before	After	Average	Before	After	Average	Before	After	Average
39.5	44.5	42.0	4.4	5.4	4.9	30.1	30.5	30.3

Table 2. Material properties of the concrete.

Maximum Aggregate Size	Water-Cement Ratio	Slump	Sand Aggregate Size	Air Content	Quantity of material per unit volume of concrete(kg/m^3)				Air entrainig and high-range water reducing agent
(mm)	(%)	(cm)	(%)	(%)	Water	Cement	Fine aggregate	Coarse Aggregate	(kg/m^3)
20	65	8.5	49	1.8	168	258	944	998	1.29

2 TESTS TO OBTAIN BONDING FORCE AND FRICTION COEFFICIENT

2.1 Test Summary and Specimens

We made eleven specimens whose steel plate does not have any hole, and is sandwiched with two concrete blocks at the both sides such as Figure 1. At that condition, the concrete and the steel plate are adhered each other by the bonding force only. After applying the lateral force up to a value by using PC rods set at the both sides, the slip force is subjected at the top of the steel plate as shown in Figure 2. Then, the maximum slip forces at slipping are obtained under several lateral forces.

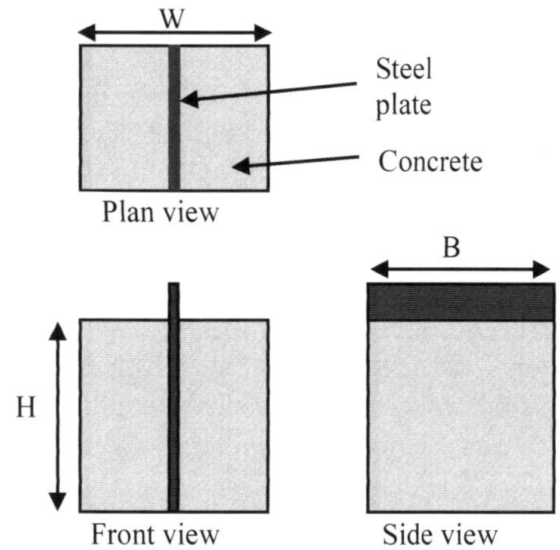

Figure 1. Test specimen.

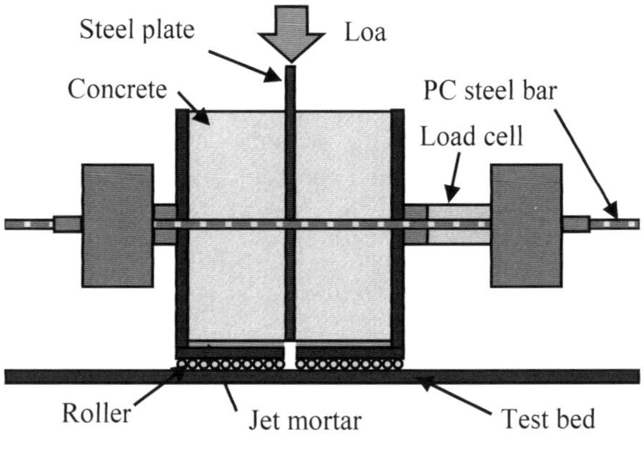

Figure 2. Loading condition.

2.2 Test Results

Test results are shown in Table 3 and a typical example (I-P_L150 test specimen) of the load vs. slip curve is also shown in Figure 3. From the figure, it is noticed that the slip between concrete and steel plate appears even when the slip load is small. If steel and concrete are rigid body, the slip will not appear until the slip load reaches a value. Though we cannot clarify whether the slip at small slip load is caused by the deformation of concrete and steel or not, regarding the maximum force as the slip load, the relationship between slip load and lateral force is plotted in Figure 4. It is noticed from the figure that as lateral force becomes larger, the slip loads increase approximately linearly, where the coefficient of correlation is 0.93. Based on the test results, the regression line between the slip load and lateral force can be obtained. The intercept and the slope of the regression line mean the bonding force and the frictional coefficient, respectively. These values are shown in Table 4.

Table 3. Dimensions of specimen and test results.

The year of the tests conducted	Name of specimen	Width of specimen	Depth of specimen	Height of specimen	Thickness of plate	Initial lateral force	Lateral force at Slipping	Load at slipping
		W (mm)	B (mm)	H (mm)	t (mm)	P_L (kN)	P_L' (kN)	Vu (kN/surface)
2016	I-P_L0-A	200	200	300	9	0.00	0.00	4.15
	I-P_L0-B					0.00	0.00	1.03
	I-P_L50-A					50.00	52.00	28.53
	I-P_L50-B					50.00	89.50	89.79
	I-P_L100-A					100.00	100.00	36.41
	I-P_L100-B					100.00	114.50	81.86
	I-P_L180-A					180.00	197.33	112.63
	I-P_L180-B					180.00	203.67	141.96
2017	I-P_L50-C					50.00	52.00	31.83
	I-P_L100-C					100.00	107.73	67.63
	I-P_L150					150.00	165.33	97.13

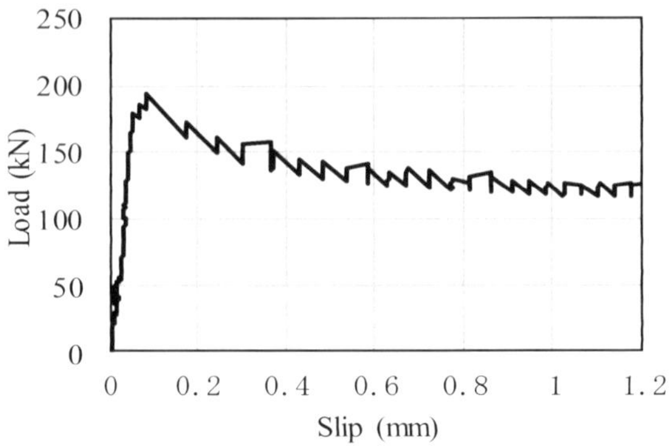

Figure 3. Load-Slip curve.

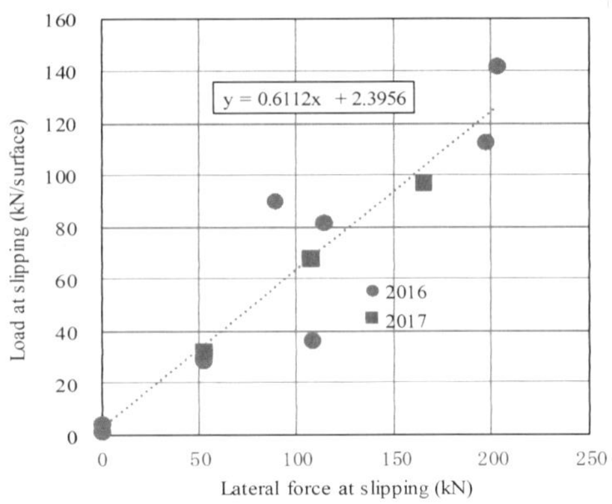

Figure 4. Relation between slip load and lateral force.

Table 4. The results of bonding force and frictional coefficient.

Bonding force	Coefficient of friction
C' (kN/m^2)	μ
40.0	0.61

3 TESTS TO OBTAIN THE SLIP STRENGTH OF PBL

3.1 Test summary and Specimens

Push out test is conducted for eight push-out specimens in order to investigate the effect of frictional force and bonding force on slip strength, changing the initial lateral force and the height of concrete block. During the tests, the lateral force is subjected to the both sides of concrete parallel to the PBL plates, as illustrated in Figure 6.

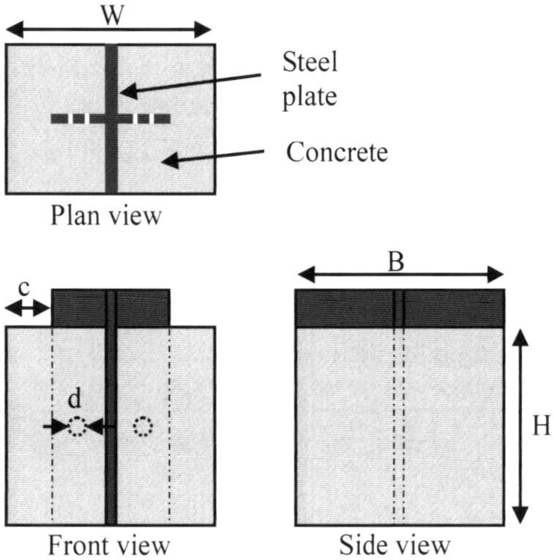

Figure 5. Test specimens.

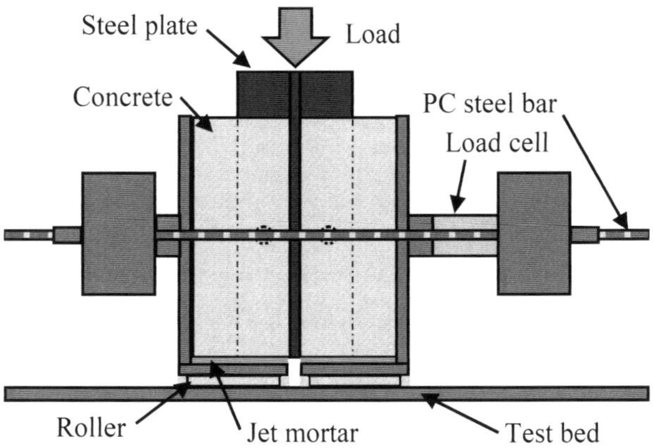

Figure 6. Loading condition.

3.2 Test Results

Test results are summarized in Table 5. "Lateral forces at slipping" and "Slip strength" show the values per one PBL hole. As the slip force increases, the lateral force varies and becomes larger from the initial lateral force, as shown in Table 5. This phenomenon is clearly due to the surface irregularity at the boundary between steel and concrete.

At the fracture of PBL, the slip force reaches the maximum; and concrete crack, which separates a concrete block to two blocks, appears; then the slip force decreases. After loading test, the concrete in the hole remained with two fracture surfaces parallel to PBL plate.

Table 5. The size of specimens and results.

Name of specimen	Diameter of PBL hole	Thickness of cover concrete	Width of specimen	Depth of specimen	Thickness of PBL plate	Height of specimen	Initial lateral force	Lateral force at Slipping	Slip strength
	d (mm)	c (mm)	W (mm)	B (mm)	t (mm)	H (mm)	P_L (kN)	P_L'(kN)	Vu (kN/hole)
J-H300-P_L0	60	100	420	400	12	300	0	0.00	77.39
J-H500-P_L0						500		0.00	100.41
J-H300-P_L20						300	20	34.08	130.44
J-H500-P_L20						500		28.00	152.46
J-H300-P_L50						300	50	61.75	138.27
J-H500-P_L50						500		68.33	193.38
J-H300-P_L80						300	80	81.92	203.27
J-H500-P_L80						500		87.75	203.30

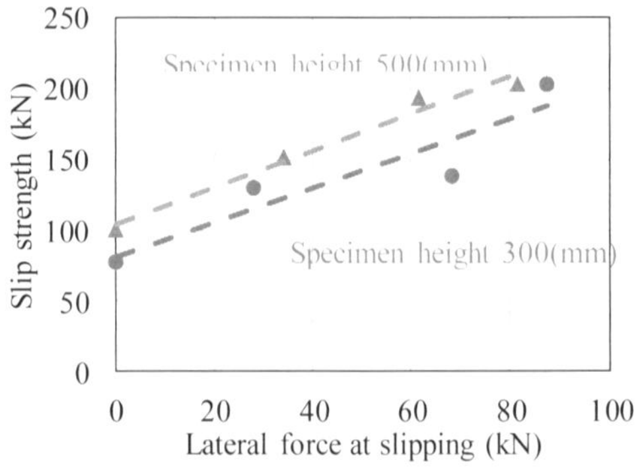

Figure 7. Relationship between lateral force and slip strength.

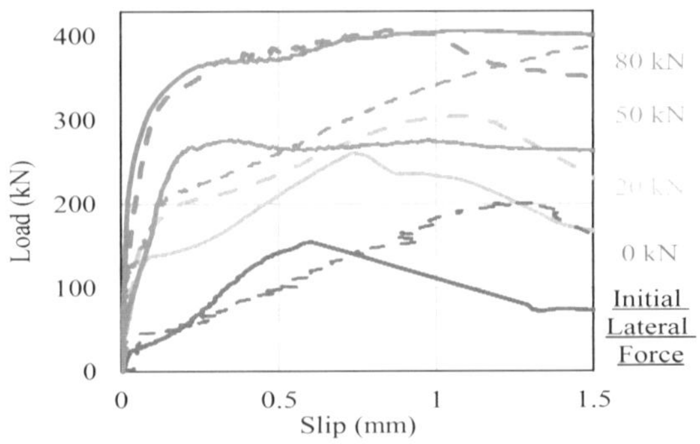

Figure 8. Slip load vs. slip curve.

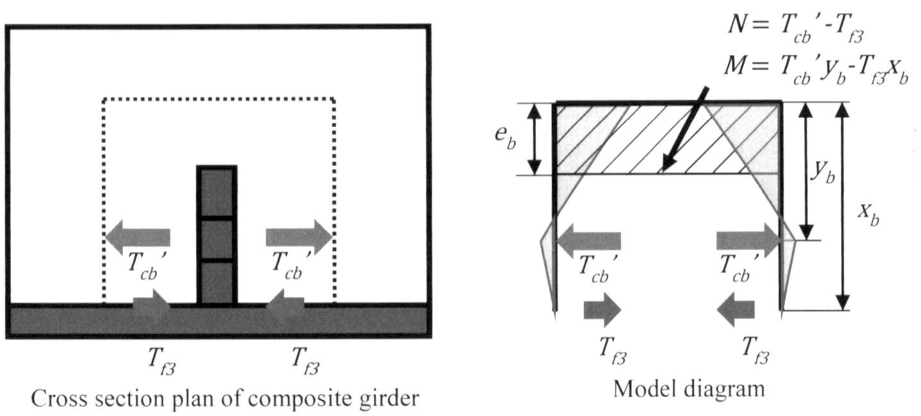

Figure 9. Rigid frame model.

4 CONSIDERATION

4.1 Effects of Bonding and Frictional Force

The relationship between the slip strength of one PBL hole and the lateral force is shown in Figure 7. From this figure, it can be noticed that the slip strength rises as the lateral force increases. This increase of the slip strength is clearly due to the bonding force and the friction force appearing at the interface of concrete and steel plate.

Slip load vs. slip curve for each lateral force is shown in Figure 8. Here, slip means the relative slip-value between steel and concrete. As is shown in Figure 8, it can be also noticed that lateral force can make enhance not only the slip strength, but also the slip rigidity.

4.2 Evaluation Formula of Slip Strength

In the followings, we try to improve the slip strength evaluation formula proposed by Fujii *et al.* (2014), considering with the bonding force and the friction force at the interface of steel and concrete.

Taking the bonding force and friction force into account, the slip strength evaluation formula proposed by Fujii *et al.* (2014) can be rewritten as the following equations. At this time, the assumption for the concrete fracture is the quite same as that by Fujii *et al.* (2014). Namely, the fracture of PBL occurs when the normal stress of concrete at the top of PBL plate reach to the tensile stress.

$$V_u = V_{\text{int}} + 2.5T_{cb}' + T_{f2} \tag{1}$$

$$V_{\text{int}} = \frac{\pi d^2}{2} \times f_{ct} \tag{2}$$

$$T_{cb}' = \frac{f_{ct} + T_{f3}\left(\dfrac{e_b x_b}{I_{cb}} + \dfrac{1}{A_{cb}}\right)}{\dfrac{e_b y_b}{I_{cb}} \times \dfrac{1}{A_{cb}}} \tag{3}$$

The evaluated slip strengths obtained from the modified formula are plotted in Figure 10, comparing with experimental results. In the figure, those based on Fujii *et al.*(2014) are also shown. As is shown in Figure 10, it can be concluded that the modified formula can evaluate well the experimental results with coefficient of correlation of 0.95, though the values from Fujii *et al.* (2014) are not in accordance.

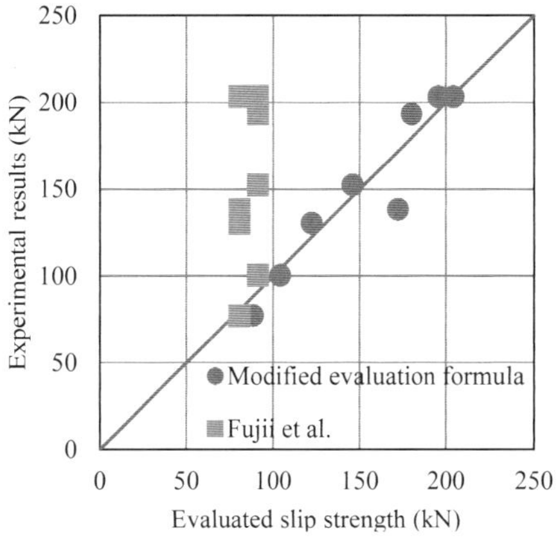

Figure 10. Relationship between evaluated strength and test results.

5 CONCLUSIONS

The results obtained from this study can be summarized as below:

1. According to the bonding force and the friction force, the slip strength and the slip rigidity increase, and they become larger as the lateral force increases.
2. Though the bonding and frictional forces have been usually ignored in the previous researches, it is clarified that they cannot be ignored because the slip strength of PBL is not a little influenced by them.
3. The modified evaluation formula indicated here in is able to evaluate the slip strength well in consideration with the bonding force and the frictional force.

References

Leonhardt, F., Andrä, W., Andrä, H., and Harre, W., *Neues, vorteilhaftes Verbundmittel für Stahlverbund-Tragwerke mit hoher Dauerfestigkeit, Beton-und Stahlbetonbau,* 325-331, December, 1987.

Fujii, K., Dokan,Y., Iwasaki, H., Himukai, M., Mori, K. and Yamaguchi, S., *Ultimate Shear Strength of Perfobond Strip*, JSCE Journal of Hybrid Structures, Vol. 70, No. 5, pp. II-53-II-68, 2014.

Streamlining Information Transfer between Construction and Structural Engineering
Edited by Shiau, J., Vimonsatit, V., Yazdani, S., and Singh, A.

ISBN: 978-0-9960437-7-9

ULTIMATE SLIP STRENGTH OF PERFOBOND STRIP CONFINED BY CONCRETE COVER

SHINTARO TANI[1], KATASHI FUJII[2], and RYO ISUDA[2]

[1]*Kyokuto Kowa Corporation, Hiroshima, Japan*
[2]*Dept of Civil and Environmental Engineering, Graduate School of Engineering, Hiroshima University, Higashi-Hiroshima, Japan*

The ultimate slip strength of a perfobond strip (called "PBL" hereafter) depends significantly on the concrete confinement conditions around PBL hole as well as concrete shear fractures at the hole. If the hole is small, the effect of concrete confinement will still more strongly appear. In this paper, focusing on the concrete confinement due to concrete cover, the ultimate slip strength of PBL with a small hole is investigated experimentally through push-out tests, in which the diameter of PBL hole is set at 30 mm and the thickness of the concrete cover is varied from 0 to 100 mm. Test results show that the concrete separates into two blocks with concrete crack parallel to the PBL plate and that the ultimate slip strength of a 50-mm-thick concrete cover is 34.4 kN, which is about 10 times of the slip strength without concrete cover (3.5 kN). Thus, it can be concluded that thicker concrete cover makes the slip strength remarkably larger, in the case of small-hole PBLs.

Keywords: Push-out tests, Confinement, PBL.

1 INTRODUCTION

PBL proposed by Leonhardt *et al.* (1987), is a shear connector to resist the slippage between different materials in composite structures. In order to design PBL, the evaluation formula of the ultimate slip strength was proposed by Leonhardt *et al.* (1987) etc. and also indicated by the Japanese Standard Specifications for Hybrid Structures (2015). These formulas are formulated by only two parameters, the diameter of the PBL hole and the compressive strength of concrete. Consequently, the ultimate slip strength of PBL is determined by the concrete strength and the area of PBL hole, in the case of without rebar. However, Nakajima *et al.* (2011), and Fujii *et al.* (2014) showed that the ultimate slip strength of PBL depends on the confinement condition of the concrete surrounding PBL hole.

Although PBLs have been used for preventing slip between steel girders and concrete floor slabs in composite girders, in recent years, their use is spreading widely to various composite structures, such as composite slab, the connection between pre-cast members, and so on. In particular, for application to thin composite slabs, PBL with a small hole will be used. Then, its ultimate slip strength should be clarified considering the effect of concrete confinement for economical design.

In this paper, push-out tests of PBL with small-hole of 30 mm diameter are conducted, changing the thickness of concrete cover in order to clarify the effect of the confinement due to concrete cover confinement on the ultimate PBL slip strength.

2 OVERVIEW OF PUSH-OUT TESTS

We made two types of specimens, in which one (A series) has a concrete block inserted a PBL plate with a hole, as shown in Figure 1, the other (B series) has two concrete blocks sandwiching a cross-shape steel column whose two plates have a hole respectively, as shown in Figure 2 (Furukawa *et al.* 2011).

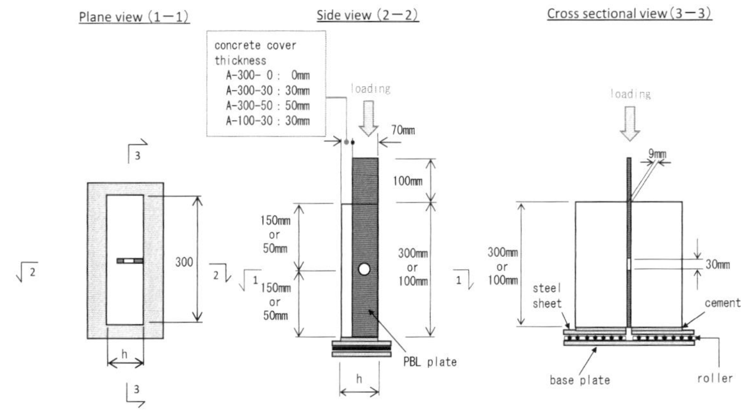

Figure 1. Specimen shape (A series).

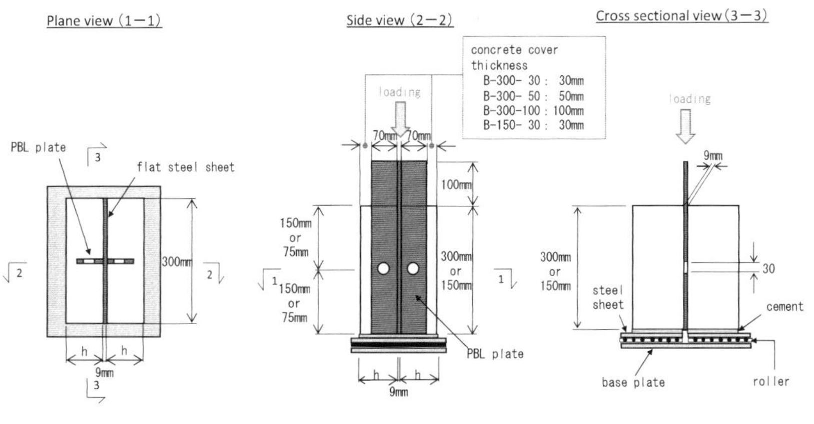

Figure 2. Specimen shape (B series).

Dimensions and material properties of specimens are indicated in Table 1. In the table, concrete cover thickness means the distance from the PBL-plate tip to the concrete surface as shown in Figure 1 and 2. Focusing on the confined effect of concrete cover, in this paper, all specimens have a different concrete cover (0, 30, 50 and 100 mm), respectively, whereas the diameter of PBL hole is the same as 30mm. For the size of the concrete block, the height of almost specimens is set 300mm, but adding to these, the specimens of A-100-30 and B-150-30 are also made, in order to investigate the effect of concrete confinement due to the height of the concrete block. Moreover, the PBL plate surfaces are not processed to remove bonding between concrete and PBL plate, though the bonding force between concrete and flat steel plate at B-series is removed by coating grease.

Before loading, a specimen is placed on the roller support, as shown in Figure 1 and 2, in order to remove the frictional force between concrete block and test bed at the bottom. Then slip load is subjected to the top of PBL plate and increased up to the maximum load. At the fracture, the concrete is divided into two blocks as shown in Figure 3, then, the load decreases rapidly.

If the specimen is set directly on the layer such as gypsum or mortar laid on the test bed, the frictional force appears. Consequently, the frictional force, which works to confine the concrete around the hole, makes the slip strength of PBL enhance significantly (Fujii *et al.* 2014). Readers will notice that this frictional force never appears in the actual structure, except the case of slip test such like a push-out test.

Table 1. Dimensions and material properties.

series	test sample	concrete					PBL plate	
		length (mm)	heigth (mm)	concrete cover thickness (mm)	compressive strength (N/mm^2)	Aggregate diameter (mm)	plate thickness (mm)	yield strength (N/mm^2)
A	A-300-0	70	300	0	62.5	20	9	more than 245
	A-300-30	100	300	30	59.5			
	A-300-50	120	300	50	59.5			
	A-100-30	100	100	30	59.5			
B	B-300-30	100	300	30	72.8	20	9	more than 245
	B-300-50	120	300	50	72.8			
	B-300-100	170	300	100	72.8			
	B-150-30	100	150	30	72.8			

3 RESULTS AND DISCUSSION

3.1 Ultimate Slip Strength

A-300-30

B-300-30

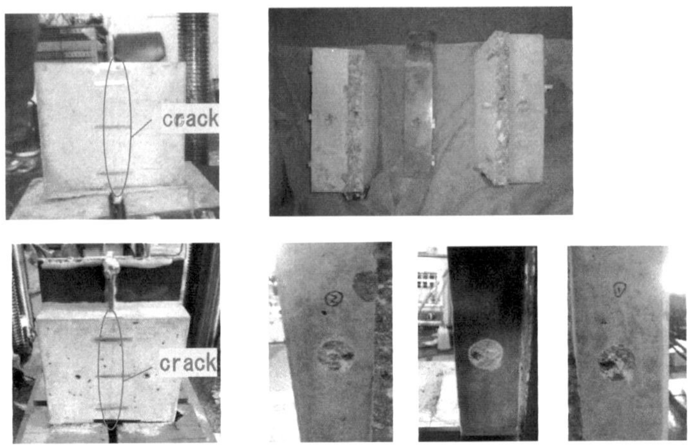

Figure 3. Specimen appearance after failure.

Every specimen were failed by the occurrence of concrete cracks at PBL hole and also of concrete crack parallel to PBL plate in the concrete cover. Consequently, a concrete block sandwiching PBL plate was divided into two pieces, as shown in Figure 3. Ultimate slip strength of PBL obtained from the tests is indicated in Table 2, also those based on three evaluation formulas, where the evaluation formula of Eq. (1) in the table was proposed by Leonhardt *et al.* (1987), Eq. (2) is now being adopted in JSSHS (Japanese Standard Specifications for Hybrid Structures 2015) and Eq. (3) was proposed by Fujii *et al.* (2014).

Table 2. Ultimate slip strength.

series	test sample	[A] experimentally maximum load (kN)	calculated ulitimate slip strength (evaluation formula)						concrete properties (N/mm^2)	
			Eq. (1)		Eq. (2)		Eq. (3)			
			① strength (kN)	[A]/①	② strength (kN)	[A]/②	③ strength (kN)	[A]/③	tensile strength	Young's modulus
A	A-300-0	3.5	98.4	0.04	90.0	0.04	3.6	0.97	2.5	2.6×10^4
	A-300-30	18.5	93.7	0.20	85.7	0.22	9.3	1.99	2.5	2.6×10^4
	A-300-50	34.4	93.7	0.37	85.7	0.40	16.1	2.14	2.5	2.6×10^4
	A-100-30	12.2	93.7	0.13	85.7	0.14	5.5	2.22	2.5	2.6×10^4
B	B-300-30	※ 38.3	114.7	0.33	104.9	0.37	12.3	3.11	3.4	4.2×10^4
	B-300-50	※ 42.9	114.7	0.37	104.9	0.41	21.3	2.01	3.4	4.2×10^4
	B-300-100	※ 59.7	114.7	0.52	104.9	0.57	48.1	1.24	3.4	4.2×10^4
	B-150-30	※ 14.7	114.7	0.13	104.9	0.14	8.5	1.73	3.4	4.2×10^4

※ half of the loading load

Leonhardt *et al.* (1987):

$$V=1.4\times d^2\times\beta_{wn} \tag{1}$$

where, V: ultimate slip strength [N], d: diameter of PBL hole [mm], and β_{wn}: compressive strength obtained from cubic concrete specimen based on ISO 1920-3 [N/mm^2]. Since the compressive strength f'_c is usually used based on the test result of a cylindrical specimen according to JIS A 1132 in Japan, the compressive strength β_{wn} is given by changing from cylindrical to cubic compressive strength using the relationship of $f'_c\,/\,\beta_{wn}$=0.8.

JSSHS (2015):

$$V=\frac{1.6\times d^2\times f'_c}{\gamma_b} \tag{2}$$

where f'_c: compressive strength of a cylindrical specimen [N/mm^2], and γ_b: member factor (=1.0).

Fujii *et al.* (2014):

$$V=V_{int}+2.5\times T_c \tag{3a}$$

$$V_{int}=2\times(\pi\times\frac{d^2}{4}+(n-1)\times A_s)\times\tau_{ct} \tag{3b}$$

In Eq. (3), V_{int}: the pure shear strength under the condition without concrete confinement [N], this condition corresponds to the specimen without concrete cover in this paper. T_c: the maximum confined force caused by the concrete cover, that is the maximum resistance force resisting the splitting force which causes concrete block dividing into two blocks. n: the ratio of the elastic modulus $E_s/\,E_c$ for concrete E_c and rebar E_s, A_s: the cross-sectional area of rebar set through -PBL hole [mm^2] (=0 in this paper), and τ_{ct}: shear strength of concrete, which Fujii *et al.* (2014) adopts tensile strength of concrete [N/mm^2].

Exactly speaking, Fujii *et al.* (2014) indicated confinement factors not only the cover but also other factors, that is: rebar set through PBL hole, rebar set perpendicular to PBL plate in the cover, and the friction force along the boundary between block and test bed at the bottom as mentioned above, in the case of push-out test. By the way, Fujii *et al.* (2014) made the evaluation formula Eq. (3) of ultimate slip strength based on the following assumptions:

(1) When slip force (shear force) V is subjected to PBL, the slip force will produce the splitting force T at the hole, as shown in Figure 4, as well as a shear force which corresponds to V_{int} in Eq. (3). Then, the splitting force will be larger as the slip force increases.

(2) The splitting force will make concrete fracture at the hole, consequently, the concrete is broken to two pieces, as shown in Figure 4.

(3) If confinement force, which prevents the above concrete fracture, exists, it will make the ultimate slip strength enhance up to when the splitting force exceeds the confinement force.

For the confinement force due to the concrete cover in the push-out test, the concrete sandwiching the PBL plate can be regarded as a rigid frame consisted of the cover and both sides concrete, as shown in Figure 3. The maximum confinement force T_c can be calculated as the fiber stress reaches to the tensile strength of the concrete at the top of the PBL plate in Figure 3, by using beam theory. Then, the ultimate slip strength becomes larger as the concrete height becomes higher by the reason of bigger inertial moment and cross-sectional area of the cover concrete.

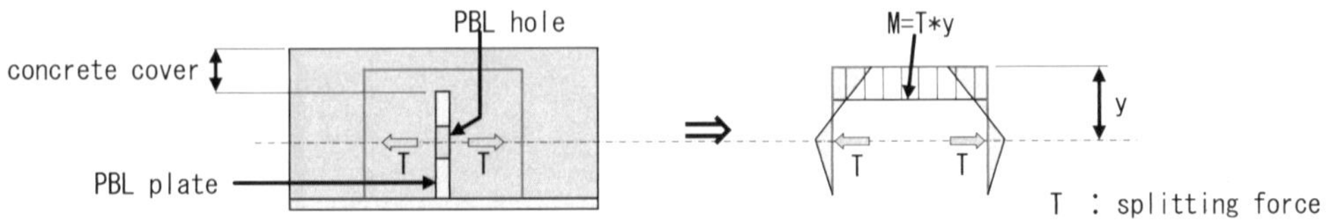

Figure 4. PBL model used by Fujii *et al.* (2014).

Ultimate slip strengths obtained from tests are indicated in Table 2, and also calculated by each evaluation formula are shown to compare. From the test results, it is noticed that the ultimate strength of PBL grows big, as the concrete cover becomes thicker. Moreover, when the concrete cover exists, the higher height of the concrete block gives bigger ultimate slip strength, comparing A-300-30 with A-100-30 or B-300-30 with B-150-30. On the other hand, when the cover does not exist such as the specimen A-300-0, ultimate slip strength does not change and is almost the same as A-300-0 even if the concrete height varies, although this fact could not enough show here. These phenomena are obviously caused by the confinement effect depending on the concrete cover. As for Eq. (1) and Eq. (2), it is noticed that the ultimate slip strength is determined only by the diameter of PBL and compressive strength of concrete. Therefore, the ultimate slip strength becomes consequently the same value for each series in Table 2, because these equations cannot take the confinement effect into account. Adding to this, also noticed that the ultimate slip strengths based on both equations are significantly larger than test results. This reason is probably that these evaluation equations were built by regression analysis based on the experimental data, which must have contained several confinement effects. In contrast, Eq. (3) can show the tendency of the phenomena due to confined effect, as indicated in Table 2. However, the value of ultimate slip strength has significant difference between test results and the evaluated values by Eq. (3), without the case of without concrete cover.

Thus, the confinement force effects largely on the ultimate slip strength of PBL, for example, when the specimens of concrete strength at the same level is compared, the ultimate slip strength of the specimen A-300-50 (cover-thickness=50 mm) was approximately 10 times higher than that of A-300-0 (without cover). Consequently, it can be concluded that the confined effect must be evaluated correctly in the evaluation formula.

4 CONCLUSIONS

Focusing on only the concrete confinement of the concrete cover, ultimate slip strength of PBL with a small hole was investigated by conducting push-out tests. Since the ultimate slip strength depends clearly on the confinement condition of concrete, the friction force at the bottom, one of the confinement factors, was removed completely by the roller in this paper. Test results were compared with three evaluation formulas having been proposed. Conclusions are as below:

(1) Ultimate slip strength becomes larger as concrete cover becomes thicker. This fact is clearly due to the confined effect caused by the cover. When the cover is 50 mm, the slip strength (34.4 kN) is about 10 times of that without cover (3.5 kN). Thus, ultimate slip strength is significantly affected by the concrete confinement.

(2) For ultimate slip strength, test results are far smaller than that obtained by both Leonhardt *et al.* (1987) and Japanese Standard Specifications for Hybrid Structures. Especially in the case without concrete cover, the test result is about 1/25 of them. This reason is that their evaluation formulas are tacitly taken confined effects into account because they are derived from a regression analysis of experimental data, many of which contain several concrete confinements.

(3) When the specimen has no concrete cover, the ultimate slip strengths are almost the same between the test and Fujii *et al.* (2014). But, in the case with concrete cover, the remarkable difference is recognized between them. Though Fujii *et al.* (2014) can show the tendency of the confinement phenomena due to the cover, their evaluation formula should be improved more in the future.

(4) Ultimate slip strength increased as the concrete height of specimen becomes larger. This phenomenon is also caused by the concrete confinement of the cover and can be explained also by the concept of Fujii *et al.* (2014).

References

Fujii, K., Doukan, Y., Iwasaki,H., Himukai, M., Mori, K., and Yamaguchi,S., Ultimate Shear Strength of Perfobond Strip, *Journal of Japan Society of Civil Engineers*, Japan Society of Civil Engineers, A1, 70(5), 53-68, May, 2014.

Furukawa, Y., Fujii, K., Doukan, Y., and Yamaguchi,S., *On Pushes Tests Slip Strength of PBL Under Dimensional Parameters,* The 9th Symposium on Research and Application of Hybrid and Composite Structures, Japan Society of Civil Engineers, 58-65, November, 2011.

Japanese Standard Specifications for Hybrid Structures, *2014 Design,* Japan Society of Civil Engineers, 74-77, 581-584, March, 2015.

Leonhardt, F., Andrä, W., Andrä, H., and Harre, W., Neues, Vorteilhaftes Verbundmittel für Stahlverbund-Tragwerke mit hoher Dauerfestigkeit, *Beton-und Stahlbetonbau*, 325-331, December, 1987.

Nakajima, A., Hashimoto,M., Koseki, S., and Suzuki, Y., *Evalution of Shear Strength of Perfobond Strip Based on Restrained Push-Out Test*, The 9th Symposium on Research and Application of Hybrid and Composite Structures, Japan Society of Civil Engineers, 49-57, November, 2011.

Streamlining Information Transfer between Construction and Structural Engineering
Edited by Shiau, J., Vimonsatit, V., Yazdani, S., and Singh, A.

ISBN: 978-0-9960437-7-9

THREE FACE ATTACHMENT RETROFIT AGAINST FATIGUE CRACKING AT STEEL GIRDER WEB PENETRATION

CHIHIRO SAKAMOTO[1], MASAHIRO SAKANO[1], HIDEYUKI KONISHI[2], and TAKASHI FUJII[3]

[1] *Kansai University, Osaka, Japan*
[2] *Japan Bridge Association, Osaka, Japan*
[3] *Ministry of Land, Infrastructure and Transport, Himeji, Japan*

Fatigue cracking in steel girder web penetration details is so dangerous that it can break steel girders. Since a number of highway bridges have such web penetration details in Japan, it is of urgent importance to grasp these fatigue strength properties. In this study, we investigate stress reduction effects of three face attachment retrofit through fatigue tests using a large girder specimen with web penetration details where cross beam lower flanges are connected to the lower surface of a slot by welding. As a result, there is very little difference between two and three face attachments about stress reduction effects, while they are more effective than one face attachment. The upper side attachment is more effective than the lower side attachment, while both side attachment is best. Two and three face both side attachments can reduce about 40% of stress concentration, while two and three face upper side attachments can reduce 50–60%.

Keywords: Fatigue test, Retrofit, Highway steel bridge, Cracks.

1 INTRODUCTION

Fatigue cracking in steel girder web penetration details is so dangerous that it can break steel girders (Fisher 1984). The one-meter-long crack was detected in Yamazoe Bridge in 2006 (Nara National Highway Office 2006). Since a number of highway bridges have such web penetration details in Japan, it is of urgent importance to grasp these fatigue strength properties. However, few fatigue tests have been reported on steel girder web penetration details (Sakano *et al.* 1995).

In previous studies, we investigated stress distributions around web penetration details and fatigue cracking behavior through fatigue tests using girder specimens with web penetration details (Yoshida *et al.* 2014) and verified the effect of retrofitting methods against fatigue cracking in web penetration details (Sakamoto *et al.* 2018). However, these methods cannot prevent fatigue crack initiation perfectly.

In this study, we investigate the stress reduction effect by three face attachment retrofit through fatigue tests using a large girder specimen with web penetration details where cross beam lower flanges are connected to lower surface of a slot by welding.

2 EXPERIMENTAL PROCEDURE

2.1 Specimen

Figure 1 shows configurations and dimensions of the specimen and location of strain gauges. Dimensions of the specimen are increased from the previous one (Yoshida *et al.* 2014), so that three face attachments can be fixed to the space between the cross beam flange and the main beam flange. Figure 2 shows the details of three face attachment retrofit.

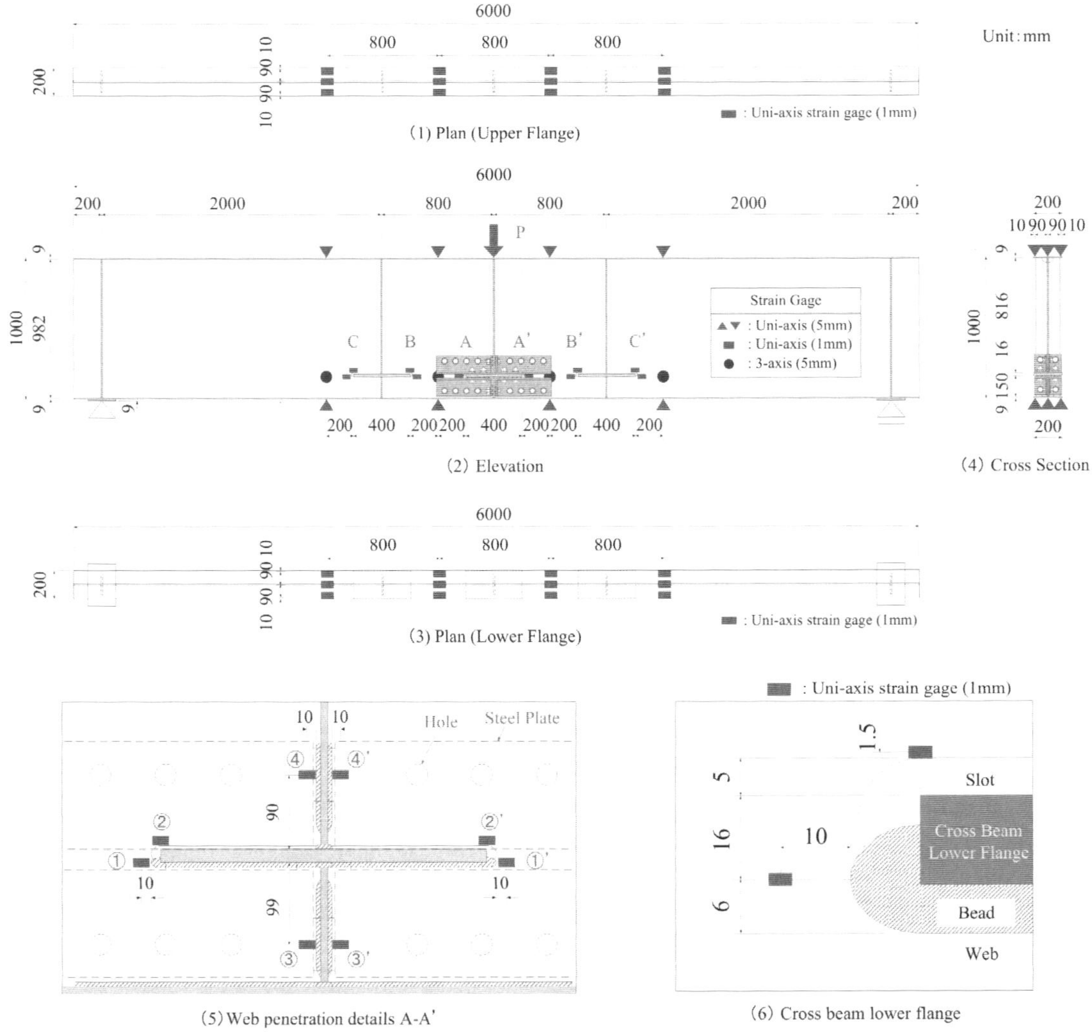

Figure 1. Configurations and dimensions of the specimen, and location of strain gauges.

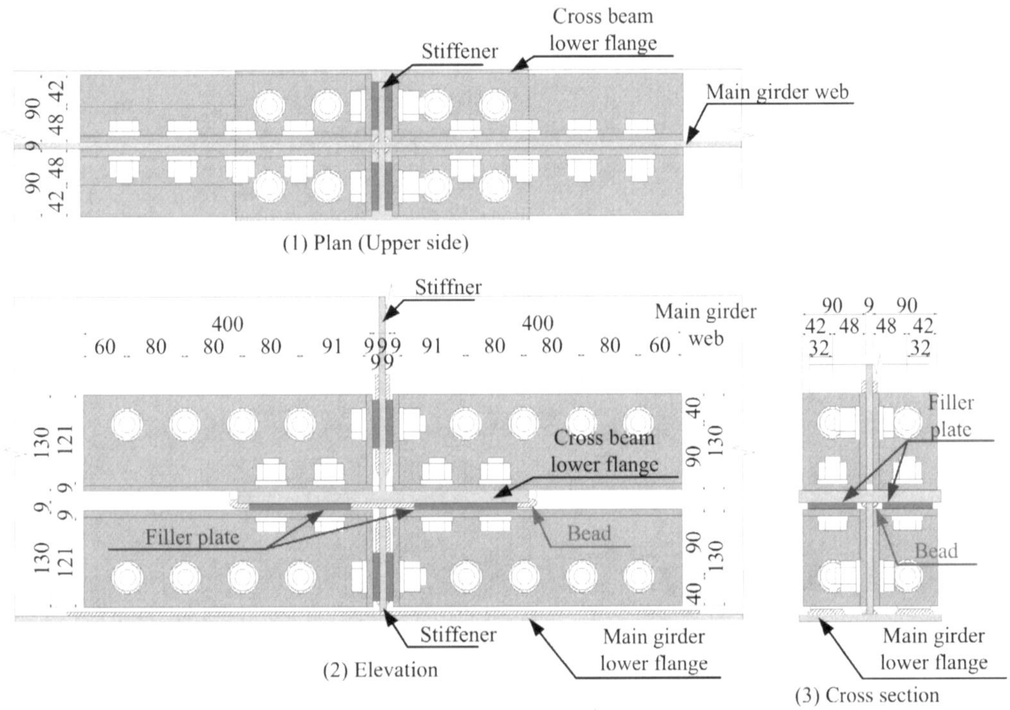

Figure 2. Three face attachment retrofit.

2.2 Preventive Measure

Figure 3 shows ten cases of the retrofitting methods. Case one is no retrofit, Cases from two to four are upper and lower side attachments, Cases from five to seven are upper side attachments, Cases from eight to ten are lower side attachments. While, Cases two, five and eight are three face attachments, Cases three, six and nine are two face attachments, and Cases four, seven and ten are one face attachments.

2.3 Static Loading Test Procedure

We conducted static loading tests in the three-point bending condition to grasp the stress distributions around web penetration details and to investigate the stress reduction effect. Figure 4 shows the loading conditions. The load range was set to 100kN so that the maximum tensile stress of about 40MPa could be generated in the bottom flange similarly as the previous study (Yoshida *et al.* 2014).

3 EXPERIMENTAL RESULTS

Figure 5 shows the stress change at all gages in the case of upper and lower attachment. At gages ① and ①', stress decreases remarkably after attachments. Two and three face attachments are more effective than one face attachment, and there is very little difference between them.

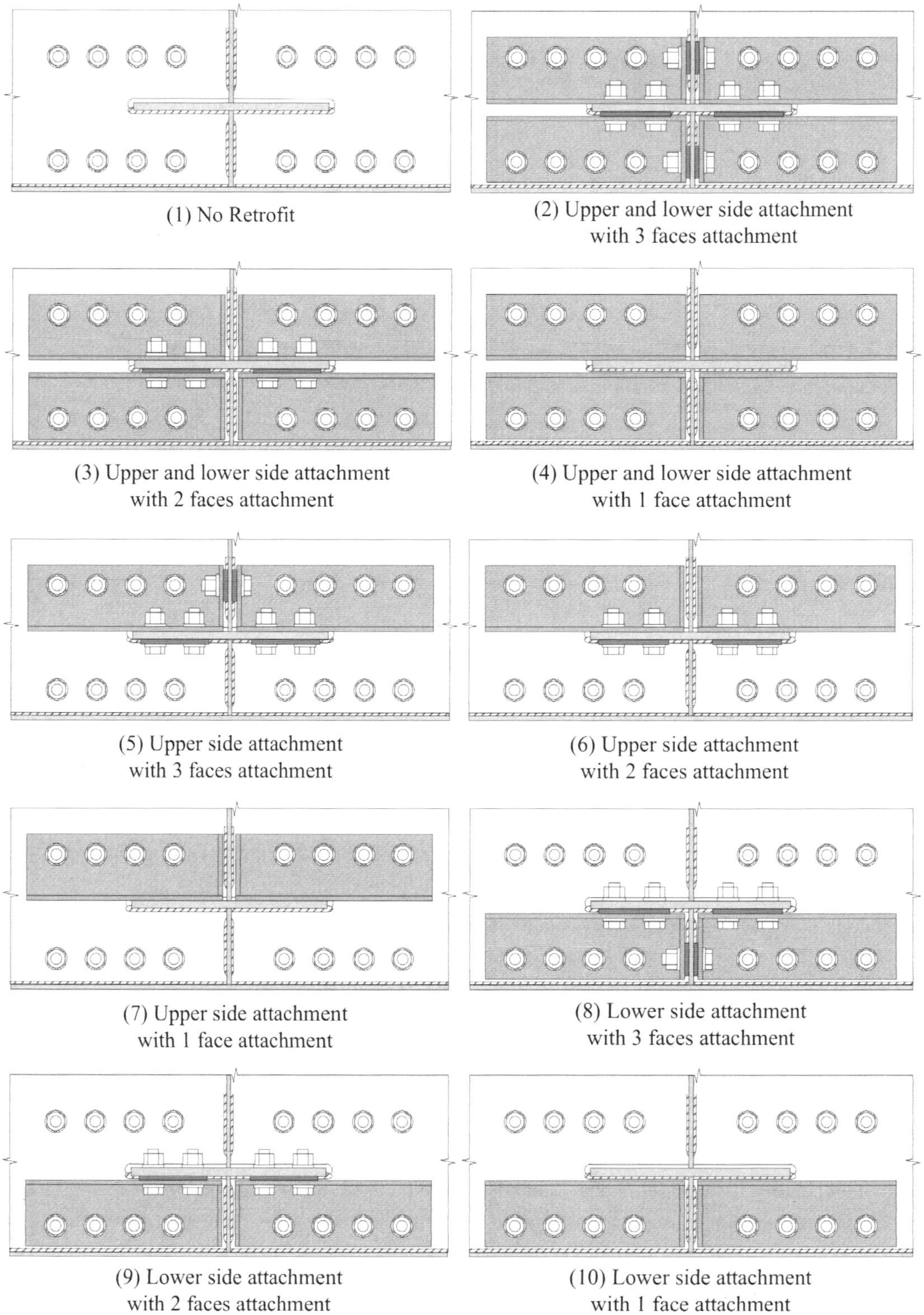

Figure 3. Ten cases of attachments.

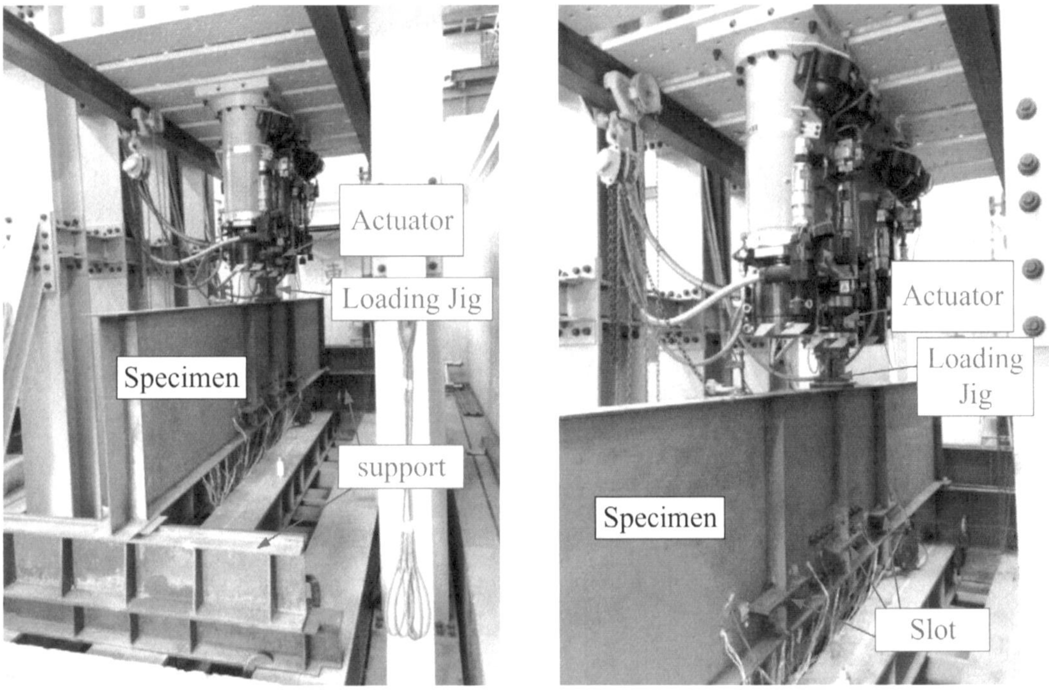

Figure 4. Loading conditions.

Figure 6 shows the stress change at all gages in the case of three face attachment. At gages ① and ①' where high tensile stress occurred before attachments, the upper side attachment is more effective than the lower side attachment, while both side attachment is best.

Figure 7 shows the stress change at gages ① and ①' in all cases of attachments. Two and three face both side attachments can reduce about 40% stress concentration, while two and three face upper side attachment can reduce 50–60%.

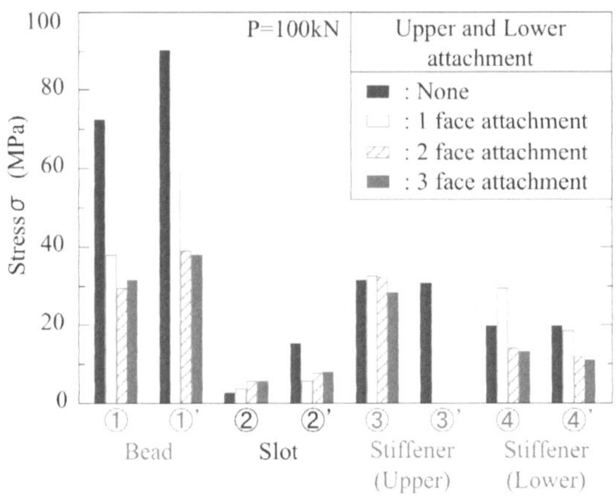

Figure 5. Stress change at all gages in the case of upper and lower attachment.

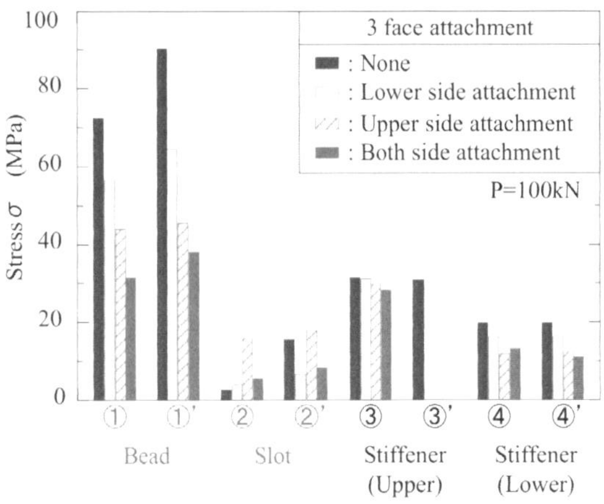

Figure 6. Stress change at all gages in the case of three face attachment.

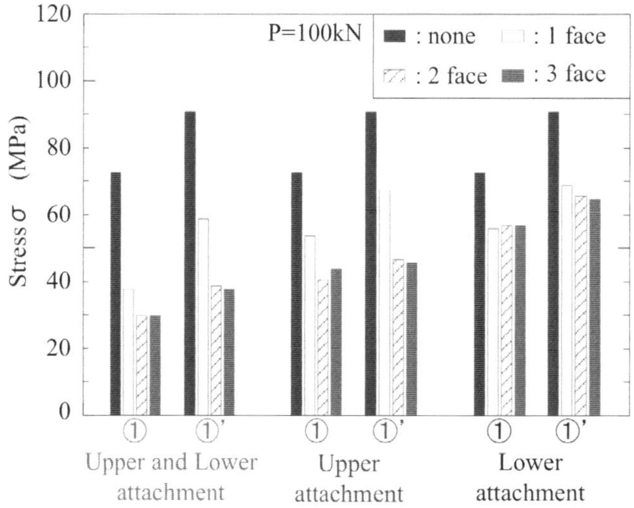

Figure 7. Stress change at gages ① and ①' in all cases of attachments.

4 SUMMARY

(1) Two and three face attachments are more effective than one face attachment about stress reduction effect, while there is very little difference between them.
(2) The upper side attachment is more effective than the lower side attachment, while both side attachment is best.
(3) Two and three face both side attachments can reduce about 40% of stress concentration, while two and three face upper side attachments can reduce 50–60%.

References

Fisher, J. W., *Fatigue and Fracture in Steel Bridges,* Chapter 6, Wiley-Interscience, 1984.

Nara National Highway Office, (in Japanese), 2006. Retrieved from http://www.kkr.mlit.go.jp/nara/ on September 2018.

Sakamoto, C., Sakano, M., Konishi, H., and Fujii, T., *Retrofit Against Fatigue Cracking at Web Penetration Details with a Slot in Steel Girder,* 9th International Symposium on Steel Bridges, Prague, Czech Republic, (to be presented in September) 2018.

Sakano, M., Mikami, I., Murayama, H., and Moriwaki, K., Fatigue Strength of Railway Plate Girder with Web Penetration (in Japanese), *Journal of Structural Engineering*, 41A, 965-973, 1995.

Yoshida, N., Sakano, M., Konishi, H., and Fujii, T., *Fatigue Test of Steel Girder Web Penetration Details with a Slit,* Sustainable Solutions in Structural Engineering and Construction, ASEA-SEC-2, Bangkok, Thailand. doi: 10.14455/ISEC.res.2014.11. Nov 3-7, 2014.

Streamlining Information Transfer between Construction and Structural Engineering
Edited by Shiau, J., Vimonsatit, V., Yazdani, S., and Singh, A.

ISBN: 978-0-9960437-7-9

A UNIFIED INELASTICITY AND DAMAGE MECHANICS FOR RUBBER-LIKE MATERIALS

SIAMAK YAZDANI[1] and ASHKAN SABOORI[2]

[1]*Heyer Engineering, Fargo, USA*
[2]*Dept of Civil Engineering, University of California, Davis, USA*

The nonlinearities observed in the behavior of rubber and polymeric materials are influenced not only by the far-field stress conditions but also by the morphology, microstructures, and changes at the meso-level such as kinks, crosslinking, and micro tearing. Considering constants temperature applications, the occurrence of micro-tearing has a significant contribution on the performance of polymers, soft tissues and other rubber-like materials. In this paper, a unified damage mechanics and nonlinear elasticity approach is presented to model nonlinear behavior of rubber-like materials under constant temperatures. The theory is cast within the general framework of the internal variable theory of thermodynamics with large deformations where the Clausius-Duhem inequality is provoked to develop general damage potential. The strain energy density function is formulated in terms of an effective Lagrangian strain tensor that evolves with cumulative damage as cracking and micro-tearing take places. Piola-Kirchhoff (PK) stress tensor is presented and a new form of the damage response tensor is proposed. The model prediction is illustrated against experimental results with good agreement.

Keywords: Continuum mechanics, Response tensor, Plasticity, Large deformation.

1 INTRODUCTION

Rubbery materials such as polymers, general elastomers, and soft tissues exhibit certain deformational characteristics that require the use of finite deformation theories in their modeling. These materials are generally considered incompressible, materially nonlinear, and non-homogenous. Experiments have shown that they exhibit hysteresis under cyclic loading and could experience microstructural tearing under higher levels of stresses. The micro-tearing causes the material to be more compliant and leads to further anisotropy. Even though a considerable literature can be found on the constitutive modeling of rubber-like materials (Mullins 1969, Simo 1987, Oktay 1993, Le Cam 2010, Volokh 2013, Wang *et al.* 2014, Armstrong and Spontak 2017) the development of a unified approach to rubber inelasticity with damage is surprisingly lagging the other modeling efforts.

The nonlinear deformational characteristics of rubber like materials stem from their microstructural properties. These properties exhibit a threshold by which when exceeded, micro-tearing and hence damage is initiated rendering the material more compliant. It is also shown that when rubber is stretched to certain strain level, the material cannot be damaged any further until the maximum previously experienced strain is reached again. This behavior is best demonstrated in the cyclic loading shown here schematically in Figure 1. The load-unloading curve shows the phenomenon known as the "Mullin's effect".

In this paper, a unified damage mechanics and nonlinear elasticity approach is combined and presented within the internal variable theory of continuum thermodynamics with finite deformations. The Lagrangian strain tensor along with a cumulative damage parameter representing the effects of micro-tearing is utilized. Using the spectral decomposition of tensors, the Lagrangian strain tensor is then decomposed into its positive and negative cones. This would allow for the decoupling of damage in tensile and compression modes. A form of strain energy density function is then postulated serving as a potential function for the Piola-Kirchhoff (PK) stress tensor. Finally, the paper proposes new anisotropic response tenors for certain elastomers to account for potential cross micro-tearing during large deformations.

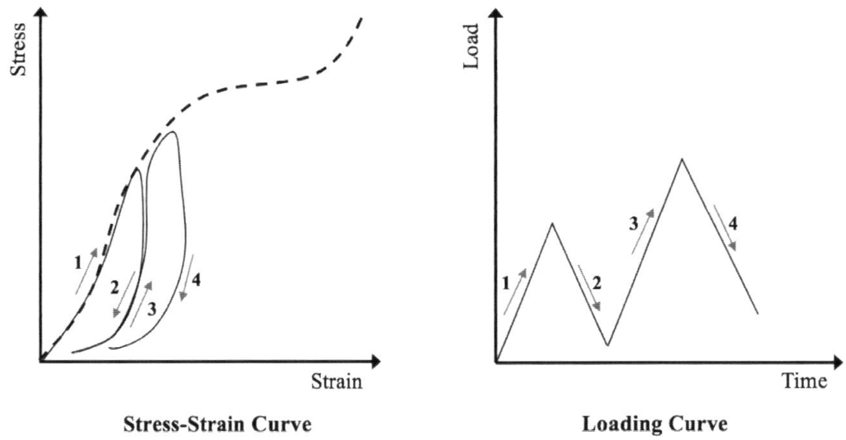

Figure 1. Schematic representation of Mullin's effect in uniaxial tension.

2 FORMULATION

For isothermal processes and utilizing the internal variable theory of thermodynamics, the Clausius-Duhem inequality can be shown to yield the following dissipation inequality (Eq. (1)) as (Lubliner 1972):

$$\frac{\partial}{\partial D}\psi(\boldsymbol{E}, D)\dot{D} \geq 0 \tag{1}$$

where, D represents a cumulative damage parameter, the superdot "." signifies the time rate of deformations, and the function $\psi(\boldsymbol{E}, D)$ represents the strain energy density function for the material. The Lagrangian strain tensor is given by $\boldsymbol{E}$. This is a standard approach in solid mechanics where Lagrangian strain measure is used for large deformations. The dependence of the strain energy density function on $\boldsymbol{E}$ and $\boldsymbol{D}$ allows for the description of the current state of the material as deformation evolves with damage. It is further postulated that the current state of the deformation is related to the undamaged material state through energy equivalence (Eq. (2)) as:

$$\psi(\boldsymbol{E}, D) = \psi^0\{\boldsymbol{E}^*(\boldsymbol{E}, D)\} \tag{2}$$

where, $\boldsymbol{E}^*$ is defined as an effective Lagrangian strain tensor and Ψ^0 represents the equivalent undamaged strain energy density function. Specific forms of $\boldsymbol{E}^*$ would define different models in the literature. The standard thermodynamic argument leads to obtaining Piola-Kirchhoff stress tensor, $\boldsymbol{S}$, as thermodynamic forces conjugate to fluxes $\dot{\boldsymbol{E}}$ as:

$$\boldsymbol{S} = \frac{\partial \psi(\boldsymbol{E}, D)}{\partial \boldsymbol{E}} \tag{3}$$

and,

$$\boldsymbol{S} = \frac{\partial \psi^0\{\boldsymbol{E}^*(\boldsymbol{E}, D)\}}{\partial \boldsymbol{E}} \tag{4}$$

Eq. (3) and Eq. (4) are the same but in terms of current and undamaged elastic strain energy density functions, respectively. To simplify the analysis, it is customary to decouple damage that may occur into tension and compression modes. There is an elegant mathematical decomposition to accomplish this known as the "spectral decomposition" and can be found elsewhere in the literature. With this, the Lagrangian strain tensor can be decomposed into its positive and negative cones (Eq. (5)) as:

$$\boldsymbol{E} = \boldsymbol{E}^+ + \boldsymbol{E}^- \tag{5}$$

where, $\boldsymbol{E}^+$ and $\boldsymbol{E}^-$ are the positive and the negative cones of the strain tensor $\boldsymbol{E}$. The nature of the decomposition is such that all the eigen-values of $\boldsymbol{E}^+$ are positive and that all the eigen-values of $\boldsymbol{E}^-$ are negative. This decoupling would significantly simplify damage formulations. To progress further, it is assumed that the damage is irreversible, i.e. $\dot{D} \geq 0$. The dissipation inequality is then expressed (Eq. (6)) as:

$$\frac{\partial \psi}{\partial \boldsymbol{E}} : \dot{\boldsymbol{E}} + \frac{\partial \psi}{\partial D} \dot{D} \geq 0 \tag{6}$$

It is accepted in the literature and evidenced by experimental work that damage in rubber-like materials would only occur under tensile stress states and not in compression. With this and using the positive cone of the strain tensor, the general form of the damage surface $\boldsymbol{\phi}(\boldsymbol{E}^+, \boldsymbol{D})$ is obtained (Eq. (7)) by postulating a positive square function $\boldsymbol{G}^2(\boldsymbol{D})$ such that,

$$\phi(\boldsymbol{E}^+, D) = \boldsymbol{E}^+ : \boldsymbol{R} : \boldsymbol{E}^+ - G^2(D) = 0 \tag{7}$$

where G is considered as the hardening-softening function and must be determined experimentally for specific materials and where $\boldsymbol{R}$ is defined as the response tensor. The response tensor $\boldsymbol{R}$ records the directionality of damage and could lead to anisotropic or isotropic formations. If R is formulated to be an isotropic fourth-order tensor, then the process will remain isotropic. Isotropic formulations are popular in the literature because it would simplify the analysis. However, it is not realistic. Micro-tearing and cracking have a strong directionality and hence requires an anisotropic approach. In this paper a new damage function is postulated as outlined below. Let, the maximum eigen-value of $\boldsymbol{E}^+$ be given by α^* and the associated eigen-vector as $\boldsymbol{q}^*$. With this, a modified Lagrangian strain tensor, as a sub-set of total Lagrangian strain tensor, is identified (Eq. (8)) as

$$\tilde{\mathbf{E}}^+ = \alpha^* \mathbf{q}^* \otimes \mathbf{q}^* \tag{8}$$

With this, a fourth-order and new anisotropic response tensor is postulated to be:

$$\boldsymbol{R} = \frac{\tilde{\boldsymbol{E}}^+ \otimes \tilde{\boldsymbol{E}}^+}{\tilde{\boldsymbol{E}}^+ : \tilde{\boldsymbol{E}}^+} \tag{9}$$

Eq. (9) would predict that under the multi-axial stress states, the damage would occur only in the direction of maximum stress as supported by experiments. It is of interest to note that under equal triaxial extension where all three stress components in the eigen-space are equal, the formulation would assign equal damage levels to each direction. No damage would be predicted under any compressive stress state as discussed before. Using the Ogden strain energy function for the first two terms utilizing stretch ratios, $\boldsymbol{\lambda_i}$, the strain function is expressed (Eq. (10)) as:

$$\psi = \sum_{n=1}^{2} \frac{\mu_n}{\alpha_n} \cdot \left(\lambda_1^{\alpha_n} + \lambda_2^{\alpha_n} + \lambda_3^{\alpha_n} - 3\right) \tag{10}$$

To complete the theory, a specific damage function $\boldsymbol{G(D)}$ is required. In this paper we have used the damage function as proposed by (Hokanson and Yazdani 1997) and given by Eq. (11):

$$g(D) = 1.0 - \left(\alpha . \exp\left(\frac{\beta G(D)}{\varepsilon_f}\right)\right) \tag{11}$$

where, α and β are material parameter and ϵ_f represents a strain measure below which no damage takes place.

The model with the new damage response tensor is tested for bovine artery soft tissue data as shown below in Figure 2. Values of $\boldsymbol{\alpha}$ = 0.004, $\boldsymbol{\beta} = \mathbf{5.5}$, and $\boldsymbol{\varepsilon_f} = \mathbf{0.60}$ were used. In the Ogden function, values of $\boldsymbol{\alpha_1} = -\boldsymbol{\alpha_2} = \mathbf{7.6}$ and $\boldsymbol{\mu_1} = -\boldsymbol{\mu_2} = \mathbf{0.65}$ were utilized. The general agreement between the experiment and theoretical results are well considering the simplicity of the new damage response tensor used.

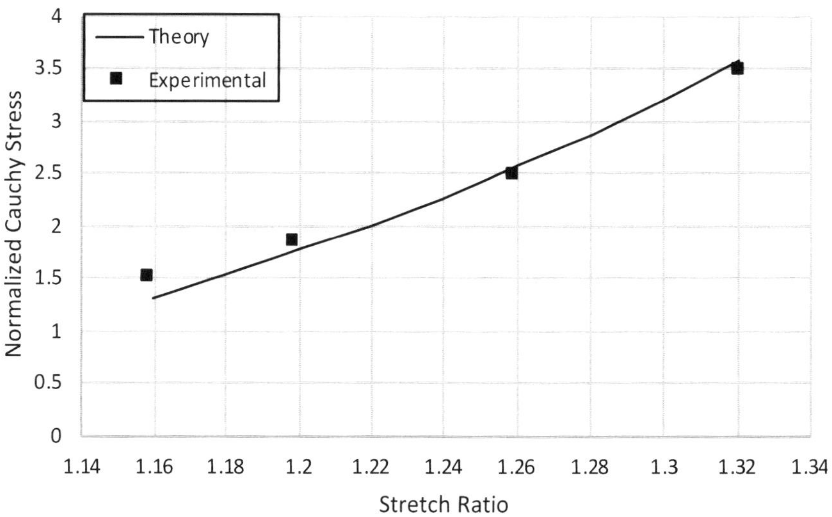

Figure 2. Experimental and theoretical results for Bovine artery soft tissue data.

3 CONCLUSION

It was shown in this paper that a unified damage and nonlinear elasticity can effectively be used for rubber-like materials with large deformations. The model represents a rigorous mathematical formulation with dissipation inequality to form a damage surface. The decomposition of the damage modes into negative and positive cones allows for the decoupling of the damage modes which simplifies the analysis. At the same time, it allows for the formulation of damage functions in both tensile and compression modes, if warranted. For polymeric materials, the compression stress state does not generally lead to any damage and hence validates the assumptions made in this paper. The model is robust and could easily be incorporated in computational codes.

References

Armstrong, D. P., and Spontak, R. J., Designing Dielectric Elastomers Over Multiple Length Scales for 21st Century Soft Materials Technologies, *Rubber Chemistry and Technology,* 90(2), 207-224, 2017.

Hokanson, J., and Yazdani, S., A Constitutive Model of the Artery with Damage, *Mechanics Research Communications*, 24(2), 151-159, 1997.

Le Cam, J. B., A Review of Volume Changes in Rubbers: The Effect of Stretching, *Rubber Chemistry and Technology*, 83(3), 247-269, 2010.

Lubliner, J., On the Thermodynamic Foundations of Nonlinear Solid Mechanics, *International Journal of Nonlinear Mechanics*, 7, 237-254, 1972.

Mullins, L., Softening of Rubber by Deformation, *Rubber Chemistry and Technology,* 42(1), 339-362, 1969.

Oktay, H. S., *Continuum Damage Mechanics of Balloon Angioplasty,* Dissertation, University of Maryland, 1993.

Simo, J. C., On a Fully Three-Dimensional Finite-Strain Viscoelastic Damage Model: Formulation And Computational Aspects, *Computer Methods in Applied Mechanics and Engineering,* 60(2), 153-173, 1987.

Volokh, K. Y., Review of The Energy Limiters Approach to Modeling Failure of Rubber, *Rubber Chemistry and Technology,* 86(3), 470-487, 2013.

Wang, X. M., Li, H., Yin, Z., and Xiao, H., Multiaxial Strain Energy Functions of Rubberlike Materials: An Explicit Approach Based on Polynomial Interpolation, *Rubber Chemistry and Technology,* 87(1), 168-183, 2014.

Streamlining Information Transfer between Construction and Structural Engineering
Edited by Shiau, J., Vimonsatit, V., Yazdani, S., and Singh, A.

ISBN: 978-0-9960437-7-9

STRUCTURAL ANALYSIS OF WOODEN FRAMES USED FOR MODERN WOODEN SCHOOL BUILDINGS

MITSUHIRO MIYAMOTO and MISAKI EBISUOKA

Dept of Safety Systems Construction Eng., Kagawa University, Takamatsu, Japan

Japanese school buildings have been changed from modern timber structures to reinforced concrete structures after the Building Standard Law was established in 1950. Therefore, only few modern wooden school buildings exist currently. Ensuring the seismic resilience of these wooden school buildings is essential for their preservation. The objective of this study is to develop an analysis model for the seismic performance evaluation of modern wooden school buildings. The fracture modes and the relationship between the lateral load and deformation of wooden frames were examined based on the results of full-scale static lateral loading tests performed in a previous study. The analysis models of these specimens were developed and compared with test results to examine their validity. Analysis results were in good agreement with test results.

Keywords: Single brace, Cross brace, Knee brace, Analysis model, Static pushover analysis, Static loading test.

1 INTRODUCTION

Japanese school buildings have been changed from modern timber structures to reinforced concrete structures after the Building Standard Law was established in 1950. Therefore, only few modern wooden school buildings exist currently. Ensuring the seismic resilience of these wooden school buildings is essential for their preservation.

A large number of studies in Japan focused on the seismic performance of wooden brace frames (Okamoto *et al.* 2010). However, only a few studies have focused on the large section brace used for modern wooden school buildings. In addition, a small number of studies have addressed wooden knee brace frames (Kamiya and Isoka 2012). Thus, almost no experimental data are available for evaluating the seismic performance of modern wooden school buildings.

The objective of this study is to develop an analysis model for the seismic performance evaluation of modern wooden school buildings. In a previous study, full-scale static lateral loading tests were conducted for wooden frame, single brace, cross brace, and knee brace specimens to examine the fracture modes and the relationship between the lateral load and deformation (Ebisuoka and Miyamoto 2018). These four specimens were designed based on the site investigation of existing modern wooden school buildings. In the present study, static pushover analysis was carried out for these specimens using a 2D frame model to simulate the relationship between lateral load and deformation. The analysis models were developed and compared with test results to examine their validity.

2 OUTLINE OF MODERN WOODEN SCHOOL BUILDINGS

The north and east school buildings of Yashiro Junior High School are wooden structures located in Ehime Prefecture. These buildings were built in 1960 and 1961. The buildings were dismantled recently; however, we were presented with an opportunity to investigate their structural characteristics before their dismantlement. Figure 1 shows the north school building. The main seismic elements are large section braces and knee braces (Figures 2 and 3). The dimensions of the column, groundsill, and brace are 135 × 135 mm, and those of the beam and knee brace are 135 × 135 mm or 150 × 150 mm and 90 × 90 mm, respectively. All braces and knee braces were fixed on the column using a bolt with a diameter of 12 mm. The column was connected to the beam or groundsill using metal connectors on each side of a connection.

Figure 1. School building of Yashiro Junior High School.

(a) Brace (b) Knee brace

Figure 2. Main seismic elements.

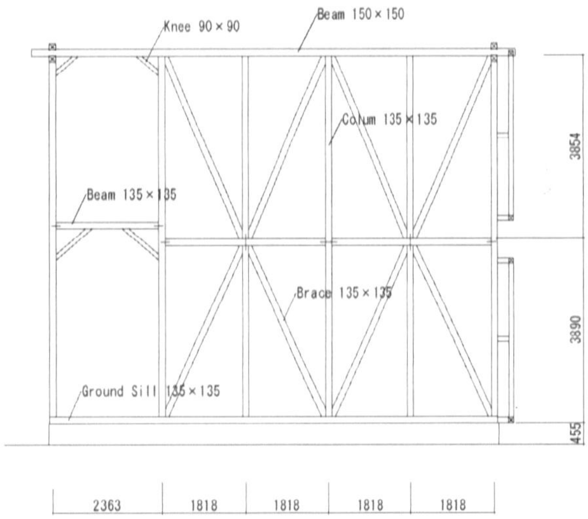

Unit: mm

Figure 3. Cross section (transverse direction).

3 PROPOSED ANALYSIS MODEL

The simulation analysis was executed using a 2D frame model to estimate the relationship between the load and deformation angle of each specimen.

3.1 Outline of Specimen

Four specimens were used for the tests, as shown in Figure 4. In addition to a wooden frame specimen, single brace, cross brace, and knee brace specimens were used because they are the main seismic elements of wooden school buildings. One brace in a cross brace specimen was separated at the intersection. The height and width of each specimen were 2752.5 mm and 1820 mm, respectively. The dimensions of the column, groundsill, and brace were 135 × 135 mm and those of the beam and knee brace were 135 × 180 mm and 90 × 90 mm, respectively. The column and brace were made of cedar, the beam and knee brace were made of Douglas fir, and the groundsill was made of Japanese cypress. All connections were designed to be the same as those of actual wooden school buildings based on site investigation. In the column-to-brace connection, the brace was connected to the beam or groundsill with a depth of 15 mm. The column was connected to the beam or groundsill using a short tenon with a height, width, and thickness of 67.5 mm, 90 mm, and 30 mm, respectively. Similarly, in the column-to-knee brace connection, the knee brace was connected to the column with a depth of 15 mm. In addition, metal connectors were fixed on each side of the connection using six nails. The height, width, and thickness of the metal connectors were 270 mm, 40 mm, and 5 mm, respectively. The through brace, separate brace and knee brace were fixed on the column using a bolt with a diameter of 12 mm. These specimens were subjected to cyclic lateral loads, with the real shear deformation angle being gradually increased symmetrically from 1/450, 1/300, 1/200, 1/150, 1/100, 1/75, 1/50, 1/30, 1/20 to 1/15rad. Three cyclic loadings were applied. Finally, the specimen was loaded to 1/10rad at one end.

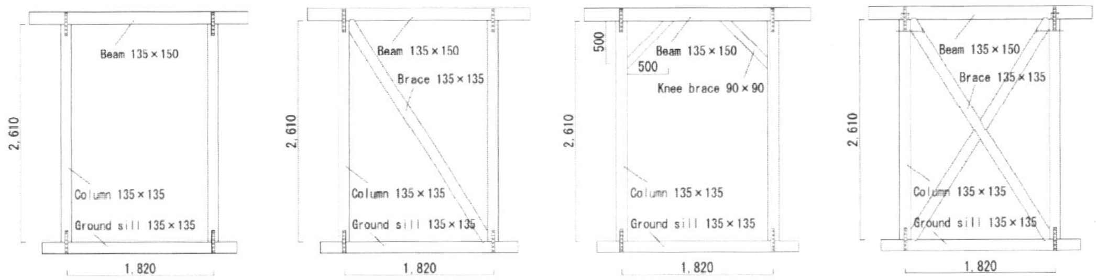

(a) Specimen A (b) Specimen B (c) Specimen C (d) Specimen D

Figure 4. Outline of specimens (Unit: mm).

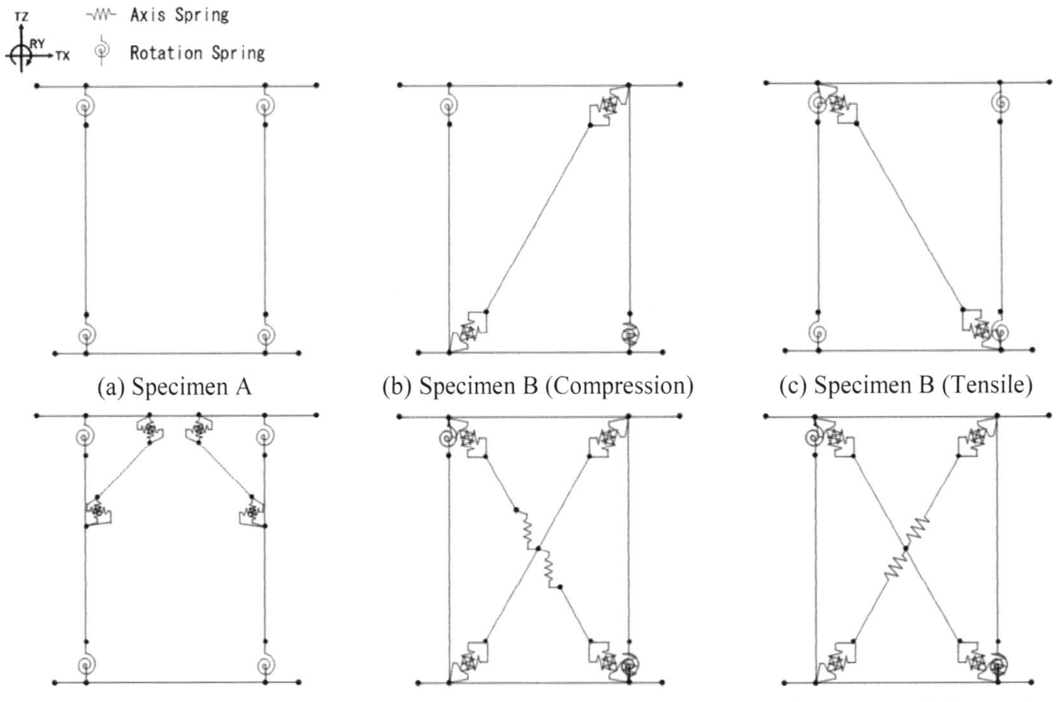

(a) Specimen A (b) Specimen B (Compression) (c) Specimen B (Tensile)

(d) Specimen C (e) Specimen D (Compression) (f) Specimen D (Tensile)

Figure 5. Model of each specimen.

Table 1. Resistance elements and direction.

Resistance element	Specimen	A	B (Com)	B (Ten)	C	D (Com)	D (Ten)
Metal connector	Bending strength	RY	RY	RY	RY	RY	RY
Short tenon	Bending strength	RY	RY	RY	RY	RY	RY
	Embedment to beam	*	TX	*	*	TX	TX
	Embedment from groundsill	*	TX	*	*	TX	TX
Bolt	Shear	—	—	TZ	TX, TZ	TZ	TZ, Axis
	Tensile	—	—	TX	TX, TZ	TX	TX
Brace	Embedment to column	—	TX	—	—	TX	TX
	Embedment to beam	—	TZ	—	—	TZ	TZ
	Embedment to groundsill	—	TZ	—	—	TZ	TZ
	Embedment to brace	—	—	—	—	—	Axis
Knee brace	Embedment to column	—	—	—	TX	—	—
	Embedment to beam	—	—	—	TZ	—	—
	Embedment from lack of column	—	—	—	TZ	—	—
	Embedment from lack of beam	—	—	—	TX	—	—
Column	Up lift of column bottom	—	TZ	—	—	TZ	TZ

* Included in bending strengh of short tenon

3.2 Outline of Analysis Model

Figure 5 shows the analysis model used for the static pushover analysis. The analysis models of specimens B and D were assumed to be different because the resistance element of a through brace was different for tensile and compressive braces. The bending resistance at the connection of the metal connector and short tenon was expressed as a rotation spring. The rotation springs of these connections with a through brace on specimens B (compression) and D were excluded because there was a brace to prevent the connection from rotating. The uplift behavior at one side of the column-to-groundsill connection on specimens B (compression) and D was expressed as an axis spring (Appendix A). The embedment resistance at the connection of the through brace, separate brace, or knee brace and the beam, groundsill, or column was expressed as an axis spring. The shear and tensile resistance of a bolt were expressed as an axis spring. A bilinear hysteresis model was used for each rotation spring and axis spring. The initial stiffness and yield moment or strength were calculated based on a previous study (Architectural Institute of Japan 2006, Architectural Institute of Japan 2009, Sakata *et al.* 2012). The material property was estimated through the results of destructive tests for each wooden member and based on a previous study (Architectural Institute of Japan 2006, Architectural Institute of Japan 2009). Table 1 summarizes the type and direction of each resistance element. The direction "TX", "TZ" and "RY" are defined in Figure 5.

3.3 Comparison of Test and Analysis Results

The static pushover analysis was performed using general purpose analysis software, *i.e.*, SNAP (Ver. 7). Figure 6 shows the comparison of the relationship between the load and deformation angle of each specimen as obtained from the test and analysis results. The analysis results are in good agreement with the test results for each specimen. The load to the range of large deformation angle on almost all specimens obtained through the analysis is smaller than that obtained in the tests because the design equations were used for calculation.

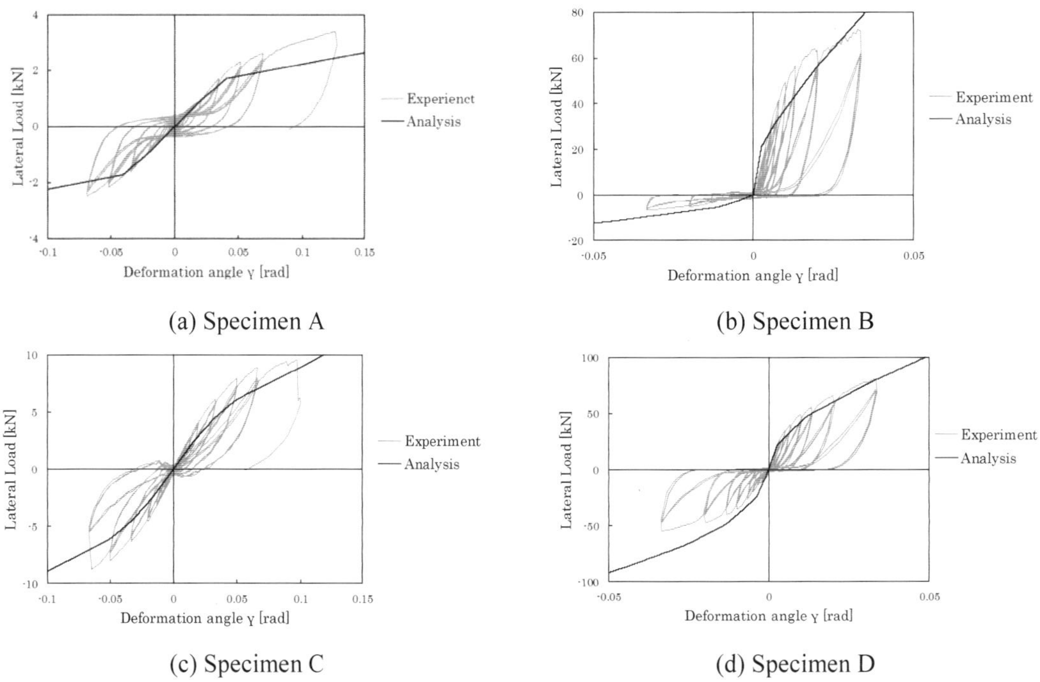

Figure 6. Comparison of experimental and analysis results.

4 CONCLUSIONS

In this study, an analysis model was developed for the seismic performance evaluation of modern wooden school buildings. The results obtained using the model were compared with test results to examine the validity of the model. Four specimens were used for the tests, *i.e.*, wooden frame, single brace, cross brace, and knee brace specimens. This was because they are the main seismic elements of wooden school buildings. Analysis results exhibited good agreement with the test results.

Acknowledgments

This research was partially supported by the Asahi Glass Foundation. We gratefully acknowledge this support.

Appendix A. Tensile Test Results of Column-to-Groundsill Connection

The tensile test for column-to-groundsill connection was conducted as shown in Appendix figures 1–3. The specimen size was the same as that for the full-scale static lateral loading tests. The number of specimens was five.

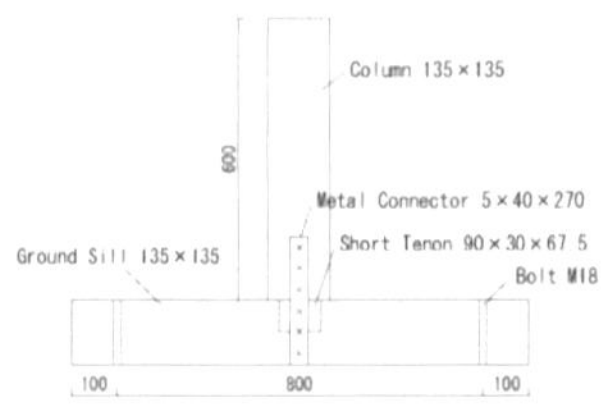

Appendix Figure 1. Outline of specimen.

Appendix Figure 2. Loading instrument.

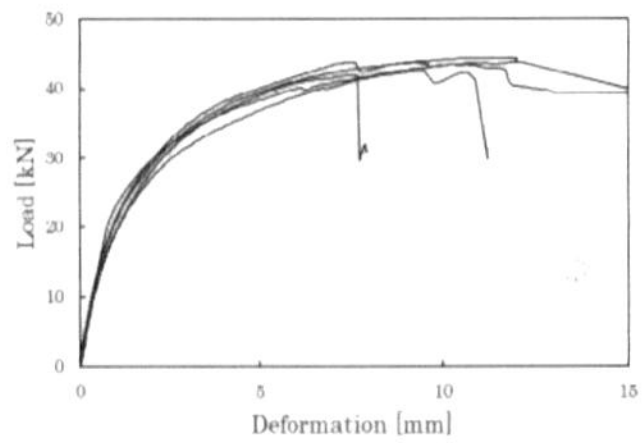

Appendix Figure 3. Relationship between load and deformation.

References

Architectural Institute of Japan (edited), *Design Manual for Engineered Timber Joints*, Maruzen, Japan, 2009.

Architectural Institute of Japan (edited), *Standard for Structural Design of Timber Structures*, Maruzen, Japan, 2006.

Ebisuoka, M., and Miyamoto, M., Study on Seismic Performance of Timber Frame Used for Modern Timber School Building, *AIJ Journal of Structural Engineering*, 64B, 453-461, March, 2018.

Kamiya, S., and Isoda, H., Lateral Load Capacity and Stiffness of Post-Beam Frame with Knee Brace Connecting to Member Surface, *AIJ Journal of Structural Engineering*, 58B, 259-264, March, 2012.

Okamoto, S., Murakami, M., and Inayama, M., Prediction and Verification of Shear Force-Drift Angle Relations for Compressive Bracing Shear Wall with Arbitrary Wall Length, *AIJ Journal of Structural and Construction Engineering*, 75(657), 2019-2026, November, 2010.

Sakata, H., Yamazaki, Y., Udagawa, H., and Ohashi, Y., Experimental Study on Flexural-Shear Behavior of Mortise-Tenon Joint with Dowel, *AIJ Journal of Structural and Construction Engineering*, 77(671), 45-54, January, 2012.

AUTHOR INDEX

AUTHOR INDEX

ISEC PRESS
PO Box 1334
Fargo, North Dakota 58107
USA

ISBN 978-1-5108-9998-8

CURRAN ASSOCIATES INC.
proceedings.com

23rd International Workshop on Rare Earth and Future Permanent Magnets and Their Applications (REPM2014)

Annapolis, Maryland, USA
17 - 21 August 2014